Conversion Factors

Density:

$$1.00 \, \text{kg/m}^3 = 1.9404 \times 10^{-3} \, \text{slug/ft}^3 = 6.2430 \times 10^{-2} \, \text{lbm/ft}^3$$
$$1.00 \, \text{lbm/ft}^3 = 3.1081 \times 10^{-2} \, \text{slug/ft}^3 = 16.018 \, \text{kg/m}^3$$

Energy or work:

$$1.00 \, \text{cal} = 4.187 \, \text{J} = 4.187 \, \text{N} \cdot \text{m}$$
$$1.00 \, \text{Btu} = 778.2 \, \text{ft lbf} = 0.2520 \, \text{kcal} = 1055 \, \text{J}$$
$$1.00 \, \text{kW h} = 3.600 \times 10^6 \, \text{J}$$

Flow rates:

$$1.00 \, \text{gal/min} = 6.309 \times 10^{-5} \, \text{m}^3/\text{s} = 2.228 \times 10^{-3} \, \text{ft}^3/\text{s}$$

Force:

$$1.00 \, \text{N} \qquad = 10^5 \, \text{dyne} = 0.2248 \, \text{lbf}$$
$$1.00 \, \text{lbf} \qquad = 4.4482 \, \text{N}$$
$$1.00 \, \text{lbf} \qquad = 16.0 \, \text{oz}$$
$$1.00 \, \text{U.S. ton} = 2000 \, \text{lbf} = 2.0 \, \text{kip}$$

Heat flux:

$$1.00 \, \text{W/cm}^2 = 0.2388 \, \text{cal/s} \cdot \text{cm}^2 = 0.8806 \, \text{Btu/ft}^2 \cdot \text{s}$$
$$= 3.170 \times 10^3 \, \text{But/ft}^2 \cdot \text{h}$$

Length:

$$1.00 \, \text{m} \quad = 3.2808 \, \text{ft} = 39.37 \, \text{in.}$$
$$1.00 \, \text{km} \quad = 0.6214 \, \text{mile} = 1093.6 \, \text{yd}$$
$$1.00 \, \text{ft} \quad = 0.3048 \, \text{m} = 30.48 \, \text{cm}$$
$$1.00 \, \text{ft} \quad = 12 \, \text{in} = 0.333 \, \text{yd}$$
$$1.00 \, \text{mile} = 5280 \, \text{ft} = 1760 \, \text{yd} = 1609.344 \, \text{m}$$

Mass:

$$1.00 \, \text{kg} \quad = 1000 \, \text{g} = 2.2047 \, \text{lbm}$$
$$1.00 \, \text{slug} = 32.174 \, \text{lbm} = 14.593 \, \text{kg}$$

Power:

$$1.00 \, \text{kW} \quad = 1000.0 \, \text{W} = 1000.0 \, \text{N} \cdot \text{m/s} = 0.2388 \, \text{kcal/s}$$
$$1.00 \, \text{kW} \quad = 1.341 \, \text{hp}$$
$$1.00 \, \text{hp} \quad = 550 \, \text{ft} \cdot \text{lbf/s} = 0.7457 \, \text{kW}$$
$$1.00 \, \text{Btu/s} \quad = 1.415 \, \text{hp}$$
$$1.00 \, \text{Btu/h} \quad = 0.29307 \, \text{W}$$

Pressure:

$$1.00 \text{ N/m}^2 = 1.4504 \times 10^{-4} \text{ lbf/in.}^2 = 2.0886 \times 10^{-2} \text{ lbf/ft}^2$$
$$1.00 \text{ bar} = 10^5 \text{ N/m}^2 = 10^5 \text{ Pa}$$
$$1.00 \text{ lbf/in.}^2 = 6.8947 \times 10^3 \text{ N/m}^2$$
$$1.00 \text{ lbf/in.}^2 = 144.00 \text{ lbf/ft}^2$$

Specific enthalpy:

$$1.00 \text{ N} \cdot \text{m/kg} = 1.00 \text{ J/kg} = 1.00 \text{ m}^2/\text{s}^2$$
$$1.00 \text{ N} \cdot \text{m/kg} = 4.3069 \times 10^{-4} \text{ Btu/lbm} = 1.0781 \text{ ft lbf/slug}$$
$$1.00 \text{ Btu/lbm} = 2.3218 \times 10^3 \text{ N} \cdot \text{m/kg}$$

Specific heat:

$$1.00 \text{ N} \cdot \text{m/kg} \cdot \text{K} = 1.00 \text{ J/kg} \cdot \text{K} = 2.388 \times 10^{-4} \text{ Btu/lbm} \cdot {}^\circ\text{R}$$
$$= 5.979 \text{ ft} \cdot \text{lbf/slug} \cdot {}^\circ\text{R}$$
$$1.00 \text{ Btu/lbm} \cdot {}^\circ\text{R} = 32.174 \text{ Btu/slug} \cdot {}^\circ\text{R} = 4.1879 \times 10^3 \text{ N} \cdot \text{m/kg} \cdot \text{K}$$
$$= 4.1879 \times 10^3 \text{ J/kg} \cdot \text{K}$$

Temperature:

The temperature of the ice point is 273.15 K (491.67°R).

$$1.00 \text{ K} = 1.80°\text{R}$$
$$\text{K} = °\text{C} + 273.15$$
$$°\text{R} = °\text{F} + 459.67$$
$$T°\text{F} = 1.8(T°\text{C}) + 32$$
$$T°\text{R} = 1.8(T \text{ K})$$

Velocity:

$$1.00 \text{ m/s} = 3.60 \text{ km/h}$$
$$1.00 \text{ km/h} = 0.2778 \text{ m/s} = 0.6214 \text{ mile/h} = 0.9113 \text{ ft/s}$$
$$1.00 \text{ ft/s} = 0.6818 \text{ mile/h} = 0.59209 \text{ knot}$$
$$1.00 \text{ mile/h} = 1.467 \text{ ft/s} \simeq 1.609 \text{ km/h} = 0.4470 \text{ m/s}$$
$$1.00 \text{ knot} = 1.15155 \text{ mile/h}$$

Viscosity:

$$1.00 \text{ kg/m} \cdot \text{s} = 0.67197 \text{ lbm/ft} \cdot \text{s} = 2.0886 \times 10^{-2} \text{ lbf} \cdot \text{s/ft}^2$$
$$1.00 \text{ lbm/ft} \cdot \text{s} = 3.1081 \times 10^{-2} \text{ lbf} \cdot \text{s/ft}^2 = 1.4882 \text{ kg/m} \cdot \text{s}$$
$$1.00 \text{ lbf} \cdot \text{s/ft}^2 = 47.88 \text{ N} \cdot \text{s/m}^2 = 47.88 \text{ Pa} \cdot \text{s}$$
$$1.00 \text{ centipoise} = 0.001 \text{ Pa} \cdot \text{s} = 0.001 \text{ kg/m} \cdot \text{s} = 6.7197 \times 10^{-4} \text{ lbm/ft} \cdot \text{s}$$
$$1.00 \text{ stoke} = 1.00 \times 10^{-4} \text{ m}^2/\text{s(kinematic viscosity)}$$

Volume:

$$1.00 \text{ liter} = 1000.0 \text{ cm}^3$$
$$1.00 \text{ barrel} = 5.6146 \text{ ft}^3$$
$$1.00 \text{ ft}^3 = 1728 \text{ in.}^3 = 0.03704 \text{ yd}^3 = 7.481 \text{ gal}$$
$$= 28.32 \text{ liters}$$
$$1.00 \text{ gal} = 3.785 \text{ liters} = 3.785 \times 10^{-3} \text{ m}^3$$
$$1.00 \text{ bushel} = 3.5239 \times 10^{-2} \text{ m}^3$$

AERODYNAMICS FOR ENGINEERS

Fifth Edition

JOHN J. BERTIN

Professor Emeritus, United States Air Force Academy

and

RUSSELL M. CUMMINGS

Professor, United States Air Force Academy

PEARSON

Prentice
Hall

Pearson Education International

Vice President and Editorial Director, ECS: *Marcia J. Horton*
Acquisitions Editor: *Tacy Quinn*
Associate Editor: *Dee Bernhard*
Managing Editor: *Scott Disanno*
Production Editor: *Rose Kernan*
Art Director: *Kenny Beck*
Art Editor: *Greg Dulles*
Cover Designer: *Kristine Carney*
Senior Operations Supervisor: *Alexis Heydt-Long*
Operations Specialist: *Lisa McDowell*
Marketing Manager: *Tim Galligan*

Cover images left to right: Air Force B-1B Lancer | Gregg Stansbery photo | Boeing Graphic based on photo and data provided by Lockheed Martin Aeronautics F-18 | NASA.

© 2009, 2002, 1998, 1989, 1979 by Pearson Education, Inc.
Pearson Prentice-Hall
Upper Saddle River, NJ 07458

ISBN-10: 0-13-235521-3
ISBN-13: 978-0-13-235521-6
Printed in the United States of America
10 9 8 7 6 5 4 3 2 1

Pearson Education Ltd., London
Pearson Education Singapore, Pte. Ltd
Pearson Education Canada, Inc.
Pearson Education–Japan
Pearson Education Australia PTY, Limited
Pearson Education North Asia, Ltd., Hong Kong
Pearson Educación de Mexico, S.A. de C.V.
Pearson Education Malaysia, Pte. Ltd.
Pearson Education Upper Saddle River, New Jersey

The authors would like to dedicate this book to their teachers and mentors.

To my parents, Andrew and Yolanda,
to my wife, Ruth E. Bertin,
and to my children, Thomas, Randolph, Elizabeth, and Michael Bertin.

To my parents, Bud and Ann,
to my wife, Signe G. Balch,
and to my children, Cornelia and Carl Cummings.

[Photograph from the collection of Capt. Dale M. McKinney, USAF (ret.).]

To the brave airmen who have sacrificed so much in service to their countries.

Contents

Preface to the Fifth Edition

There were two main goals for writing the Fifth Edition of *Aerodynamics for Engineers*: 1) to provide readers with a motivation for studying aerodynamics in a more casual, enjoyable, and readable manner, and 2) to update the technical innovations and advancements that have taken place in aerodynamics since the writing of the previous edition.

To help achieve the first goal we provided readers with background for the true purpose of aerodynamics. Namely, we believe that the goal of aerodynamics is to predict the forces and moments that act on an airplane in flight in order to better understand the resulting performance benefits of various design choices. In order to better accomplish this, Chapter 1 begins with a fun, readable, and motivational presentation on aircraft performance using material on Specific Excess Power (a topic which is taught to all cadets at the U.S. Air Force Academy). This new introduction should help to make it clear to students and engineers alike that understanding aerodynamics is crucial to understanding how an airplane performs, and why one airplane may be better than another at a specific task.

Throughout the remainder of the fifth edition we have added new and emerging aircraft technologies that relate to aerodynamics. These innovations include detailed discussion about: laminar flow and low Reynolds number airfoils, as well as modern high-lift systems (Chapter 6); micro UAV and high altitude/long endurance wing geometries (Chapter 7); the role of experimentation in determining aerodynamics, including the impact of scaling data for full-scale aircraft (Chapter 8); slender-body theory and sonic boom reduction (Chapter 11); hypersonic transition (Chapter 12); and wing-tip devices, as well as modern wing planforms (Chapter 13). Significant new material on practical methods for estimating aircraft drag have also been incorporated into Chapters 4 and 5, including methods for estimating skin friction, form factor, roughness effects, and the impact of boundary-layer transition. Of special interest in the fifth edition is a description of the aerodynamic design of the F-35, now included in Chapter 13.

In addition, there are 32 new figures containing updated and new information, as well as numerous, additional up-to-date references throughout the book. New problems have been added to almost every chapter, as well as example problems showing students how the theoretical concepts can be applied to practical problems. Users of the fourth edition of the book will find that all material included in that edition is still included in the fifth edition, with the new material added throughout the book to bring a real-world flavor to the concepts being developed. We hope that readers will find the inclusion of all of this additional material helpful and informative.

In order to help accomplish these goals a new co-author, Professor Russell M. Cummings of the U.S. Air Force Academy, has been added for the fifth edition of *Aerodynamics for Engineers*. Based on his significant contributions to both the writing and presentation of new and updated material, he makes a welcome addition to the quality and usefulness of the book.

Finally, no major revision of a book like *Aerodynamics for Engineers* can take place without the help of many people. The authors are especially indebted to everyone who aided in collecting new materials for the fifth edition. We want to especially thank Doug McLean, John McMasters, and their associates at Boeing; Rick Baker, Mark Buchholz, and their associates from Lockheed Martin; Charles Boccadoro, David Graham, and their associates of Northrop Grumman; Mark Drela, Massachusetts Institute of Technology; Michael Selig, University of Illinois; and Case van Dam, University of California, Davis. In addition, we are very grateful for the excellent suggestions and comments made by the reviewers of the fifth edition: Doyle Knight of Rutgers University, Hui Hu of Iowa State University, and Gabriel Karpouzian of the U.S. Naval Academy. Finally, we also want to thank Shirley Orlofsky of the U.S. Air Force Academy for her unfailing support throughout this project.

Preface to the Fourth Edition

This text is designed for use by undergraduate students in intermediate and advanced classes in aerodynamics and by graduate students in mechanical engineering and aerospace engineering. Basic fluid mechanic principles are presented in the first four chapters. Fluid properties and a model for the standard atmosphere are discussed in Chapter 1, "Fluid Properties." The equations governing fluid motion are presented in Chapter 2, "Fundamentals of Fluid Mechanics." Differential and integral forms of the continuity equation (based on the conservation of mass), the linear momentum equation (based on Newton's law of motion), and the energy equation (based on the first law of thermodynamics) are presented. Modeling inviscid, incompressible flows is the subject of Chapter 3, "Dynamics of an Incompressible, Inviscid Flow Field." Modeling viscous boundary layers, with emphasis on incompressible flows, is the subject of Chapter 4, "Viscous Boundary Layers." Thus, Chapters 1 through 4 present material that covers the principles upon which the aerodynamic applications are based. For the reader who already has had a course (or courses) in fluid mechanics, these four chapters provide a comprehensive review of fluid mechanics and an introduction to the nomenclature and style of the present text.

At this point, the reader is ready to begin material focused on aerodynamic applications. Parameters that characterize the geometry of aerodynamic configurations and parameters that characterize aerodynamic performance are presented in Chapter 5, "Characteristic Parameters for Airfoil and Wing Aerodynamics." Techniques for modeling the aerodynamic performance of two-dimensional airfoils and of finite-span wings at low speeds (where variations in density are negligible) are presented in Chapters 6 and 7, respectively. Chapter 6 is titled "Incompressible Flows around Wings of Infinite Span," and Chapter 7 is titled "Incompressible Flow about Wings of Finite Span."

The next five chapters deal with compressible flow fields. To provide the reader with the necessary background for high-speed aerodynamics, the basic fluid mechanic principles for compressible flows are discussed in Chapter 8, "Dynamics of a Compressible Flow Field." Thus, from a pedagogical point of view, the material presented in Chapter 8 complements the material presented in Chapters 1 through 4. Techniques for modeling high-speed flows (where density variations cannot be neglected) are presented in Chapters 9 through 12. Aerodynamic performance for compressible, subsonic flows through transonic speeds is the subject of Chapter 9, "Compressible Subsonic Flows and Transonic Flows." Supersonic aerodynamics for two-dimensional airfoils is the subject of Chapter 10, "Two-Dimensional Supersonic Flows about Thin Airfoils" and for finite-span wings in Chapter 11, "Supersonic Flows over Wings and Airplane Configurations." Hypersonic flows are the subject of Chapter 12.

At this point, chapters have been dedicated to the development of basic models for calculating the aerodynamic performance parameters for each of the possible speed ranges. The assumptions and, therefore, the restrictions incorporated into the development of the theory are carefully noted. The applications of the theory are illustrated by working one or more problems. Solutions are obtained using numerical techniques in order to apply the theory for those flows where closed-form solutions are impractical or impossible. In each of the chapters, the computed aerodynamic parameters are compared with experimental data from the open literature to illustrate both the validity of the theoretical analysis and its limitations (or, equivalently, the range of conditions for which the theory is applicable). One objective is to use the experimental data to determine the limits of applicability for the proposed models.

Extensive discussions of the effects of viscosity, compressibility, shock/boundary-layer interactions, turbulence modeling, and other practical aspects of contemporary aerodynamic design are also presented. Problems at the end of each chapter are designed to complement the material presented within the chapter and to develop the student's understanding of the relative importance of various phenomena. The text emphasizes practical problems and the techniques through which solutions to these problems can be obtained. Because both the International System of Units (Système International d'Unitès, abbreviated SI) and English units are commonly used in the aerospace industry, both are used in this text. Conversion factors between SI units and English units are presented on the inside covers.

Advanced material relating to design features of aircraft over more than a century and to the tools used to define the aerodynamic parameters are presented in Chapters 13 and14. Chapter 13 is titled "Aerodynamic Design Considerations," and Chapter 14 is titled "Tools for Defining the Aerodynamic Environment." Chapter 14 presents an explanation of the complementary role of experiment and of computation in defining the aerodynamic environment. Furthermore, the advantages, limitations, and roles of computational techniques of varying degrees of rigor are discussed. The material presented in Chapters 13 and14 not only should provide interesting reading for the student but, should be useful to professionals long after they have completed their formal academic training.

COMMENTS ON THE FIRST THREE EDITIONS

The author would like to thank Michael L. Smith for his significant contributions to *Aerodynamics for Engineers*. Michael Smith's contributions helped establish the quality of the text from the outset and the foundation upon which the subsequent editions have been based. For these contributions, he was recognized as coauthor of the first three editions.

The author is indebted to his many friends and colleagues for their help in preparing the first three editions of this text. I thank them for their suggestions, their support, and for copies of photographs, illustrations, and reference documents. The author is indebted to L. C. Squire of Cambridge University; V. G. Szebehely of the University of Texas at Austin; F. A. Wierum of the Rice University; T. J. Mueller of the University of Notre Dame; R. G. Bradley and C. Smith of General Dynamics; G. E. Erickson of Northrop; L. E. Ericsson of Lockheed Missiles and Space; L. Lemmerman and A. S. W. Thomas of Lockheed Georgia; J. Periaux of Avions Marcel Dassault; H. W. Carlson, M. L. Spearman, and P. F. Covell of the Langley Research Center; D. Kanipe of the Johnson Space Center; R. C. Maydew, S. McAlees, and W. H. Rutledge of the Sandia National Labs; M. J. Nipper of the Lockheed Martin Tactical Aircraft Systems; H. J. Hillaker (formerly) of General Dynamics; R. Chase of the ANSER Corporation; and Lt. Col. S. A. Brandt, Lt. Col. W. B. McClure, and Maj. M. C. Towne of the U.S. Air Force Academy. F. R. DeJarnette of North Carolina State University, and J. F. Marchman III, of Virginia Polytechnic Institute and State University provided valuable comments as reviewers of the third edition.

Not only has T. C. Valdez served as the graphics artist for the first three editions of this text, but he has regularly located interesting articles on aircraft design that have been incorporated into the various editions.

THE FOURTH EDITION

Rapid advances in software and hardware have resulted in the ever-increasing use of computational fluid dynamics (CFD) in the design of aerospace vehicles. The increased reliance on computational methods has led to three changes unique to the fourth edition.

1. Some very sophisticated numerical solutions for high alpha flow fields (Chapter 7), transonic flows around an NACA airfoil (Chapter 9), and flow over the SR-71 at three high-speed Mach numbers (Chapter 11) appear for the first time in *Aerodynamics for Engineers*. Although these results have appeared in the open literature, the high-quality figures were provided by Cobalt Solutions, LLC, using the postprocessing packages Fieldview and EnSight. Captain J. R. Forsythe was instrumental in obtaining the appropriate graphics.

2. The discussion of the complementary use of experiment and computation as tools for defining the aerodynamic environment was the greatest single change to the text. Chapter 14 was a major effort, intended to put in perspective the strengths and limitations of the various tools that were discussed individually throughout the text.

3. A CD with complementary homework problems and animated graphics is available to adopters. Please contact the author at USAFA.

Major D. C. Blake, Capt. J. R. Forsythe, and M. C. Towne were valuable contributors to the changes that have been made to the fourth edition. They served as sounding boards before the text was written, as editors to the modified text, and as suppliers of graphic art. Since it was the desire of the author to reflect the current role of computations (limitations, strengths, and usage) and to present some challenging applications, the author appreciates the many contributions of Maj. Blake, Capt. Forsythe, and Dr. Towne, who are active experts in the use and in the development of CFD in aerodynamic design.

The author would also like to thank M. Gen. E. R. Bracken for supplying information and photographs regarding the design and operation of military aircraft. G. E. Peters of the Boeing Company and M. C. Towne of Lockheed Martin Aeronautics served as points of contact with their companies in providing material new to the fourth edition.

The author would like to thank John Evans Burkhalter of Auburn University, Richard S. Figliola of Clemson University, Marilyn Smith of the Georgia Institute of Technology, and Leland A. Carlson of Texas A & M University, who, as reviewers of a draft manuscript, provided comments that have been incorporated either into the text or into the corresponding CD.

The author would also like to thank the American Institute of Aeronautics and Astronautics (AIAA), the Advisory Group for Aerospace Research and Development, North Atlantic Treaty Organization (AGARD/NATO),[1] the Boeing Company, and the Lockheed Martin Tactical Aircraft System for allowing the author to reproduce significant amounts of archival material. This material not only constitutes a critical part of the fourth edition, but it also serves as an excellent foundation upon which the reader can explore new topics.

Finally, thank you Margaret Baker and Shirley Orlofsky.

JOHN J. BERTIN
United States Air Force Academy

---

[1]The AGARD/NATO material was first published in the following publications: *Conf. Proc. High Lift System Aerodynamics*, CP-515, Sept. 1993; *Conf. Proc. Validation of Computational Fluid Dynamics*, CP-437, vol. 1, Dec. 1988; Report, *Special Course on Aerothermodynamics of Hypersonic Vehicles*, R-761, June 1989.

1 WHY STUDY AERODYNAMICS?

1.1 THE ENERGY-MANEUVERABILITY TECHNIQUE

Early in the First World War, fighter pilots (at least those good enough to survive their first engagements with the enemy) quickly developed tactics that were to serve them throughout the years. German aces, such as Oswald Boelcke and Max Immelman, realized that, if they initiated combat starting from an altitude that was greater than that of their adversary, they could dive upon their foe, trading potential energy (height) for kinetic energy (velocity). Using the greater speed of his airplane to close from the rear (i.e., from the target aircraft's "six o'clock position"), the pilot of the attacking aircraft could dictate the conditions of the initial phase of the air-to-air combat. Starting from a superior altitude and converting potential energy to kinetic energy, the attacker might be able to destroy his opponent on the first pass. These tactics were refined, as the successful fighter aces gained a better understanding of the nuances of air combat by building an empirical data base through successful air-to-air battles. A language grew up to codify these tactics: "Check your six."

This data base of tactics learned from successful combat provided an empirical understanding of factors that are important to aerial combat. Clearly, the sum of the potential energy plus the kinetic energy (i.e., the total energy) of the aircraft is one of the factors.

EXAMPLE 1.1:   The total energy

Compare the total energy of a B-52 that weighs 450,000 pounds and that is cruising at a true air speed of 250 knots at an altitude of 20,000 feet with the total energy of an F-5 that weighs 12,000 pounds and that is cruising at a true air speed of 250 knots at an altitude of 20,000 feet. The equation for the total energy is

$$E = 0.5\, m\, V^2 + m\, g\, h \qquad\qquad (1.1)$$

Solution: To have consistent units, the units for velocity should be feet per second rather than knots. A knot is a nautical mile per hour and is equal to 1.69 feet per second. Thus, 250 knots is equal to 422.5 ft/s. Since the mass is given by the equation,

$$m = \frac{W}{g} \qquad\qquad (1.2)$$

Note that the units of mass could be grams, kilograms, lbm, slugs, or $lbf \cdot s^2/ft$. The choice of units often will reflect how mass appears in the application. The mass of the "Buff" (i.e., the B-52) is $13,986\ lbf \cdot s^2/ft$ or 13,986 slugs, while that for the F-5 is $373\ lbf \cdot s^2/ft$. Thus, the total energy for the B-52 is

$$E = 0.5 \left(13{,}986 \frac{lbf \cdot s^2}{ft} \right) \left(422.5 \frac{ft}{s} \right)^2 + (450{,}000\ lbf)\,(20{,}000\ ft)$$

$$E = 1.0248 \times 10^{10}\ ft \cdot lbf$$

Similarly, the total energy of the F-5 fighter is

$$E = 0.5 \left(373 \frac{lbf \cdot s^2}{ft} \right) \left(422.5 \frac{ft}{s} \right)^2 + (12{,}000\ lbf)\,(20{,}000\ ft)$$

$$E = 2.7329 \times 10^{8}\ ft \cdot lbf$$

The total energy of the B-52 is 37.5 times the total energy of the F-5. Even though the total energy of the B-52 is so very much greater than that for the F-5, it just doesn't seem likely that a B-52 would have a significant advantage in air-to-air combat with an F-5. Note that the two aircraft are cruising at the same flight condition (velocity/altitude combination). Thus, the difference in total energy is in direct proportion to the difference in the weights of the two aircraft. Perhaps the specific energy (i.e., the energy per unit weight) is a more realistic parameter when trying to predict which aircraft would have an edge in air-to-air combat.

EXAMPLE 1.2:   The energy height

Since the weight specific energy also has units of height, it will be given the symbol H_e and is called the energy height. Dividing the terms in equation (1.1) by the weight of the aircraft ($W = m\,g$).

$$H_e = \frac{E}{W} = \frac{V^2}{2\,g} + h \qquad (1.3)$$

Compare the energy height of a B-52 flying at 250 knots at an altitude of 20,000 feet with that of an F-5 cruising at the same altitude and at the same velocity.

Solution: The energy height of the B-52 is

$$H_e = 0.5 \frac{\left(422.5 \frac{\text{ft}}{\text{s}}\right)^2}{32.174 \frac{\text{ft}}{\text{s}^2}} + 20000 \text{ ft}$$

$$H_e = 22774 \text{ ft}$$

Since the F-5 is cruising at the same altitude and at the same true air speed as the B-52, it has the same energy height (i.e., the same weight specific energy). If we consider only this weight specific energy, the B-52 and the F-5 are equivalent. This is obviously an improvement over the factor of 37.5 that the "Buff" had over the F-5, when the comparison was made based on the total energy. However, the fact that the energy height is the same for these two aircraft indicates that further effort is needed to provide a more realistic comparison for air-to-air combat.

Thus, there must be some additional parameters that are relevant when comparing the one-on-one capabilities of two aircraft in air-to-air combat. Captain Oswald Boelcke developed a series of rules based on his combat experience as a forty-victory ace by October 19, 1916. Boelcke specified seven rules, or "dicta" [Werner (2005)]. The first five, which deal with tactics, are

1. Always try to secure an advantageous position before attacking. Climb before and during the approach in order to surprise the enemy from above, and dive on him swiftly from the rear when the moment to attack is at hand.

2. Try to place yourself between the sun and the enemy. This puts the glare of the sun in the enemy's eyes and makes it difficult to see you and impossible to shoot with any accuracy.

3. Do not fire the machine guns until the enemy is within range and you have him squarely within your sights.

4. Attack when the enemy least expects it or when he is preoccupied with other duties, such as observation, photography, or bombing.

5. Never turn your back and try to run away from an enemy fighter. If you are surprised by an attack on your tail, turn and face the enemy with your guns.

Although Boelcke's dicta were to guide fighter pilots for decades to come, they were experienced-based empirical rules. The first dictum deals with your total energy, the sum of the potential energy plus the kinetic energy. We learned from the first two example calculations that predicting the probable victor in one-on-one air-to-air combat is not based on energy alone.

Note that the fifth dictum deals with maneuverability. ***Energy AND Maneuverability!*** The governing equations should include maneuverability as well as the specific energy.

It wasn't until almost half a century later that a Captain in the U.S. Air Force brought the needed complement of talents to bear on the problem [Coram (2002)]. Captain John R. Boyd was an aggressive and talented fighter pilot who had an insatiable intellectual curiosity for understanding the scientific equations that had to be the basis of the "Boelcke dicta". John R. Boyd was driven to understand the physics that was the foundation of the tactics that, until that time, had been learned by experience for the fighter pilot lucky enough to survive his early air-to-air encounters with an enemy. In his role as Director of Academics at the U.S. Air Force Fighter Weapons School, it became not only his passion, but his job.

Air combat is a dynamic ballet of move and countermove that occurs over a continuum of time. Thus, Boyd postulated that perhaps the time derivatives of the energy height are more relevant than the energy height itself. How fast can we, in the target aircraft, with an enemy on our six, quickly dump energy and allow the foe to pass? Once the enemy has passed, how quickly can we increase our energy height and take the offensive? John R. Boyd taught these tactics in the Fighter Weapons School. Now he became obsessed with the challenge of developing the science of fighter tactics.

1.1.1 Specific Excess Power

If the pilot of the 12,000 lbf F-5 that is flying at a velocity of 250 knots (422.5 ft/s) and at an altitude of 20,000 feet is to gain the upper hand in air-to-air combat, his aircraft must have sufficient power either to out accelerate or to out climb his adversary. Consider the case where the F-5 is flying at a constant altitude. If the engine is capable of generating more thrust than the drag acting on the aircraft, the acceleration of the aircraft can be calculated using Newton's Law:

$$\Sigma F = m\,a$$

which for an aircraft accelerating at a constant altitude becomes

$$T - D = \frac{W}{g}\frac{dV}{dt} \tag{1.4}$$

Multiplying both sides of the Equation (1.4) by V and dividing by W gives

$$\frac{(T-D)\,V}{W} = \frac{V}{g}\frac{dV}{dt} \tag{1.5}$$

EXAMPLE 1.3: The specific excess power and acceleration

The left-hand side of equation (1.5) is excess power per unit weight, or specific excess power, P_s. Use equation (1.5) to calculate the maximum acceleration for a 12,000-lbf F-5 that is flying at 250 knots (422.5 ft/s) at 20,000 feet.

Solution: Performance charts for an F-5 that is flying at these conditions indicate that it is capable of generating 3550 lbf thrust (T) with the afterburner lit, while the total drag (D) acting on the aircraft is 1750 lbf. Thus, the specific excess power (P_s) is

$$P_s = \frac{(T - D)V}{W} = \frac{[(3550 - 1750)\,\text{lbf}]\,422.5\,\text{ft/s}}{12000\,\text{lbf}} = 63.38\,\text{ft/s}$$

Rearranging Equation (1.5) to solve for the acceleration gives

$$\frac{dV}{dt} = P_s\frac{g}{V} = (63.38\,\text{ft/s})\frac{32.174\,\text{ft/s}^2}{422.5\,\text{ft/s}} = 4.83\,\text{ft/s}^2$$

1.1.2 Using Specific Excess Power to Change the Energy Height

Taking the derivative with respect to time of the two terms in equation (1.3), one obtains

$$\frac{dH_e}{dt} = \frac{V}{g}\frac{dV}{dt} + \frac{dh}{dt} \tag{1.6}$$

The first term on the right-hand side of equation (1.6) represents the rate of change of kinetic energy (per unit weight). It is a function of the rate of change of the velocity as seen by the pilot $\left(\frac{dV}{dt}\right)$. The significance of the second term is even less cosmic. It is the rate of change of the potential energy (per unit weight). Note also that $\left(\frac{dh}{dt}\right)$ is the vertical component of the velocity [i.e., the rate of climb (ROC)] as seen by the pilot on his altimeter. Air speed and altitude—these are parameters that fighter pilots can take to heart.

Combining the logic that led us to equations (1.5) and (1.6) leads us to the conclusion that the specific excess power is equal to the time-rate-of-change of the energy height. Thus,

$$P_s = \frac{(T - D)V}{W} = \frac{dH_e}{dt} = \frac{V}{g}\frac{dV}{dt} + \frac{dh}{dt} \tag{1.7}$$

Given the specific excess power calculated in Example 1.3, one could use equation (1.7) to calculate the maximum rate-of-climb (for a constant velocity) for the 12,000-lbf F-5 as it passes through 20,000 feet at 250 knots.

$$\frac{dh}{dt} = P_s = 63.38\,\text{ft/s} = 3802.8\,\text{ft/min}$$

Clearly, to be able to generate positive values for the terms in equation (1.7), we need an aircraft with excess power (i.e., one for which the thrust exceeds the drag). Weight is another important factor, since the lighter the aircraft, the greater the benefits of the available excess power.

"Boyd, as a combat pilot in Korea and as a tactics instructor at Nellis AFB in the Nevada desert, observed, analyzed, and assimilated the relative energy states of his aircraft and those of his opponent's during air combat engagements ... He also noted that, when in a position of advantage, his energy was higher than that of his opponent and that he lost that advantage when he allowed his energy to decay to less than that of his opponent."

"He knew that, when turning from a steady–state flight condition, the airplane under a given power setting would either slow down, lose altitude, or both. The result meant he was losing energy (the drag exceeded the thrust available from the engine). From these observations, he concluded that maneuvering for position was basically an energy problem. Winning required the proper management of energy available at the conditions existing at any point during a combat engagement." [Hillaker (1997)]

In the mid 1960s, Boyd had gathered energy-maneuverability data on all of the fighter aircraft in the U.S. Air Force inventory and on their adversaries. He sought to understand the intricacies of maneuvering flight. What was it about the airplane that would limit or prevent him from making it do what he wanted it to do?

1.1.3 John R. Boyd Meet Harry Hillaker

The relation between John R. Boyd and Harry Hillaker "dated from an evening in the mid-1960s when a General Dynamics engineer named Harry Hillaker was sitting in the Officer's Club at Eglin AFB, Florida, having an after dinner drink. Hillaker's host introduced him to a tall, blustery pilot named John R. Boyd, who immediately launched a frontal attack on GD's F-111 fighter. Hillaker was annoyed but bantered back." [Grier (2004)] Hillaker countered that the F-111 was designated a fighter-bomber.

"A few days later, he (Hillaker) received a call—Boyd had been impressed by Hillaker's grasp of aircraft conceptual design and wanted to know if Hillaker was interested in more organized meetings."

"Thus was born a group that others in the Air Force dubbed the 'fighter mafia.' Their basic belief was that fighters did not need to overwhelm opponents with speed and size. Experience in Vietnam against nimble Soviet-built MiGs had convinced them that technology had not yet turned air-to-air combat into a long-range shoot-out." [Grier (2004)]

The fighter mafia knew that a small aircraft could enjoy a high thrust-to-weight ratio. Small aircraft have less drag. "The original F-16 design had about one-third the drag of an F-4 in level flight and one-fifteenth the drag of an F-4 at a high angle-of-attack."

1.2 SOLVING FOR THE AEROTHERMODYNAMIC PARAMETERS

A fundamental problem facing the aerodynamicist is to predict the aerodynamic forces and moments and the heat-transfer rates acting on a vehicle in flight. In order to predict these aerodynamic forces and moments with suitable accuracy, it is necessary to be able to describe the pattern of flow around the vehicle. The resultant flow pattern depends on the geometry of the vehicle, its orientation with respect to the undisturbed free stream, and the altitude and speed at which the vehicle is traveling. In analyzing the various flows that an aerodynamicist may encounter, assumptions about the fluid properties may be introduced. In some applications, the temperature variations are so small that they do not affect the velocity field. In addition, for those applications where the temperature variations have a negligible effect on the flow field, it is often assumed that the density is essentially constant. However, in analyzing high-speed flows, the density variations cannot be neglected. Since density is a function of pressure and temperature, it may be expressed in terms of these two parameters. In fact, for a gas in thermodynamic equilibrium, any thermodynamic property may be expressed as a function of two other independent, thermodynamic properties. Thus, it is possible to formulate the governing equations using the enthalpy and the entropy as the flow properties instead of the pressure and the temperature.

1.2.1 Concept of a Fluid

From the point of view of fluid mechanics, matter can be in one of two states, either solid or fluid. The technical distinction between these two states lies in their response to an applied shear, or tangential, stress. A solid can resist a shear stress by a static deformation; a fluid cannot. A *fluid* is a substance that deforms continuously under the action of shearing forces. An important corollary of this definition is that there can be no shear stresses acting on fluid particles if there is no relative motion within the fluid; that is, such fluid particles are not deformed. Thus, if the fluid particles are at rest or if they are all moving at the same velocity, there are no shear stresses in the fluid. This zero shear stress condition is known as the *hydrostatic stress condition*.

   A fluid can be either a liquid or a gas. A liquid is composed of relatively closely packed molecules with strong cohesive forces. As a result, a given mass of liquid will occupy a definite volume of space. If a liquid is poured into a container, it assumes the shape of the container up to the volume it occupies and will form a free surface in a gravitational field if unconfined from above. The upper (or free) surface is planar and perpendicular to the direction of gravity. Gas molecules are widely spaced with relatively small cohesive forces. Therefore, if a gas is placed in a closed container, it will expand until it fills the entire volume of the container. A gas has no definite volume. Thus, if it is unconfined, it forms an atmosphere that is essentially hydrostatic.

1.2.2 Fluid as a Continuum

When developing equations to describe the motion of a system of fluid particles, one can either define the motion of each and every molecule or one can define the average behavior of the molecules within a given elemental volume. The size of the elemental volume is important, but only in relation to the number of fluid particles contained in the volume and to the physical dimensions of the flow field. Thus, the elemental volume should be large compared with the volume occupied by a single molecule so that it contains a large number of molecules at any instant of time. Furthermore, the number of molecules within the volume will remain essentially constant even though there is a continuous flux of molecules through the boundaries. If the elemental volume is too large, there could be a noticeable variation in the fluid properties determined statistically at various points in the volume.

In problems of interest to this text, our primary concern is not with the motion of individual molecules, but with the general behavior of the fluid. Thus, we are concerned with describing the fluid motion in spaces that are very large compared to molecular dimensions and that, therefore, contain a large number of molecules. The fluid in these problems may be considered to be a continuous material whose properties can be determined from a statistical average for the particles in the volume, that is, a macroscopic representation. The assumption of a continuous fluid is valid when the smallest volume of fluid that is of interest contains so many molecules that statistical averages are meaningful.

The number of molecules in a cubic meter of air at room temperature and at sea-level pressure is approximately 2.5×10^{25}. Thus, there are 2.5×10^{10} molecules in a cube 0.01 mm on a side. The mean free path at sea level is 6.6×10^{-8} m. There are sufficient molecules in this volume for the fluid to be considered a continuum, and the fluid properties can be determined from statistical averages. However, at an altitude of 130 km, there are only 1.6×10^{17} molecules in a cube 1 m on a side. The mean free path at this altitude is 10.2 m. Thus, at this altitude the fluid cannot be considered a continuum.

A parameter that is commonly used to identify the onset of low-density effects is the Knudsen number, which is the ratio of the mean free path to a characteristic dimension of the body. Although there is no definitive criterion, the continuum flow model starts to break down when the Knudsen number is roughly of the order of 0.1.

1.2.3 Fluid Properties

By employing the concept of a continuum, we can describe the gross behavior of the fluid motion using certain observable, macroscopic properties. Properties used to describe a general fluid motion include the temperature, the pressure, the density, the viscosity, and the speed of sound.

Temperature. We are all familiar with *temperature* in qualitative terms; that is, an object feels hot (or cold) to the touch. However, because of the difficulty in quantitatively defining the temperature, we define the equality of temperature. Two bodies have equality of temperature when no change in any observable property occurs when they are in thermal contact. Further, two bodies respectively equal in temperature to a third body must be equal in temperature to each other. It follows that an arbitrary scale of temperature can be defined in terms of a convenient property of a standard body.

Pressure. Because of the random motion due to their thermal energy, the individual molecules of a fluid would continually strike a surface that is placed in the fluid. These collisions occur even though the surface is at rest relative to the fluid. By Newton's second law, a force is exerted on the surface equal to the time rate of change of the momentum of the rebounding molecules. *Pressure* is the magnitude of this force per unit area of surface. Since a fluid that is at rest cannot sustain tangential forces, the pressure

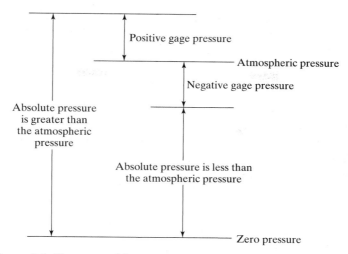

Figure 1.1 Terms used in pressure measurements.

on the surface must act in the direction perpendicular to that surface. Furthermore, the pressure acting at a point in a fluid at rest is the same in all directions.

Standard atmospheric pressure at sea level is defined as the pressure that can support a column of mercury 760 mm in length when the density of the mercury is 13.5951 g/cm^3 and the acceleration due to gravity is the standard value. The standard atmospheric pressure at sea level is $1.01325 \times 10^5 \, N/m^2$. In English units, the standard atmospheric pressure at sea level is 14.696 lbf/in^2 or 2116.22 lb/ft^2.

In many aerodynamic applications, we are interested in the difference between the absolute value of the local pressure and the atmospheric pressure. Many pressure gages indicate the difference between the absolute pressure and the atmospheric pressure existing at the gage. This difference, which is referred to as *gage pressure*, is illustrated in Fig. 1.1.

Density. The *density* of a fluid at a point in space is the mass of the fluid per unit volume surrounding the point. As is the case when evaluating the other fluid properties, the incremental volume must be large compared to molecular dimensions yet very small relative to the dimensions of the vehicle whose flow field we seek to analyze. Thus, provided that the fluid may be assumed to be a continuum, the density at a point is defined as

$$\rho = \lim_{\delta(\text{vol}) \to 0} \frac{\delta(\text{mass})}{\delta(\text{vol})} \tag{1.8}$$

The dimensions of density are $(\text{mass})/(\text{length})^3$.

In general, the density of a gas is a function of the composition of the gas, its temperature, and its pressure. The relation

$$\rho(\text{composition}, T, p) \tag{1.9}$$

is known as an *equation of state*. For a thermally perfect gas, the equation of state is

$$\rho = \frac{p}{RT} \tag{1.10}$$

The gas constant R has a particular value for each substance. The gas constant for air has the value 287.05 N · m/kg · K in SI units and 53.34 ft · lbf/lbm · °R or 1716.16 ft^2/s^R · °R in English units. The temperature in equation (1.10) should be in absolute units. Thus, the temperature is either in K or in °R, but never in °C or in °F.

EXAMPLE 1.4:   Density in SI units

Calculate the density of air when the pressure is 1.01325×10^5 N/m^2 and the temperature is 288.15 K. Since air at this pressure and temperature behaves as a perfect gas, we can use equation (1.10).

Solution:

$$\rho = \frac{1.01325 \times 10^5 \text{ N/m}^2}{(287.05 \text{ N} \cdot \text{m/kg} \cdot \text{K})(288.15 \text{ K})}$$

$$= 1.2250 \text{ kg/m}^3$$

EXAMPLE 1.5:   Density in English units

Calculate the density of air when the pressure is 2116.22 lbf/ft^2 and the temperature is 518.67°R. Since air at this pressure and temperature behaves as a perfect gas, we can use equation (1.10). Note that throughout the remainder of this book, air will be assumed to behave as a perfect gas unless specifically stated otherwise.

Solution:

$$\rho = \frac{2116.22 \dfrac{\text{lbf}}{\text{ft}^2}}{\left(53.34 \dfrac{\text{ft} \cdot \text{lbf}}{\text{lbm} \cdot °\text{R}}\right)(518.67°\text{R})} = 0.07649 \frac{\text{lbm}}{\text{ft}^3}$$

Alternatively,

$$\rho = \frac{2116.22 \dfrac{\text{lbf}}{\text{ft}^2}}{\left(1716.16 \dfrac{\text{ft}^2}{\text{s}^2 \cdot °\text{R}}\right)(518.67°\text{R})} = 0.002377 \frac{\text{lbf} \cdot \text{s}^2}{\text{ft}^4}$$

The unit lbf · s^2/ft^4 is often written as slugs/ft^3, where slugs are alternative units of mass in the English system. One slug is the equivalent of 32.174 lbm.

For vehicles that are flying at approximately 100 m/s (330 ft/s), or less, the density of the air flowing past the vehicle is assumed constant when obtaining a solution for the flow field. Rigorous application of equation (1.10) would require that the pressure and the temperature remain constant (or change proportionally) in order for the density to remain constant throughout the flow field. We know that the pressure around the vehicle is not constant, since the aerodynamic forces and moments in which we are interested are the result of pressure variations associated with the flow pattern. However, the assumption of constant density for velocities below 100 m/s is a valid approximation because the pressure changes that occur from one point to another in the flow field are small relative to the absolute value of the pressure.

Viscosity. In all real fluids, a shearing deformation is accompanied by a shearing stress. The fluids of interest in this text are *Newtonian* in nature; that is, the shearing stress is proportional to the rate of shearing deformation. The constant of proportionality is called the *coefficient of viscosity*, μ. Thus,

$$\text{shear stress} = \mu \times \text{transverse gradient of velocity} \quad \textbf{(1.11)}$$

There are many problems of interest to us in which the effects of viscosity can be neglected. In such problems, the magnitude of the coefficient of viscosity of the fluid and of the velocity gradients in the flow field are such that their product is negligible relative to the inertia of the fluid particles and to the pressure forces acting on them. We shall use the term *inviscid flow* in these cases to emphasize the fact that it is the character both of the flow field and of the fluid which allows us to neglect viscous effects. No real fluid has a zero coefficient of viscosity.

The viscosity of a fluid relates to the transport of momentum in the direction of the velocity gradient (but opposite in sense). Therefore, viscosity is a transport property. In general, the coefficient of viscosity is a function of the composition of the gas, its temperature, and its pressure. For temperatures below 3000 K, the viscosity of air is independent of pressure. In this temperature range, we could use Sutherland's equation to calculate the coefficient of viscosity:

$$\mu = 1.458 \times 10^{-6} \frac{T^{1.5}}{T + 110.4} \quad \textbf{(1.12a)}$$

Here T is the temperature in K and the units for μ are kg/s · m.

EXAMPLE 1.6: Viscosity in SI units

Calculate the viscosity of air when the temperature is 288.15 K.

Solution:

$$\mu = 1.458 \times 10^{-6} \frac{(288.15)^{1.5}}{288.15 + 110.4}$$

$$= 1.7894 \times 10^{-5} \text{ kg/s · m}$$

For temperatures below 5400°R, the viscosity of air is independent of pressure. In this temperature range, Sutherland's equation for the viscosity of air in English units is

$$\mu = 2.27 \times 10^{-8} \frac{T^{1.5}}{T + 198.6} \qquad \textbf{(1.12b)}$$

where T is the temperature in °R and the units for μ are lbf · s/ft^2.

EXAMPLE 1.7:   Viscosity in English units

Calculate the viscosity of air when the temperature is 59.0°F.

Solution: First, convert the temperature to the absolute scale for English units, °R, 59.0°F + 459.67 = 518.67°R.

$$\mu = 2.27 \times 10^{-8} \frac{(518.67)^{1.5}}{518.67 + 198.6}$$

$$= 3.7383 \times 10^{-7} \frac{\text{lbf} \cdot \text{s}}{\text{ft}^2}$$

Equations used to calculate the coefficient of viscosity depend on the model used to describe the intermolecular forces of the gas molecules, so that it is necessary to define the potential energy of the interaction of the colliding molecules. Svehla (1962) noted that the potential for the Sutherland model is described physically as a rigid, impenetrable sphere, surrounded by an inverse-power attractive force. This model is qualitatively correct in that the molecules attract one another when they are far apart and exert strong repulsive forces upon one another when they are close together.

Chapman and Cowling (1960) note that equations (1.12a) and (1.12b) closely represent the variation of μ with temperature over a "fairly" wide range of temperatures. They caution, however, that the success of Sutherland's equation in representing the variation of μ with temperature for several gases does not establish the validity of Sutherland's molecular model for those gases. "In general it is not adequate to represent the core of a molecule as a rigid sphere, or to take molecular attractions into account to a first order only. The greater rapidity of the experimental increase of μ with T, as compared with that for non-attracting rigid spheres, has to be explained as due partly to the 'softness' of the repulsive field at small distances, and partly to attractive forces which have more than a first-order effect. The chief value of Sutherland's formula seems to be as a simple interpolation formula over restricted ranges of temperature."

The Lennard-Jones model for the potential energy of an interaction, which takes into account both the softness of the molecules and their mutual attraction at large distances, has been used by Svehla (1962) to calculate the viscosity and the thermal conductivity of gases at high temperatures. The coefficients of viscosity for air as tabulated by Svehla are compared with the values calculated using equation (1.12a) in Table 1.1. These comments are made to emphasize the fact that even the basic fluid properties may involve approximate models that have a limited range of applicability.

TABLE 1.1 Comparison of the Coefficient of Viscosity for Air as Tabulated by Svehla (1962) and as Calculated Using Sutherland's Equation [Equation (1.12a)]

T (K)	$\mu \times 10^5$ (kg/m·s)*	$\mu \times 10^5$ (kg/m·s)†
200	1.360	1.329
400	2.272	2.285
600	2.992	3.016
800	3.614	3.624
1000	4.171	4.152
1200	4.695	4.625
1400	5.197	5.057
1600	5.670	5.456
1800	6.121	5.828
2000	6.553	6.179
2200	6.970	6.512
2400	7.373	6.829
2600	7.765	7.132
2800	8.145	7.422
3000	8.516	7.702
3200	8.878	7.973
3400	9.232	8.234
3600	9.579	8.488
3800	9.918	8.734
4000	10.252	8.974
4200	10.580	9.207
4400	10.902	9.435
4600	11.219	9.657
4800	11.531	9.874
5000	11.838	10.087

*From Svehla (1962)
†Calculated using equation (1.12a)

Kinematic Viscosity. The aerodynamicist may encounter many applications where the ratio μ/ρ has been replaced by a single parameter. Because this ratio appears frequently, it has been given a special name, the kinematic viscosity. The symbol used to represent the kinematic viscosity is ν:

$$\nu = \frac{\mu}{\rho}$$

(1.13)

In this ratio, the force units (or, equivalently, the mass units) cancel. Thus, ν has the dimensions of L^2/T (e.g., square meters per second or square feet per second).

EXAMPLE 1.8: Kinematic Viscosity in English units

Using the results of Examples 1.5 and 1.7, calculate the kinematic viscosity of air when the temperature is 518.67°R and the pressure is 2116.22 lbf/ft².

Solution: From Example 1.5, $\rho = 0.07649 \text{ lbm/ft}^3 = 0.002377 \text{ lbf} \cdot \text{s}^2/\text{ft}^4$; while, from Example 1.7, $\mu = 3.7383 \times 10^{-7} \text{ lbf} \cdot \text{s/ft}^2$. Thus,

$$\nu = \frac{\mu}{\rho} = \frac{3.7383 \times 10^{-7} \dfrac{\text{lbf} \cdot \text{s}}{\text{ft}^2}}{0.002377 \dfrac{\text{lbf} \cdot \text{s}^2}{\text{ft}^4}} = 1.573 \times 10^{-4} \frac{\text{ft}^2}{\text{s}}$$

If we use the alternative units for the density, we must employ the factor g_c, which is equal to 32.174 ft $\cdot$ lbm/lbf $\cdot$ s², to arrive at the appropriate units.

$$\nu = \frac{\mu}{\rho} = \frac{3.7383 \times 10^{-7} \dfrac{\text{lbf} \cdot \text{s}}{\text{ft}^2}}{0.07649 \dfrac{\text{lbm}}{\text{ft}^3}} \left(32.174 \frac{\text{ft} \cdot \text{lbm}}{\text{lbf} \cdot \text{s}^2} \right)$$

$$= 1.573 \times 10^{-4} \text{ ft}^2/\text{s}$$

Speed of Sound. The speed at which a disturbance of infinitesimal proportions propagates through a fluid that is at rest is known as the *speed of sound*, which is designated in this book as a. The speed of sound is established by the properties of the fluid. For a perfect gas $a = \sqrt{\gamma R T}$, where γ is the ratio of specific heats (see Chapter 8) and R is the gas constant. For the range of temperature over which air behaves as a perfect gas, $\gamma = 1.4$ and the speed of sound is given by

$$a = 20.047\sqrt{T} \tag{1.14a}$$

where T is the temperature in K and the units for the speed of sound are m/s. In English units

$$a = 49.02\sqrt{T} \tag{1.14b}$$

where T is the temperature in °R and the units for the speed of sound are ft/s.

1.2.4 Pressure Variation in a Static Fluid Medium

In order to compute the forces and moments or the heat-transfer rates acting on a vehicle or to determine the flight path (i.e., the trajectory) of the vehicle, the engineer will often develop an analytic model of the atmosphere instead of using a table, such as Table 1.2.

TABLE 1.2A U.S. Standard Atmosphere, 1976 SI Units

Geometric Altitude (km)	Pressure (p/p_{SL})	Temperature (K)	Density (ρ/ρ_{SL})	Viscosity (μ/μ_{SL})	Speed of Sound (m/s)
0	1.0000 E+00	288.150	1.0000 E+00	1.00000	340.29
1	8.8700 E−01	281.651	9.0748 E−01	0.98237	336.43
2	7.8461 E−01	275.154	8.2168 E−01	0.96456	332.53
3	6.9204 E−01	268.659	7.4225 E−01	0.94656	328.58
4	6.0854 E−01	262.166	6.6885 E−01	0.92836	324.59
5	5.3341 E−01	255.676	6.0117 E−01	0.90995	320.55
6	4.6600 E−01	249.187	5.3887 E−01	0.89133	316.45
7	4.0567 E−01	242.700	4.8165 E−01	0.87249	312.31
8	3.5185 E−01	236.215	4.2921 E−01	0.85343	308.11
9	3.0397 E−01	229.733	3.8128 E−01	0.83414	303.85
10	2.6153 E−01	223.252	3.3756 E−01	0.81461	299.53
11	2.2403 E−01	216.774	2.9780 E−01	0.79485	295.15
12	1.9145 E−01	216.650	2.5464 E−01	0.79447	295.07
13	1.6362 E−01	216.650	2.1763 E−01	0.79447	295.07
14	1.3985 E−01	216.650	1.8601 E−01	0.79447	295.07
15	1.1953 E−01	216.650	1.5898 E−01	0.79447	295.07
16	1.0217 E−01	216.650	1.3589 E−01	0.79447	295.07
17	8.7340 E−02	216.650	1.1616 E−01	0.79447	295.07
18	7.4663 E−02	216.650	9.9304 E−02	0.79447	295.07
19	6.3829 E−02	216.650	8.4894 E−02	0.79447	295.07
20	5.4570 E−02	216.650	7.2580 E−02	0.79447	295.07
21	4.6671 E−02	217.581	6.1808 E−02	0.79732	295.70
22	3.9945 E−02	218.574	5.2661 E−02	0.80037	296.38
23	3.4215 E−02	219.567	4.4903 E−02	0.80340	297.05
24	2.9328 E−02	220.560	3.8317 E−02	0.80643	297.72
25	2.5158 E−02	221.552	3.2722 E−02	0.80945	298.39
26	2.1597 E−02	222.544	2.7965 E−02	0.81247	299.06
27	1.8553 E−02	223.536	2.3917 E−02	0.81547	299.72
28	1.5950 E−02	224.527	2.0470 E−02	0.81847	300.39
29	1.3722 E−02	225.518	1.7533 E−02	0.82147	301.05
30	1.1813 E−02	226.509	1.5029 E−02	0.82446	301.71

Reference values: $p_{SL} = 1.01325 \times 10^5 \, N/m^2$; $T_{SL} = 288.150 \, K$

$\rho_{SL} = 1.2250 \, kg/m^3$; $\mu_{SL} = 1.7894 \times 10^{-5} \, kg/s \cdot m$

(continues on next page)

TABLE 1.2B   U.S. Standard Atmosphere, 1976 English Units

Geometric Altitude (kft)	Pressure (p/p_{SL})	Temperature (°R)	Density (ρ/ρ_{SL})	Viscosity (μ/μ_{SL})	Speed of Sound (ft/s)
0	1.0000 E+00	518.67	1.0000 E+00	1.0000 E+00	1116.44
2	9.2981 E−01	511.54	9.4278 E−01	9.8928 E−01	1108.76
4	8.6368 E−01	504.41	8.8811 E−01	9.7849 E−01	1100.98
6	8.0142 E−01	497.28	8.3590 E−01	9.6763 E−01	1093.18
8	7.4286 E−01	490.15	7.8609 E−01	9.5670 E−01	1085.33
10	6.8783 E−01	483.02	7.3859 E−01	9.4569 E−01	1077.40
12	6.3615 E−01	475.90	6.9333 E−01	9.3461 E−01	1069.42
14	5.8767 E−01	468.78	6.5022 E−01	9.2346 E−01	1061.38
16	5.4224 E−01	461.66	6.0921 E−01	9.1223 E−01	1053.31
18	4.9970 E−01	454.53	5.7021 E−01	9.0092 E−01	1045.14
20	4.5991 E−01	447.42	5.3316 E−01	8.8953 E−01	1036.94
22	4.2273 E−01	440.30	4.9798 E−01	8.7806 E−01	1028.64
24	3.8803 E−01	433.18	4.6462 E−01	8.6650 E−01	1020.31
26	3.5568 E−01	426.07	4.3300 E−01	8.5487 E−01	1011.88
28	3.2556 E−01	418.95	4.0305 E−01	8.4315 E−01	1003.41
30	2.9754 E−01	411.84	3.7473 E−01	8.3134 E−01	994.85
32	2.7151 E−01	404.73	3.4795 E−01	8.1945 E−01	986.22
34	2.4736 E−01	397.62	3.2267 E−01	8.0746 E−01	977.53
36	2.2498 E−01	390.51	2.9883 E−01	7.9539 E−01	968.73
38	2.0443 E−01	389.97	2.7191 E−01	7.9447 E−01	968.08
40	1.8576 E−01	389.97	2.4708 E−01	7.9447 E−01	968.08
42	1.6880 E−01	389.97	2.2452 E−01	7.9447 E−01	968.08
44	1.5339 E−01	389.97	2.0402 E−01	7.9447 E−01	968.08
46	1.3939 E−01	389.97	1.8540 E−01	7.9447 E−01	968.08
48	1.2667 E−01	389.97	1.6848 E−01	7.9447 E−01	968.08
50	1.1511 E−01	389.97	1.5311 E−01	7.9447 E−01	968.08
52	1.0461 E−01	389.97	1.3914 E−01	7.9447 E−01	968.08
54	9.5072 E−02	389.97	1.2645 E−01	7.9447 E−01	968.08
56	8.6402 E−02	389.97	1.1492 E−01	7.9447 E−01	968.08
58	7.8524 E−02	389.97	1.0444 E−01	7.9447 E−01	968.08
60	7.1366 E−02	389.97	9.4919 E−02	7.9447 E−01	968.08
62	6.4861 E−02	389.97	8.6268 E−02	7.9447 E−01	968.08
64	5.8951 E−02	389.97	7.8407 E−02	7.9447 E−01	968.08
66	5.3580 E−02	390.07	7.1246 E−02	7.9463 E−01	968.21
68	4.8707 E−02	391.16	6.4585 E−02	7.9649 E−01	969.55
70	4.4289 E−02	392.25	5.8565 E−02	7.9835 E−01	970.90

(continues on next page)

TABLE 1.2B U.S. Standard Atmosphere, 1976 English Units (continued)

Geometric Altitude (kft)	Pressure (p/p_{SL})	Temperature (°R)	Density (ρ/ρ_{SL})	Viscosity (μ/μ_{SL})	Speed of Sound (ft/s)
72	4.0284 E−02	393.34	5.3121 E−02	8.0020 E−01	972.24
74	3.6651 E−02	394.43	4.8197 E−02	8.0205 E−01	973.59
76	3.3355 E−02	395.52	4.3742 E−02	8.0390 E−01	974.93
78	3.0364 E−02	396.60	3.9710 E−02	8.0575 E−01	976.28
80	2.7649 E−02	397.69	3.6060 E−02	8.0759 E−01	977.62
82	2.5183 E−02	398.78	3.2755 E−02	8.0943 E−01	978.94
84	2.2943 E−02	399.87	2.9761 E−02	8.1127 E−01	980.28
86	2.0909 E−02	400.96	2.7048 E−02	8.1311 E−01	981.63
88	1.9060 E−02	402.05	2.4589 E−02	8.1494 E−01	982.94
90	1.7379 E−02	403.14	2.2360 E−02	8.1677 E−01	984.28
92	1.5850 E−02	404.22	2.0339 E−02	8.1860 E−01	985.60
94	1.4460 E−02	405.31	1.8505 E−02	8.2043 E−01	986.94
96	1.3195 E−02	406.40	1.6841 E−02	8.2225 E−01	988.25
98	1.2044 E−02	407.49	1.5331 E−02	8.2407 E−01	989.57
100	1.0997 E−02	408.57	1.3960 E−02	8.2589 E−01	990.91

Reference values: $p_{SL} = 2116.22$ lbf/ft^2
$T_{SL} = 518.67$°R
$\rho_{SL} = 0.002377$ slugs/ft^3
$\mu_{SL} = 1.2024 \times 10^{-5}$ lbm/ft·s
$\qquad = 3.740 \times 10^{-7}$ lbf·s/ft^2

To do this, let us develop the equations describing the pressure variation in a static fluid medium. If fluid particles, when viewed as a continuum, are either all at rest or all moving with the same velocity, the fluid is said to be a *static medium*. Thus, the term *static fluid properties* may be applied to situations in which the elements of the fluid are moving, provided that there is no relative motion between fluid elements. Since there is no relative motion between adjacent layers of the fluid, there are no shear forces. Thus, with no relative motion between fluid elements, the viscosity of the fluid is of no concern. For these inviscid flows, the only forces acting on the surface of the fluid element are pressure forces.

Consider the small fluid element whose center is defined by the coordinates x, y, z as shown in Fig. 1.2. A first-order Taylor's series expansion is used to evaluate the pressure at each face. Thus, the pressure at the back face of the element is $p - (\partial p/\partial x)(\Delta x/2)$; that at the front face is $p + (\partial p/\partial x)(\Delta x/2)$. If the fluid is not accelerating, the element must be in equilibrium. For equilibrium, the sum of the forces in any direction must be zero. Thus,

$$-\left(p + \frac{\partial p}{\partial x}\frac{\Delta x}{2}\right)\Delta y\,\Delta z + \left(p - \frac{\partial p}{\partial x}\frac{\Delta x}{2}\right)\Delta y\,\Delta z = 0 \qquad \textbf{(1.15a)}$$

$$-\left(p + \frac{\partial p}{\partial y}\frac{\Delta y}{2}\right)\Delta x\,\Delta z + \left(p - \frac{\partial p}{\partial y}\frac{\Delta y}{2}\right)\Delta x\,\Delta z = 0 \qquad \textbf{(1.15b)}$$

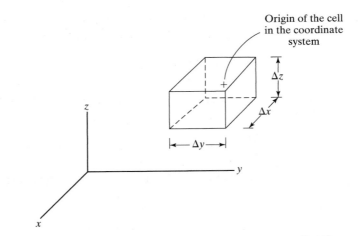

Figure 1.2 Derivation of equations (1.15) through (1.17).

$$-\left(p + \frac{\partial p}{\partial z}\frac{\Delta z}{2}\right)\Delta x\,\Delta y + \left(p - \frac{\partial p}{\partial z}\frac{\Delta z}{2}\right)\Delta x\,\Delta y - \rho g\,\Delta x\,\Delta y\,\Delta z = 0 \quad \textbf{(1.15c)}$$

Note that the coordinate system has been chosen such that gravity acts in the negative z direction. Combining terms and dividing by $\Delta x\,\Delta y\,\Delta z$ gives us

$$\frac{\partial p}{\partial x} = 0 \qquad\qquad\qquad\qquad \textbf{(1.16a)}$$

$$\frac{\partial p}{\partial y} = 0 \qquad\qquad\qquad\qquad \textbf{(1.16b)}$$

$$\frac{\partial p}{\partial z} = -\rho g \qquad\qquad\qquad\qquad \textbf{(1.16c)}$$

The three equations can be written as one using vector notation:

$$\nabla p = \rho\vec{f} = -\rho g\hat{k} \qquad\qquad\qquad \textbf{(1.17)}$$

where $\vec{f}$ represents the body force per unit mass and ∇ is the gradient operator. For the cases of interest in this book, the body force is gravity.

These equations illustrate two important principles for a nonaccelerating, hydrostatic, or shear-free, flow: (1) There is no pressure variation in the horizontal direction, that is, the pressure is constant in a plane perpendicular to the direction of gravity; and (2) the vertical pressure variation is proportional to gravity, density, and change in depth. Furthermore, as the element shrinks to zero volume (i.e., as $\Delta z \rightarrow 0$), it can be seen that the pressure is the same on all faces. That is, pressure at a point in a static fluid is independent of orientation.

Since the pressure varies only with z, that is, it is not a function of x or y, an ordinary derivative may be used and equation (1.16c) may be written

$$\frac{dp}{dz} = -\rho g \qquad\qquad\qquad\qquad \textbf{(1.18)}$$

Let us assume that the air behaves as a perfect gas. Thus, the expression for density given by equation (1.10) can be substituted into equation (1.18) to give

$$\frac{dp}{dz} = -\rho g = -\frac{pg}{RT} \tag{1.19}$$

In those regions where the temperature can be assumed to constant, separating the variables and integrating between two points yields

$$\int \frac{dp}{p} = \ln \frac{p_2}{p_1} = -\frac{g}{RT} \int dz = -\frac{g}{RT}(z_2 - z_1)$$

where the integration reflects the fact that the temperature has been assumed constant. Rearranging yields

$$p_2 = p_1 \exp \left[\frac{g(z_1 - z_2)}{RT} \right] \tag{1.20}$$

The pressure variation described by equation (1.20) is a reasonable approximation of that in the atmosphere near the earth's surface.

An improved correlation for pressure variation in the earth's atmosphere can be obtained if one accounts for the temperature variation with altitude. The earth's mean atmospheric temperature decreases almost linearly with z up to an altitude of nearly 11,000 m. That is,

$$T = T_0 - Bz \tag{1.21}$$

where T_0 is the sea-level temperature (absolute) and B is the lapse rate, both of which vary from day to day. The following standard values will be assumed to apply from 0 to 11,000 m:

$$T_0 = 288.15 \text{ K} \quad \text{and} \quad B = 0.0065 \text{ K/m}$$

Substituting equation (1.21) into the relation

$$\int \frac{dp}{p} = -\int \frac{g \, dz}{RT}$$

and integrating, we obtain

$$p = p_0 \left(1 - \frac{Bz}{T_0} \right)^{g/RB} \tag{1.22}$$

The exponent g/RB, which is dimensionless, is equal to 5.26 for air.

1.2.5 The Standard Atmosphere

In order to correlate flight-test data with wind-tunnel data acquired at different times at different conditions or to compute flow fields, it is important to have agreed-upon standards of atmospheric properties as a function of altitude. Since the earliest days of aeronautical research, "standard" atmospheres have been developed based on the knowledge of the atmosphere at the time. The one used in this text is the 1976 U.S. Standard Atmosphere. The atmospheric properties most commonly used in the analysis and design of flight vehicles, as taken from the U.S. Standard Atmosphere (1976), are reproduced in Table 1.2. These are the properties used in the examples in this text.

The basis for establishing a standard atmosphere is a defined variation of temperature with altitude. This atmospheric temperature profile is developed from measurements obtained from balloons, from sounding rockets, and from aircraft at a variety of locations at various times of the year and represents a mean expression of these measurements. A reasonable approximation is that the temperature varies linearly with altitude in some regions and is constant in other altitude regions. Given the temperature profile, the hydrostatic equation, equation (1.17), and the perfect-gas equation of state, equation (1.10), are used to derive the pressure and the density as functions of altitude. Viscosity and the speed of sound can be determined as functions of altitude from equations such as equation (1.12), Sutherland's equation, and equation (1.14), respectively. In reality, variations would exist from one location on earth to another and over the seasons at a given location. Nevertheless, a standard atmosphere is a valuable tool that provides engineers with a standard when conducting analyses and performance comparisons of different aircraft designs.

EXAMPLE 1.9:   Properties of the standard atmosphere at 10 km

Using equations (1.21) and (1.22), calculate the temperature and pressure of air at an altitude of 10 km. Compare the tabulated values with those presented in Table 1.2.

Solution: The ambient temperature at 10,000 m is

$$T = T_0 - Bz = 288.15 - 0.0065(10^4) = 223.15 \text{ K}$$

The tabulated value from Table 1.2 is 223.252 K. The calculated value for the ambient pressure is

$$p = p_0 \left(1 - \frac{Bz}{T_0} \right)^{g/RB}$$

$$= 1.01325 \times 10^5 \left[1 - \frac{0.0065(10^4)}{288.15} \right]^{5.26}$$

$$= 1.01325 \times 10^5 (0.26063) = 2.641 \times 10^4 \text{ N/m}^2$$

The comparable value in Table 1.2 is $2.650 \times 10^4 \text{ N/m}^2$.

EXAMPLE 1.10:   Properties of the standard atmosphere in English units

Develop equations for the pressure and for the density as a function of altitude from 0 to 65,000 ft. The analytical model of the atmosphere should make use of the hydrostatic equations [i.e., equation (1.18)], for which the density is eliminated through the use of the equation of state for a thermally perfect gas. Assume that the temperature of air from 0 to 36,100 ft is given by

$$T = 518.67 - 0.003565z$$

and that the temperature from 36,100 to 65,000 ft is constant at 389.97°R.

Solution: From 0 to 36,000 ft, the temperature varies linearly as described in general by equation (1.21). Specifically,

$$T = 518.67 - 0.003565z$$

Thus, $T_0 = 518.67°R$ and $B = 0.003565°R/ft$. Using English unit terms in equation (1.22) gives us

$$p = p_0 \left(1 - \frac{Bz}{T_0} \right)^{g/RB}$$

$$= 2116.22(1.0 - 6.873 \times 10^{-6}z)^{5.26}$$

For a thermally perfect gas, the density is

$$\rho = \frac{p}{RT} = \frac{2116.22(1.0 - 6.873 \times 10^{-6}z)^{5.26}}{53.34(518.67 - 0.003565z)}$$

Dividing by ρ_0, the value of the density at standard sea-level conditions,

$$\rho_0 = \frac{p_0}{RT_0} = \frac{2116.22}{(53.34)(518.67)}$$

one obtains the nondimensionalized density:

$$\frac{\rho}{\rho_0} = (1.0 - 6.873 \times 10^{-6}z)^{4.26}$$

Since the temperature is constant from 36,100 to 65,000 ft, equation (1.20) can be used to express the pressure, with the values at 36,100 ft serving as the reference values p_1 and z_1:

$$p_{36,100} = 2116.22(1.0 - 6.873 \times 10^{-6}z)^{5.26} = 472.19 \, \text{lbf/ft}^2$$

Thus,

$$p = 472.9 \exp \left[\frac{g(36,100 - z)}{RT} \right]$$

In English units,

$$\frac{g}{RT} = \frac{32.174 \dfrac{\text{ft}}{\text{s}^2}}{\left(53.34 \dfrac{\text{ft} \cdot \text{lbf}}{\text{lbm} \cdot °R} \right)(389.97°R)}$$

However, to have the correct units, multiply by $(1/g_c)$, so that

$$\frac{g}{RT} = \frac{32.174 \dfrac{\text{ft}}{\text{s}^2}}{\left(53.34 \dfrac{\text{ft} \cdot \text{lbf}}{\text{lbm} \cdot °R} \right)(389.97°R)\left(32.174 \dfrac{\text{ft} \cdot \text{lbm}}{\text{lbf} \cdot \text{s}^2} \right)} = 4.8075 \times 10^{-5}/\text{ft}$$

Thus,

$$\frac{p}{p_0} = 0.2231 \exp\left(1.7355 - 4.8075 \times 10^{-5} z\right)$$

The nondimensionalized density is

$$\frac{\rho}{\rho_0} = \frac{p}{p_0} \frac{T_0}{T}$$

Since $T = 389.97°R = 0.7519 T_0$,

$$\frac{\rho}{\rho_0} = 0.2967 \exp\left(1.7355 - 4.8075 \times 10^{-5} z\right)$$

1.3 SUMMARY

Specific values and equations for fluid properties (e.g., viscosity, density, and speed of sound) have been presented in this chapter. The reader should note that it may be necessary to use alternative relations for calculating fluid properties. For instance, for the relatively high temperatures associated with hypersonic flight, it may be necessary to account for real-gas effects (e.g., dissociation). Numerous references present the thermodynamic properties and transport properties of gases at high temperatures and pressures [e.g., Moeckel and Weston (1958), Hansen (1957), and Yos (1963)].

PROBLEMS

Problems 1.1 through 1.5 deal with the Energy-Maneuverability Technique for a T-38A that is powered by two J85-GE-5A engines. Presented in Fig. P1.1 are the thrust available and the thrust required for the T-38A that is cruising at 20,000 feet. The thrust available is presented as a function of Mach number for the engines operating at military power ("Mil") or operating with the afterburner ("Max"). More will be said about such curves in Chapter 5. With the aircraft cruising at a constant altitude (of 20,000 feet), the speed of sound is constant for Fig. P1.1 and the Mach number could be replaced by the velocity, i.e., by the true air speed.

When the vehicle is cruising at a constant altitude and at a constant attitude, the total drag is equal to the thrust required and the lift balances the weight. As will be discussed in Chapter 5, the total drag is the sum of the induced drag, the parasite drag, and the wave drag. Therefore, when the drag (or thrust required) curves are presented for aircraft weights of 8,000 lbf, 10,000 lbf, and 12,000 lbf, they reflect the fact that the induced drag depends on the lift. But the lift is equal to the weight. Thus, at the lower velocities, where the induced drag dominates, the drag is a function of the weight of the aircraft.

1.1. The maximum velocity at which an aircraft can cruise occurs when the thrust available with the engines operating with the afterburner lit ("Max") equals the thrust required, which are represented by the bucket shaped curves. What is the maximum cruise velocity that a 10,000-lbf T-38A can sustain at 20,000 feet?

As the vehicle slows down, the drag acting on the vehicle (which is equal to the thrust required to cruise at constant velocity and altitude) reaches a minimum (D_{min}). The lift-to-drag ratio

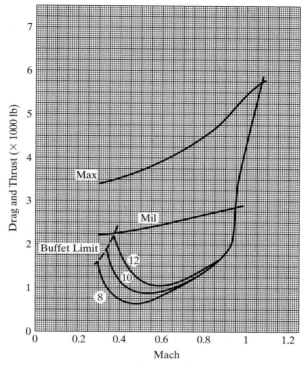

Figure P1.1

is, therefore, a maximum $[(L/D)_{max}]$. What is the maximum value of the lift-to-drag ratio $[(L/D)_{max}]$ for our 10,000-lbf T-38A cruising at 20,000 ft? What is the velocity at which the vehicle cruises, when the lift-to-drag ratio is a maximum? As the vehicle slows to speeds below that for $[L/D_{min}]$, which is equal to $[(L/D)_{max}]$, it actually requires more thrust (i. e., more power) to fly slower. You are operating the aircraft in the region of reverse command. More thrust is required to cruise at a slower speed. Eventually, one of two things happens: either the aircraft stalls (which is designated by the term "Buffet Limit" in Fig. P1.1) or the drag acting on the aircraft exceeds the thrust available. What is the minimum velocity at which a 10,000-lbf T-38A can cruise at 20,000 ft? Is this minimum velocity due to stall or is it due to the lack of sufficient power?

1.2. What are the total energy, the energy height, and the specific excess power, if our 10,000-lbf T-38A is using "Mil" thrust to cruise at a Mach number of 0.65 at 20,000 ft?

1.3. What is the maximum acceleration that our 10,000-lbf T-38A can achieve using "Mil" thrust, while passing through Mach 0.65 at a constant altitude of 20,000 ft? What is the maximum rate-of-climb that our 10,000-lbf T-38A can achieve at a constant velocity (specifically, at a Mach number of 0.65), when using "Mil" thrust while climbing through 20,000 ft?

1.4. Compare the values of $(L/D)_{max}$ for aircraft weights of 8,000 lbf, 10,000 lbf, and 12,000 lbf, when our T-38A aircraft cruises at 20,000 ft. Compare the velocity that is required to cruise at $(L/D)_{max}$ for each of the three aircraft weights.

1.5. Compare the specific excess power for a 10,000-lbf T-38A cruising at the Mach number required for $(L/D)_{max}$ while operating at "Mil" thrust with that for the aircraft cruising at a Mach number of 0.35 and with that for the aircraft cruising at a Mach number of 0.70.

1.6. Nitrogen is often used in wind tunnels as the test gas substitute for air. Compare the value of the kinematic viscosity

$$\nu = \frac{\mu}{\rho} \tag{1.6}$$

for nitrogen at a temperature of 350°F and at a pressure of 150 psia with that for air at the same conditions.

The constants for Sutherland's equation to calculate the coefficient of viscosity, i.e., Eqn. 1.12b, are:

$$C_1 = 2.27 \times 10^{-8} \frac{\text{lbf} \cdot \text{s}}{\text{ft}^2 \cdot {}^{\circ}\text{R}^{0.5}} \text{ and } C_2 = 198.6{}^{\circ}\text{R}$$

for air. Similarly,

$$C_1 = 2.16 \times 10^{-8} \frac{\text{lbf} \cdot \text{s}}{\text{ft}^2 \cdot {}^{\circ}\text{R}^{0.5}} \text{ and } C_2 = 183.6{}^{\circ}\text{R}$$

for nitrogen. The gas constant, which is used in the calculation of the density for a thermally perfect gas,

$$\rho = \frac{p}{RT} \tag{1.10}$$

is equal to $53.34 \frac{\text{ft} \cdot \text{lbf}}{\text{lbm} \cdot {}^{\circ}\text{R}}$ for air and to $55.15 \frac{\text{ft} \cdot \text{lbf}}{\text{lbm} \cdot {}^{\circ}\text{R}}$ for nitrogen.

1.7. Compare the value of the kinematic viscosity for nitrogen in a wind-tunnel test, where the free-stream static pressure is $586 \frac{\text{N}}{\text{m}^2}$ and the free-stream static temperature is 54.3 K, with the value for air at the same conditions.

The constants for Sutherland's equation to calculate the coefficient of viscosity, i.e., Eqn. 1.12a, are:

$$C_1 = 1.458 \times 10^{-6} \frac{\text{kg}}{\text{s} \cdot \text{m} \cdot \text{K}^{0.5}} \text{ and } C_2 = 110.4\text{K}$$

for air. Similarly,

$$C_1 = 1.39 \times 10^{-6} \frac{\text{kg}}{\text{s} \cdot \text{m} \cdot \text{K}^{0.5}} \text{ and } C_2 = 102K$$

for nitrogen. The gas constant, which is used in the calculation of the density for a thermally perfect gas,

$$\rho = \frac{p}{RT} \tag{1.10}$$

is equal to $287.05 \dfrac{N \cdot m}{kg \cdot K}$ for air and to $297 \dfrac{N \cdot m}{kg \cdot K}$ for nitrogen. What would be the advantage(s) of using nitrogen as the test gas instead of air?

1.8. A perfect gas undergoes a process whereby the pressure is doubled and its density is decreased to three-quarters of its original value. If the initial temperature is 200°F, what is the final temperature in °F? in °C?

1.9. The isentropic expansion of perfect helium takes place such that $p/\rho^{1.66}$ is a constant. If the pressure decreases to one-half of its original value, what happens to the temperature? If the initial temperature is 20° C, what is the final temperature?

1.10. Using the values for the pressure and for the temperature given in Table 1.2, calculate the density [equation (1.10)] and the viscosity [equation (1.12a)] at 20 km. Compare the calculated values with those given in Table 1.2. What is the kinematic viscosity at this altitude [equation (1.13)]?

1.11. Using the values for the pressure and for the temperature given in Table 1.2, calculate the density [equation (1.10)] and the viscosity [equation (1.12b)] at 35,000 ft. Compare the calculated values with those given in Table 1.2. What is the kinematic viscosity at this altitude [equation (1.13)]?

1.12. The conditions in the reservoir (or stagnation chamber) of Tunnel B at the Arnold Engineering Development Center (AEDC) are that the pressure (p_{t1}) is 5.723×10^6 N/m^2 and the temperature (T_A) is 750 K. Using the perfect-gas relations, what are the density and the viscosity in the reservoir?

1.13. The air in Tunnel B is expanded through a convergent/divergent nozzle to the test section, where the Mach number is 8, the free-stream static pressure is 586 N/m^2, and the free-stream temperature is 54.3 K. Using the perfect-gas relations, what are the corresponding values for the test-section density, viscosity, and velocity? Note that $U_\infty = M_\infty a_\infty$.

1.14. The conditions in the reservoir (or stagnation chamber) of Aerothermal Tunnel C at the Arnold Engineering Development Center (AEDC) are that the pressure (p_{t1}) is 24.5 psia and the temperature (T_t) is 1660°R. Using the perfect-gas relations, what are the density and the viscosity in the reservoir?

1.15. The air in Tunnel C accelerates through a convergent/divergent nozzle until the Mach number is 4 in the test section. The corresponding values for the free-stream pressure and the free-stream temperature in the test section are 23.25 lbf/ft^2 abs. and −65°F, respectively. What are the corresponding values for the free-stream density, viscosity, and velocity in the test section? Note that $M_\infty = U_\infty/a_\infty$. Using the values for the static pressure given in Table 1.2, what is the pressure altitude simulated in the wind tunnel by this test condition?

1.16. The pilot announces that you are flying at a velocity of 470 knots at an altitude of 35,000 ft. What is the velocity of the airplane in km/h? In ft/s?

1.17. Using equations (1.21) and (1.22), calculate the temperature and pressure of the atmosphere at 7000 m. Compare the tabulated values with those presented in Table 1.2.

1.18. Using an approach similar to that used in Example 1.7, develop metric-unit expressions for the pressure, the temperature, and the density of the atmosphere from 11,000 to 20,000 m. The temperature is constant and equal to 216.650 K over this range of altitude.

1.19. Using the expressions developed in Problem 1.18, what are the pressure, the density, the viscosity, and the speed of sound for the ambient atmospheric air at 18 km? Compare these values with the corresponding values tabulated in Table 1.2.

1.20. Using the expressions developed in Example 1.7, calculate the pressure, the temperature, and the density for the ambient atmospheric air at 10,000, 30,000 and 65,000 ft. Compare these values with the corresponding values presented in Table 1.2.

Problems 1.21 and 1.22 deal with standard atmosphere usage. The properties of the standard atmosphere are frequently used as the free-stream reference conditions for aircraft performance predictions. It is common to refer to the free-stream properties by the altitude in the atmosphericmodel at which those conditions occur. For instance, if the density of the free-stream flow is 0.00199 slugs/ft^{3}, then the density altitude (h_ρ) would be 6000 ft.

1.21. One of the design requirements for a multirole jet fighter is that it can survive a maximum sustained load factor of 9g at 15,000 *ft* MSL (mean sea level). What are the atmospheric values of the pressure, of the temperature, and of the density, that define the free-stream properties that you would use in the calculations that determine if a proposed design can meet this requirement?

1.22. An aircraft flying at geometric altitude of 20,000 ft has instrument readings of $p = 900\,\text{lbf/ft}^2$ and $T = 460°\text{R}$.

    (a) Find the values for the pressure altitude (h_p), the temperature altitude (h_T), and the density altitude (h_ρ) to the nearest 500 ft.

    (b) If the aircraft were flying in a standard atmosphere, what would be the relationship among h_p, h_T, and h_ρ.

1.23. If the water in a lake is everywhere at rest, what is the pressure as a function of the distance from the surface? The air above the surface of the water is at standard sea-level atmospheric conditions. How far down must one go before the pressure is 1 atm greater than the pressure at the surface? Use equation (1.18).

1.24. A U-tube manometer is used to measure the pressure at the stagnation point of a model in a wind tunnel. One side of the manometer goes to an orifice at the stagnation point; the other side is open to the atmosphere (Fig. P1.24). If there is a difference of 2.4 cm in the mercury levels in the two tubes, what is the pressure difference in N/m^2.

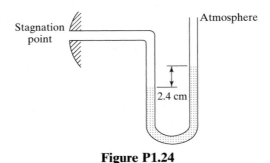

Figure P1.24

1.25. Consult a reference that contains thermodynamic charts for the properties of air, e.g., Hansen (1957), and delineate the temperature and pressure ranges for which air behaves:

(1) as a thermally perfect gas, i.e., $\dfrac{p}{\rho RT} = 1$ and

(2) as a calorically perfect gas, i.e., $h = c_p T$, where c_p is a constant.

1.26. A fairing for an optically perfect window for an airborne telescope is being tested in a wind tunnel. A manometer is connected to two pressure ports, one on the inner side of the window and the other port on the outside. The manometer fluid is water. During the testing at the maximum design airspeed, the column of water in the tube that is connected to the outside pressure port is 30 cm higher that the column of water in the tube that is connected to the inside port.

(a) What is the difference between the pressure that is acting on the inner surface of the window relative to the pressure acting on the outer surface of the window?

(b) If the window has a total area of $0.5 m^2$, what is the total force acting on the window due to the pressure difference?

REFERENCES

Chapman S, Cowling TG. 1960. *The Mathematical Theory of Non-uniform Gases*. Cambridge: Cambridge University Press

Coram R. 2002. *Boyd, The Fighter Pilot Who Changed the Art of War*. Boston: Little Brown

Grier P. 2004. The Viper revolution. *Air Force Magazine* 87(1):64–69

Hansen CF. 1957. Approximations for the thermodynamic properties of air in chemical equilibrium. *NACA Tech. Report R-50*

Hillaker H. 1997. Tribute to John R. Boyd. *Code One Magazine* 12(3)

Moeckel WE, Weston KC. 1958. Composition and thermodynamic properties of air in chemical equilibrium. *NACA Tech. Note 4265*

Svehla RA.1962. Estimated viscosities and thermal conductivities of gases at high temperatures. *NASA Tech.Report R-132*

1976. *U. S. Standard Atmosphere*. Washington, DC: U.S. Government Printing Office

Werner J. 2005. *Knight of Germany: Oswald Boelcke – German Ace*. Mechanicsburg: Stackpole Books

Yos JM. 1963. Transport properties of nitrogen, hydrogen, oxygen, and air to 30,000 K. *AVCO Corp. RAD-TM-63-7*

2 FUNDAMENTALS OF FLUID MECHANICS

As noted in Chapter 1, to predict accurately the aerodynamic forces and moments that act on a vehicle in flight, it is necessary to be able to describe the pattern of flow around the configuration. The resultant flow pattern depends on the geometry of the vehicle, its orientation with respect to the undisturbed free stream, and the altitude and speed at which the vehicle is traveling. The fundamental physical laws used to solve for the fluid motion in a general problem are

1. Conservation of mass (or the continuity equation)
2. Conservation of linear momentum (or Newton's second law of motion)
3. Conservation of energy (or the first law of thermodynamics)

Because the flow patterns are often very complex, it may be necessary to use experimental investigations as well as theoretical analysis to describe the resultant flow. The theoretical descriptions may utilize simplifying approximations in order to obtain any solution at all. The validity of the simplifying approximations for a particular application should be verified experimentally. Thus, it is important that we understand the fundamental laws that govern the fluid motion so that we can relate the theoretical solutions obtained using approximate flow models with the experimental results, which usually involve scale models.

2.1 INTRODUCTION TO FLUID DYNAMICS

To calculate the aerodynamic forces acting on an airplane, it is necessary to solve the equations governing the flow field about the vehicle. The flow-field solution can be formulated from the point of view of an observer on the ground or from the point of view of the pilot. Provided that the two observers apply the appropriate boundary conditions to the governing equations, both observers will obtain the same values for the aerodynamic forces acting on the airplane.

To an observer on the ground, the airplane is flying into a mass of air substantially at rest (assuming there is no wind). The neighboring air particles are accelerated and decelerated by the airplane and the reaction of the particles to the acceleration results in a force on the airplane. The motion of a typical air particle is shown in Fig. 2.1. The particle, which is initially at rest well ahead of the airplane, is accelerated by the passing airplane. The description of the flow field in the ground-observer-fixed coordinate system must represent the time-dependent motion (i.e., a nonsteady flow).

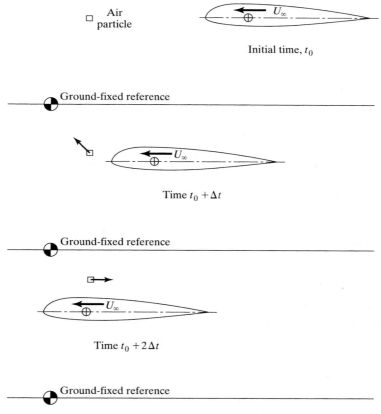

Figure 2.1 (Nonsteady) airflow around a wing in the ground-fixed coordinate system.

As viewed by the pilot, the air is flowing past the airplane and moves in response to the geometry of the vehicle. If the airplane is flying at constant altitude and constant velocity, the terms of the flow-field equations that contain partial derivatives with respect to time are zero in the vehicle-fixed coordinate system. Thus, as shown in Fig. 2.2, the velocity and the flow properties of the air particles that pass through a specific location relative to the vehicle are independent of time. The flow field is steady relative to a set of axes fixed to the vehicle (or pilot). Therefore, the equations are usually easier to solve in the vehicle (or pilot)-fixed coordinate system than in the ground-observer-fixed coordinate system. Because of the resulting simplification of the mathematics through the Gallilean transformation from the ground-fixed-reference coordinate system to the vehicle-fixed-reference coordinate system, many problems in aerodynamics are formulated as the flow of a stream of fluid past a body at rest. Note that the subsequent locations of the air particle which passed through our control volume at time t_0 are included for comparison with Fig. 2.2.

In this text we shall use the vehicle (or pilot)-fixed coordinate system. Thus, instead of describing the fluid motion around a vehicle flying through the air, we will examine air

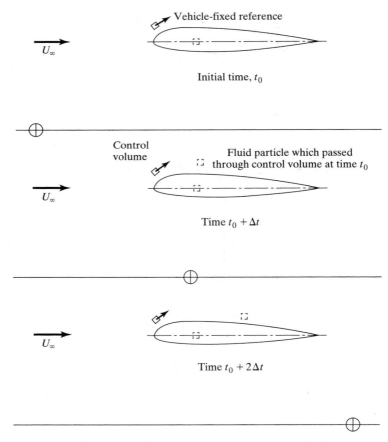

Figure 2.2 (Steady) airflow around a wing in a vehicle-fixed coordinate system.

flowing around a fixed vehicle. At points far from the vehicle (i.e., the undisturbed free stream), the fluid particles are moving toward the vehicle with the velocity U_∞ (see Fig. 2.2), which is in reality the speed of the vehicle (see Fig. 2.1). The subscript ∞ or 1 will be used to denote the undisturbed (or free-stream) flow conditions (i.e., those conditions far from the vehicle). Since all the fluid particles in the free stream are moving with the same velocity, there is no relative motion between them, and, hence, there are no shearing stresses in the free-stream flow. When there is no relative motion between the fluid particles, the fluid is termed a *static medium*, see Section 1.2.4. The values of the static fluid properties (e.g., pressure and temperature) are the same for either coordinate system.

2.2 CONSERVATION OF MASS

Let us apply the principle of conservation of mass to a small volume of space (a control volume) through which the fluid can move freely. For convenience, we shall use a Cartesian coordinate system (x, y, z). Furthermore, in the interest of simplicity, we shall treat a two-dimensional flow, that is, one in which there is no flow along the z axis. Flow patterns are the same for any xy plane. As indicated in the sketch of Fig. 2.3, the component of the fluid velocity in the x direction will be designated by u, and that in the y direction by v. The net outflow of mass through the surface surrounding the volume must be equal to the decrease of mass within the volume. The mass-flow rate through a surface bounding the element is equal to the product of the density, the velocity component normal to the surface, and the area of that surface. Flow out of the volume is considered positive. A first-order Taylor's series expansion is used to evaluate the flow properties at the faces of the element, since the properties are a function of position. Referring to Fig. 2.3, the net outflow of mass per unit time per unit depth (into the paper) is

$$\left[\rho u + \frac{\partial(\rho u)}{\partial x} \frac{\Delta x}{2} \right] \Delta y + \left[\rho v + \frac{\partial(\rho v)}{\partial y} \frac{\Delta y}{2} \right] \Delta x$$
$$- \left[\rho u - \frac{\partial(\rho u)}{\partial x} \frac{\Delta x}{2} \right] \Delta y - \left[\rho v - \frac{\partial(\rho v)}{\partial y} \frac{\Delta y}{2} \right] \Delta x$$

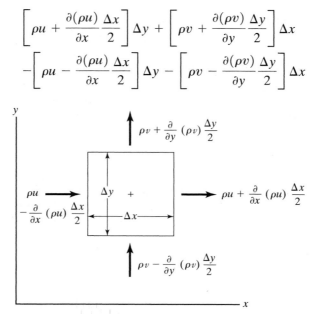

Figure 2.3 Velocities and densities for the mass-flow balance through a fixed volume element in two dimensions.

which must equal the rate at which the mass contained within the element decreases:

$$-\frac{\partial \rho}{\partial t}\, \Delta x\, \Delta y$$

Equating the two expressions, combining terms, and dividing by $\Delta x\, \Delta y$, we obtain

$$\frac{\partial \rho}{\partial t} + \frac{\partial}{\partial x}(\rho u) + \frac{\partial}{\partial y}(\rho v) = 0$$

If the approach were extended to include flow in the z direction, we would obtain the general differential form of the continuity equation:

$$\frac{\partial \rho}{\partial t} + \frac{\partial}{\partial x}(\rho u) + \frac{\partial}{\partial y}(\rho v) + \frac{\partial}{\partial z}(\rho w) = 0 \qquad \textbf{(2.1)}$$

In vector form, the equation is

$$\frac{\partial \rho}{\partial t} + \nabla \cdot (\rho \vec{V}) = 0 \qquad \textbf{(2.2)}$$

As has been discussed, the pressure variations that occur in relatively low-speed flows are sufficiently small, so that the density is essentially constant. For these incompressible flows, the continuity equation becomes

$$\frac{\partial u}{\partial x} + \frac{\partial v}{\partial y} + \frac{\partial w}{\partial z} = 0 \qquad \textbf{(2.3)}$$

In vector form, this equation is

$$\nabla \cdot \vec{V} = 0 \qquad \textbf{(2.4)}$$

Using boundary conditions, such as the requirement that there is no flow through a solid surface (i.e., the normal component of the velocity is zero at a solid surface), we can solve equation (2.4) for the velocity field. In so doing, we obtain a detailed picture of the velocity as a function of position.

EXAMPLE 2.1:   Incompressible boundary layer

Consider the case where a steady, incompressible, uniform flow whose velocity is U_∞ (i.e., a free-stream flow) approaches a flat plate. In the viscous region near the surface, which is called the boundary layer and is discussed at length in Chapter 4, the streamwise component of velocity is given by

$$u = U_\infty \left(\frac{y}{\delta}\right)^{1/7}$$

where δ, the boundary-layer thickness at a given station, is a function of x. Is a horizontal line parallel to the plate and a distance Δ from the plate (where Δ is equal to δ at the downstream station) a streamline? (See Fig. 2.4.)

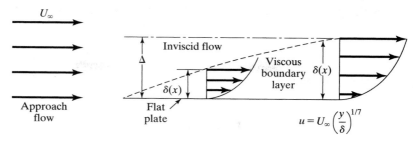

Figure 2.4 Flow diagram for Example 2.1.

Solution: By continuity for this steady, incompressible flow,

$$\frac{\partial u}{\partial x} + \frac{\partial v}{\partial y} = 0$$

Since $u = U_\infty(y/\delta)^{1/7}$ and $\delta(x)$,

$$\frac{\partial v}{\partial y} = -\frac{\partial u}{\partial x} = \frac{U_\infty}{7} \frac{y^{1/7}}{\delta^{8/7}} \frac{d\delta}{dx}$$

Integrating with respect to y yields

$$v = \frac{U_\infty}{8} \frac{y^{8/7}}{\delta^{8/7}} \frac{d\delta}{dx} + C$$

where C, the constant of integration, can be set equal to zero since $v = 0$ when $y = 0$ (i.e., there is no flow through the wall). Thus, when $y = \Delta$,

$$v_e = \frac{U_\infty}{8} \left(\frac{\Delta}{\delta}\right)^{8/7} \frac{d\delta}{dx}$$

Since v is not equal to zero, there is flow across the horizontal line which is Δ above the surface, and this line is not a streamline.

If the details of the flow are not of concern, the mass conservation principle can be applied directly to the entire region. Integrating equation (2.2) over a fixed finite volume in our fluid space (see Fig. 2.5) yields

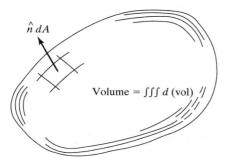

Figure 2.5 Nomenclature for the integral form of the continuity equation.

$$\iiint_{\text{vol}} \frac{\partial \rho}{\partial t} d(\text{vol}) + \iiint_{\text{vol}} \nabla \cdot (\rho \vec{V}) d(\text{vol}) = 0$$

The second volume integral can be transformed into a surface integral using Gauss's theorem, which is

$$\iiint_{\text{vol}} \nabla \cdot (\rho \vec{V}) d(\text{vol}) = \oiint_{A} \hat{n} \cdot \rho \vec{V} \, dA$$

where $\hat{n} \, dA$ is a vector normal to the surface dA which is positive when pointing outward from the enclosed volume and which is equal in magnitude to the surface area. The circle through the integral sign for the area indicates that the integration is to be performed over the entire surface bounding the volume. The resultant equation is the general integral expression for the conservation of mass:

$$\frac{\partial}{\partial t} \iiint_{\text{vol}} \rho d(\text{vol}) + \oiint_{A} \rho \vec{V} \cdot \hat{n} \, dA = 0 \qquad \textbf{(2.5)}$$

In words, the time rate of change of the mass within the volume plus the net efflux (outflow) of mass through the surface bounding the volume must be zero.

The volumetric flux Q is the flow rate through a particular surface and is equal to $\iint \vec{V} \cdot \hat{n} \, dA$.

For a sample problem using the integral form of the continuity equation, see Example 2.3.

2.3 CONSERVATION OF LINEAR MOMENTUM

The equation for the conservation of linear momentum is obtained by applying Newton's second law: The net force acting on a fluid particle is equal to the time rate of change of the linear momentum of the fluid particle. As the fluid element moves in space, its velocity, density, shape, and volume may change, but its mass is conserved. Thus, using a coordinate system that is neither accelerating nor rotating, which is called an *inertial coordinate system*, we may write

$$\vec{F} = m \frac{d\vec{V}}{dt} \qquad \textbf{(2.6)}$$

The velocity $\vec{V}$ of a fluid particle is, in general, an explicit function of time t as well as of its position x, y, z. Furthermore, the position coordinates x, y, z of the fluid particle are themselves a function of time. Since the time differentiation of equation (2.6) follows a given particle in its motion, the derivative is frequently termed the *particle, total*, or *substantial derivative* of $\vec{V}$. Since $\vec{V}(x, y, z, t)$ and $x(t), y(t)$, and $z(t)$,

$$\frac{d\vec{V}}{dt} = \frac{\partial \vec{V}}{\partial x} \frac{dx}{dt} + \frac{\partial \vec{V}}{\partial y} \frac{dy}{dt} + \frac{\partial \vec{V}}{\partial z} \frac{dz}{dt} + \frac{\partial \vec{V}}{\partial t} \qquad \textbf{(2.7)}$$

(The reader should note that some authors use D/Dt instead of d/dt to represent the substantial derivative.) However,

$$\frac{dx}{dt} = u \qquad \frac{dy}{dt} = v \qquad \frac{dz}{dt} = w$$

Therefore, the acceleration of a fluid particle is

$$\frac{d\vec{V}}{dt} = \frac{\partial \vec{V}}{\partial t} + u\frac{\partial \vec{V}}{\partial x} + v\frac{\partial \vec{V}}{\partial y} + w\frac{\partial \vec{V}}{\partial z} \tag{2.8}$$

or

$$\frac{d\vec{V}}{dt} = \frac{\partial \vec{V}}{\partial t} + (\vec{V} \cdot \nabla)\vec{V} \left(= \frac{D\vec{V}}{Dt} \right)$$
$$\text{total} \quad \text{local} \quad \text{convective} \quad \text{total} \tag{2.9}$$

Thus, the substantial derivative is the sum of the local, time-dependent changes that occur at a point in the flow field and of the changes that occur because the fluid particle moves around in space. Problems where the local, time-dependent changes are zero,

$$\frac{\partial \vec{V}}{\partial t} = 0$$

are known as *steady-state flows*. Note that even for a steady-state flow where $\partial\vec{V}/\partial t$ is equal to zero, fluid particles can accelerate due to the unbalanced forces acting on them. This is the case for an air particle that accelerates as it moves from the stagnation region to the low-pressure region above the airfoil. The convective acceleration of a fluid particle as it moves to different points in space is represented by the second term in equation (2.9).

The principal forces with which we are concerned are those which act directly on the mass of the fluid element, the *body forces*, and those which act on its surface, the *pressure forces* and *shear forces*. The stress system acting on an element of the surface is illustrated in Fig. 2.6. The stress components τ acting on the small cube are assigned

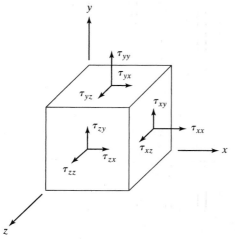

Figure 2.6 Nomenclature for the normal stresses and the shear stresses acting on a fluid element.

subscripts. The first subscript indicates the direction of the normal to the surface on which the stress acts and the second indicates the direction in which the stress acts. Thus, τ_{xy} denotes a stress acting in the y direction on the surface whose normal points in the x direction. Similarly, τ_{xx} denotes a normal stress acting on that surface. The stresses are described in terms of a right-hand coordinate system in which the outwardly directed surface normal indicates the positive direction.

The properties of most fluids have no preferred direction in space; that is, fluids are isotropic. As a result,

$$\tau_{xy} = \tau_{yx} \qquad \tau_{yz} = \tau_{zy} \qquad \tau_{zx} = \tau_{xz} \tag{2.10}$$

as shown in Schlichting (1968).

In general, the various stresses change from point to point. Thus, they produce net forces on the fluid particle, which cause it to accelerate. The forces acting on each surface are obtained by taking into account the variations of stress with position by using the center of the element as a reference point. To simplify the illustration of the force balance on the fluid particle we shall again consider a two-dimensional flow, as indicated in Fig. 2.7. The resultant force in the x direction (for a unit depth in the z direction) is

$$\rho f_x \, \Delta x \, \Delta y + \frac{\partial}{\partial x}(\tau_{xx}) \, \Delta x \, \Delta y + \frac{\partial}{\partial y}(\tau_{yx}) \, \Delta y \, \Delta x$$

where f_x is the body force per unit mass in the x direction. The body force for the flow fields of interest to this text is gravity.

Including flow in the z direction, the resultant force in the x direction is

$$F_x = \rho f_x \, \Delta x \, \Delta y \, \Delta z + \frac{\partial}{\partial x}(\tau_{xx}) \, \Delta x \, \Delta y \, \Delta z + \frac{\partial}{\partial y}(\tau_{yx}) \, \Delta y \, \Delta x \, \Delta z$$
$$+ \frac{\partial}{\partial z}(\tau_{zx}) \, \Delta z \, \Delta y \, \Delta x$$

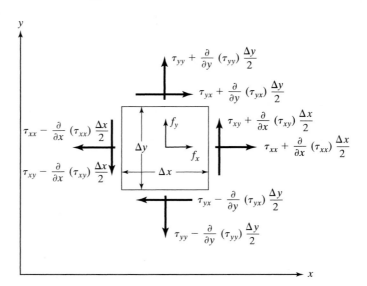

Figure 2.7 Stresses acting on a two-dimensional element of fluid.

which, by equation (2.6), is equal to

$$ma_x = \rho\, \Delta x\, \Delta y\, \Delta z\, \frac{du}{dt} = \rho\, \Delta x\, \Delta y\, \Delta z \left[\frac{\partial u}{\partial t} + (\vec{V} \cdot \nabla)u \right]$$

Equating the two and dividing by the volume of the fluid particle $\Delta x\, \Delta y \Delta z$ yields the linear momentum equation for the x direction:

$$\rho\frac{du}{dt} = \rho f_x + \frac{\partial}{\partial x}\tau_{xx} + \frac{\partial}{\partial y}\tau_{yx} + \frac{\partial}{\partial z}\tau_{zx} \qquad \textbf{(2.11a)}$$

Similarly, we obtain the equation of motion for the y direction:

$$\rho\frac{dv}{dt} = \rho f_y + \frac{\partial}{\partial x}\tau_{xy} + \frac{\partial}{\partial y}\tau_{yy} + \frac{\partial}{\partial z}\tau_{zy} \qquad \textbf{(2.11b)}$$

and for the z direction:

$$\rho\frac{dw}{dt} = \rho f_z + \frac{\partial}{\partial x}\tau_{xz} + \frac{\partial}{\partial y}\tau_{yz} + \frac{\partial}{\partial z}\tau_{zz} \qquad \textbf{(2.11c)}$$

Next, we need to relate the stresses to the motion of the fluid. For a fluid at rest or for a flow for which all the fluid particles are moving at the same velocity, there is no shearing stress, and the normal stress is in the nature of a pressure. For fluid particles, the stress is related to the rate of strain by a physical law based on the following assumptions:

1. Stress components may be expressed as a linear function of the components of the rate of strain. The friction law for the flow of a Newtonian fluid where $\tau = \mu(\partial u/\partial y)$ is a special case of this linear stress/rate-of-strain relation. The viscosity μ is more precisely called the first viscosity coefficient. In a more rigorous development, one should include the second viscosity coefficient (λ), which would appear in the normal stress terms. The term involving λ disappears completely when the flow is incompressible, since $\nabla \cdot \vec{V} = 0$ by continuity. For other flows, Stokes's hypothesis $\left(\lambda = -\frac{2}{3}\mu\right)$ is presumed to apply. See Schlichting (1968). The second viscosity coefficient is of significance in a few specialized problems, such as the analysis of the shockwave structure, where extremely large changes in pressure and in temperature take place over very short distances.

2. The relations between the stress components and the rate-of-strain components must be invariant to a coordinate transformation consisting of either a rotation or a mirror reflection of axes, since a physical law cannot depend upon the choice of the coordinate system.

3. When all velocity gradients are zero (i.e., the shear stress vanishes), the stress components must reduce to the hydrostatic pressure, p.

For a fluid that satisfies these criteria,

$$\tau_{xx} = -p - \frac{2}{3}\mu\nabla\cdot\vec{V} + 2\mu\frac{\partial u}{\partial x}$$

$$\tau_{yy} = -p - \frac{2}{3}\mu\nabla\cdot\vec{V} + 2\mu\frac{\partial v}{\partial y}$$

$$\tau_{zz} = -p - \frac{2}{3}\mu\nabla\cdot\vec{V} + 2\mu\frac{\partial w}{\partial z}$$

$$\tau_{xy} = \tau_{yx} = \mu\left(\frac{\partial u}{\partial y} + \frac{\partial v}{\partial x}\right)$$

$$\tau_{xz} = \tau_{zx} = \mu\left(\frac{\partial u}{\partial z} + \frac{\partial w}{\partial x}\right)$$

$$\tau_{yz} = \tau_{zy} = \mu\left(\frac{\partial v}{\partial z} + \frac{\partial w}{\partial y}\right)$$

With the appropriate expressions for the surface stresses substituted into equation (2.11), we obtain

$$\rho\frac{\partial u}{\partial t} + \rho(\vec{V} \cdot \nabla)u = \rho f_x - \frac{\partial p}{\partial x} + \frac{\partial}{\partial x}\left(2\mu\frac{\partial u}{\partial x} - \frac{2}{3}\mu\nabla\cdot\vec{V}\right)$$
$$+ \frac{\partial}{\partial y}\left[\mu\left(\frac{\partial u}{\partial y} + \frac{\partial v}{\partial x}\right)\right] + \frac{\partial}{\partial z}\left[\mu\left(\frac{\partial w}{\partial x} + \frac{\partial u}{\partial z}\right)\right] \quad \textbf{(2.12a)}$$

$$\rho\frac{\partial v}{\partial t} + \rho(\vec{V} \cdot \nabla)v = \rho f_y + \frac{\partial}{\partial x}\left[\mu\left(\frac{\partial u}{\partial y} + \frac{\partial v}{\partial x}\right)\right]$$
$$- \frac{\partial p}{\partial y} + \frac{\partial}{\partial y}\left(2\mu\frac{\partial v}{\partial y} - \frac{2}{3}\mu\nabla\cdot\vec{V}\right)$$
$$+ \frac{\partial}{\partial z}\left[\mu\left(\frac{\partial w}{\partial y} + \frac{\partial v}{\partial z}\right)\right] \quad \textbf{(2.12b)}$$

$$\rho\frac{\partial w}{\partial t} + \rho(\vec{V} \cdot \nabla)w = \rho f_z + \frac{\partial}{\partial x}\left[\mu\left(\frac{\partial w}{\partial x} + \frac{\partial u}{\partial z}\right)\right]$$
$$+ \frac{\partial}{\partial y}\left[\mu\left(\frac{\partial v}{\partial z} + \frac{\partial w}{\partial y}\right)\right] - \frac{\partial p}{\partial z}$$
$$+ \frac{\partial}{\partial z}\left(2\mu\frac{\partial w}{\partial z} - \frac{2}{3}\mu\nabla\cdot\vec{V}\right) \quad \textbf{(2.12c)}$$

These general, differential equations for the conservation of linear momentum are known as the *Navier-Stokes equations*. Note that the viscosity μ is considered to be dependent on the spatial coordinates. This is done since, for a compressible flow, the changes in velocity and pressure, together with the heat due to friction, bring about considerable temperature variations. The temperature dependence of viscosity in the general case should, therefore, be incorporated into the governing equations.

For a general application, the unknown parameters that appear in the Navier-Stokes equations are the three velocity components (u, v, and w), the pressure (p), the density (ρ), and the viscosity (μ). As we discussed in Chapter 1, for a fluid of known composition that is in equilibrium, the density and the viscosity are unique functions of pressure and temperature. Thus, there are five primary (or primitive) variables for a general flow problem: the three velocity components, the pressure, and the temperature. However, at present we have only four equations: the continuity equation, equation (2.2), and the three components of the momentum equation, equations (2.12a) through (2.12c). To solve for a general flow involving all five variables, we would need to introduce the energy equation.

Since equations (2.12a) through (2.12c) are the general differential equations for the conservation of linear momentum, the equations for a static medium can be obtained by neglecting the terms relating to the acceleration of the fluid particles and to the viscous forces. Neglecting these terms in equations (2.12a) through (2.12c) and assuming that the body force is gravity and that it acts in the z direction, the reader would obtain equations (1.16a) through (1.16c).

The integral form of the momentum equation can be obtained by returning to Newton's law. The sum of the forces acting on a system of fluid particles is equal to the rate of change of momentum of the fluid particles. Thus, the sum of the body forces and of the surface forces equals the time rate of change of momentum within the volume plus the net efflux of momentum through the surface bounding the volume. In vector form,

$$\vec{F}_{\text{body}} + \vec{F}_{\text{surface}} = \frac{\partial}{\partial t} \iiint_{\text{vol}} \rho\vec{V}\, d(\text{vol}) + \oiint_{A} \vec{V}(\rho\vec{V}\cdot\hat{n}\, dA) \tag{2.13}$$

This equation can also be obtained by integrating equation (2.12) over a volume and using Gauss's Theorem.

2.4 APPLICATIONS TO CONSTANT-PROPERTY FLOWS

For many flows, temperature variations are sufficiently small that the density and viscosity may be assumed constant throughout the flow field. Such flows will be termed *constant-property* flows in this text. The terms *low-speed* and/or *incompressible* flows will also be used in the description of these flows. A gas flow is considered incompressible if the Mach number is less than 0.3 to 0.5, depending upon the application. For these flows, there are only four unknowns: the three velocity components (u, v, and w), and the pressure (p). Thus, we have a system of four independent equations that can be solved for the four unknowns; that is, the energy equation is not needed to obtain the velocity components and the pressure of a constant-property flow.

Let us consider two constant-property flows, one for which the solution will be obtained using differential equations and one for which the integral equations are used.

EXAMPLE 2.2: Poiseuille flow

Consider a steady, low-speed flow of a viscous fluid in an infinitely long, two-dimensional channel of height h (Fig. 2.8). This is known as *Poiseuille flow*. Since the flow is low speed, we will assume that the viscosity and the density are constant. Because the channel is infinitely long, the velocity components do not change in the x direction. In this text, such a flow is termed a *fully developed flow*. Let us assume that the body forces are negligible. We are to determine the velocity profile and the shear-stress distribution.

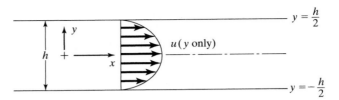

Figure 2.8 Flow diagram for Example 2.2.

Solution: For a two-dimensional flow, $w \equiv 0$ and all the derivatives with respect to z are zero. The continuity equation for this steady-state, constant-property flow yields

$$\frac{\partial u}{\partial x} + \frac{\partial v}{\partial y} = 0$$

Since the velocity components do not change in the x direction, $\dfrac{\partial u}{\partial x} = 0$ and, hence,

$$\frac{\partial v}{\partial y} = 0$$

Further, since $v = 0$ at both walls (i.e., there is no flow through the walls or, equivalently, the walls are streamlines) and v does not depend on x or z, $v \equiv 0$ everywhere. Thus, the flow is everywhere parallel to the x axis.

At this point, we know that $v \equiv 0$ everywhere, $w = 0$ everywhere, and u is a function of y only, since it does not depend on x, z, or t. We can also neglect the body forces. Thus, all the terms in equations (2.12b) and (2.12c) are zero and need not be considered further. Expanding the acceleration term of equation (2.12a), we obtain

$$\rho\frac{\partial u}{\partial t} + \rho u\frac{\partial u}{\partial x} + \rho v\frac{\partial u}{\partial y} + \rho w\frac{\partial u}{\partial z} = \rho f_x - \frac{\partial p}{\partial x} + \frac{\partial}{\partial x}\left(2\mu\frac{\partial u}{\partial x} - \frac{2}{3}\mu\nabla\cdot\vec{V}\right)$$

$$+ \frac{\partial}{\partial y}\left[\mu\left(\frac{\partial u}{\partial y} + \frac{\partial v}{\partial x}\right)\right] + \frac{\partial}{\partial z}\left[\mu\left(\frac{\partial u}{\partial z} + \frac{\partial w}{\partial x}\right)\right] \quad \textbf{(2.14)}$$

Because we are considering low-speed of a simple fluid, μ is constant throughout the flow field. Thus, we can rewrite the viscous terms of this equation as follows:

$$\frac{\partial}{\partial x}\left(2\mu\frac{\partial u}{\partial x} - \frac{2}{3}\mu\nabla\cdot\vec{V}\right) + \frac{\partial}{\partial y}\left[\mu\left(\frac{\partial u}{\partial y} + \frac{\partial v}{\partial x}\right)\right] + \frac{\partial}{\partial z}\left[\mu\left(\frac{\partial u}{\partial z} + \frac{\partial w}{\partial x}\right)\right]$$

$$= \mu\frac{\partial^2 u}{\partial x^2} + \mu\frac{\partial}{\partial x}\left(\frac{\partial u}{\partial x}\right) - \frac{2}{3}\mu\frac{\partial}{\partial x}(\nabla\cdot\vec{V}) + \mu\frac{\partial^2 u}{\partial y^2}$$

$$+ \mu\frac{\partial}{\partial x}\left(\frac{\partial v}{\partial y}\right) + \mu\frac{\partial^2 u}{\partial z^2} + \mu\frac{\partial}{\partial x}\left(\frac{\partial w}{\partial z}\right)$$

$$= \mu\left(\frac{\partial^2 u}{\partial x^2} + \frac{\partial^2 u}{\partial y^2} + \frac{\partial^2 u}{\partial z^2}\right) + \mu\frac{\partial}{\partial x}\left(\frac{\partial u}{\partial x} + \frac{\partial v}{\partial y} + \frac{\partial w}{\partial z}\right) - \frac{2}{3}\mu\frac{\partial}{\partial x}(\nabla\cdot\vec{V})$$

Noting that

$$\frac{\partial u}{\partial x} + \frac{\partial v}{\partial y} + \frac{\partial w}{\partial z} = \nabla \cdot \vec{V}$$

and that $\nabla \cdot \vec{V} = 0$ (since these are two ways of writing the continuity equation for a constant density flow), we can write equation (2.14) as

$$\rho \frac{\partial u}{\partial t} + \rho u \frac{\partial u}{\partial x} + \rho v \frac{\partial u}{\partial y} + \rho w \frac{\partial u}{\partial z} = \rho f_x - \frac{\partial p}{\partial x} + \mu \left(\frac{\partial^2 u}{\partial x^2} + \frac{\partial^2 u}{\partial y^2} + \frac{\partial^2 u}{\partial z^2} \right) \quad \textbf{(2.15)}$$

However, we can further simplify equation (2.15) by eliminating terms whose value is zero.

$$\rho \frac{\partial u}{\partial t} = 0 \qquad\qquad \text{because the flow is steady}$$

$$\rho u \frac{\partial u}{\partial x} = 0 \qquad\qquad \textit{because } u = u(y \text{ only})$$

$$\rho v \frac{\partial u}{\partial y} = 0 \qquad\qquad \text{because } v \equiv 0$$

$$\rho w \frac{\partial u}{\partial z} = \mu \frac{\partial^2 u}{\partial x^2} = \mu \frac{\partial^2 u}{\partial z^2} = 0 \qquad \text{because } u = u(y \text{ only})$$

$$\rho f_x = 0 \qquad\qquad \text{because body forces are negligible}$$

Thus,

$$0 = -\frac{\partial p}{\partial x} + \mu \frac{\partial^2 u}{\partial y^2}$$

$$0 = -\frac{\partial p}{\partial y}$$

$$0 = -\frac{\partial p}{\partial z}$$

These three equations require that the pressure is a function of x only. Recall that u is a function of y only. These two statements can be true only if

$$\mu \frac{d^2 u}{dy^2} = \frac{dp}{dx} = \text{constant}$$

Integrating twice gives

$$u = \frac{1}{2\mu} \frac{dp}{dx} y^2 + C_1 y + C_2$$

To evaluate the constants of integration, we apply the viscous-flow boundary condition that the fluid particles at a solid surface move with the same speed as the surface (i.e., do not slip relative to the surface). Thus,

$$\text{At } y = -\frac{h}{2} \qquad u = 0$$

$$\text{At } y = +\frac{h}{2} \qquad u = 0$$

When we do this, we find that

$$C_1 = 0$$

$$C_2 = -\frac{1}{2\mu}\frac{dp}{dx}\frac{h^2}{4}$$

so that

$$u = +\frac{1}{2\mu}\frac{dp}{dx}\left(y^2 - \frac{h^2}{4}\right)$$

The velocity profile is parabolic, with the maximum velocity at the center of the channel. The shear stress at the two walls is

$$\tau = \mu\frac{du}{dy}\bigg|_{y=\pm h/2} = +\frac{h}{2}\frac{dp}{dx}$$

The pressure must decrease in the x direction (i.e., dp/dx must be negative) to have a velocity in the direction shown. The negative, or favorable, pressure gradient results because of the viscous forces.

Examination of the integral momentum equation (2.13) verifies that a change in pressure must occur to balance the shear forces. Let us verify this by using equation (2.13) on the control volume shown in Fig. 2.9.

Since the flow is fully developed, the velocity profile at the upstream station (i.e., station 1) is identical to that at the downstream station (i.e., station 2). Thus, the positive momentum efflux at station 2 is balanced by the negative momentum influx at station 1:

$$\oiint \vec{V}(\rho\vec{V}\cdot\hat{n}\,dA) = 0$$

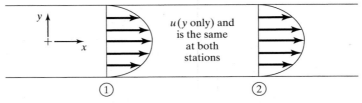

Figure 2.9 Another flow diagram for Example 2.2.

Thus, for this steady flow with negligible body forces,

$$\vec{F}_{surface} = 0$$

or

$$p_1 h - p_2 h - 2\tau \, \Delta x = 0$$

where the factor of 2 accounts for the existence of shear forces at the upper and lower walls. Finally, as shown in the approach using the differential equation,

$$\tau = -\frac{p_2 - p_1}{\Delta x}\frac{h}{2} = -\frac{dp}{dx}\frac{h}{2}$$

Note, as discussed previously, that there are subtle implications regarding the signs in these terms. For the velocity profile shown, the shear acts to retard the fluid motion and the pressure must decrease in the x direction (i.e., $dp/dx < 0$).

EXAMPLE 2.3: Drag on a flat-plate airfoil

A steady, uniform ($\vec{V} = U_\infty \hat{\imath}$), low-speed flow approaches a very thin, flat-plate "airfoil" whose length is c. Because of viscosity, the flow near the plate slows, such that velocity measurements at the trailing edge of the plate indicate that the x component of the velocity (above the plate) varies as

$$u = U_\infty \left(\frac{y}{\delta}\right)^{1/7}$$

Below the plate, the velocity is a mirror image of this profile. The pressure is uniform over the entire control surface. Neglecting the body forces, what is the drag coefficient for this flow if $\delta = 0.01c$? The drag coefficient is the drag per unit span (unit depth into the page) divided by the free-stream dynamic pressure $\left(\frac{1}{2}\rho_\infty U_\infty^2\right)$ times the reference area per unit span (which is the chord length c):

$$C_d = \frac{d}{\frac{1}{2}\rho_\infty U_\infty^2 c}$$

Solution: Let us apply the integral form of the momentum equation [i.e., equation (2.13)] to this flow. Noting that the momentum equation is a vector equation, let us consider the x component only, since that is the direction in which the drag force acts. Furthermore, since the flow is planar symmetric, only the control volume above the plate will be considered (i.e., from $y = 0$ to $y = \delta$). Since the pressure is uniform over the control surface and since the body forces are negligible, the only force acting on the fluid in the control volume is the retarding force of the plate on the fluid, which is $-d/2$.

Because the flow is steady, the first term on the right-hand side of equation (2.13) is zero. Thus,

$$-\frac{d}{2} = \oiint (\rho\vec{V}\cdot\hat{n}\,dA)V_x$$

Noting that $V_x \equiv u$ and using the values of which are shown in Fig. 2.10, we obtain

$$-\frac{d}{2} = \int_0^\delta [\rho(U_\infty \hat{i}) \cdot (-\hat{i}\, dy)] U_\infty + \int_0^c [\rho(U_\infty \hat{i} + v_e \hat{j}) \cdot (\hat{j}\, dx)] U_\infty$$

$$\overleftarrow{\hspace{1.5cm}} (1) \overrightarrow{\hspace{1.5cm}} \qquad \overleftarrow{\hspace{1.5cm}} (2) \overrightarrow{\hspace{1.5cm}}$$

$$+ \int_0^\delta \left\{ \rho \left[U_\infty \left(\frac{y}{\delta}\right)^{1/7} \hat{i} + v\hat{j} \right] \cdot (\hat{i}\, dy) \right\} U_\infty \left(\frac{y}{\delta}\right)^{1/7}$$

$$\overleftarrow{\hspace{2.5cm}} (3) \overrightarrow{\hspace{2.5cm}}$$

Let us comment about each of the terms describing the momentum efflux from each surface of the control volume. (1) Since $\hat{n}\, dA$ is a vector normal to the surface dA which is positive when pointing outward from the enclosed volume, $\hat{n}$ is in the $-x$ direction and dA per unit span is equal to dy. (2) Because viscosity slows the air particles near the plate, line 2 is not a streamline, and the velocity vector along this surface is

$$\vec{V} = U_\infty \hat{i} + v_e \hat{j}$$

The volumetric efflux (per unit depth) across this surface is $\int_0^c v_e dx$, which will be represented by the symbol Q_2. (3) Because of viscosity, the velocity vector has a y component, which is a function of y. However, this y component of velocity does not transport fluid across the area $\hat{i}\, dy$. (4) There is no flow across this surface of the control volume, since it is at a solid wall. Furthermore, because the flow is low speed, the density will be assumed constant, and the momentum equation becomes

$$-\frac{d}{2} = -\rho U_\infty^2 \delta + \rho U_\infty Q_2 + \frac{7}{9} \rho U_\infty^2 \delta$$

To obtain an expression for Q_2, let us use the integral form of the continuity equation, equation (2.5), for the flow of Fig. 2.10.

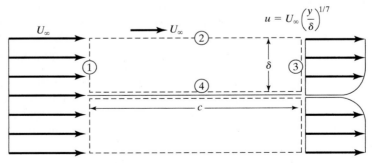

Figure 2.10 Flow diagram for Example 2.3.

$$\oiint \rho \vec{V} \cdot \hat{n}\, dA = \int_0^\delta [\rho(U_\infty \hat{i}) \cdot (-\hat{i}\, dy)]$$

$$\overleftarrow{\qquad (1) \qquad}\rightarrow$$

$$+ \int_0^c [\rho(U_\infty \hat{i} + v_e \hat{j}) \cdot (\hat{j}\, dx)] + \int_0^\delta \left\{ \rho \left[U_\infty \left(\frac{y}{\delta}\right)^{1/7} \hat{i} + v\hat{j} \right] \cdot (\hat{i}\, dy) \right\} = 0$$

$$\overleftarrow{\qquad (2) \qquad}\rightarrow \qquad\qquad \overleftarrow{\qquad (3) \qquad}\rightarrow$$

$$-\rho U_\infty \delta + \rho \int_0^c v_e dx + \rho U_\infty \left(\tfrac{7}{8}\delta\right) = 0 \qquad\qquad \int_0^c v_e dx = Q_2 = \tfrac{1}{8} U_\infty \delta$$

Substituting this expression into the momentum equation yields

$$-\frac{d}{2} = -\rho U_\infty^2 \delta + \frac{1}{8}\rho U_\infty^2 \delta + \frac{7}{9}\rho U_\infty^2 \delta$$

$$d = \left(2 - \frac{1}{4} - \frac{14}{9}\right)\rho U_\infty^2 \delta = \frac{7}{36}\rho U_\infty^2 \delta$$

$$C_d = \frac{d}{\frac{1}{2}\rho_\infty U_\infty^2 c} = \frac{\frac{7}{36}\rho U_\infty^2 (0.01c)}{\frac{1}{2}\rho_\infty U_\infty^2 c} = 0.00389$$

2.5 REYNOLDS NUMBER AND MACH NUMBER AS SIMILARITY PARAMETERS

Because of the difficulty of obtaining theoretical solutions of the flow field around a vehicle, numerous experimental programs have been conducted to measure directly the parameters that define the flow field. Some of the objectives of such test programs are as follows:

1. To obtain information necessary to develop a flow model that could be used in numerical solutions
2. To investigate the effect of various geometric parameters on the flow field (such as determining the best location for the engines on a supersonic transport)
3. To verify numerical predictions of the aerodynamic characteristics for a particular configuration
4. To measure directly the aerodynamic characteristics of a complete vehicle

Usually, either scale models of the complete vehicle or large-scale simulations of elements of the vehicle (such as the wing section) have been used in these wind-tunnel programs. Furthermore, in many test programs, the free-stream conditions (such as the velocity, the static pressure, etc.) for the wind-tunnel tests were not equal to the values for the flight condition that was to be simulated.

It is important, then, to determine under what conditions the experimental results obtained for one flow are applicable to another flow which is confined by boundaries

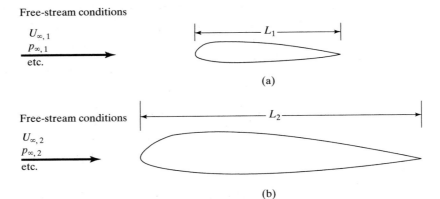

Figure 2.11 Flow around geometrically similar (but different size) configurations: (a) first flow; (b) second flow.

that are geometrically similar (but of different size). To do this, consider the x-momentum equation as applied to the two flows of Fig. 2.11. For simplicity, let us limit ourselves to constant-property flows. Since the body-force term is usually negligible in aerodynamic problems, equation (2.15) can be written

$$\rho\frac{\partial u}{\partial t} + \rho u\frac{\partial u}{\partial x} + \rho v\frac{\partial u}{\partial y} + \rho w\frac{\partial u}{\partial z} = -\frac{\partial p}{\partial x} + \mu\frac{\partial^2 u}{\partial x^2} + \mu\frac{\partial^2 u}{\partial y^2} + \mu\frac{\partial^2 u}{\partial z^2} \qquad \textbf{(2.16)}$$

Let us divide each of the thermodynamic properties by the value of that property at a point far from the vehicle (i.e., the free-stream value of the property) for each of the two flows. Thus, for the first flow,

$$p_1^* = \frac{p}{p_{\infty,1}} \qquad \rho_1^* = \frac{\rho}{\rho_{\infty,1}} \qquad \mu_1^* = \frac{\mu}{\mu_{\infty,1}}$$

and for the second flow,

$$p_2^* = \frac{p}{p_{\infty,2}} \qquad \rho_2^* = \frac{\rho}{\rho_{\infty,2}} \qquad \mu_2^* = \frac{\mu}{\mu_{\infty,2}}$$

Note that the free-stream values for all three nondimensionalized (*) thermodynamic properties are unity for both cases. Similarly, let us divide the velocity components by the free-stream velocity. Thus, for the first flow,

$$u_1^* = \frac{u}{U_{\infty,1}} \qquad v_1^* = \frac{v}{U_{\infty,1}} \qquad w_1^* = \frac{w}{U_{\infty,1}}$$

and for the second flow,

$$u_2^* = \frac{u}{U_{\infty,2}} \qquad v_2^* = \frac{v}{U_{\infty,2}} \qquad w_2^* = \frac{w}{U_{\infty,2}}$$

With the velocity components thus nondimensionalized, the free-stream boundary conditions are the same for both flows: that is, at points far from the vehicle

$$u_1^* = u_2^* = 1 \quad \text{and} \quad v_1^* = v_2^* = w_1^* = w_2^* = 0$$

A characteristic dimension L is used to nondimensionalize the independent variables. L/U_∞ is a characteristic time.

$$x_1^* = \frac{x}{L_1} \qquad y_1^* = \frac{y}{L_1} \qquad z_1^* = \frac{z}{L_1} \qquad t_1^* = \frac{tU_{\infty,1}}{L_1}$$

and

$$x_2^* = \frac{x}{L_2} \qquad y_2^* = \frac{y}{L_2} \qquad z_2^* = \frac{z}{L_2} \qquad t_2^* = \frac{tU_{\infty,2}}{L_2}$$

In terms of these dimensionless parameters, the x-momentum equation (2.16) becomes

$$\rho_1^* \frac{\partial u_1^*}{\partial t_1^*} + \rho_1^* u_1^* \frac{\partial u_1^*}{\partial x_1^*} + \rho_1^* v_1^* \frac{\partial u_1^*}{\partial y_1^*} + \rho_1^* w_1^* \frac{\partial u_1^*}{\partial z_1^*}$$

$$= -\left(\frac{p_{\infty,1}}{\rho_{\infty,1} U_{\infty,1}^2}\right) \frac{\partial p_1^*}{\partial x_1^*} + \left(\frac{\mu_{\infty,1}}{\rho_{\infty,1} U_{\infty,1} L_1}\right)\left(\mu_1^* \frac{\partial^2 u_1^*}{\partial x_1^{*2}} + \mu_1^* \frac{\partial^2 u_1^*}{\partial y_1^{*2}} + \mu_1^* \frac{\partial^2 u_1^*}{\partial z_1^{*2}}\right) \quad \textbf{(2.17a)}$$

for the first flow. For the second flow,

$$\rho_2^* \frac{\partial u_2^*}{\partial t_2^*} + \rho_2^* u_2^* \frac{\partial u_2^*}{\partial x_2^*} + \rho_2^* v_2^* \frac{\partial u_2^*}{\partial y_2^*} + \rho_2^* w_2^* \frac{\partial u_2^*}{\partial z_2^*}$$

$$= -\left(\frac{p_{\infty,2}}{\rho_{\infty,2} U_{\infty,2}^2}\right) \frac{\partial p_2^*}{\partial x_2^*} + \left(\frac{\mu_{\infty,2}}{\rho_{\infty,2} U_{\infty,2} L_2}\right)\left(\mu_2^* \frac{\partial^2 u_2^*}{\partial x_2^{*2}} + \mu_2^* \frac{\partial^2 u_2^*}{\partial y_2^{*2}} + \mu_2^* \frac{\partial^2 u_2^*}{\partial z_2^{*2}}\right) \quad \textbf{(2.17b)}$$

Both the dependent variables and the independent variables have been nondimensionalized, as indicated by the * quantities. The dimensionless *boundary-condition values* for the dependent variables are the same for the two flows around geometrically similar configurations. As a consequence, the solutions of the two problems in terms of the dimensionless variables will be identical provided that the differential equations are identical. The differential equations will be identical if the parameters in the parentheses have the same values for both problems. In this case, the flows are said to be dynamically similar as well as geometrically similar.

Let us examine the first similarity parameter from equation (2.17),

$$\frac{p_\infty}{\rho_\infty U_\infty^2} \tag{2.18}$$

Recall that for a perfect gas, the equation of state is

$$p_\infty = \rho_\infty R T_\infty$$

and the free-stream speed of sound is given by

$$a_\infty = \sqrt{\gamma R T_\infty}$$

Substituting these relations into equation (2.18) yields

$$\frac{p_\infty}{\rho_\infty U_\infty^2} = \frac{R T_\infty}{U_\infty^2} = \frac{a_\infty^2}{\gamma U_\infty^2} = \frac{1}{\gamma M_\infty^2} \qquad (2.19)$$

since $M_\infty = U_\infty/a_\infty$. Thus, the first dimensionless similarity parameter can be interpreted in terms of the free-stream Mach number.

The inverse of the second similarity parameter is written

$$\text{Re}_{\infty,L} = \frac{\rho_\infty U_\infty L}{\mu_\infty} \qquad (2.20)$$

which is the *Reynolds number*, a measure of the ratio of inertia forces to viscous forces.

As has been discussed, the free-stream values of the fluid properties, such as the static pressure and the static temperature, are a function of altitude. Thus, once the velocity, the altitude, and the characteristic dimension of the vehicle are defined, the free-stream Mach number and the free-stream Reynolds number can be calculated as a function of velocity and altitude. This has been done using the values presented in Table 1.2. The free-stream Reynolds number is defined by equation (2.20) with the characteristic length L (e.g., the chord of the wing or the diameter of the missile) chosen to be 1.0 m for the correlations of Fig. 2.12. The correlations represent altitudes up to 30 km (9.84×10^4 ft) and velocities up to 2500 km/h (1554 mi/h or 1350 knots). Note that 1 knot $\equiv$ 1 nautical mile per hour.

EXAMPLE 2.4: Calculating the Reynolds number

An airplane is flying at a Mach number of 2 at an altitude of 40,000 ft. If the characteristic length for the aircraft is 14 ft, what is the velocity in mi/h, and what is the Reynolds number for this flight condition?

Solution: The density, the viscosity, and the speed of sound of the free-stream flow at 40,000 ft can be found in Table 1.2.

$$\mu_\infty = 0.79447\mu_{SL} = 2.9713 \times 10^{-7} \frac{\text{lbf} \cdot \text{s}}{\text{ft}^2}$$

$$\rho_\infty = 0.2471\rho_{SL} = 5.8711 \times 10^{-4} \frac{\text{slug}}{\text{ft}^3}$$

$$a_\infty = 968.08 \text{ ft/s}$$

Since the Mach number is 2.0,

$$U_\infty = M_\infty a_\infty = 2.0\left(968.08\frac{\text{ft}}{\text{s}}\right) = 1936.16\frac{\text{ft}}{\text{s}}\left(\frac{3600\frac{\text{s}}{\text{h}}}{5280\frac{\text{ft}}{\text{mi}}}\right) = 1320.11\frac{\text{mi}}{\text{h}}$$

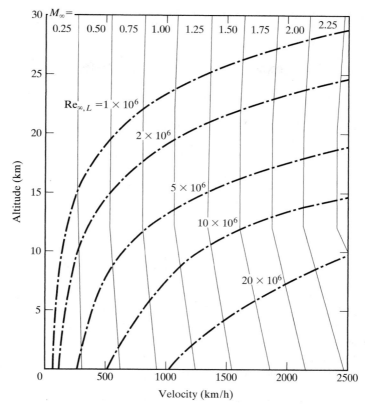

Figure 2.12 Reynolds number/Mach number correlations as a function of velocity and altitude for U.S. Standard Atmosphere.

The corresponding Reynolds number is

$$\text{Re}_{\infty, L} = \frac{\rho_\infty U_\infty L}{\mu_\infty} = \frac{\left(5.8711 \times 10^{-4} \frac{\text{lbf} \cdot \text{s}^2}{\text{ft}^4}\right)\left(1936.16 \frac{\text{ft}}{\text{s}}\right)(14 \text{ ft})}{2.9713 \times 10^{-7} \frac{\text{lbf} \cdot \text{s}}{\text{ft}^2}}$$

$$= 5.3560 \times 10^7$$

2.6 CONCEPT OF THE BOUNDARY LAYER

For many high-Reynolds-number flows (such as those of interest to the aerodynamicist), the flow field may be divided into two regions: (1) a viscous *boundary layer* adjacent to the surface of the vehicle and (2) the essentially inviscid flow outside the boundary layer. The velocity of the fluid particles increases from a value of zero (in a vehicle-fixed coordinate system) at the wall to the value that corresponds to the external "frictionless"

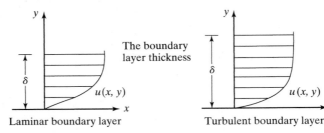

- Relatively thin layer with limited mass transfer
- Relatively low velocity gradient near the wall
- Relatively low skin friction

- Thicker layer with considerable mass transport
- Higher velocities near the surface
- Higher skin friction

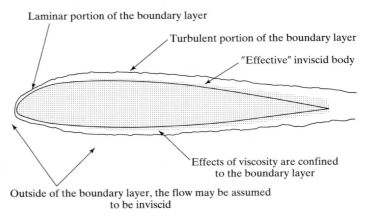

Figure 2.13 Viscous boundary layer on an airfoil.

flow outside the boundary layer, whose edge is represented by the solid lines in Fig. 2.13. Because of the resultant velocity gradients, the shear forces are relatively large in the boundary layer. Outside the boundary layer, the velocity gradients become so small that the shear stresses acting on a fluid element are negligible. Thus, the effect of the viscous terms may be ignored in the solution for the flow field external to the boundary layer. To generate a solution for the inviscid portion of the flow field, we require that the velocity of the fluid particles at the surface be parallel to the surface (but not necessarily of zero magnitude). This represents the physical requirement that there is no flow through a solid surface. The analyst may approximate the effects of the boundary layer on the inviscid solution by defining the geometry of the surface to be that of the actual surface plus a displacement due to the presence of the boundary layer (as represented by the shaded area in Fig. 2.13). The "effective" inviscid body (the actual configuration plus the displacement thickness) is represented by the shaded area of Fig. 2.13. The solution of the boundary-layer equations and the subsequent determination of a corresponding displacement thickness are dependent on the velocity at the edge of the boundary layer (which is, in effect, the velocity at the surface that corresponds to the inviscid solution). The process of determining the interaction of the solutions provided by the inviscid-flow equations with those for the boundary-layer equations requires a thorough understanding of the problem [e.g., refer to Brune et al., (1974)].

For many problems involving flow past streamlined shapes such as airfoils and wings (at low angles of attack), the presence of the boundary layer causes the actual pressure distribution to be only negligibly different from the inviscid pressure distribution. Let us consider this statement further by studying the x and y components of equations (2.12) for a two-dimensional incompressible flow. The resultant equations, which define the flow in the boundary layer shown in Fig. 2.13, are

$$\rho \frac{\partial u}{\partial t} + \rho u \frac{\partial u}{\partial x} + \rho v \frac{\partial u}{\partial y} = -\frac{\partial p}{\partial x} + \mu \frac{\partial^2 u}{\partial x^2} + \mu \frac{\partial^2 u}{\partial y^2}$$

and

$$\rho \frac{\partial v}{\partial t} + \rho u \frac{\partial v}{\partial x} + \rho v \frac{\partial v}{\partial y} = -\frac{\partial p}{\partial y} + \mu \frac{\partial^2 v}{\partial x^2} + \mu \frac{\partial^2 v}{\partial y^2}$$

where the x coordinate is measured parallel to the airfoil surface and the y coordinate is measured perpendicular to it. Solving for the pressure gradients gives us

$$-\frac{\partial p}{\partial x} = \left(\rho \frac{\partial}{\partial t} + \rho u \frac{\partial}{\partial x} + \rho v \frac{\partial}{\partial y} - \mu \frac{\partial^2}{\partial x^2} - \mu \frac{\partial^2}{\partial y^2} \right) u$$

$$-\frac{\partial p}{\partial y} = \left(\rho \frac{\partial}{\partial t} + \rho u \frac{\partial}{\partial x} + \rho v \frac{\partial}{\partial y} - \mu \frac{\partial^2}{\partial x^2} - \mu \frac{\partial^2}{\partial y^2} \right) v$$

Providing that the boundary layer near the solid surface is thin, the normal component of velocity is usually much less than the streamwise component of velocity (i.e., $v < u$). Thus, the terms on the right-hand side of the equation in the second line are typically smaller than the corresponding term in the first line. We conclude, then, that

$$\frac{\partial p}{\partial y} < \frac{\partial p}{\partial x}$$

As a result, the pressure gradient normal to the surface is negligible:

$$\frac{\partial p}{\partial y} \approx 0$$

which is verified by experiment. Since the static pressure variation across the boundary layer is usually negligible, the pressure distribution around the airfoil is essentially that of the inviscid flow (accounting for the displacement effect of the boundary layer).

The assumption that the static pressure variation across the boundary layer is negligible breaks down for turbulent boundary layers at very high Mach numbers. Bushnell et al., (1977) cite data for which the wall pressure is significantly greater than the edge value for turbulent boundary layers where the edge Mach number is approximately 20. The characteristics distinguishing laminar and turbulent boundary layers are discussed in Chapter 4.

When the combined action of an adverse pressure gradient and the viscous forces causes the boundary layer to separate from the vehicle surface (which may occur for blunt bodies or for streamlined shapes at high angles of attack), the flow field is very sensitive to the Reynolds number. The Reynolds number, therefore, also serves as an indicator of how much of the flow can be accurately described by the inviscid-flow equations. For detailed discussions of the viscous portion of the flow field, the reader is referred to Chapter 4 and to Schlichting (1979), White (2005), Schetz (1993), and Wilcox (1998).

2.7   CONSERVATION OF ENERGY

There are many flows that involve sufficient temperature variations so that convective heat transfer is important, but for which the constant-property assumption is reasonable. An example is flow in a heat exchanger. For such flows, the temperature field is obtained by solving the energy equation after the velocity field has been determined by solving the continuity equation and the momentum equation. This is because, for this case, the continuity equation and the momentum equation are independent of the energy equation, but not vice versa.

We must also include the energy equation in the solution algorithm for compressible flows. Compressible flows are those in which the pressure and temperature variations are sufficiently large that we must account for changes in the other fluid properties (e.g., density and viscosity). For compressible flows, the continuity equation, the momentum equation, and the energy equation must be solved simultaneously. Recall the discussion relating to equations (2.12a) through (2.12c).

In the remainder of this chapter, we derive the energy equation and discuss its application to various flows.

2.8   FIRST LAW OF THERMODYNAMICS

Consider a system of fluid particles. Everything outside the group of particles is called the *surroundings* of the system. The *first law of thermodynamics* results from the fundamental experiments of James Joule. Joule found that, for a cyclic process, that is, one in which the initial state and the final state of the fluid are identical,

$$\oint \delta q - \oint \delta w = 0 \qquad \textbf{(2.21)}$$

Thus, Joule has shown that the heat transferred from the surroundings to the system less the work done by the system on its surroundings during a cyclic process is zero. In equation (2.21), we have adopted the convention that heat transfer to the system is positive and that work done by the system is positive. The use of lower case symbols to represent the parameters means that we are considering the magnitude of the parameter per unit mass of the fluid. We use the symbols δq and δw to designate that the incremental heat transfer to the system and the work done by the system are not exact differentials but depend on the process used in going from state 1 to state 2. Equation (2.21) is true for any and all cyclic processes. Thus, if we apply it to a process that takes place between any two states (1 and 2), then

$$\delta q - \delta w = de = e_2 - e_1 \qquad \textbf{(2.22)}$$

where e is the total energy per unit mass of the fluid. Note that de is an exact differential and the energy is, therefore, a property of the fluid. The energy is usually divided into three components: (1) kinetic energy, (2) potential energy, and (3) all other energy. The internal energy of the fluid is part of the third component. In this book, we will be concerned only with kinetic, potential, and internal energies. Chemical, nuclear, and other forms of energy are normally not relevant to the study of aerodynamics. Since we are normally only concerned with changes in energy rather than its absolute value, an arbitrary zero energy (or datum) state can be assigned.

In terms of the three energy components, equation (2.22) becomes

$$\delta q - \delta w = dke + dpe + du_e \qquad (2.23)$$

Note that u_e is the symbol used for specific internal energy (i.e., the internal energy per unit mass).

Work.   In mechanics, work is defined as the effect that is produced by a system on its surroundings when the system moves the surroundings in the direction of the force exerted by the system on its surroundings. The magnitude of the effect is measured by the product of the displacement times the component of the force in the direction of the motion. Thermodynamics deals with phenomena considerably more complex than covered by this definition from mechanics. Thus, we may say that work is done by a system on its surroundings if we can postulate a process in which the system passes through the same series of states as in the original process, but in which the sole effect on the surroundings is the raising of a weight.

In an inviscid flow, the only forces acting on a fluid system (providing we neglect gravity) are the pressure forces. Consider a small element of the surface dA of a fluid system, as shown in Fig. 2.14. The force acting on dA due to the fluid in the system is $p\,dA$. If this force displaces the surface a differential distance ds in the direction of the force, the work done is $p\,dA\,ds$. Differential displacements are assumed so that the process is reversible; that is, there are no dissipative factors such as friction and/or heat transfer. But the product of dA times ds is just $d(\text{vol})$, the change in volume of the system. Thus, the work per unit mass is

$$\delta w = +p\,dv \qquad (2.24a)$$

where v is the volume per unit mass (or specific volume). It is, therefore, the reciprocal of the density. The reader should not confuse this use of the symbol v with the y component of velocity. Equivalently,

$$w = +\int_1^2 p\,dv \qquad (2.24b)$$

where the work done by the system on its surroundings in going from state 1 to state 2 (a finite process), as given by equation (2.24b), is positive when dv represents an increase in volume.

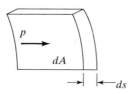

Figure 2.14 Incremental work done by the pressure force, which acts normal to the surface.

2.9 DERIVATION OF THE ENERGY EQUATION

Having discussed the first law and its implications, we are now ready to derive the differential form of the energy equation for a viscous, heat-conducting compressible flow. Consider the fluid particles shown in Fig. 2.15. Writing equation (2.23) in rate form, we can describe the energy balance on the particle as it moves along in the flow:

$$\rho\dot{q} - \rho\dot{w} = \rho\frac{d}{dt}(e) = \rho\frac{d}{dt}(ke) + \rho\frac{d}{dt}(pe) + \rho\frac{d}{dt}(u_e) \qquad \textbf{(2.25)}$$

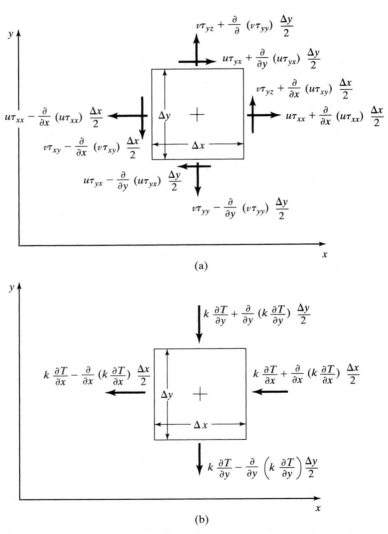

(a)

(b)

Figure 2.15 Heat-transfer and flow-work terms for the energy equation for a two-dimensional fluid element: (a) work done by stresses acting on a two-dimensional element; (b) heat transfer to a two-dimensional element.

where the overdot notation denotes differentiation with respect to time. Recall that the substantial (or total) derivative is

$$\frac{d}{dt} = \frac{\partial}{\partial t} + \vec{V} \cdot \nabla$$

and therefore represents the local, time-dependent changes, as well as those due to convection through space.

To simplify the illustration of the energy balance on the fluid particle, we shall again consider a two-dimensional flow, as shown in Fig. 2.15. The rate at which work is done by the system on its surrounding is equal to the negative of the product of the forces acting on a boundary surface times the flow velocity (i.e., the displacement per unit time) at that surface. The work done by the body forces is not included in this term. It is accounted for in the potential energy term, see equation (2.27). Thus, using the nomenclature of Fig. 2.15a, we can evaluate the rate at which work is done by the system (per unit depth):

$$-\dot{W} = \frac{\partial}{\partial x}(u\tau_{xx})\,\Delta x\,\Delta y + \frac{\partial}{\partial x}(v\tau_{xy})\,\Delta x\,\Delta y + \frac{\partial}{\partial y}(u\tau_{yx})\,\Delta y\,\Delta x + \frac{\partial}{\partial y}(v\tau_{yy})\,\Delta y\,\Delta x$$

Using the constitutive relations for $\tau_{xx}, \tau_{xy}, \tau_{yx}$, and τ_{yy} given earlier in this chapter and dividing by $\Delta x\,\Delta y$, we obtain

$$-\rho\dot{w} = 2\mu\left[\left(\frac{\partial u}{\partial x}\right)^2 + \left(\frac{\partial v}{\partial y}\right)^2\right] - p\left(\frac{\partial u}{\partial x} + \frac{\partial v}{\partial y}\right) - \frac{2}{3}\mu(\nabla \cdot \vec{V})^2$$

$$+ \mu\left[\left(\frac{\partial u}{\partial y} + \frac{\partial v}{\partial x}\right)^2\right] + u\frac{\partial\tau_{xx}}{\partial x} + v\frac{\partial\tau_{yy}}{\partial y} + v\frac{\partial\tau_{xy}}{\partial x} + u\frac{\partial\tau_{yx}}{\partial y} \quad \textbf{(2.26a)}$$

From the component momentum equations (2.11),

$$u\frac{\partial\tau_{xx}}{\partial x} + u\frac{\partial\tau_{yx}}{\partial y} = u\rho\frac{du}{dt} - u\rho f_x \quad\quad\quad \textbf{(2.26b)}$$

$$v\frac{\partial\tau_{xy}}{\partial x} + v\frac{\partial\tau_{yy}}{\partial y} = v\rho\frac{dv}{dt} - v\rho f_y \quad\quad\quad \textbf{(2.26c)}$$

From Fourier's law of heat conduction,

$$\dot{Q} = -k\hat{n}A \cdot \nabla T$$

we can evaluate the rate at which heat is added to the system (per unit depth). Note that the symbol T will be used to denote temperature, the symbol t, time, and $\dot{Q}$, the total heat flux rate. Referring to Fig. 2.15b and noting that, if the temperature is increasing in the outward direction, heat is added to the particle (which is positive by our convention),

$$\dot{Q} = +\frac{\partial}{\partial x}\left(k\frac{\partial T}{\partial x}\right)\Delta x\,\Delta y + \frac{\partial}{\partial y}\left(k\frac{\partial T}{\partial y}\right)\Delta y\,\Delta x$$

Therefore,

$$\rho\dot{q} = \frac{\partial}{\partial x}\left(k\frac{\partial T}{\partial x}\right) + \frac{\partial}{\partial y}\left(k\frac{\partial T}{\partial y}\right) \tag{2.26d}$$

Substituting equations (2.26) into equation (2.25), we obtain

$$\frac{\partial}{\partial x}\left(k\frac{\partial T}{\partial x}\right) + \frac{\partial}{\partial y}\left(k\frac{\partial T}{\partial y}\right) + 2\mu\left[\left(\frac{\partial u}{\partial x}\right)^2 + \left(\frac{\partial v}{\partial y}\right)^2\right] - p\nabla\cdot\vec{V}$$

$$- \frac{2}{3}\mu(\nabla\cdot\vec{V})^2 + \mu\left[\left(\frac{\partial u}{\partial y} + \frac{\partial v}{\partial x}\right)^2\right] + \rho u\frac{du}{dt} - \rho u f_x + \rho v\frac{dv}{dt} - \rho v f_x$$

$$= \rho\frac{d[(u^2 + v^2)/2]}{dt} + \rho\frac{d(pe)}{dt} + \rho\frac{d(u_e)}{dt} \tag{2.27}$$

From the continuity equation,

$$\nabla\cdot\vec{V} = -\frac{1}{\rho}\frac{d\rho}{dt}$$

and by definition

$$\rho\frac{d(p/\rho)}{dt} = \frac{dp}{dt} - \frac{p}{\rho}\frac{d\rho}{dt}$$

Thus,

$$-p\nabla\cdot\vec{V} = -\rho\frac{d(p/\rho)}{dt} + \frac{dp}{dt} \tag{2.28a}$$

For a conservative force field,

$$\rho\frac{d(pe)}{dt} = \rho\vec{V}\cdot\nabla F = -\rho u f_x - \rho v f_y \tag{2.28b}$$

where F is the body-force potential and $\nabla F = -\bar{f}$, as introduced in equation (3.3). Substituting equations (2.28) into equation (2.27), we obtain

$$\frac{\partial}{\partial x}\left(k\frac{\partial T}{\partial x}\right) + \frac{\partial}{\partial y}\left(k\frac{\partial T}{\partial y}\right) + 2\mu\left[\left(\frac{\partial u}{\partial x}\right)^2 + \left(\frac{\partial v}{\partial y}\right)^2\right] - \frac{2}{3}\mu(\nabla\cdot\vec{V})^2$$

$$+ \mu\left[\left(\frac{\partial u}{\partial y} + \frac{\partial v}{\partial x}\right)^2\right] - \rho\frac{d(p/\rho)}{dt} + \frac{dp}{dt} + \rho u\frac{du}{dt} + \rho v\frac{dv}{dt} - \rho u f_x - \rho v f_y$$

$$= \rho u\frac{du}{dt} + \rho v\frac{dv}{dt} - \rho u f_x - \rho v f_y + \rho\frac{d(u_e)}{dt} \tag{2.29}$$

Since the terms u_e and p/ρ appear as a sum in many flow applications, it is convenient to introduce a symbol for this sum. Let us introduce the definition that

$$h = u_e + \frac{p}{\rho} \tag{2.30}$$

where h is called the specific enthalpy. Using equation (2.30) and combining terms, we can write equation (2.29) as

$$\frac{\partial}{\partial x}\left(k\frac{\partial T}{\partial x}\right) + \frac{\partial}{\partial y}\left(k\frac{\partial T}{\partial y}\right) + 2\mu\left[\left(\frac{\partial u}{\partial x}\right)^2 + \left(\frac{\partial v}{\partial y}\right)^2\right]$$

$$-\frac{2}{3}\mu(\nabla\cdot\vec{V})^2 + \mu\left[\left(\frac{\partial u}{\partial y} + \frac{\partial v}{\partial x}\right)^2\right] = \rho\frac{dh}{dt} - \frac{dp}{dt} \tag{2.31}$$

This is the energy equation for a general, compressible flow in two dimensions. The process can be extended to a three-dimensional flow field to yield

$$\rho\frac{dh}{dt} - \frac{dp}{dt} = \nabla\cdot(k\nabla T) + \phi \tag{2.32a}$$

where

$$\phi = -\frac{2}{3}\mu(\nabla\cdot\vec{V})^2 + 2\mu\left[\left(\frac{\partial u}{\partial x}\right)^2 + \left(\frac{\partial v}{\partial y}\right)^2 + \left(\frac{\partial w}{\partial z}\right)^2\right]$$

$$+ \mu\left[\left(\frac{\partial u}{\partial y} + \frac{\partial v}{\partial x}\right)^2 + \left(\frac{\partial v}{\partial z} + \frac{\partial w}{\partial y}\right)^2 + \left(\frac{\partial w}{\partial x} + \frac{\partial u}{\partial z}\right)^2\right] \tag{2.32b}$$

Equation (2.32b) defines the dissipation function ϕ, which represents the rate at which work is done by the viscous forces per unit volume.

2.9.1 Integral Form of the Energy Equation

The integral form of the energy equation is

$$\dot{Q} - \dot{W} = \iiint \frac{\partial}{\partial t}(\rho e)d(\text{vol}) + \oiint e\rho\vec{V}\cdot\hat{n}\,dA \tag{2.33}$$

That is, the net rate heat is added to the system less the net rate work is done by the system is equal to the time rate of change of energy within the control volume plus the net efflux of energy across the system boundary. Note that the heat added to the system is positive. So, too, is the work done by the system. Conversely, heat transferred from the system or work done on the system is negative by this convention.

2.9.2 Energy of the System

As noted earlier, the energy of the system can take a variety of forms. They are usually grouped as follows:

1. *Kinetic energy (ke)*: energy associated with the directed motion of the mass
2. *Potential energy (pe)*: energy associated with the position of the mass in the external field
3. *Internal energy (u_e)*: energy associated with the internal fields and the random motion of the molecules

Thus, the energy of the system may be written as

$$e = ke + pe + u_e \tag{2.34a}$$

Let us further examine the terms that comprise the energy of the system. The kinetic energy per unit mass is given by

$$ke = \tfrac{1}{2}V^2 \tag{2.34b}$$

Note that the change in kinetic energy during a process clearly depends only on the initial velocity and final velocity of the system of fluid particles. Assuming that the external force field is that of gravity, the potential energy per unit mass is given by

$$pe = gz \tag{2.34c}$$

Note that the change in the potential energy depends only on the initial and final elevations. Furthermore, the change in internal energy is a function of the values at the endpoints only.

Substituting equations (2.34) into equation (2.33), we obtain

$$\dot{Q} - \dot{W} = \frac{\partial}{\partial t} \iiint \rho \left(\frac{V^2}{2} + gz + u_e \right) d(\text{vol})$$

$$+ \oiint \rho \left(\frac{V^2}{2} + gz + u_e \right) \vec{V} \cdot \hat{n}\, dA \tag{2.35}$$

It should be noted that, whereas the changes in the energy components are a function of the states, the amount of heat transferred and the amount of work done during a process are path dependent. That is, the changes depend not only on the intial and final states but on the process that takes place between these states.

Let us consider further the term for the rate at which work is done, $\dot{W}$. For convenience, the total work rate is divided into flow work rate ($\dot{W}_f$), viscous work rate ($\dot{W}_v$), and shaft work rate ($\dot{W}_s$).

2.9.3 Flow Work

Flow work is the work done by the pressure forces on the surroundings as the fluid moves through space. Consider flow through the streamtube shown in Fig. 2.16. The

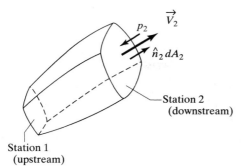

Station 1
(upstream)

Figure 2.16 Streamtube for derivation of equation (2.36).

pressure p_2 acts over the differential area $\hat{n}_2\, dA_2$ at the right end (i.e., the downstream end) of the control volume. Recall that the pressure is a compressive force acting on the system of particles. Thus, the force acting on the right end surface is $-p_2\hat{n}_2\, dA_2$. In moving the surrounding fluid through the distance $\vec{V}_2\,\Delta t$ for the velocity shown in Fig. 2.16, the system does work on the surroundings (which is positive by our sign convention). Thus, the work done is the dot product of the force times the distance:

$$p_2\hat{n}_2\, dA_2 \cdot \vec{V}_2\,\Delta t$$

$$\dot{W}_{f,2} = p_2\vec{V}_2 \cdot \hat{n}_2\, dA_2 = \frac{p_2}{\rho_2}\rho_2\vec{V}_2 \cdot \hat{n}_2\, dA_2 \qquad \textbf{(2.36a)}$$

The positive sign is consistent with the assumed directions of Fig. 2.16; that is, the velocity and the area vectors are in the same direction. Thus, the dot product is consistent with the convention that work done by the system on the surroundings is positive.

In a similar manner, it can be shown that the flow work done on the surrounding fluid at the upstream end (station 1) is

$$\dot{W}_{f,1} = -\frac{p_1}{\rho_1}\rho_1\vec{V}_1 \cdot \hat{n}_1\, dA_1 \qquad \textbf{(2.36b)}$$

The negative sign results because the pressure force is compressive (acts on the system of particles), and the assumed velocity represents movement of the fluid particles in that direction. Thus, since work is done by the surroundings on the system at the upstream end, it is negative by our sign convention.

2.9.4 Viscous Work

Viscous work is similar to flow work in that it is the result of a force acting on the surface bounding the fluid system. In the case of viscous work, however, the pressure is replaced by the viscous shear. The rate at which viscous work is done by the system over some incremental area of the system surface (dA) is

$$\dot{W}_v = -\bar{\tau} \cdot \vec{V}\, dA \qquad \textbf{(2.36c)}$$

2.9.5 Shaft Work

Shaft work is defined as any other work done by the system other than the flow work and the viscous work. This usually enters or leaves the system through the action of a shaft (from which the term originates), which either takes energy out of or puts energy into the system. Since a turbine extracts energy from the system, the system does work on the surroundings and $\dot{W}_s$ is positive. In the case where the shaft is that of a pump, the surroundings are doing work on the system and $\dot{W}_s$ is negative.

2.9.6 Application of the Integral Form of the Energy Equation

Thus, the energy equation can be written

$$\dot{Q} - \dot{W}_v - \dot{W}_s = \frac{\partial}{\partial t} \iiint \rho\left(\frac{V^2}{2} + gz + u_e\right) d(\text{vol})$$

$$+ \oiint \rho\left(\frac{V^2}{2} + gz + u_e + \frac{p}{\rho}\right) \vec{V} \cdot \hat{n} \, dA \qquad \text{(2.37)}$$

Note that the flow work, as represented by equation (2.36), has been incorporated into the second integral of the right-hand side of equation (2.37).

For a steady, adiabatic flow ($\dot{Q} = 0$) with no shaft work ($\dot{W}_s = 0$) and with no viscous work ($\dot{W}_v = 0$), equation (2.37) can be written

$$\oiint \rho\left(\frac{V^2}{2} + gz + h\right) \vec{V} \cdot \hat{n} \, dA = 0 \qquad \text{(2.38)}$$

where the definition for the enthalpy has been used.

EXAMPLE 2.5:   A flow where the energy equation is Bernoulli's equation

Consider the steady, inviscid, one-dimensional flow of water in the curved pipe shown in Fig. 2.17. If water drains to the atmosphere at station 2 at the rate of 0.001π m^3/s what is the static pressure at station 1? There is no shaft work or heat transfer, and there are no perceptible changes in the internal energy.

Solution: We will apply equation (2.37) to the control volume that encloses the fluid in the pipe between stations 1 and 2. Applying the conditions and assumptions in the problem statement,

No heat transfer $\dot{Q} = 0$

No shaft work $\dot{W}_s = 0$

No viscous work $\dot{W}_v = 0$

Steady flow $\dfrac{\partial}{\partial t} = 0$

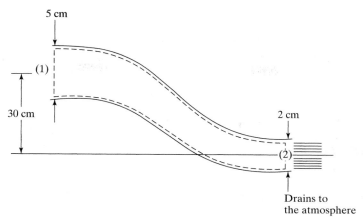

Figure 2.17 Pipe flow for Example 2.5.

Thus,

$$\oiint \left(\frac{V^2}{2} + gz + u_e + \frac{p}{\rho} \right) \rho \vec{V} \cdot \hat{n} \, dA = 0$$

Since the properties for the inviscid, one-dimensional flow are uniform over the plane of each station and since the velocities are perpendicular to the cross-sectional area, the integral can readily be evaluated. It is

$$\left(\frac{V_1^2}{2} + gz_1 + u_{e1} + \frac{p_1}{\rho_1} \right) \rho_1 V_1 A_1 = \left(\frac{V_2^2}{2} + gz_2 + u_{e2} + \frac{p_2}{\rho_2} \right) \rho_2 V_2 A_2$$

By continuity, $\rho_1 V_1 A_1 = \rho_2 V_2 A_2$. Since water is incompressible, $\rho_1 = \rho_2$. Furthermore, we are told that there are no perceptible changes in the internal energy (i.e., $u_{e1} = u_{e2}$). Thus,

$$\frac{V_1^2}{2} + gz_1 + \frac{p_1}{\rho} = \frac{V_2^2}{2} + gz_2 + \frac{p_2}{\rho}$$

Note that the resultant form of the energy equation for this flow is Bernoulli's equation.

Thus, for an incompressible, steady, nondissipative flow, we have a mechanical energy equation which simply equates the flow work with the sum of the changes in potential energy and in kinetic energy.

$$V_2 = \frac{Q}{A_2} = \frac{0.001\pi}{[\pi(0.02)^2]/4} = 10 \, \text{m/s}$$

$$V_1 = \left(\frac{D_2}{D_1} \right)^2 V_2 = 0.16(10) = 1.6 \, \text{m/s}$$

Thus,

$$\frac{(1.6)^2}{2} + 9.8066(0.3) + \frac{p_1}{1000} = \frac{10^2}{2} + \frac{p_{atm}}{1000}$$

$$p_1 = p_{atm} + (50 - 1.28 - 2.94)1000$$

$$= 4.58 \times 10^4 \, \text{N/m}^2, \text{gage}$$

2.10 SUMMARY

Both the differential and integral forms of the equations of motion for a compressible, viscous flow have been developed in this chapter. Presented were the continuity equation, equation (2.2) or (2.5); the momentum equation, equation (2.12) or (2.13); and the energy equation, equation (2.32) or (2.33). The dependent variables, or unknowns, in these equations include pressure, temperature, velocity, density, viscosity, thermal conductivity, internal energy, and enthalpy. For many applications, simplifying assumptions can be introduced to eliminate one or more of the dependent variables. For example, a common assumption is that the variation in the fluid properties as the fluid moves through the flow field are very small when the Mach number is less than 0.3. Thus, assuming that the density, the viscosity, and the thermal conductivity are constant, we can eliminate them as variables and still obtain solutions of suitable accuracy. Examples of constant-property flows will be worked out in subsequent chapters.

PROBLEMS

2.1. Derive the continuity equation in cylindrical coordinates, starting with the general vector form

$$\frac{\partial \rho}{\partial t} + \nabla \cdot (\rho \vec{V}) = 0$$

where

$$\nabla = \hat{e}_r \frac{\partial}{\partial r} + \frac{\hat{e}_\theta}{r} \frac{\partial}{\partial \theta} + \hat{e}_z \frac{\partial}{\partial z}$$

in cylindrical coordinates. Note also that $\dfrac{\partial \hat{e}_r}{\partial \theta}$ and $\dfrac{\partial \hat{e}_\theta}{\partial \theta}$ are not zero.

$$\frac{1}{r} \frac{\partial (\rho r v_r)}{\partial r} + \frac{1}{r} \frac{\partial (\rho v_\theta)}{\partial \theta} + \frac{\partial (\rho v_z)}{\partial z} = 0$$

2.2. Which of the following flows are physically possible, that is, satisfy the continuity equation? Substitute the expressions for density and for the velocity field into the continuity equation to substantiate your answer.

(a) Water, which has a density of $1.0 \, \text{g/cm}^3$, is flowing radially outward from a source in a plane such that $\vec{V} = (K/2\pi r)\hat{e}_r$. Note that $v_\theta = v_z = 0$. Note also that, in cylindrical coordinates,

$$\nabla = \hat{e}_r \frac{\partial}{\partial r} + \frac{\hat{e}_\theta}{r} \frac{\partial}{\partial \theta} + \hat{e}_z \frac{\partial}{\partial z}$$

(b) A gas is flowing at relatively low speeds (so that its density may be assumed constant) where

$$u = -\frac{2xyz}{(x^2 + y^2)^2}U_\infty L$$

$$v = \frac{(x^2 - y^2)z}{(x^2 + y^2)^2}U_\infty L$$

$$w = \frac{y}{x^2 + y^2}U_\infty L$$

Here U_∞ and L are a reference velocity and a reference length, respectively.

2.3. Two of the three velocity components for an incompressible flow are:

$$u = x^2 + 2xz \qquad v = y^2 + 2yz$$

What is the general form of the velocity component $w(x,y,z)$ that satisfies the continuity equation?

2.4. A two-dimensional velocity field is given by

$$u = -\frac{Ky}{x^2 + y^2} \qquad v = +\frac{Kx}{x^2 + y^2}$$

where K is a constant. Does this velocity field satisfy the continuity equation for incompressible flow? Transform these velocity components into the polar components v_r and v_θ in terms of r and θ. What type of flow might this velocity field represent?

2.5. The velocity components for a two-dimensional flow are

$$u = \frac{C(y^2 - x^2)}{(x^2 + y^2)^2} \qquad v = \frac{-2Cxy}{(x^2 + y^2)^2}$$

where C is a constant. Does this flow satisfy the continuity equation?

2.6. For the two-dimensional flow of incompressible air near the surface of a flat plate, the steamwise (or x) component of the velocity may be approximated by the relation

$$u = a_1 \frac{y}{\sqrt{x}} - a_2 \frac{y^3}{x^{1.5}}$$

Using the continuity equation, what is the velocity component v in the y direction? Evaluate the constant of integration by noting that $v = 0$ at $y = 0$.

2.7. Consider a one-dimensional steady flow along a streamtube. Differentiate the resultant integral continuity equation to show that

$$\frac{d\rho}{\rho} + \frac{dA}{A} + \frac{dV}{V} = 0$$

For a low-speed, constant-density flow, what is the relation between the change in area and the change in velocity?

2.8. Water flows through a circular pipe, as shown in Fig. P2.8, at a constant volumetric flow rate of $0.5 \text{ m}^3/\text{s}$. Assuming that the velocities at stations 1, 2, and 3 are uniform across the cross section (i.e., that the flow is one dimensional), use the integral form of the continuity equation to calculate the velocities, V_1, V_2, and V_3. The corresponding diameters are $d_1 = 0.4 \text{ m}$, $d_2 = 0.2 \text{ m}$, and $d_3 = 0.6 \text{ m}$.

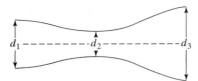

Figure P2.8

2.9. A long pipe (with a reducer section) is attached to a large tank, as shown in Fig. P2.9. The diameter of the tank is 5.0 m; the diameter to the pipe is 20 cm at station 1 and 10 cm at station 2. The effects of viscosity are such that the velocity (u) may be considered constant across the cross section at the surface (s) and at station 1, but varies with the radius at station 2 such that

$$u = U_0\left(1 - \frac{r^2}{R_2^2}\right)$$

where U_0 is the velocity at the centerline, R_2 the radius of the pipe at station 2, and r the radial coordinate. If the density is 0.85 g/cm^3 and the mass flow rate is 10 kg/s, what are the velocities at s and 1, and what is the value of U_0?

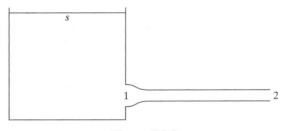

Figure P2.9

A note for Problems 2.10 through 2.13 and 2.27 through 2.31. The drag force acting on an airfoil can be calculated by determining the change in the momentum of the fluid as it flows past the airfoil. As part of this exercise, one measures the velocity distribution well upstream of the airfoil and well downstream of the airfoil. The integral equations of motion can be applied to either a rectangular control volume or a control volume bounded by streamlines.

2.10. Velocity profiles are measured at the upstream end (surface 1) and at the downstream end (surface 2) of a rectangular control volume, as shown in Fig. P2.10. If the flow is incompressible, two dimensional, and steady, what is the total volumetric flow rate ($\iint \vec{V} \cdot \hat{n}\, dA$) across the horizontal surfaces (surfaces 3 and 4)?

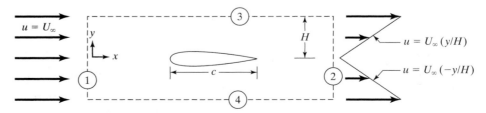

Figure P2.10

2.11. Velocity profiles are measured at the upstream end (surface 1) and at the downstream end (surface 2) of the control volume shown in Fig. P2.11. The flow is incompressible, two dimensional, and steady. If surfaces 3 and 4 are streamlines, what is the vertical dimension of the upstream station (H_U)?

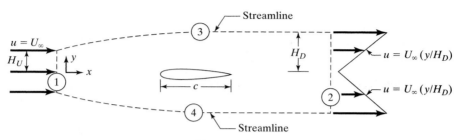

Figure P2.11

2.12. Velocity profiles are measured at the upstream end (surface 1) and at the downstream end (surface 2) of a rectangular control volume, as shown in Fig. P2.12. If the flow is incompressible, two dimensional, and steady, what is the total volumetric flow rate ($\iint \vec{V} \cdot \hat{n}\, dA$) across the horizontal surfaces (surfaces 3 and 4)?

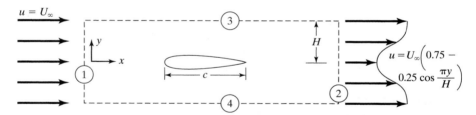

Figure P2.12

2.13. Velocity profiles are measured at the upstream end (surface 1) and at the downstream end (surface 2) of the control volume shown in Fig. P2.13. The flow is incompressible, two dimensional, and steady. If surfaces 3 and 4 are streamlines, what is the vertical dimension of the upstream station (H_U)?

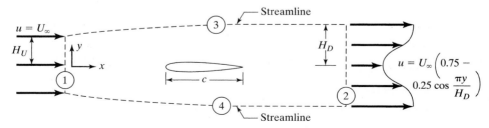

Figure P2.13

2.14. One cubic meter per second of water enters a rectangular duct as shown in Fig. P2.14. Two of the surfaces of the duct are porous. Water is added through the upper surface at a rate shown by the parabolic curve, while it leaves through the front face at a rate that decreases linearly with the distance from the entrance. The maximum value of both flow rates, shown in the sketch, are given in cubic meters per second per unit length along the duct. What is the average velocity of water leaving the duct if it is 1.0 m long and has a cross section of 0.1 m^2?

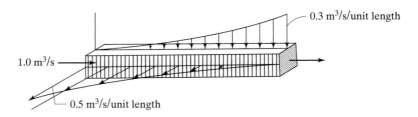

Figure P2.14

2.15. For the conditions of Problem 2.14, determine the position along the duct where the average velocity of flow is a minimum. What is the average velocity at this station?

2.16. As shown in Fig. P2.16, 1.5 m^3/s of water leaves a rectangular duct. Two of the surfaces of the duct are porous. Water leaves through the upper surface at a rate shown by the parabolic curve, while it enters the front face at a rate that decreases linearly with distance from the entrance. The maximum values of both flow rates, shown in the sketch, are given in cubic meters per second per unit length along the duct. What is the average velocity of the water entering the duct if it is 1.0 m long and has a cross section of 0.1 m^2?

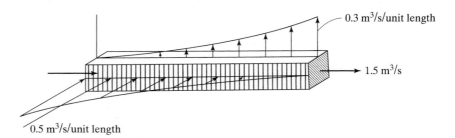

Figure P2.16

2.17. Consider the velocity field

$$\vec{V} = -\frac{x}{2t}\hat{i}$$

in a compressible flow where $\rho = \rho_0 xt$. Using equation (2.8), what is the total acceleration of a particle at $(1, 1, 1)$ at time $t = 10$?

2.18. Given the velocity field

$$\vec{V} = (6 + 2xy + t^2)\hat{i} - (xy^2 + 10t)\hat{j} + 25\hat{k}$$

what is the acceleration of a particle at $(3, 0, 2)$ at time $t = 1$?

2.19. Consider steady two-dimensional flow about a cylinder of radius R (Fig. P2.19). Using cylindrical coordinates, we can express the velocity field for steady, inviscid, incompressible flow around the cylinder as

$$\vec{V}(r,\theta) = U_\infty\left(1 - \frac{R^2}{r^2}\right)\cos\theta\,\hat{e}_r - U_\infty\left(1 + \frac{R^2}{r^2}\right)\sin\theta\,\hat{e}_\theta$$

where U_∞ is the velocity of the undisturbed stream (and is, therefore, a constant). Derive the expression for the acceleration of a fluid particle at the surface of the cylinder (i.e., at points where $r = R$). Use equation (2.8) and the definition that

$$\nabla = \hat{e}_r\frac{\partial}{\partial r} + \frac{\hat{e}_\theta}{r}\frac{\partial}{\partial\theta} + \hat{e}_z\frac{\partial}{\partial z}$$

and

$$\vec{V} = v_r\hat{e}_r + v_\theta\hat{e}_\theta + v_z\hat{e}_z$$

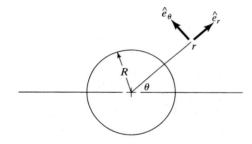

Figure P2.19

2.20. Consider the one-dimensional motion of a fluid particle moving on the centerline of the converging channel, as shown in Fig. P2.20. The vertical dimension of the channel (and, thus, the area per unit depth) varies as

$$2y = 2h - h\sin\left(\frac{\pi}{2}\frac{x}{L}\right)$$

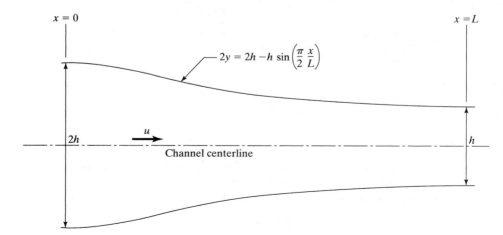

Figure P2.20

Assume that the flow is steady and incompressible. Use the integral form of the continuity equation to describe the velocity along the channel centerline. Also determine the corresponding axial acceleration. If u at $x = 0$ is 2 m/s, h is 1 m, and $L = 1$ m, calculate the acceleration when $x = 0$ and when $x = 0.5L$.

2.21. You are relaxing on an international flight when a terrorist leaps up and tries to take over the airplane. The crew refuses the demands of the terrorist and he fires his pistol, shooting a small hole in the airplane. Panic strikes the crew and other passengers. But you leap up and shout, "Do not worry! I am an engineering student and I know that it will take _____ seconds for the cabin pressure to drop from 0.5×10^5 N/m^2 to 0.25×10^5 N/m^2." Calculate how long it will take the cabin pressure to drop. Make the following assumptions:

(i) The air in the cabin behaves as a perfect gas: $\rho_c = \dfrac{p_c}{RT_c}$ where the subscript c stands for the cabin. $R = 287.05 \ N \cdot m/kg \cdot K$. Furthermore, $T_c = 22°C$ and is constant for the whole time.

(ii) The volume of air in the cabin is 71.0 m^3. The bullet hole is 0.75 cm in diameter.

(iii) Air escapes through the bullet hole according to the equation: $\dot{m}_c = -0.040415 \dfrac{p_c}{\sqrt{T_c}} [A_{hole}]$ where p_c is in N/m^2, T_c is in K, A_{hole} is in m^2, and $\dot{m}_c$ is in kg/s.

2.22. The crew refuses the demands of a terrorist and he fires the pistol, shooting a small hole in the airplane. Panic strikes the crew and other passengers. But you leap up and shout, "Do not worry! I am an engineering student and I know that it will take _____ seconds for the cabin pressure to drop by a factor of two from 7.0 *psia* to 3.5 *psia*." Calculate how long it will take the cabin pressure to drop. Make the following assumptions:

(i) The air in the cabin behaves as a perfect gas: $\rho_c = \dfrac{p_c}{RT_c}$ where the subscript c stands for the cabin. $R = 53.34 \ \dfrac{\text{ft. lbf}}{\text{lbm °R}}$. Furthermore, $T_c = 80°F$ and is constant for the whole time.

(ii) The volume of air in the cabin is 2513 ft^3. The bullet hole is 0.3 in. diameter.

(iii) Air escapes through the bullet hole according to the equation: $\dot{m}_c = -0.5318 \dfrac{p_c}{\sqrt{T_c}} [A_{hole}]$ where p_c is in lbf/ft^2, T_c is in °R, A_{hole} is in ft^2, and $\dot{m}_c$ is in lbm/s.

2.23. Oxygen leaks slowly through a small orifice from an oxygen bottle. The volume of the bottle is 0.1 mm and the diameter of the orifice is 0.1 mm. Assume that the temperature in the tank remains constant at 18°C and that the oxygen behaves as a perfect gas. The mass flow rate is given by

$$\dot{m}_{O_2} = -0.6847 \frac{p_{O_2}}{\sqrt{R_{O_2} T_{O_2}}} [A_{hole}]$$

(The units are those of Prob. 2.21). How long does it take for the pressure in the tank to decrease from 10 to 5 MPa?

2.24. Consider steady, low-speed flow of a viscous fluid in an infinitely long, two-dimensional channel of height h (i.e., the flow is fully developed; Fig. P2.24). Since this is a low-speed flow, we will assume that the viscosity and the density are constant. Assume the body forces to be negligible. The upper plate (which is at $y = h$) moves in the x direction at the speed V_0, while the lower plate (which is at $y = 0$) is stationary.

(a) Develop expressions for u, v, and w (which satisfy the boundary conditions) as functions of U_0, h, μ, dp/dx, and y.
(b) Write the expression for dp/dx in terms of μ, U_0, and h, if $u = 0$ at $y = h/2$.

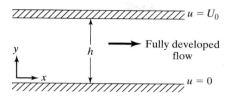

Figure P2.24

2.25. Consider steady, laminar, incompressible flow between two parallel plates, as shown in Fig. P2.25. The upper plate moves at velocity U_0 to the right and the lower plate is stationary. The pressure gradient is zero. The lower half of the region between the plates (i.e., $0 \leq y \leq h/2$) is filled with fluid of density ρ_1 and viscosity μ_1, and the upper half ($h/2 \leq y \leq h$) is filled with fluid of density ρ_2 and viscosity μ_2.
(a) State the condition that the shear stress must satisfy for $0 < y < h$.
(b) State the conditions that must be satisfied by the fluid velocity at the walls and at the interface of the two fluids.
(c) Obtain the velocity profile in each of the two regions and sketch the result for $\mu_1 > \mu_2$.
(d) Calculate the shear stress at the lower wall.

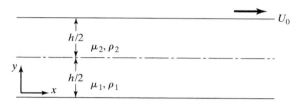

Figure P2.25

2.26. Consider the fully developed flow in a circular pipe, as shown in Fig. P2.26. The velocity u is a function of the radial coordinate only:

$$u = U_{C.L.}\left(1 - \frac{r^2}{R^2}\right)$$

where $U_{C.L.}$ is the magnitude of the velocity at the centerline (or axis) of the pipe. Use the integral form of the momentum equation [i.e., equation (2.13)] to show how the pressure drop per unit length dp/dx changes if the radius of the pipe were to be doubled while the mass flux through the pipe is held constant at the value $\dot{m}$. Neglect the weight of the fluid in the control volume and assume that the fluid properties are constant.

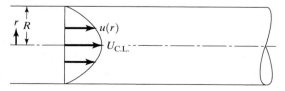

Figure P2.26

2.27. Velocity profiles are measured at the upstream end (surface 1) and at the downstream end (surface 2) of a rectangular control volume, as shown in Fig. P2.27. If the flow is incompressible, two dimensional, and steady, what is the drag coefficient for the airfoil? The vertical dimension H is $0.025c$ and

$$C_d = \frac{d}{\frac{1}{2}\rho_\infty U_\infty^2 c}$$

The pressure is p_∞ (a constant) over the entire surface of the control volume. (This problem is an extension of Problem 2.10.)

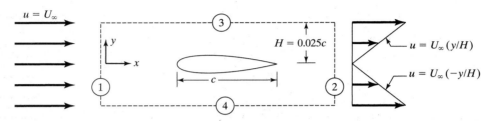

Figure P2.27

2.28. Velocity profiles are measured at the upstream end (surface 1) and at the downstream end (surface 2) of the control volume shown in Fig. P2.28. Surfaces 3 and 4 are streamlines. If the flow is incompressible, two dimensional, and steady, what is the drag coefficient for the airfoil? The vertical dimension H_D is $0.025c$. You will need to calculate the vertical dimension of the upstream station (H_U). The pressure is p_∞ (a constant) over the entire surface of the control volume. (This problem is an extension of Problem 2.11.)

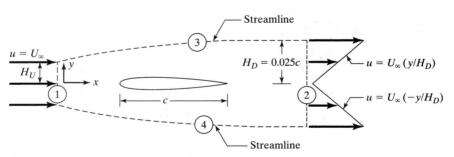

Figure P2.28

2.29. Velocity profiles are measured at the upstream end (surface 1) and at the downstream end (surface 2) of a rectangular control volume, as shown in Fig. P2.29. If the flow is incompressible, two dimensional, and steady, what is the drag coefficient? The vertical dimension $H = 0.025c$. The pressure is p_∞ (a constant) over the entire surface of the control volume. (This problem is a variation of Problem 2.27.) At the upstream end (surface 1), $\vec{V} = U_\infty \hat{\imath}$. At the downstream end of the control volume (surface 2),

$$0 \leq y \leq H \qquad \vec{V} = \frac{U_\infty y}{H}\hat{\imath} + v\hat{\jmath}$$

$$H \leq y \leq 2H \qquad \vec{V} = U_\infty \hat{\imath} + v_0\hat{\jmath}$$

$$-H \le y \le 0 \qquad \vec{V} = -\frac{U_\infty y}{H}\hat{i} - v\hat{j}$$

$$-2H \le y \le -H \qquad \vec{V} = U_\infty\hat{i} - v_0\hat{j}$$

where $v(x, y)$ and $v_0(x)$ are y components of the velocity which are not measured.

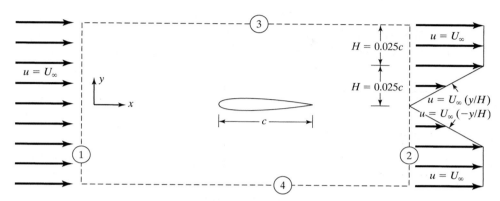

Figure P2.29

2.30. Velocity profiles are measured at the upstream end (surface 1) and at the downstream end (surface 2) of a rectangular control volume, as shown in Fig. P2.30. If the flow is incompressible, two dimensional, and steady, what is the drag coefficient for the airfoil? The vertical dimension H is 0.022c. The pressure is p_∞ (a constant) over the entire surface of the control volume. (This is an extension of Problem 2.12.)

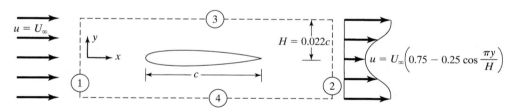

Figure P2.30

2.31. Velocity profiles are measured at the upstream end (surface 1) and at the downstream end (surface 2) of the control volume shown in Fig. P2.31. Surfaces 3 and 4 are streamlines. If the flow is incompressible, two dimensional, and steady, what is the drag coefficient for the

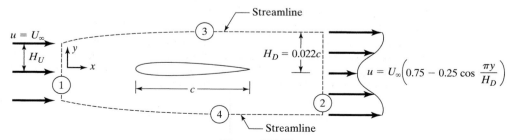

Figure P2.31

airfoil? The vertical dimension at the downstream station (station 2) $H_D = 0.023c$. The pressure is p_∞ (a constant) over the entire surface of the control volume. (This problem is an extension of Problem 2.13.)

2.32. What are the free-stream Reynolds number [as given by equation (2.20)] and the free-stream Mach number [as given by equation (2.19)] for the following flows?

    (a) A golf ball, whose characteristic length (i.e., its diameter) is 1.7 inches, moves through the standard sea-level atmosphere at 200 ft/s.

    (b) A hypersonic transport flies at a Mach number of 6.0 at an altitude of 30 km. The characteristic length of the transport is 32.8 m.

2.33. **(a)** An airplane has a characteristic chord length of 10.4 m. What is the free-stream Reynolds number for the Mach 3 flight at an altitude of 20 km?

    (b) What is the characteristic free-stream Reynolds number of an airplane flying 160 mi/h in a standard sea-level environment? The characteristic chord length is 4.0 ft.

2.34. To illustrate the point that the two integrals in equation (2.21) are path dependent, consider a system consisting of air contained in a piston/cylinder arrangement (Fig. P2.34). The system of air particles is made to undergo two cyclic processes. Note that all properties (p, T, ρ, etc.) return to their original value (i.e., undergo a net change of zero), since the processes are cyclic.

    (a) Assume that both cycles are reversible and determine (1) $\oint \delta q$ and (2) $\oint \delta w$ for each cycle.

    (b) Describe what occurs physically with the piston/cylinder/air configuration during each leg of each cycle.

    (c) Using the answers to part (a), what is the value of $\left(\oint \delta q - \oint \delta w \right)$ for each cycle?

    (d) Is the first law satisfied for this system of air particles?

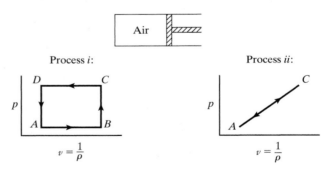

Figure P2.34

2.35. In Problem 2.25, the entropy change in going from A to C directly (i.e., following process ii) is

$$s_C - s_A$$

Going via B (i.e., following process i), the entropy change is

$$s_C - s_A = (s_C - s_B) + (s_B - s_A)$$

    (a) Is the net entropy change $(s_C - s_A)$ the same for both paths?

    (b) Processes AC and ABC were specified to be reversible. What is $s_C - s_A$ if the processes are irreversible? Does $s_C - s_A$ depend on the path if the process is irreversible?

2.36. What assumptions were made in deriving equation (2.32)?

2.37. Show that for an adiabatic, inviscid flow, equation (2.32a) can be written as $ds/dt = 0$.

2.38. Consider the wing-leading edge of a Cessna 172 flying at 130 mi/h through the standard atmosphere at 10,000 ft. Using the integral form of the energy equation for a steady, one-dimensional, adiabatic flow,

$$H_t = h_\infty + \tfrac{1}{2}U_\infty^2$$

For a perfect gas,

$$h_\infty = c_p T_\infty \quad \text{and} \quad H_t = c_p T_t$$

where T_∞ is the free-stream static temperature and T_t is the total (or stagnation point) temperature. If $c_p = 0.2404$ Btu/lbm · °R, compare the total temperature with the free-stream static temperature for this flow. Is convective heating likely to be a problem for this aircraft?

2.39. Consider the wing-leading edge of an SR-71 flying at Mach 3 at 80,000 ft. Using the integral form of the energy equation for a steady, one-dimensional, adiabatic flow,

$$H_t = h_\infty + \tfrac{1}{2}U_\infty^2$$

For a perfect gas,

$$h_\infty = c_p T_\infty \quad \text{and} \quad H_t = c_p T_t$$

where T_∞ is the free-stream static temperature and T_t is the total (or stagnation point) temperature. If $c_p = 0.2404$ Btu/lbm · °R, compare the total temperature with the free-stream static temperature for this flow. Is convective heating likely to be a problem for this aircraft?

2.40. Start with the integral form of the energy equation for a one-dimensional, steady, adiabatic flow:

$$H_t = h + \tfrac{1}{2}U^2$$

and the equation for the entropy change for a perfect gas:

$$s - s_t = c_p \ln \frac{T}{T_t} - R \ln \frac{p}{p_t}$$

and develop the expression relating the local pressure to the stagnation pressure:

$$\frac{p}{p_{t1}} = \left(1 + \frac{\gamma - 1}{2} M^2 \right)^{-\gamma(\gamma-1)}$$

Carefully note the assumptions made at each step of the derivation. Under what conditions is this valid?

REFERENCES

Brune GW, Rubbert PW, Nark TC. 1974. *A new approach to inviscid flow/boundary layer matching*. Presented at Fluid and Plasma Dynamics Conf., 7th, AIAA Pap. 74-601, Palo Alto, CA

Bushnell DM, Cary AM, Harris JE. 1977. Calculation methods for compressible turbulent boundary layers. *NASA SP-422*

Schetz JA. 1993. *Boundary Layer Analysis*. Englewood Cliffs, NJ: Prentice-Hall

Schlichting H. 1979. *Boundary Layer Theory*. 7th Ed. New York: McGraw-Hill

White FM. 2005. *Viscous Fluid Flow*. 3rd Ed. New York: McGraw-Hill

Wilcox DC. 1998. *Turbulence Modeling for CFD*. 2nd Ed. La Cañada, CA: DCW Industries

3 DYNAMICS OF AN INCOMPRESSIBLE, INVISCID FLOW FIELD

As will be discussed in Chapter 14, for many applications, solutions of the inviscid region can provide important design information. Furthermore, once the inviscid flow field has been defined, it can be used as boundary conditions for a thin, viscous boundary layer adjacent to the surface. For the majority of this text, the analysis of the flow field will make use of a two-region flow model (one region in which the viscous forces are negligible, i.e., the inviscid region, and one in which viscous forces cannot be neglected, i.e., the viscous boundary layer near the surface).

3.1 INVISCID FLOWS

As noted in Chapter 2, the shearing stresses may be expressed as the product of the viscosity μ times the shearing stress velocity gradient. There are no real fluids for which the viscosity is zero. However, there are many situations where the product of the viscosity times the shearing velocity gradient is sufficiently small that the shear-stress terms may be neglected when compared to the other terms in the governing equations. Let us use the term *inviscid flow* to describe the flow in those regions of the flow field where the viscous shear-stresses are negligibly small. By using the term *inviscid flow* instead of *inviscid fluid*, we emphasize that the viscous shear stresses are small because the combined product of viscosity and the shearing velocity gradients has a small effect on the flow field and not that the fluid viscosity is zero. In fact, once the solution for the

94

inviscid flow field is obtained, the engineer may want to solve the boundary-layer equations and calculate the skin friction drag on the configuration.

In regions of the flow field where the viscous shear stresses are negligibly small (i.e., in regions where the flow is *inviscid*), equation (2.12) becomes

$$\rho \frac{du}{dt} = \rho f_x - \frac{\partial p}{\partial x} \tag{3.1a}$$

$$\rho \frac{dv}{dt} = \rho f_y - \frac{\partial p}{\partial y} \tag{3.1b}$$

$$\rho \frac{dw}{dt} = \rho f_z - \frac{\partial p}{\partial z} \tag{3.1c}$$

In vector form, the equation is

$$\frac{d\vec{V}}{dt} = \frac{\partial \vec{V}}{\partial t} + (\vec{V} \cdot \nabla)\vec{V} = \vec{f} - \frac{1}{\rho}\nabla p \tag{3.2}$$

No assumption has been made about density, so these equations apply to a compressible flow as well as to an incompressible one. These equations, derived in 1755 by Euler, are called the *Euler equations*.

In this chapter we develop fundamental concepts for describing the flow around configurations in a low-speed stream. Let us assume that the viscous boundary layer is thin and therefore has a negligible influence on the inviscid flow field. (The effect of violating this assumption will be discussed when we compare theoretical results with data.) We will seek the solution for the inviscid portion of the flow field (i.e., the flow outside the boundary layer). The momentum equation is Euler's equation.

3.2 BERNOULLI'S EQUATION

As has been discussed, the density is essentially constant when the gas particles in the flow field move at relatively low speeds or when the fluid is a liquid. Further, let us consider only body forces that are conservative (such as is the case for gravity),

$$\vec{f} = -\nabla F \tag{3.3}$$

and flows that are steady (or steady state):

$$\frac{\partial \vec{V}}{\partial t} = 0$$

Using the vector identity that

$$(\vec{V} \cdot \nabla)\vec{V} = \nabla\left(\frac{U^2}{2}\right) - \vec{V} \times (\nabla \times \vec{V})$$

equation (3.2) becomes (for these assumptions)

$$\nabla\left(\frac{U^2}{2}\right) + \nabla F + \frac{1}{\rho}\nabla p - \vec{V} \times (\nabla \times \vec{V}) = 0 \tag{3.4}$$

In these equations U is the scalar magnitude of the velocity $\vec{V}$. Consideration of the subsequent applications leads us to use U (rather than V).

Let us calculate the change in the magnitude of each of these terms along an arbitrary path whose length and direction are defined by the vector $\vec{ds}$. To do this, we take the dot product of each term in equation (3.4) and the vector $\vec{ds}$. The result is

$$d\left(\frac{U^2}{2}\right) + dF + \frac{dp}{\rho} - \vec{V} \times (\nabla \times \vec{V}) \cdot \vec{ds} = 0 \tag{3.5}$$

Note that, since $\vec{V} \times (\nabla \times \vec{V})$ is a vector perpendicular to $\vec{V}$, the last term is zero (1) for any displacement $\vec{ds}$ if the flow is irrotational (i.e., where $\nabla \times \vec{V} = 0$), or (2) for a displacement along a streamline if the flow is rotational. Thus, for a flow that is

1. Inviscid,
2. Incompressible,
3. Steady, and
4. Irrotational (or, if the flow is rotational, we consider only displacements along a streamline), and for which
5. The body forces are conservative,

the first integral of Euler's equation is

$$\int d\left(\frac{U^2}{2}\right) + \int dF + \int \frac{dp}{\rho} = \text{constant} \tag{3.6}$$

Since each term involves an exact differential,

$$\frac{U^2}{2} + F + \frac{p}{\rho} = \text{constant} \tag{3.7}$$

The force potential most often encountered is that due to gravity. Let us take the z axis to be positive when pointing upward and normal to the surface of the earth. The force per unit mass due to gravity is directed downward and is of magnitude g. Therefore, referring to equation (3.3),

$$\vec{f} = -\frac{\partial F}{\partial z}\hat{k} = -g\hat{k}$$

so

$$F = gz \tag{3.8}$$

The momentum equation becomes

$$\frac{U^2}{2} + gz + \frac{p}{\rho} = \text{constant} \tag{3.9}$$

Equation (3.9) is known as *Bernoulli's equation*.

Because the density has been assumed constant, it is not necessary to include the energy equation in the procedure to solve for the velocity and pressure fields. In fact, equation (3.9), which is a form of the momentum equation, can be derived from equation (2.37), which is the integral form of the energy equation. As indicated in Example 2.5, there is an interrelation between the total energy of the flow and the flow work. Note that, in deriving equation (3.9), we have assumed that dissipative mechanisms do not significantly affect the flow. As a corollary, Bernoulli's equation is valid only for flows where there is no mechanism for dissipation, such as viscosity. In thermodynamics, the flow process would be called *reversible*.

Note that, if the acceleration is zero throughout the entire flow field, the pressure variation in a static fluid as given by equation (3.9) is identical to that given by equation (1.17). This is as it should be, since the five conditions required for Bernoulli's equation are valid for the static fluid.

For aerodynamic problems, the changes in potential energy are negligible. Neglecting the change in potential energy, equation (3.9) may be written

$$p + \tfrac{1}{2}\rho U^2 = \text{constant} \tag{3.10}$$

This equation establishes a direct relation between the static pressure and the velocity. Thus, if either parameter is known, the other can be uniquely determined provided that the flow does not violate the assumptions listed previously. The equation can be used to relate the flow at various points around the vehicle; for example, (1) a point far from the vehicle (i.e., the free stream), (2) a point where the velocity relative to the vehicle is zero (i.e., a stagnation point), and (3) a general point just outside the boundary layer. The nomenclature for these points is illustrated in Fig. 3.1.

Recall from the discussion associated with Fig. 2.1 that the flow around a wing in a ground-fixed coordinate system is unsteady. Thus, we can not apply Bernoulli's equation to the flow depicted in Fig. 3.1a. However, the flow can be made steady through the Gallilean transformation to the vehicle-fixed coordinate system of Fig. 2.2. In the

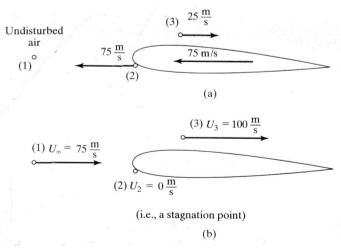

Fgure 3.1 Velocity field around an airfoil: (a) ground-fixed coordinate system; (b) vehicle-fixed coordinate system.

vehicle-fixed coordinate system of Figure 3.1b, we can apply Bernoulli's equation to points (1), (2), and (3).

$$p_\infty + \tfrac{1}{2}\rho_\infty U_\infty^2 = p_t = p_3 + \tfrac{1}{2}\rho_\infty U_3^2 \qquad (3.11)$$

Note that at point (2), the static pressure is equal to the total pressure since the velocity at this point is zero. The *stagnation* (or total) *pressure*, which is the constant of equation (3.10), is the sum of the free-stream static pressure (p_∞) and the free-stream *dynamic pressure* $\left(\tfrac{1}{2}\rho_\infty U_\infty^2\right)$, which is designated by the symbol q_∞. This statement is not true, if the flow is compressible.

EXAMPLE 3.1:   Calculations made using Bernoulli's equation

The airfoil of Fig. 3.1a moves through the air at 75 m/s at an altitude of 2 km. The fluid at point 3 moves downstream at 25 m/s relative to the ground-fixed coordinate system. What are the values of the static pressure at points (1), (2), and (3)?

Solution: To solve this problem, let us superimpose a velocity of 75 m/s to the right so that the airfoil is at rest in the transformed coordinate system. In this vehicle-fixed coordinate system, the fluid "moves" past the airfoil, as shown in Fig. 3.1b. The velocity at point 3 is 100 m/s relative to the stationary airfoil. The resultant flow is steady. p_∞ is found directly in Table 1.2.

$$\text{Point 1: } p_\infty = 0.7846 p_{\text{SL}}$$
$$= 79{,}501 \text{ N/m}^2$$

$$\text{Point 2: } p_t = p_\infty + \tfrac{1}{2}\rho_\infty U_\infty^2$$
$$= 79{,}501 \text{ N/m}^2 + \tfrac{1}{2}(1.0066 \text{ kg/m}^3)(75 \text{ m/s})^2$$
$$= 82{,}332 \text{ N/m}^2$$

$$\text{Point 3: } p_3 + \tfrac{1}{2}\rho_\infty U_3^2 = p_\infty + \tfrac{1}{2}\rho_\infty U_\infty^2$$
$$p_3 = 82{,}332 \text{ N/m}^2 - \tfrac{1}{2}(1.0066 \text{ kg/m}^3)(100 \text{ m/s})^2$$
$$= 77{,}299 \text{ N/m}^2$$

3.3   USE OF BERNOULLI'S EQUATION TO DETERMINE AIRSPEED

Equation (3.11) indicates that a Pitot-static probe (see Fig. 3.2) can be used to obtain a measure of the vehicle's airspeed. The Pitot head has no internal flow velocity, and the pressure in the Pitot tube is equal to the total pressure of the airstream (p_t). The purpose of the static ports is to sense the true static pressure of the free stream (p_∞). When the aircraft is operated through a large angle of attack range, the surface pressure may vary markedly, and, as a result, the pressure sensed at the static port may be significantly different from the free-stream static pressure. The total-pressure and the static-pressure lines can be attached to a differential pressure gage in order to determine

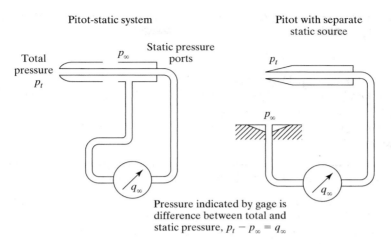

Pressure indicated by gage is difference between total and static pressure, $p_t - p_\infty = q_\infty$

Figure 3.2 Pitot-static probes that can be used to "measure" air speed.

the airspeed using the value of the free-stream density for the altitude at which the vehicle is flying:

$$U_\infty = \sqrt{\frac{2(p_t - p_\infty)}{\rho_\infty}}$$

As indicated in Fig. 3.2, the measurements of the local static pressure are often made using an orifice flush-mounted at the vehicle's surface. Although the orifice opening is located on the surface beneath the viscous boundary layer, the static pressure measurement is used to calculate the velocity at the (outside) edge of the boundary layer (i.e., the velocity of the inviscid stream). Nevertheless, the use of Bernoulli's equation, which is valid only for an inviscid, incompressible flow, is appropriate. It is appropriate because (as discussed in Chapter 2) the analysis of the y-momentum equation reveals that the static pressure is essentially constant across a thin boundary layer. As a result, the value of the static pressure measured at the wall is essentially equal to the value of the static pressure in the inviscid stream (immediately outside the boundary layer).

There can be many conditions of flight where the airspeed indicator may not reflect the actual velocity of the vehicle relative to the air. For instance, when the aircraft is operated through a large angle of attack range, the surface pressure may vary markedly, and, as a result, the pressure sensed at the static port may be significantly different from the free-stream static pressure. The definitions for various terms associated with airspeed are as follows:

1. *Indicated airspeed (IAS).* Indicated airspeed is equal to the Pitot-static airspeed indicator reading as installed in the airplane without correction for airspeed indicator system errors but including the sea-level standard adiabatic compressible flow correction. (The latter correction is included in the calibration of the airspeed instrument dials.)

2. *Calibrated airspeed (CAS).*   CAS is the result of correcting IAS for errors of the instrument and errors due to position or location of the installation. The instrument error may be small by design of the equipment and is usually negligible in equipment that is properly maintained and cared for. The position error of the installation must be small in the range of airspeed involving critical performance conditions. Position errors are most usually confined to the static source in that the actual static pressure sensed at the static port may be different from the free airstream static pressure.

3. *Equivalent airspeed (EAS).*   Equivalent airspeed is equal to the airspeed indicator reading corrected for position error, instrument error, and for adiabatic compressible flow for the particular altitude. The equivalent airspeed (EAS) is the flight speed in the standard sea-level air mass that would produce the same free-stream dynamic pressure as flight at the true airspeed at the correct density altitude.

4. *True airspeed (TAS).*   The true airspeed results when the EAS is corrected for density altitude. Since the airspeed indicator is calibrated for the dynamic pressures corresponding to air speeds at standard sea-level conditions, we must account for variations in air density. To relate EAS and TAS requires consideration that the EAS coupled with standard sea-level density produces the same dynamic pressure as the TAS coupled with the actual air density of the flight condition. From this reasoning, it can be shown that

$$\text{TAS} = \text{EAS}\sqrt{\frac{\rho_{\text{SL}}}{\rho}}$$

where

$$\text{TAS} = \text{true airspeed}$$
$$\text{EAS} = \text{equivalent airspeed}$$
$$\rho = \text{actual air density}$$
$$\rho_{\text{SL}} = \text{standard sea-level air density.}$$

The result shows that the EAS is a function of TAS and density altitude. Table 3.1 presents the EAS and the dynamic pressure as a function of TAS and altitude. The free-stream properties are those of the U.S. standard atmosphere U.S. Standard Atmosphere (1976).

TABLE 3.1   Dynamic Pressure and EAS as a Function of Altitude and TAS

	Altitude					
	Sea level $(\rho = 1.0000\rho_{\text{SL}})$		10,000 m $(\rho = 0.3376\rho_{\text{SL}})$		20,000 m $(\rho = 0.0726\rho_{\text{SL}})$	
TAS (km/h)	$q\infty$ (N/m^2)	EAS (km/h)	$q\infty$ (N/m^2)	EAS (km/h)	$q\infty$ (N/m^2)	EAS (km/h)
200	1.89×10^3	200	6.38×10^2	116.2	1.37×10^2	53.9
400	7.56×10^3	400	2.55×10^3	232.4	5.49×10^2	107.8
600	1.70×10^4	600	5.74×10^3	348.6	1.23×10^3	161.6
800	3.02×10^4	800	1.02×10^4	464.8	2.20×10^3	215.5
1000	4.73×10^4	1000	1.59×10^3	581.0	3.43×10^3	269.4

3.4 THE PRESSURE COEFFICIENT

The engineer often uses experimental data or theoretical solutions for one flow condition to gain insight into the flow field which exists at another flow condition. Wind-tunnel data, where scale models are exposed to flow conditions that simulate the design flight environment, are used to gain insight to describe the full-scale flow field at other flow conditions. Therefore, it is most desirable to present (experimental or theoretical) correlations in terms of dimensionless coefficients which depend only upon the configuration geometry and upon the angle of attack. One such dimensionless coefficient is the *pressure coefficient*, which is defined as

$$C_p = \frac{p - p_\infty}{\frac{1}{2}\rho_\infty U_\infty^2} = \frac{p - p_\infty}{q_\infty} \tag{3.12}$$

In flight tests and wind-tunnel tests, pressure orifices, which are located flush mounted in the surface, sense the local static pressure at the wall [p in equation (3.12)]. Nevertheless, these experimentally determined static pressures, which are located beneath the viscous boundary layer, can be presented as the dimensionless pressure coefficient using equation (3.12).

If we consider those flows for which Bernoulli's equation applies, we can express the pressure coefficient in terms of the nondimensionalized local velocity. Rearranging equation (3.11),

$$p_3 - p_\infty = \frac{1}{2}\rho_\infty U_\infty^2 \left[1 - \frac{U_3^2}{U_\infty^2} \right]$$

Treating point 3 as a general point in the flow field, we can write the pressure coefficient as

$$C_p = \frac{p - p_\infty}{\frac{1}{2}\rho_\infty U_\infty^2} = 1 - \frac{U^2}{U_\infty^2} \tag{3.13}$$

Thus, at the stagnation point, where the local velocity is zero, $C_p = C_{p,t} = 1.0$ for an incompressible flow. Note that the stagnation-point value is independent of the free-stream flow conditions or the configuration geometry.

EXAMPLE 3.2: Flow in an open test-section wind tunnel

Consider flow in a subsonic wind tunnel with an open test section; that is, the region where the model is located is open to the room in which the tunnel is located. Thus, as shown in Fig. 3.3, the air accelerates from a reservoir (or stagnation chamber) where the velocity is essentially zero, through a converging nozzle, exhausting into the room (i.e., the test section) in a uniform, parallel stream of velocity U_∞. Using a barometer located on the wall in the room where the tunnel is located, we know that the barometric pressure in the room is 29.5 in Hg.

The model is a cylinder of infinite span (i.e., the dimension normal to the paper is infinite). There are two pressure orifices flush with the wind-tunnel walls and two orifices flush with the model surface, as shown in Figure 3.3. The pressure sensed at orifice 3, which is at the stagnation point

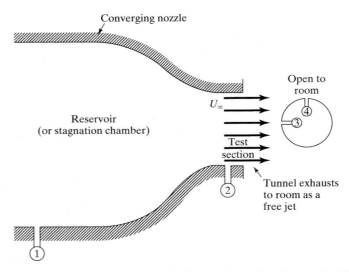

Figure 3.3 Open test-section wind tunnel used in Example 3.2.

of the cylindrical model, is +2.0 in of water, gage. Furthermore, we know that the pressure coefficient for point 4 is −1.2.

What is the pressure sensed by orifice 1 in the stagnation chamber? What is the pressure sensed by orifice 2, located in the exit plane of the wind-tunnel nozzle? What is the free-stream velocity, U_∞? What is the static pressure at point 4? What is the velocity of an air particle just outside the boundary layer at point 4?

In working this example, we will assume that the variation in the static pressure across the boundary layer is negligible. Thus, as will be discussed in Section 4.1, the static pressure at the wall is approximately equal to the static pressure at the edge of the boundary layer.

Solution: As discussed in Chapter 1, the standard atmospheric pressure at sea level is defined as that pressure which can support a column of mercury of 760 mm high. It is also equal to 2116.22 lbf/ft². Thus, an equivalency statement may be written

$$760 \text{ mm Hg} = 29.92 \text{ in Hg} = 2116.22 \text{ lbf/ft}^2$$

Since the barometric pressure is 29.5 in Hg, the pressure in the room is

$$p_{\text{room}} = 29.5 \text{ in Hg}\left(\frac{2116.22 \text{ lbf/ft}^2/\text{atm}}{29.92 \text{ in Hg/atm}}\right) = 2086.51 \text{ lbf/ft}^2$$

Furthermore, since the nozzle exhausts into the room at subsonic speeds and since the streamlines are essentially straight (so that there is no pressure variation normal to the streamlines), the pressure in the room is equal to the free-stream static pressure for the test section (p_∞) and is equal to the static pressure in the exit plane of the nozzle (p_2).

$$p_{room} = p_\infty = p_2 = 2086.51 \text{ lbf/ft}^2$$

We will assume that the temperature changes are negligible and, therefore, the free-stream density is reduced proportionally from the standard atmosphere's sea-level value:

$$\rho_\infty = (2086.51 \text{ lbf/ft}^2)\left(\frac{0.002376 \text{ slug/ft}^3}{2116.22 \text{ lbf/ft}^2}\right) = 0.00234\frac{\text{slug}}{\text{ft}^3}$$

Since the pressure measurement sensed by orifice 3 is given as a gage pressure, it is the difference between the stagnation pressure and the free-stream pressure, that is,

$$p_3 - p_\infty = 2 \text{ in H}_2\text{O, gage}\left(\frac{2116.22 \text{ lbf/ft}^2/\text{atm}}{407.481 \text{ in H}_2\text{O/atm}}\right)$$

$$= 10.387 \text{ lbf/ft}^2, \text{gage}$$

where 407.481 in H_2O is the column of water equivalent to 760 mm Hg if the density of water is 1.937 slugs/ft^3. But since $p_3 = p_t$, we can rearrange Bernoulli's equation:

$$p_3 - p_\infty = p_t - p_\infty = \tfrac{1}{2}\rho_\infty U_\infty^2$$

Equating these expressions and solving for U_∞,

$$U_\infty = \sqrt{\frac{2(10.387 \text{ lbf/ft}^2)}{0.00234 \text{ lbf} \cdot \text{s}^2/\text{ft}^4}} = 94.22 \text{ ft/s}$$

Since $\tfrac{1}{2}\rho_\infty U_\infty^2$ is the dynamic pressure, we can rearrange the definition for the pressure coefficient to find the static pressure at point 4:

$$p_4 = p_\infty + C_{p4}q_\infty = 2086.51 + (-1.2)(10.387)$$

$$= 2074.05 \text{ lbf/ft}^2$$

Since we seek the velocity of the air particles just outside the boundary layer above orifice 4, Bernoulli's equation is applicable, and we can use equation (3.13):

$$\frac{U_4^2}{U_\infty^2} = 1 - C_{p4} = 2.2$$

Thus, $U_4 = 139.75$ ft/s.

3.5   CIRCULATION

The *circulation* is defined as the line integral of the velocity around any closed curve. Refering to the closed curve C of Fig. 3.4, the circulation is given by

$$-\Gamma = \oint_C \vec{V} \cdot \vec{ds} \tag{3.14}$$

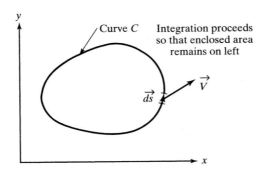

Figure 3.4 Concept of circulation.

where $\vec{V} \cdot \vec{ds}$ is the scalar product of the velocity vector and the differential vector length along the path of integration. As indicated by the circle through the integral sign, the integration is carried out for the complete closed path. The path of the integration is counterclockwise, so the area enclosed by the curve C is always on the left. A negative sign is used in equation (3.14) for convenience in the subsequent application to lifting-surface aerodynamics.

Consider the circulation around a small, square element in the xy plane, as shown in Fig. 3.5a. Integrating the velocity components along each of the sides and proceeding counterclockwise (i.e., keeping the area on the left of the path),

$$-\Delta\Gamma = u\,\Delta x + \left(v + \frac{\partial v}{\partial x}\Delta x\right)\Delta y - \left(u + \frac{\partial u}{\partial y}\Delta y\right)\Delta x - v\,\Delta y$$

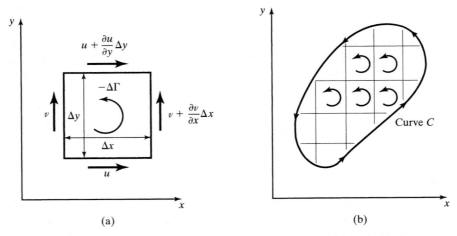

(a)

(b)

Figure 3.5 Circulation for elementary closed curves: (a) rectangular element; (b) general curve C.

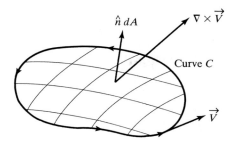

Figure 3.6 Nomenclature for Stokes's theorem.

Simplifying yields

$$-\Delta\Gamma = \left(\frac{\partial v}{\partial x} - \frac{\partial u}{\partial y}\right)\Delta x\,\Delta y$$

This procedure can be extended to calculate the circulation around a general curve C in the xy plane, such as that of Fig. 3.5b. The result for this general curve in the xy plane is

$$-\Gamma = \oint_C (u\,dx + v\,dy) = \iint_A \left(\frac{\partial v}{\partial x} - \frac{\partial u}{\partial y}\right) dx\,dy \qquad (3.15)$$

Equation (3.15) represents *Green's lemma* for the transformation from a line integral to a surface integral in two-dimensional space. The transformation from a line integral to a surface integral in three-dimensional space is governed by *Stokes's theorem*:

$$\oint_C \vec{V}\cdot\vec{ds} = \iint_A (\nabla\times\vec{V})\cdot\hat{n}\,dA \qquad (3.16)$$

where $\hat{n}\,dA$ is a vector normal to the surface, positive when pointing outward from the enclosed volume, and equal in magnitude to the incremental surface area. (See Fig. 3.6.) Note that equation (3.15) is a planar simplification of the more general vector equation, equation (3.16). In words, the integral of the normal component of the curl of the velocity vector over any surface A is equal to the line integral of the tangential component of the velocity around the curve C which bounds A. Stokes's theorem is valid when A represents a simply connected region in which $\vec{V}$ is continuously differentiable. (A simply connected region is one where any closed curve can be shrunk to a point without leaving the simply connected region.) Thus, equation (3.16) is not valid if the area A contains regions where the velocity is infinite.

3.6   IRROTATIONAL FLOW

By means of Stokes's theorem, it is apparent that, if the curl of $\vec{V}$ (i.e., $\nabla\times\vec{V}$) is zero at all points in the region bounded by C, then the line integral of $\vec{V}\cdot\vec{ds}$ around the closed path is zero. If

$$\nabla\times\vec{V} \equiv 0 \qquad (3.17)$$

and the flow contains no singularities, the flow is said to be *irrotational*. Stokes's theorem leads us to the conclusion that

$$\oint_C \vec{V} \cdot \vec{ds} = -\Gamma = 0$$

For this irrotational velocity field, the line integral

$$\int \vec{V} \cdot \vec{ds}$$

is independent of path. A necessary and sufficient condition that

$$\int \vec{V} \cdot \vec{ds}$$

be independent of path is that the curl of $\vec{V}$ is everywhere zero. Thus, its value depends only on its limits. However, a line integral can be independent of the path of integration only if the integrand is an exact differential. Therefore,

$$\vec{V} \cdot \vec{ds} = d\phi \qquad \textbf{(3.18)}$$

where $d\phi$ is an exact differential. Expanding equation (3.18) in Cartesian coordinates,

$$u\,dx + v\,dy + w\,dz = \frac{\partial \phi}{\partial x}\,dx + \frac{\partial \phi}{\partial y}\,dy + \frac{\partial \phi}{\partial z}\,dz$$

It is apparent that

$$\vec{V} = \nabla\phi \qquad \textbf{(3.19)}$$

Thus, a *velocity potential* $\phi(x, y, z)$ exists for this flow such that the partial derivative of ϕ in any direction is the velocity component in that direction. That equation (3.19) is a valid representation of the velocity field for an irrotational flow can be seen by noting that

$$\nabla \times \nabla\phi \equiv 0 \qquad \textbf{(3.20)}$$

That is, the curl of any gradient is necessarily zero. Thus, an irrotational flow is termed a *potential flow*.

3.7   KELVIN'S THEOREM

Having defined the necessary and sufficient condition for the existence of a flow that has no circulation, let us examine a theorem first demonstrated by Lord Kelvin. For an inviscid, barotropic flow with conservative body forces, the circulation around a closed fluid line remains constant with respect to time. A barotropic flow (sometimes called a homogeneous flow) is one in which the density depends only on the pressure.

The time derivative of the circulation along a closed fluid line (i.e., a fluid line that is composed of the same fluid particles) is

$$-\frac{d\Gamma}{dt} = \frac{d}{dt}\left(\oint_C \vec{V} \cdot \overrightarrow{ds}\right) = \oint_C \frac{\overrightarrow{dV}}{dt} \cdot \overrightarrow{ds} + \oint_C \vec{V} \cdot \frac{d}{dt}(\overrightarrow{ds}) \tag{3.21}$$

Again, the negative sign is used for convenience, as discussed in equation (3.14). Euler's equation, equation (3.2), which is the momentum equation for an inviscid flow, yields

$$\frac{\overrightarrow{dV}}{dt} = \vec{f} - \frac{1}{\rho}\nabla p$$

Using the constraint that the body forces are conservative (as is true for gravity, the body force of most interest to the readers of this text), we have

$$\vec{f} = -\nabla F$$

and

$$\frac{\overrightarrow{dV}}{dt} = -\nabla F - \frac{1}{\rho}\nabla p \tag{3.22}$$

where F is the body-force potential. Since we are following a particular fluid particle, the order of time and space differentiation does not matter.

$$\frac{d}{dt}(\overrightarrow{ds}) = d\left(\frac{\overrightarrow{ds}}{dt}\right) = \overrightarrow{dV} \tag{3.23}$$

Substituting equations (3.22) and (3.23) into equation (3.21) yields

$$\frac{d}{dt}\oint_C \vec{V} \cdot \overrightarrow{ds} = -\oint_C dF - \oint_C \frac{dp}{\rho} + \oint_C \vec{V} \cdot \overrightarrow{dV} \tag{3.24}$$

Since the density is a function of the pressure only, all the terms on the right-hand side involve exact differentials. The integral of an exact differential around a closed curve is zero. Thus,

$$\frac{d}{dt}\left(\oint_C \vec{V} \cdot \overrightarrow{ds}\right) = 0 \tag{3.25}$$

Or, as given in the statement of Kelvin's theorem, the circulation remains constant along the closed fluid line for the conservative flow.

3.7.1 Implication of Kelvin's Theorem

If the fluid starts from rest, or if the velocity of the fluid in some region is uniform and parallel, the rotation in this region is zero. Kelvin's theorem leads to the important conclusion that the entire flow remains irrotational in the absence of viscous forces and of discontinuities provided that the flow is barotropic and the body forces can be described by a potential function.

In many flow problems (including most of those of interest to the readers of this text), the undisturbed, free-stream flow is a uniform parallel flow in which there are no shear stresses. Kelvin's theorem implies that, although the fluid particles in the subsequent flow patterns may follow curved paths, the flow remains irrotational except in those regions where the dissipative viscous forces are an important factor.

3.8 INCOMPRESSIBLE, IRROTATIONAL FLOW

Kelvin's theorem states that for an inviscid flow having a conservative force field, the circulation must be constant around a path that moves so as always to touch the same particles and which contains no singularities. Thus, since the free-stream flow is irrotational, a barotropic flow around the vehicle will remain irrotational provided that viscous effects are not important. For an irrotational flow, the velocity may be expressed in terms of a potential function;

$$\vec{V} = \nabla\phi \tag{3.19}$$

For relatively low speed flows (i.e., incompressible flows), the continuity equation is

$$\nabla \cdot \vec{V} = 0 \tag{2.4}$$

Note that equation (2.4) is valid for a three-dimensional flow, as well as a two-dimensional flow.

Combining equations (2.4) and (3.19), one finds that for an incompressible, irrotational flow,

$$\nabla^2\phi = 0 \tag{3.26}$$

Thus, the governing equation, which is known as *Laplace's equation*, is a linear, second-order partial differential equation of the elliptic type.

3.8.1 Boundary Conditions

Within certain constraints on geometric slope continuity, a bounded, simply connected velocity field is uniquely determined by the distribution on the flow boundaries either of the normal component of the total velocity $\nabla\phi \cdot \hat{n}$ or of the total potential ϕ. These boundary-value problems are respectively designated Neumann or Dirichlet problems. For applications in this book, the Neumann formulation will be used since most practical cases involve prescribed normal velocity boundary conditions. Specifically, the flow tangency requirement associated with inviscid flow past a solid body is expressed mathematically as

$$\nabla\phi \cdot \hat{n} = 0$$

because the velocity component normal to the surface is zero at a solid surface.

3.9 STREAM FUNCTION IN A TWO-DIMENSIONAL, INCOMPRESSIBLE FLOW

Just as the condition of irrotationality is the necessary and sufficient condition for the existence of a velocity potential, so the equation of continuity for an incompressible, two-dimensional flow is the necessary and sufficient condition for the existence of a stream function. The flow need be two dimensional only in the sense that it requires only two spatial coordinates to describe the motion. Therefore, stream functions exist both for plane flow and for axially symmetric flow. The reader might note, although it is not relevant to this chapter, that stream functions exist for compressible, two-dimensional flows, if they are steady.

Examining the continuity equation for an incompressible, two-dimensional flow in Cartesian coordinates,

$$\nabla \cdot \vec{V} = \frac{\partial u}{\partial x} + \frac{\partial v}{\partial y} = 0$$

It is obvious that the equation is satisfied by a stream function ψ, for which the velocity components can be calculated as

$$u = \frac{\partial \psi}{\partial y} \tag{3.27a}$$

$$v = -\frac{\partial \psi}{\partial x} \tag{3.27b}$$

A corollary to this is the existence of a stream function is a necessary condition for a physically possible flow (i.e., one that satisfies the continuity equation).

Since ψ is a point function,

$$d\psi = \frac{\partial \psi}{\partial x} dx + \frac{\partial \psi}{\partial y} dy$$

so that

$$d\psi = -v \, dx + u \, dy \tag{3.28a}$$

Since a streamline is a curve whose tangent at every point coincides with the direction of the velocity vector, the definition of a streamline in a two-dimensional flow is

$$\frac{dx}{u} = \frac{dy}{v}$$

Rearranging, it can be seen that

$$u \, dy - v \, dx = 0 \tag{3.28b}$$

along a streamline. Equating equations (3.28a) and (3.28b), we find that

$$d\psi = 0$$

along a streamline. Thus, the change in ψ is zero along a streamline or, equivalently, ψ is a constant along a streamline. A corollary statement is that lines of constant ψ are streamlines of the flow. It follows, then, that the volumetric flow rate (per unit depth) between any two points in the flow is the difference between the values of the stream function at the two points of interest.

Referring to Fig. 3.7, it is clear that the product $v(-dx)$ represents the volumetric flow rate per unit depth across AO and the product $u \, dy$ represents the volumetric flow rate per unit depth across OB. By continuity, the fluid crossing lines AO and OB must cross the

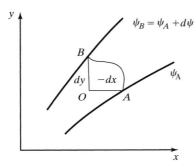

Figure 3.7 The significance of the stream function.

curve AB. Therefore, $d\psi$ is a measure of the volumetric flow rate per unit depth across AB. A line can be passed through A for which $\psi = \psi_A$ (a constant), while a line can be passed through B for which $\psi = \psi_B = \psi_A + d\psi$ (a different constant). The difference $d\psi$ is the volumetric flow rate (per unit depth) between the two streamlines. The fact that the flow is always tangent to a streamline and has no component of velocity normal to it has an important consequence. Any streamline in an inviscid flow can be replaced by a solid boundary of the same shape without affecting the remainder of the flow pattern.

The velocity components for a two-dimensional flow in cylindrical coordinates can also be calculated using a stream function as

$$v_r = \frac{1}{r}\frac{\partial \psi}{\partial \theta} \tag{3.29a}$$

and

$$v_\theta = -\frac{\partial \psi}{\partial r} \tag{3.29b}$$

If the flow is irrotational,

$$\nabla \times \vec{V} = 0$$

Then writing the velocity components in terms of the stream function, as defined in equation (3.27), we obtain

$$\nabla^2 \psi = 0 \tag{3.30}$$

Thus, for an irrotational, two-dimensional, incompressible flow, the stream function is also governed by Laplace's equation.

3.10 RELATION BETWEEN STREAMLINES AND EQUIPOTENTIAL LINES

If a flow is incompressible, irrotational, and two dimensional, the velocity field may be calculated using either a potential function or a stream function. Using the potential function, the velocity components in Cartesian coordinates are

$$u = \frac{\partial \phi}{\partial x} \qquad v = \frac{\partial \phi}{\partial y}$$

For a potential function,

$$d\phi = \frac{\partial \phi}{\partial x} dx + \frac{\partial \phi}{\partial y} dy = u\, dx + v\, dy$$

Therefore, for lines of constant potential ($d\phi = 0$),

$$\left(\frac{dy}{dx}\right)_{\phi=C} = -\frac{u}{v} \qquad\qquad\qquad \textbf{(3.31)}$$

Since a streamline is everywhere tangent to the local velocity, the slope of a streamline, which is a line of constant ψ, is

$$\left(\frac{dy}{dx}\right)_{\psi=C} = \frac{v}{u} \qquad\qquad\qquad \textbf{(3.32)}$$

Refer to the discussion associated with equation (3.28a).

Comparing equations (3.31) and (3.32) yields

$$\left(\frac{dy}{dx}\right)_{\phi=C} = -\frac{1}{(dy/dx)_{\psi=C}} \qquad\qquad \textbf{(3.33)}$$

The slope of an equipotential line is the negative reciprocal of the slope of a streamline. Therefore, streamlines (ψ = constant) are perpendicular to equipotential lines (ϕ = constant), except at stagnation points, where the components vanish simultaneously.

EXAMPLE 3.3:   Equipotential lines and streamlines for a corner flow

Consider the incompressible, irrotational, two-dimensional flow, where the stream function is

$$\psi = 2xy$$

(a) What is the velocity at $x = 1$, $y = 1$? At $x = 2$, $y = \frac{1}{2}$? Note that both points are on the same streamline, since $\psi = 2$ for both points.

(b) Sketch the streamline pattern and discuss the significance of the spacing between the streamlines.

(c) What is the velocity potential for this flow?

(d) Sketch the lines of constant potential. How do the lines of equipotential relate to the streamlines?

Solution:

(a) The stream function can be used to calculate the velocity components:

$$u = \frac{\partial \psi}{\partial y} = 2x \qquad v = -\frac{\partial \psi}{\partial x} = -2y$$

Therefore,

$$\vec{V} = 2x\hat{i} - 2y\hat{j}$$

At $x = 1$, $y = 1$, $\vec{V} = 2\hat{i} - 2\hat{j}$, and the magnitude of the velocity is

$$U = 2.8284$$

At $x = 2$, $y = \frac{1}{2}$, $\vec{V} = 4\hat{i} - \hat{j}$, and the magnitude of the velocity is

$$U = 4.1231$$

(b) A sketch of the streamline pattern is presented in Fig. 3.8. Results are presented only for the first quadrant (x positive, y positive). Mirror-image patterns would exist in other quadrants. Note that the $x = 0$ and the $y = 0$ axes represent the $\psi = 0$ "streamline."

Since the flow is incompressible and steady, the integral form of the continuity equation (2.5) indicates that the product of the velocity times the distance between the streamlines is a constant. That is, since $\rho = $ constant,

$$\oiint \vec{V} \cdot \hat{n} \, dA = 0$$

Therefore, the distance between the streamlines decreases as the magnitude of the velocity increases.

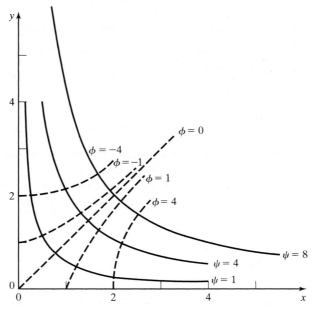

Figure 3.8 Equipotential lines and streamlines for Example 3.3.

(c) Since $u = \partial\phi/\partial x$ and $v = \partial\phi/\partial y$,

$$\phi = \int u \, dx + g(y) = \int 2x \, dx + g(y)$$

Also,

$$\phi = \int v \, dy + f(x) = -\int 2y \, dy + f(x)$$

The potential function which would satisfy both of these equations is

$$\phi = x^2 - y^2 + C$$

where C is an arbitrary constant.

(d) The equipotential lines are included in Fig. 3.8, where C, the arbitrary constant, has been set equal to zero. The lines of equipotential are perpendicular to the streamlines.

3.11 SUPERPOSITION OF FLOWS

Since equation (3.26) for the potential function and equation (3.30) for the stream function are linear, functions that individually satisfy these (Laplace's) equations may be added together to describe a desired, complex flow. The boundary conditions are such that the resultant velocity is equal to the free-stream value at points far from the solid surface and that the component of the velocity normal to the solid surface is zero (i.e., the surface is a streamline). There are numerous two-dimensional and axisymmetric solutions available through "inverse" methods. These inverse methods do not begin with a prescribed boundary surface and directly solve for the potential flow, but instead assume a set of known singularities in the presence of an onset flow. The total potential function (or stream function) for the singularities and the onset flow are then used to determine the streamlines, any one of which may be considered to be a "boundary surface." If the resultant boundary surface corresponds to the shape of interest, the desired solution has been obtained. The singularities most often used in such approaches, which were suggested by Rankine in 1871, include a source, a sink, a doublet, and a vortex.

For a constant-density potential flow, the velocity field can be determined using only the continuity equation and the condition of irrotationality. Thus, the equation of motion is not used, and the velocity may be determined independently of the pressure. Once the velocity field has been determined, Bernoulli's equation can be used to calculate the corresponding pressure field. It is important to note that the pressures of the component flows cannot be superimposed (or added together), since they are nonlinear functions of the velocity. Referring to equation (3.10), the reader can see that the pressure is a quadratic function of the velocity.

3.12 ELEMENTARY FLOWS

3.12.1 Uniform Flow

The simplest flow is a *uniform stream* moving in a fixed direction at a constant speed. Thus, the streamlines are straight and parallel to each other everywhere in the flow field

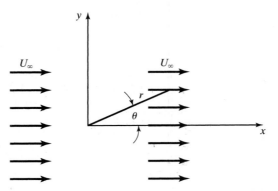

Figure 3.9 Streamlines for a uniform flow parallel to the *x* axis.

(see Fig. 3.9). Using a cylindrical coordinate system, the potential function for a uniform flow moving parallel to the *x* axis is

$$\phi = U_\infty r \cos \theta \qquad (3.34)$$

where U_∞ is the velocity of the fluid particles. Using a Cartesian coordinate system, the potential function for the uniform stream of Figure 3.9 is

$$\phi = U_\infty x \qquad (3.35a)$$

For a uniform stream inclined relative to the *x* axis by the angle α, the potential function is

$$\phi = U_\infty(x \cos \alpha + y \sin \alpha) \qquad (3.35b)$$

3.12.2 Source or Sink

A *source* is defined as a point from which fluid issues and flows radially outward (see Fig. 3.10) such that the continuity equation is satisfied everywhere but at the singularity that exists at the source's center. The potential function for the two-dimensional (planar) source centered at the origin is

$$\phi = \frac{K}{2\pi} \ln r \qquad (3.36)$$

where *r* is the radial coordinate from the center of the source and *K* is the source strength. Such a two-dimensional source is sometimes referred to as a line source, because its axis extends infinitely far out of and far into the page. The resultant velocity field in cylindrical coordinates is

$$\vec{V} = \nabla\phi = \hat{e}_r \frac{\partial \phi}{\partial r} + \frac{\hat{e}_\theta}{r} \frac{\partial \phi}{\partial \theta} \qquad (3.37)$$

since

$$\vec{V} = \hat{e}_r v_r + \hat{e}_\theta v_\theta$$

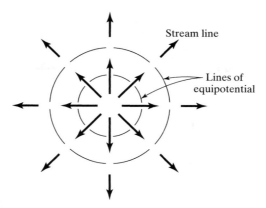

Figure 3.10 Equipotential lines and streamlines for flow from a two-dimensional source.

$$v_r = \frac{\partial \phi}{\partial r} = \frac{K}{2\pi r}$$

and

$$v_\theta = \frac{1}{r}\frac{\partial \phi}{\partial \theta} = 0$$

Note that the resultant velocity has only a radial component and that this component varies inversely with the radial distance from the source.

A sink is a negative source. That is, fluid flows into a sink along radial streamlines. Thus, for a sink of strength K centered at the origin,

$$\phi = -\frac{K}{2\pi}\ln r \tag{3.38}$$

Note that the dimensions of K are (length)^2/(time).

EXAMPLE 3.4:   A two-dimensional source

Show that the flow rate passing through a circle of radius r is proportional to K, the strength of the two-dimensional source, and is independent of the radius.

Solution:

$$\dot{m} = \iint \rho \vec{V} \cdot \hat{n}\, dA$$

$$= \int_0^{2\pi} \rho\left(\frac{K}{2\pi r}\right) r\, d\theta$$

$$= K\rho$$

3.12.3 Doublet

A *doublet* is defined to be the singularity resulting when a source and a sink of equal strength are made to approach each other such that the product of their strengths (K) and their distance apart (a) remains constant at a preselected finite value in the limit as the distance between them approaches zero. The line along which the approach is made is called the *axis of the doublet* and is considered to have a positive direction when oriented from sink to source. The potential for a two-dimensional (line) doublet for which the flow proceeds out from the origin in the negative x direction (see Fig. 3.11) is

$$\phi = \frac{B}{r} \cos \theta \tag{3.39a}$$

where B is a constant. In general, the potential function of a line doublet whose axis is at an angle α relative to the positive x axis is

$$\phi = -\frac{B}{r} \cos \alpha \cos \theta \tag{3.39b}$$

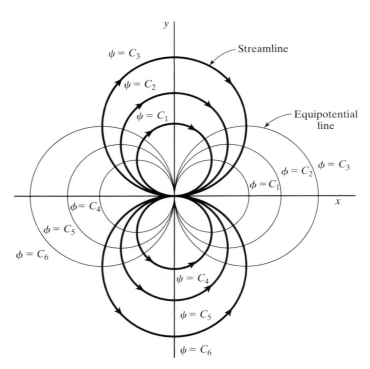

Figure 3.11 Equipotential lines and streamlines for a doublet (flow proceeds out from the origin in the negative x direction).

3.12.4 Potential Vortex

A *potential vortex* is defined as a singularity about which fluid flows with concentric streamlines (see Fig. 3.12). The potential for a vortex centered at the origin is

$$\phi = -\frac{\Gamma\theta}{2\pi} \tag{3.40}$$

where Γ is the strength of the vortex. We have used a minus sign to represent a vortex with clockwise circulation. Differentiating the potential function, one finds the velocity distribution about an isolated vortex to be

$$v_r = \frac{\partial\phi}{\partial r} = 0$$

$$v_\theta = \frac{1}{r}\frac{\partial\phi}{\partial\theta} = -\frac{\Gamma}{2\pi r}$$

Thus, there is no radial velocity component and the circumferential component varies with the reciprocal of the radial distance from the vortex. Note that the dimensions of Γ are $(\text{length})^2/(\text{time})$.

    The curl of the velocity vector for the potential vortex can be found using the definition for the curl of $\vec{V}$ in cylindrical coordinates

$$\nabla \times \vec{V} = \frac{1}{r}\begin{vmatrix} \hat{e}_r & r\hat{e}_\theta & \hat{e}_z \\ \dfrac{\partial}{\partial r} & \dfrac{\partial}{\partial\theta} & \dfrac{\partial}{\partial z} \\ v_r & rv_\theta & v_z \end{vmatrix}$$

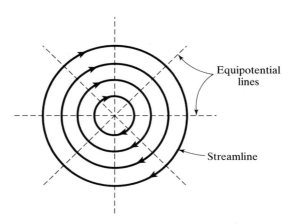

Figure 3.12 Equipotential lines and streamlines for a potential vortex.

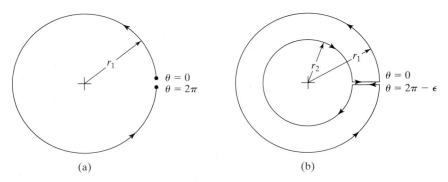

(a)   (b)

Figure 3.13 Paths for the calculation of the circulation for a potential vortex: (a) closed curve C_1, which encloses origin; (b) closed curve C_2, which does not enclose the origin.

We find that

$$\nabla \times \vec{V} = 0$$

Thus, although the flow is irrotational, we must remember that the velocity is infinite at the origin (i.e., when $r = 0$).

Let us calculate the circulation around a closed curve C_1 which encloses the origin. We shall choose a circle of radius r_1, as shown in Figure 3.13a. Using equation (3.14), the circulation is

$$-\Gamma_{C_1} = \oint_{C_1} \vec{V} \cdot \vec{ds} = \int_0^{2\pi} \left(-\frac{\Gamma}{2\pi r_1} \hat{e}_\theta \right) \cdot r_1 \, d\theta \, \hat{e}_\theta$$

$$= \int_0^{2\pi} (-)\frac{\Gamma}{2\pi} \, d\theta = -\Gamma$$

Recall that Stokes's theorem [equation (3.16)] is not valid if the region contains points where the velocity is infinite.

However, if we calculate the circulation around a closed curve C_2, which does not enclose the origin, such as that shown in Figure 3.13b, we find that

$$-\Gamma_{C_2} = \oint_{C_2} \vec{V} \cdot \vec{ds} = \int_0^{2\pi-\epsilon} (-)\frac{\Gamma}{2\pi r_1} r_1 \, d\theta + \int_{2\pi-\epsilon}^0 (-)\frac{\Gamma}{2\pi r_2} r_2 \, d\theta$$

or

$$-\Gamma_{C_2} = 0$$

Thus, the circulation around a closed curve not containing the origin is zero.

The reader may be familiar with the rotation of a two-dimensional, solid body about its axis, such as the rotation of a wheel. For solid-body rotation,

$$v_r = 0$$

$$v_\theta = r\omega$$

where ω is the angular velocity. Substituting these velocity components into the definition

$$\nabla \times \vec{V} = \frac{1}{r}\left[\frac{\partial(rv_\theta)}{\partial r} - \frac{\partial v_r}{\partial \theta}\right]\hat{e}_z$$

we find that

$$\nabla \times \vec{V} = 2\omega\hat{e}_z$$

We see that the velocity field which describes two-dimensional solid-body rotation is not irrotational and, therefore, cannot be defined using a potential function.

Vortex lines (or filaments) will have an important role in the study of the flow around wings. Therefore, let us summarize the vortex theorems of Helmholtz. For a barotropic (homogeneous) inviscid flow acted upon by conservative body forces, the following statements are true:

1. The circulation around a given vortex line (i.e., the strength of the vortex filament) is constant along its length.

2. A vortex filament cannot end in a fluid. It must form a closed path, end at a boundary, or go to infinity. Examples of these three kinds of behavior are a smoke ring, a vortex bound to a two-dimensional airfoil that spans from one wall to the other in a wind tunnel (see Chapter 6), and the downstream ends of the horseshoe vortices representing the loading on a three-dimensional wing (see Chapter 7).

3. No fluid particle can have rotation, if it did not originally rotate. Or, equivalently, in the absence of rotational external forces, a fluid that is initially irrotational remains irrotational. In general, we can conclude that vortices are preserved as time passes. Only through the action of viscosity (or some other dissipative mechanism) can they decay or disappear.

Two vortices which are created by the rotational motion of jet engines can be seen in the photograph of Fig. 3.14, which is taken from Campbell and Chambers (1994). One vortex can be seen entering the left engine. Because a vortex filament cannot end in a fluid, the vortex axis turns sharply and the vortex quickly goes to the ground. The right-hand vortex is in the form of a horseshoe vortex.

3.12.5 Summary of Stream Functions and of Potential Functions

Table 3.2 summarizes the potential functions and the stream functions for the elementary flows discussed previously.

3.13 ADDING ELEMENTARY FLOWS TO DESCRIBE FLOW AROUND A CYLINDER

3.13.1 Velocity Field

Consider the case where a uniform flow is superimposed on a doublet whose axis is parallel to the direction of the uniform flow and is so oriented that the direction of the

Figure 3.14 Ground vortices between engine intake and ground and between adjacent engines for Asuka STOL research aircraft during static tests. [Taken from Campbell and Chambers (1994).]

efflux opposes the uniform flow (see Fig. 3.15). Substituting the potential function for a uniform flow [equation (3.34)] and that for the doublet [equation (3.39a)] into the expression for the velocity field [equation (3.37)], one finds that

$$v_\theta = \frac{1}{r}\frac{\partial \phi}{\partial \theta} = -U_\infty \sin\theta - \frac{B}{r^2}\sin\theta$$

TABLE 3.2   Stream Functions and Potential Functions for Elementary Flows

Flow	Ψ	ϕ
Uniform flow	$U_\infty r \sin\theta$	$U_\infty r \cos\theta$
Source	$\dfrac{K\theta}{2\pi}$	$\dfrac{K}{2\pi}\ln r$
Doublet	$-\dfrac{B}{r}\sin\theta$	$\dfrac{B}{r}\cos\theta$
Vortex (with clockwise circulation)	$\dfrac{\Gamma}{2\pi}\ln r$	$-\dfrac{\Gamma\theta}{2\pi}$
90° corner flow	Axy	$\frac{1}{2}A(x^2-y^2)$
Solid-body rotation	$\frac{1}{2}\omega r^2$	Does not exist

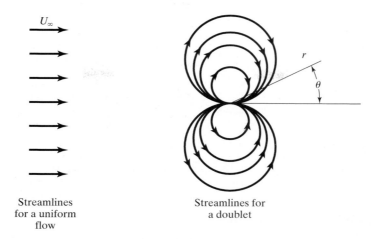

Streamlines
for a uniform
flow

Streamlines for
a doublet

Figure 3.15 Streamlines for the two elementary flows which, when superimposed, describe the flow around a cylinder.

and

$$v_r = \frac{\partial \phi}{\partial r} = U_\infty \cos \theta - \frac{B}{r^2} \cos \theta$$

Note that $v_r = 0$ at every point where $r = \sqrt{B/U_\infty}$, which is a constant. Since the velocity is always tangent to a streamline, the fact that velocity component (v_r) perpendicular to a circle of $r = R = \sqrt{B/U_\infty}$ is zero means that the circle may be considered as a streamline of the flow field. Replacing B by $R^2 U_\infty$ allows us to write the velocity components as

$$v_\theta = -U_\infty \sin \theta \left(1 + \frac{R^2}{r^2} \right) \tag{3.41a}$$

$$v_r = U_\infty \cos \theta \left(1 - \frac{R^2}{r^2} \right) \tag{3.41b}$$

The velocity field not only satisfies the surface boundary condition that the inviscid flow is tangent to a solid wall, but the velocity at points far from the cylinder is equal to the undisturbed free-stream velocity U_∞. Streamlines for the resultant inviscid flow field are illustrated in Fig. 3.16. The resultant two-dimensional, irrotational (inviscid), incompressible flow is that around a cylinder of radius R whose axis is perpendicular to the free-stream direction.

Setting $r = R$, we see that the velocity at the surface of the cylinder is equal to

$$v_\theta = -2U_\infty \sin \theta \tag{3.42}$$

Of course, as noted earlier, $v_r = 0$. Since the solution is for the inviscid model of the flow field, it is not inconsistent that the fluid particles next to the surface move relative to the surface (i.e., violate the no-slip requirement). When $\theta = 0$ or π (points B and A,

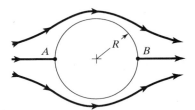

Figure 3.16 Two-dimensional, inviscid flow around a cylinder with zero circulation.

respectively, of Figure 3.16), the fluid is at rest with respect to the cylinder (i.e., $(v_r = v_\theta = 0)$. These points are, therefore, termed *stagnation points*.

3.13.2 Pressure Distribution

Because the velocity at the surface of the cylinder is a function of θ, the local static pressure will also be a function of θ. Once the pressure distribution has been defined, it can be used to determine the forces and the moments acting on the configuration. Using Bernoulli's equation [equation (3.10)], we obtain the expression for the θ-distribution of the static pressure using dimensional parameters:

$$p = p_\infty + \tfrac{1}{2}\rho_\infty U_\infty^2 - 2\rho_\infty U_\infty^2 \sin^2\theta \qquad (3.43)$$

Expressing the pressure in terms of the dimensionless pressure coefficient, which is presented in equation (3.12), we have:

$$C_\rho = 1 - 4\sin^2\theta \qquad (3.44)$$

The pressure coefficients, thus calculated, are presented in Fig. 3.17 as a function of θ. Recall that, in the nomenclature of this chapter, $\theta = 180°$ corresponds to the plane of symmetry for the windward surface or forebody (i.e., the surface facing the free stream). Starting with the undisturbed free-stream flow and following the streamline that wets the surface, the flow is decelerated from the free-stream velocity to zero velocity at the (windward) stagnation point in the plane of symmetry. The flow then accelerates along the surface of the cylinder, reaching a maximum velocity equal in magnitude to twice the free-stream velocity. From these maxima (which occur at $\theta = 90°$ and at 270°), the flow tangent to the leeward surface decelerates to a stagnation point at the surface in the leeward plane of symmetry (at $\theta = 0°$).

However, even though the viscosity of air is relatively small, the actual flow field is radically different from the inviscid solution described in the previous paragraphs. When the air particles in the boundary layer, which have already been slowed by the action of viscosity, encounter the relatively large adverse pressure gradient associated with the deceleration of the leeward flow for this blunt configuration, boundary-layer separation occurs. The photograph of the smoke patterns for flow around a cylinder presented in Fig. 3.18 clearly illustrates the flow separation. Note that separation results when the fluid particles in the boundary layer (already slowed by viscosity) encounter an adverse pressure gradient that they cannot overcome. However, not all boundary layers separate when they encounter an adverse pressure gradient. There is a

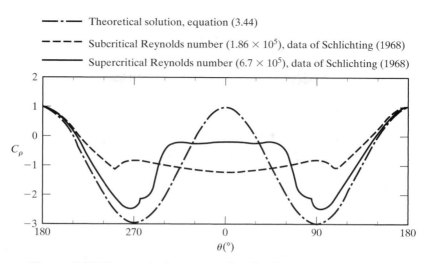

Figure 3.17 Theoretical pressure distribution around a circular cylinder, compared with data for a subcritical Reynolds number and that for a supercritical Reynolds number. [From Boundary Layer Theory by H. Schlichting (1968), used with permission of McGraw-Hill Book Company.]

relation between the characteristics of the boundary layer and the magnitude of the adverse pressure gradient that is required to produce separation. A turbulent boundary layer, which has relatively fast moving particles near the wall, would remain attached longer than a laminar boundary layer, which has slower-moving particles near the wall for the same value of the edge velocity. (Boundary layers are discussed in more detail in Chapter 4.) Therefore, the separation location, the size of the wake, and the surface pressure in the wake region depend on the character of the forebody boundary layer.

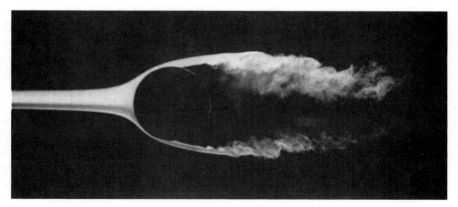

Figure 3.18 Smoke visualization of flow pattern for subcritical flow around a cylinder. (Photograph by F. N. M. Brown, courtesy of T. J. Mueller of University of Notre Dame.)

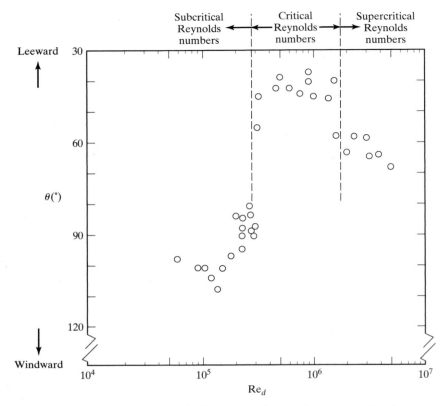

Figure 3.19 Location of the separation points on a circular cylinder as a function of the Reynolds number. [Data from Achenbach (1968).]

The experimentally determined separation locations for a circular cylinder as reported by Achenbach (1968) are presented as a function of Reynolds number in Fig. 3.19. As discussed in Chapter 2, the Reynolds number is a dimensionless parameter (in this case, $Re_d = \rho_\infty U_\infty d / \mu_\infty$) that relates to the viscous characteristics of the flow. At subcritical Reynolds numbers (i.e., less than approximately 3×10^5), the boundary layer on the windward surface (or forebody) is laminar and separation occurs for $\theta \sim 100°$, that is, 80° from the windward stagnation point. Note that the occurrence of separation so alters the flow that separation actually occurs on the windward surface, where the inviscid solution, as given by equation (3.44) and presented in Fig. 3.17, indicates that there still should be a favorable pressure gradient (i.e., one for which the static pressure decreases in the streamwise direction). If the pressure were actually decreasing in the streamwise direction, separation would not occur. Thus, the occurrence of separation alters the pressure distribution on the forebody (windward surface) of the cylinder. Above the critical Reynolds number, the forebody boundary layer is turbulent. Due to the higher levels of energy for the fluid particles near the surface in a turbulent boundary layer, the flow is able to run longer against the adverse pressure gradient. In the critical region, Achenbach observed an intermediate "separation bubble" with final separation not occurring until $\theta = 40°$ (i.e., 140° from the stagnation point). For $\mathrm{Re}_d > 1.5 \times 10^6$, the separation bubble no longer occurs, indicating that the supercritical state of flow has been reached. For supercritical Reynolds numbers, separation occurs in the range $60° < \theta < 70°$ (The reader should note that the critical Reynolds number is sensitive both to the turbulence level in the free stream and to surface roughness.)

Experimental pressure distributions are presented in Fig. 3.17 for the cases where the forebody boundary layer is laminar (a subcritical Reynolds number) and where the forebody boundary layer is turbulent (a supercritical Reynolds number). The subcritical pressure-coefficient distribution is essentially unchanged over a wide range of Reynolds numbers below the critical Reynolds numbers. Similarly, the supercritical pressure-coefficient distribution is independent of Reynolds numbers over a wide range of Reynolds numbers above the critical Reynolds number. For the flow upstream of the separation location, the boundary layer is thin, and the pressure-coefficient distribution is essentially independent of the character of the boundary layer for the cylinder. However, because the character of the attached boundary layer affects the separation location, it affects the pressure in the separated region. If the attached boundary layer is turbulent, separation is delayed and the pressure in the separated region is higher and closer to the inviscid level. See Figure 3.17

3.13.3 Lift and Drag

The motion of the air particles around the cylinder produces forces that may be viewed as a normal (or pressure) component and a tangential (or shear) component. It is conventional to resolve the resultant force on the cylinder into a component perpendicular to the free-stream velocity direction (called the *lift*) and a component parallel to the free-stream velocity direction (called the *drag*). The nomenclature is illustrated in Fig. 3.20.

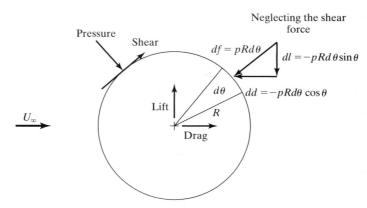

Figure 3.20 Forces acting on a cylinder whose axis is perpendicular to the free-stream flow.

   Since the expressions for the velocity distribution [equation (3.42)] and for the pressure distribution [equation (3.43) or (3.44)] were obtained for an inviscid flow, we shall consider only the contribution of the pressure to the lift and to the drag. As shown in Figure 3.20, the lift per unit span of the cylinder is

$$l = -\int_0^{2\pi} p \sin \theta \, R \, d\theta \tag{3.45}$$

Using equation (3.43) to define the static pressure as a function of θ, one finds that

$$l = 0 \tag{3.46}$$

It is not surprising that there is zero lift per unit span of the cylinder, since the pressure distribution is symmetric about a horizontal plane.

   Instead of using equation (3.43), which is the expression for the static pressure, the aerodynamicist might be more likely to use equation (3.44), which is the expression for the dimensionless pressure coefficient. To do this, note that the net force in any direction due to a constant pressure acting on a closed surface is zero. As a result,

$$\int_0^{2\pi} p_\infty \sin \theta \, R \, d\theta = 0 \tag{3.47}$$

Adding equations (3.45) and (3.47) yields

$$l = -\int_0^{2\pi} (p - p_\infty) \sin \theta \, R \, d\theta$$

Dividing both sides of this equation by the product $q_\infty 2R$, which is (dynamic pressure) times (area per unit span in the x plane), yields

$$\frac{l}{q_\infty 2R} = -0.5 \int_0^{2\pi} C_p \sin \theta \, d\theta \tag{3.48}$$

Both sides of the equation (3.48) are dimensionless. The expression of the left-hand side is known as the *section lift coefficient* for a cylinder:

$$C_l = \frac{l}{q_\infty 2R} \tag{3.49}$$

Using equation (3.44) to define C_p as a function of θ,

$$C_l = -0.5 \int_0^{2\pi} (1 - 4 \sin^2 \theta) \sin \theta \, d\theta = 0$$

which, of course, is the same result as was obtained by integrating the pressure directly.

   Referring to Figure 3.20 and following a similar procedure, we can calculate the drag per unit span for the cylinder in an inviscid flow. Thus, the drag per unit span is

$$d = -\int_0^{2\pi} p \cos \theta \, R \, d\theta \tag{3.50}$$

Substituting equation (3.43) for the local pressure,

$$d = -\int_0^{2\pi} \left(p_\infty + \tfrac{1}{2}\rho_\infty U_\infty^2 - 2\rho_\infty U_\infty^2 \sin^2 \theta \right) \cos \theta \, R \, d\theta$$

we find that

$$d = 0 \qquad\qquad\qquad (3.51)$$

A drag of zero is an obvious contradiction to the reader's experience (and is known as *d'Alembert's paradox*). Note that the actual pressure in the separated, wake region near the leeward plane of symmetry (in the vicinity of $\theta = 0$ in Fig. 3.17) is much less than the theoretical value. It is the resultant difference between the high pressure acting near the windward plane of symmetry (in the vicinity of $\theta = 180°$, i.e., the stagnation point) and the relatively low pressures acting near the leeward plane of symmetry which produces the large drag component.

A drag force that represents the streamwise component of the pressure force integrated over the entire configuration is termed *pressure* (or *form*) *drag*. The drag force that is obtained by integrating the streamwise component of the shear force over the vehicle is termed *skin friction drag*. Note that in the case of real flow past a cylinder, the skin friction drag is small. However, significant form drag results because of the action of viscosity, which causes the boundary layer to separate and therefore radically alters the pressure field. The pressure near the leeward plane of symmetry is higher (and closer to the inviscid values) when the forebody boundary layer is turbulent. Thus, the difference between the pressure acting on the foreward surface and that acting on the leeward surface is less in the turbulent case. As a result, the form drag for a turbulent boundary layer is markedly less than the corresponding value for a laminar (forebody) boundary layer.

The *drag coefficient* per unit span for a cylinder is

$$C_d = \frac{d}{q_\infty 2R} \qquad\qquad\qquad (3.52)$$

Experimental drag coefficients for a smooth circular cylinder in a low-speed stream [Schlichting (1968)] are presented as a function of Reynolds number in Fig. 3.21. For Reynolds numbers below 300,000, the drag coefficient is essentially constant (approximately 1.2), independent of Reynolds number. Recall that when we were discussing the experimental values of C_p presented in Fig. 3.17, it was noted that the subcritical pressure-coefficient distribution is essentially unchanged over a wide range of Reynolds number. For blunt bodies, the pressure (or form) drag is the dominant drag component. Since the pressure coefficient distribution for a circular cylinder is essentially independent of Reynolds number below the critical Reynolds number, it follows that the drag coefficient would be essentially independent of the Reynolds number. (For streamlined bodies at small angles of attack, the dominant component of drag is skin friction, which is Reynolds-number dependent.) Thus,

$$C_d \approx -0.5 \int_0^{2\pi} C_p \cos \theta \, d\theta$$

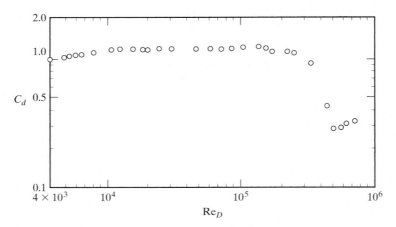

Figure 3.21 Drag coefficient for a smooth circular cylinder as a function of the Reynolds number. [From *Boundary Layer Theory* by H. Schlichting (1968), used with permission of McGraw-Hill Book Company.]

for the blunt body. Above the critical Reynolds number (when the forebody boundary layer is turbulent), the drag coefficient is significantly lower. Reviewing the supercritical pressure distribution, we recall that the pressure in the separated region is closer to the inviscid level. In a situation where the Reynolds number is subcritical, it may be desirable to induce boundary-layer transition by roughening the surface. Examples of such transition-promoting roughness elements are the dimples on a golf ball. The dimples on a golf ball are intended to reduce drag by reducing the form (or pressure) drag with only a slight increase in the friction drag.

3.14 LIFT AND DRAG COEFFICIENTS AS DIMENSIONLESS FLOW-FIELD PARAMETERS

The formulas for the drag coefficient [equation (3.52)] and for the lift coefficient for a cylinder [equation (3.49)] have the same elements. Thus, we can define a *force coefficient* as

$$C_F = \frac{\text{force}}{\underbrace{\left(\tfrac{1}{2}\rho_\infty U_\infty^2\right)}_{\substack{\text{dynamic} \\ \text{pressure}}}(\text{reference area})} \tag{3.53}$$

Note that, for a configuration of infinite span, a force per unit span would be divided by the reference area per unit span. Ideally, the force coefficient would be independent of size and would be a function of configuration geometry and of attitude only. However, the effects of viscosity and compressibility cause variations in the force coefficients. These effects can be correlated in terms of parameters, such as the Reynolds number and the Mach number. Such variations are evident in the drag coefficient measurements presented in this chapter.

From equation (3.53), it is clear that an aerodynamic force is proportional to the square of the free-stream velocity, to the free-stream density, to the size of the object, and to the force coefficient. An indication of the effect of configuration geometry on the total drag and on the drag coefficient is given in Figs. 3.22 and 3.23, which are taken from Talay (1975). The actual drag for several incompressible, flow condition/configuration geometry combinations are presented in Fig. 3.22. Compare the results for configurations (a), (b), and (c), which are configurations having the same dimension and exposed

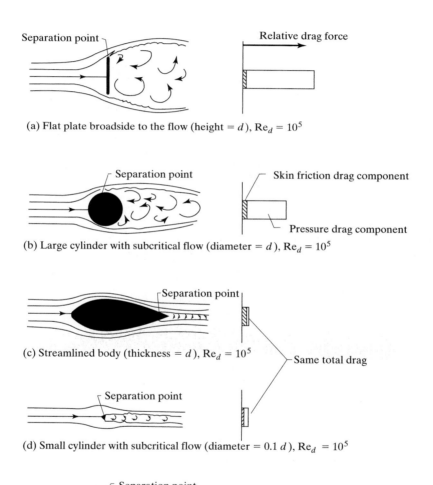

(a) Flat plate broadside to the flow (height $= d$), $\mathrm{Re}_d = 10^5$

(b) Large cylinder with subcritical flow (diameter $= d$), $\mathrm{Re}_d = 10^5$

(c) Streamlined body (thickness $= d$), $\mathrm{Re}_d = 10^5$

(d) Small cylinder with subcritical flow (diameter $= 0.1\,d$), $\mathrm{Re}_d = 10^5$

(e) Large cylinder with supercritical flow (diameter $= d$), $\mathrm{Re}_d = 10^7$

Figure 3.22 Comparison of the drag components for various shapes and flows. [From Talay (1975).]

(a) Flat plate broadside to the flow (height $= d$), $Re_d = 10^5$

(b) Large cylinder with subcritical flow (diameter $= d$), $Re_d = 10^5$

(c) Streamlined body (thickness $= d$), $Re_d = 10^5$

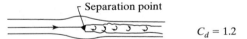

(d) Small cylinder with subcritical flow (diameter $= 0.1d$), $Re_d = 10^5$

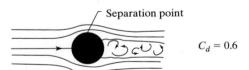

(e) Large cylinder with supercritical flow (diameter $= d$), $Re_d = 10^7$

Figure 3.23 Comparison of section drag coefficients for various shapes and flows. [From Talay (1975).]

to the same Reynolds number stream. Streamlining produces dramatic reductions in the pressure (or form) drag with only a slight increase in skin friction drag at this Reynolds number. Thus, streamlining reduces the drag coefficient.

Note that the diameter of the small cylinder is one-tenth that of the other configurations. Therefore, in the same free-stream flow as configuration (b), the small cylinder operates at a Reynolds number of 10^4. Because the size is reduced, the drag forces for (d) are an order of magnitude less than for (b). However, over this range of Reynolds number, the drag coefficients are essentially equal (see Fig. 3.21). As shown in Fig. 3.22, the total drag of the small cylinder is equal to that of the much thicker streamlined shape. The student can readily imagine how much additional drag was produced by the interplane wire bracing of the World War I biplanes.

When the Reynolds number is increased to 10^7 (corresponding to the supercritical flow of Fig. 3.21), the pressure drag is very large. However, the drag coefficient for this condition (e) is only 0.6, which is less than the drag coefficient for the subcritical flow (b) even though the pressure drag is significantly greater. Note that since the cylinder diameter is the same for both (b) and (e), the two order of magnitude increase in Reynolds number is accomplished by increasing the free-stream density and the free-stream velocity. Therefore, the denominator of equation (3.52) increases more than the numerator. As a result, even though the dimensional force is increased, the nondimensionalized force coefficient is decreased.

There are a variety of sources of aerodynamic data. However, Hoerner (1958) and Hoerner (1975) offer the reader unique and entertaining collections of data. In these volumes the reader will find aerodynamic coefficients for flags, World War II airplanes, and vintage automobiles as well as more classical configurations.

EXAMPLE 3.5:   Forces on a (semi-cylinder) quonset hut

You are to design a quonset hut to serve as temporary housing near the seashore. The quonset hut may be considered to be a closed (no leaks) semicylinder, whose radius is 5 m, mounted on tie-down blocks, as shown in Fig. 3.24. Neglect viscous effects and assume that the flow field over the top of the hut is identical to the flow over the cylinder for $0 \leq \theta \leq \pi$. When calculating the flow over the upper surface of the hut, neglect the presence of the air space under the hut. The air under the hut is at rest and the pressure is equal to the stagnation pressure, p_t.

What is the net lift force acting on the quonset hut? The wind speed is 50 m/s and the static free-stream properties are those for standard sea-level conditions.

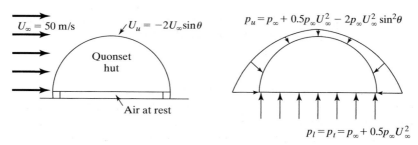

Figure 3.24 Inviscid flow model for quonset hut of Example 3.5.

Solution: Since we are to assume that the flow over the upper surface of the quonset hut is identical to an inviscid flow over a cylinder, the velocity is given by equation (3.42):

$$U_u = v_\theta = -2U_\infty \sin \theta$$

and the pressure is given by equation (3.43):

$$p_u = p_\infty + \tfrac{1}{2}\rho_\infty U_\infty^2 - 2\rho_\infty U_\infty^2 \sin^2 \theta$$

The pressure on the lower surface (under the hut) is

$$p_l = p_t = p_\infty + \tfrac{1}{2}\rho_\infty U_\infty^2$$

Equation (3.45) and Figure 3.20 can be used to calculate the lifting contribution of the upper surface but not the lower surface, since it is a "flat plate" rather than a circular arc. The lift per unit depth of the quonset hut is

$$l = -\int_0^\pi p_u \sin \theta \, R \, d\theta + p_t(2R)$$

$$= -R \int_0^\pi \left(p_\infty \sin \theta + \tfrac{1}{2}\rho_\infty U_\infty^2 \sin \theta - 2\rho_\infty U_\infty^2 \sin^3 \theta \right) d\theta$$

$$+ \, p_\infty 2R + \tfrac{1}{2}\rho_\infty U_\infty^2 2R$$

$$= \, +p_\infty R \cos \theta \Big|_0^\pi + \tfrac{1}{2}\rho_\infty U_\infty^2 R \cos \theta \Big|_0^\pi + 2\rho_\infty U_\infty^2 R \left(-\cos \theta + \tfrac{1}{3}\cos^3 \theta \right) \Big|_0^\pi$$

$$+ \, p_\infty 2R + \tfrac{1}{2}\rho_\infty U_\infty^2 2R$$

$$= \tfrac{8}{3}\rho_\infty U_\infty^2 R$$

Note that the lift coefficient is

$$C_l = \frac{l}{\tfrac{1}{2}\rho_\infty U_\infty^2(2R)} = \frac{\tfrac{8}{3}\rho_\infty U_\infty^2 R}{\rho_\infty U_\infty^2 R} = \frac{8}{3}$$

Since we have assumed that the flow is inviscid and incompressible, the lift coefficient is independent of the Mach number and of the Reynolds number. The actual lift force is

$$l = \frac{8}{3}\left(1.225 \frac{\text{kg}}{\text{m}^3} \right)\left(50 \frac{\text{m}}{\text{s}} \right)^2 (5 \text{ m}) = 40{,}833 \text{ N/m}$$

By symmetry, the drag is zero, which reflects the fact that we neglected the effects of viscosity.

3.15   FLOW AROUND A CYLINDER WITH CIRCULATION

3.15.1   Velocity Field

Let us consider the flow field that results if a vortex with clockwise circulation is super-imposed on the doublet/uniform-flow combination discussed above. The resultant potential function is

$$\phi = U_\infty r \cos \theta + \frac{B}{r} \cos \theta - \frac{\Gamma \theta}{2\pi} \tag{3.54}$$

Thus,

$$v_r = \frac{\partial \phi}{\partial r} = U_\infty \cos \theta - \frac{B \cos \theta}{r^2} \tag{3.55a}$$

and

$$v_\theta = \frac{1}{r} \frac{\partial \phi}{\partial \theta} = \frac{1}{r} \left(-U_\infty r \sin \theta - \frac{B \sin \theta}{r} - \frac{\Gamma}{2\pi} \right) \tag{3.55b}$$

Again, $v_r = 0$ at every point where $r = \sqrt{B/U_\infty}$, which is a constant and will be designated as R. Since the velocity is always tangent to streamline, the fact that the velocity component (v_r) perpendicular to a circle of radius R is zero means that the circle may be considered as a streamline of the flow field. Thus, the resultant potential function also represents flow around a cylinder. For this flow, however, the streamline pattern away from the surface is not symmetric about the horizontal plane. The velocity at the surface of the cylinder is equal to

$$v_\theta = -U = -2U_\infty \sin \theta - \frac{\Gamma}{2\pi R} \tag{3.56}$$

The resultant irrotational flow about the cylinder is uniquely determined once the magnitude of the circulation around the body is specified. Using the definition for the pressure coefficient [equation (3.13)], we obtain

$$C_p = 1 - \frac{U^2}{U_\infty^2} = 1 - \frac{1}{U_\infty^2} \left[4U_\infty^2 \sin^2 \theta + \frac{2\Gamma U_\infty \sin \theta}{\pi R} + \left(\frac{\Gamma}{2\pi R} \right)^2 \right] \tag{3.57}$$

3.15.2   Lift and Drag

If the expression for the pressure distribution is substituted into the expression for the drag force per unit span of the cylinder,

$$d = - \int_0^{2\pi} p(\cos \theta) R \, d\theta = 0$$

The prediction of zero drag may be generalized to apply to any general, two-dimensional body in an irrotational, steady, incompressible flow. In any real two-dimensional flow, a drag force does exist. For incompressible flow, drag is due to viscous effects, which produce the shear force at the surface and which may also produce significant changes in the pressure field (causing form drag).

Integrating the pressure distribution to determine the lift force per unit span for the cylinder, one obtains

$$l = -\int_0^{2\pi} p(\sin\theta)R\,d\theta = \rho_\infty U_\infty \Gamma \qquad (3.58)$$

Thus, the lift per unit span is directly related to the circulation about the cylinder. This result, which is known as the *Kutta-Joukowski theorem*, applies to the potential flow about closed cylinders of arbitrary cross section. To see this, consider the circulating flow field around the closed configuration to be represented by the superposition of a uniform flow and a unique set of sources, sinks, and vortices within the body. For a closed body, continuity requires that the sum of the source strengths be equal to the sum of the sink strengths. When one considers the flow field from a point far from the surface of the body, the distance between the sources and sinks becomes negligible and the flow field appears to be that generated by a single doublet with circulation equal to the sum of the vortex strengths within the body. Thus, in the limit, the forces acting are independent of the shape of the body and

$$l = \rho_\infty U_\infty \Gamma$$

The locations of the stagnation points (see Fig. 3.25) also depend on the circulation. To locate the stagnation points, we need to find where

$$v_r = v_\theta = 0$$

Since $v_r = 0$ at every point on the cylinder, the stagnation points occur when $v_\theta = 0$. Therefore,

$$-2U_\infty \sin\theta - \frac{\Gamma}{2\pi R} = 0$$

or

$$\theta = \sin^{-1}\left(-\frac{\Gamma}{4\pi R U_\infty}\right) \qquad (3.59)$$

If $\Gamma < 4\pi R U_\infty$, there are two stagnation points on the surface of the cylinder. They are symmetrically located about the y axis and both are below the x axis (see Fig. 3.25). If $\Gamma = 4\pi U_\infty R$, only one stagnation point exists on the cylinder and it exists at $\theta = 270°$. For this magnitude of the circulation, the lift per unit span is

$$l = \rho_\infty U_\infty \Gamma = \rho_\infty U_\infty^2 R 4\pi \qquad (3.60)$$

The lift coefficient per unit span of the cylinder is

$$C_l = \frac{l}{q_\infty 2R}$$

Thus,

$$C_l = \frac{\rho_\infty U_\infty^2 R 4\pi}{\frac{1}{2}\rho_\infty U_\infty^2 2R} = 4\pi \qquad (3.61)$$

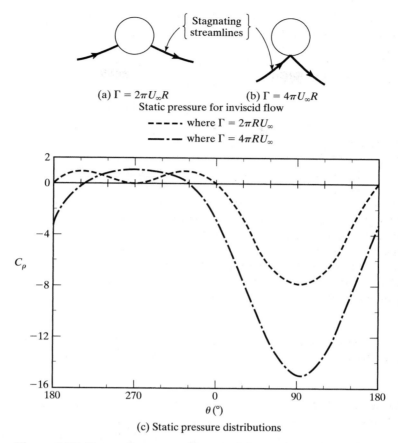

(c) Static pressure distributions

Figure 3.25 Stagnating streamlines and the static pressure distribution for a two-dimensional circulating flow around a cylinder. (a) $\Gamma = 2\pi U_\infty R$; (b) $\Gamma = 4\pi U_\infty R$.

The value 4π represents the maximum lift coefficient that can be generated for a circulating flow around a cylinder unless the circulation is so strong that no stagnation point exists on the body.

3.16 SOURCE DENSITY DISTRIBUTION ON THE BODY SURFACE

Thus far, we have studied fundamental fluid phenomena, such as the Kutta-Joukowski theorem, using the inverse method. Flow fields for other elementary configurations, such as axisymmetric shapes in a uniform stream parallel to the axis of symmetry, can be represented by a source distribution along the axis of symmetry. An "exact" solution for the flow around an arbitrary configuration can be approached using a direct method in a variety of ways, all of which must finally become numerical and make use of a computing machine. The reader is referred to Hess and Smith (1966) for an extensive review of the problem.

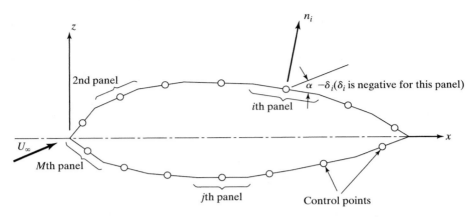

Figure 3.26 Source density distribution of the body surface.

Consider a two-dimensional configuration in a uniform stream, such as shown in Fig. 3.26. The coordinate system used in this section (i.e., x in the chordwise direction and y in the spanwise direction) will be used in subsequent chapters on wing and airfoil aerodynamics. The configuration is represented by a finite number (M) of linear segments, or panels. The effect of the jth panel on the flow field is characterized by a distributed source whose strength is uniform over the surface of the panel. Referring to equation (3.36), a source distribution on the jth panel causes an induced velocity whose potential at a point (x, z) is given by

$$\phi(x, z) = \int \frac{k_j \, ds_j}{2\pi} \ln r \qquad (3.62)$$

where k_j is defined as the volume of fluid discharged per unit area of the panel and the integration is carried out over the length of the panel ds_j. Note also that

$$r = \sqrt{(x - x_j)^2 + (z - z_j)^2} \qquad (3.63)$$

Since the flow is two dimensional, all calculations are for a unit length along the y axis, or span.

Each of the M panels can be represented by similar sources. To determine the strengths of the various sources k_j, we need to satisfy the physical requirement that the surface must be a streamline. Thus, we require that the sum of the source-induced velocities and the free-stream velocity is zero in the direction normal to the surface of the panel at the surface of each of the M panels. The points at which the requirement that the resultant flow is tangent to the surface will be numerically satisfied are called the *control points*. The control points are chosen to be the midpoints of the panels, as shown in Fig. 3.26.

At the control point of the ith panel, the velocity potential for the flow resulting from the superposition of the M source panels and the free-stream flow is

$$\phi(x_i, z_i) = U_\infty x_i \cos \alpha + U_\infty z_i \sin \alpha + \sum_{j=1}^{M} \frac{k_j}{2\pi} \int \ln r_{ij} \, ds_j \qquad (3.64)$$

where r_{ij} is the distance from the control point of the ith panel to a point on the jth panel.

$$r_{ij} = \sqrt{(x_i - x_j)^2 + (z_i - z_j)^2} \tag{3.65}$$

Note that the source strength k_j has been taken out of the integral, since it is constant over the jth panel. Each term in the summation represents the contribution of the jth panel (integrated over the length of the panel) to the potential at the control point of the ith panel.

The boundary conditions require that the resultant velocity normal to the surface be zero at each of the control points. Thus,

$$\frac{\partial}{\partial n_i} \phi_i(x_i, z_i) = 0 \tag{3.66}$$

must be satisfied at each and every control point. Care is required in evaluating the spatial derivatives of equation (3.64), because the derivatives become singular when the contribution of the ith panel is evaluated. Referring to equation (3.65), we have

$$r_{ij} = 0$$

where $j = i$. A rigorous development of the limiting process is given by Kellogg (1953). Although the details will not be repeated here, the resultant differentiation indicated in equation (3.66) yields

$$\frac{k_i}{2} + \sum_{\substack{j=1 \\ (j \neq i)}}^{M} \frac{k_j}{2\pi} \int \frac{\partial}{\partial n_i} (\ln r_{ij}) \, ds_j = -U_\infty \sin(\alpha - \delta_i) \tag{3.67}$$

where δ_i is the slope of the ith panel relative to the x axis. Note that the summation is carried out for all values of j except $j = i$. The two terms of the left side of equation (3.67) have a simple interpretation. The first term is the contribution of the source density of the ith panel to the outward normal velocity at the point (x_i, z_i), that is, the control point of the ith panel. The second term represents the contribution of the remainder of the boundary surface to the outward normal velocity at the control point of the ith panel.

Evaluating the terms of equation (3.67) for a particular ith control point yields a linear equation in terms of the unknown source strengths k_j (for $j = 1$ to M, including $j = i$). Evaluating the equation for all values of i (i.e., for each of the M control points) yields a set of M simultaneous equations which can be solved for the source strengths. Once the panel source strengths have been determined, the velocity can be determined at any point in the flow field using equations (3.64) and (3.65). With the velocity known, Bernoulli's equation can be used to calculate the pressure field.

Lift can be introduced by including vortex or doublet distributions and by introducing the Kutta condition; see Chapters 6 and 7.

EXAMPLE 3.6:   Application of the source density distribution

Let us apply the surface source density distribution to describe the flow around a cylinder in a uniform stream, where the free-stream velocity is U_∞.

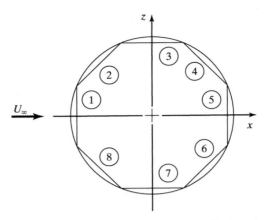

Figure 3.27 Representation of flow around a cylinder of unit radius by eight surface source panels.

The radius of the cylinder is unity. The cylinder is represented by eight, equal-length linear segments, as shown in Fig. 3.27. The panels are arranged such that panel 1 is perpendicular to the undisturbed stream.

Solution: Let us calculate the contribution of the source distribution on panel 2 to the normal velocity at the control point of panel 3. A detailed sketch of the two panels involved in this sample calculation is presented in Fig. 3.28. Referring to equation (3.67), we are to evaluate the integral:

$$\int \frac{\partial}{\partial n_i}(\ln r_{ij})ds_j$$

where $i = 3$ and $j = 2$. We will call this integral I_{32}. Note that

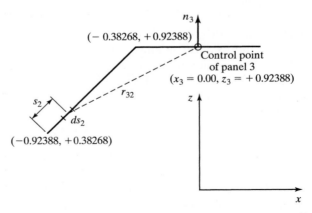

Figure 3.28 Detailed sketch for calculation of the contribution of the source distribution on panel 2 to the normal velocity at the control point of panel 3.

$$\frac{\partial}{\partial n_3}(\ln r_{32}) = \frac{1}{r_{32}}\frac{\partial r_{32}}{\partial n_3}$$

$$= \frac{(x_3 - x_2)\dfrac{\partial x_3}{\partial n_3} + (z_3 - z_2)\dfrac{\partial z_3}{\partial n_3}}{(x_3 - x_2)^2 + (z_3 - z_2)^2} \tag{3.68}$$

where $x_3 = 0.00$ and $z_3 = 0.92388$ are the coordinates of the control point of panel 3. Note also that

$$\frac{\partial x_3}{\partial n_3} = 0.00, \qquad \frac{\partial z_3}{\partial n_3} = 1.00$$

Furthermore, for the source line represented by panel 2,

$$x_2 = -0.92388 + 0.70711 s_2$$

$$z_2 = +0.38268 + 0.70711 s_2$$

and the length of the panel is

$$l_2 = 0.76537$$

Combining these expressions, we obtain

$$I_{32} = \int_0^{0.76537} \frac{(0.92388 - 0.38268 - 0.70711 s_2)\, ds_2}{(0.92388 - 0.70711 s_2)^2 + (0.92388 - 0.38268 - 0.70711 s_2)^2}$$

This equation can be rewritten

$$I_{32} = 0.54120 \int_0^{0.76537} \frac{ds_2}{1.14645 - 2.07195 s_2 + 1.00002 s_2^2}$$

$$-0.70711 \int_0^{0.76537} \frac{s_2\, ds_2}{1.14645 - 2.07195 s_2 + 1.00002 s_2^2}$$

Using the integral tables to evaluate these expressions, we obtain

$$I_{32} = -0.70711\left[\tfrac{1}{2}\ln(s_2^2 - 2.07195 s_2 + 1.14645)\right]_{s_2=0}^{s_2=0.76537}$$

$$-0.70711\left[\tan^{-1}\left(\frac{2s_2 - 2.07195}{\sqrt{0.29291}}\right)\right]_{s_2=0}^{s_2=0.76537}$$

Thus,

$$I_{32} = 0.3528$$

In a similar manner, we could calculate the contributions of source panels 1, 4, 5, 6, 7, and 8 to the normal velocity at the control point of panel 3. Substituting the values of these integrals into equation (3.67), we obtain a linear equation of the form

$$I_{31}k_1 + I_{32}k_2 + \pi k_3 + I_{34}k_4 + I_{35}k_5 + I_{36}k_6 + I_{37}k_7 + I_{38}k_8 = 0.00 \tag{3.69}$$

The right-hand side is zero since $\alpha = 0$ and $\delta_3 = 0$.

Repeating the process for all eight control points, we would obtain a set of eight linear equations involving the eight unknown source strengths. Solving the system of equations, we would find that

$$k_1 = 2\pi U_\infty(+0.3765)$$
$$k_2 = 2\pi U_\infty(+0.2662)$$
$$k_3 = 0.00$$
$$k_4 = 2\pi U_\infty(-0.2662)$$
$$k_5 = 2\pi U_\infty(-0.3765)$$
$$k_6 = 2\pi U_\infty(-0.2662)$$
$$k_7 = 0.00$$
$$k_8 = 2\pi U_\infty(+0.2662)$$

Note there is a symmetrical pattern in the source distribution, as should be expected. Also,

$$\sum k_i = 0 \tag{3.70}$$

as must be true since the sum of the strengths of the sources and sinks (negative sources) must be zero if we are to have a closed configuration.

3.17 INCOMPRESSIBLE, AXISYMMETRIC FLOW

The irrotational flows discussed thus far have been planar, two dimensional. That is, the flow field that exists in the plane of the paper will exist in any and every plane parallel to the plane of the paper. Thus, although sketches of the flow field defined by equations (3.41) though (3.61) depict the flow around a circle of radius R, in reality they represent the flow around a cylinder whose axis is perpendicular to the plane of the paper. For these flows, $w \equiv 0$ and $\partial/\partial z \equiv 0$.

Let us now consider another type of "two-dimensional" flow: an axisymmetric flow. The coordinate system is illustrated in Fig. 3.29. There are no circumferential variations in an axisymmetric flow; that is,

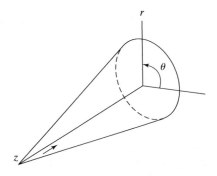

Figure 3.29 Coordinate system for an axisymmetric flow.

$$v_\theta \equiv 0 \quad \text{and} \quad \frac{\partial}{\partial \theta} \equiv 0$$

Thus, the incompressible, continuity equation becomes

$$\frac{\partial v_r}{\partial r} + \frac{v_r}{r} + \frac{\partial v_z}{\partial z} = 0$$

Noting that r and z are the independent coordinates (i.e., variables), we can rewrite this expression as

$$\frac{\partial}{\partial r}(rv_r) + \frac{\partial}{\partial z}(rv_z) = 0 \tag{3.71}$$

As has been discussed, a stream function will exist for an incompressible, two-dimensional flow. The flow need be two dimensional only in the sense that it requires only two spatial coordinates to describe the motion. The stream function that identically satisfies equation (3.71) is

$$\frac{\partial \psi}{\partial z} = rv_r \quad \text{and} \quad \frac{\partial \psi}{\partial r} = -rv_z$$

Thus, in the coordinate system of Fig. 3.29,

$$v_r = \frac{1}{r}\frac{\partial \psi}{\partial z} \quad \text{and} \quad v_z = -\frac{1}{r}\frac{\partial \psi}{\partial r} \tag{3.72}$$

Note that $\psi = $ constant defines a stream surface.

3.17.1   Flow around a Sphere

To describe a steady, inviscid, incompressible flow around a sphere, we will add the axisymmetric potential functions for a uniform flow and for a point doublet. We will first introduce the necessary relations in spherical coordinates. For a spherical coordinate system,

$$v_r = \frac{\partial \phi}{\partial r} \qquad v_\omega = \frac{1}{r}\frac{\partial \phi}{\partial \omega} \qquad v_\theta = \frac{1}{r \sin \omega}\frac{\partial \phi}{\partial \theta} \tag{3.73}$$

for an irrotational flow where $\vec{V} = \nabla \phi$. In equation (3.73), ϕ represents the potential function, and r, θ, and ω represent the independent coordinates. By symmetry,

$$v_\theta = 0 \quad \text{and} \quad \frac{\partial}{\partial \theta} = 0$$

The velocity potential for an axisymmetric doublet is

$$\phi = +\frac{B}{4\pi r^2}\cos \omega$$

where the doublet is so oriented that the source is placed upstream and the doublet axis is parallel to the uniform flow. The potential function for a uniform flow is

$$\phi = U_\infty r \cos \omega$$

Thus, the sum of the potential functions is

$$\phi = U_\infty r \cos \omega + \frac{B}{4\pi r^2} \cos \omega \tag{3.74}$$

The velocity components for this potential function are

$$v_r = \frac{\partial \phi}{\partial r} = U_\infty \cos \omega - \frac{B}{2\pi r^3} \cos \omega \tag{3.75a}$$

and

$$v_\omega = \frac{1}{r}\frac{\partial \phi}{\partial \omega} = -U_\infty \sin \omega - \frac{B}{4\pi r^3} \sin \omega \tag{3.75b}$$

As we did when modeling the inviscid flow around a cylinder, we note that

$$v_r = 0$$

when

$$r^3 = \frac{B}{2\pi U_\infty} = \text{constant} = R^3$$

Thus, if $B = 2\pi U_\infty R^3$, we can use the potential function described by equation (3.74) to describe steady, inviscid, incompressible flow around a sphere of radius R. For this flow,

$$v_r = U_\infty \left(1 - \frac{R^3}{r^3}\right) \cos \omega \tag{3.76a}$$

and

$$v_\omega = -U_\infty \left(1 + \frac{R^3}{2r^3}\right) \sin \omega \tag{3.76b}$$

On the surface of the sphere (i.e., for $r = R$), the resultant velocity is given by

$$U = |\vec{V}| = v_\omega = -\tfrac{3}{2}U_\infty \sin \omega \tag{3.77}$$

The static pressure acting at any point on the sphere can be calculated using equation (3.77) to represent the local velocity in Bernoulli's equation:

$$p = p_\infty + \tfrac{1}{2}\rho_\infty U_\infty^2 - \tfrac{1}{2}\rho_\infty U_\infty^2 \left(\tfrac{9}{4}\sin^2 \omega\right) \tag{3.78}$$

Rearranging the terms, we obtain the expression for the pressure coefficient for steady, inviscid, incompressible flow around a sphere:

$$C_p = 1 - \tfrac{9}{4}\sin^2 \omega \tag{3.79}$$

Compare this expression with equation (3.44) for flow around a cylinder of infinite span whose axis is perpendicular to the free-stream flow:

$$C_p = 1 - 4\sin^2 \theta$$

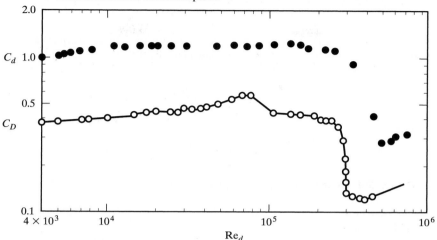

● Measurements for a smooth cylinder (see Fig. 3.21)
○ Measurements for a smooth sphere

Figure 3.30 Drag coefficient for a sphere as a function of the Reynolds number. [From Schlichting, *Boundary-Layer Theory* (1968), with permission of McGraw-Hill.]

Note that both θ and ω represent the angular coordinate relative to the axis, one for the two-dimensional flow, the other for axisymmetric flow. Thus, although the configurations have the same cross section in the plane of the paper (a circle) and both are described in terms of two coordinates, the flows are significantly different.

The drag coefficients for a sphere, as reported in Schlichting (1968), are presented as a function of the Reynolds number in Figure 3.30. The drag coefficient for a sphere is defined as

$$C_D = \frac{\text{drag}}{q_\infty(\pi d^2/4)} \tag{3.80}$$

The Reynolds number dependence of the drag coefficient for a smooth sphere is similar to that for a smooth cylinder. Again, a significant reduction in drag occurs as the critical Reynolds number is exceeded and the windward boundary layer becomes turbulent.

3.18 SUMMARY

In most flow fields of interest to the aerodynamicist, there are regions where the product of the viscosity times the shearing velocity gradient is sufficiently small that we may neglect the shear stress terms in our analysis. The momentum equation for these inviscid flows is known as Euler's equation. From Kelvin's theorem, we know that a flow remains irrotational in the absence of viscous forces and discontinuities provided that the flow is barotropic and the body forces are conservative. Potential functions can be used to describe the velocity field for such flows. If we assume further that the flow is incompressible (i.e., low speed), we can linearly add potential functions to obtain the velocity field for complex configurations and use Bernoulli's equation to determine the corresponding pressure distribution. The inviscid flow field solutions, thus obtained, form the outer (edge) boundary conditions for the thin viscous (boundary) layer adjacent to the wall. The characteristics of the boundary layer and techniques for analyzing it are described in the next chapter.

PROBLEMS

3.1. A truck carries an open tank, that is 6 m long, 2 m wide, and 3 m deep. Assuming that the driver will not accelerate or decelerate the truck at a rate greater than 2 m/s², what is the maximum depth to which the tank may be filled so that the water will not be spilled?

3.2. A truck carries an open tank that is 20 ft long, 6 ft wide, and 10 ft deep. Assuming that the driver will not accelerate or decelerate the truck at a rate greater than 6.3 ft/s², what is the maximum depth to which the tank may be filled so that the water will not be spilled?

3.3. What conditions must be satisfied before we can use Bernoulli's equation to relate the flow characteristics between two points in the flow field?

3.4. Water fills the circular tank (which is 20.0 ft in diameter) shown in Fig. P3.4. Water flows out of a hole which is 1.0 in. in diameter and which is located in the side of the tank, 15.0 ft from the top and 15.0 ft from the bottom. Consider the water to be inviscid. $\rho_{H_2O} = 1.940 \text{ slug/ft}^3$.

(a) Calculate the static pressure and the velocity at points 1, 2, and 3. For these calculations you can assume that the fluid velocities are negligible at points more than 10.0 ft from the opening.

(b) Having calculated U_3 in part (a), what is the velocity U_1? Was the assumption of part (a) valid?

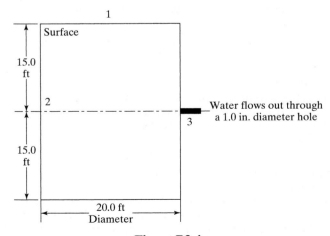

Figure P3.4

3.5. Consider a low-speed, steady flow around the thin airfoil shown in Fig. P3.5. We know the velocity and altitude at which the vehicle is flying. Thus, we know p_∞ (i.e., p_1) and U_∞. We have obtained experimental values of the local static pressure at points 2 through 6. At which of these points can we use Bernoulli's equation to determine the local velocity? If we cannot, why not?

Point 2: at the stagnation point of airfoil

Point 3: at a point in the inviscid region just outside the laminar boundary layer

Point 4: at a point in the laminar boundary layer

Point 5: at a point in the turbulent boundary layer

Point 6: at a point in the inviscid region just outside the turbulent boundary layer.

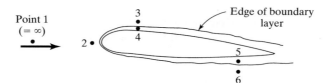

Figure P3.5

3.6. Assume that the airfoil of problem 3.5 is moving at 300 km/h at an altitude of 3 km. The experimentally determined pressure coefficients are

Point	2	3	4	5	6
C_p	1.00	−3.00	−3.00	+0.16	+0.16

(a) What is the Mach number and the Reynolds number for this configuration? Assume that the characteristic dimension for the airfoil is 1.5 m.

(b) Calculate the local pressure in N/m² and in lbf/in.² at all five points. What is the percentage change in the pressure relative to the free-stream value? That is, what is $(p_{\text{local}} - p_\infty)/p_\infty$? Was it reasonable to assume that the pressure changes are sufficiently small that the density is approximately constant?

(c) Why are the pressures at points 3 and 4 equal and at points 5 and 6 equal?

(d) At those points where Bernoulli's equation can be used validly, calculate the local velocity.

3.7. A Pitot-static probe is used to determine the airspeed of an airplane that is flying at an altitude of 6000 m. If the stagnation pressure is 4.8540×10^4 N/m², what is the airspeed? What is the pressure recorded by a gage that measures the difference between the stagnation pressure and the static pressure (such as that shown in Fig. 3.2)? How fast would an airplane have to fly at sea level to produce the same reading on this gage?

3.8. A high-rise office building located in Albuquerque, New Mexico, at an altitude of 1 mi is exposed to a wind of 40 mi/h. What is the static pressure in the airstream away from the influence of the building? What is the maximum pressure acting on the building? Pressure measurements indicate that a value of $C_p = -5$ occurs near the corner of the wall parallel to the wind direction. If the internal pressure is equal to the free-stream static pressure, what is the total force on a pane of glass 3 ft × 8 ft located in this region?

3.9. A high-rise office building located in a city at sea level is exposed to a wind of 75 km/h. What is the static pressure of the airstream away from the influence of the building? What is the maximum pressure acting on the building? Pressure measurements indicate that a value of $C_p = -4$ occurs near the corner of the wall parallel to the wind direction. If the internal pressure equals to the free-stream static pressure, what is the total force on the pane of glass 1 m × 3 m located in this region?

3.10. You are working as a flight-test engineer at the Dryden Flight Research Center. During the low-speed phase of the test program for the X-37, you know that the plane is flying at an altitude of 8 km. The pressure at gage 1 is 1550 N/m², gage; the pressure at gage 2 is −3875 N/m², gage.

     (a) If gage 1 is known to be at the stagnation point, what is the velocity of the airplane? What is its Mach number?

     (b) What is the free-stream dynamic pressure for this test condition?

     (c) What is the velocity of the air at the edge of the boundary layer at the second point relative to the airplane? What is the velocity relative to the ground? What is C_p for this gage?

3.11. Air flows through a converging pipe section, as shown in Fig. P3.11. Since the centerline of the duct is horizontal, the change in potential energy is zero. The Pitot probe at the upstream station provides a measure of the total pressure (or stagnation pressure). The downstream end of the U-tube provides a measure of the static pressure at the second station. Assuming the density of air to be 0.00238 slug/ft³ and neglecting the effects of viscosity, compute the volumetric flow rate in ft³/s. The fluid in the manometer is unity weight oil $(\rho_{oil} = 1.9404 \text{ slug/ft}^3)$.

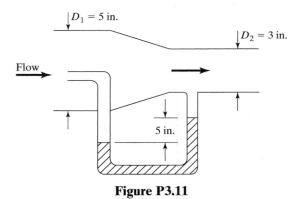

$|D_1 = 5$ in.

$|D_2 = 3$ in.

Flow

5 in.

Figure P3.11

3.12. An in-draft wind tunnel (Fig. P3.12) takes air from the quiescent atmosphere (outside the tunnel) and accelerates it in the converging section, so that the velocity of the air at a point in the test section but far from the model is 60 m/s. What is the static pressure at this point? What is the pressure at the stagnation point on a model in the test section? Use Table 1.2 to obtain the properties of the ambient air, assuming that the conditions are those for the standard atmosphere at sea level.

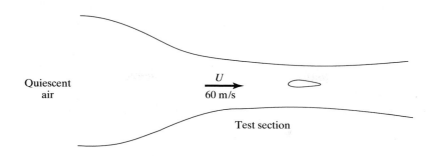

Figure P3.12.

3.13. A venturi meter is a device that is inserted into a pipeline to measure incompressible flow rates. As shown in Fig. P.3.13, it consists of a convergent section that reduces the diameter to between one-half to one-fourth of the pipe diameter. This is followed by a divergent section through which the flow is returned to the original diameter. The pressure difference between a location just before the venturi and one at the throat of the venturi is used to determine the volumetric flow rate (Q). Show that

$$Q = C_d \left[\frac{A_2}{\sqrt{1 - (A_2/A_1)^2}} \sqrt{\frac{2g(p_1 - p_2)}{\gamma}} \right]$$

where C_d is the coefficient of discharge, which takes into account the frictional effects and is determined experimentally or from handbook tabulations.

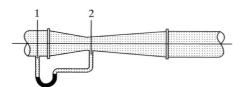

Figure P3.13

3.14. You are in charge of the pumping unit used to pressurize a large water tank on a fire truck. The fire that you are to extinguish is on the sixth floor of a building, 70 ft higher than the truck hose level, as shown in Fig. P3.14.

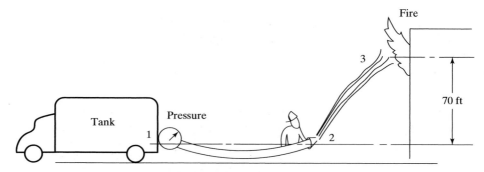

Figure P3.14.

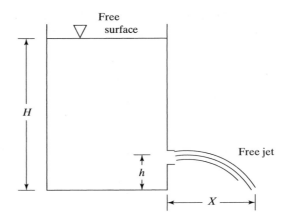

Figure P3.15.

(a) What is the minimum pressure in the large tank for the water to reach the fire? Neglect pressure losses in the hose.

(b) What is the velocity of the water as it exits the hose? The diameter of the nozzle is 3.0 in. What is the flow rate in gallons per minute? Note that 1 gal/min equals 0.002228 ft^3/s.

3.15. A free jet of water leaves the tank horizontally, as shown in Fig. P3.15. Assuming that the tank is large and the losses are negligible, derive an expression for the distance X (from the tank to the point where the jet strikes the floor) as a function of h and H? What is X, if the liquid involved was gasoline for which $\sigma = 0.70$?

3.16. (a) What conditions are necessary before you can use a stream function to solve for the flow field?

(b) What conditions are necessary before you can use a potential function to solve for the flow field?

(c) What conditions are necessary before you can apply Bernoulli's equation to relate two points in a flow field?

(d) Under what conditions does the circulation around a closed fluid line remain constant with respect to time?

3.17. What is the circulation around a circle of constant radius R_1 for the velocity field given as

$$\vec{V} = \frac{\Gamma}{2\pi r}\hat{e}_\theta$$

3.18. The velocity field for the fully developed viscous flow discussed in Example 2.2 is

$$u = \frac{1}{2\mu}\frac{dp}{dx}\left(y^2 - \frac{h^2}{4}\right)$$

$$v = 0$$

$$w = 0$$

Is the flow rotational or irrotational? Why?

3.19. Find the integral along the path $\vec{s}$ between the points $(0,0)$ and $(1,2)$ of the component of $\vec{V}$ in the direction of $\vec{s}$ for the following three cases:

(a) $\vec{s}$ a straight line.

(b) $\vec{s}$ a parabola with vertex at the origin and opening to the right.

(c) $\vec{s}$ a portion of the x axis and a straight line perpendicular to it.
The components of $\vec{V}$ are given by the expressions

$$u = x^2 + y^2$$
$$v = 2xy^2$$

3.20. Consider the velocity field given in Problem 3.12:

$$\vec{V} = (x^2 + y^2)\hat{\imath} + 2xy^2\hat{\jmath}$$

Is the flow rotational or irrotational? Calculate the circulation around the right triangle shown in Fig. P3.20.

$$\oint \vec{V} \cdot \vec{ds} = ?$$

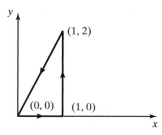

x **Figure P3.20**

What is the integral of the component of the curl $\vec{V}$ over the surface of the triangle? That is,

$$\iint (\nabla \times \vec{V}) \cdot \hat{n}\, \vec{dA} = ?$$

Are the results consistent with Stokes's theorem?

3.21. The absolute value of velocity and the equation of the potential function lines in a two-dimensional velocity field (Fig. P3.21) are given by the expressions

$$|\vec{V}| = \sqrt{4x^2 + 4y^2}$$
$$\phi = x^2 - y^2 + C$$

Evaluate both the left-hand side and the right-hand side of equation (3.16) to demonstrate the validity of Stokes's theorem of this irrotational flow.

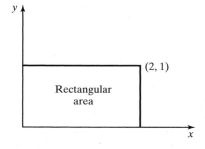

x **Figure P3.21**

3.22. Consider the incompressible, irrotational two-dimensional flow where the potential function is

$$\phi = K \ln \sqrt{x^2 + y^2}$$

where K is an arbitrary constant.

(a) What is the velocity field for this flow? Verify that the flow is irrotational. What is the magnitude and direction of the velocity at $(2, 0)$, at $(\sqrt{2}, \sqrt{2})$, and at $(0, 2)$?

(b) What is the stream function for this flow? Sketch the streamline pattern.

(c) Sketch the lines of constant potential. How do the lines of equipotential relate to the streamlines?

3.23. The stream function of a two-dimensional, incompressible flow is given by

$$\psi = \frac{\Gamma}{2\pi} \ln r$$

(a) Graph the streamlines.

(b) What is the velocity field represented by this stream function? Does the resultant velocity field satisfy the continuity equation?

(c) Find the circulation about a path enclosing the origin. For the path of integration, use a circle of radius 3 with a center at the origin. How does the circulation depend on the radius?

3.24. The absolute value of the velocity and the equation of the streamlines in a velocity field are given by

$$|\vec{V}| = \sqrt{4x^2 - 4xy + 5y^2}$$
$$4xy - y^2 = y^2 + 2xy = \text{constant}$$

Find u and v.

3.25. The absolute value of the velocity and the equation of the streamlines in a two-dimensional velocity field (Fig. P3.25) are given by the expressions

$$|\vec{V}| = \sqrt{5y^2 + x^2 + 4xy}$$
$$\psi = xy + y^2 = C$$

Find the integral over the surface shown of the normal component of curl $\vec{V}$ by two methods.

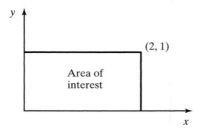

Figure P3.25

3.26. Given an incompressible, steady flow, where the velocity is

$$\vec{V} = (x^2 y - xy^2)\hat{\imath} + \left(\frac{y^3}{3} - xy^2\right)\hat{\jmath}$$

(a) Does the velocity field satisfy the continuity equation? Does a stream function exist? If a stream function exists, what is it?

(b) Does a potential function exist? If a potential function exists, what is it?

(c) For the region shown in Fig. P3.26, evaluate

$$\iint (\nabla \times \vec{V}) \cdot \hat{n}\, \overrightarrow{dA} = ?$$

and

$$\oint \vec{V} \cdot \overrightarrow{ds} = ?$$

to demonstrate that Stokes's theorem is valid.

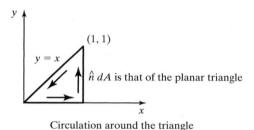

Circulation around the triangle

Figure P3.26

3.27. Consider the superposition of a uniform flow and a source of strength K. If the distance from the source to the stagnation point is R, calculate the strength of the source in terms of U_∞ and R.

(a) Determine the equation of the streamline that passes through the stagnation point. Let this streamline represent the surface of the configuration of interest.

(b) Noting that

$$v_r = \frac{1}{r}\frac{\partial \psi}{\partial \theta} \qquad v_\theta = -\frac{\partial \psi}{\partial r}$$

complete the following table for the surface of the configuration.

θ	$\dfrac{r}{R}$	$\dfrac{U}{U_\infty}$	C_p
30°			
45°			
90°			
135°			
150°			
180°			

3.28. A two-dimensional free vortex is located near an infinite plane at a distance h above the plane (Fig. P3.28). The pressure at infinity is p_∞ and the velocity at infinity is U_∞ parallel to the plane. Find the total force (per unit depth normal to the paper) on the plane if the pressure on the underside of the plane is p_∞. The strength of the vortex is Γ. The fluid is incompressible and perfect. To what expression does the force simplify if h becomes very large?

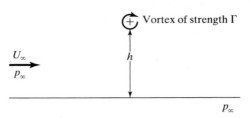

Figure P3.28

3.29. A perfect, incompressible irrotational fluid is flowing past a wall with a sink of strength K per unit length at the origin (Fig. P3.29). At infinity the flow is parallel and of uniform velocity U_∞. Determine the location of the stagnation point x_0 in terms of U_∞ and K. Find the pressure distribution along the wall as a function of x. Taking the free-stream static pressure at infinity to be p_∞, express the pressure coefficient as a function of x/x_0. Sketch the resulting pressure distribution.

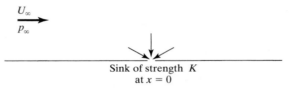

Sink of strength K
at $x = 0$

Figure P3.29

3.30. What is the stream function that represents the potential flow about a cylinder whose radius is 1 m and which is located in an air stream where the free-stream velocity is 50 m/s? What is the change in pressure from the free-stream value to the value at the top of the cylinder (i.e., $\theta = 90°$)? What is the change in pressure from the free-stream value to that at the stagnation point (i.e., $\theta = 180°$)? Assume that the free-stream conditions are those of the standard atmosphere at sea level.

3.31. Consider the flow formed by a uniform flow superimposed on a doublet whose axis is parallel to the direction of the uniform flow and is so oriented that the direction of the efflux opposes the uniform flow. This is the flow field of Section 3.13.1. Using the stream functions for these two elementary flows, show that a circle of radius R, where

$$R = \sqrt{\frac{B}{U_\infty}}$$

is a streamline in the flow field.

3.32. Consider the flow field that results when a vortex with clockwise circulation is superimposed on the doublet/uniform-flow combination discussed in Problem 3.31. This is the flow field of Section 3.15.1. Using the stream functions for these three elementary flows, show that a circle of radius R, where

$$R = \sqrt{\frac{B}{U_\infty}}$$

is a streamline in the flow field.

3.33. A cylindrical tube with three radially drilled orifices, as shown in Fig. P3.33, can be used as a flow-direction indicator. Whenever the pressure on the two side holes is equal, the pressure at the center hole is the stagnation pressure. The instrument is called a *direction-finding Pitot tube*, or a *cylindrical yaw probe*.

(a) If the orifices of a direction-finding Pitot tube were to be used to measure the free-stream static pressure, where would they have to be located if we use our solution for flow around a cylinder?

(b) For a direction-finding Pitot tube with orifices located as calculated in part (a), what is the sensitivity? Let the sensitivity be defined as the pressure change per unit angular change (i.e., $\partial p/\partial \theta$).

Figure P3.33

3.34. An infinite-span cylinder (two-dimensional) serves as a plug between the two airstreams , as shown in Fig. P3.34. Both air flows may be considered to be steady, inviscid, and incompressible, Neglecting the body forces in the air and the weight of the cylinder, in which direction does the plug more (i.e., due to the airflow)?

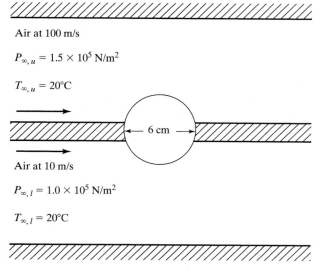

Air at 100 m/s

$P_{\infty, u} = 1.5 \times 10^5 \ N/m^2$

$T_{\infty, u} = 20°C$

← 6 cm →

Air at 10 m/s

$P_{\infty, l} = 1.0 \times 10^5 \ N/m^2$

$T_{\infty, l} = 20°C$

Figure P3.34.

3.35. Using the data of Fig. 3.30 calculate the force and the overturning moment exerted by a 4 m/s wind on a cylindrical smokestack that has a diameter of 3 m and a height of 50 m. Neglect variations in the velocity of the wind over the height of the smokestack. The temperature of the air is 30°C; its pressure is 99 kPa. What is the Reynolds number for this flow?

3.36. Calculate the force and the overturning moment exerted by a 45-mph wind on a cylindrical flagpole that has a diameter of 6 in. and a height of 15 ft. Neglect variations in the velocity of the wind over the height of the flagpole. The temperature of the air is 85°F; its pressure is 14.4 psi. What is the Reynolds number of this flow?

3.37. A cylinder 3 ft in diameter is placed in a stream of air at 68°F where the free-stream velocity is 120 ft/s. What is the vortex strength required in order to place the stagnation points at $\theta = 30°$ and $\theta = 150°$? If the free-stream pressure is 2000 lbf/ft², what is the pressure at the stagnation points? What will be the velocity and the static pressure at $\theta = 90°$? at $\theta = 270°$? What will be the theoretical value of the lift per spanwise foot of the cylinder?

3.38. Consider the flow around the quonset hut shown in Fig. P3.38 to be represented by super-imposing a uniform flow and a doublet. Assume steady, incompressible, potential flow. The ground plane is represented by the plane of symmetry and the hut by the upper half of the cylinder. The free-stream velocity is 175 km/h; the radius R_0 of the hut is 6 m. The door is not well sealed, and the static pressure inside the hut is equal to that on the outer surface of the hut, where the door is located.
   (a) If the door to the hut is located at ground level (i.e., at the stagnation point), what is the net lift acting on the hut? What is the lift coefficient?
   (b) Where should the door be located (i.e., at what angle θ_0 relative to the ground) so that the net force on the hut will vanish?
   For both parts of the problem, the opening is very small compared to the radius R_0. Thus, the pressure on the door is essentially constant and equal to the value of the angle θ_0 at which the door is located. Assume that the wall is negligibly thin.

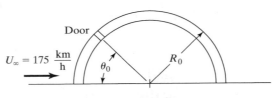

Figure P3.38

3.39. Consider an incompressible flow around a semicylinder, as shown in Figure. P3.39. Assume that velocity distribution for the windward surface of the cylinder is given by the inviscid solution

$$\vec{V} = -2U_\infty \sin \theta \hat{e}_\theta$$

Calculate the lift and drag coefficients if the base pressure (i.e., the pressure on the flat, or leeward, surface) is equal to the pressure at the separation point, p_{corner}.

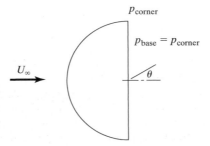

Figure P3.39

3.40. A semicylindrical tube, as shown in Fig. P3.39, is submerged in a stream of air where $\rho_\infty = 1.22 \, \text{kg/m}^3$ and $U_\infty = 75 \, \text{m/s}$. The radius is 0.3 m. What are the lift and drag forces acting on the tube using the equations developed in Problem 3.39.

3.41. You are to design quonset huts for a military base in the mideast. The design wind speed is 100 ft/s. The static free-stream properties are those for standard sea-level conditions. The quonset hut may be considered to be a closed (no leaks) semicylinder, whose radius is 15 ft, mounted on tie-down blocks, as shown in Example 3.5 The flow is such that the velocity distribution and, thus, the pressure distribution over the top of the hut (the semicircle of the sketch) is represented by the potential function

$$\phi = U_\infty r \cos\theta + \frac{B}{r}\cos\theta$$

When calculating the flow over the hut, neglect the presence of the air space under the hut. The air under the hut is at rest and the pressure is equal to stagnation pressure, $p_t \left(= p_\infty + \frac{1}{2}\rho_\infty U_\infty^2 \right)$.

(a) What is the value of B for the 15-ft-radius (R) quonset hut?
(b) What is the net lift force acting on the quonset hut?
(c) What is the net drag force acting on the quonset hut?

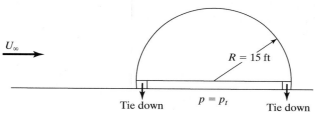

Figure P3.41 See Example 3.5

3.42. Using equation (3.56) to define the surface velocity distribution for inviscid flow around a cylinder with circulation, derive the expression for the local static pressure as a function of θ. Substitute the pressure distribution into the expression for the lift to verify that equation (3.58) gives the lift force per unit span. Using the definition that

$$C_l = \frac{l}{q_\infty 2R}$$

(where l is the lift per unit span), what is the section lift coefficient?

3.43. Combining equations (3.45) and (3.49), it has been shown that the section lift coefficient for inviscid flow around a cylinder is

$$C_l = -\frac{1}{2}\int_0^{2\pi} C_p \sin\theta \, d\theta \qquad\qquad (3.48)$$

Using equation (3.57) to define the pressure coefficient distribution for inviscid flow with circulation, calculate the section lift coefficient for this flow.

3.44. There were early attempts in the development of the airplane to use rotating cylinders as airfoils. Consider such a cylinder having a diameter of 1 m and a length of 10 m. If this cylinder is rotated at 100 rpm while the plane moves at a speed of 100 km/h through the air at 2 km standard atmosphere, estimate the maximum lift that could be developed, disregarding end effects.

3.45. Using the procedures illustrated in Example 3.6, calculate the contribution of the source distribution on panel 3 to the normal velocity at the control point of panel 4. The configuration geometry is illustrated in Fig. 3.27.

3.46. Consider the pressure distribution shown in Fig. P3.46 for the windward and leeward surfaces of a thick disk whose axis is parallel to the free-stream flow. What is the corresponding drag coefficient?

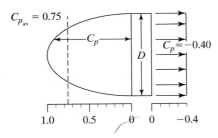

Figure P3.46

3.47. Consider an incompressible flow around a hemisphere, as shown in Fig. P3.47. Assume that the velocity distribution for the windward surface of the cylinder is given by the inviscid solution

$$V = -\tfrac{3}{2}U_\infty \sin \omega \tag{3.77}$$

Calculate the lift and drag coefficients if the base pressure (i.e., the pressure on the flat, or leeward surface) is equal to the pressure at the separation point, p_{corner}. How does the drag coefficient for a hemisphere compare with that for a hemicylinder (i.e., Problem 3.47.)

3.48. A hemisphere, as shown in Fig. P3.48, is submerged in an airstream where $\rho_\infty = 0.002376 \text{ slug/ft}^3$ and $U_\infty = 200 \text{ ft/s}$. The radius is 1.0 ft. What are the lift and drag forces on the hemisphere using the equations developed in Problem 3.47.)

3.49. Consider air flowing past a hemisphere resting on a flat surface, as shown in Fig. P3.49. Neglecting the effects of viscosity, if the internal pressure is p_i, find an expression for the pressure force on the hemisphere. At what angular location should a hole be cut in the surface of the hemisphere so that the net pressure force will be zero?

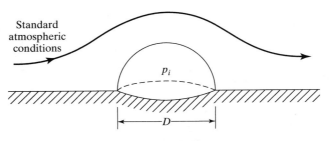

Figure P3.49

3.50. A major league pitcher is accused of hiding sandpaper in his back pocket in order to scuff up an otherwise "smooth" baseball. Why would he do this? To estimate the Reynolds number of the baseball, assume its speed to be 90 mi/h and its diameter 2.75 in.

3.51. Derive the stream functions for the elementary flows of Table 3.2.

3.52. What condition(s) must prevail in order for a velocity potential to exist? For a stream function to exist?

REFERENCES

Achenbach E. 1968. Distributions of local pressure and skin friction around a circular cylinder in cross flow up to Re $= 5 \times 10^6$. *J. Fluid Mechanics* 34:625–639

Campbell J, Chambers JR. 1994. Patterns in the sky: natural visualization of aircraft flow fields. *NASA SP-514*

Hess JL, Smith AMO. 1967. Calculations of potential flow about arbitrary bodies. *Progr. Aeronaut. Sci.* 8:1-138

Hoerner SF. 1958. *Fluid Dynamic Drag*. Midland Park, NJ: published by the author

Hoerner SF, Borst HV. 1975. *Fluid Dynamic Lift*. Midland Park, NJ: published by the authors

Kellogg OD. 1953. *Fundamentals of Potential Theory*. New York: Dover

Schlichting H. 1968. *Boundary Layer Theory*. 6th Ed. New York: McGraw-Hill

Talay TA. 1975. Introduction to the aerodynamics of flight. *NASA SP-367*

1976. *U.S. Standard Atmosphere*. Washington, DC: U.S. Government Printing Office

4 VISCOUS BOUNDARY LAYERS

The equation for the conservation of linear momentum was developed in Chapter 2 by applying Newton's law, which states that the net force acting on a fluid particle is equal to the time rate of change of the linear momentum of the fluid particle. The principal forces considered were those that act directly on the mass of the fluid element (i.e., the body forces) and those that act on its surface (i.e., the pressure forces and shear forces). The resultant equations are known as the Navier-Stokes equations. Even today, there are no general solutions for the complete Navier-Stokes equations. Nevertheless, reasonable approximations can be introduced to describe the motion of a viscous fluid if the viscosity is either very large or very small. The latter case is of special interest to us, since two important fluids, water and air, have very small viscosities. Not only is the viscosity of these fluids very small, but the velocity for many of the practical applications relevant to this text is such that the Reynolds number is very large.

Even in the limiting case where the Reynolds number is large, it is not permissible simply to omit the viscous terms completely, because the solution of the simplified equation could not be made to satisfy the complete boundary conditions. However, for many high Reynolds number flows, the flow field may be divided into two regions: (1) a viscous boundary layer adjacent to the surface of the vehicle and (2) the essentially inviscid flow outside the boundary layer. The velocity of the fluid particles increases from a value of zero (in a vehicle-fixed coordinate system) at the wall to the value that corresponds to the external "frictionless" flow outside the boundary layer. Outside the boundary layer, the transverse velocity gradients become so small that the shear stresses acting on a fluid element are negligibly small. Thus, the effect of the viscous terms may be ignored when solving for the flow field external to the boundary layer.

When using the two-region flow model to solve for the flow field, the first step is to solve for the inviscid portion of the flow field. The solution for the inviscid portion of the flow field must satisfy the boundary conditions: (1) that the velocity of the fluid particles far from the body be equal to the free-stream value and (2) that the velocity of the fluid particles adjacent to the body are parallel to the "surface." The second boundary condition represents the physical requirement that there is no flow through a solid surface. However, since the flow is inviscid, the velocity component parallel to the surface does not have to be zero. Having solved for the inviscid flow field, the second step is to calculate the boundary layer using the inviscid flow as the outer boundary condition.

If the boundary layer is relatively thick, it may be necessary to use an iterative process for calculating the flow field. To start the second iteration, the inviscid flow field is recalculated, replacing the actual configuration by the "effective" configuration, which is determined by adding the displacement thickness of the boundary layer from the first iteration to the surface coordinate of the actual configuration. See Fig. 2.13. The boundary layer is recalculated using the second-iterate inviscid flow as the boundary condition. As discussed in DeJarnette and Ratcliffe (1996), the iterative procedure required to converge to a solution requires an understanding of each region of the flow field and their interactions.

In Chapter 3, we generated solutions for the inviscid flow field for a variety of configurations. In this chapter we examine the viscous region in detail, assuming that the inviscid flow field is known.

4.1 EQUATIONS GOVERNING THE BOUNDARY LAYER FOR A STEADY, TWO-DIMENSIONAL, INCOMPRESSIBLE FLOW

In this chapter, we discuss techniques by which we can obtain engineering solutions when the boundary layer is either laminar or turbulent. Thus, for the purpose of this text, we shall assume that we know whether the boundary layer is laminar or turbulent. The transition process through which the boundary layer "transitions" from a laminar state to a turbulent state is quite complex and depends on many parameters (e.g., surface roughness, surface temperature, pressure gradient, and Mach number). A brief summary of the factors affecting transition is presented later in this chapter. For a more detailed discussion of the parameters that affect transition, the reader is referred to Schlichting (1979) and White (2005).

To simplify the development of the solution techniques, we will consider the flow to be steady, two-dimensional, and constant property (or, equivalently, incompressible for a gas flow). By restricting ourselves to such flows, we can concentrate on the development of the solution techniques themselves. As shown in Fig. 4.1, the coordinate system is fixed to the surface of the body. The x coordinate is measured in the streamwise direction along the surface of the configuration. The stagnation point (or the leading edge if the configuration is a sharp object) is at $x = 0$. The y coordinate is perpendicular to the surface. This coordinate system is used throughout this chapter.

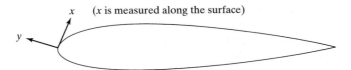

Figure 4.1 Coordinate system for the boundary-layer equations.

Referring to equation (2.3), the differential form of the continuity equation for this flow is

$$\frac{\partial u}{\partial x} + \frac{\partial v}{\partial y} = 0 \tag{4.1}$$

Referring to equation (2.16) and neglecting the body forces, the x component of momentum for this steady, two-dimensional flow is

$$\rho u \frac{\partial u}{\partial x} + \rho v \frac{\partial u}{\partial y} = -\frac{\partial p}{\partial x} + \mu \frac{\partial^2 u}{\partial x^2} + \mu \frac{\partial^2 u}{\partial y^2} \tag{4.2}$$

Similarly, the y component of momentum is

$$\rho u \frac{\partial v}{\partial x} + \rho v \frac{\partial v}{\partial y} = -\frac{\partial p}{\partial y} + \mu \frac{\partial^2 v}{\partial x^2} + \mu \frac{\partial^2 v}{\partial y^2} \tag{4.3}$$

Note that, if the boundary layer is thin and the streamlines are not highly curved, then $u \gg v$. Thus, if we compare each term in equation (4.3) with the corresponding term in equation (4.2), we conclude that

$$\rho u \frac{\partial u}{\partial x} > \rho u \frac{\partial v}{\partial x} \qquad \rho v \frac{\partial u}{\partial y} > \rho v \frac{\partial v}{\partial y} \qquad \mu \frac{\partial^2 u}{\partial x^2} > \mu \frac{\partial^2 v}{\partial x^2} \qquad \mu \frac{\partial^2 u}{\partial y^2} > \mu \frac{\partial^2 v}{\partial y^2}$$

Thus, as discussed in Chapter 2, we conclude that

$$\frac{\partial p}{\partial x} > \frac{\partial p}{\partial y}$$

The essential information supplied by the y component of the momentum equation is that the static pressure variation in the y direction may be neglected for most boundary layer flows. This is true whether the boundary layer is laminar, transitional, or turbulent. It is not true in wake flows, that is, separated regions in the lee side of blunt bodies such as those behind cylinders, which were discussed in Chapter 3. The assumption that the static pressure variation across a thin boundary layer is negligible only breaks down for turbulent boundary layers at very high Mach numbers. The common assumption for thin boundary layers also may be written as

$$\frac{\partial p}{\partial y} \approx 0 \tag{4.4}$$

Thus, the local static pressure is a function of x only and is determined from the solution of the inviscid portion of the flow field. As a result, Euler's equation for a steady flow with negligible body forces, which relates the streamwise pressure gradient to the velocity gradient for the inviscid flow, can be used to evaluate the pressure gradient in the viscous region, also.

$$-\frac{\partial p}{\partial x} = -\frac{dp_e}{dx} = \rho_e u_e \frac{du_e}{dx} \tag{4.5}$$

Substituting equation (4.5) into equation (4.2), and noting that $\mu(\partial^2 u/\partial x^2) < \mu(\partial^2 u/\partial y^2)$, we obtain

$$\rho u \frac{\partial u}{\partial x} + \rho v \frac{\partial u}{\partial y} = \rho_e u_e \frac{du_e}{dx} + \mu \frac{\partial^2 u}{\partial y^2} \tag{4.6}$$

Let us examine equations (4.1) and (4.6). The assumption that the flow is constant property (or incompressible) implies that fluid properties, such as density ρ and viscosity μ, are constants. For low-speed flows of gases, the changes in pressure and temperature through the flow field are sufficiently small that the corresponding changes in ρ and μ have a negligible effect on the flow field. By limiting ourselves to incompressible flows, it is not necessary to include the energy equation in the formulation of our solution. For compressible (or high-speed) flows, the temperature changes in the flow field are sufficiently large that the temperature dependence of the viscosity and of the density must be included. As a result, the analysis of a compressible boundary layer involves the simultaneous solution of the continuity equation, the x momentum equation, and the energy equation. For a detailed treatment of compressible boundary layers, the reader is referred to other sources [e.g., Schlichting (1979) and Dorrance (1962)] and to Chapter 8.

When the boundary layer is laminar, the transverse exchange of momentum (i.e., the momentum transfer in a direction perpendicular to the principal flow direction) takes place on a molecular (or microscopic) scale. As a result of the molecular movement, slower moving fluid particles from the lower layer (or lamina) of fluid move upward, slowing the particles in the upper layer. Conversely, when the faster-moving fluid particles from the upper layer migrate downward, they tend to accelerate the fluid particles in that layer. This molecular interchange of momentum for a laminar flow is depicted in Fig. 4.2a. Thus, the shear stress at a point in a Newtonian fluid is that given by constitutive relations preceding equations (2.12).

For a turbulent boundary layer, there is a macroscopic transport of fluid particles, as shown in Fig. 4.2b. Thus, in addition to the laminar shear stress described in the preceding paragraph, there is an effective *turbulent shear stress* that is due to the transverse transport of momentum and that is very large. Because slower-moving fluid particles near the wall are transported well upward, the turbulent boundary layer is relatively thick. Because faster-moving fluid particles (which are normally located near the edge of the boundary layer) are transported toward the wall, they produce relatively high velocities for the fluid particles near the surface. Thus, the shear stress at the wall for a turbulent boundary layer is larger than that for a laminar boundary layer. Because the macroscopic transport of fluid introduces large localized variations in the flow at any instant, the values of the fluid properties and the velocity components are (in general) the sum of the "average" value and a fluctuating component.

We could introduce the fluctuating characteristics of turbulent flow at this point and treat both laminar and turbulent boundary layers in a unified fashion. For a laminar

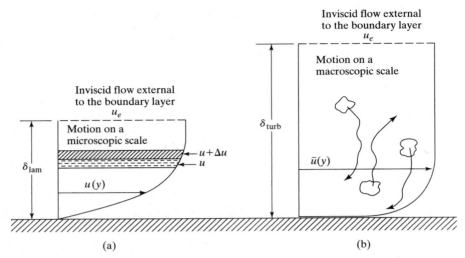

Figure 4.2 Momentum-transport models: (a) laminar boundary layer; (b) turbulent boundary layer.

boundary layer the fluctuating components of the flow would be zero. However, to simplify the discussion, we will first discuss laminar flows and their analysis and then turbulent boundary layers and their analysis.

4.2 BOUNDARY CONDITIONS

Let us now consider the boundary conditions that we must apply in order to obtain the desired solutions. Since we are considering that portion of the flow field where the viscous forces are important, the condition of no slip on the solid boundaries must be satisfied. That is, at $y = 0$,

$$u(x,0) = 0 \tag{4.7a}$$

At a solid wall, the normal component of velocity must be zero. Thus,

$$v(x,0) = 0 \tag{4.7b}$$

The velocity boundary conditions for porous walls through which fluid can flow are treated in the problems at the end of the chapter. Furthermore, at points far from the wall (i.e., at y large), we reach the edge of the boundary layer where the streamwise component of the velocity equals that given by the inviscid solution. In equation form,

$$u(x, y \text{ large}) = u_e(x) \tag{4.8}$$

Note that throughout this chapter, the subscript e will be used to denote parameters evaluated at the edge of the boundary layer (i.e., those for the inviscid solution).

4.3 INCOMPRESSIBLE, LAMINAR BOUNDARY LAYER

In this section we analyze the boundary layer in the region from the stagnation point (or from the leading edge of a sharp object) to the onset of transition (i.e., that "point" at which the boundary layer becomes turbulent). The reader should note that, in reality, the boundary layer does not go from a laminar state to a turbulent state at a point but that the transition process takes place over a distance. The length of the transition zone may be as long as the laminar region. Typical velocity profiles for the laminar boundary layer are presented in Fig. 4.3. The streamwise (or x) component of velocity is presented as a function of distance from the wall (the y coordinate). Instead of presenting the dimensional parameter u, which is a function both of x and y, let us seek a dimensionless velocity parameter that perhaps can be written as a function of a single variable. Note that, at each station, the velocity varies from zero at $y = 0$ (i.e., at the wall) to u_e for the inviscid flow outside of the boundary layer. The local velocity at the edge of the boundary layer u_e is a function of x only. Thus, a logical dimensionless velocity parameter is u/u_e.

Instead of using the dimensional y coordinate, we will use a dimensionless coordinate η, which is proportional to y/δ for these incompressible, laminar boundary layers. The boundary-layer thickness δ at any x station depends not only on the magnitude

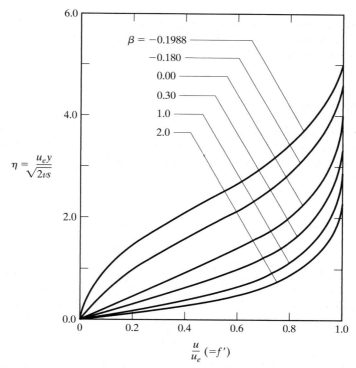

Figure 4.3 Solutions for the dimensionless streamwise velocity for the Falkner-Skan, laminar, similarity flows.

of x but on the kinematic viscosity, the local velocity at the edge of the boundary layer, and the velocity variation from the origin to the point of interest. Thus, we will introduce the coordinate transformation for η:

$$\eta = \frac{u_e y}{\sqrt{2\nu s}} \tag{4.9a}$$

where ν is the kinematic viscosity, as defined in Chapter 1, and where

$$s = \int u_e \, dx \tag{4.9b}$$

Note that for flow past a flat plate, u_e is a constant (independent of x) and is equal to the free-stream velocity upstream of the plate U_∞.

$$\eta = y\sqrt{\frac{u_e}{2\nu x}} \tag{4.10}$$

Those readers who are familiar with the transformations used in more complete treatments of boundary layers will recognize that this definition for η is consistent with that commonly used to transform the incompressible laminar boundary layer on a flat plate [White (2005)]. The flat-plate solution is the classical Blasius solution. This transformation is also consistent with more general forms used in the analysis of a compressible laminar flow [Dorrance (1962)]. By using this definition of s as the transformed x coordinate, we can account for the effect of the variation in u_e on the streamwise growth of the boundary layer.

Note that we have two equations [equations (4.1) and (4.6)] with two unknowns, the velocity components: u and v. Since the flow is two dimensional and the density is constant, the necessary and sufficient conditions for the existence of a stream function are satisfied. (*Note*: Because of viscosity, the boundary-layer flow cannot be considered as irrotational. Therefore, potential functions cannot be used to describe the flow in the boundary layer.) We shall define the stream function such that

$$u = \left(\frac{\partial \psi}{\partial y}\right)_x \quad \text{and} \quad v = -\left(\frac{\partial \psi}{\partial x}\right)_y$$

By introducing the stream function, the continuity equation (4.1) is automatically satisfied. Thus, we need to solve only one equation, the x component of the momentum equation, in terms of only one unknown, the stream function.

Let us now transform our equations from the x, y coordinate system to the s, η coordinate system. To do this, note that

$$\left(\frac{\partial}{\partial y}\right)_x = \left(\frac{\partial \eta}{\partial y}\right)_x \left(\frac{\partial}{\partial \eta}\right)_s = \frac{u_e}{\sqrt{2\nu s}}\left(\frac{\partial}{\partial \eta}\right)_s \tag{4.11a}$$

$$\left(\frac{\partial}{\partial x}\right)_y = \left(\frac{\partial s}{\partial x}\right)_y\left[\left(\frac{\partial \eta}{\partial s}\right)_y\left(\frac{\partial}{\partial \eta}\right)_s + \left(\frac{\partial}{\partial s}\right)_\eta\right] \tag{4.11b}$$

Thus, the streamwise component of velocity may be written in terms of the stream function as

$$u = \left(\frac{\partial \psi}{\partial y}\right)_x = \frac{u_e}{\sqrt{2\nu s}}\left(\frac{\partial \psi}{\partial \eta}\right)_s \tag{4.12a}$$

Let us introduce a transformed stream function f, which we define so that

$$u = u_e\left(\frac{\partial f}{\partial \eta}\right)_s \tag{4.12b}$$

Comparing equations (4.12a) and (4.12b), we see that

$$f = \frac{1}{\sqrt{2\nu s}}\psi \tag{4.13}$$

Similarly, we can develop an expression for the transverse component of velocity:

$$v = -\left(\frac{\partial \psi}{\partial x}\right)_y$$

$$= -u_e\sqrt{2\nu s}\left[\left(\frac{\partial \eta}{\partial s}\right)_y\left(\frac{\partial f}{\partial \eta}\right)_s + \left(\frac{\partial f}{\partial s}\right)_\eta + \left(\frac{f}{2s}\right)\right] \tag{4.14}$$

In equations (4.12b) and (4.14), we have written the two velocity components, which were the unknowns in the original formulation of the problem, in terms of the transformed stream function. We can rewrite equation (4.6) using the differentials of the variables in the s, η coordinate system. For example,

$$\frac{\partial^2 u}{\partial y^2} = \frac{\partial}{\partial y}\left[\frac{\partial \eta}{\partial y}\frac{\partial(u_e f')}{\partial \eta}\right] = \left(\frac{\partial \eta}{\partial y}\right)^2 u_e\frac{\partial^2 f'}{\partial \eta^2}$$

$$= \frac{u_e^2}{2\nu s}u_e f'''$$

where the prime ($'$) denotes differentiation with respect to η. Using these substitutions, the momentum equation becomes

$$ff'' + f''' + [1 - (f')^2]\frac{2s}{u_e}\frac{du_e}{ds} = 2s\left[f'\left(\frac{\partial f'}{\partial s}\right)_y - f''\left(\frac{\partial f}{\partial s}\right)_y\right] \tag{4.15}$$

As discussed earlier, by using a stream function we automatically satisfy the continuity equation. Thus, we have reduced the formulation to one equation with one unknown.

For many problems, the parameter $(2s/u_e)(du_e/ds)$, which is represented by the symbol β, is assumed to be constant. The assumption that β is a constant implies that the s derivatives of f and f' are zero. As a result, the transformed stream function and its derivatives are functions of η only, and equation (4.15) becomes the ordinary differential equation:

$$ff'' + f''' + [1 - (f')^2]\beta = 0 \tag{4.16}$$

Because the dimensionless velocity function f' is a function of η only, the velocity profiles at one s station are the same as those at another. Thus, the solutions are called *similar solutions*. Note that the Reynolds number does not appear as a parameter when the momentum equation is written in the transformed coordinates. It will appear when our solutions are transformed back into the x, y coordinate system. There are no analytical solutions to this third-order equation, which is known as the *Falkner-Skan equation*. Nevertheless, there are a variety of well-documented numerical techniques available to solve it.

Let us examine the three boundary conditions necessary to solve the equation. Substituting the definition that

$$f' = \frac{u}{u_e}$$

into the boundary conditions given by equations (4.7) and (4.8), the wall,

$$f'(s, 0) = 0 \qquad\qquad \textbf{(4.17a)}$$

and far from the wall,

$$f'(s, \eta_{\text{large}}) = 1.0 \qquad\qquad \textbf{(4.17b)}$$

Using equations (4.14) and (4.7), the boundary condition that the transverse velocity be zero at the wall becomes

$$f(s, 0) = 0 \qquad\qquad \textbf{(4.17c)}$$

Since f is the transformed stream function, this third boundary condition states that the stream function is constant along the wall (i.e., the surface is streamline). This is consistent with the requirement that $v(x, 0) = 0$, which results because the component of velocity normal to a streamline is zero.

4.3.1 Numerical Solutions for the Falkner-Skan Problem

Numerical solutions of equation (4.16) that satisfy the boundary conditions represented by equation (4.17) have been generated for $-0.1988 \leq \beta \leq +2.0$. The resultant velocity profiles are presented in Fig. 4.3 and Table 4.1. Since

$$\beta = \frac{2s}{u_e} \frac{du_e}{ds}$$

these solutions represent a variety of inviscid flow fields and, therefore, represent the flow around different configurations. Note that when $\beta = 0$, $u_e = $ constant, and the solution is that for flow past a flat plate (known as the Blasius solution). Negative values of β correspond to cases where the inviscid flow is decelerating, which corresponds to an adverse pressure gradient [i.e., $(dp/dx) > 0$]. The positive values of β correspond to an accelerating inviscid flow, which results from a favorable pressure gradient [i.e., $(dp/dx) < 0$].

As noted in the discussion of flow around cylinder in Chapter 3, when the air particles in the boundary layer encounter a relatively large adverse pressure gradient, boundary-layer separation may occur. Separation results because the fluid particles in the viscous layer have been slowed to the point that they cannot overcome the adverse pressure gradient. The effect of an adverse pressure gradient is evident in the velocity profiles presented in Fig. 4.3.

TABLE 4.1 Numerical Values of the Dimensionless Streamwise Velocity $f'(\eta)$ for the Falkner-Skan, Laminar, Similarity Flows

				β		
η	−0.1988	−0.180	0.000	0.300	1.000	2.000
0.0	0.0000	0.0000	0.0000	0.0000	0.0000	0.0000
0.1	0.0010	0.0138	0.0470	0.0760	0.1183	0.1588
0.2	0.0040	0.0293	0.0939	0.1489	0.2266	0.2979
0.3	0.0089	0.0467	0.1408	0.2188	0.3252	0.4185
0.4	0.0159	0.0658	0.1876	0.2857	0.4145	0.5219
0.5	0.0248	0.0867	0.2342	0.3494	0.4946	0.6096
0.6	0.0358	0.1094	0.2806	0.4099	0.5663	0.6834
0.7	0.0487	0.1337	0.3265	0.4671	0.6299	0.7450
0.8	0.0636	0.1597	0.3720	0.5211	0.6859	0.7959
0.9	0.0804	0.1874	0.4167	0.5717	0.7351	0.8377
1.0	0.0991	0.2165	0.4606	0.6189	0.7779	0.8717
1.2	0.1423	0.2790	0.5452	0.7032	0.8467	0.9214
1.4	0.1927	0.3462	0.6244	0.7742	0.8968	0.9531
1.6	0.2498	0.4169	0.6967	0.8325	0.9323	0.9727
1.8	0.3127	0.4895	0.7611	0.8791	0.9568	0.9845
2.0	0.3802	0.5620	0.8167	0.9151	0.9732	0.9915
2.2	0.4510	0.6327	0.8633	0.9421	0.9839	0.9955
2.4	0.5231	0.6994	0.9011	0.9617	0.9906	
2.6	0.5946	0.7605	0.9306	0.9755	0.9946	
2.8	0.6635	0.8145	0.9529	0.9848		
3.0	0.7277	0.8606	0.9691	0.9909		
3.2	0.7858	0.8985	0.9804	0.9947		
3.4	0.8363	0.9285	0.9880			
3.6	0.8788	0.9514	0.9929			
3.8	0.9131	0.9681	0.9959			
4.0	0.9398	0.9798				
4.2	0.9597	0.9876				
4.4	0.9740	0.9927				
4.6	0.9838	0.9959				
4.8	0.9903					
5.0	0.9944					

When $\beta = -0.1988$, not only is the streamwise velocity zero at the wall, but the velocity gradient $\partial u/\partial y$ is also zero at the wall. If the adverse pressure gradient were any larger, the laminar boundary layer would separate from the surface, and flow reversal would occur.

For the accelerating flows (i.e., positive β), the velocity increases rapidly with distance from the wall. Thus, $\partial u/\partial y$ at the wall is relatively large. Referring to equation (1.11), one would expect that the shear force at the wall would be relatively large. To calculate the shear force at the wall,

$$\tau = \left(\mu \frac{\partial u}{\partial y}\right)_{y=0} \tag{4.18}$$

let us introduce the transformation presented in equation (4.11a). Thus, the shear is

$$\tau = \frac{\mu u_e^2}{\sqrt{2\nu s}} f''(0) \tag{4.19}$$

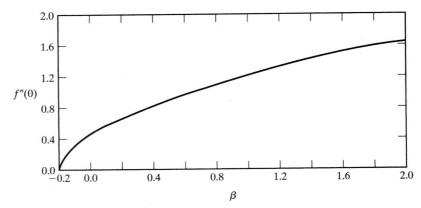

Figure 4.4 Transformed shear function at the wall for laminar boundary layers as a function of β.

Because of its use in equation (4.19), we will call f'' the *transformed shear function*. Theoretical values of $f''(0)$ are presented in Fig. 4.4 and in Table 4.2. Note that $f''(0)$ is a unique function of β for these incompressible, laminar boundary layers. The value does not depend on the stream conditions, such as velocity or Reynolds number.

When $\beta = 0, f''(0) = 0.4696$. Thus, for the laminar boundary layer on a flat plate,

$$\tau = 0.332\sqrt{\frac{\rho\mu u_e^3}{x}} \tag{4.20}$$

As noted earlier, for flow past a flat plate, velocity at the edge of the boundary layer (u_e) is equal to the free-stream value (U_∞). We can express the shear in terms of the dimensionless skin-friction coefficient

$$C_f = \frac{\tau}{\frac{1}{2}\rho_\infty U_\infty^2} = \frac{0.664}{\sqrt{Re_x}} \tag{4.21}$$

where

$$Re_x = \frac{\rho u_e x}{\mu} = \frac{\rho U_\infty x}{\mu} \tag{4.22}$$

Mentally substituting the values of $f''(0)$ presented in Fig. 4.4, we see that the shear is zero when $\beta = -0.1988$. Thus, this value of β corresponds to the onset of separation. Conversely, when the inviscid flow is accelerating, the shear is greater than that for a zero pressure gradient flow.

TABLE 4.2 Theoretical Values of the Transformed Shear Function at the Wall for Laminar Boundary Layers as a Function of β

β	−0.1988	−0.180	0.000	0.300	1.000	2.000
$f''(0)$	0.000	0.1286	0.4696	0.7748	1.2326	1.6872

TABLE 4.3   Solution for the Laminar Boundary Layer on a Flat Plate

η	f	f'	f''
0.0	0.0000	0.0000	0.4696
0.1	0.0023	0.0470	0.4696
0.2	0.0094	0.0939	0.4693
0.3	0.0211	0.1408	0.4686
0.4	0.0375	0.1876	0.4673
0.5	0.0586	0.2342	0.4650
0.6	0.0844	0.2806	0.4617
0.7	0.1147	0.3265	0.4572
0.8	0.1497	0.3720	0.4512
0.9	0.1891	0.4167	0.4436
1.0	0.2330	0.4606	0.4344
1.2	0.3336	0.5452	0.4106
1.4	0.4507	0.6244	0.3797
1.6	0.5829	0.6967	0.3425
1.8	0.7288	0.7610	0.3005
2.0	0.8868	0.8167	0.2557
2.2	1.0549	0.8633	0.2106
2.4	1.2315	0.9010	0.1676
2.6	1.4148	0.9306	0.1286
2.8	1.6032	0.9529	0.0951
3.0	1.7955	0.9691	0.0677
3.2	1.9905	0.9804	0.0464
3.4	2.1874	0.9880	0.0305
3.5	2.2863	0.9907	0.0244
4.0	2.7838	0.9978	0.0069
4.5	3.2832	0.9994	0.0015

The transformed stream function (f), the dimensionless streamwise velocity (f'), and the shear function (f'') are presented as a function of η for a laminar boundary layer on a flat plate in Table 4.3. Note that as η increases (i.e., as y increases) the shear goes to zero and the function f' tends asymptotically to 1.0. Let us define the boundary-layer thickness δ as that distance from the wall for which $u = 0.99u_e$. We see that the value of η corresponding to the boundary-layer thickness

$$\eta_\delta = 3.5$$

independent of the specific flow properties of the free stream. Converting this to a physical distance, the corresponding boundary-layer thickness (δ) is

$$\delta = y_\delta = \eta_\delta \sqrt{\frac{2\nu x}{u_e}}$$

or

$$\frac{\delta}{x} = \frac{5.0}{\sqrt{\text{Re}_x}} \qquad \textbf{(4.23)}$$

Thus, the thickness of a laminar boundary layer is proportional to $\sqrt{x}$ and is inversely proportional to the square root of the Reynolds number.

Although the transverse component of velocity at the wall is zero, it is not zero at the edge of the boundary layer. Referring to equation (4.14), we can see that

$$\frac{v_e}{u_e} = \frac{1}{\sqrt{2}}\sqrt{\frac{\nu}{u_e x}}\left[\eta_e(f')_e - f_e\right] \tag{4.24}$$

Using the values given in Table 4.3,

$$\frac{v_e}{u_e} = \frac{0.84}{\sqrt{Re_x}} \tag{4.25}$$

This means that at the outer edge there is an outward flow, which is due to the fact that the increasing boundary-layer thickness causes the fluid to be displaced from the wall as it flows along it. There is no boundary-layer separation for flow past a flat plate, since the streamwise pressure gradient is zero.

Since the streamwise component of the velocity in the boundary layer asymptotically approaches the local free-stream value, the magnitude of δ is very sensitive to the ratio of u/u_e, which is chosen as the criterion for the edge of the boundary layer [e.g., 0.99 was the value used to develop equation (4.23)]. A more significant measure of the boundary layer is the displacement thickness δ^*, which is the distance by which the external streamlines are shifted due to the presence of the boundary layer. Referring to Fig. 4.5,

$$\rho_e u_e \delta^* = \int_0^\delta \rho(u_e - u)\,dy$$

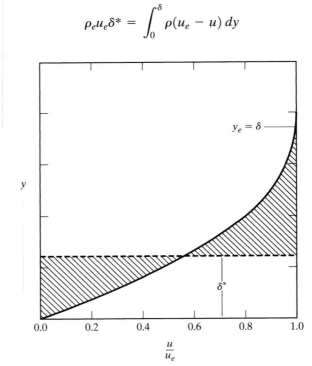

Figure 4.5 Velocity profile for a laminar boundary layer on a flat plate illustrating the boundary-layer thickness δ and the displacement thickness δ^*.

Thus, for any incompressible boundary layer,

$$\delta^* = \int_0^\delta \left(1 - \frac{u}{u_e}\right) dy \tag{4.26}$$

Note that, since the integrand is zero for any point beyond δ, the upper limit for the integration does not matter providing it is equal to (or greater than) δ. Substituting the transformation of equation (4.10) for the laminar boundary layer on a flat plate,

$$\delta^* = \sqrt{\frac{2\nu x}{u_e}} \int_0^\infty (1 - f') \, d\eta$$

Using the values presented in Table 4.3, we obtain

$$\frac{\delta^*}{x} = \frac{\sqrt{2}(\eta_e - f_e)}{\sqrt{Re_x}} = \frac{1.72}{\sqrt{Re_x}} \tag{4.27}$$

Thus, for a flat plate at zero incidence in a uniform stream, the displacement thickness δ^* is on the order of one-third the boundary-layer thickness δ.

The momentum thickness, θ, for an incompressible boundary layer is given by

$$\theta = \int_0^\infty \frac{u}{u_e}\left(1 - \frac{u}{u_e}\right) dy \tag{4.28}$$

The momentum thickness represents the height of the free-stream flow which would be needed to make up the deficiency in momentum flux within the boundary layer due to the shear force at the surface. For an incompressible, laminar boundary layer,

$$\frac{\theta}{x} = \frac{0.664}{\sqrt{Re_x}} \tag{4.29}$$

Another convenient formulation for skin-friction on a flat plate is found by integrating the "local" skin-friction coefficient, C_f, found in equation (4.21) to obtain a "total" or "average" skin-friction drag coefficient on the flat plate [White (2005)]. The use of the total skin-friction drag coefficient avoids performing the same integration numerous times with different flat-plate lengths accounting for different results. The total skin-friction coefficient is defined as

$$\overline{C}_f \equiv \frac{D_f}{q_\infty S_{\text{wet}}} \tag{4.30}$$

where D_f is the friction drag on the plate and S_{wet} is the wetted area of the plate (the wetted area is the area of the plate in contact with the fluid—for one side of the plate, $S_{\text{wet}} = Lb$). The total skin-friction coefficient for laminar boundary layers becomes

$$\overline{C}_f = \frac{1}{q_\infty S_{\text{wet}}} b \int_0^L \tau \, dx = \frac{b}{q_\infty Lb} \int_0^L C_f(x) q_\infty \, dx$$

$$= \frac{1}{L} \int_0^L \frac{0.664}{\sqrt{Re_x}} dx = 2C_f(L) = \frac{2\theta}{L} \tag{4.31}$$

which is just twice the value of the local skin-friction coefficient evaluated at $x = L$. The total skin-friction coefficient for laminar flow simply becomes

$$\overline{C}_f = \frac{1.328}{\sqrt{Re_L}} \tag{4.32}$$

where Re_L is the Reynolds number evaluated at $x = L$, which is the end of the flat plate.

Since drag coefficients are normally nondimensionalized by a reference area rather than a wetted area [see equation (3.53)], the drag coefficient due to skin-friction is obtained from equation (4.30) as

$$C_D \equiv \frac{D_f}{q_\infty S_{ref}} = \overline{C}_f \frac{S_{wet}}{S_{ref}} \tag{4.33}$$

It can be tempting to add together the total skin-friction coefficients for various flat plates in order to obtain a total skin-friction drag—this must never be done! Since each total skin-friction coefficient is defined with a different wetted area, doing this would result in an incorrect result. In other words,

$$\overline{C}_{f_{total}} \neq \sum_{i=1}^{N} \overline{C}_{f_i}$$

Always convert total skin-friction coefficients into drag coefficients (based on a single reference area) and then add the drag coefficients to obtain a total skin-friction drag coefficient:

$$C_D = \sum_{i=1}^{N} C_{D_i} \tag{4.34}$$

EXAMPLE 4.1:

A rectangular plate, whose streamwise dimension (or chord c) is 0.2 m and whose width (or span b) is 1.8 m, is mounted in a wind tunnel. The free-stream velocity is 40 m/s. The density of the air is 1.2250 kg/m³, and the absolute viscosity is 1.7894×10^{-5} kg/m·s. Graph the velocity profiles at $x = 0.0$ m, $x = 0.05$ m, $x = 0.10$ m, and $x = 0.20$ m. Calculate the chordwise distribution of the skin-friction coefficient and the displacement thickness. What is the drag coefficient for the plate?

Solution: Since the span (or width) of the plate is 9.0 times the chord (or streamwise dimension), let us assume that the flow is two dimensional (i.e., it is independent of the spanwise coordinate). The maximum value of the local Reynolds number, which occurs when $x = c$, is

$$Re_c = \frac{(1.225 \text{ kg/m}^3)(40 \text{ m/s})(0.2 \text{ m})}{(1.7894 \times 10^{-5} \text{ kg/m} \cdot \text{s})} = 5.477 \times 10^5$$

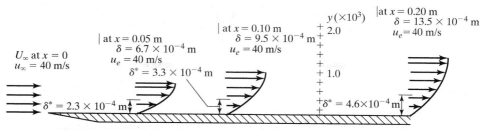

Figure 4.6 Velocity profile for the flat-plate laminar boundary layer $\text{Re}_c = 5.477 \times 10^5$.

This Reynolds number is close enough to the transition criteria for a flat plate that we will assume that the boundary layer is laminar for its entire length. Thus, we will use the relations developed in this section to calculate the required parameters.

Noting that

$$y = \sqrt{\frac{2\nu x}{u_e}}\,\eta = 8.546 \times 10^{-4}\sqrt{x}\,\eta$$

we can use the results presented in Table 4.3 to calculate the velocity profiles. The resultant profiles are presented in Fig. 4.6. At the leading edge of the flat plate (i.e., at $x = 0$), the velocity is constant (independent of y). The profiles at the other stations illustrate the growth of the boundary layer with distance from the leading edge. Note that the scale of the y coordinate is greatly expanded relative to that for the x coordinate. Even though the streamwise velocity at the edge of the boundary layer (u_e) is the same at all stations, the velocity within the boundary layer is a function of x and y. However, if the dimensionless velocity (u/u_e) is presented as a function of η, the profile is the same at all stations. Specifically, the profile is that for $\beta = 0.0$ in Fig. 4.3. Since the dimensionless profiles are similar at all x stations, the solutions are termed *similarity solutions*.

The displacement thickness in meters is

$$\delta^* = \frac{1.72x}{\sqrt{\text{Re}_x}} = 1.0394 \times 10^{-3}\sqrt{x}$$

The chordwise (or streamwise) distribution of the displacement thickness is presented in Fig. 4.6. These calculations verify the validity of the common assumption that the boundary layer is thin. Therefore, the inviscid solution obtained neglecting the boundary layer altogether and that obtained for the effective geometry (the actual surface plus the displacement thickness) are essentially the same.

The skin-friction coefficient is

$$C_f = \frac{0.664}{\sqrt{\text{Re}_x}} = \frac{4.013 \times 10^{-4}}{\sqrt{x}}$$

Let us now calculate the drag coefficient for the plate. Obviously, the pressure contributes nothing to the drag. Therefore, the drag force acting on the flat plate is due only to skin friction. Using general notation, we see that

$$D = 2b \int_0^c \tau \, dx \tag{4.35}$$

We need integrate only in the x direction, since by assuming the flow to be two dimensional, we have assumed that there is no spanwise variation in the flow. In equation (4.35), the integral, which represents the drag per unit width (or span) of the plate, is multiplied by b (the span) and by 2 (since friction acts on both the top and bottom surfaces of the plate). Substituting the expression for the laminar shear forces, given in equation (4.20),

$$D = 0.664 b \sqrt{\rho \mu u_e^3} \int_0^c \frac{dx}{\sqrt{x}}$$

$$= 1.328 b \sqrt{c \rho \mu u_e^3} \tag{4.36}$$

Since the edge velocity (u_e) is equal to the free-stream velocity (U_∞), the drag coefficient for the plate is, therefore,

$$C_D = \frac{D}{q_\infty c b} = \frac{2.656}{\sqrt{\mathrm{Re}_c}} \tag{4.37}$$

For the present problem, $C_D = 3.589 \times 10^{-3}$.

Alternatively, using the total skin-friction coefficient equation (4.32) and computing drag on the top and bottom of the plate,

$$C_D = \overline{C_f} \frac{S_{\text{wet}}}{S_{\text{ref}}} = \frac{1.328}{\sqrt{\mathrm{Re}_L}} \frac{2Lb}{Lb} = 3.589 \times 10^{-3}$$

EXAMPLE 4.2:

The streamwise velocity component for a laminar boundary layer is sometimes assumed to be roughly approximated by the linear relation

$$u = \frac{y}{\delta} u_e$$

where $\delta = 1.25 \times 10^{-2} \sqrt{x}$. Assume that we are trying to approximate the flow of air at standard sea-level conditions past a flat plate where $u_e = 2.337$ m/s. Calculate the streamwise distribution of the displacement thickness (δ^*), the velocity at the edge of the boundary layer (v_e), and the skin-friction coefficient (C_f). Compare the values obtained assuming a linear velocity profile with the more exact solutions presented in this chapter.

Solution: As given in Table 1.2, the standard atmospheric conditions at sea level include

$$\rho_\infty = 1.2250 \text{ kg/m}^3 \quad \text{and} \quad \mu_\infty = 1.7894 \times 10^{-5} \text{ kg/s} \cdot \text{m}$$

Thus, for constant-property flow past a flat plate,

$$\mathrm{Re}_x = \frac{\rho_\infty u_e x}{\mu_\infty} = 1.60 \times 10^5 x$$

Using the definition for the displacement thickness of an incompressible boundary layer, equation (4.26),

$$\delta^* = \int_0^\delta \left(1 - \frac{u}{u_e}\right) dy = \delta \int_0^1 \left(1 - \frac{u}{u_e}\right) d\left(\frac{y}{\delta}\right)$$

Notice that, since we have u/u_e in terms of y/δ, we have changed our independent variable from y to y/δ. We must also change the upper limit on our integral from δ to 1. Thus, since

$$\frac{u}{u_e} = \frac{y}{\delta} \quad \text{and} \quad \delta = 1.25 \times 10^{-2} \sqrt{x}$$

then

$$\delta^* = 1.25 \times 10^{-2} \sqrt{x} \int_0^1 \left(1 - \frac{y}{\delta}\right) d\left(\frac{y}{\delta}\right) = 0.625 \times 10^{-2} \sqrt{x}$$

for the linear profile.

Using the equation for the more exact formulation [equation (4.27)], and noting that $\mathrm{Re}_x = 1.60 \times 10^5 x$, we find that

$$\delta^* = 0.430 \times 10^{-2} \sqrt{x}$$

Using the continuity equation, we would find that the linear approximation gives a value for v_e of

$$v_e = \frac{3.125 \times 10^{-3}}{\sqrt{x}} u_e$$

Using the more exact formulation of equation (4.25) yields

$$v_e = \frac{0.84}{\sqrt{\mathrm{Re}_x}} u_e = \frac{2.10 \times 10^{-3}}{\sqrt{x}} u_e$$

Finally, we find that the skin friction for the linear velocity approximation is given by

$$\tau = \mu\left(\frac{\partial u}{\partial y}\right)_{y=0} = \frac{\mu u_e}{\delta}$$

Thus, the skin-friction coefficient is

$$C_f = \frac{\tau}{\frac{1}{2}\rho_\infty u_e^2} = \frac{2\mu_\infty}{\rho_\infty u_e \delta} = \frac{2}{1.60 \times 10^5 (1.25 \times 10^{-2} \sqrt{x})}$$

$$= \frac{1.00 \times 10^{-3}}{\sqrt{x}}$$

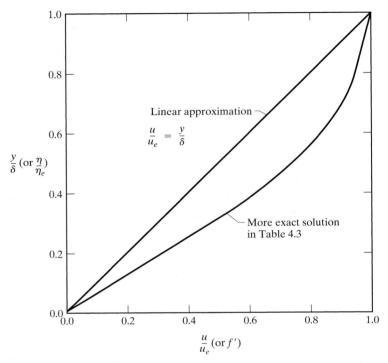

Figure 4.7 Comparison of velocity profiles for a laminar boundary layer on a flat plate.

For the more exact formulation,

$$C_f = \frac{0.664}{\sqrt{Re_x}} = \frac{1.66 \times 10^{-3}}{\sqrt{x}}$$

Summarizing these calculations provides the following comparison:

	Linear approximation	*More exact solution*
δ^*	$0.625 \times 10^{-2} \sqrt{x}$	$0.430 \times 10^{-2} \sqrt{x}$
v_e	$(3.125 \times 10^{-3} u_e)/\sqrt{x}$	$(2.10 \times 10^{-3} u_e)/\sqrt{x}$
C_f	$(1.00 \times 10^{-3})/\sqrt{x}$	$(1.66 \times 10^{-3})/\sqrt{x}$

Comparing the velocity profiles, which are presented in Fig. 4.7, the reader should be able to use physical reasoning to determine that these relationships are intuitively correct. That is, if one uses a linear profile, the shear would be less than that for the exact solution, where δ^* and v_e would be greater for the linear profile.

In this example, we assumed that the boundary-layer thickness δ was $1.25 \times 10^{-2} \sqrt{x}$, which is the value obtained using the more exact formulation [i.e., equation (4.23)]. However, if we had used the integral approach to determine the value of δ for a linear profile, we would have obtained

$$\delta = \frac{3.464x}{\sqrt{Re_x}} = 0.866 \times 10^{-2} \sqrt{x}$$

Although this is considerably less than the assumed (more correct) value, the values of the other parameters (e.g., δ^* and C_f) would be in closer agreement with those given by the more exact solution.

Although the linear profile for the streamwise velocity component is a convenient approximation to use when demonstrating points about the continuity equation or about Kelvin's theorem, it clearly does not provide reasonable values for engineering parameters, such as δ^* and C_f. A more realistic approximation for the streamwise velocity component in a laminar boundary layer would be

$$\frac{u}{u_e} = \frac{3}{2}\left(\frac{y}{\delta}\right) - \frac{1}{2}\left(\frac{y}{\delta}\right)^3 \qquad (4.38)$$

EXAMPLE 4.3:   Calculate the velocity gradient, β

Calculate the velocity gradient parameter β, which appears in the Falkner-Skan form of the momentum equation [equation (4.16)] for the NACA 65-006 airfoil. The coordinates of this airfoil section, which are given in Table 4.4, are given in terms of the coordinate system used in Fig. 4.8. Note that the maximum thickness is located relatively far aft in order to maintain a favorable pressure gradient, which tends to delay transition. The β distribution is required as an input to obtain the local similarity solutions for a laminar boundary layer.

Solution: Using the definition for β gives us

$$\beta = \frac{2s}{u_e}\frac{du_e}{ds} = \frac{2\int u_e\,dx}{u_e}\frac{du_e}{dx}\frac{dx}{ds}$$

But

$$\frac{dx}{ds} = \frac{1}{u_e}$$

$$\frac{u_e}{U_\infty} = (1 - C_p)^{0.5}$$

Therefore, at any chordwise location for a thin airfoil

$$\beta = -\frac{\int_0^{\widetilde{x}} (1 - C_p)^{0.5}\,d\widetilde{x}}{(1 - C_p)^{1.5}}\frac{dC_p}{d\widetilde{x}}$$

where $\widetilde{x} = x/c$.

TABLE 4.4 Pressure Distribution for the NACA 65–006

$\widetilde{x}(= x/c)$	$\widetilde{z}(= z/c)$	C_p
0.000	0.0000	1.000
0.005	0.0048	−0.044
0.025	0.0096	−0.081
0.050	0.0131	−0.100
0.100	0.0182	−0.120
0.150	0.0220	−0.134
0.200	0.0248	−0.143
0.250	0.0270	−0.149
0.300	0.0285	−0.155
0.350	0.0295	−0.159
0.400	0.0300	−0.163
0.450	0.0298	−0.166
0.500	0.0290	−0.165
0.550	0.0274	−0.145
0.600	0.0252	−0.124
0.650	0.0225	−0.100
0.700	0.0194	−0.073
0.750	0.0159	−0.044
0.800	0.0123	−0.013
0.850	0.0087	+0.019
0.900	0.0051	+0.056
0.950	0.0020	+0.098
1.000	0.0000	+0.142

Source: I. H. Abbott and A. E. von Doenhoff, *Theory of Wing Sections.* New York: Dover Publications, 1949 [Abbott and von Doenhoff (1949).]

The resultant β distribution is presented in Fig. 4.9. Note that a favorable pressure gradient acts over the first half of the airfoil. For $\widetilde{x} \geq 0.6$, the negative values of β exceed that required for separation of a similar laminar boundary layer. Because of the large streamwise variations in β, the nonsimilar

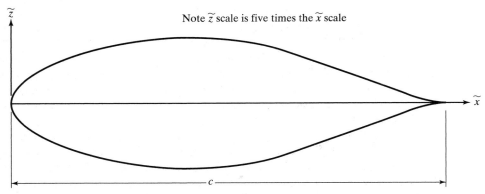

Note $\widetilde{z}$ scale is five times the $\widetilde{x}$ scale

Figure 4.8 Cross section for symmetric NACA 65-006 airfoil of Example 4.3.

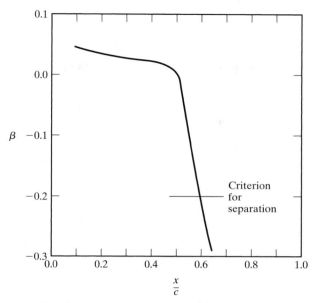

Figure 4.9 Distribution for NACA 65-606 airfoil (assuming that the boundary layer does not separate).

character of the boundary layer should be taken into account when establishing a separation criteria. Nevertheless, these calculations indicate that, if the boundary layer were laminar along its entire length, it would separate, even for this airfoil at zero angle of attack. Boundary-layer separation would result in significant changes in the flow field. However, the experimental measurements of the pressure distribution indicate that the actual flow field corresponds closely to the inviscid flow field. Thus, boundary-layer separation apparently does not occur at zero angle of attack. The reason that separation does not occur is as follows. At the relatively high Reynolds numbers associated with airplane flight, the boundary layer is turbulent over a considerable portion of the airfoil. As discussed previously, a turbulent boundary layer can overcome an adverse pressure gradient longer, and separation is not as likely to occur.

The fact that a turbulent boundary layer can flow without separation into regions of much steeper adverse pressure gradients than can a laminar boundary layer is illustrated in Fig. 4.10. Incompressible boundary-layer solutions were generated for symmetrical Joukowski airfoils at zero angle of attack. The edge velocity and therefore the corresponding inviscid pressure distributions are shown in Fig. 4.10. At the conditions indicated, boundary-layer separation will occur for any Joukowski airfoil that is thicker than 4.6% if the flow is entirely laminar. However, if the boundary layer is turbulent, separation will not occur until a thickness of about 31% has been exceeded. The boundary layer effectively thickens the airfoil, especially near the trailing edge, since δ^* in creases with distance. This thickening alleviates the adverse pressure gradients, which in turn permits somewhat thicker sections before separation occurs. To

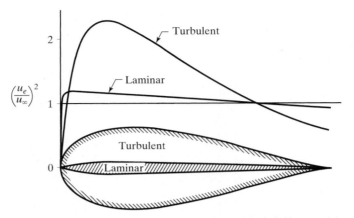

Figure 4.10 Thickest symmetrical Joukowski airfoils capable of supporting fully attached laminar and turbulent flows. The angle of attack is 0°, and the Mach number is 0. For turbulent flow, transition is assumed to occur at the velocity peak. The turbulent case is calculated for $Re_c = 10^7$. Results for laminar flow are independent of Reynolds number. Maximum thickness for laminar flow is about 4.6%, for turbulent flow, 31%. If displacement-thickness effects on pressure distribution were included, the turbulent airfoil would increase to about 33%. The change in the laminar case would be negligible. [From Cebeci and Smith (1974).]

ensure that boundary-layer transition occurs and, thus, delay or avoid separation altogether, one might use vortex generators or other forms of surface roughness, such as shown in Fig. 4.11.

4.4 BOUNDARY-LAYER TRANSITION

As the boundary layer develops in the streamwise direction, it is subjected to numerous disturbances. The disturbances may be due to surface roughness, temperature irregularities, background noise, and so on. For some flows, these disturbances are damped and the flow remains laminar. For other flows, the disturbances amplify and the boundary layer becomes turbulent. The onset of transition from a laminar boundary layer to a turbulent layer (if it occurs at all) depends on many parameters, such as the following:

1. Pressure gradient
2. Surface roughness
3. Compressibility effects (usually related to the Mach number)
4. Surface temperature
5. Suction or blowing at the surface
6. Free-stream turbulence

Figure 4.11 Vortex generators, which can be seen in front of the ailerons and near the wing leading edge of an A-4, are an effective, but not necessarily an aerodynamically efficient, way of delaying separation (from Ruth Bertin's collection).

Obviously, no single criterion for the onset of transition can be applied to a wide variety of flow conditions. However, as a rule of thumb, adverse pressure gradients, surface roughness, blowing at the surface, and free-stream turbulence promote transition, that is, cause it to occur early. Conversely, favorable pressure gradients, increased Mach numbers, suction at the surface, and surface cooling delay transition. Although the parameters used and the correlation formula for the onset of transition depend on the details of the application, transition criteria incorporate a Reynolds number. For incompressible flow past a flat plate, a typical transition criterion is

$$\mathrm{Re}_{x,\mathrm{tr}} = 500,000 \tag{4.39}$$

Thus, the location for the onset of boundary-layer transition would occur at

$$x_{\mathrm{tr}} = \frac{\mathrm{Re}_{x,\mathrm{tr}}}{\rho u_e / \mu} \tag{4.40}$$

Once the critical Reynolds number is exceeded, the flat-plate boundary layer would contain regions with the following characteristics as it transitioned from the laminar state to a fully turbulent flow:

1. Stable, laminar flow near the leading edge
2. Unstable flow containing two-dimensional Tollmien-Schlichting (T-S) waves

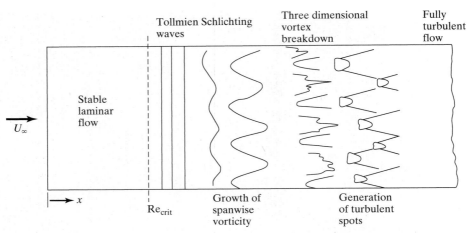

Figure 4.12 Idealized sketch of the transition process on a flat plate. (Based on the sketch from *Viscous Fluid Flow* by F. M. White. Copyright © 1974 by McGraw-Hill Book Company.)

3. A region where three-dimensional unstable waves and hairpin eddies develop
4. A region where vortex breakdown produces locally high shear
5. Fluctuating, three-dimensional flow due to cascading vortex breakdown
6. A region where turbulent spots form
7. Fully turbulent flow

A sketch of the idealized transition process is presented in Fig. 4.12.

Stability theory predicts and experiment verifies that the initial instability is in the form of two-dimensional T-S waves that travel in the mean flow direction. Even though the mean flow is two dimensional, three-dimensional unstable waves and hairpin eddies soon develop as the T-S waves begin to show spanwise variations. The experimental verification of the transition process is illustrated in the photograph of Fig. 4.13. A vibrating ribbon perturbs the low-speed flow upstream of the left margin of the photograph. Smoke accumulation in the small recirculation regions associated with the T-S waves can be seen at the left edge of the photograph. The sudden appearance of three dimensionality is associated with the nonlinear growth region of the laminar instability. In the advanced stages of the transition process, intense local fluctuations occur at various times and locations in the viscous layer. From these local intensities, true turbulence bursts forth and grows into a turbulent spot. Downstream of the region where the spots first form, the flow becomes fully turbulent.

Transition-promoting phenomena, such as an adverse pressure gradient and finite surface roughness, may short circuit the transition process, eliminating one or more of the five transitional regions described previously. When one or more of the transitional regions are by-passed, we term the cause (e.g., roughness) a by-pass mechanism.

Figure 4.13 Flow visualization of the transition process on a flat plate. (Photograph supplied by A. S. W. Thomas, Lockheed Aeronautical Systems Company, Georgia Division.)

4.5 INCOMPRESSIBLE, TURBULENT BOUNDARY LAYER

Let us now consider flows where transition has occurred and the boundary layer is fully turbulent. A turbulent flow is one in which irregular fluctuations (mixing or eddying motions) are superimposed on the mean flow. Thus, the velocity at any point in a turbulent boundary layer is a function of time. The fluctuations occur in the direction of the mean flow and at right angles to it, and they affect macroscopic lumps of fluid.

Therefore, even when the inviscid (mean) flow is two dimensional, a turbulent boundary layer will be three dimensional because of the three-dimensional character of the fluctuations. However, whereas momentum transport occurs on a microscopic (or molecular) scale in a laminar boundary layer, it occurs on a macroscopic scale in a turbulent boundary layer. It should be noted that, although the velocity fluctuations may be only several percent of the local streamwise values, they have a decisive effect on the overall motion. The size of these macroscopic lumps determines the scale of turbulence.

The effects caused by the fluctuations are as if the viscosity were increased by a factor of 10 or more. As a result, the shear forces at the wall and the skin-friction component of the drag are much larger when the boundary layer is turbulent. However, since a turbulent boundary layer can negotiate an adverse pressure gradient for a longer distance, boundary-layer separation may be delayed or even avoided altogether. Delaying (or avoiding) the onset of separation reduces the pressure component of the drag (i.e., the form drag). For a blunt body or for a slender body at angle of attack, the reduction in form drag usually dominates the increase in skin friction drag.

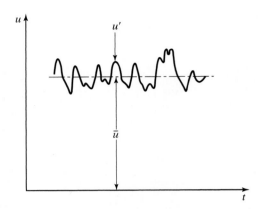

Figure 4.14 Histories of the mean component ($\bar{u}$) and the fluctuating component (u') of the streamwise velocity u for a turbulent boundary layer.

When describing a turbulent flow, it is convenient to express the local velocity components as the sum of a mean motion plus a fluctuating, or eddying, motion. For example, as illustrated in Fig. 4.14,

$$u = \bar{u} + u' \tag{4.41}$$

where $\bar{u}$ is the time-averaged value of the u component of velocity, and u' is the time-dependent magnitude of the fluctuating component. The time-averaged value at a given point in space is calculated as

$$\bar{u} = \frac{1}{\Delta t} \int_{t_0}^{t_0 + \Delta t} u \, dt \tag{4.42}$$

The integration interval Δt should be much larger than any significant period of the fluctuation velocity u'. As a result, the mean value for a steady flow is independent of time, as it should be. The integration interval depends on the physics and geometry of the problem. Referring to equation (4.42), we see that $\overline{u'} = 0$, by definition. The time average of any fluctuating parameter or its derivative is zero. The time average of products of fluctuating parameters and their derivatives is not zero. For example, $\overline{v'} = 0, \left[\partial\left(\overline{v'}\right)\right]/\partial x = 0$; but $\overline{u'v'} \neq 0$. Of fundamental importance to turbulent motion is the way in which the fluctuations u', v', and w' influence the mean motion $\bar{u}$, $\bar{v}$, and $\bar{w}$.

4.5.1 Derivation of the Momentum Equation for Turbulent Boundary Layer

Let us now derive the x (or streamwise) momentum equation for a steady, constant-property, two-dimensional, turbulent boundary layer. Since the density is constant, the continuity equation is

$$\frac{\partial(\bar{u} + u')}{\partial x} + \frac{\partial(\bar{v} + v')}{\partial y} = 0 \tag{4.43}$$

Expanding yields

$$\frac{\partial \bar{u}}{\partial x} + \frac{\partial \bar{v}}{\partial y} + \frac{\partial u'}{\partial x} + \frac{\partial v'}{\partial y} = 0 \tag{4.44}$$

Let us take the time-averaged value for each of these terms. The first two terms already are time-averaged values. As noted when discussing equation (4.42), the time-averaged value of a fluctuating component is zero; that is,

$$\frac{\overline{\partial u'}}{\partial x} = \frac{\overline{\partial v'}}{\partial y} = 0$$

Thus, for a turbulent flow, we learn from the continuity equation that

$$\frac{\partial \bar{u}}{\partial x} + \frac{\partial \bar{v}}{\partial y} = 0 \tag{4.45a}$$

and that

$$\frac{\partial u'}{\partial x} + \frac{\partial v'}{\partial y} = 0 \tag{4.45b}$$

Substituting the fluctuating descriptions for the velocity components into the x momentum equation (4.6), we have

$$\rho(\bar{u} + u')\frac{\partial(\bar{u} + u')}{\partial x} + \rho(\bar{v} + v')\frac{\partial(\bar{u} + u')}{\partial y} = \rho_e u_e \frac{du_e}{dx} + \mu \frac{\partial^2(\bar{u} + u')}{\partial y^2}$$

Expanding gives us

$$\rho\bar{u}\frac{\partial \bar{u}}{\partial x} + \rho\bar{v}\frac{\partial \bar{u}}{\partial y} + \rho u'\frac{\partial \bar{u}}{\partial x} + \rho v'\frac{\partial \bar{u}}{\partial y} + \rho u'\frac{\partial u'}{\partial x} + \rho v'\frac{\partial u'}{\partial y} + \rho\bar{u}\frac{\partial u'}{\partial x} + \rho\bar{v}\frac{\partial u'}{\partial y}$$

$$= \rho_e u_e \frac{du_e}{dx} + \mu\frac{\partial^2 \bar{u}}{\partial y^2} + \mu\frac{\partial^2 u'}{\partial y^2}$$

Taking the time average of the terms in this equation, the terms that contain only one fluctuating parameter vanish, since their time-averaged value is zero. However, the time average of terms involving the product of fluctuating terms is not zero. Thus, we obtain

$$\rho\bar{u}\frac{\partial \bar{u}}{\partial x} + \rho\bar{v}\frac{\partial \bar{u}}{\partial y} + \overline{\rho u'\frac{\partial u'}{\partial x}} + \overline{\rho v'\frac{\partial u'}{\partial y}} = \rho_e u_e \frac{du_e}{dx} + \mu\frac{\partial^2 \bar{u}}{\partial y^2} \tag{4.46}$$

Let us now multiply the fluctuating portion of the continuity equation (4.45b) by $\rho(\bar{u} + u')$. We obtain

$$\rho\bar{u}\frac{\partial u'}{\partial x} + \rho u'\frac{\partial u'}{\partial x} + \rho\bar{u}\frac{\partial v'}{\partial y} + \rho u'\frac{\partial v'}{\partial y} = 0$$

Taking the time average of these terms, we find that

$$\overline{\rho u'\frac{\partial u'}{\partial x}} + \overline{\rho u'\frac{\partial v'}{\partial y}} = 0 \tag{4.47}$$

Adding equation (4.47) to (4.46) and rearranging the terms, we obtain

$$\rho\bar{u}\frac{\partial\bar{u}}{\partial x} + \rho\bar{v}\frac{\partial\bar{u}}{\partial y} = \rho_e u_e\frac{du_e}{dx} + \mu\frac{\partial^2\bar{u}}{\partial y^2} - \rho\frac{\partial}{\partial y}\left(\overline{u'v'}\right) - \rho\frac{\partial}{\partial x}\overline{(u')^2} \qquad \textbf{(4.48)}$$

We will neglect the streamwise gradient of the time-averaged value of the square of the fluctuating velocity component, that is, $(\partial/\partial x)\overline{(u')^2}$ as compared to the transverse gradient. Thus, the momentum equation becomes

$$\rho\bar{u}\frac{\partial\bar{u}}{\partial x} + \rho\bar{v}\frac{\partial\bar{u}}{\partial y} = \rho_e u_e\frac{du_e}{dx} + \mu\frac{\partial^2\bar{u}}{\partial y^2} - \rho\frac{\partial}{\partial y}\left(\overline{u'v'}\right) \qquad \textbf{(4.49)}$$

Let us further examine the last two terms,

$$\frac{\partial}{\partial y}\left(\mu\frac{\partial\bar{u}}{\partial y} - \rho\overline{u'v'}\right) \qquad \textbf{(4.50)}$$

Recall that the first term is the laminar shear stress. To evaluate the second term, let us consider a differential area dA such that the normal to dA is parallel to the y axis, and the directions x and z are in the plane of dA. The mass of fluid passing through this area in time dt is given by the product $(\rho v)(dA)(dt)$. The flux of momentum in the x direction is given by the product $(u)(\rho v)(dA)(dt)$. For a constant density flow, the time-averaged flux of momentum per unit time is

$$\overline{\rho uv}\,dA = \rho\left(\overline{uv} + \overline{u'v'}\right)dA$$

Since the flux of momentum per unit time through an area is equivalent to an equal-and-opposite force exerted on the area by the surroundings, we can treat the term $-\rho\overline{u'v'}$ as equivalent to a "turbulent" shear stress. This "apparent," or Reynolds, stress can be added to the stresses associated with the mean flow. Thus, we can write

$$\tau_{xy} = \mu\left(\frac{\partial\bar{u}}{\partial y}\right) - \rho\overline{u'v'} \qquad \textbf{(4.51)}$$

Mathematically, then, the turbulent inertia terms behave as if the total stress on the system were composed of the Newtonian viscous stress plus an apparent turbulent stress.

The term $-\rho\overline{u'v'}$ is the source of considerable difficulties in the analysis of a turbulent boundary layer because its analytical form is not known a priori. It is related not only to physical properties of the fluid but also to the local flow conditions (velocity, geometry, surface roughness, upstream history, etc.). Furthermore, the magnitude of $-\rho\overline{u'v'}$ depends on the distance from the wall. Because the wall is a streamline, there is no flow through it. Thus, $\bar{v}$ and v' go to zero at the wall, and the flow for $y < 0.02\delta$ is basically laminar.

The term $-\rho\overline{u'v'}$ represents the turbulent transport of momentum and is known as the *turbulent shear stress* or *Reynolds stress*. At points away from the wall, $-\rho\overline{u'v'}$ is the dominant term. The determination of the turbulent shear-stress term is the critical problem in the analysis of turbulent shear flows. However, this new variable can be defined only through an understanding of the detailed turbulent structure. These terms are related not only to physical fluid properties but also to local flow conditions. Since there are no further physical laws available to evaluate these terms, empirically based correlations are introduced to model them. There is a hierarchy of techniques for *closure*, which have been developed from models of varying degrees of rigor.

4.5.2 Approaches to Turbulence Modeling

For the flow fields of practical interest to this text, the boundary layer is usually turbulent. As discussed by Spalart (2000), current approaches to turbulence modeling include direct numerical simulations (DNS), large-eddy simulations (LES), and Reynolds-averaged Navier-Stokes (RANS). The direct numerical simulation approach attempts to resolve all scales of turbulence. Because DNS must model all scales from the largest to the smallest, the grid resolution requirements are very stringent and increase dramatically with Reynolds number. Large-eddy simulations attempt to model the smaller, more homogeneous scales while resolving the larger, energy containing scales. This makes LES grid requirements less stringent than those for DNS. The RANS approach attempts to solve for the time-averaged flow, such as that described in equation (4.49). This means that all scales of turbulence must be modeled. Recently, hybrid approaches that combine RANS and LES have been proposed in an attempt to combine the best features of these two approaches.

For the RANS approach, quantities of interest are time averaged. When this averaging process is applied to the Navier-Stokes equations (such as was done in Section 4.5.1), the result is an equation for the mean quantities with extra term(s) involving the fluctuating quantities, the Reynolds stress tensor (e.g., $-\rho\overline{u'v'}$). The Reynolds stress tensor takes into account the transfer of momentum by the turbulent fluctuations. It is often written that the Reynolds stress tensor is proportional to the mean strain-rate tensor [i.e., $\mu_t(\partial\overline{u}/\partial y)$]. This is known as the Boussinesq eddy-viscosity approximation, where μ_t is the unknown turbulent eddy viscosity. The turbulent eddy viscosity is determined using a turbulence model. The development of correlations in terms of known parameters is usually termed as the *closure problem*. Closure procedures for the turbulent eddy viscosity are generally categorized by the number of partial differential equations that are solved, with zero-equation, one-equation, and two-equation models being the most popular.

Zero-equation models use no differential equations and are commonly known as algebraic models. Zero-equation models are well adapted to simple, attached flows where local turbulence equilibrium exists (i.e., the local production of turbulence is balanced by the local dissipation of turbulence). Smith (1991) noted, "For solutions of external flows around full aircraft configurations, algebraic turbulence models remain the most popular choice due to their simplicity." However, he further noted, "In general, while algebraic turbulence models are computationally simple, they are more difficult for the user to apply. Since algebraic models are accurate for a narrow range of flows, different algebraic models must be applied to the different types of turbulent flows in a single flow problem. The user must define in advance which model applies to which region, or complex logic must be implemented to automate this process .... Different results can be obtained with different implementation of the same turbulence model."

By solving one or more differential equations, the transport of turbulence can be included. That is, the effect of flow history on the turbulence can be modeled. One example where modeling the flow history of turbulence is crucial is in the calculation of turbulent flow over a multielement airfoil. One equation models are perhaps the simplest way to model this effect. Wilcox (1998), when discussing the one-equation model of Spalart-Allmaras (1992), noted that "it is especially attractive for airfoil and wing applications, for which it has been calibrated." As a result, the Spalart-Allmaras model, which solves a single partial differential equation for a variable that is related to the turbulent kinematic eddy viscosity, is one of the most popular turbulence models.

In the two-equation models, one transport equation is used for the computation of the specific turbulence kinetic energy (k) and a second transport equation is used to determine the turbulent length scale (or dissipation length scale). A variety of transport equations have been proposed for determining the turbulent length scale, including k-ε, k-ω, and k-kl models. Again, the reader is referred to Wilcox (1998) for detailed discussions of these models.

Smith (1991) notes, " For a two equation model, the normal Reynolds stress components are assumed to be equal, while for algebraic models the normal stresses are entirely neglected. Experimental results show the streamwise Reynolds stress component to be two to three times larger than the normal component. For shear flows with only gradual variations in the streamwise direction, the Reynolds shear stress is the dominant stress term in the momentum equations and the two equation models are reasonably accurate. For more complex strain fields, the errors can be significant."

Neumann (1989) notes, "Turbulence models employed in computational schemes to specify the character of turbulent flows are just that ... models, nonphysical ways of describing the character of the physical situation of turbulence. The models are the result of generalizing and applying fundamental experimental observations; they are not governed by the physical principles of turbulence and they are not unique."

In evaluating computations for the flow over aircraft at high angles of attack, Smith (1991) notes that "at higher angles of attack, turbulence modeling becomes more of a factor in the accuracy of the solution." The advantage of the RANS models is that they are relatively cheap to compute and can provide accurate solutions to many engineering flows. However, RANS models lack generality. The coefficients in the various models are usually determined by matching the computations to simple building-block experimental flows (e.g., zero-pressure-gradient (flat-plate) boundary layers.) Therefore, when deciding which turbulence model to use, the user should take care to insure that the selected turbulence model has been calibrated using measurements from relevant flow fields. Furthermore, the model should have sufficient accuracy and suitable numerical efficiency for the intended applications.

For an in-depth review of turbulence models and their applications, the reader is referred to Wilcox (1998).

4.5.3 Turbulent Boundary Layer for a Flat Plate

Since u_e is a constant for a flat plate, the pressure gradient term is zero. Even with this simplification, there is no exact solution for the turbulent boundary layer. Very near the wall, the viscous shear dominates. Ludwig Prandtl deduced that the mean velocity in this region must depend on the wall shear stress, the fluid's physical properties, and the distance y from the wall. Thus, $\bar{u}$ is a function of (τ_w, ρ, μ, y). To a first order, the velocity profile in this region is linear; that is, $\bar{u}$ is proportional to y. Thus,

$$\tau_w = \mu \frac{\partial \bar{u}}{\partial y} = \mu \frac{\bar{u}}{y} \tag{4.52}$$

Let us define

$$u^+ = \frac{\bar{u}}{u^*} \tag{4.53a}$$

and

$$y^+ = \frac{yu^*}{\nu} \tag{4.53b}$$

where u^* is called the *wall-friction velocity* and is defined as

$$u^* = \sqrt{\frac{\tau_w}{\rho}} \tag{4.53c}$$

Note that y^+ has the form of a Reynolds number.

Substituting these definitions into equation (4.52), we obtain

$$\tau_w = \mu \frac{u^+ u^*}{(y^+ \nu)/u^*} = (u^+/y^+)\rho u^{*2}$$

Introducing the definition of the wall-friction velocity, it is clear that

$$u^+ = y^+ \tag{4.54}$$

for the laminar sublayer. In the laminar sublayer, the velocities are so small that viscous forces dominate and there is no turbulence. The edge of the laminar sublayer corresponds to a y^+ of 5 to 10.

In 1930, Theodor von Kármán deduced that, in the outer region of a turbulent boundary layer, the mean velocity $\bar{u}$ is reduced below the free-stream value (u_e) in a manner that is independent of the viscosity but is dependent on the wall shear stress and the distance y over which its effect has diffused. Thus, the velocity defect $(u_e - \bar{u})$ for the outer region is a function of $(\tau_w, \rho, y, \delta)$. For the outer region, the velocity-defect law is given by

$$\frac{u_e - \bar{u}}{u^*} = g\left(\frac{y}{\delta}\right) \tag{4.55}$$

The outer region of a turbulent boundary layer contains 80 to 90% of the boundary-layer thickness δ.

In 1933, Prandtl deduced that the mean velocity in the inner region must depend on the wall shear stress, the fluid physical properties, and the distance y from the wall. Thus, $\bar{u}$ is a function of (τ_w, ρ, μ, y). Specifically,

$$\frac{\bar{u}}{u^*} = f\left[\left(\frac{y}{\delta}\right)\left(\frac{\delta u^*}{\nu}\right)\right] \tag{4.56}$$

for the inner region.

Since the velocities of the two regions must match at their interface,

$$\frac{\bar{u}}{u^*} = f\left[\left(\frac{y}{\delta}\right)\left(\frac{\delta u^*}{\nu}\right)\right] = \frac{u_e}{u^*} - g\left(\frac{y}{\delta}\right)$$

As a result, the velocity in the inner region is given by

$$\frac{\bar{u}}{u^*} = \frac{1}{\kappa}\ln\frac{yu^*}{\nu} + B \qquad\qquad \textbf{(4.57a)}$$

or, in terms of u^+, y^+, the equation can be written as

$$u^+ = \frac{1}{\kappa}\ln y^+ + B \qquad\qquad \textbf{(4.57b)}$$

This velocity correlation is valid only in regions where the laminar shear stress can be neglected in comparison with the turbulent stress. Thus, the flow in this region (i.e., $70 < y^+ < 400$) is fully turbulent.

The velocity in the outer region is given by

$$\frac{u_e - \bar{u}}{u^*} = -\frac{1}{\kappa}\ln\frac{y}{\delta} + A \qquad\qquad \textbf{(4.58)}$$

where κ, A, and B are dimensionless parameters. For incompressible flow past a flat plate,

$$\kappa \simeq 0.40 \quad\text{or}\quad 0.41$$
$$A \simeq 2.35$$
$$B \simeq 5.0 \quad\text{to}\quad 5.5$$

The resultant velocity profile is presented in Fig. 4.15.

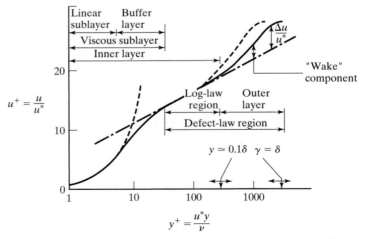

Figure 4.15 Turbulent boundary layer illustrating wall-layer nomenclature.

   The computation of the turbulent skin-friction drag for realistic aerodynamic applications presents considerable challenges to the analyst, both because of grid generation considerations and because of the need to develop turbulence models of suitable accuracy for the complex flow-field phenomena that may occur (e.g., viscous/inviscid interactions). In order to determine accurately velocity gradients near the wall, the computational grid should include points in the laminar sublayer. Referring to Fig. 4.15, the computational grid should, therefore, contain points at a y^+ of 5, or less. While there are many turbulence models of suitable engineering accuracy available in the literature, the analysts should calibrate the particular model to be used in their codes against a relevant data base to insure that the model provides results of suitable accuracy for the applications of interest.

4.6 EDDY VISCOSITY AND MIXING LENGTH CONCEPTS

In the late nineteenth century, Boussinesq introduced the concept of eddy viscosity to model the Reynolds shear stress. It was assumed that the Reynolds stresses act like the viscous (laminar) shear stresses and are proportional to the transverse gradient of the (mean) streamwise velocity component. The coefficient of the proportionality is called the *eddy viscosity* (ε_m) and is defined as

$$-\rho\overline{u'v'} = \rho\varepsilon_m\left(\frac{\partial\bar{u}}{\partial y}\right) \tag{4.59}$$

Having introduced the concept of eddy viscosity, equation (4.51) for the total shear stress may be written

$$\tau = \tau_l + \tau_t = \rho\nu\frac{\partial\bar{u}}{\partial y} + \rho\epsilon_m\left(\frac{\partial\bar{u}}{\partial y}\right) \tag{4.60}$$

Like the kinematic viscosity ν, ε_m has the units of L^2/T. However, whereas ν is a property of the fluid and is defined once the pressure and the temperature are known, ε_m is a function of the flow field (including such factors as surface roughness, pressure gradients, etc.).

   In an attempt to obtain a more generally applicable relation, Prandtl proposed the mixing length concept, whereby the shear stress is given as

$$-\rho\overline{u'v'} = \rho l^2\left|\frac{\partial\bar{u}}{\partial y}\right|\frac{\partial\bar{u}}{\partial y} \tag{4.61}$$

Equating the expressions for the Reynolds stress, given by equations (4.59) and (4.61), we can now write a relation between the eddy viscosity and the mixing length:

$$\varepsilon_m = l^2\left|\frac{\partial\bar{u}}{\partial y}\right| \tag{4.62}$$

From this point on, we will use only the time-average (or mean-flow) properties. Thus, we will drop the overbar notation in the subsequent analysis.

   The distributions of ε_m and of l across the boundary layer are based on experimental data. Because the eddy viscosity and the mixing length concepts are based on local equilibrium ideas, they provide only rough approximations to the actual flow and are said to lack generality. In fact, the original derivations included erroneous physical arguments. However, they are relatively simple to use and provide reasonable values of the shear stress for many engineering applications.

A general conclusion that is drawn from the experimental evidence is that the turbulent boundary layer should be treated as a composite layer consisting of an inner region and an outer region. For the inner region, the mixing length is given as

$$l_i = \kappa y[(1 - \exp(-y/A)] \tag{4.63}$$

where $\kappa = 0.41$ (as discussed earlier) and A, the van Driest damping parameter, is

$$A = \frac{26\nu}{Nu^*} \tag{4.64a}$$

where u^* is the wall friction velocity as defined by equation (4.53c)

$$N = (1 - 11.8p^+)^{0.5} \tag{4.64b}$$

and

$$p^+ = \frac{\nu u_e}{(u^*)^3} \frac{du_e}{dx} \tag{4.64c}$$

Thus, the eddy viscosity for the inner region becomes

$$(\varepsilon_m)_i = (\kappa y)^2 [1 - \exp(-y/A)]^2 \left| \frac{\partial u}{\partial y} \right| \tag{4.65}$$

For the outer region, the eddy viscosity is given as

$$(\varepsilon_m)_0 = \alpha u_e \delta^* \tag{4.66}$$

where δ^* is the displacement thickness as defined by equation (4.26),

$$\alpha = \frac{0.02604}{1 + \Pi} \tag{4.67a}$$

$$\Pi = 0.55[1 - \exp(-0.243\sqrt{z_1} - 0.298z_1)] \tag{4.67b}$$

and

$$z_1 = \frac{Re_\theta}{425} - 1 \tag{4.67c}$$

where Re_θ is the Reynolds number based on momentum thickness, that is, the characteristic dimension in the expression for Re_θ is the momentum thickness, see equation (4.28).

The y coordinate of the interface between the inner region and the outer region is determined by the requirement that the y distribution of the eddy viscosity be continuous. Thus, the inner region expression, equation (4.65), is used to calculate the eddy viscosity until its value becomes equal to that given by the outer region expression, equation (4.66). The y coordinate, where the two expressions for the eddy viscosity are equal [i.e., when $(\varepsilon_m)_i = (\varepsilon_m)_0$], is the interface value, y_c. For $y \geq y_c$, the eddy viscosity is calculated using the outer region expression, equation (4.66).

Recall that the transition process occurs over a finite length; that is, the boundary layer does not instantaneously change from a laminar state to a fully turbulent profile. For most practical boundary-layer calculations, it is necessary to calculate the viscous flow along its entire length. That is, for a given pressure distribution (inviscid flow field) and for a given transition criterion, the boundary-layer calculation is started at the leading edge or at the forward stagnation point of the body (where the boundary layer is laminar) and proceeds downstream through the transitional flow into the fully turbulent region. To treat the boundary layer in the transition zone, the expressions for the eddy viscosity are multiplied by an intermittency factor, γ_{tr}. The expression for γ_{tr} is

$$\gamma_{tr} = 1 - \exp\left[-G(x - x_{tr}) \int_{x_{tr}}^{x} \frac{dx}{u_e} \right] \tag{4.68a}$$

where x_{tr} is the x coordinate for the onset of transition and

$$G = 8.35 \times 10^{-4}\left(\frac{u_e^3}{\nu^2}\right)(\mathrm{Re}_{x,tr})^{-1.34} \tag{4.68b}$$

The intermittency factor varies from 0 (in the laminar region and at the onset of transition) to 1 (at the end of the transition zone and for fully turbulent flow). Thus, in the transition zone,

$$(\varepsilon_m)_i = (\kappa y)^2[1 - \exp(-y/A)]^2\left|\frac{\partial u}{\partial y}\right|\gamma_{tr} \tag{4.69a}$$

and

$$(\varepsilon_m)_0 = \alpha u_e \delta^* \gamma_{tr} \tag{4.69b}$$

Solutions have been obtained using these equations to describe the laminar, transitional, and turbulent boundary layer for flow past a flat plate. The results are presented in Fig. 4.16.

4.7 INTEGRAL EQUATIONS FOR A FLAT-PLATE BOUNDARY LAYER

The eddy viscosity concept or one of the higher-order methods are used in developing turbulent boundary-layer solutions using the differential equations of motion. Although approaches using the differential equations are most common in computational fluid dynamics, the integral approach can also be used to obtain approximate solutions for a turbulent boundary layer. Following a suggestion by Prandtl, the turbulent velocity profile will be represented by a power-law approximation.

We shall use the mean-flow properties in the integral form of the equation of motion to develop engineering correlations for the skin-friction coefficient and the boundary-layer thickness for an incompressible, turbulent boundary layer on a flat plate. Since we will use only the time-averaged (or mean-flow) properties in this section, we will drop the overbar notation. Consider the control volume shown in Fig. 4.17. Note that the free-stream velocity of the flow approaching the plate (U_∞) and the velocity of the flow outside of the boundary layer adjacent to the plate (u_e) are equal and are

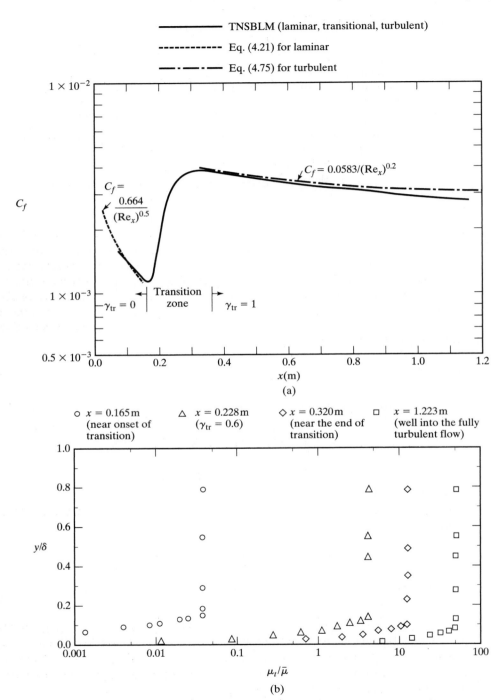

Figure 4.16 Sample of computed boundary layer for incompressible flow past a flat plate, $u_e = 114.1 \, \text{ft/s}$, $T_e = 542°R$, $P_e = 2101.5 \, \text{psf}$, $T_w = 540°R$: (a) skin friction distribution; (b) turbulent viscosity/thermodynamic viscosity ratio.

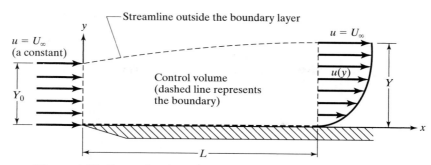

Figure 4.17 Control volume used to analyze the boundary layer on a flat plate.

used interchangeably. The wall (which is, of course, a streamline) is the inner boundary of the control volume. A streamline outside the boundary layer is the outer boundary. Any streamline that is outside the boundary layer (and, therefore, has zero shear force acting across it) will do. Because the viscous action retards the flow near the surface, the outer boundary is not parallel to the wall. Thus, the streamline is a distance Y_0 away from the wall at the initial station and is a distance Y away from the wall at the downstream station, with $Y > Y_0$. Since $\vec{V} \cdot \hat{n} \, dA$ is zero for both boundary streamlines, the continuity equation (2.5) yields

$$\int_0^Y u \, dy - u_e Y_0 = 0 \tag{4.70}$$

But also,

$$\int_0^Y u \, dy = \int_0^Y [u_e + (u - u_e)] \, dy$$

$$= u_e Y + \int_0^Y (u - u_e) \, dy \tag{4.71}$$

Combining these two equations and introducing the definition for the displacement thickness,

$$\delta^* = \int_0^\delta \left(1 - \frac{u}{u_e}\right) dy$$

we find that

$$Y - Y_0 = \delta^* \tag{4.72}$$

Thus, we have derived the expected result that the outer streamline is deflected by the transverse distance δ^*. In developing this relation, we have used both δ and Y as the upper limit for the integration. Since the integrand goes to zero for $y \geq \delta$, the integral is independent of the upper limit of integration, provided that it is at, or beyond, the edge of the boundary layer.

Similarly, application of the integral form of the momentum equation (2.13) yields

$$-d = \int_0^Y u(\rho u \, dy) - \int_0^{Y_0} u_e(\rho u_e \, dy)$$

Note that the lowercase letter designates the drag per unit span (d). Thus,

$$d = \rho u_e^2 Y_0 - \int_0^Y (\rho u^2 \, dy) \tag{4.73}$$

Using equation (4.70), we find that

$$d = \rho u_e \int_0^Y u \, dy - \int_0^Y \rho u^2 \, dy$$

This equation can be rewritten in terms of the section drag coefficient as

$$
\begin{aligned}
C_d &= \frac{d}{\frac{1}{2}\rho_\infty U_\infty^2 L} \\
&= \frac{2}{L}\left(\int_0^Y \frac{u}{u_e}\, dy - \int_0^Y \frac{u^2}{u_e^2}\, dy \right)
\end{aligned} \tag{4.74}
$$

Recall that the momentum thickness for an incompressible flow is

$$\theta = \int_0^\delta \frac{u}{u_e}\left(1 - \frac{u}{u_e}\right) dy \tag{4.28}$$

Note that the result is independent of the upper limit of integration provided that the upper limit is equal to or greater than the boundary-layer thickness. Thus, the drag coefficient (for one side of a flat plate of length L) is

$$C_d = \frac{2\theta}{L} \tag{4.75}$$

The equations developed in this section are valid for incompressible flow past a flat plate whether the boundary layer is laminar or turbulent. The value of the integral technique is that it requires only a "reasonable" approximation for the velocity profile [i.e., $u(y)$] in order to achieve "fairly accurate" drag predictions, because the integration often averages out positive and negative deviations in the assumed velocity function.

4.7.1 Application of the Integral Equations of Motion to a Turbulent, Flat-Plate Boundary Layer

Now let us apply these equations to develop correlations for a turbulent boundary layer on a flat plate. As discussed earlier, an analytical form for the turbulent shear is not known a priori. Therefore, we need some experimental information. Experimental measurements have shown that the time-averaged velocity may be represented by the power law,

$$\frac{u}{u_e} = \left(\frac{y}{\delta}\right)^{1/7} \tag{4.76}$$

when the local Reynolds number Re_x is in the range 5×10^5 to 1×10^7. However, note that the velocity gradient for this profile,

$$\frac{\partial u}{\partial y} = \frac{u_e}{7} \frac{1}{\delta^{1/7}} \frac{1}{y^{6/7}}$$

goes to infinity at the wall. Thus, although the correlation given in equation (4.76) provides a reasonable representation of the actual velocity profile, we need another piece of experimental data: a correlation for the shear at the wall. Blasius found that the skin friction coefficient for a turbulent boundary layer on a flat plate where the local Reynolds number is in the range 5×10^5 to 1×10^7 is given by

$$C_f = \frac{\tau}{\frac{1}{2}\rho u_e^2} = 0.0456 \left(\frac{\nu}{u_e \delta} \right)^{0.25} \tag{4.77}$$

Differentiating equation (4.69) gives us

$$C_f = -2 \frac{d}{dx} \left[\delta \int_0^1 \frac{u}{u_e} \left(\frac{u}{u_e} - 1 \right) d\left(\frac{y}{\delta} \right) \right] \tag{4.78}$$

Substituting equations (4.76) and (4.77) into equation (4.78), we obtain

$$0.0456 \left(\frac{\nu}{u_e \delta} \right)^{0.25} = -2 \frac{d}{dx} \left\{ \delta \int_0^1 \left[\left(\frac{y}{\delta} \right)^{2/7} - \left(\frac{y}{\delta} \right)^{1/7} \right] d\left(\frac{y}{\delta} \right) \right\}$$

which becomes

$$\delta^{0.25} d\delta = 0.2345 \left(\frac{\nu}{u_e} \right)^{0.25} dx$$

If we assume that the boundary-layer thickness is zero when $x = 0$, we find that

$$\delta = 0.3747 \left(\frac{\nu}{u_e} \right)^{0.2} (x)^{0.8}$$

Rearranging, the thickness of a turbulent boundary layer on a flat plate is given by

$$\frac{\delta}{x} = \frac{0.3747}{(\mathrm{Re}_x)^{0.2}} \tag{4.79}$$

Comparing the turbulent correlation given by equation (4.79) with the laminar correlation given by equation (4.23), we see that a turbulent boundary layer grows at a faster rate than a laminar boundary layer subject to the same conditions. Furthermore, at a given x station, a turbulent boundary layer is thicker than a laminar boundary layer for the same stream conditions.

Substitution of equation (4.79) into equation (4.77) yields

$$C_f = \frac{0.0583}{(\mathrm{Re}_x)^{0.2}} \tag{4.80}$$

As with the laminar skin-friction coefficient found in Sec. 4.3.1, a total skin-friction coefficient can be found for turbulent flow by integrating equation (4.80) over the length of a flat plate:

$$\overline{C}_f = \frac{1}{L} \int_0^L C_f(x) dx = \frac{1}{L} \int_0^L \frac{0.0583}{(\mathrm{Re}_x)^{0.2}} dx$$

$$\overline{C}_f = \frac{0.074}{(\mathrm{Re}_L)^{0.2}} \tag{4.81}$$

This formula, known as the Prandtl formula, is an exact theoretical representation of the turbulent skin-friction drag. However, when compared with experimental data, it is found to be only $\pm 25\%$ accurate. A number of other empirical and semi-empirical turbulent skin-friction coefficient relations also have been developed, some of which are considerably more accurate than the Prandtl formula [White (2005)]:

Prandtl-Schlichting:

$$\overline{C}_f \equiv \frac{0.455}{(\log_{10} \mathrm{Re}_L)^{2.58}} \quad \pm 3\% \text{ accurate} \tag{4.82}$$

Karman-Schoenherr:

$$\frac{1}{\sqrt{\overline{C}_f}} \equiv 4.13 \log_{10}(\mathrm{Re}_L \, \overline{C}_f) \quad \pm 2\% \text{ accurate} \tag{4.83}$$

Schultz-Grunow:

$$\overline{C}_f \equiv \frac{0.427}{(\log_{10} \mathrm{Re}_L - 0.407)^{2.64}} \quad \pm 7\% \text{ accurate} \tag{4.84}$$

While the Karman-Schoenherr relation is the most accurate of these relationships, it requires an iterative solution method to obtain a result, since the drag coefficient is not explicitly represented. Therefore, the most accurate relation which is also straight-forward to use is the Prandtl-Schlichting relation, which should usually be used instead of the Prandtl theoretical relation, equation (4.81).

The calculation of the skin-friction drag for a flat plate with transition theoretically would require using the local skin-friction coefficients, equation (4.21) for the laminar portion of the flow and equation (4.80) for the turbulent part of the flow, taking care to only integrate each relation over the laminar and turbulent lengths, respectively.

$$C_D = \frac{1}{L} \left\{ \int_0^{x_{tr}} C_{f_{\mathrm{lam}}} \, dx + \int_{x_{tr}}^{L} C_{f_{\mathrm{turb}}} \, dx \right\} \tag{4.85}$$

Another approach is to use the total skin-friction coefficients, which are equation (4.32) for the laminar portion of the flow and equation (4.82) for the turbulent part of the flow. Care must be taken, however, when using the total skin-friction coefficients, since they already represent the integrated skin friction over the entire plate from $x = 0$ to $x = L$. In order to properly simulate a flat plate with transitional flow present, the process shown in Fig. 4.18 should be used. Since you want to simulate the plate with both laminar and turbulent boundary layers present, you start by evaluating the entire plate by assuming that the boundary layer is turbulent along the entire length of the plate. Since the distance from the leading edge of the plate to the transition location should be evaluated with laminar flow, that portion of the plate should be evaluated with the turbulent-flow skin-friction relation and also with the laminar-flow skin-friction relation by subtracting the turbulent-flow drag from the total plate turbulent results and adding the laminar-flow portion. In equation form, this would be:

$$C_D = \overline{C}_{f_{\mathrm{turb}}} \frac{Lb}{S_{\mathrm{ref}}} - \overline{C}_{f_{\mathrm{turb}}} \frac{x_{\mathrm{tr}} b}{S_{\mathrm{ref}}} + \overline{C}_{f_{\mathrm{lam}}} \frac{x_{\mathrm{tr}} b}{S_{\mathrm{ref}}} \tag{4.86}$$

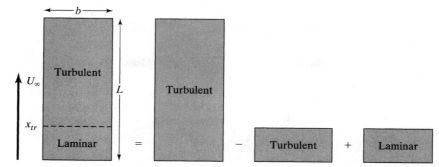

Figure 4.18 Calculation of skin-friction drag coefficient using total skin-friction coefficients.

A more straightforward approach to model transitional flow is to use an empirical correction to the Prandtl-Schlichting turbulent skin-friction relation [Dommasch, et al. (1967)], in equation (4.82):

$$\overline{C}_f \equiv \frac{0.455}{(\log_{10} \text{Re}_L)^{2.58}} - \frac{A}{\text{Re}_L} \qquad (4.87)$$

where the correction term reduces the skin friction since laminar boundary layers produce less skin friction than turbulent boundary layers. The experimentally determined constant, A, varies depending on the transition Reynolds number, as shown in Table 4.5. The value $A = 1700$ represents the laminar correction for a transition Reynolds number of $\text{Re}_{x,\text{tr}} = 500{,}000$. It can be seen from this formulation that if the Reynolds number at the end of the plate is very high, then the laminar correction term plays a fairly insignificant role in the total skin-friction drag on the plate. A good rule of thumb is to assume that if transition takes place at less than 10% of the length of the plate, then the laminar correction usually can be ignored, since it is relatively small.

Figure 4.19 shows how the total skin-friction coefficient varies from the laminar value in equation (4.32), through transition, and finally to the fully turbulent value, in equation (4.82).

TABLE 4.5 Empirical Relations for Transition Correction [Schlichting (1979).]

$Re_{x,\,tr}$	A
300,000	1050
500,000	1700
1,000,000	3300
3,000,000	8700

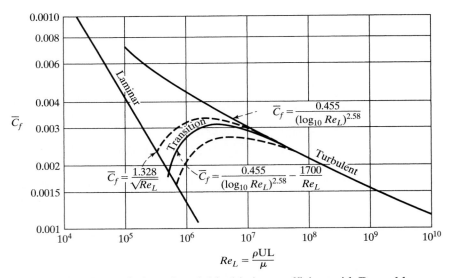

Figure 4.19 Variation of total skin-friction coefficient with Reynolds number for a smooth, flat plate. [From Dommasch, et al. (1967).]

EXAMPLE 4.4:   Computing the velocity profiles at the "transition point"

Air at standard sea-level atmospheric pressure and 5°C flows at 200 km/h across a flat plate. Compare the velocity distribution for a laminar boundary layer and for a turbulent boundary layer at the transition point, assuming that the transition process is completed instantaneously at that location.

Solution: For air at atmospheric pressure and 5°C,

$$\rho_\infty = \frac{1.01325 \times 10^5 \, \text{N/m}^2}{(287.05 \, \text{N} \cdot \text{m/kg} \cdot \text{K})(278.15 \, \text{K})} = 1.2691 \, \text{kg/m}^3$$

$$\mu_\infty = 1.458 \times 10^{-6} \frac{(278.15)^{1.5}}{278.15 + 110.4} = 1.7404 \times 10^{-5} \, \text{kg/s} \cdot \text{m}$$

and

$$U_\infty = \frac{(200 \, \text{km/h})(1000 \, \text{m/km})}{3600 \, \text{s/h}} = 55.556 \, \text{m/s}$$

We will assume that the transition Reynolds number for this incompressible flow past a flat plate is 500,000. Thus,

$$x_{\text{tr}} = \frac{\text{Re}_{x,\text{tr}}}{\rho u_e / \mu} = 0.12344 \, \text{m}$$

The thickness of a laminar boundary layer at this point is

$$\delta_{\text{lam}} = \frac{5.0x}{\sqrt{\text{Re}_x}} = 8.729 \times 10^{-4}\,\text{m}$$

For comparison, we will calculate the thickness of the turbulent boundary layer at this point for this Reynolds number, assuming that the boundary layer is turbulent all the way from the leading edge. Thus,

$$\delta_{\text{turb}} = \frac{0.3747x}{(\text{Re}_x)^{0.2}} = 3.353 \times 10^{-3}\,\text{m}$$

In reality, the flow is continuous at the transition location and the boundary-layer thickness does not change instantaneously. Furthermore, since we are at the transition location, it is not realistic to use the assumption that the boundary layer is turbulent all the way from the leading edge. (This assumption would be reasonable far downstream of the transition location so that $x \gg x_{\text{tr}}$.) Nevertheless, the object of these calculations is to illustrate the characteristics of the turbulent boundary layer relative to a laminar boundary layer at the same conditions.

The resultant velocity profiles are compared in Table 4.6 and Fig. 4.20. Note that the streamwise velocity component u increases much more rapidly with y near the wall for the turbulent boundary layer. Thus, the shear at the wall is greater for the turbulent boundary layer even though this layer is much thicker than the laminar boundary layer for the same conditions at a given x station. The macroscopic transport of fluid in the y direction causes both increased shear and increased thickness of the boundary layer.

TABLE 4.6 Velocity Profiles for Example 4.4

y (m)	u_{lam} (m/s)	u_{turb} (m/s)
0.00000	0.00	0.00
0.00017	17.78	36.21
0.00034	33.33	39.98
0.00067	52.50	44.14
0.00101	Inviscid flow	46.78
0.00134		48.74
0.00168		50.32
0.00201		51.64
0.00235		52.79
0.00268		53.81
0.00302		54.71
0.00335		55.56

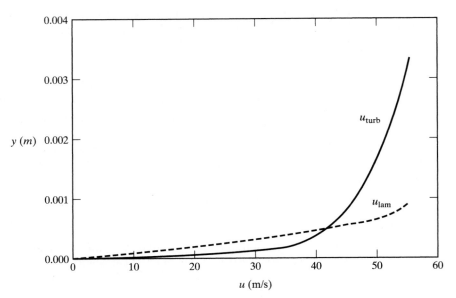

Figure 4.20 Velocity profiles for Example 4.4.

4.7.2 Integral Solutions for a Turbulent Boundary Layer with a Pressure Gradient

If we apply the integral equations of motion to a flow with a velocity gradient external to the boundary layer, we obtain

$$\frac{d\theta}{dx} + (2 + H)\frac{\theta}{u_e}\frac{du_e}{dx} = \frac{C_f}{2} \tag{4.88}$$

where θ, the momentum thickness, was defined in equation (4.28). H, the momentum shape factor, is defined as

$$H = \frac{\delta^*}{\theta} \tag{4.89}$$

where δ^*, the displacement thickness, is defined in equation (4.26). Equation (4.88) contains three unknown parameters, θ, H and C_f, for a given external velocity distribution. For a turbulent boundary layer, these parameters are interrelated in a complex way. Head (1969) assumed that the rate of entertainment is given by

$$\frac{d}{dx}(u_e\theta H_1) = u_e F \tag{4.90}$$

where H_1 is defined as

$$H_1 = \frac{\delta - \delta^*}{\theta} \tag{4.91}$$

Head also assumed that H_1 is a function of the shape factor H, that is, $H_1 = G(H)$. Correlations of several sets of experimental data that were developed by Cebeci and Bradshaw (1979) yielded

$$F = 0.0306(H_1 - 3.0)^{-0.6169} \tag{4.92}$$

and

$$G = \begin{cases} 0.8234(H - 1.1)^{-1.287} + 3.3 & \text{for } H \leq 1.6 \\ 1.5501(H - 0.6778)^{-3.064} + 3.3 & \text{for } H \geq 1.6 \end{cases} \tag{4.93}$$

Equations (4.90) through (4.93) provide a relationship between θ and H. A relation between C_f, θ, and H is needed to complete our system of equations. A curvefit formula given in White (2005) is

$$C_f = \frac{0.3e^{-1.33H}}{(\log \mathrm{Re}_\theta)^{(1.74 + 0.31H)}} \tag{4.94}$$

where Re_θ is the Reynolds number based on the momentum thickness:

$$\mathrm{Re}_\theta = \frac{\rho u_e \theta}{\mu} \tag{4.95}$$

We can numerically solve this system of equations for a given inviscid flow field. To start the calculations at some initial streamwise station, such as the transition location, values for two of the three parameters, θ, H, and C_f, must be specified at this station. The third parameter is then calculated using equation (4.94). Using this method, the shape factor H can be used as a criterion for separation. Although it is not possible to define an exact value of H corresponding to the separation, the value of H for separation is usually in the range 1.8 to 2.8.

4.8 THERMAL BOUNDARY LAYER FOR CONSTANT-PROPERTY FLOWS

As noted earlier, there are many constant-property flows for which we are interested in calculating the convective heat transfer. Thus, the temperature variations in the flow field are sufficiently large that there is heat transfer to or from a body in the flow but are small enough that the corresponding variations in density and viscosity can be neglected. Let us examine one such flow, the thermal boundary layer for a steady, low-speed flow past a flat plate. We will consider flows where the boundary layer is laminar. The solution for the velocity field for this flow has been described earlier in this chapter; see equations (4.6) through (4.27).

We will now solve the energy equation (2.32), in order to determine the temperature distribution. For a low-speed, constant-property, laminar boundary layer, the viscous dissipation is negligible (i.e., $\phi = 0$). For flow past a flat plate, $dp/dt = 0$. Thus, for a calorically perfect gas, which will be defined in Chapter 8, equation (2.32) becomes

$$\rho u c_p \frac{\partial T}{\partial x} + \rho v c_p \frac{\partial T}{\partial y} = k \frac{\partial^2 T}{\partial y^2} \tag{4.96}$$

Note that we have already neglected $k(\partial^2 T/\partial x^2)$ since it is small compared to $k(\partial^2 T/\partial y^2)$. We made a similar assumption about the corresponding velocity gradients when working with the momentum equation; see equation (4.6).

Let us now change the dependent variable from T to the dimensionless parameter θ, where

$$\theta = \frac{T - T_w}{T_e - T_w}$$

Note that $\theta = 0$ at the wall (i.e., at $y = 0$), and $\theta = 1$ at the edge of the thermal boundary layer. Using θ as the dependent variable, the energy equation becomes

$$\rho u \frac{\partial \theta}{\partial x} + \rho v \frac{\partial \theta}{\partial y} = \frac{k}{c_p} \frac{\partial^2 \theta}{\partial y^2} \tag{4.97}$$

Since the pressure is constant along the flat plate, the velocity at the edge of the boundary layer (u_e) is constant, and the momentum equation becomes

$$\rho u \frac{\partial u}{\partial x} + \rho v \frac{\partial u}{\partial y} = \mu \frac{\partial^2 u}{\partial y^2} \tag{4.98}$$

Let us replace u in the derivatives by the dimensionless parameter, u^*, where

$$u^* = \frac{u}{u_e}$$

Thus, equation (4.98) becomes

$$\rho u \frac{\partial u^*}{\partial x} + \rho v \frac{\partial u^*}{\partial y} = \mu \frac{\partial^2 u^*}{\partial y^2} \tag{4.99}$$

Note that $u^* = 0$ at the wall (i.e., at $y = 0$), and $u^* = 1$ at the edge of the velocity boundary layer.

Compare equations (4.97) and (4.98). Note that the equations are identical if $k/c_p = \mu$. Furthermore, the boundary conditions are identical: $\theta = 0$ and $u^* = 0$ at the wall, and $\theta = 1$ and $u^* = 1$ at the edge of the boundary layer. Thus, if

$$\frac{\mu c_p}{k} = 1$$

the velocity and the thermal boundary layers are identical. This ratio is called the Prandtl number (Pr) in honor of the German scientist:

$$\text{Pr} = \frac{\mu c_p}{k} \tag{4.100}$$

The Prandtl number is an important dimensionless parameter for problems involving convective heat transfer where one encounters both fluid motion and heat conduction.

4.8.1 Reynolds Analogy

The shear at the wall is defined as

$$\tau = \left(\mu \frac{\partial u}{\partial y} \right)_{y=0}$$

Therefore, the skin-friction coefficient for a flat plate is

$$C_f = \frac{\tau}{\frac{1}{2}\rho u_e^2} = \frac{2\mu}{\rho u_e} \frac{\partial u^*}{\partial y} \tag{4.101}$$

The rate at which heat is transferred to the surface $(\dot{q})$ is defined as

$$\dot{q} = \left(k \frac{\partial T}{\partial y} \right)_{y=0} \tag{4.102}$$

The Stanton number (designated by the symbols St or C_h), which is a dimensionless heat-transfer coefficient, is defined as

$$St \equiv C_h = \frac{\dot{q}}{\rho u_e c_p (T_e - T_w)} \tag{4.103}$$

Combining these last two expressions, we find that the Stanton number is

$$St = \frac{k}{\rho u_e c_p} \frac{\partial \theta}{\partial y} \tag{4.104}$$

Relating the Stanton number, as given by equation (4.104), to the skin-friction coefficient, as defined by equation (4.101), we obtain the ratio

$$\frac{C_f}{St} = \frac{2\mu c_p}{k} \frac{\partial u^*/\partial y}{\partial \theta/\partial y} \tag{4.105}$$

Note that if $\mu c_p/k \equiv Pr = 1$, then

$$\frac{\partial u^*}{\partial y} = \frac{\partial \theta}{\partial y}$$

Thus, if the Prandtl number is 1,

$$St = \frac{C_f}{2} \tag{4.106}$$

This relation between the heat-transfer coefficient and the skin-friction coefficient is known as the Reynolds analogy.

EXAMPLE 4.5:   Calculating the thermal properties of air

The thermal conductivity of air can be calculated using the relation

$$k = 4.76 \times 10^{-6} \frac{T^{1.5}}{T + 112} \, \text{cal/cm} \cdot \text{s} \cdot \text{K} \qquad (4.107)$$

over the range of temperatures below those for which oxygen dissociates, that is, approximately 2000 K at atmospheric pressure. What is the Prandtl number for air at $15°C$, that is, at 288.15 K?

Solution: Using the results from Example 1.3, the viscosity is 1.7894×10^{-5} kg/s · m. The specific heat is 1004.7 J/kg · K. Using the equation to calculate the thermal conductivity,

$$k = 4.76 \times 10^{-6} \frac{(288.15)^{1.5}}{400.15} = 5.819 \times 10^{-5} \, \text{cal/cm} \cdot \text{s} \cdot \text{K}$$

Noting that there are 4.187 J/cal, the thermal conductivity is

$$k = 2.436 \times 10^{-2} \, \text{J/m} \cdot \text{s} \cdot \text{K}$$

Thus,

$$\text{Pr} = \frac{\mu c_p}{k} = \frac{(1.7894 \times 10^{-5} \, \text{kg/s} \cdot \text{m})(1004.7 \, \text{J/kg} \cdot \text{K})}{2.436 \times 10^{-2} \, \text{J/m} \cdot \text{s} \cdot \text{K}} = 0.738$$

Note that, as a rule of thumb, the Prandtl number for air is essentially constant (approximately 0.7) over a wide range of flow conditions.

4.8.2   Thermal Boundary Layer for Pr ≠ 1

To solve for the temperature distribution for the laminar, flat-plate boundary layer, let us introduce the transformation of equation (4.10):

$$\eta = y\sqrt{\frac{u_e}{2\nu x}}$$

Using the transformed stream function f, as defined by equation (4.13), equation (4.85) becomes

$$\theta'' + (\text{Pr})f\theta' = 0 \qquad (4.108a)$$

where the $'$ denotes differentiation with respect to η. But we have already obtained the solution for the stream function. Referring to equation (4.16) for a flat plate (i.e., $\beta = 0$), we have

$$f = -\frac{f'''}{f''} \qquad (4.108b)$$

Combining equations (4.108a) and (4.108b) and rearranging, we obtain

$$\frac{\theta''}{\theta'} = (\text{Pr})\frac{f'''}{f''}$$

Integrating twice yields

$$\theta = C \int_0^{\eta} (f'')^{\text{Pr}} d\eta + \theta_0 \tag{4.109}$$

where C and θ_0 are constant of integration. They can be evaluated by applying the boundary conditions (1) at $\eta = 0, \theta = 0$, and (2) for $\eta \to$ large, $\theta = 1$.

$$\theta = \frac{T - T_w}{T_e - T_w} = 1 - \frac{\int_{\eta}^{\infty} (f'')^{\text{Pr}} d\eta}{\int_0^{\infty} (f'')^{\text{Pr}} d\eta} \tag{4.110}$$

The rate at which heat is transferred to the wall $(\dot{q})$ can be calculated using

$$\dot{q} = \left(k\frac{\partial T}{\partial y} \right)_{y=0} = k(T_e - T_w)\frac{\partial \eta}{\partial y}\left(\frac{\partial \theta}{\partial \eta} \right)_{\eta=0}$$

Using the values of Pohlhausen, we find that

$$\left(\frac{\partial \theta}{\partial \eta} \right)_{\eta=0} = 0.4696(\text{Pr})^{0.333}$$

Combining these two relations, the rate at which heat is transferred from a laminar boundary layer to the wall is given by the relation

$$\dot{q} = 0.332k(T_e - T_w)(\text{Pr})^{0.333}\sqrt{\frac{u_e}{\nu x}} \tag{4.111}$$

The heat transfer can be expressed in terms of the Stanton number using equation (4.103). Thus,

$$\text{St} = 0.332\frac{k}{\mu c_p}\sqrt{\frac{\mu}{\rho u_e x}}(\text{Pr})^{0.333}$$

Using the definitions for the Reynolds number and Prandtl number, the Stanton number is

$$\text{St} = \frac{0.332}{(\text{Pr})^{0.667}(\text{Re}_x)^{0.5}} \tag{4.112}$$

Another popular dimensionless heat-transfer parameter is the Nusselt number. The Nusselt number is defined as

$$\text{Nu}_x = \frac{hx}{k} \tag{4.113a}$$

In this equation, h is the local heat-transfer coefficient, which is defined as

$$h = \frac{\dot{q}}{T_e - T_w} \tag{4.113b}$$

Combining this definition with equation (4.111) and (4.113a) gives us

$$\mathrm{Nu}_x = 0.332(\mathrm{Re}_x)^{0.5}(\mathrm{Pr})^{0.333} \tag{4.114}$$

By dividing the expression for the Stanton number, equation (4.112), by that for the skin-friction coefficient, equation (4.21), we obtain

$$\mathrm{St} = \frac{C_f}{2(\mathrm{Pr})^{0.667}} \tag{4.115}$$

Because of the similarity between this equation and equation (4.106), we shall call this the *modified Reynolds analogy.*

EXAMPLE 4.6:　Calculating the heat-transfer rate for a turbulent boundary layer on a flat plate

Using the modified Reynolds analogy, develop relations for the dimensionless heat-transfer parameters, St and Nu_x, for a turbulent flat-plate boundary layer.

Solution:　Referring to the discussion of turbulent boundary layers, we note that

$$C_f = \frac{0.0583}{(\mathrm{Re}_x)^{0.2}} \tag{4.80}$$

Thus, using equation (4.115) for the modified Reynolds analogy, we can approximate the Stanton number as

$$\mathrm{St} = \frac{0.0292}{(\mathrm{Re}_x)^{0.2}(\mathrm{Pr})^{0.667}} \tag{4.116}$$

Comparing equations (4.112) and (4.114), we can see that the Nu_x is given as

$$\mathrm{Nu}_x = (\mathrm{St})(\mathrm{Pr})(\mathrm{Re}_x)$$

Thus, the Nusselt number for turbulent flow past a flat plate can be approximated as

$$\mathrm{Nu}_x = 0.0292(\mathrm{Re}_x)^{0.8}(\mathrm{Pr})^{0.333} \tag{4.117}$$

EXAMPLE 4.7:　Calculating the heat transfer

The radiator systems on many of the early racing aircraft were flush mounted on the external surface of the airplane. Let us assume that the local heat-transfer rate can be estimated using the flat-plate relations. What is the local heating rate for $x = 3.0$ m when the airplane is flying at 468 km/h at an altitude of 3 km? The surface temperature is 330 K.

Solution:　Using Table 1.2 to find the free-stream flow properties,

$$\rho_\infty = 7.012 \times 10^4\,\mathrm{N/m^2} \qquad T_\infty = 268.659\,\mathrm{K}$$

$$\rho_\infty = 0.9092\,\mathrm{kg/m^3} \qquad \mu_\infty = 1.6938 \times 10^{-5}\,\mathrm{kg/s \cdot m}$$

Since we have assumed that the flow corresponds to that for a flat plate, these values are also the local properties at the edge of the boundary layer at $x = 3.0$ m. Note that

$$u_e = U_\infty = 468 \text{ km/h} = 130 \text{ m/s}$$

To determine whether the boundary layer is laminar or turbulent, let us calculate the local Reynolds number:

$$\text{Re}_x = \frac{\rho_e u_e x}{\mu_e} = \frac{(0.9092)(130)(3.0)}{1.6938 \times 10^{-5}} = 2.093 \times 10^7$$

This is well above the transition value. In fact, if the transition Reynolds number is assumed to be 500,000, transition would occur at a point

$$x_{\text{tr}} = \frac{500{,}000}{(\rho_e u_e)/\mu_e} = 0.072 \text{ m}$$

from the leading edge. Thus, the calculation of the heating will be based on the assumption that the boundary layer is turbulent over its entire length.

Combining equations (4.113a) and (4.113b),

$$\dot{q} = \frac{\text{Nu}_x k(T_e - T_w)}{x}$$

where the Nu_x is given by equation (4.117). Thus,

$$\dot{q} = \frac{0.0292(\text{Re}_x)^{0.8}(\text{Pr})^{0.333} k(T_e - T_w)}{x}$$

To calculate the thermal conductivity for air,

$$k = 4.76 \times 10^{-6} \frac{T^{1.5}}{T + 112} = 5.506 \times 10^{-5} \text{ cal/cm} \cdot \text{s} \cdot \text{K}$$

$$= 2.306 \times 10^{-2} \text{ J/m} \cdot \text{s} \cdot \text{K}$$

Since $1 \text{ W} = 1 \text{ J/s}$,

$$k = 2.306 \times 10^{-2} \text{ W/m} \cdot \text{K}$$

Furthermore, the Prandtl number is

$$\text{Pr} = \frac{\mu c_p}{k} = 0.738$$

Thus,

$$\dot{q} = \frac{(0.0292)(2.093 \times 10^7)^{0.8}(0.738)^{0.333}(2.306 \times 10^{-2} \text{ W/m} \cdot \text{K})(268.659 - 330) \text{ K}}{3.0 \text{ m}}$$

$$= -8.944 \times 10^3 \text{ W/m}^2 = -8.944 \text{ kW/m}^2$$

The minus sign indicates that heat is transferred from the surface to the air flowing past the aircraft. This is as it should be, since the surface is hotter than the adjacent air. Furthermore, since the problem discusses a radiator, proper performance would produce cooling. Since there are 1.341 hp/kW, the heat transfer rate is equivalent to 1.114 hp/ft^2.

4.9   SUMMARY

In this chapter we have developed techniques by which we can obtain solutions for a thin, viscous boundary layer near the surface. Techniques have been developed both for a laminar and for a turbulent boundary layer using both integral and differential approaches. We now have reviewed the basic concepts of fluid mechanics in Chapters 1 through 4 and are ready to apply them specifically to aerodynamic problems.

PROBLEMS

4.1. A very thin, "flat-plate" wing of a model airplane moves through the air at standard sea-level conditions at a velocity of 15 m/s. The dimensions of the plate are such that its chord (streamwise dimension) is 0.5 m and its span (length perpendicular to the flow direction) is 5 m. What is the Reynolds number at the trailing edge ($x = 0.5$ m)? Assume that the boundary layer is laminar in answering the remaining questions. What are the boundary-layer thickness and the displacement thickness at the trailing edge? What are the local shear at the wall and the skin-friction coefficient at $x = 0.5$ m? Calculate the total drag on the wing (both sides). Prepare a graph of $\tilde{u}$ as a function of y, where $\tilde{u}$ designates the x component of velocity relative to a point on the ground, at $x = 0.5$ m.

4.2. Assume that the inviscid external flow over a configuration is given by

$$u_e = Ax$$

Thus, the stagnation point occurs at the leading edge of the configuration (i.e., at $x = 0$). Obtain the expression for β. Using Fig. 4.4 and assuming that the boundary layer is laminar, determine the value of $f''(0)$, that is, the value of the shear function at the wall. What is the relation between the shear at a given value of x for this flow and that for a flat plate?

4.3. Consider two-dimensional, incompressible flow over a cylinder. For ease of use with the nomenclature of the current chapter, we will assume that the windward plane of symmetry (i.e., the stagnation point) is $\theta = 0$ and that θ increases in the streamwise direction. Thus,

$$u_e = 2U_\infty \sin \theta \quad \text{and} \quad x = R\theta$$

Determine the values of β, at $\theta = 30°$, at $\theta = 45°$, and at $\theta = 90°$.

4.4. Assume that the wall is porous so that there can be flow through the wall; that is, $v(x,0) = v_w \neq 0$. Using equation (4.14), show that

$$\frac{v_w}{u_e} = -\frac{f(0)}{\sqrt{2\text{Re}_x}}$$

in order to have similarity solutions; that is, $(\partial f/\partial s)_\eta = 0$ for steady, incompressible flow past a flat plate.

4.5. We plan to use suction through a porous wall as a means of boundary-layer control. Using the equation developed in Problem 4.4, determine $f(0)$ if $v_w = -0.001u_e$ for steady flow past a flat plate where $u_e = 10$ m/s at standard sea-level conditions. What are the remaining two boundary conditions?

4.6. Transpiration (or injecting gas through a porous wall into the boundary layer) is to be used to reduce the skin-friction drag for steady, laminar flow past a flat plate. Using the equation developed in Problem 4.4, determine v_w if $f(0) = -0.25$. The inviscid velocity (u_e) is 50 ft/s with standard atmospheric conditions.

4.7. When we derived the integral equations for a flat-plate boundary layer, the outer boundary of our control volume was a streamline outside the boundary layer (see Fig. 4.17). Let us now apply the integral equations to a rectangular control volume to calculate the sectional drag coefficient for incompressible flow past a flat plate of length L. Thus, as shown in Fig. P4.7, the outer boundary is a line parallel to the wall and outside the boundary layer at all x stations. Owing to the growth of the boundary layer, fluid flows through the upper boundary with a velocity v_e which is a function of x. How does the resultant expression compare with equation (4.70)?

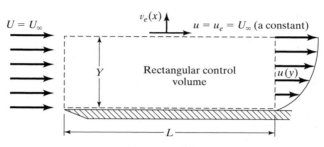

Figure P4.7

4.8. Use the integral momentum analysis and the assumed velocity profile for a laminar boundary layer:

$$\frac{u}{u_e} = \frac{3}{2}\left(\frac{y}{\delta}\right) - \frac{1}{2}\left(\frac{y}{\delta}\right)^3$$

where δ is the boundary-layer thickness, to describe the incompressible flow past a flat plate. For this profile, compute (a) $(\delta/x)\sqrt{Re_x}$, (b) $(\delta^*/x)\sqrt{Re_x}$, (c) $(v_e/u_e)\sqrt{Re_x}$, (d) $C_f\sqrt{Re_x}$, and (e) $C_d\sqrt{Re_x}$. Compare these values with those presented in the text, which were obtained using the more exact differential technique [e.g., $(\delta/x)\sqrt{Re_x} = 5.0$]. Prepare a graph comparing this approximate velocity profile and that given in Table 4.3. For the differential solution, use $\eta = 3.5$ to define δ when calculating y/δ.

4.9. Use the integral momentum analysis and a linear velocity profile for a laminar boundary layer

$$\frac{u}{u_e} = \frac{y}{\delta}$$

where δ is the boundary layer thickness. If the viscous flow is incompressible, calculate $(\delta/x)\sqrt{Re_x}$, $(\delta^*/x)\sqrt{Re_x}$, and $C_f\sqrt{Re_x}$. Compare these values with those presented in the chapter that were obtained using the more exact differential technique [e.g., $(\delta/x)\sqrt{Re_x} = 5.0$].

4.10. Let us represent the wing of an airplane by a flat plate. The airplane is flying at standard sea-level conditions at 180 mph. The dimensions of the wing are chord = 5 ft and span = 30 ft. What is the total friction drag acting on the wing? What is the drag coefficient?

4.11. A flat plate at zero angle of attack is mounted in a wind tunnel where

$$p_\infty = 1.01325 \times 10^5 \, \text{N/m}^2 \qquad U_\infty = 100 \, \text{m/s}$$
$$\mu_\infty = 1.7894 \times 10^{-5} \, \text{kg/m} \cdot \text{s} \qquad p_\infty = 1.2250 \, \text{kg/m}^3$$

A Pitot probe is to be used to determine the velocity profile at a station 1.0 m from the leading edge (Fig. P4.11).

(a) Using a transition criterion that $Re_{x,tr} = 500{,}000$, where does transition occur?

(b) Use equation (4.79) to calculate the thickness of the turbulent boundary layer at a point 1.00 m from the leading edge.

(c) If the streamwise velocity varies as the $\frac{1}{7}$th power law [i.e., $u/u_e = (y/\delta)^{1/7}$], calculate the pressure you should expect to measure with the Pitot probe $p_t(y)$ as a function of y. Present the predicted values as

    (1) The difference between that sensed by the Pitot probe and that sensed by the static port in the wall [i.e., y versus $p_t(y) - p_{\text{static}}$]

    (2) The pressure coefficient

$$y \text{ versus } C_p(y) = \frac{p_t(y) - p_\infty}{\frac{1}{2}\rho_\infty U_\infty^2}$$

Note that for part (c) we can use Bernoulli's equation to relate the static pressure and the velocity on the streamline just ahead of the probe and the stagnation pressure sensed by the probe. Even though this is in the boundary layer, we can use Bernoulli's equation, since we relate properties on a streamline and since we calculate these properties at "point." Thus, the flow slows down isentropically to zero velocity over a very short distance at the mouth of the probe.

(d) Is the flow described by this velocity function rotational or irrotational?

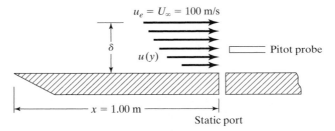

Figure P4.11

4.12. Air at atmospheric pressure and 100°C flows at 100 km/h across a flat plate. Compare the streamwise velocity as a function of y for a laminar boundary layer and for a turbulent boundary layer at the transition point, assuming that the transition process is completed instantaneously at that location. Use Table 4.3 to define the laminar profile and the one-seventh power law to describe the turbulent profile.

4.13. A thin symmetric airfoil section is mounted at zero angle of attack in a low-speed wind tunnel. A Pitot probe is used to determine the velocity profile in the viscous region downstream of the airfoil, as shown in Fig. P4.13. The resultant velocity distribution in the region $-w \le z \le +w$

$$u(z) = U_\infty - \frac{U_\infty}{2}\cos\frac{\pi z}{2w}$$

If we apply the integral form of the momentum equation [equation (2.13)] to the flow between the.two streamlines bounding this wake, we can calculate the drag force acting on the airfoil section. The integral continuity equation [equation (2.5)] can be used to relate the spacing between the streamlines in the undisturbed flow ($2h$) to their spacing ($2w$) at the x location where the Pitot profile was obtained. If $w = 0.009c$, what is the section drag coefficient C_d?

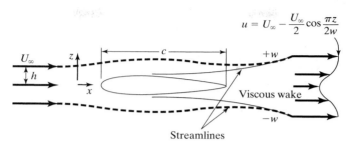

Figure P4.13

4.14. For the wing dimensions and flow properties of Problem 4.1, find the total skin friction coefficient and the total drag on the wing (both sides) using the method of equation (4.86) and the Prandtl-Schlichting turbulent skin friction relation. Perform the estimation again using the approximate method of equation (4.87). Comment on the difference between the two approaches and the accuracy of the approximate method.

4.15. Assume the flow over the flat plate of Problem 4.11 is at $U_\infty = 80 \ m/s$ (all other conditions are the same). Find the total skin friction drag of the flat plate (both sides) using the Prandtl-Schlichting relation (initially assume that the flow is completely turbulent along the length of the plate). Now find the skin friction drag for the plate using the approximate formula of equation (4.87) which takes into account transition effects. How accurate was the initial assumption of fully turbulent flow?

4.16. Derive equation (4.86) to find the total drag coefficient for the flat plate shown in Fig. 4.18. Convert the resulting relation into a drag relation for the flat plate and simplify the results.

4.17. Using equation (4.95), calculate the thermal conductivity of air at 2000 K. What is the Prandtl number of perfect air at this temperature?

4.18. The boundary conditions that were used in developing the equation for the laminar thermal boundary layer were that the temperature is known at the two limits (1) $\theta = 0$ at $\eta = 0$ and (2) $\theta = 1$ at $\eta \to$ large. What would be the temperature distribution if the boundary conditions were (1) an adiabatic wall (i.e., $\theta' = 0$ at $\eta = 0$), and (2) $\theta = 1$ at $\eta \to$ large? *Hint*: From equation (4.97),

$$\theta' = C(f'')^{\text{Pr}}$$

4.19. Represent the wing of an airplane by a flat plate. The airplane is flying at standard sea-level conditions at 180 mi/h. The dimensions of the wing are chord $= 5$ ft and span $= 30$ ft. What is the total heat transferred to the wing if the temperature of the wing is 45°F?

4.20. A wind tunnel has a 1-m², 6-m-long test section in which air at standard sea-level conditions moves at 70 m/s. It is planned to let the walls diverge slightly (slant outward) to compensate for the growth in boundary-layer displacement thickness, and thus maintain a constant area for the inviscid flow. This allows the free-stream velocity to remain constant. At what angle should the walls diverge to maintain a constant velocity between $x = 1.5$ m and $x = 6$ m?

REFERENCES

Abbott IH, von Doenhoff AE. 1949. *Theory of Wing Sections*. New York: Dover

Cebeci T, Bradshaw P. 1979. *Momentum Transfer in Boundary Layers*. New York: McGraw-Hill

Cebeci T, Smith AMO. 1974. *Analysis of Turbulent Boundary Layers*. Orlando: Academic Press

DeJarnette FR, Ratcliffe RA. 1996. Matching inviscid/boundary layer flowfields. *AIAA J.* 34:35–42

Dommasch DO, Sherby SS, Connolly TF. 1967. *Airplane Aerodynamics*. 4th Ed. New York: Pitman

Dorrance WH. 1962. *Viscous Hypersonic Flow*. New York: McGraw-Hill

Head MR. 1969. Cambridge work on entrainment. In *Proceedings, Computation of Turbulent Boundary Layers — 1968 AFOSR-IFP-Stanford Conference*, Vol. 1. Stanford: Stanford University Press

Neumann RD. 1989. Defining the aerothermodynamic environment. In *Hypersonics, Vol I: Defining the Hypersonic Environment*, Ed. Bertin JJ, Glowinski R, Periaux J. Boston: Birkhauser Boston

Schlichting H. 1979. *Boundary Layer Theory*. 7th Ed. New York: McGraw-Hill

Smith BR. 1991. Application *of turbulence modeling to the design of military aircraft*. Presented at AIAA Aerosp. Sci. Meet., 29th, AIAA Pap. 91-0513, Reno, NV

Spalart PR, Allmaras SR. 1992. *A one-equation turbulence model for aerodynamic flows*. Presented at AIAA Aerosp. Sci. Meet., 30th, AIAA Pap. 92-0439, Reno, NV

Spalart PR. 2000. *Trends in turbulence treatments*. Presented at Fluids 2000 Conf., AIAA Pap. 2000-2306, Denver, CO

White FM. 2005. *Viscous Fluid Flow*. 3rd Ed. New York: McGraw-Hill

Wilcox DC. 1998. *Turbulence Modeling for CFD*. 2nd Ed. LaCañada, CA: DCW Industries

5 CHARACTERISTIC PARAMETERS FOR AIRFOIL AND WING AERODYNAMICS

5.1 CHARACTERIZATION OF AERODYNAMIC FORCES AND MOMENTS

5.1.1 General Comments

The motion of air around the vehicle produces pressure and velocity variations through the flow field. Although viscosity is a fluid property and, therefore, acts throughout the flow field, the viscous forces acting on the vehicle depend on the velocity gradients near the surface as well as the viscosity itself. The normal (pressure) forces and the tangential (shear) forces, which act on the surface due to the motion of air around the vehicle, are shown in Fig. 5.1. The pressures and the shear forces can be integrated over the surface

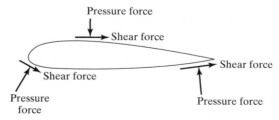

Figure 5.1 Normal (or pressure) and tangential (or shear) forces on an airfoil surface.

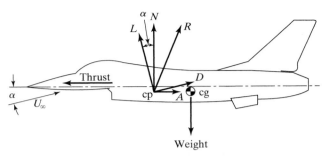

Figure 5.2 Nomenclature for aerodynamic forces in the pitch plane.

on which they act in order to yield the resultant aerodynamic force (R), which acts at the center of pressure (cp) of the vehicle.

For convenience, the total force vector is usually resolved into components. Body-oriented force components are used when the application is primarily concerned with the vehicle response (e.g., the aerodynamics or the structural dynamics). Let us first consider the forces and the moments in the plane of symmetry (i.e., the pitch plane). For the pitch-plane forces depicted in Fig. 5.2, the body-oriented components are the axial force, which is the force parallel to the vehicle axis (A), and the normal force, which is the force perpendicular to the vehicle axis (N).

For applications such as trajectory analysis, the resultant force is divided into components taken relative to the velocity vector (i.e., the flight path). Thus, for these applications, the resultant force is divided into a component parallel to the flight path, the drag (D), and a component perpendicular to the flight path, the lift (L), as shown in Fig. 5.2. As the airplane moves through the earth's atmosphere, its motion is determined by its weight, the thrust produced by the engine, and the aerodynamic forces acting on the vehicle. Consider the case of steady, unaccelerated level flight in a horizontal plane. This condition requires (1) that the sum of the forces along the flight path is zero and (2) that the sum of the forces perpendicular to the flight path is zero. Let us consider only cases where the angles are small (e.g., the component of the thrust parallel to the free-stream velocity vector is only slightly less than the thrust itself). Summing the forces along the flight path (parallel to the free-stream velocity), the equilibrium condition requires that the thrust must equal the drag acting on the airplane. Summing the forces perpendicular to the flight path leads to the conclusion that the weight of the aircraft is balanced by the lift.

Consider the case where the lift generated by the wing/body configuration acts ahead of the center of gravity, as shown in Fig. 5.3. Because the lift force generated by the wing/body configuration acts ahead of the center of gravity, it will produce a nose-up (positive) pitching moment about the center of gravity (cg). The aircraft is said to be trimmed, when the sum of the moments about the cg is zero,

$$\Sigma M_{cg} = 0$$

Thus, a force from a control surface located aft of the cg (e.g., a tail surface) is needed to produce a nose-down (negative) pitching moment about the cg, which could balance the

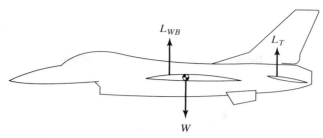

Figure 5.3 Moment balance to trim an aircraft.

moment produced by the wing/body lift. The tail-generated lift force is indicated in Fig. 5.3. The orientation of the tail surface which produces the lift force depicted in Fig. 5.3 also produces a drag force, which is known as the trim drag. Typically, the trim drag may vary from 0.5 to 5% of the total cruise drag for the airplane. The reader should note that the trim drag is associated with the lift generated to trim the vehicle, but does not include the tail profile drag (which is included in the total drag of the aircraft at zero lift conditions).

In addition to the force components which act in the pitch plane (i.e., the lift, which acts upward perpendicular to the undisturbed free-stream velocity, and the drag, which acts in the same direction as the free-stream velocity) there is a side force. The side force is the component of force in a direction perpendicular both to the lift and to the drag. The side force is positive when acting toward the starboard wing (i.e., the pilot's right).

As noted earlier, the resulting aerodynamic force usually will not act through the origin of the airplane's axis system (i.e., the center of gravity). The moment due to the resultant force acting at a distance from the origin may be divided into three components, referred to the airplane's reference axes. The three moment components are the pitching moment, the rolling moment, and the yawing moment, as shown in Fig. 5.4.

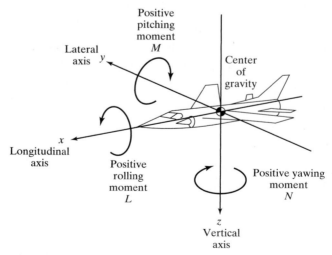

Figure 5.4 Reference axes of the airplane and the corresponding aerodynamic moments.

1. *Pitching moment.* The moment about the lateral axis (the y axis of the airplane-fixed coordinate system) is the pitching moment. The pitching moment is the result of the lift and the drag forces acting on the vehicle. A positive pitching moment is in the nose-up direction.

2. *Rolling moment.* The moment about the longitudinal axis of the airplane (the x axis) is the rolling moment. A rolling moment is often created by a differential lift, generated by some type of ailerons or spoilers. A positive rolling moment causes the right, or starboard, wingtip to move downward.

3. *Yawing moment.* The moment about the vertical axis of the airplane (the z axis) is the yawing moment. A positive yawing moment tends to rotate the nose to the pilot's right.

5.1.2 Parameters That Govern Aerodynamic Forces

The magnitude of the forces and of the moments that act on a vehicle depend on the combined effects of many different variables. Personal observations of the aerodynamic forces acting on an arm extended from a car window or on a ball in flight demonstrate the effect of velocity and of configuration. Pilot manuals advise that a longer length of runway is required if the ambient temperature is relatively high or if the airport elevation is high (i.e., the ambient density is relatively low). The parameters that govern the magnitude of aerodynamic forces and moments include the following:

1. Configuration geometry
2. Angle of attack (i.e., vehicle attitude in the pitch plane relative to the flight direction)
3. Vehicle size or model scale
4. Free-stream velocity
5. Density of the undisturbed air
6. Reynolds number (as it relates to viscous effects)
7. Mach number (as it relates to compressibility effects)

The calculation of the aerodynamic forces and moments acting on a vehicle often requires that the engineer be able to relate data obtained at other flow conditions to the conditions of interest. Thus, the engineer often uses data from the wind tunnel, where scale models are exposed to flow conditions that simulate the design environment or data from flight tests at other flow conditions. So that one can correlate the data for various free-stream conditions and configuration scales, the measurements are usually presented in dimensionless form. Ideally, once in dimensionless form, the results would be independent of all but the first two parameters listed, configuration geometry and angle of attack. In practice, flow phenomena, such as boundary-layer separation, shock-wave/boundary-layer interactions, and compressibility effects, limit the range of flow conditions over which the dimensionless force and moment coefficients remain constant. For such cases, parameters such as the Reynolds number and the Mach number appear in the correlations for the force coefficient and for the moment coefficients.

5.2 AIRFOIL GEOMETRY PARAMETERS

If a horizontal wing is cut by a vertical plane parallel to the centerline of the vehicle, the resultant section is called the *airfoil section*. The generated lift and the stall characteristics of the wing depend strongly on the geometry of the airfoil sections that make up the wing. Geometric parameters that have an important effect on the aerodynamic characteristics of an airfoil section include (1) the leading-edge radius, (2) the mean camber line, (3) the maximum thickness and the thickness distribution of the profile, and (4) the trailing-edge angle. The effect of these parameters, which are illustrated in Fig. 5.5, will be discussed after a brief introduction to airfoil-section nomenclature.

5.2.1 Airfoil-Section Nomenclature

Quoting from Abbott and von Doenhoff (1949), "The gradual development of wing theory tended to isolate the wing-section problems from the effects of planform and led to a more systematic experimental approach. The tests made at Göttingen during World War I contributed much to the development of modern types of wing sections. Up to about World War II, most wing sections in common use were derived from more or less direct extensions of the work at Göttingen. During this period, many families of wing sections were tested in the laboratories of various countries, but the work of the NACA was outstanding. The NACA investigations were further systematized by separation of the effects of camber and thickness distribution, and the experimental work was performed at higher Reynolds number than were generally obtained elsewhere."

As a result, the geometry of many airfoil sections is uniquely defined by the NACA designation for the airfoil. There are a variety of classifications, including NACA four-digit wing sections, NACA five-digit wing sections, and NACA 6 series wing sections. As an example, consider the NACA four-digit wing sections. The first integer indicates the

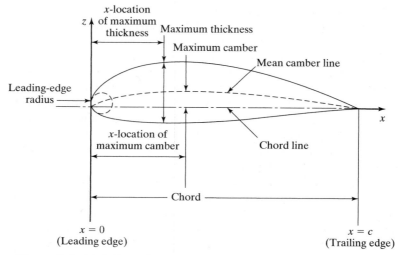

Figure 5.5 Airfoil-section geometry and its nomenclature.

maximum value of the mean camber-line ordinate (see Fig. 5.5) in percent of the chord. The second integer indicates the distance from the leading edge to the maximum camber in tenths of the chord. The last two integers indicate the maximum section thickness in percent of the chord. Thus, the NACA 0010 is a symmetric airfoil section whose maximum thickness is 10% of the chord. The NACA 4412 airfoil section is a 12% thick airfoil which has a 4% maximum camber located at 40% of the chord.

A series of "standard" modifications are designated by a suffix consisting of a dash followed by two digits. These modifications consist essentially of (1) changes of the leading-edge radius from the normal value and (2) changes of the position of maximum thickness from the normal position (which is at $0.3c$). Thus,

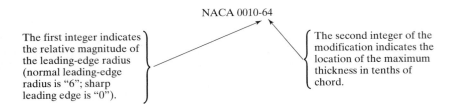

NACA 0010-64

The first integer indicates the relative magnitude of the leading-edge radius (normal leading-edge radius is "6"; sharp leading edge is "0").

The second integer of the modification indicates the location of the maximum thickness in tenths of chord.

However, because of the rapid improvements, both in computer hardware and computer software, and because of the broad use of sophisticated numerical codes, one often encounters airfoil sections being developed that are not described by the standard NACA geometries.

5.2.2 Leading-Edge Radius and Chord Line

The *chord line* is defined as the straight line connecting the leading and trailing edges. The leading edge of airfoils used in subsonic applications is rounded, with a radius that is on the order of 1% of the chord length. The leading-edge radius of the airfoil section is the radius of a circle centered on a line tangent to the leading-edge camber connecting tangency points of the upper and the lower surfaces with the leading edge. The center of the leading-edge radius is located such that the cambered section projects slightly forward of the leading-edge point. The magnitude of the leading-edge radius has a significant effect on the stall (or boundary-layer separation) characteristics of the airfoil section.

The geometric angle of attack is the angle between the chord line and the direction of the undisturbed, "free-stream" flow. For many airplanes the chord lines of the airfoil sections are inclined relative to the vehicle axis.

5.2.3 Mean Camber Line

The locus of the points midway between the upper surface and the lower surface, as measured perpendicular to the chord line, defines the *mean camber line*. The shape of the mean camber line is very important in determining the aerodynamic characteristics of an airfoil section. As will be seen in the theoretical solutions and in the experimental data that will be presented in this book, cambered airfoils in a subsonic flow generate lift even when the section angle of attack is zero. Thus, an effect of camber is a change in the zero-lift angle of attack, α_{0l}. While the symmetric sections have zero lift at zero angle of attack, zero lift results for sections with positive camber when they are at negative angles of attack.

Furthermore, camber has a beneficial effect on the maximum value of the section lift coefficient. If the maximum lift coefficient is high, the stall speed will be low, all other factors being the same. It should be noted, however, that the high thickness and camber necessary for high maximum values for the section lift coefficient produce low critical Mach numbers (see Chapter 9) and high twisting moments at high speeds. Thus, one needs to consider the trade-offs in selecting a design value for a particular parameter.

5.2.4 Maximum Thickness and Thickness Distribution

The maximum thickness and the thickness distribution strongly influence the aerodynamic characteristics of the airfoil section as well. The maximum local velocity to which a fluid particle accelerates as it flows around an airfoil section increases as the maximum thickness increases (see the discussion associated with Fig. 4.10). Thus, the minimum pressure value is smallest for the thickest airfoil. As a result, the adverse pressure gradient associated with the deceleration of the flow from the location of this pressure minimum to the trailing edge is greatest for the thickest airfoil. As the adverse pressure gradient becomes larger, the boundary layer becomes thicker (and is more likely to separate producing relatively large values for the form drag). Thus, the beneficial effects of increasing the maximum thickness are limited.

Consider the maximum section lift coefficients for several different thickness-ratio airfoils presented in this table. The values are taken from the figures presented in Abbott and von Doenhoff (1949).

Airfoil Section	$C_{l, max}$
NACA 2408	1.5
NACA 2410	1.65
NACA 2412	1.7
NACA 2415	1.63
NACA 2418	1.48
NACA 2424	1.3

For a very thin airfoil section (which has a relatively small leading-edge radius), boundary-layer separation occurs early, not far from the leading edge of the upper (leeward) surface. As a result, the maximum section lift coefficient for a very thin airfoil section is relatively small. The maximum section lift coefficient increases as the thickness ratio increases from 8% of the chord to 12% of the chord. The separation phenomena described in the previous paragraph causes the maximum section-lift coefficients for the relatively thick airfoil sections (i.e., those with a thickness ratio of 18% of the chord and of 24% of the chord) to be less than those for medium thickness airfoil sections.

The thickness distribution for an airfoil affects the pressure distribution and the character of the boundary layer. As the location of the maximum thickness moves aft, the velocity gradient (and hence the pressure gradient) in the midchord region decreases. The resultant favorable pressure gradient in the midchord region promotes boundary-layer stability and increases the possibility that the boundary layer remains laminar. Laminar boundary layers produce less skin friction drag than turbulent boundary layers but are also more likely to separate under the influence of an adverse pressure gradient. This will be discussed in more detail later in this chapter.

In addition, the thicker airfoils benefit more from the use of high lift devices but have a lower critical Mach number.

5.2.5 Trailing-Edge Angle

The trailing-edge angle affects the location of the aerodynamic center (which is defined later in this chapter). The aerodynamic center of thin airfoil sections in a subsonic stream is theoretically located at the quarter-chord.

5.3 WING-GEOMETRY PARAMETERS

By placing the airfoil sections discussed in the preceding section in spanwise combinations, wings, horizontal tails, vertical tails, canards, and/or other lifting surfaces are formed. When the parameters that characterize the wing planform are introduced, attention must be directed to the existence of flow components in the spanwise direction. In other words, airfoil section properties deal with flow in two dimensions while planform properties relate to the resultant flow in three dimensions.

In order to fully describe the planform of a wing, several terms are required. The terms that are pertinent to defining the aerodynamic characteristics of a wing are illustrated in Fig. 5.6.

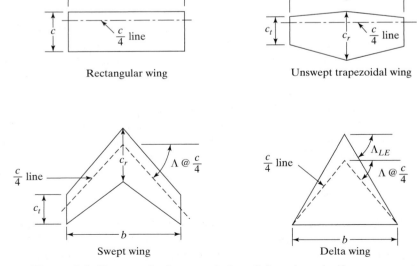

Figure 5.6 Geometric characteristics of the wing planform.

1. The *wing area, S*, is simply the plan surface area of the wing. Although a portion of the area may be covered by fuselage or nacelles, the pressure carryover on these surfaces allows legitimate consideration of the entire plan area.

2. The *wing span, b*, is measured tip to tip.

3. The *average chord, c̄*, is determined from the equation that the product of the span and the average chord is the wing area ($b \times \bar{c} = S$).

4. The *aspect ratio*, AR, is the ratio of the span and the average chord. For a rectangular wing, the aspect ratio is simply

$$AR = \frac{b}{c}$$

For a nonrectangular wing,

$$AR = \frac{b^2}{S}$$

The aspect ratio is a fineness ratio of the wing and is useful in determining the aerodynamic characteristics and structural weight. Typical aspect ratios vary from 35 for a high-performance sailplane to 2 for a supersonic jet fighter.

5. The *root chord, c_r*, is the chord at the wing centerline, and the *tip chord, c_t*, is measured at the tip.

6. Considering the wing planform to have straight lines for the leading and trailing edges, the *taper ratio*, λ, is the ratio of the tip chord to the root chord:

$$\lambda = \frac{c_t}{c_r}$$

The taper ratio affects the lift distribution and the structural weight of the wing. A rectangular wing has a taper ratio of 1.0 while the pointed tip delta wing has a taper ratio of 0.0.

7. The *sweep angle* Λ is usually measured as the angle between the line of 25% chord and a perpendicular to the root chord. Sweep angles of the leading edge or of the trailing edge are often presented with the parameters, since they are of interest for many applications. The sweep of a wing causes definite changes in the maximum lift, in the stall characteristics, and in the effects of compressibility.

8. The *mean aerodynamic chord* (mac) is used together with S to nondimensionalize the pitching moments. Thus, the mean aerodynamic chord represents an average chord which, when multiplied by the product of the average section moment coefficient, the dynamic pressure, and the wing area, gives the moment for the entire wing. The mean aerodynamic chord is given by

$$mac = \frac{1}{S} \int_{-0.5b}^{+0.5b} [c(y)]^2 \, dy$$

9. The *dihedral angle* is the angle between a horizontal plane containing the root chord and a plane midway between the upper and lower surfaces of the wing. If the wing lies below the horizontal plane, it is termed an *anhedral angle*. The dihedral angle affects the lateral stability characteristics of the airplane.

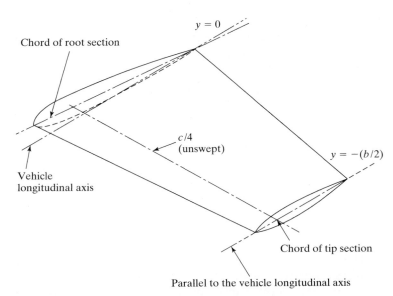

Figure 5.7 Unswept, tapered wing with geometric twist (wash out).

10. *Geometric twist* defines the situation where the chord lines for the spanwise distribution of airfoil sections do not all lie in the same plane. Thus, there is a spanwise variation in the geometric angle of incidence for the sections. The chord of the root section of the wing shown in the sketch of Fig. 5.7 is inclined 4° relative to the vehicle axis. The chord of the tip section, however, is parallel to the longitudinal axis of the vehicle. In this case, where the incidence of the airfoil sections relative to the vehicle axis decrease toward the tip, the wing has "wash out." The wings of numerous subsonic aircraft have wash out to control the spanwise lift distribution and, hence, the boundary-layer separation (i.e., stall) characteristics. If the angle of incidence increases toward the tip, the wing has "wash in."

    The airfoil section distribution, the aspect ratio, the taper ratio, the twist, and the sweep back of a planform are the principal factors that determine the aerodynamic characteristics of a wing and have an important bearing on its stall properties. These same quantities also have a definite influence on the structural weight and stiffness of a wing. Values of these parameters for a variety of aircraft have been taken from *Jane's All the World's Aircraft* (1973, 1966, and 1984). and are summarized in Table 5.1. Data are presented for four-place single-engine aircraft, commercial jetliners and transports, and high-speed military aircraft. Note how the values of these parameters vary from one group of aircraft to another.

    The data presented in Table 5.1 indicate that similar designs result for similar applications, regardless of the national origin of the specific design. For example, the aspect ratio for the four-seat, single-engine civil aviation designs is approximately 7, whereas the aspect ratio for the supersonic military aircraft is between 1.7 and 3.8. In fact, as noted in Stuart (1978), which was a case study in aircraft design for the Northrop F-5, "the selection of wing aspect ratio represents an interplay between a large value for low drag-due-to-lift and a small value for reduced wing weight."

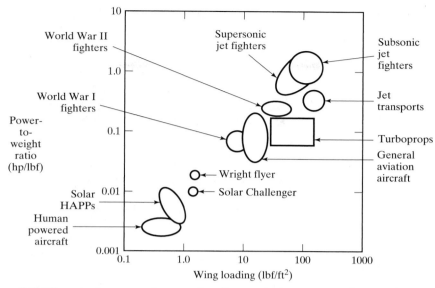

Figure 5.8 Historical ranges of power loading and wing loading. [From Hall (1985).]

There are other trends exhibited in parameters relating to aircraft performance. Notice the grouping by generic classes of aircraft for the correlation between the power-to-weight ratio and the wing loading that is presented in Fig. 5.8. There is a tendency for airplanes to get larger, heavier, and more complex with technological improvements. The trend toward larger, heavier aircraft is evident in this figure. Note that a fully loaded B-17G Flying Fortress, a heavy bomber of World War II, weighed 29,700 kg (65,500 lb) with a wing span of 31.62 m (103.75 ft), whereas the F15, a modern fighter aircraft, has a maximum takeoff weight of 30,845 kg (68,000 lb) with a wing span of 13.05 m (42.81 ft). However, the successful human-powered aircraft fall in the lower left corner of Fig. 5.8, with wing loadings (the ratio of takeoff gross weight to wing area) less than 1 lbf/ft². It is in this same region that Lockheed's solar high-altitude powered platform (Solar HAPP) operates.

EXAMPLE 5.1: Develop an expression for the aspect ratio

Develop an expression for the aspect ratio of a delta wing in terms of the leading edge sweep angle (Λ_{LE}).

Solution: Referring to the sketch of the delta wing in Fig. 5.6, the wing area is

$$S = \frac{bc_r}{2}$$

and the tan Λ_{LE} is given by

$$\tan \Lambda_{LE} = \frac{c_r}{(b/2)}$$

TABLE 5.1 Wing-Geometry Parameters

Type	Wing span [m(ft)]	Aspect ratio, AR	Sweep angle	Dihedral	Airfoil section	Speed [km/h (mi/h)]
a. Four-Place Single-Engine Aircraft						
Socata Rallye (France)	9.61 (31.52)	7.57	None	7°	63A414(mod), 63A416, inc. 4°	173–245 (108–152)
Ambrosini NF 15 (Italy)	9.90 (32.5)	7.37	None	6°	64–215, inc. 4°	325 (202)
Beechcraft Bonanza V35B	10.20 (33.46)	6.30	None	6°	23016.5 at root, 23012 at tip, inc. 4° at root, 1° at tip	298–322 (185–200)
Beechcraft Sierra	9.98 (32.75)	7.35	None	6°30′	63_2A415, inc. 3° at root, 1° at tip	211–281 (131–162)
Cessna 172	10.92 (35.83)	7.32	None	1°44′	NACA 2412, inc. 1°30′ at root, −1°30′ at tip	211 (131)
Piper Commanche	10.97 (36.0)	7.28	2°30′ forward	5°	64_2A215, inc. 2°	298 (185)
Bellanca, Model 25	10.67 (35.0)	6.70	None	2°	NACA 63_2–215, inc. 2°	458–499 (285–310)
Piper Warrior II	10.67 (35.0)	7.24	None	7°	NACA 65_2–415, inc. 2° at root, −1° at tip	191–235 (119–146)
b. Commercial Jetliners and Transports						
Caravelle 210 (France)	34.3 (112.5)	8.02	20° at c/4	3°	NACA $65_1$212	790 (490)
BAC 111 (UK)	26.97 (88.5)	8.00	20° at c/4	2°	NACA cambered section (mod.), t/c = 0.125 at root, 0.11 at tip, inc. 2°30′	815 (507)
Tupolev 154 (USSR)	37.55 (123.2)	7.03	35° at c/4	—	—	975 (605)
McDonnell-Douglas DC9	27.25 (89.42)	8.25	24° at c/4	—	t/c = 0.116 (av.)	903 (561)
Boeing 727	32.92 (108.0)	7.67	32° at c/4	3°	Special Boeing sections t/c = 0.09 to 0.13, inc. 2°	975 (605)
Boeing 737	28.35 (93.0)	8.83	25° at c/4	6°	t/c = 0.129 (av.)	848 (527)

Aircraft	Span, m (ft)	Aspect ratio	Sweep	Dihedral	Airfoil t/c	Max speed, knots (mph)
Boeing 747	59.64 (195.7)	6.95	37°30' at $c/4$	7°	t/c = 0.134 (inboard), 0.078 (midspan), 0.080 (outboard), inc. 2°	958 (595)
Lockheed C-5A	67.88 (222.8)	7.75	25° at $c/4$	Anhedral 5°30'	NACA 0011 (mod.) near midspan, inc. 3°30'	815 (507)
Airbus A310 (International)	43.89 (144.0)	8.8	28° at $c/4$	11°8' (inboard at the trailing edge)	t/c = 0.152 at root, t/c = 0.108 at tip, inc. 5°3' at root	667–895 (414–556)
Ilyushin IL76 (USSR)	50.50 (165.7)	8.5	25° at $c/4$	Anhedral with wing mounted above the fuselage		750–850 (466–528)
c. High-Speed Military Aircraft						
SAAB-35 Draken (Sweden)	9.40 (30.8)	1.77	Central: leading edge 80°, outer: leading edge 57°	—	t/c = 0.05	Mach 1.4–2.0
Dassault Mirage III (France)	8.22 (27.0)	1.94	Leading edge 60°34'	Anhedral 1°	t/c = 0.045–0.035	Mach 2.2
Northrop F-5E	8.13 (26.67)	3.82	24° at $c/4$	None	65A004.8 (mod.), t/c = 0.048	Mach 1.23
McDonnell-Douglas F4	11.70 (38.4)	2.78	45°	Outer panel 12°	t/c = 0.051 (av.)	Over Mach 2.0
LTV F-8	10.81 (35.7)	3.39	35°	Anhedral 5°	Thin, laminar flow section	Nearly Mach 2 1123 (698)
LTV A-7	11.80 (38.75)	4.0	35° at $c/4$	Anhedral 5°	65A007, inc. −1°	Mach 1.6
Mitsubishi T2 (Japan)	7.88 (25.85)	2.93	35°47' at $c/4$	Anhedral 9°	NACA 65 series (mod.), t/c = 0.0466	Mach 1.6
General Dynamics F-16	9.14 (30.0)	3.0	40° on leading edges	—	NACA 64A-204	Mach 2.0+

Source: Data from Jane's All the World's Aircraft (1973, 1966, and 1984).

Solving this second expression for c_r and substituting it into the expression for the wing area, we obtain

$$S = \frac{b^2}{4} \tan \Lambda_{LE}$$

Substituting this expression for the wing area into the expression for the aspect ratio,

$$AR = \frac{b^2}{S} = \frac{4}{\tan \Lambda_{LE}} \tag{5.1}$$

EXAMPLE 5.2:   Calculate the wing-geometry parameters for the Space Shuttle Orbiter

To calculate the wing-geometry parameters for the Space Shuttle *Orbiter*, the complex shape of the actual wing is replaced by a swept, trapezoidal wing, as shown in Fig. 5.9. For the reference wing of the *Orbiter*, the root chord (c_r), is 57.44 ft, the tip chord (c_t) is 11.48 ft, and the span (b) is 78.056 ft. Using these values which define the reference wing, calculate (a) the wing area (S), (b) the aspect ratio (AR), (c) the taper ratio (λ), and (d) the mean aerodynamic chord (mac).

Solution:

   (a) The area for the trapezoidal reference wing is

$$S = \left(\frac{c_t + c_r}{2} \right) \frac{b}{2} 2 = 2690 \text{ ft}^2$$

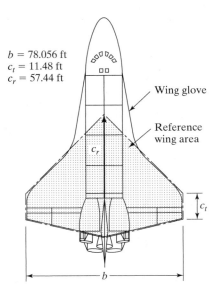

$b = 78.056$ ft
$c_t = 11.48$ ft
$c_r = 57.44$ ft

Wing glove

Reference wing area

c_r

c_t

b

Figure 5.9 Sketch of Space Shuttle *Orbiter* geometry for Example 5.2.

(b) The aspect ratio for this swept, trapezoidal wing is

$$AR = \frac{b^2}{S} = \frac{(78.056)^2}{2690} = 2.265$$

(c) The taper ratio is

$$\lambda = \frac{c_t}{c_r} = \frac{11.48}{57.44} = 0.20$$

(d) To calculate the mean aerodynamic chord, we will first need an expression for the chord as a function of the distance from the plane of symmetry [i.e., $c(y)$]. The required expression is

$$c(y) = c_r + \frac{11.48 - 57.44}{39.028} y = 57.44 - 1.1776y$$

Substituting this expression for the chord as a function of y into the equation for the mean aerodynamic chord, we obtain

$$mac = \frac{2}{S} \int_0^{0.5b} [c(y)]^2 \, dy = \frac{2}{2690} \int_0^{39.028} (57.44 - 1.1776y)^2 \, dy$$

Integrating this expression, we obtain

$$mac = 39.57 \text{ ft}$$

5.4 AERODYNAMIC FORCE AND MOMENT COEFFICIENTS

5.4.1 Lift Coefficient

Let us develop the equation for the normal force coefficient to illustrate the physical significance of a dimensionless force coefficient. We choose the normal (or z) component of the resultant force since it is relatively simple to calculate and it has the same relation to the pressure and the shear forces as does the lift. For a relatively thin airfoil section at a relatively low angle of attack, it is clear from Fig. 5.1 that the lift (and similarly the normal force) results primarily from the action of the pressure forces. The shear forces act primarily in the chordwise direction (i.e., contribute primarily to the drag). Therefore, to calculate the force in the z direction, we need consider only the pressure contribution, which is illustrated in Fig. 5.10. The pressure force acting on a differential

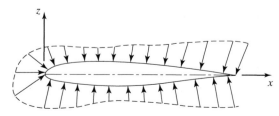

Figure 5.10 Pressure distribution for a lifting airfoil section.

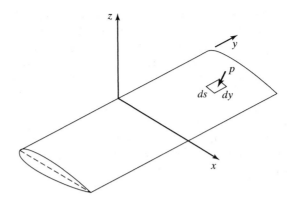

Figure 5.11 Pressure acting on an elemental surface area.

area of the vehicle surface is $dF = p\,ds\,dy$, as shown in Fig. 5.11. The elemental surface area is the product of ds, the wetted length of the element in the plane of the cross section, times dy, the element's length in the direction perpendicular to the plane of the cross section (or spanwise direction). Since the pressure force acts normal to the surface, the force component in the z direction is the product of the pressure times the projected planform area:

$$dF_z = p\,dx\,dy \tag{5.2}$$

Integrating over the entire wing surface (including the upper and the lower surfaces), the net force in the z direction is given by

$$F_z = \oiint p\,dx\,dy \tag{5.3}$$

Note that the resultant force in any direction due to a constant pressure over a closed surface is zero. Thus, the force in the z direction due to a uniform pressure p_∞ acting over the entire wing area is zero.

$$\oiint p_\infty\,dx\,dy = 0 \tag{5.4}$$

Combining equations (5.3) and (5.4), the resultant force component is

$$F_z = \oiint (p - p_\infty)\,dx\,dy \tag{5.5}$$

To nondimensionalize the factors on the right-hand side of equation (5.5), divide by the product $q_\infty cb$, which has the units of force.

$$\frac{F_z}{q_\infty cb} = \oiint \frac{p - p_\infty}{q_\infty} d\left(\frac{x}{c}\right) d\left(\frac{y}{b}\right)$$

Since the product cb represents the planform area S of the rectangular wing of Fig. 5.11,

$$\frac{F_z}{q_\infty S} = \oiint C_p\, d\left(\frac{x}{c}\right) d\left(\frac{y}{b}\right) \tag{5.6}$$

When the boundary layer is thin, the pressure distribution around the airfoil is essentially that of an inviscid flow. Thus, the pressure distribution is independent of Reynolds number and does not depend on whether the boundary layer is laminar or turbulent. When the boundary layer is thin, the pressure coefficient at a particular location on the surface given by the dimensionless coordinates $(x/c, y/b)$ is independent of vehicle scale and of the flow conditions. Over the range of flow conditions for which the pressure coefficient is a unique function of the dimensionless coordinates $(x/c, y/b)$, the value of the integral in equation (5.6) depends only on the configuration geometry and on the angle of attack. Thus, the resulting dimensionless force parameter, or force coefficient (in this case, the normal force coefficient), is independent of model scale and of flow conditions.

A similar analysis can be used to calculate the lift coefficient, which is defined as

$$C_L = \frac{L}{q_\infty S} \tag{5.7}$$

Data are presented in Fig. 5.12 for a NACA 23012 airfoil; that is, the wind-tunnel model represented a wing of infinite span. The lift acting on a wing of infinite span does not vary in the y direction. For this two-dimensional flow, we are interested in determining the lift acting on a unit width of the wing [i.e., the lift per unit span (l)]. Thus, the lift measurements are presented in terms of the section lift coefficient C_l. The section lift coefficient is the lift per unit span (l) divided by the product of the dynamic pressure times the plan area per unit span, which is the chord length (c).

$$C_l = \frac{l}{q_\infty c} \tag{5.8}$$

The data from Abbott and von Doenhoff (1949) were obtained in a wind tunnel that could be operated at pressures up to 10 atm. As a result, the Reynolds number ranged from 3×10^6 to 9×10^6 at Mach numbers less than 0.17. In addition to the measurements obtained with a smooth model, data are presented for a model that had "standard" surface roughness applied near the leading edge. Additional comments will be made about surface roughness later in this chapter.

The experimental section lift coefficient is independent of Reynolds number and is a linear function of the angle of attack from $-10°$ to $+10°$. The slope of this linear portion of the curve is called the *two-dimensional lift-curve slope*. Using the experimental data for this airfoil,

$$\frac{dC_l}{d\alpha} = C_{l,\alpha} = 0.104 \text{ per degree}$$

Note that equations to be developed in Chapter 6 will show that the theoretical value for the two-dimensional lift-curve slope is 2π per radian (0.1097 per degree).

Since the NACA 23012 airfoil section is cambered (the maximum camber is approximately 2% of the chord length), lift is generated at zero angle of attack. In fact, zero lift is obtained at $-1.2°$, which is designated α_{0l} or the section angle of attack for zero lift. As the angle of attack is increased above $10°$, the section lift coefficient continues to increase (but not linearly with angle of attack) until a maximum value, $C_{l,\max}$, is reached. Referring to Fig. 5.12, $C_{l,\max}$ is 1.79 and occurs at an angle of attack of $18°$. Partly because

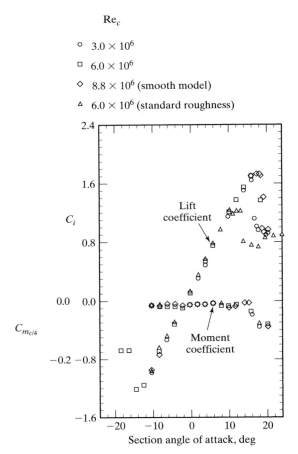

Figure 5.12 Section lift coefficient and section moment coefficient (with respect to $c/4$) for an NACA 23012 airfoil. [Data from Abbott and von Doenhoff (1949).]

of this relatively high value of $C_{l,max}$, the NACA 23012 section has been used on many aircraft (e.g., the Beechcraft Bonanza; see Table 5.1, and the Brewster Buffalo).

At angles of attack in excess of 10°, the section lift coefficients exhibit a Reynolds-number dependence. Note that the adverse pressure gradient which the air particles encounter as they move toward the trailing edge of the upper surface increases as the angle of attack increases. At these angles of attack, the air particles which have been slowed by the viscous forces cannot overcome the relatively large adverse pressure gradient, and the boundary layer separates. The separation location depends on the character (laminar or turbulent) of the boundary layer and its thickness, and therefore on the Reynolds number. As will be discussed, boundary-layer separation has a profound effect on the drag acting on the airfoil.

The study of airfoil lift as a function of incidence has shown that, in many instances, the presence of a separation bubble near the leading edge of the airfoil results in laminar section stall. Experimental data on two-dimensional "peaky" airfoil sections indicate that $C_{l,\max}$, as limited by laminar stall, is strongly dependent on the leading-edge shape and on the Reynolds number. If the laminar boundary layer, which develops near the leading edge of the upper surface of the airfoil, is subjected to a relatively high adverse pressure gradient, it will separate because the relatively low kinetic energy level of the air particles near the wall is insufficient to surmount the "pressure hill" of the adverse pressure gradient. The separated shear layer that is formed may curve back onto the surface within a very short distance. This is known as *short bubble separation*. The separated viscous layer near the pressure peak may not reattach to the surface at all, or it may reattach within 0.3 chord length or more downstream. This extended separation region is known as *long bubble separation*.

The separation bubble may not appear at all on relatively thick, strongly cambered profiles operating at high Reynolds numbers. The reason for this is that the Reynolds number is then large enough for a natural transition to a turbulent boundary layer to occur upstream of the strong pressure rise. The relatively high kinetic energy of the air particles in a turbulent boundary layer permits them to climb the "pressure hill," and boundary-layer separation occurs only a short distance upstream of the trailing edge (trailing-edge stall). The separation point moves upstream continuously with increasing angle of attack, and the lift does not drop abruptly after $C_{l,\max}$ but decreases gradually.

EXAMPLE 5.3: Calculate the lift per unit span on a NACA 23012 airfoil section

Consider tests of an unswept wing that spans the wind tunnel and whose airfoil section is NACA 23012. Since the wing model spans the test section, we will assume that the flow is two dimensional. The chord of the model is 1.3 m. The test section conditions simulate a density altitude of 3 km. The velocity in the test section is 360 km/h.

What is the lift per unit span (in N/m) that you would expect to measure when the angle of attack is 4°? What would be the corresponding section lift coefficient?

Solution: First, we need to calculate the section lift coefficient. We will assume that the lift is a linear function of the angle of attack and that it is independent of the Reynolds number (i.e., the viscous effects are negligible) at these test conditions. That these are reasonable assumptions can be seen by referring to Fig. 5.12. Thus, the section lift coefficient is

$$C_l = C_{l,\alpha}(\alpha - \alpha_{0l}) \qquad \textbf{(5.9)}$$

Using the values presented in the discussion associated with Fig. 5.12,

$$C_l = 0.104(4.0 - (-1.2)) = 0.541$$

At an angle of attack of 4°, the experimental values of the section lift coefficient for an NACA 23012 airfoil section range from 0.50 to 0.57.

To calculate the corresponding lift force per unit span, we rearrange equation (5.8) to obtain

$$l = C_l q_\infty c$$

To calculate the dynamic pressure (q_∞), we need the velocity in m/s and the density in kg/m^3.

$$U_\infty = 360 \frac{\text{km}}{\text{h}} \frac{1000\,\text{m}}{\text{km}} \frac{\text{h}}{3600\,\text{s}} = 100 \frac{\text{m}}{\text{s}}$$

Given the density altitude is 3 km, we refer to Table 1.2 to find that

$$\frac{\rho}{\rho_{\text{SL}}} = 0.74225$$

(*Note:* The fact that we are given the density altitude as 3 km does not provide specific information either about the temperature or about the pressure.)

$$\rho = (0.74225)1.2250 \frac{\text{kg}}{\text{m}^3} = 0.9093 \frac{\text{kg}}{\text{m}^3}$$

Thus, the lift per unit span is

$$l = (0.541)\left[0.5\left(0.9093 \frac{\text{kg}}{\text{m}^3} \right)\left(100 \frac{\text{m}}{\text{s}} \right)^2 \right](1.3\,\text{m})$$

$$l = (0.541)\left[4546.5 \frac{\text{N}}{\text{m}^2} \right](1.3\,\text{m}) = 3197.6 \frac{\text{N}}{\text{m}}$$

5.4.2 Moment Coefficient

The moment created by the aerodynamic forces acting on a wing (or airfoil) is determined about a particular reference axis. The reference axis may be the leading edge, the quarter-chord location, the aerodynamic center, and so on. The significance of these reference axes in relation to the coefficients for thin airfoils will be discussed in subsequent chapters.

The procedure used to nondimensionalize the moments created by the aerodynamic forces is similar to that used to nondimensionalize the lift. To demonstrate this nondimensionalization, the pitching moment about the leading edge due to the pressures acting on the surface will be calculated (refer again to Fig. 5.11). The contribution of the chordwise component of the pressure force and of the skin friction force to the pitching moment is small and is neglected. Thus, the pitching moment about the leading edge due to the pressure force acting on the surface element whose area is ds times dy and which is located at a distance x from the leading edge is

$$dM_0 = p\,dx\,dy\,x \tag{5.10}$$

where $dx\,dy$ is the projected area. Integrating over the entire wing surface, the net pitching moment is given by

$$M_0 = \oiint px\,dx\,dy \tag{5.11}$$

When a uniform pressure acts on any closed surface, the resultant pitching moment due to this constant pressure is zero. Thus,

$$\oiint p_\infty x \, dx \, dy = 0 \tag{5.12}$$

Combining equations (5.11) and (5.12), the resulting pitching moment about the leading edge is

$$M_0 = \oiint (p - p_\infty) x \, dx \, dy \tag{5.13}$$

To nondimensionalize the factors on the right-hand side of equation (5.13), divide by $q_\infty c^2 b$, which has the units of force times length:

$$\frac{M_0}{q_\infty c^2 b} = \oiint \frac{p - p_\infty}{q_\infty} \frac{x}{c} d\left(\frac{x}{c}\right) d\left(\frac{y}{b}\right)$$

Since the product of cb represents the planform area of the wing S,

$$\frac{M_0}{q_\infty S c} = \oiint C_p \frac{x}{c} d\left(\frac{x}{c}\right) d\left(\frac{y}{b}\right) \tag{5.14}$$

Thus, the dimensionless moment coefficient is

$$C_{M_0} = \frac{M_0}{q_\infty S c} \tag{5.15}$$

Since the derivation of equation (5.15) was for the rectangular wing of Fig. 5.11, the chord c is used. However, as noted previously in this chapter, the mean aerodynamic chord is used together with S to nondimensionalize the pitching moment for a general wing.

The section moment coefficient is used to represent the dimensionless moment per unit span (m_0):

$$C_{m_0} = \frac{m_0}{q_\infty c c} \tag{5.16}$$

since the surface area per unit span is the chord length c. The section pitching moment coefficient depends on the camber and on the thickness ratio. Section pitching moment coefficients for a NACA 23012 airfoil section with respect to the quarter chord and with respect to the aerodynamic center are presented in Figs. 5.12 and 5.13, respectively. The *aerodynamic center* is that point about which the section moment coefficient is independent of the angle of attack. Thus, the aerodynamic center is that point along the chord where all changes in lift effectively take place. Since the moment about the aerodynamic center is the product of a force (the lift that acts at the center of pressure) and a lever arm (the distance from the aerodynamic center to the center of pressure), the center of pressure must move toward the aerodynamic center as the lift increases. The quarter-chord location is significant, since it is the theoretical aerodynamic center for incompressible flow about a two-dimensional airfoil.

Note that the pitching moment coefficient is independent of the Reynolds number (for those angles of attack where the lift coefficient is independent of the Reynolds number), since the pressure coefficient depends only on the dimensionless space coordinates $(x/c, y/b)$ [see equation (5.14)]. One of the features of the NACA 23012 section is a relatively high $C_{l,\max}$ with only a small $C_{m_{ac}}$.

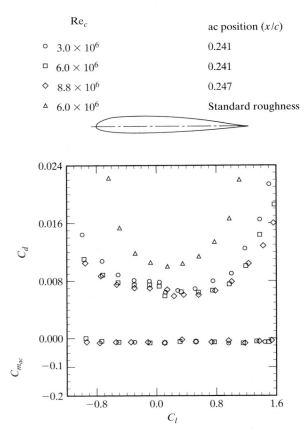

	Re_c	ac position (x/c)
○	3.0×10^6	0.241
□	6.0×10^6	0.241
◇	8.8×10^6	0.247
△	6.0×10^6	Standard roughness

Figure 5.13 Section drag coefficient and section moment coefficient (with respect to the ac) for a NACA 23012 airfoil. [Data from Abbott and von Doenhoff (1949).]

The characteristic length (or moment arm) for the rolling moment and for the yawing moment is the wing span b (instead of the chord). Therefore, the rolling moment coefficient is

$$C_{\mathscr{L}} = \frac{\mathscr{L}}{q_\infty S b} \tag{5.17}$$

and the yawing moment coefficient is

$$C_N = \frac{N}{q_\infty S b} \tag{5.18}$$

5.4.3 Drag Coefficient

The drag force on a wing is due in part to skin friction and in part to the integrated effect of pressure. If $\vec{t}$ denotes the tangential shear stress at a point on the body surface,

p the static pressure, and $\hat{n}$ the external normal to the element of surface dS, the drag can be formally expressed as

$$D = \oiint \vec{\tau} \cdot \hat{e}_\infty dS - \oiint p\hat{n} \cdot \hat{e}_\infty dS \tag{5.19}$$

where $\hat{e}_\infty$ is a unit vector parallel to the free stream and the integration takes place over the entire wetted surface. The first integral represents the friction component and the second integral represents the pressure drag.

The most straightforward approach to calculating the pressure drag is to perform the numerical integration indicated by the second term in equation (5.19). This approach is known as the near-field method of drag computation. Unfortunately, this can be a relatively inaccurate procedure for streamlined configurations at small angles of attack. The inaccuracy results because the pressure drag integral is the difference between the integration on forward-facing and rearward-facing surface elements, this difference being a second-order quantity for slender bodies. Furthermore, the reader should realize that subtle differences between the computed pressure distribution and the actual pressure distribution can have a significant effect on the validity of the drag estimates, depending upon where the differences occur. If the pressure difference is near the middle of the aerodynamic configuration, where the local slope is roughly parallel to the free-stream direction, it will have a relatively small effect on the validity of the estimated drag. However, if the pressure difference is near the aft end of the configuration (for instance, on a nozzle boattail), even a small difference between the computed pressure and the actual pressure can have a significant effect on the accuracy of the predicted drag.

In Chapter 3 we learned that zero drag results for irrotational, steady, incompressible flow past a two-dimensional body. For an airfoil section (i.e., a two-dimensional geometry) which is at a relatively low angle of attack so that the boundary layer is thin and does not separate, the pressure distribution is essentially that for an inviscid flow. Thus, skin friction is a major component of the chordwise force per unit span (f_x). Referring to Figs. 5.1 and 5.10, sf_x can be approximated as

$$sf_x \approx \oint \tau\, dx \tag{5.20}$$

where sf_x is the chordwise force per unit span due to skin friction. Dividing both sides of equation (5.20) by the product $q_\infty c$ yields an expression for the dimensionless force coefficient:

$$\frac{sf_x}{q_\infty c} \approx \oint C_f d\left(\frac{x}{c}\right) \tag{5.21}$$

C_f, the skin friction coefficient, is defined as

$$C_f = \frac{\tau}{\frac{1}{2}\rho_\infty U_\infty^2} \tag{5.22}$$

As was stated in the general discussion of the boundary-layer characteristics in Chapter 4 (see Fig. 4.18), skin friction for a turbulent boundary layer is much greater than that for a laminar boundary layer for given flow conditions. Equations for calculating the skin friction coefficient were developed in Chapter 4. However, let us introduce the correlations for the skin friction coefficient for incompressible flow past a flat plate to gain insight into the force coefficient of equation (5.21). Of course, the results for a flat plate only approximate those for an airfoil. The potential function given by equation (3.35a) shows that the velocity at the edge of the boundary layer and, therefore, the local static pressure, is constant along the plate. Such is not the case for an airfoil section, for which the flow accelerates from a forward stagnation point to a maximum velocity, then decelerates to the trailing edge. Nevertheless, the analysis will provide useful insights into the section drag coefficient,

$$C_d = \frac{d}{q_\infty c} \tag{5.23}$$

for an airfoil at relatively low angles of attack.

Referring to Chapter 4, when the boundary layer is laminar,

$$C_f = \frac{0.664}{(\text{Re}_x)^{0.5}} \tag{5.24}$$

If the boundary layer is turbulent,

$$C_f = \frac{0.0583}{(\text{Re}_x)^{0.2}} \tag{5.25}$$

For equations (5.24) and (5.25), the local Reynolds number is defined as

$$\text{Re}_x = \frac{\rho_\infty U_\infty x}{\mu_\infty} \tag{5.26}$$

Also, as was shown in Chapter 4, total skin-friction coefficients can be defined. The total skin-friction coefficient for laminar flow is given by

$$\overline{C}_f = \frac{1.328}{\sqrt{\text{Re}_L}} \tag{5.27}$$

and the total skin-friction coefficient for turbulent flow is

$$\overline{C}_f = \frac{0.074}{(\text{Re}_L)^{0.2}} \tag{5.28}$$

which is the Prandtl formulation; although the Prandtl-Schlichting formulation is more accurate:

$$\overline{C}_f \equiv \frac{0.455}{(\log_{10}\text{Re}_L)^{2.58}} \tag{5.29}$$

These total skin-friction coefficients use the length-based Reynolds number given by

$$\text{Re}_L = \frac{\rho_\infty U_\infty L}{\mu_\infty} \tag{5.30}$$

where L is the length of a flat plate.

EXAMPLE 5.4:   Calculate the local skin friction

Calculate the local skin friction at a point 0.5 m from the leading edge of a flat-plate airfoil flying at 60 m/s at a height of 6 km.

Solution: Refer to Table 1.2 to obtain the static properties of undisturbed air at 6 km:

$$\rho_\infty = 0.6601 \text{ kg/m}^3$$

$$\mu_\infty = 1.5949 \times 10^{-5} \text{ kg/s} \cdot \text{m}$$

Thus, using equation (5.26),

$$\text{Re}_x = \frac{(0.6601 \text{ kg/m}^3)(60 \text{ m/s})(0.5 \text{ m})}{1.5949 \times 10^{-5} \text{ kg/s} \cdot \text{m}}$$

$$= 1.242 \times 10^6$$

If the boundary layer is laminar,

$$C_f = \frac{0.664}{(\text{Re}_x)^{0.5}} = 5.959 \times 10^{-4}$$

$$\tau = C_f\left(\tfrac{1}{2}\rho_\infty U_\infty^2\right) = 0.708 \text{ N/m}^2$$

If the boundary layer is turbulent,

$$C_f = \frac{0.0583}{(\text{Re}_x)^{0.2}} = 3.522 \times 10^{-3}$$

$$\tau = C_f\left(\tfrac{1}{2}\rho_\infty U_\infty^2\right) = 4.185 \text{ N/m}^2$$

As discussed in the text of Chapter 2 and in the homework problems of Chapters 2 and 4, the integral form of the momentum equation can be used to determine the drag acting on an airfoil section. This approach is known as the far-field method of drag determination. Wing-section profile-drag measurements have been made for the Boeing 727 in flight using the Boeing Airborne Traversing Probe, examples of which are presented in Fig. 5.14. The probe consists of four main components: (1) flow sensors, (2) a rotating arm, (3) the drive unit, and (4) the mounting base. As reported by Bowes (1974), "The measured minimum section profile drag at $M = .73$ was about 15 percent higher than predicted from wind-tunnel test data for a smooth airfoil. The wind-tunnel data used in this correlation were also from wake surveys on the 727 wing. The data were adjusted to fully turbulent flow and extrapolated to flight Reynolds numbers. This quite sizeable difference between the measured and extrapolated values of $C_{d,\min}$ has been attributed to surface roughness and excrescences on the airplane wing, although the 15-percent increase in wing-section profile drag is larger than traditionally allotted in airplane drag estimates. The wing section where this survey was performed was inspected and had numerous steps and bumps due to control devices and manufacturing tolerances which would account for this local level of excrescence drag. This is not representative of the entire wing surface."

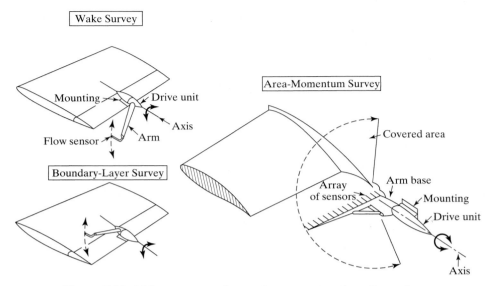

Figure 5.14 Airborne traversing probe concept and configurations.
[From Bowes (1974).]

5.4.4   Boundary-Layer Transition

It is obvious that the force coefficient of equation (5.21) depends on the Reynolds number. The Reynolds number not only affects the magnitude of C_f, but it is also used as an indicator of whether the boundary layer is laminar or turbulent. Since the calculation of the force coefficient requires integration of C_f over the chord length, we must know at what point, if any, the boundary layer becomes turbulent (i.e., where transition occurs).

Near the forward stagnation point on an airfoil, or on a wing, or near the leading edge of a flat plate, the boundary layer is laminar. As the flow proceeds downstream, the boundary layer thickens and the viscous forces continue to dissipate the energy of the airstream. Disturbances to the flow in the growing viscous layer may be caused by surface roughness, a temperature variation in the surface, pressure pulses, and so on. If the Reynolds number is low, the disturbances will be damped by viscosity and the boundary layer will remain laminar. At higher Reynolds numbers, the disturbances may grow. In such cases, the boundary layer may become unstable and, eventually, turbulent (i.e., transition will occur). The details of the transition process are quite complex and depend on many parameters.

The engineer who must develop a transition criterion for design purposes usually uses the Reynolds number. For instance, if the surface of a flat plate is smooth and if the external airstream has no turbulence, transition "occurs" at a Reynolds number (Re_x) of approximately 500,000. However, experience has shown that the Reynolds number at which the disturbances will grow and the length over which the transition process takes place depends on the magnitude of the disturbances and on the flow field. Let us consider briefly the effect of surface roughness, of the surface temperature, of a pressure gradient in the inviscid flow, and of the local Mach number on transition.

1. *Surface roughness.* Since transition is the amplification of disturbances to the flow, the presence of surface roughness significantly promotes transition (i.e., causes transition to occur at relatively low Reynolds numbers).

2. *Surface temperature.* The boundary-layer thickness decreases as the surface temperature is decreased. Cooling the surface usually delays transition. However, for supersonic flows, there is a complex relationship between boundary-layer transition and surface cooling.

3. *Pressure gradient.* A favorable pressure gradient (i.e., the static pressure decreases in the streamwise direction or, equivalently, the inviscid flow is accelerating) delays transition. Conversely, an adverse pressure gradient promotes transition.

4. *Mach number.* The transition Reynolds number is usually higher when the flow is compressible (i.e., as the Mach number is increased).

Stability theory [e.g., Mack (1984).] can be an important tool that provides insights into the importance of individual parameters without introducing spurious effects that might be due to flow disturbances peculiar to a test facility (e.g., "noise" in a wind tunnel). For a more detailed discussion of transition, the reader is referred to a textbook on boundary-layer theory [e.g., Schlichting (1968).]

If the skin friction is the dominant component of the drag, transition should be delayed as long as possible to obtain a low-drag section. To delay transition on a low-speed airfoil section, the point of maximum thickness could be moved aft so that the boundary layer is subjected to a favorable pressure gradient over a longer run. Consider the NACA 0009 section and the NACA 66–009 section. Both are symmetric, having a maximum thickness of $0.09c$. The maximum thickness for the NACA 66–009 section is at $0.45c$, while that for the NACA 0009 section is at $0.3c$ (see Fig. 5.15). As a result, the minimum pressure coefficient occurs at $x = 0.6c$ for the NACA 66–009 and a favorable pressure gradient acts to stabilize the boundary layer to this point. For the NACA 0009, the minimum pressure occurs near $x = 0.1c$. The lower local velocities near the leading edge and the extended region of favorable pressure gradient cause transition to be farther aft on the NACA 66–009. Since the drag for a streamlined airfoil at low angles of attack is primarily due to skin friction, use of equation (5.21) would indicate that the drag is lower for the NACA 66–009. This is verified by the data from Abbott and von Doenhoff (1949) which are reproduced in Fig. 5.16. The subsequent reduction in the friction drag creates a *drag bucket* for the NACA 66–009 section. Note that the section drag curve varies only slightly with C_l for moderate excursions in angle of attack, since the skin friction coefficient varies little with angle of attack. At the very high Reynolds numbers that occur at some flight conditions, it is difficult to maintain a long run of laminar boundary layer, especially if surface roughnesses develop during the flight operations. However, a *laminar-flow section*, such as the NACA 66–009, offers additional benefits. Comparing the cross sections presented in Fig. 5.15, the cross section of the NACA 66–009 airfoil provides more flexibility for carrying fuel and for accommodating the load-carrying structure.

For larger angles of attack, the section drag coefficient depends both on Reynolds number and on angle of attack. As the angle of attack and the section lift coefficient increase, the minimum pressure coefficient decreases. The adverse pressure gradient that results as the flow decelerates toward the trailing edge increases. When the air particles in the boundary layer, already slowed by viscous action, encounter the relatively

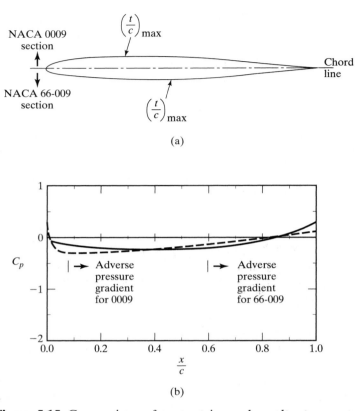

Figure 5.15 Comparison of geometries and resultant pressure distributions for a "standard" airfoil section (NACA 0009) and for a laminar airfoil section (NACA 66–009): (a) comparison of cross section for an NACA 0009 airfoil with that for an NACA 66–009 airfoil; (b) static pressure distribution.

strong adverse pressure gradient, the boundary layer thickens and separates. Because the thickening boundary layer and its separation from the surface cause the pressure distribution to be significantly different from the inviscid model at the higher angles of attack, form drag dominates. Note that, at the higher angles of attack (where form drag is important), the drag coefficient for the NACA 66–009 is greater than that for the NACA 0009. See Fig. 5.16.

Thus, when the viscous effects are secondary, we see that the lift coefficient and the moment coefficient depend only on vehicle geometry and angle of attack for low-speed flows. However, the drag coefficient exhibits a Reynolds number dependence both at the low angles of attack, where the boundary layer is thin (and the transition location is important), and at high angles of attack, where extensive regions of separated flow exist.

The section drag coefficient for a NACA 23012 airfoil is presented as a function of the section lift coefficient in Fig. 5.13. The data illustrate the dependence on Reynolds number and on angle of attack, which has been discussed. Note that measurements, which are taken from Abbott and von Doenhoff (1949), include data for a *standard roughness*.

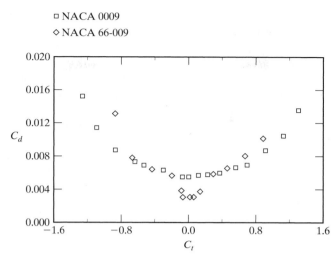

Figure 5.16 Section drag coefficients for NACA 0009 airfoil and the NACA 66–009 airfoil, $Re_c = 6 \times 10^6$. [Data from Abbott and von Doenhoff (1949).]

5.4.5   Effect of Surface Roughness on the Aerodynamic Forces

As discussed in Chapter 2, the Reynolds number is an important parameter when comparing the viscous character of two fields. If one desires to reproduce the Reynolds number for a flight condition in the wind tunnel, then

$$\left(\frac{\rho_\infty U_\infty c}{\mu_\infty}\right)_{\text{wt}} = \left(\frac{\rho_\infty U_\infty c}{\mu_\infty}\right)_{\text{ft}} \tag{5.31}$$

where the subscripts wt and ft designate wind-tunnel and flight conditions, respectively. In many low-speed wind tunnels, the free-stream values for density and for viscosity are roughly equal to the atmospheric values. Thus,

$$(U_\infty c)_{\text{wt}} \simeq (U_\infty c)_{\text{ft}} \tag{5.32}$$

If the wind-tunnel model is 0.2 scale, the wind-tunnel value for the free-stream velocity would have to be 5 times the flight value. As a result, the tunnel flow would be transonic or supersonic, which obviously would not be a reasonable simulation. Thus, since the maximum Reynolds number for this "equal density" subsonic wind-tunnel simulation is much less than the flight value, controlled surface roughness is often added to the model to fix boundary-layer transition at the location at which it would occur naturally in flight.

   Abbott and von Doenhoff (1949) present data on the effect of surface condition on the aerodynamic forces. "The standard leading-edge roughness selected by the NACA for 24-in chord models consisted of 0.011-in carborundum grains applied to the surface of the model at the leading edge over a surface length of 0.08c measured from the leading edge on both surfaces. The grains were thinly spread to cover 5 to 10% of the area. This

standard roughness is considerably more severe than that caused by the usual manufacturing irregularities or deterioration in service, but it is considerably less severe than
that likely to be encountered in service as a result of accumulation of ice, mud, or
damage in military combat." The data for the NACA 23012 airfoil (Fig. 5.12) indicate that
the angle of zero lift and the lift-curve slope are practically unaffected by the standard
leading-edge roughness. However, the maximum lift coefficient is affected by surface
roughness. This is further illustrated by the data presented in Fig. 5.17.

When there is no appreciable separation of the flow, the drag on the airfoil is
caused primarily by skin friction. Thus, the value of the drag coefficient depends on the
relative extent of the laminar boundary layer. A sharp increase in the drag coefficient
results when transition is suddenly shifted forward. If the wing surface is sufficiently
rough to cause transition near the wing leading edge, large increases in drag are observed,
as is evident in the data of Fig. 5.13 for the NACA 23012 airfoil section. In other test
results presented in Abbott and von Doenhoff (1949), the location of the roughness
strip was systematically varied. The minimum drag increased progressively with forward
movement of the roughness strip.

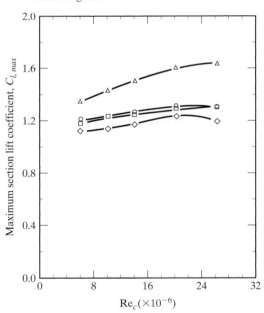

Figure 5.17 Effect of roughness near the leading edge on the
maximum section lift for the NACA, 63(420)-422 airfoil. [Data
from Abbott and von Doenhoff (1949).]

Scaling effects between model simulations and flight applications (as they relate to the viscous parameters) are especially important when the flow field includes an interaction between a shock wave and the boundary layer. The transonic flow field for an airfoil may include a shock-induced separation, a subsequent reattachment to the airfoil surface, and another boundary-layer separation near the trailing edge. According to Pearcey, et al. (1968), the prime requirements for correct simulation of these transonic shock/boundary-layer interactions include that the boundary layer is turbulent at the point of interaction and that the thickness of the turbulent boundary layer for the model flow is not so large in relation to the full-scale flow that a rear separation would occur in the simulation that would not occur in the full-scale flow.

Braslow, et al. (1966) provide some general guidelines for the use of grit-type boundary-layer transition trips. Whereas it is possible to fix boundary-layer transition far forward on wind-tunnel models at subsonic speeds using grit-type transition trips having little or no grit drag, the roughness configurations that are required to fix transition in a supersonic flow often cause grit drag. Fixing transition on wind-tunnel models becomes increasingly difficult as the Mach number is increased. Since roughness heights several times larger than the boundary-layer thickness can be required to fix transition in a hypersonic flow, the required roughness often produces undesirable distortions of the flow. Sterret, et al. (1966) provides some general guidelines for the use of boundary-layer trips for hypersonic flows. Data are presented to illustrate problems that can arise from poor trip designs.

The comments regarding the effects of surface roughness were presented for two reasons. (1) The reader should note that, since it is impossible to match the Reynolds number in many scale-model simulations, surface roughness (in the form of boundary-layer trips) is often used to fix transition and therefore the relative extent of the laminar boundary layer. (2) When surface roughness is used, considerable care should be taken to properly size and locate the roughness elements in order to properly simulate the desired flow.

The previous discussion in this section has focused on the effects of roughness elements that have been intentionally placed on the surface to fix artificially the location of transition. As discussed, the use of boundary-layer trips is intended to compensate for the inability to simulate the Reynolds number in ground-test facilities. However, surface roughness produced by environmental "contamination" may have a significant (and unexpected) effect on the transition location. As noted by van Dam and Holmes (1986), loss of laminar flow can be caused by surface contamination such as insect debris, ice crystals, moisture due to mist or rain, surface damage, and "innocent" modifications such as the addition of a spanwise paint stripe.

As noted by van Dam and Holmes (1986), "The surface roughness caused by such contamination can lead to early transition near the leading edge. A turbulent boundary layer which originates near the leading edge of an airfoil, substantially ahead of the point of minimum pressure, will produce a thicker boundary layer at the onset of the pressure recovery as compared to the conditions produced by a turbulent boundary layer which originates further downstream. With sufficiently steep pressure gradients in the recovery region, a change in the turbulent boundary layer conditions ... can lead to premature turbulent separation ..., thus affecting the aerodynamic characteristics and the effectiveness of trailing-edge control surfaces. ... Also, forward movement of transition location and turbulent separation produce a large increase in section drag."

5.4.6   Method for Predicting Aircraft Parasite Drag

As you can imagine, total aircraft parasite drag is a complex combination of aircraft configuration, skin friction, pressure distribution, interference among aircraft components, and flight conditions (among other things). Accurately predicting parasite drag would seem an almost impossible task, especially compared with predicting lift. To make matters even more difficult, there are a variety of terms associated with drag, all of which add confusion about drag and predicting drag. Some of the common terms used to describe drag are [McCormick (1979)]:

- *Induced* (or *vortex*) *drag*—drag due to the trailing vortex system
- *Skin friction drag*—due to viscous stress acting on the surface of the body
- *Form* (or *pressure*) *drag*—due to the integrated pressure acting on the body, caused by flow separation
- *Interference drag*—due to the proximity of two (or more) bodies (e.g., wing and fuselage)
- *Trim drag*—due to aerodynamic forces required to trim the airplane about the center of gravity
- *Profile drag*—the sum of skin-friction and pressure drag for an airfoil section
- *Parasite drag*—the sum of skin-friction and pressure drag for an aircraft
- *Base drag*—the pressure drag due to a blunt base or afterbody
- *Wave drag*—due to shockwave energy losses

In spite of these complexities, numerous straightforward estimation methods exist for predicting the parasite drag of aircraft, most of which use a combination of theoretical, empirical, and semi-empirical approaches.

While the building block methods for predicting skin-friction drag have been presented in Section 4.7.1 and 5.4.3, there are a variety of methods for applying skin-friction prediction methods to determine total aircraft drag. Every aerodynamics group at each aircraft manufacturer has different methods for estimating subsonic aircraft drag. The basic approaches, however, are probably quite similar:

1. Estimate an equivalent flat-plate skin-friction coefficient for each component of the aircraft (wing, fuselage, stabilizers, etc.)
2. Correct the skin-friction coefficient for surface roughness
3. Apply a form factor correction to each component's skin-friction coefficient to take into account supervelocities (velocities greater than free stream around the component) as well as pressure drag due to flow separation to obtain a parasite-drag coefficient
4. Convert each corrected skin-friction coefficient into an aircraft drag coefficient for that component
5. Sum all aircraft parasite-drag coefficients to obtain a total aircraft drag coefficient

Of course, this approach does not take into account a variety of sources of aircraft drag, including:

- Interference drag
- Excrescence drag (the drag due to various small drag-producing protuberances, including rivets, bolts, wires, etc.)
- Engine installation drag
- Drag due to control surface gaps
- Drag due to fuselage upsweep
- Landing gear drag

The total aircraft drag coefficient is defined as

$$C_D \equiv \frac{D_f}{q_\infty S_{\text{ref}}} \tag{5.33}$$

where S_{ref} is usually the wing planform area for an airplane. When the airplane drag coefficient is defined in this way, the term "drag count" refers to a drag coefficient of $C_D = 0.0001$ (a drag coefficient of $C_D = 0.0100$ would be 100 drag counts); many aerodynamicists refer to drag counts rather than the drag coefficient.

The following approach to determining subsonic aircraft parasite drag is due to Shevell (1989) and also is presented in Schaufele (2000). This approach assumes that each component of the aircraft contributes to the total drag without interfering with each other. While this is not true, the approach provides a good starting point for the estimation of drag. The zero-lift drag coefficient for subsonic flow is obtained by:

$$C_{D_0} = \sum_{i=1}^{N} \frac{K_i \overline{C}_{f_i} S_{\text{wet}_i}}{S_{\text{ref}}} \tag{5.34}$$

where N is the total number of aircraft components making up the aircraft (wing, fuselage, stabilizers, nacelles, pylons, etc.), K_i is the form factor for each component, $\overline{C}_{f_i}$ is the total skin-friction coefficient for each component, S_{wet_i} is the wetted area of each component, and S_{ref} is the aircraft reference area (there is only one reference area for the entire aircraft, which is usually the wing planform area).

Most aircraft components fall into one of two geometric categories: (1) wing-like shapes and (2) body-like shapes. Because of this, there are two basic ways to find the equivalent flat-plate skin-friction coefficient for the various aircraft components.

Wing Method. A wing with a trapezoidal planform (as shown in Fig. 5.18) can be defined by a root chord c_r, a tip chord c_t, a leading-edge sweep Λ, and a semi-span, $b/2$. The difficulty comes in applying the flat-plate skin-friction analysis to a wing with variable chord lengths along the span. Since the total skin-friction coefficient is a function of the Reynolds number at the "end of the plate", each spanwise station would have a different Reynolds number, and hence, a different total skin-friction coefficient.

It would be possible to perform a double integration along the chord and span of the wing to obtain a total skin-friction coefficient for the wing, however, an easier approach

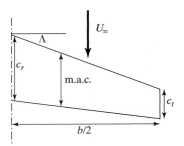

Figure 5.18 Geometry of a wing with a trapezoidal planform.

is to find an equivalent rectangular flat plate using the mean aerodynamic chord (m.a.c.) of the wing as the appropriate flat-plate length. The mean aerodynamic chord, described in Section 5.3, can be calculated using the following formulation which is valid for a trapezoidal wing

$$\text{m.a.c.} = \frac{2}{3}\left(c_r + c_t - \frac{c_r c_t}{c_r + c_t}\right) = \frac{2}{3}c_r\left(\frac{\lambda^2 + \lambda + 1}{\lambda + 1}\right) \tag{5.35}$$

where $\lambda = c_t/c_r$ is the taper ratio of the wing. The mean aerodynamic chord can then be used to define a "mean" Reynolds number for the wing:

$$\text{Re}_L = \frac{\rho_\infty U_\infty \text{m.a.c.}}{\mu_\infty} \tag{5.36}$$

It is important to remember that if a portion of the "theoretical" wing is submerged in the fuselage of the aircraft, then that portion of the wing should not be included in the calculation—the mean aerodynamic chord should be calculated using the root chord at the side of the fuselage!

Now the total skin-friction coefficient for the wing can be found using the Prandtl-Schlichting formula, including the correction for laminar flow (assuming transition takes place at $\text{Re}_{x,tr} = 500{,}000$), from equation (4.87):

$$\overline{C}_f = \frac{0.455}{(\log_{10}\text{Re}_L)^{2.58}} - \frac{1700}{\text{Re}_L} \tag{5.37}$$

Before proceeding any farther, the skin-friction coefficient should be corrected for surface roughness and imperfections. Various approaches exist for making this correction, some of which include small imperfections in the wing surface, such as rivets, seams, gaps, etc. In general, there is no straightforward method for correcting for surface roughness, so an empirical correction is often used, based on the actual flight test data of aircraft compared with the drag prediction using the approach outlined in this section. Most subsonic aircraft have a 6 to 9% increase in drag due to surface roughness, rivets, etc. [Kroo (2003)]. However, Kroo reports that, "carefully built laminar flow, composite aircraft may achieve a lower drag associated with roughness, perhaps as low as 2 to 3%."

The key factor in correcting for surface roughness is the relative height of the imperfections in the surface compared with the size of the viscous sublayer of the boundary layer (see Figure 4.15 for details about the viscous sublayer). The viscous sublayer

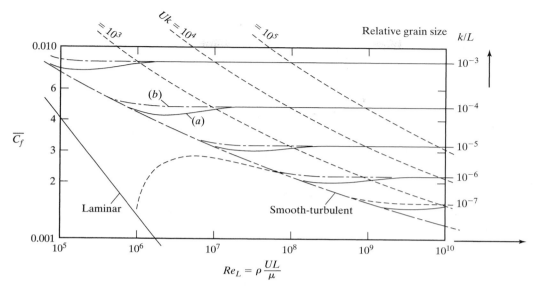

Figure 5.19 Effect of surface roughness on skin-friction drag. [From Hoerner (1965)].

usually is contained within a distance from the wall of $y^+ \approx 10$. So, if the equivalent "sand" grain roughness of the surface is contained within the viscous sublayer, than the surface is aerodynamically "smooth". As the sand grain roughness, k, increases in size, the skin friction will increase accordingly, as shown in Fig. 5.19, and eventually remains at a constant value with increasing Reynolds number. Since the thickness of the viscous sublayer actually decreases with Reynolds number, the impact of surface roughness is increased at higher Reynolds numbers, since the roughness will emerge from the sublayer and begin to impact the characteristics of the turbulent boundary layer [Hoerner (1965)]. Notice that, as the relative grain size increases, the skin-friction coefficient can deviate from the smooth turbulent value by factors as high as 300%. Keeping aerodynamic surfaces as smooth as possible is essential to reducing skin-friction drag! Equivalent sand grain roughness for different surfaces vary from approximately $k = 0.06 \times 10^{-3}$ in. for a polished metal surface, to $k = 2 \times 10^{-3}$ in. for mass production spray paint and to $k = 6 \times 10^{-3}$ in. for galvanized metal [Blake (1998)].

The form factor for the wing can be found from Fig. 5.20. This figure is based on empirical information and shows the correction to the skin-friction coefficient to take into account supervelocities (flow acceleration over the wing which alters the boundary layer properties which are assumed to be based on freestream levels) and pressure drag due to flow separation. Thicker wings have higher form factors and hence higher drag, while thinner wings have lower form factors and lower drag. An increase in wing sweep also tends to reduce the form factor and the drag coefficient.

Finally, the wetted area of the wing can be calculated using the following relationship [Kroo (2003)]:

$$S_{\text{wet}} \approx 2.0(1 + 0.2t/c)S_{\text{exposed}} \tag{5.38}$$

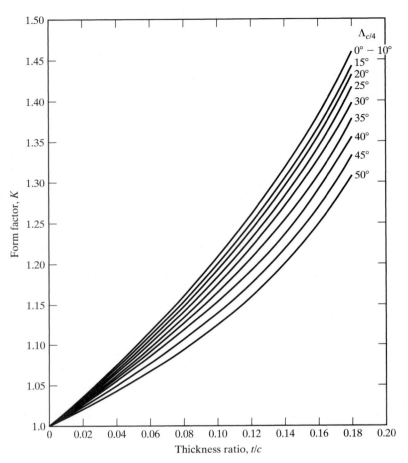

Figure 5.20 Wing form factor as a function of wing thickness ratio and quarter-chord sweep angle. [From Shevell (1989).]

where S_{exposed} is the portion of the wing planform that is not buried within the fuselage. The thickness factor in the previous equation takes into account the slight increase in flat-plate area due to the fact that the wing thickness increases the arc length of the wing chord. The exposed area is doubled to take into account the top and bottom of the wing.

The wing parasite-drag coefficient now can be calculated from equation (5.34)

$$C_{D_0} = K\overline{C}_f S_{\text{wet}} / S_{\text{ref}} \qquad (5.34)$$

Fuselage Method. Since the fuselage has a single length (as shown in Fig. 5.21), the calculation of the skin-friction coefficient is simpler than for a wing. First, find the Reynolds number at the end of the fuselage as

$$\text{Re}_L = \frac{\rho_\infty U_\infty L}{\mu_\infty} \qquad (5.30)$$

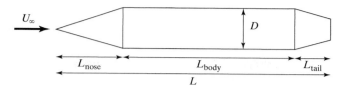

Figure 5.21 Geometry of a fuselage with circular cross sections.

and calculate the total skin-friction coefficient using equation (5.37). The skin-friction coefficient also can be corrected for surface roughness using the same approximation as discussed for the wing.

The fuselage form factor is a function of the fineness ratio of the body, which is defined as the length of the fuselage divided by the maximum diameter of the fuselage, L/D. The form factor can be found in Fig. 5.22 and shows that the more long and slender the fuselage, the smaller the form factor, and the smaller the drag coefficient increment due to separation. Conversely, a short, bluff body would have a high form factor due to large amounts of flow separation and hence a higher drag coefficient.

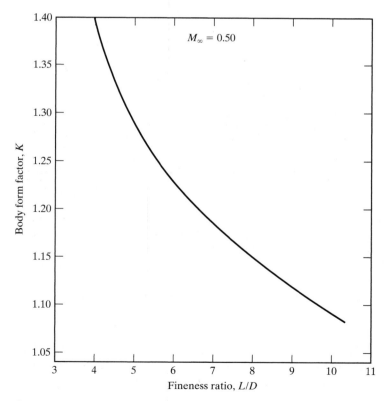

Figure 5.22 Body form factor as a function of fuselage fineness ratio. [From Shevell (1989).]

The total wetted area of the fuselage can be calculated as

$$S_{\text{wet}} \approx S_{\text{wet}_{\text{nose}}} + S_{\text{wet}_{\text{body}}} + S_{\text{wet}_{\text{tail}}} \tag{5.39}$$

where the wetted areas can be found assuming that the various portions of the fuselage can be approximated as cones, cylinders, and conical sections [from Kroo (2003)].

$$S_{\text{wet}_{\text{nose}}} = 0.75\pi DL_{\text{nose}}$$
$$S_{\text{wet}_{\text{body}}} = \pi DL_{\text{body}} \tag{5.40}$$
$$S_{\text{wet}_{\text{tail}}} = 0.72\pi DL_{\text{tail}}$$

These formulas do not double the exposed area of the fuselage, since air only flows over one side of the fuselage. (Namely, hopefully, the outside of the fuselage!)

Total Aircraft Parasite Drag. Now that the drag coefficient for the wing and fuselage have been calculated, the remaining components of the aircraft must be included, as shown in Fig. 5.23. Most of the remaining components can be approximated as either wing surfaces or fuselage surfaces, using the same methods as have been described previously. For example,

- *Vertical* and *horizontal stabilizers* — use wing method
- *Engine pylons* — use wing method
- *Engines nacelles* — use fuselage method
- *Antenna* — use wing method

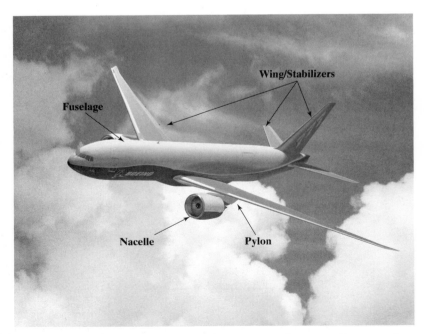

Figure 5.23 Aircraft showing the major components that contribute to drag. [(Boeing 777 photograph courtesy of The Boeing Company.)]

This approach can be continued for the vast majority of the aircraft. Now the total aircraft zero-lift drag coefficient can be found by using equation (5.34) as

$$C_{D_0} = \sum_{i=1}^{N} \frac{K_i \overline{C}_{f_i} S_{\text{wet}_i}}{S_{\text{ref}}}$$

The remaining drag components are due to smaller affects, such as excrescence drag, base drag, and interference drag. Some of these drag components would be extremely difficult (and time consuming) to calculate. In spite of this, semi-empirical and empirical methods for estimating the drag of rivets, bolts, flap gaps, and other small protuberances can be found in Hoerner's book on drag [Hoerner (1965)]. It is probably more expedient to use an empirical correction for the remainder of the drag, which is often done at aircraft manufacturers based on historical data from previous aircraft [Shevell (1989)].

EXAMPLE 5.5: **Estimate the subsonic parasite-drag coefficient**

This example will show how the subsonic parasite-drag coefficient for the F-16 can be estimated at a specific altitude. It is important to note that, unlike other aerodynamic coefficients, the subsonic drag coefficient is a function of altitude and Mach number, since the Reynolds number over the surfaces of the aircraft will vary with altitude and Mach. This example will assume that the aircraft is flying at an altitude of 30,000 ft and has a Mach number of 0.4 (to match available flight test data). The theoretical wing area of the F-16 is 300 ft^2, which will serve as the reference area for the aircraft: $S_{\text{ref}} = 300$ ft^2

Solution: The first task is to estimate the wetted area of the various surfaces of the aircraft. A good estimate of these areas has been completed in Brandt et al. (2004) and is reproduced here in Fig. 5.24 and Tables 5.2 and 5.3. The aircraft is approximated with a series of wing-like and fuselage-like shapes, as shown in Fig. 5.24. The wetted area for these simplified geometric surfaces is approximated using equations (5.38 and 5.40).

This simplified approximation yields a total wetted area of 1418 ft^2, which is very close to the actual wetted area of the F-16, which is 1495 ft^2 [Brandt et al.(2004)]. Now that these wetted areas have been obtained, the parasite-drag coefficient for each surface can be estimated.

TABLE 5.2 F-16 Wing-Like Surface Wetted Area Estimations					
Surface	*Span*, ft	c_r, ft	c_t, ft	*t/c*	S_{wet}, ft^2
Wing (1 and 2)	12	14	3.5	0.04	419.4
Horizontal tail (3 and 4)	6	7.8	2	0.04	117.5
Strake (5 and 6)	2	9.6	0	0.06	38.6
Inboard vertical tail (7)	1.4	12.5	6	0.10	26.3
Outboard vertical tail (8)	7	8	3	0.06	77.3
Dorsal fins (9 and 10)*	1.5	5	3	0.03	23.9

*not shown in Fig. 5.24

Surface	Length, ft	Height, ft	Width, ft	S_{wet}, ft²	Net S_{wet}, ft²
Fuselage (cylinder 1)	39	2.5	5	551.3	551.3
Nose (cone 1)	6	2.5	5	42.4	42.4
Boattail (cylinder 2)	4	6	6	62.8	62.8
Side (half cylinder 1 and 2)	24	0.8	1	67.9	29.5
Canopy (half cylinder 3)	5	2	2	15.7	5.7
Engine (half cylinder 4)	30	2.5	5	180	32.1
Canopy front (half cone 1)	2	2	2	3.1	1.1
Canopy rear (half cone 2)	4	2	2	6.3	2.3

TABLE 5.3 F-16 Fuselage-Like Surface Wetted Area Estimations

Wing. First, estimate the mean aerodynamic chord of the wing as

$$\text{m.a.c.} = \frac{2}{3}\left(c_r + c_t - \frac{c_r c_t}{c_r + c_t}\right) = \frac{2}{3}\left(14 + 3.5 - \frac{(14)(3.5)}{14 + 3.5}\right) = 9.800 \text{ ft}$$

and use the m.a.c. to calculate the Reynolds number for the equivalent rectangular wing as

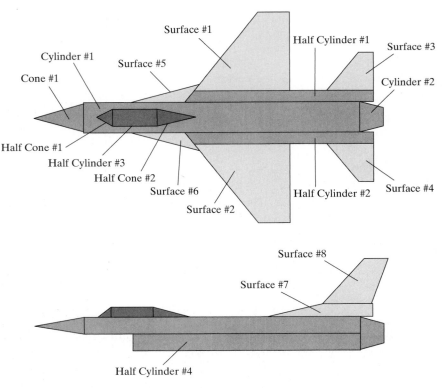

Figure 5.24 F-16 geometry approximated by simple shapes. [From Brandt et al. (2004).]

$$\text{Re}_L = \frac{\rho_\infty U_\infty \text{m.a.c.}}{\mu_\infty} = \frac{(0.000891 \text{ slug/ft}^3)(397.92 \text{ ft/s})(9.800 \text{ ft})}{3.107 \times 10^{-7} \text{lb-s/ft}^2}$$

$$= 11.18 \times 10^6$$

Finally, the total skin-friction coefficient for the wing is calculated as

$$\overline{C}_f = \frac{0.455}{(\log_{10} \text{Re}_L)^{2.58}} - \frac{1700}{\text{Re}_L} = \frac{0.455}{(\log_{10} 11.18 \times 10^6)^{2.58}} - \frac{1700}{11.18 \times 10^6}$$

$$= 0.00280$$

Assume for the sake of this example that the wing is aerodynamically smooth, so no roughness correction will be applied. However, a form factor correction should be performed. From Fig. 5.20, for a leading edge sweep of $40°$ and a thickness ratio of 0.04, $K = 1.06$ and the parasite drag coefficient is

$$C_{D_0} = \frac{K\overline{C}_f S_{\text{wet}}}{S_{\text{ref}}} = \frac{(1.06)(0.00280)(419.4 \text{ ft}^2)}{300 \text{ ft}^2} = 0.00415$$

The other wing-like components of the F-16 have had similar analysis performed, resulting in the zero-lift drag predictions presented in Table 5.4.

Fuselage.

$$L = L_{\text{nose}} + L_{\text{fuselage}} + L_{\text{boattail}} = 6 + 39 + 4 = 49.0 \text{ ft}$$

$$\text{Re}_L = \frac{(0.000891 \text{ slug/ft}^3)(397.92 \text{ ft/s})(49.0 \text{ ft})}{3.107 \times 10^{-7} \text{ lb-s/ft}^2} = 55.91 \times 10^6$$

$$\overline{C}_f = \frac{0.455}{(\log_{10} 55.91 \times 10^6)^{2.58}} - \frac{1700}{55.91 \times 10^6} = 0.00228$$

From Fig. 5.22, for a fineness ratio of $L/D = 49.0/0.5*(2.5 + 5.0) = 13.067$, $K = 1.05$ (requires some extrapolation from the graph) and the parasite drag coefficient is

$$C_{D_0} = \frac{K\overline{C}_f S_{\text{wet}}}{S_{\text{ref}}} = \frac{(1.05)(0.00228)(656.5 \text{ ft}^2)}{300 \text{ ft}^2} = 0.00524$$

TABLE 5.4 F-16 Wing-Like Surface Zero-Lift Drag Estimations

Surface	m.a.c., ft	$\text{Re}_L(\times 10^{-6})$	$\overline{C}_f$	K	C_{D_0}
Wing (1 and 2)	9.800	11.18	0.00280	1.06	0.00415
Horizontal tail (3 and 4)	5.472	6.568	0.00296	1.06	0.00123
Strake (5 and 6)	6.400	7.303	0.00293	1.04	0.00039
Inboard vertical tail (7)	9.631	10.99	0.00280	1.04	0.00026
Outboard vertical tail (8)	5.879	6.708	0.00295	1.08	0.00082
Dorsal Fins (9 and 10)	4.083	4.660	0.00304	1.04	0.00025
Total					0.00710

TABLE 5.5   F-16 Fuselage-Like Surface Zero-Lift Drag Estimations

Surface	Length, ft	$Re_L (\times 10^{-6})$	$\bar{C}_f$	K	C_{D_0}
Fuselage + nose + boattail	49.0	55.91	0.00228	1.05	0.00524
Side (half cylinder 1 and 2)	24.0	27.39	0.00251	1.01	0.00025
Canopy (front + center + rear)	11.0	12.55	0.00276	1.25	0.00011
Engine (half cylinder 4)	30.0	34.23	0.00244	1.15	0.00030
Total					0.00590

The other fuselage-like components of the F-16 have had similar analysis performed, resulting in the zero-lift drag predictions presented in Table 5.5.

The total aircraft zero-lift drag coefficient (assuming aerodynamically smooth surfaces) is

$$C_{D_0} = C_{D_0 \text{ (Wings)}} + C_{D_0 \text{ (Fuselages)}} = 0.00710 + 0.00590 = 0.01300$$

Since the total wetted-area estimate from this analysis was 5.4% lower than the actual wetted area of the F-16 (something which could be improved with a better representation of the aircraft surfaces, such as from a CAD geometry), it would be reasonable to increase the zero-lift drag value by 5.4% to take into account the simplicity of the geometry model. This would result in a zero-lift drag coefficient of $C_{D_0} = 0.01370$.

Again, this result assumes that the surfaces are aerodynamically smooth, that there are no drag increments due to excrescence or base drag, and that there is no interference among the various components of the aircraft. If we assume that the other components of drag account for an additional 10%, then our final estimate for the zero-lift drag coefficient would be $C_{D_0} = 0.0151$. Initial flight test data for the F-16 showed that the subsonic zero-lift drag coefficient varied between $C_{D_0} = 0.0160$ and $C_{D_0} = 0.0190$ after correcting for engine effects and the presence of missiles in the flight test data [Webb, et al. (1977)]. These results should be considered quite good for a fairly straightforward method that can be used easily on a spreadsheet.

5.5   WINGS OF FINITE SPAN

Much of the information about aerodynamic coefficients presented thus far are for two-dimensional airfoils (i.e., configurations of infinite span). For a wing of finite span that is generating lift, the pressure differential between the lower surface and the upper surface causes a spanwise flow. As will be discussed in Chapter 7, the spanwise variation of lift for the resultant three-dimensional flow field produces a corresponding distribution of streamwise vortices. These streamwise vortices in turn create a downwash, which has the effect of "tilting" the undisturbed air, reducing the effective angle of attack. As a result of the induced downwash velocity, the lift generated by the airfoil section of a

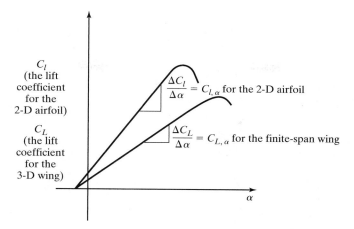

Figure 5.25 Comparison of the lift-curve slope of a two-dimensional airfoil with that for a finite-span wing.

finite-span wing which is at the geometric angle of attack α is less than that for the same airfoil section of an infinite-span airfoil configuration at the same angle of attack. Furthermore, the trailing vortex system produces an additional component of drag, known as the *vortex drag* (or the *induced drag*). Incompressible flows for three-dimensional wings will be discussed in detail in Chapter 7.

5.5.1 Lift

The typical lift curve for a three-dimensional wing composed of a given airfoil section is compared in Fig. 5.25 with that for a two-dimensional airfoil having the same airfoil section. Note that the lift-curve slope for the three-dimensional wing (which will be represented by the symbol $C_{L,\alpha}$) is considerably less than the lift-curve slope for an unswept, two-dimensional airfoil (which will be represented by the symbol $C_{l,\alpha}$). Recall that a lift-curve slope of approximately 0.1097 per degree is typical for an unswept, two-dimensional airfoil, as discussed in Section 5.4.1. The lift-curve slope for three-dimensional unswept wing ($C_{L,\alpha}$) can be approximated as

$$C_{L,\alpha} = \frac{C_{l,\alpha}}{1 + \dfrac{57.3 C_{l,\alpha}}{\pi e AR}} \tag{5.41}$$

where e is the efficiency factor, typical values for which are between 0.6 and 0.95.

EXAMPLE 5.6:   An F-16C in steady, level, unaccelerated flight

The pilot of an F-16 wants to maintain a constant altitude of 30,000 ft flying at idle power. Recall from the discussion at the start of the chapter, for flight in a horizontal plane (i.e., one of constant altitude), where the angles are

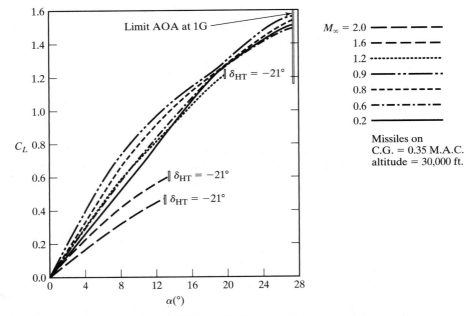

Figure 5.26 Trimmed lift coefficient as a function of the angle of attack for the F-16C. δ_{HT} is deflection of the horizontal tail. [Provided by the General Dynamics Staff (1976).]

small, the lift must balance the weight of the aircraft, which is 23,750 pounds. Therefore, as the vehicle slows down, the pilot must increase the angle of attack of the aircraft in order to increase the lift coefficient to compensate for the decreasing dynamic pressure. Use the lift curves for the F-16 aircraft, which were provided by the General Dynamics Staff (1976) and are presented in Fig. 5.26 for several Mach numbers. Assume that the lift curve for $M = 0.2$ is typical of that for incompressible flow, which for the purposes of this problem will be Mach numbers of 0.35, or less. Prepare a graph of the angle of attack as a function of the air speed in knots (nautical miles per hour) as the aircraft decelerates from a Mach number of 1.2 until it reaches its minimum flight speed. The minimum flight speed (i.e., the stall speed), is the velocity at which the vehicle must fly at its limit angle of attack in order to generate sufficient lift to balance the aircraft's weight. The wing (reference) area S is 300 ft^2.

Solution: Let us first calculate the lower limit for the speed range specified (i.e., the stall velocity). Since the aircraft is to fly in a horizontal plane (one of constant altitude), the lift is equal to the weight. Thus,

$$W = L = C_L(0.5\rho_\infty U_\infty^2)S \qquad \textbf{(5.42a)}$$

In order to compensate for the low dynamic pressure that occurs when flying at the stall speed (U_{stall}), the lift coefficient must be the maximum attainable value, $C_{L,max}$. Thus,

$$W = L = C_{L,max}(0.5\rho_\infty(U_{stall})^2)S$$

Solving for U_{stall},

$$U_{stall} = \sqrt{\frac{2W}{\rho_\infty C_{L,max}S}}$$

Referring to Fig. 5.26, the maximum value of the lift coefficient (assuming it occurs at an incompressible flow condition) is 1.57, which corresponds to the limit angle of attack. From Table 1.2b, the free-stream density at 30,000 ft is 0.0008907 slugs/ft^3, or 0.0008907 lbf s^2/ft^4. Thus,

$$U_{stall} = \sqrt{\frac{2(23750\,\text{lbf})}{(0.0008907\,\text{lbf s}^2/\text{ft}^4)(1.57)(300\,\text{ft}^2)}}$$

$$U_{stall} = 336.5\,\text{ft/s} = 199.2\,\text{knots}$$

Thus, the minimum velocity at which the weight of the F-16 is balanced by the lift at 30,000 ft is 199.2 knots with the aircraft at an angle of attack of 27.5°. The corresponding Mach number is

$$M_{stall} = \frac{U_{stall}}{a} = \frac{336.5\,\text{ft/s}}{994.85\,\text{ft/s}} = 0.338$$

To calculate the lift coefficient as a function of Mach number from $M_\infty = 1.2$ down to the stall value for the Mach number,

$$C_L = \frac{2W}{\rho_\infty(M_\infty a_\infty)^2 S} \tag{5.42b}$$

The velocity in knots is given by

$$U_\infty = M_\infty a_\infty(0.59209)$$

where U_∞ has the units of knots. The values obtained using these equations and using the lift curves presented in Fig. 5.26 to determine the angle of attack required to produce the required lift coefficient at a given Mach number are presented in the following table.

M_∞ (—)	C_L (—)	α (°)	U_∞ (ft/s)	U_∞ (knots)
1.2	0.125	1.9	1193.82	706.8
0.9	0.222	2.4	895.37	530.1
0.8	0.281	3.5	795.88	471.2
0.6	0.499	6.2	596.91	353.4
0.338	1.57	27.5	336.50	199.2

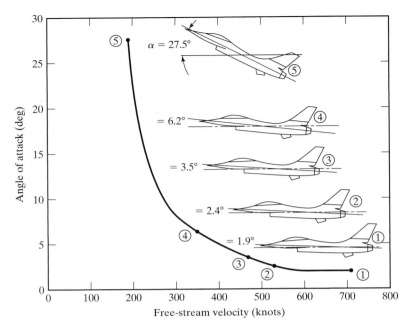

Figure 5.27 Correlation between the angle of attack and the velocity to maintain an F-16C in steady, level unaccelerated flight.

The correlation between the angle of attack and the velocity (in knots), as determined in this example and as presented in the table, is presented in Fig. 5.27. Note how rapidly the angle of attack increases toward the stall angle attack in order to generate a lift coefficient sufficient to maintain the altitude as the speed of the F-16 decreases toward the stall velocity. The angle of attack varies much more slowly with velocity at transonic Mach numbers.

5.5.2   Drag

As noted at the start of this section, the three-dimensional flow past a finite-span wing produces an additional component of drag associated with the trailing vortex system. This drag component is proportional to the square of the lift coefficient. In fact, a general expression for the drag acting on a finite-span wing (for which the flow is three dimensional) or on a complete aircraft is

$$C_D = C_{D,\min} + k'C_L^2 + k''(C_L - C_{L,\min})^2 \tag{5.43}$$

In equation (5.43), k' is a coefficient for the inviscid drag due to lift (which is also known as the induced drag or as the vortex drag). Similarly, k'' is a coefficient for the viscous drag due to lift (which includes the skin friction drag and the pressure drag associated with the viscous-induced changes to the pressure distribution, such as those due to separation).

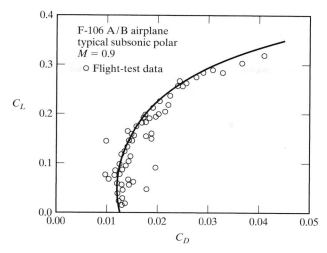

Figure 5.28 Flight data for a drag polar for F-106A/B aircraft at a Mach number of 0.9. [Taken from Piszkin et al. (1961)]

Data for a subsonic drag polar are presented in Fig. 5.28 for F-106A/B aircraft at a Mach number of 0.9. Equation (5.43) correlates these data, which are taken from Piszkin, et al. (1961). Note that the minimum drag occurs when C_L is approximately 0.07. Thus, $C_{L,\text{min}}$ is approximately 0.07.

Expanding the terms in equation (5.31), we obtain

$$C_D = (k' + k'')C_L^2 - (2k''C_{L,\text{min}})C_L$$
$$+ (C_{D,\text{min}} + k''C_{L,\text{min}}^2) \tag{5.44}$$

We can rewrite equation (5.44) as

$$C_D = k_1C_L^2 + k_2C_L + C_{D0} \tag{5.45}$$

where

$$k_1 = k' + k''$$

$$k_2 = -2k''C_{L,\text{min}}$$

and

$$C_{D0} = C_{D,\text{min}} + k''C_{L,\text{min}}^2$$

Referring to equation (5.45), C_{D0} is the drag coefficient when the lift is zero. C_{D0} is also known as the parasite drag coefficient. As noted when discussing Fig. 5.28, $C_{L,\text{min}}$ is relatively small. Thus, k_2 is often neglected. In such cases, one could also assume,

$$C_{D0} \cong C_{D,\text{min}}$$

Incorporating these comments into equation (5.45), we can write

$$C_D = C_{D0} + kC_L^2 \tag{5.46}$$

The reader will note that k_1 has been replaced by k (since there is only one constant left). How one accounts for both the inviscid and the viscous contributions to the lift-related drag will be discussed under item 2 in a subsequent list [(see equation (5.48)].

Let us introduce an additional term to account for the contributions of the compressibility effects to the drag, ΔC_{DM}. Thus, one obtains the expression for the total drag acting on an airplane:

$$C_D = C_{D0} + kC_L^2 + \Delta C_{DM} \tag{5.47}$$

Thus, the total drag for an airplane may be written as the sum of (1) of parasite drag, which is independent of the lift coefficient and therefore exists when the configuration generates zero lift (C_{D0} or ΔC_{Dp}), (2) the drag associated with lift (kC_L^2), and (3) the compressibility related effects that are correlated in terms of the Mach number and are known as wave drag.

Although the relative importance of the different drag sources depends on the aircraft type and the mission to be flown, approximate breakdowns (by category) for large, subsonic transports are presented in Figs. 5.29 and 5.30. According to Thomas (1985), the greatest contribution arises from turbulent skin friction drag. The next most significant contribution arises from lift-induced drag, which, when added to skin friction drag, accounts for about 85% of the total drag of a typical transport aircraft (see Fig. 5.29). Thomas cited the pressure drag due to open separation in the afterbody and other regions, interference effects between aerodynamic components, wave drag due to the compressibility effect at near-sonic flight conditions, and miscellaneous effects, such as roughness effects and leakage, as constituting the remaining 15%.

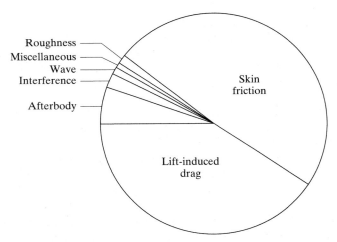

Figure 5.29 Contributions of different drag sources for a typical transport aircraft. [From Thomas (1985).]

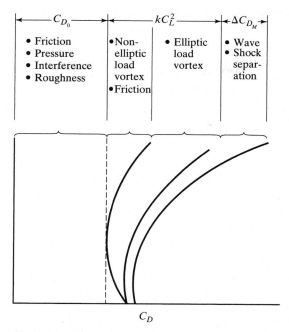

Figure 5.30 Lift/drag polar for a large, subsonic transport. [From Bowes (1974). The original version of this material was first published by the Advisory Group for Aerospace Research and Development, North Atlantic Treaty Organization (AGARD, NATO) in Lecture Series 67, May 1974.]

Using slightly different division of the drag-contribution sources, Bowes (1974) presented the lift/drag polar which is reproduced in Fig. 5.30. The majority of the lift-related drag is the vortex drag for an elliptic load distribution at subcritical speeds (i.e., the entire flow is subsonic). Bowes notes that a good wing design should approach an elliptic loading at the design condition.

1. *Zero-lift drag.* Skin friction and form drag components can be calculated for the wing, tail, fuselage, nacelles, and so on. When evaluating the zero-lift drag, one must consider interactions such as how the growth of the boundary layer on the fuselage reduces the boundary-layer velocities on the wing-root surface. Because of the interaction, the wing-root boundary layer is more easily separated in the presence of an adverse pressure gradient. Since the upper wing surface has the more critical pressure gradients, a low wing position on a circular fuselage would be sensitive to this interaction. Adequate filleting and control of the local pressure gradients can minimize the interaction effects.

    A representative value of C_{D0} would be 0.020, of which the wings may account for 50%, the fuselage and the nacelles 40%, and the tail 10%. Since the wing constitutes such a large fraction of the drag, reducing the wing area would reduce C_{D0} if all other factors are unchanged. Although other factors must be considered, the reasoning implies that an optimum airplane configuration would have a minimum wing surface area and, therefore, highest practical wing loading (N/m^2).

2. *Drag due to lift.* The drag due to lift may be represented as

$$kC_L^2 = \frac{C_L^2}{\pi(AR)e} \tag{5.48}$$

where e is the airplane efficiency factor. Typical values of the airplane efficiency factor range from 0.6 to 0.95. At high lift coefficients (near $C_{L,\max}$), e should be changed to account for increased form drag. The deviation of the actual airplane drag from the quadratic correlation, where e is a constant, is significant for airplanes with low aspect ratios and sweepback.

3. *Wave drag.* Another factor to consider is the effect of compressibility. When the free-stream Mach number is sufficiently large so that regions of supersonic flow exist in the flow field (e.g., free-stream Mach numbers of approximately 0.7, or greater), compressibility-related effects produce an additional drag component, known as wave drag. The correlation of equation (5.47) includes wave drag (i.e., the compressibility effects) as the third term, ΔC_{DM}. The aerodynamic characteristics of the F-16, which will be incorporated into Example 5.7, illustrate how the drag coefficient increases rapidly with Mach number in the transonic region. As will be discussed in Chapter 9, the designer can delay and/or reduce the compressibility drag rise by using low aspect ratio wings, by sweeping the wings, and/or by using area rule.

5.5.3 Lift/Drag Ratio

The configuration and the application of an airplane is closely related to the lift/drag ratio. Many important items of airplane performance are obtained in flight at $(L/D)_{\max}$. Performance conditions that occur at $(L/D)_{\max}$ include

1. Maximum range of propeller-driven airplanes
2. Maximum climb angle for jet-powered airplanes
3. Maximum power-off glide ratio (for jet-powered or for propeller-driven airplanes)
4. Maximum endurance for jet-powered airplanes.

Representative values of the maximum lift/drag ratios for subsonic flight speeds are as follows:

Type of airplane	$(L/D)_{\max}$
High-performance sailplane	25–40
Commercial transport	12–20
Supersonic fighter	4–12

EXAMPLE 5.7:   **Compute the drag components for an F-16 in steady, level, unacclerated flight**

The pilot of an F-16 wants to maintain steady, level (i.e., in a constant altitude, horizontal plane) unaccelerated flight. Recall from the discussion at the start of this chapter, for flight in a horizontal plane, where the angles are small, the lift must balance the weight and the thrust supplied by the engine must be sufficient to balance the total drag acting on the aircraft. For this exercise, we will assume that the total drag coefficient for this aircraft is given by equation (5.46):

$$C_{D0} + kC_L^2$$

Consider the following aerodynamic characteristics for the F-16:

M_∞ (—)	C_{D0} (—)	k (—)
0.10	0.0208	0.1168
0.84	0.0208	0.1168
1.05	0.0527	0.1667
1.50	0.0479	0.3285
1.80	0.0465	0.4211

Note that, since we are using equation (5.46) to represent the drag polar, the tabulated values of C_{D0} include both the profile drag and the compressibility effects. Thus, the values of C_{D0} presented in the preceding table include the two components of the drag not related to the lift. Had we used equation (5.47), the table would include individual values for the profile drag [the first term in equation (5.47)] and for the compressibility effects [the third term in equation (5.47)]. The parasite drag can be calculated as

$$\text{Para } D = C_{D0}(0.5\rho_\infty U_\infty^2)S$$

The induced drag can be calculated as

$$\text{Ind } D = C_{D\text{ind}}(0.5\rho_\infty U_\infty^2)S$$

where the induced-drag coefficient is given by

$$C_{D\text{ind}} = kC_L^2$$

In order for the aircraft to maintain steady, level unaccelerated flight, the lift must balance the weight, as represented by equation (5.42a). Therefore, one can solve equation (5.42b) for the lift coefficient as a function of the Mach number.

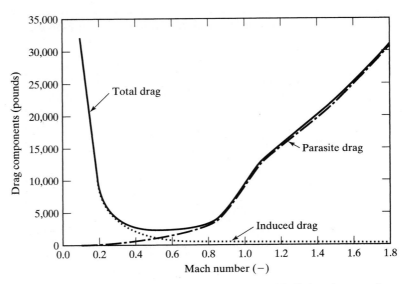

Figure 5.31 Drag components for an F-16C flying in steady, level, unaccelerated flight at 20,000 ft.

Referring to equation (5.46), the total drag is the sum of the parasite drag and the induced drag.

Additional aerodynamic characteristics for the F-16 are

$$\text{Aspect ratio } (AR) = 3.0$$

$$\text{Wing (reference) area}(S) = 300 \text{ ft}^2$$

$$\text{Airplane efficiency factor}(e) = 0.9084$$

Consider an F-16 that weighs 23,750 pounds in steady, level unaccelerated flight at an altitude of 20,000 ft (standard atmospheric conditions). Calculate the parasite drag, the induced drag, the total drag, and the lift-to-drag ratio as a function of Mach number in Mach-number increments of 0.1. Use linear fits of the tabulated values to obtain values of C_{D0} and of k at Mach numbers other than those presented in the table.

Solution: The first step is to use straight lines to generate values of C_{D0} and of k in Mach-number increments of 0.1. The results are presented in the first three columns of the accompanying table. Note, as mentioned in the problem statement, the values of C_{D0} for Mach numbers greater than 0.84 include a significant contribution of the wave drag ΔC_{DM} to the components of drag not related to the lift. The inclusion of the wave drag causes the drag coefficient C_{D0} to peak at a transonic Mach number of 1.05. Note also that

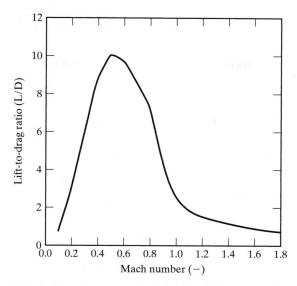

Figure 5.32 L/D ratio as a function of the Mach number for an F-16C flying at 20,000 ft.

the value of k for a Mach number of 0.10 is consistent with equation (5.48). That is,

$$k = \frac{1}{\pi e AR} = \frac{1}{\pi(0.9084)(3)} = 0.1168$$

The other tabulated values of k incorporate the effects of compressibility.

The free-stream density (0.001267 slugs/ft^3) and the free-stream speed of sound (1036.94 ft/s) for standard atmospheric conditions at 20,000 ft are taken from Table 1.2b.

The computed values of the parasite drag, of the induced drag, and of the total drag are presented in the table and in Fig. 5.31. Note that when the total drag is a minimum (which occurs at a Mach number of approximately 0.52), the parasite drag is equal to the induced drag.)

Since the lift is equal to the weight, the lift-to-drag ratio is given by

$$\frac{L}{D} = \frac{\text{Weight}}{\text{Total Drag}} = \frac{C_L}{C_{D0} + C_{D\text{ind}}} \quad \textbf{(5.49)}$$

The computed values for the lift-to-drag ratio are presented in the table and in Fig. 5.32. Since the weight of the aircraft is fixed, the maximum value of the lift-to-drag ratio occurs when the total drag is a minimum. The fact that the induced drag and the parasite drag are equal at this condition is an underlying principle to the solution of Problem 5.3.

Aircraft	F-16C
S (sq ft)	300
span (ft)	30
AR	3
e	0.9084

Mach (—)	CD,0 (—)	k (—)	a (fps) 1036.94 Vel (fps)	rho (s/ft3) 0.001267 Pdyn (psf)	ParaD (lb)	W(lb) 23750 CL (—)	CD,ind (—)	IndD (lb)	TotD (lb)	L/D (—)
0.10	0.0208	0.1168	103.69	6.81	42.50	11.6222	15.7768	32239.99	32282.50	0.7357
0.20	0.0208	0.1168	207.39	27.25	170.02	2.9056	0.9861	8060.00	8230.02	2.8858
0.30	0.0208	0.1168	311.08	61.31	382.54	1.2914	0.1948	3582.22	3964.77	5.9903
0.40	0.0208	0.1168	414.78	108.99	680.08	0.7264	0.0616	2015.00	2695.08	8.8124
0.50	0.0208	0.1168	518.47	170.29	1062.62	0.4649	0.0252	1289.60	2352.22	10.0968
0.60	0.0208	0.1168	622.16	245.22	1530.17	0.3228	0.0122	895.56	2425.73	9.7909
0.70	0.0208	0.1168	725.86	333.77	2082.74	0.2372	0.0066	657.96	2740.70	8.6657
0.80	0.0208	0.1168	829.55	435.95	2720.31	0.1816	0.0039	503.75	3224.06	7.3665
0.90	0.0300	0.1312	933.25	551.75	4965.71	0.1435	0.0027	447.10	5412.81	4.3877
1.00	0.0447	0.1542	1036.94	681.17	9134.46	0.1162	0.0021	425.63	9560.09	2.4843
1.10	0.0522	0.1832	1140.63	824.21	12907.17	0.0961	0.0017	417.92	13325.09	1.7824
1.20	0.0511	0.2195	1244.33	980.88	15036.91	0.0807	0.0014	420.75	15457.66	1.5365
1.30	0.0501	0.2558	1348.02	1151.17	17302.13	0.0688	0.0012	417.80	17719.93	1.3403
1.40	0.0490	0.2922	1451.72	1335.09	19625.80	0.0593	0.0010	411.51	20037.30	1.1853
1.50	0.0479	0.3285	1555.41	1532.63	22023.85	0.0517	0.0009	403.00	22426.85	1.0590
1.60	0.0474	0.3594	1659.10	1743.79	24796.67	0.0454	0.0007	387.52	25184.19	0.9431
1.70	0.0470	0.3902	1762.80	1968.57	27756.89	0.0402	0.0006	372.68	28129.58	0.8443
1.80	0.0465	0.4211	1866.49	2206.98	30787.41	0.0359	0.0005	358.75	31146.16	0.7625

PROBLEMS

5.1. For the delta wing of the F-106, the reference area (S) is 58.65 m² and the leading-edge sweep angle is 60°. What are the corresponding values of the aspect ratio and of the wing span?

5.2. In Example 5.1, an expression for the aspect ratio of a delta wing was developed in terms of the leading-edge sweep angle (Λ_{LE}). For this problem, develop an expression for the aspect ratio in terms of the sweep angle of the quarter chord ($\Lambda_{c/4}$).

5.3. Using equation (5.46) and treating C_{D0} and k as constants, show that the lift coefficient for maximum lift/drag ratio and the maximum lift/drag ratio for an incompressible flow are given by

$$C_{L(L/D)\max} = \sqrt{\frac{C_{D0}}{k}}$$

$$\left(\frac{L}{D}\right)_{\max} = \frac{1}{2\sqrt{kC_{D0}}}$$

5.4. A Cessna 172 is cruising at 10,000 ft on a standard day ($\rho_\infty = 0.001756$ slug/ft³) at 130 mi/h. If the airplane weighs 2300 lb, what C_L is required to maintain level flight?

5.5. Assume that the lift coefficient is a linear function of for its operating range. Assume further that the wing has a positive camber so that its zero-lift angle of attack (α_{0L}) is negative, and that the slope of the straight-line portion of the C_L versus α curve is $C_{L,\alpha}$. Using the results of Problem 5.3, derive an expression for $\alpha_{(L/D)\max}$.

5.6. Using the results of Problem 5.3, what is the drag coefficient (C_D) when the lift-to-drag ratio is a maximum? That is, what is $C_{D(L/D)\max}$?

5.7. Consider a flat plate at zero angle of attack in a uniform flow where $U_\infty = 35$ m/s in the standard sea-level atmosphere (Fig. P5.7). Assume that $Re_{x,tr} = 500,000$ defines the transition point. Determine the section drag coefficient, C_d. Neglect plate edge effects (i.e., assume two-dimensional flow). What error would be incurred if it is assumed that the boundary layer is turbulent along the entire length of the plate?

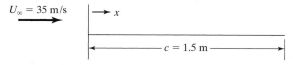

Figure P5.7

5.8. An airplane that weighs 60,000 N and has a wing area of 25.8 m² is landing at an airport.
  (a) Graph C_L as a function of the true airspeed over the airspeed range 300 to 180 km/h, if the airport is at sea level.
  (b) Repeat the problem for an airport that is at an altitude of 2000 m. For the purposes of this problem, assume that the airplane is in steady, level flight to calculate the required lift coefficients.

5.9. The lift/drag ratio of a sailplane is 30. The sailplane has a wing area of 10.0 m² and weighs 3150 N. What is C_D when the aircraft is in steady level flight at 170 km/h at an altitude of 1.0 km?

Problems (5.10) through (5.13) make use of pressure measurements taken from Pinkerton (1936) for NACA 4412 airfoil section. The measurements were taken from the centerplane of a rectangular planform wing, having a span of 30 inches and a chord of 5 inches. The test conditions included an average pressure (standard atmospheres): 21; average Reynolds number: 3,100,000. Pinkerton (1936) spent considerable effort defining the reliability of the pressure measurements. "In order to have true section characteristics (two-dimensional) for comparison with theoretical calculations, a determination must be made of the effective angle of attack, i.e., the angle that the chord of the model makes with the direction of the flow in the region of the midspan of the model." Thus, the experimentally determined pressure distributions, which are presented in Table 5.6, are presented both for the physical angle of attack (α) and the effective angle of attack (α_{eff}). However, "The determination of the effective angle of attack of the midspan section entails certain assumptions that are subject to considerable uncertainty." Nevertheless, the data allow us to develop some interesting graphs.

5.10. Pressure distribution measurements from Pinkerton (1936) are presented in Table 5.6 for the midspan section of a 76.2 cm by 12.7 cm model which had a NACA 4412 airfoil section. Graph C_p as a function of x/c for these three angles of attack. Comment on the movement of the stagnation point. Comment on changes in the magnitude of the adverse pressure gradient toward the trailing edge of the upper surface. How does this relate to possible boundary-layer separation (or stall)?

5.11. Use equation (3.13),

$$\frac{U}{U_\infty} = \sqrt{1 - C_p}$$

TABLE 5.6 Experimental Pressure Distributions for an NACA 4412 Airfoil [Abbott and von Doenhoff (1949)].[a]

x-Station (percentc c from the leading edge)	z-Ordinate (percent c above chord)	Values of the pressure coefficient, C_p		
		$\alpha = -4°$ $(\alpha_{\text{eff}} = -4°)$	$\alpha = +2°$ $(\alpha_{\text{eff}} = 1.2°)$	$\alpha = +16°$ $(\alpha_{\text{eff}} = 13.5°)$
100.00	0	0.204	0.181	0.010
97.92	−0.10	0.178	0.164	0.121
94.86	−0.16	0.151	0.154	0.179
89.90	−0.22	0.128	0.152	0.231
84.94	−0.28	0.082	0.148	0.257
74.92	−0.52	0.068	0.136	0.322
64.94	−0.84	0.028	0.120	0.374
54.48	−1.24	−0.024	0.100	0.414
49.98	−1.44	−0.053	0.091	0.426
44.90	−1.64	−0.075	0.088	0.459
39.98	−1.86	−0.105	0.071	0.485
34.90	−2.10	−0.146	0.066	0.516
29.96	−2.30	−0.190	0.048	0.551
24.90	−2.54	−0.266	0.025	0.589
19.98	−2.76	−0.365	−0.011	0.627
14.94	−2.90	−0.502	−0.053	0.713
9.96	−2.86	−0.716	−0.111	0.818
7.38	−2.72	−0.867	−0.131	0.896
4.94	−2.46	−1.106	−0.150	0.980
2.92	−2.06	−1.380	−0.098	0.993
1.66	−1.60	−1.709	0.028	0.791
0.92	−1.20	−1.812	0.254	0.264
0.36	−0.70	−1.559	0.639	−1.379
0	0	−0.296	0.989	−3.648

TABLE 5.6   Experimental Pressure Distributions for an NACA 4412 Airfoil
                [Abbott and von Doenhoff (1949)].[a]

Orifice location		Values of the pressure coefficient, C_p		
x-Station (percentc c from the leading edge)	z-Ordinate (percent c above chord)	$\alpha = -4°$ ($\alpha_{eff} = -4°$)	$\alpha = +2°$ ($\alpha_{eff} = 1.2°$)	$\alpha = +16°$ ($\alpha_{eff} = 13.5°$)
0	0.68	0.681	0.854	−6.230
0.44	1.56	0.994	0.336	−5.961
0.94	2.16	0.939	0.055	−5.210
1.70	2.78	0.782	−0.148	−4.478
2.94	3.64	0.559	−0.336	−3.765
4.90	4.68	0.333	−0.485	−3.190
7.50	5.74	0.139	−0.568	−2.709
9.96	6.56	0.017	−0.623	−2.440
12.58	7.34	−0.091	−0.676	−2.240
14.92	7.88	−0.152	−0.700	−2.149
17.44	8.40	−0.210	−0.721	−1.952
19.96	8.80	−0.262	−0.740	−1.841
22.44	9.16	−0.322	−0.769	−1.758
24.92	9.52	−0.322	−0.746	−1.640
27.44	9.62	−0.355	−0.742	−1.535
29.88	9.76	−0.364	−0.722	−1.438
34.95	9.90	−0.381	−0.693	−1.269
39.90	9.84	−0.370	−0.635	−1.099
44.80	9.64	−0.371	−0.609	−0.961
49.92	9.22	−0.329	−0.525	−0.786
54.92	8.76	−0.303	−0.471	−0.649
59.94	8.16	−0.298	−0.438	−0.551
64.90	7.54	−0.264	−0.378	−0.414
69.86	6.76	−0.225	−0.319	−0.316
74.90	5.88	−0.183	−0.252	−0.212
79.92	4.92	−0.144	−0.191	−0.147
84.88	3.88	−0.091	−0.116	−0.082
89.88	2.74	−0.019	−0.026	−0.043
94.90	1.48	−0.069	−0.076	−0.016
98.00	0.68	0.139	−0.143	−0.004

[a] *Source*: Pinkerton (1936).

to calculate the maximum value of the local velocity at the edge of the boundary layer both on the upper surface and on the lower surface for all three angles of attack. If these velocities are representative of the changes with angle of attack, how does the circulation (or lift) change with the angle of attack?

5.12. Using the small-angle approximations for the local surface inclinations, integrate the experimental chordwise pressure distributions of Table 5.6 to obtain values of the section lift coefficient for $\alpha = -4°$ and $\alpha = +2°$. Assuming that the section lift coefficient is a linear function of α (in this range of α), calculate

$$C_{l,\alpha} = \frac{dC_l}{d\alpha}$$

Does the section lift coefficient calculated using the pressure measurements for an angle of attack of 16° fall on the line? If not, why not?

5.13. Using the small-angle approximations for the local surface inclinations, integrate the experimental chordwise pressure distributions of Table 5.6 to obtain values of the section pitching moment coefficient relative to the quarter chord for each of the three angles of attack. Thus, in equation (5.10), x is replaced by $(x - 0.2c$. Recall that a positive pitching moment is in the nose-up direction.

5.14. If one seeks to maintain steady, level unaccelerated flight (SLUF), the thrust supplied by the aircraft's engine must balance the total drag acting on the aircraft. Thus, the total drag acting on the aircraft is equal to the thrust required from the power plant. As calculated in Example 5.7, the total drag of an F-16 flying at a constant altitude of 20,000 ft exhibits a minimum total drag at a Mach number of 0.52. If one wishes to fly at speeds above a Mach number of 0.52, the pilot must advance the throttle to obtain more thrust from the engine. Similarly, if one wishes to fly at speeds below a Mach number of 0.52, the pilot must initially retard the throttle to begin slowing down and then advance the throttle to obtain more thrust from the engine than would be required to cruise at a Mach number of 0.52 in order to sustain the new, slower velocity. This Mach number range is the region known as "reverse command." If the engine of the F-16 generates 10,000 pounds of thrust when flying at 20,000 ft, what is the maximum Mach number at which the F-16 can sustain SLUF? Use the results of the table and of the figures in Example 5.7. What is the minimum Mach number at which the F-16 can sustain SLUF at 20,000 ft, based on an available thrust of 10,000 pounds? If the maximum lift coefficient is 1.57 (see Fig. 5.26), at what Mach number will the aircraft stall? Use the lift coefficients presented in the table of Example 5.7.

5.15. Consider a jet in steady, level unaccelerated flight (SLUF) at an altitude of 30,000 ft (standard atmospheric conditions). Calculate the parasite drag, the induced drag, the total drag, and the lift-to-drag ratio as a function of the Mach number in Mach-number increments of 0.1, from a Mach number of 0.1 to a Mach number of 1.8. For the jet, assume the following:

Mach No. (—)	C_{D0} (—)	k (—)
0.1000	0.0215	0.1754
0.8500	0.0220	0.1760
1.0700	0.0510	0.2432
1.5000	0.0486	0.4560
1.8000	0.0425	0.5800

Other parameters for the jet include $S = 500$ ft^2; $b = 36$ ft; and $e = 0.70$. The weight of the aircraft is 32,000 pounds.

5.16. Consider the Eurofighter 2000 in steady, level unaccelerated flight (SLUF) at an altitude of 5 km (standard atmospheric conditions). Calculate the parasite drag, the induced drag, the total drag, and the lift-to-drag ratio as a function of the Mach number in Mach-number increments of 0.1, from a Mach number of 0.1 to a Mach number of 1.8. For the Eurofighter 2000, assume the following:

Mach No. (—)	C_{D0} (—)	k (—)
0.1000	0.0131	0.1725
0.8600	0.0131	0.1725
1.1100	0.0321	0.2292
1.5000	0.0289	0.3515
1.8000	0.0277	0.4417

Other parameters for the Eurofighter 2000 include $S = 50$ m^2; $b = 10.5$ m; and $e = 0.84$. The weight of the aircraft is 17,500 kg.

5.17. A finless missile is flying at sea-level at 450 mph. Estimate the parasite drag (excluding base drag) on the missile. The body has a length of 20.0 ft. and a diameter of 2.4 ft. The reference area for the missile is given by $S_{ref} = \pi d^2/4$ (the cross-sectional area of the missile). Explain why you would not need to correct your results for laminar flow.

5.18. A flying wing has a planform area of 4100 ft^2, a root chord at the airplane centerline of 36 ft, an overall taper ratio of 0.25, and a span of 180 ft. The average weighted airfoil thickness ratio is 9.9% and the wing has 38° of sweepback at the 25% chordline. The airplane is cruising at a pressure altitude of 17,000 ft on a standard day with a wing loading of 95 lb/ft^2. The cruise Mach number is 0.30. Determine the following:

(a) skin friction drag coefficient (assume a spray-painted surface)
(b) pressure drag coefficient
(c) induced drag coefficient
(d) total drag coefficient

REFERENCES

Abbott IH, von Doenhoff AE. 1949. *Theory of Wing Sections*. New York: Dover

Blake WB. 1998. *Missile DATCOM*. AFRL TR-1998-3009

Bowes GM. 1974. *Aircraft lift and drag prediction and measurement*, AGARD Lecture Series 67, 4-1 to 4-41

Brandt SA, Stiles RJ, Bertin JJ, Whitford R. 2004. *Introduction to Aeronautics: A Design Perspective*. 2nd Ed. Reston, VA: AIAA

Braslow AL, Hicks RM, Harris RV. 1966. Use of grit-type boundary-layer transition trips on wind-tunnel models. *NASA Tech. Note D-3579*

General Dynamics Staff. 1979. *F-16 Air Combat Fighter*. General Dynamics Report F-16-060

Hall DW. 1985. To fly on the wings of the sun. *Lockheed Horizons*. 18:60–68

Hoerner SR. 1965. *Fluid-Dynamic Drag*. Midland Park, NJ: published by the author

Kroo I. 2003. *Applied Aerodynamics: A Digital Textbook*. Stanford: Desktop Aeronautics

Mack LM. 1984. Boundary-layer linear stability theory. In *Special Course on Stability and Transition of Laminar Flow*, Ed. Michel R, AGARD Report 709, pp. 3–1 to 3–81

McCormick BW. 1979. *Aerodynamics, Aeronautics, and Flight Mechanics*. New York: John Wiley and Sons

Pearcey HH, Osborne J, Haines AB. 1968. The interaction between local effects at the shock and rear separation—a source of significant scale effects in wind-tunnel tests on aerofoils and wings. In *Transonic Aerodynamics*, AGARD Conference Proceedings, Vol. 35, Paper 1

Pinkerton RM. 1936. Calculated and measured pressure distributions over the midspan section of the NACA 4412 airfoil. *NACA Report 563*

Piszkin F, Walacavage R, LeClare G. 1961. *Analysis and comparison of Air Force flight test performance data with predicted and generalized flight test performance data for the F-106A and B airplanes*. Convair Technical Report AD-8-163

Schaufele RD. 2000. *The Elements of Aircraft Preliminary Design*. Santa Ana, CA: Aries Publications

Schlichting H. 1979. *Boundary Layer Theory*. 7th Ed. New York: McGraw-Hill

Shevell RS. 1989. *Fundamentals of Flight*. 2nd Ed. Englewood Cliffs, NJ: Prentice Hall

Sterret JR, Morrisette EL, Whitehead AH, Hicks RM. 1966. Transition fixing for hypersonic flow. *NASA Tech. Note D-4129*

Stuart WG. 1978. *Northrop F-5 Case Study in Aircraft Design*. Washington, DC: AIAA

Taylor JWR, ed. 1973. *Jane's All the World's Aircraft, 1973–1974*. London: Jane's Yearbooks

Taylor JWR, ed. 1966. *Jane's All the World's Aircraft, 1966–1967*. London: Jane's Yearbooks

Taylor JWR, ed. 1984. *Jane's All the World's Aircraft, 1984–1985*. London: Jane's Yearbooks

Thomas ASW. 1985. Aircraft viscous drag reduction technology. *Lockheed Horizons* 19:22–32

Van Dam CP, Holmes BJ. 1986. Boundary-layer transition effects on airplane stability and control. Presented at Atmospheric Flight Mech. Conf., AIAA Pap. 86-2229, Williamsburg, VA

Webb TS, Kent DR, Webb JB. 1977. Correlation of F-16 aerodynamics and performance predictions with early flight test results. In *Performance Prediction Methods*, AGARD Conference Proceeding, Vol. 242, Paper 19

6 INCOMPRESSIBLE FLOWS AROUND AIRFOILS OF INFINITE SPAN

6.1 GENERAL COMMENTS

Theoretical relations that describe an inviscid, low-speed flow around a thin airfoil will be developed in this chapter. To obtain the governing equations, it is assumed that the airfoil extends to infinity in both directions from the plane of symmetry (i.e., that the airfoil is a wing of infinite aspect ratio). Thus, the flow field around the airfoil is the same for any cross section perpendicular to the wing span and the flow is two dimensional.

For the rest of this book, the term *airfoil* will be used when the flow field is two dimensional. Thus, it will used for the two-dimensional flow fields that would exist when identical airfoil sections are placed side by side so that the spanwise dimension of the resultant configuration is infinite. But the term *airfoil* will also be used when a finite-span model with identical cross sections spans a wind tunnel from wall-to-wall and is perpendicular to the free-stream flow. Neglecting interactions with the tunnel-wall boundary layer, the flow around the model does not vary in the spanwise direction. For many applications (see Chapter 7), these two-dimensional airfoil flow fields will be applied to a slice of a wing flow field. That is, they approximate the flow per unit width (or per unit span) around the airfoil sections that are elements of a finite-span wing. For the rest of this book, the term *wing* will be used when the configuration is of finite span.

The flow around a two-dimensional airfoil can be idealized by superimposing a translational flow past the airfoil section, a distortion of the stream that is due to the airfoil thickness, and a circulatory flow that is related to the lifting characteristics of the airfoil.

Since it is a two-dimensional configuration, an airfoil in an incompressible stream experiences no drag force, if one neglects the effects of viscosity, as explained in Section 3.15.2. However, as discussed in Chapters 4 and 5, the viscous forces produce a velocity gradient near the surface of the airfoil and, hence, drag due to skin friction. Furthermore, the presence of the viscous flow near the surface modifies the inviscid flow field and may produce a significant drag force due to the integrated effect of the pressure field (i.e., form drag). To generate high lift, one may either place the airfoil at high angles of attack or employ leading-edge devices or trailing-edge devices. Either way, the interaction between large pressure gradients and the viscous, boundary layer produce a complex, Reynolds-number–dependent flow field.

The analytical values of aerodynamic parameters for incompressible flow around thin airfoils will be calculated in Section 6.3 through 6.5 using classical formulations from the early twentieth century. Although these formulations have long since been replaced by more rigorous numerical models (see Chapter 14), they do provide valuable information about the aerodynamic characteristics for airfoils in incompressible air streams. The comparison of the analytical values of the aerodynamic parameters with the corresponding experimental values will indicate the limits of the applicability of the analytical models. The desired characteristics and the resultant flow fields for high-lift airfoil sections will be discussed in Sections 6.6 through 6.8.

6.2 CIRCULATION AND THE GENERATION OF LIFT

For a lifting airfoil, the pressure on the lower surface of the airfoil is, on the average, greater than the pressure on the upper surface. Bernoulli's equation for steady, incompressible flow leads to the conclusion that the velocity over the upper surface is on the average, greater than that past the lower surface. Thus, the flow around the airfoil can be represented by the combination of a translational flow from left to right and a circulating flow in a clockwise direction, as shown in Fig. 6.1. This flow model assumes that the airfoil is sufficiently thin so that thickness effects may be neglected. Thickness effects can be treated using the source panel technique, discussed in Chapter 3.

If the fluid is initially at rest, the line integral of the velocity around any curve completely surrounding the airfoil is zero, because the velocity is zero for all fluid particles. Thus, the circulation around the line of Fig. 6.2a is zero. According to Kelvin's theorem for a

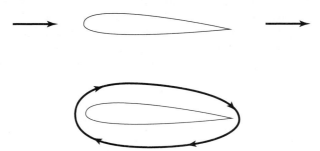

Figure 6.1 Flow around the lifting airfoil section, as represented by two elementary flows.

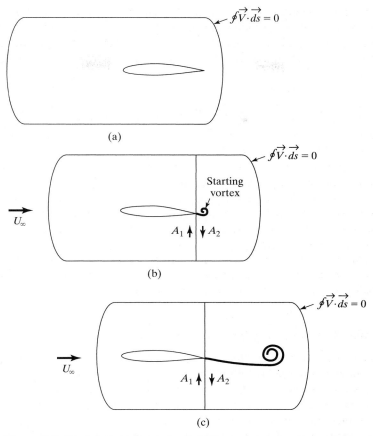

Figure 6.2 Circulation around a fluid line containing the airfoil remains zero: (a) fluid at rest; (b) fluid at time t; (c) fluid at time $t + \Delta t$.

frictionless flow, the circulation around this line of fluid particles remains zero if the fluid is suddenly given a uniform velocity with respect to the airfoil. Therefore, in Fig. 6.2b and c, the circulation around the line which encloses the lifting airfoil and which contains the same fluid particles as the line of Fig. 6.2a should be zero. However, circulation is necessary to produce lift. Thus, as explained in the following paragraphs, the circulation around each of the component curves of Fig. 6.2b and c is not zero but equal in magnitude and opposite in sign to the circulation around the other component curve.

6.2.1 Starting Vortex

When an airfoil is accelerated from rest, the circulation around it and, therefore, the lift are not produced instantaneously. At the instant of starting, the flow is a potential flow without circulation, and the streamlines are as shown in Fig. 6.3a, with a stagnation point occurring on the rear upper surface. At the sharp trailing edge, the air is required to change

Figure 6.3 Streamlines around the airfoil section: (a) zero circulation, stagnation point on the rear upper surface; (b) full circulation, stagnation point on the trailing edge.

direction suddenly. However, because of viscosity, the large velocity gradients produce large viscous forces, and the air is unable to flow around the sharp trailing edge. Instead, a surface of discontinuity emanating from the sharp trailing edge is rolled up into a vortex, which is called the *starting vortex*. The stagnation point moves toward the trailing edge, as the circulation around the airfoil and, therefore, the lift increase progressively. The circulation around the airfoil increases in intensity until the flows from the upper surface and the lower surface join smoothly at the trailing edge, as shown in Fig. 6.3b. Thus, the generation of circulation around the wing and the resultant lift are necessarily accompanied by a starting vortex, which results because of the effect of viscosity.

A line which encloses both the airfoil and the starting vortex and which always contains the same fluid particles is presented in Fig. 6.2. The total circulation around this line remains zero, since the circulation around the airfoil is equal in strength but opposite in direction to that of the starting vortex. Thus, the existence of circulation is not in contradiction to Kelvin's theorem. Referring to Fig. 6.2, the line integral of the tangential component of the velocity around the curve that encloses area A_1 must be equal and opposite to the corresponding integral for area A_2.

If either the free-stream velocity or the angle of attack of the airfoil is increased, another vortex is shed which has the same direction as the starting vortex. However, if the velocity or the angle of attack is decreased, a vortex is shed which has the opposite direction of rotation relative to the initial vortex.

A simple experiment that can be used to visualize the starting vortex requires only a pan of water and a small, thin board. Place the board upright in the water so that it cuts the surface. If the board is accelerated suddenly at moderate incidence, the starting vortex will be seen leaving the trailing edge of the "airfoil." If the board is stopped suddenly, another vortex of equal strength but of opposite rotation is generated.

6.3   GENERAL THIN-AIRFOIL THEORY

The essential assumptions of thin-airfoil theory are (1) that the lifting characteristics of an airfoil below stall are negligibly affected by the presence of the boundary layer, (2) that the airfoil is operating at a small angle of attack, and (3) that the resultant of the pressure forces (magnitude, direction, and line of action) is only slightly influenced by the airfoil thickness, since the maximum mean camber is small and the ratio of maximum thickness to chord is small.

Note that we assume that there is sufficient viscosity to produce the circulation that results in the flow depicted in Fig. 6.3b. However, we neglect the effect of viscosity as it relates to the boundary layer. The boundary layer is assumed to be thin and, therefore, does not significantly alter the static pressures from the values that correspond to those for the inviscid flow model. Furthermore, the boundary layer does not cause the flow to separate when it encounters an adverse pressure gradient. Typically, airfoil sections have a maximum thickness of approximately 12% of the chord and a maximum mean camber of approximately 2% of the chord. For thin-airfoil theory, the airfoil will be represented by its mean camber line in order to calculate the section aerodynamic characteristics.

A velocity difference across the infinitely thin profile which represents the airfoil section is required to produce the lift-generating pressure difference. A vortex sheet coincident with the mean camber line produces a velocity distribution that exhibits the required velocity jump. Therefore, the desired flow is obtained by superimposing on a uniform field of flow a field induced by a series of line vortices of infinitesimal strength which are located along the camber line, as shown in Fig. 6.4. The total circulation is the sum of the circulations of the vortex filaments

$$\Gamma = \int_0^c \gamma(s)\, ds \tag{6.1}$$

where $\gamma(s)$ is the distribution of vorticity for the line vortices. The length of an arbitrary element of the camber line is ds and the circulation is in the clockwise direction.

The velocity field around the sheet is the sum of the free-stream velocity and the velocity induced by all the vortex filaments that make up the vortex sheet. For the vortex sheet to be a streamline of the flow, it is necessary that the resultant velocity be tangent to the mean camber line at each point. Thus, the sum of the components normal to the surface for these two velocities is zero. In addition, the condition that the flows from the upper surface and the lower surface join smoothly at the trailing edge (i.e., the Kutta condition) requires that $\gamma = 0$ at the trailing edge. Ideally (i.e., for an inviscid potential flow), the circulation that forms places the rear stagnation point exactly on the sharp trailing edge. When the effects of friction are included, there is a reduction in circulation

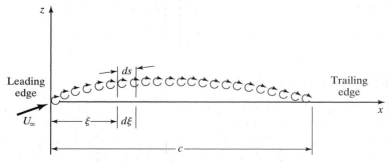

Figure 6.4 Representation of the mean camber line by a vortex sheet whose filaments are of variable strength $\gamma(s)$.

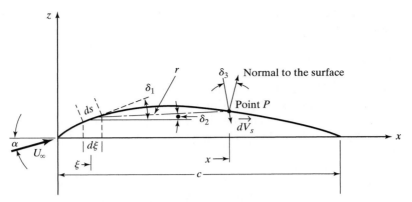

Figure 6.5 Thin-airfoil geometry parameters.

relative to the value determined for an "inviscid flow." Thus, the Kutta condition places a constraint on the vorticity distribution that is consistent with the effects of the boundary layer.

The portion of the vortex sheet designated ds in Fig. 6.5 produces a velocity at point P which is perpendicular to the line whose length is r and which joins the element ds and the point P. The induced velocity component normal to the camber line at P due to the vortex element ds is

$$dV_{s,n} = -\frac{\gamma \, ds \, \cos \delta_3}{2\pi r}$$

where the negative sign results because the circulation induces a clockwise velocity and the normal to the upper surface is positive outward. To calculate the resultant vortex-induced velocity at a particular point P, one must integrate over all the vortex filaments from the leading edge to the trailing edge. The chordwise location of the point of interest P will be designated in terms of its x coordinate. The chordwise location of a given element of vortex sheet ds will be given in terms of its ξ coordinate. Thus, to calculate the cumulative effect of all the vortex elements, it is necessary to integrate over the ξ coordinate from the leading edge ($\xi = 0$) to the trailing edge ($\xi = c$). Noting that

$$\cos \delta_2 = \frac{x - \xi}{r} \quad \text{and} \quad ds = \frac{d\xi}{\cos \delta_1}$$

the resultant vortex-induced velocity at any point P (which has the chordwise location x) is given bys

$$V_{s,n}(x) = -\frac{1}{2\pi} \int_0^c \frac{\gamma(\xi) \cos \delta_2 \cos \delta_3 \, d\xi}{(x - \xi) \cos \delta_1} \tag{6.2}$$

The component of the free-stream velocity normal to the mean camber line at P is given by

$$U_{\infty,n}(x) = U_\infty \sin(\alpha - \delta_P)$$

where α is the angle of attack and δ_P is the slope of the camber line at the point of interest P. Thus,

$$\delta_P = \tan^{-1} \frac{dz}{dx}$$

where $z(x)$ describes the mean camber line. As a result,

$$U_{\infty,n}(x) = U_\infty \sin\left(\alpha - \tan^{-1} \frac{dz}{dx} \right) \tag{6.3}$$

Since the sum of the velocity components normal to the surface must be zero at all points along the vortex sheet,

$$\frac{1}{2\pi} \int_0^c \frac{\gamma(\xi) \cos \delta_2 \cos \delta_3 \, d\xi}{(x - \xi) \cos \delta_1} = U_\infty \sin\left(\alpha - \tan^{-1} \frac{dz}{dx} \right) \tag{6.4}$$

The vorticity distribution $\gamma(\xi)$ that satisfies this integral equation makes the vortex sheet (and, therefore, the mean camber line) a streamline of the flow. The desired vorticity distribution must also satisfy the Kutta condition that $\gamma(c) = 0$.

Within the assumptions of thin-airfoil theory, the angles $\delta_1, \delta_2, \delta_3$, and α are small. Using the approximate trigonometric relations for small angles, equation (6.4) becomes

$$\frac{1}{2\pi} \int_0^c \frac{\gamma(\xi) \, d\xi}{x - \xi} = U_\infty\left(a - \frac{dz}{dx} \right) \tag{6.5}$$

6.4 THIN, FLAT-PLATE AIRFOIL (SYMMETRIC AIRFOIL)

The mean camber line of a symmetric airfoil is coincident with the chord line. Thus, when the profile is replaced by its mean camber line, a flat plate with a sharp leading edge is obtained. For subsonic flow past a flat plate even at small angles of attack, a region of dead air (or stalled flow) will exist over the upper surface. For the actual airfoil, the rounded nose allows the flow to accelerate from the stagnation point onto the upper surface without separation. Of course, when the angle of attack is sufficiently large (the value depends on the cross-section geometry), stall will occur for the actual profile. The approximate theoretical solution for a thin airfoil with two sharp edges represents an irrotational flow with finite velocity at the trailing edge but with infinite velocity at the leading edge. Because it does not account for the thickness distribution nor for the viscous effects, the approximate solution does not describe the chordwise variation of the flow around the actual airfoil. However, as will be discussed, the theoretical values of the lift coefficient (obtained by integrating the circulation distribution along the airfoil) are in reasonable agreement with the experimental values.

For the camber line of the symmetric airfoil, dz/dx is everywhere zero. Thus, equation (6.5) becomes

$$\frac{1}{2\pi} \int_0^c \frac{\gamma(\xi)}{x - \xi} \, d\xi = U_\infty \alpha \tag{6.6}$$

It is convenient to introduce the coordinate transformation

$$\xi = \frac{c}{2}(1 - \cos\theta) \tag{6.7}$$

Similarly, the x coordinate transforms to θ_0 using

$$x = \frac{c}{2}(1 - \cos\theta_0)$$

The corresponding limits of integration are

$$\xi = 0, \theta = 0 \quad \text{and} \quad \xi = c, \theta = \pi$$

Equation (6.6) becomes

$$\frac{1}{2\pi}\int_0^\pi \frac{\gamma(\theta)\sin\theta\,d\theta}{\cos\theta - \cos\theta_0} = U_\infty\alpha \tag{6.8}$$

The required vorticity distribution, $\gamma(\theta)$, must not only satisfy this integral equation, but must satisfy the Kutta condition, $\gamma(\pi) = 0$. The solution is

$$\gamma(\theta) = 2\alpha U_\infty \frac{1 + \cos\theta}{\sin\theta} \tag{6.9}$$

That this is a valid solution can be seen by substituting the expression for $\gamma(\theta)$ given by equation (6.9) into equation (6.8). The resulting equation,

$$\frac{\alpha U_\infty}{\pi}\int_0^\pi \frac{(1 + \cos\theta)\,d\theta}{\cos\theta - \cos\theta_0} = U_\infty\alpha$$

can be reduced to an identity using the relation

$$\int_0^\pi \frac{\cos n\theta\,d\theta}{\cos\theta - \cos\theta_0} = \frac{\pi\sin n\theta_0}{\sin\theta_0} \tag{6.10}$$

where n assumes only integer values. Using l'Hospital's rule, it can be readily shown that the expression for $\gamma(\theta)$ also satisfies the Kutta condition.

The Kutta-Joukowski theorem for steady flow about a two-dimensional body of any cross section shows that the force per unit span is equal to $\rho_\infty U_\infty \Gamma$ and acts perpendicular to U_∞. Thus, for two-dimensional inviscid flow, an airfoil has no drag but experiences a lift per unit span equal to the product of free-stream density, the free-stream velocity, and the total circulation. The lift per unit span is

$$l = \int_0^c \rho_\infty U_\infty \gamma(\xi)\,d\xi \tag{6.11}$$

Using the circulation distribution of equation (6.9) and the coordinate transformation of equation (6.7), the lift per unit span is

$$l = \rho_\infty U_\infty^2 \alpha c \int_0^\pi (1 + \cos\theta)\,d\theta$$

$$l = \pi\rho_\infty U_\infty^2 \alpha c \tag{6.12}$$

To determine the section lift coefficient of the airfoil, note that the reference area per unit span is the chord. The section lift coefficient is

$$C_l = \frac{l}{0.5\rho_\infty U_\infty^2 c} = 2\pi\alpha \tag{6.13}$$

where α is the angle of attack in radians. Thus, thin-airfoil theory yields a section lift coefficient for a symmetric airfoil that is directly proportional to the geometric angle of attack. The geometric angle of attack is the angle between the free-stream velocity and the chord line of the airfoil. The theoretical relation is independent of the airfoil thickness.

However, because the airfoil thickness distribution and the boundary layer affect the flow field, the actual two-dimensional lift curve slope will be less than 2π per radian. See the discussion in Section 5.4.1.

It may seem strange that the net force for an inviscid flow past a symmetric airfoil is perpendicular to the free-stream flow rather than perpendicular to the airfoil (i.e., that the resultant force has only a lift component and not both a lift and a drag component). As noted in Section 3.15.2, the prediction of zero drag may be generalized to any general, two-dimensional body in an irrotational, steady, incompressible flow. Consider an incompressible, inviscid flow about a symmetric airfoil at a small angle of attack. There is a stagnation point on the lower surface of the airfoil just downstream of the leading edge. From this stagnation point, flow accelerates around the leading edge to the upper surface. Referring to the first paragraph in this section, the approximate theoretical solution for a thin, symmetric airfoil with two sharp edges yields an infinite velocity at the leading edge. The high velocities for flow over the leading edge result in low pressures in this region, producing a component of force along the leading edge, known as the *leading-edge suction force*, which exactly cancels the streamwise component of the pressure distribution acting on the rest of the airfoil, resulting in zero drag.

As noted by Carlson and Mack (1980), "Linearized theory places no bounds on the magnitude of the peak suction pressure, which, therefore, can become much greater than practically realizable values." However, "limitations imposed by practically realizable pressures may have a relatively insignificant effect on the normal force but could, at the same time, severely limit the attainment of the thrust force."

The pressure distribution also produces a pitching moment about the leading edge (per unit span), which is given by

$$m_0 = -\int_0^c \rho_\infty U_\infty \gamma(\xi)\xi \, d\xi \tag{6.14}$$

The lift-generating circulation of an element $d\xi$ produces an upward force that acts a distance ξ downstream of the leading edge. The lift force, therefore, produces a nose-down pitching moment about the leading edge. Thus, the negative sign is used in equation (6.14) because nose-up pitching moments are considered positive. Again, using the coordinate transformation [equation (6.7)] and the circulation distribution [equation (6.9)], the pitching moment (per unit span) about the leading edge is

$$m_0 = -0.5\rho_\infty U_\infty^2 \alpha c^2 \int_0^\pi (1 - \cos^2\theta) \, d\theta$$

$$= -\frac{\pi}{4}\rho_\infty U_\infty^2 \alpha c^2 \tag{6.15}$$

The section moment coefficient is given by

$$C_{m_0} = \frac{m_0}{0.5\rho_\infty U_\infty^2 cc} \tag{6.16}$$

Note that the reference area per unit span for the airfoil is the chord and the reference length for the pitching moment is the chord. For the symmetric airfoil,

$$C_{m_0} = -\frac{\pi}{2}\alpha = -\frac{C_l}{4} \tag{6.17}$$

The center of pressure x_{cp} is the x coordinate, where the resultant lift force could be placed to produce the pitching moment m_0. Equating the moment about the leading edge [equation (6.15)] to the product of the lift [equation (6.12)] and the center of pressure yields

$$-\frac{\pi}{4}\rho_\infty U_\infty^2 \alpha c^2 = -\pi\rho_\infty U_\infty^2 \alpha c x_{cp}$$

Solving for x_{cp}, one obtains

$$x_{cp} = \frac{c}{4} \tag{6.18}$$

The result is independent of the angle of attack and is therefore independent of the section lift coefficient.

EXAMPLE 6.1: Theoretical aerodynamic coefficients for a symmetric airfoil

The theoretical aerodynamic coefficients calculated using the thin-airfoil relations are compared with the data of Abbott and von Doenhoff (1949) in Fig. 6.6. Data are presented for two different airfoil sections. One, the NACA 0009, has a maximum thickness which is 9% of chord and is located at $x = 0.3c$. The theoretical lift coefficient calculated using equation (6.13) is in excellent agreement with the data for the NACA 0009 airfoil up to an angle of attack of 12°. At higher angles of attack, the viscous effects significantly alter the flow field and hence the experimental lift coefficients. Thus, theoretical values would not be expected to agree with the data at high angles of attack. Since the theory presumes that viscous effects are small, it is valid only for angles of attack below stall. According to thin-airfoil theory, the moment about the quarter chord is zero. The measured moments for the NACA 0009 are also in excellent agreement with theory prior to stall. The correlation between the theoretical values and the experimental values is not as good for the NACA 0012–64 airfoil section. The difference in the correlation between theory and data for these two airfoil sections is attributed to viscous effects. The maximum thickness of the NACA 0012–64 is greater and located farther aft. Thus, the adverse pressure gradients that cause separation of the viscous boundary layer and thereby alter the flow field would be greater for the NACA 0012–64.

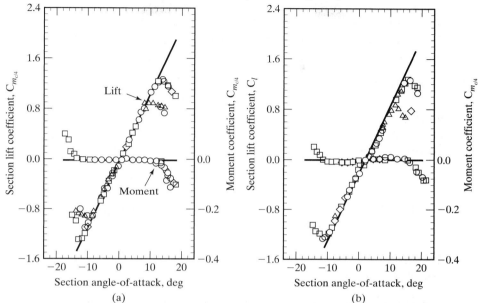

Figure 6.6 Comparison of the aerodynamic coefficients calculated using thin-airfoil theory for symmetric airfoils: (a) NACA 0009 wing section; (b) NACA 0012-64 wing section. [Data from Abbott and von Doenhoff (1949)].

6.5 THIN, CAMBERED AIRFOIL

The method of determining the aerodynamic characteristics for a cambered airfoil is similar to that followed for the symmetric airfoil. Thus, a vorticity distribution is sought which satisfies both the condition that the mean camber line is a streamline [equation (6.5)] and the Kutta condition. However, because of camber, the actual computations are more involved. Again, the coordinate transformation

$$\xi = \frac{c}{2}(1 - \cos\theta) \tag{6.7}$$

is used, so the integral equation to be solved is

$$\frac{1}{2\pi} \int_0^\pi \frac{\gamma(\theta)\sin\theta\, d\theta}{\cos\theta - \cos\theta_0} = U_\infty\left(\alpha - \frac{dz}{dx}\right) \tag{6.19}$$

Recall that this integral equation expresses the requirement that the resultant velocity for the inviscid flow is parallel to the mean camber line (which represents the airfoil).

The vorticity distribution $\gamma(\theta)$ that satisfies the integral equation makes the vortex sheet (which is coincident with the mean camber line) a streamline of the flow.

6.5.1 Vorticity Distribution

The desired vorticity distribution, which satisfies equation (6.19) and the Kutta condition, may be represented by the series involving

1. A term of the form for the vorticity distribution for a symmetric airfoil,

$$2U_\infty A_0 \frac{1 + \cos\theta}{\sin\theta}$$

2. A Fourier sine series whose terms automatically satisfy the Kutta condition,

$$2U_\infty \sum_{n=1}^{\infty} A_n \sin n\theta$$

The coefficients A_n of the Fourier series depend on the shape of the camber line. Thus,

$$\gamma(\theta) = 2U_\infty \left(A_0 \frac{1 + \cos\theta}{\sin\theta} + \sum_{n=1}^{\infty} A_n \sin n\theta \right) \tag{6.20}$$

Since each term is zero when $\theta = \pi$, the Kutta condition is satisfied [i.e., $\gamma(\pi) = 0$]. Substituting the vorticity distribution [equation (6.20)] into equation (6.19) yields

$$\frac{1}{\pi} \int_0^\pi \frac{A_0(1 + \cos\theta)\, d\theta}{\cos\theta - \cos\theta_0} + \frac{1}{\pi} \int_0^\pi \sum_{n=1}^{\infty} \frac{A_n \sin n\theta \sin\theta\, d\theta}{\cos\theta - \cos\theta_0} = \alpha - \frac{dz}{dx} \tag{6.21}$$

This integral equation can be used to evaluate the coefficients $A_0, A_1, A_2, \ldots, A_n$ in terms of the angle of attack and the mean camber-line slope, which is known for a given airfoil section. The first integral on the left-hand side of equation (6.21) can be readily evaluated using equation (6.10). To evaluate the series of integrals represented by the second term, one must use equation (6.10) and the trigonometric identity:

$$(\sin n\theta)(\sin\theta) = 0.5[\cos(n - 1)\theta - \cos(n + 1)\theta]$$

Equation (6.21) becomes

$$\frac{dz}{dx} = \alpha - A_0 + \sum_{n=1}^{\infty} A_n \cos n\theta \tag{6.22}$$

which applies to any chordwise station. Since we are evaluating both dz/dx and $\cos n\theta_0$ at the general point θ_0 (i.e., x), we have dropped the subscript 0 from equation (6.22) and from all subsequent equations. Thus, the coefficients $A_0, A_1, A_2, \ldots, A_n$ must satisfy equation (6.22) if equation (6.20) is to represent the vorticity distribution which satisfies the condition that the mean camber line is a streamline. Since the geometry of the mean camber line would be known for the airfoil of interest, the slope is a known function of θ. One can, therefore, determine the values of the coefficients.

To evaluate A_0, note that

$$\int_0^\pi A_n \cos n\theta \, d\theta = 0$$

for any value of n. Thus, by algebraic manipulation of equation (6.22),

$$A_0 = \alpha - \frac{1}{\pi} \int_0^\pi \frac{dz}{dx} \, d\theta \qquad \textbf{(6.23)}$$

Multiplying both sides of equation (6.22) by $\cos m\theta$, where m is an unspecified integer, and integrating from 0 to π, one obtains

$$\int_0^\pi \frac{dz}{dx} \cos m\theta \, d\theta = \int_0^\pi (\alpha - A_0) \cos m\theta \, d\theta + \int_0^\pi \sum_{n=1}^\infty A_n \cos n\theta \cos m\theta \, d\theta$$

The first term on the right-hand side is zero for any value of m. Note that

$$\int_0^\pi A_n \cos n\theta \cos m\theta \, d\theta = 0 \qquad \text{when } n \neq m$$

but

$$\int_0^\pi A_n \cos n\theta \cos m\theta \, d\theta = \frac{\pi}{2} A_n \qquad \text{when } n = m$$

Thus,

$$A_n = \frac{2}{\pi} \int_0^\pi \frac{dz}{dx} \cos n\theta \, d\theta \qquad \textbf{(6.24)}$$

Using equations (6.23) and (6.24) to define the coefficients, equation (6.20) can be used to evaluate the vorticity distribution for a cambered airfoil in terms of the geometric angle of attack and the shape of the mean camber line. Note that, for a symmetric airfoil, $A_0 = \alpha$, $A_1 = A_2 = \cdots = A_n = 0$. Thus, the vorticity distribution for a symmetric airfoil, as determined using equation (6.20), is

$$\gamma(\theta) = 2\alpha U_\infty \frac{1 + \cos\theta}{\sin\theta}$$

This is identical to equation (6.9). Therefore, the general expression for the cambered airfoil includes the symmetric airfoil as a special case.

6.5.2 Aerodynamic Coefficients for a Cambered Airfoil

The lift and the moment coefficients for a cambered airfoil are found in the same manner as for the symmetric airfoil. The section lift coefficient is given by

$$C_l = \frac{1}{0.5 \rho_\infty U_\infty^2 c} \int_0^c \rho_\infty U_\infty \gamma(\xi) \, d\xi$$

Using the coordinate transformation [equation (6.7)] and the expression for γ [equation (6.20)] gives us

$$C_l = 2\left[\int_0^\pi A_0(1 + \cos\theta)\,d\theta + \int_0^\pi \sum_{n=1}^\infty A_n \sin n\theta \sin\theta\,d\theta \right]$$

Note that $\int_0^\pi A_n \sin n\theta \sin\theta\,d\theta = 0$ for any value of n other than unity. Thus, upon integration, one obtains

$$C_l = \pi(2A_0 + A_1) \tag{6.25}$$

The section moment coefficient for the pitching moment about the leading edge is given by

$$C_{m_0} = \frac{-1}{0.5\rho_\infty U_\infty^2 c^2} \int_0^c \rho_\infty U_\infty \gamma(\xi)\xi\,d\xi$$

Again, using the coordinate transformation and the vorticity distribution, one obtains, upon integration,

$$C_{m_0} = -\frac{\pi}{2}\left(A_0 + A_1 - \frac{A_2}{2} \right) \tag{6.26}$$

The center of pressure relative to the leading edge is found by dividing the moment about the leading edge (per unit span) by the lift per unit span.

$$x_{\rm cp} = -\frac{m_0}{l}$$

The negative sign is used since a positive lift force with a positive moment arm $x_{\rm cp}$ results in a nose-down, or negative moment, as shown in the sketch of Fig. 6.7. Thus,

$$x_{\rm cp} = \frac{c}{4}\left(\frac{2A_0 + 2A_1 - A_2}{2A_0 + A_1} \right)$$

Noting that $C_l = \pi(2A_0 + A_1)$, the expression for the center of pressure becomes

$$x_{\rm cp} = \frac{c}{4}\left[1 + \frac{\pi}{C_l}(A_1 - A_2) \right] \tag{6.27}$$

Thus, for the cambered airfoil, the position of the center of pressure depends on the lift coefficient and hence the angle of attack. The line of action for the lift, as well as the magnitude, must be specified for each angle of attack.

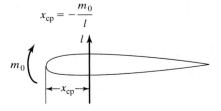

Figure 6.7 Center of pressure for a thin, cambered airfoil.

If the pitching moment per unit span produced by the pressure distribution is referred to a point $0.25c$ downstream of the leading edge (i.e., the quarter chord), the moment is given by

$$m_{0.25c} = \int_0^{0.25c} \rho_\infty U_\infty \gamma(\xi) \left(\frac{c}{4} - \xi \right) d\xi - \int_{0.25c}^c \rho_\infty U_\infty \gamma(\xi) \left(\xi - \frac{c}{4} \right) d\xi$$

Again, the signs are chosen so that a nose-up moment is positive. Rearranging the relation yields

$$m_{0.25c} = \frac{c}{4} \int_0^c \rho_\infty U_\infty \gamma(\xi) \, d\xi - \int_0^c \rho_\infty U_\infty \gamma(\xi) \xi \, d\xi$$

The first integral on the right-hand side of this equation represents the lift per unit span, while the second integral represents the moment per unit span about the leading edge. Thus,

$$m_{0.25c} = \frac{c}{4} l + m_0 \tag{6.28}$$

The section moment coefficient about the quarter-chord point is given by

$$C_{m_{0.25c}} = \frac{1}{4} C_l + C_{m_0} = \frac{\pi}{4} (A_2 - A_1) \tag{6.29}$$

Since A_1 and A_2 depend on the camber only, the section moment coefficient about the quarter-chord point is independent of the angle of attack. The point about which the section moment coefficient is independent of the angle of attack is called the *aerodynamic center of the section*. Thus, according to the theoretical relations for a thin-airfoil section, the aerodynamic center is at the quarter-chord point.

In order for the section pitching moment to remain constant as the angle of attack is increased, the product of the moment arm (relative to the aerodynamic center) and C_l must remain constant. Thus, the moment arm (relative to the aerodynamic center) decreases as the lift increases. This is evident in the expression for the center of pressure, which is given in equation (6.27). Alternatively, the aerodynamic center is the point at which all changes in lift effectively take place. Because of these factors, the center of gravity is usually located near the aerodynamic center. If we include the effects of viscosity on the flow around the airfoil, the lift due to angle of attack would not necessarily be concentrated at the exact quarter-chord point. However, for angles of attack below the onset of stall, the actual location of the aerodynamic center for the various sections is usually between the 23% chord point and the 27% chord point. Thus, the moment coefficient about the aerodynamic center, which is given the symbol $C_{m_{ac}}$, is also given by equation (6.29). If equation (6.24) is used to define A_1 and A_2, then $C_{m_{ac}}$ becomes

$$C_{m_{ac}} = \frac{1}{2} \int_0^\pi \frac{dz}{dx} (\cos 2\theta - \cos \theta) \, d\theta \tag{6.30}$$

Note, as discussed when comparing theory with data in the preceding section, the $C_{m_{ac}}$ is zero for symmetric airfoil.

EXAMPLE 6.2: Theoretical aerodynamic coefficients for a cambered airfoil

The relations developed in this section will now be used to calculate the aerodynamic coefficients for a representative airfoil section. The airfoil section selected for use in this sample problem is the NACA 2412. As discussed in Abbott and Doenhoff (1949) the first digit defines the maximum camber in percent of chord, the second digit defines the location of the maximum camber in tenths of chord, and the last two digits represent the thickness ratio (i.e., the maximum thickness in percent of chord). The equation for the mean camber line is defined in terms of the maximum camber and its location. Forward of the maximum camber position, the equation of the mean camber line is

$$\left(\frac{z}{c}\right)_{\text{fore}} = 0.125\left[0.8\left(\frac{x}{c}\right) - \left(\frac{x}{c}\right)^2\right]$$

while aft of the maximum camber position,

$$\left(\frac{z}{c}\right)_{\text{aft}} = 0.0555\left[0.2 + 0.8\left(\frac{x}{c}\right) - \left(\frac{x}{c}\right)^2\right]$$

Solution: To calculate the section lift coefficient and the section moment coefficient, it is necessary to evaluate the coefficients A_0, A_1, and A_2. To evaluate these coefficients it is necessary to integrate an expression involving the function which defines the slope of the mean camber line. Therefore, the slope of the mean camber line will be expressed in terms of the θ coordinate, which is given in equation (6.7). Forward of the maximum camber location, the slope is given by

$$\left(\frac{dz}{dx}\right)_{\text{fore}} = 0.1 - 0.25\frac{x}{c} = 0.125\cos\theta - 0.025$$

Aft of the maximum camber location, the slope is given by

$$\left(\frac{dz}{dx}\right)_{\text{aft}} = 0.0444 - 0.1110\frac{x}{c} = 0.0555\cos\theta - 0.0111$$

Since the maximum camber location serves as a limit for the integrals, it is necessary to convert the x coordinate, which is $0.4c$, to the corresponding θ coordinate. To do this,

$$\frac{c}{2}(1 - \cos\theta) = 0.4c$$

Thus, the location of the maximum camber is $\theta = 78.463° = 1.3694$ rad.
Referring to equations (6.23) and (6.24), the necessary coefficients are

$$A_0 = \alpha - \frac{1}{\pi}\left[\int_0^{1.3694}(0.125\cos\theta - 0.025)\,d\theta + \int_{1.3694}^{\pi}(0.0555\cos\theta - 0.0111)\,d\theta\right]$$

$$= \alpha - 0.004517$$

$$A_1 = \frac{2}{\pi} \left[\int_0^{1.3694} (0.125 \cos^2 \theta - 0.025 \cos \theta)\, d\theta \right.$$

$$\left. + \int_{1.3694}^{\pi} (0.0555 \cos^2 \theta - 0.0111 \cos \theta)\, d\theta \right]$$

$$= 0.08146$$

$$A_2 = \frac{2}{\pi} \left[\int_0^{1.3694} (0.125 \cos \theta \cos 2\theta - 0.025 \cos 2\theta)\, d\theta \right.$$

$$\left. + \int_{1.3694}^{\pi} (0.0555 \cos \theta \cos 2\theta - 0.0111 \cos 2\theta)\, d\theta \right]$$

$$= 0.01387$$

The section lift coefficient is

$$C_l = 2\pi \left(A_0 + \frac{A_1}{2} \right) = 2\pi\alpha + 0.2297$$

Solving for the angle of attack for zero lift, we obtain

$$\alpha_{0l} = -\frac{0.2297}{2\pi} \text{ rad} = -2.095°$$

According to thin-airfoil theory, the aerodynamic center is at the quarter-chord point. Thus, the section moment coefficient for the moment about the quarter chord is equal to that about the aerodynamic center. The two coefficients are given by

$$C_{m_{ac}} = C_{m_{0.25c}} = \frac{\pi}{4} (A_2 - A_1) = -0.05309$$

The theoretical values of the section lift coefficient and of the section moment coefficients are compared with the measured values from Abbott and Doenhoff (1949) in Figs. 6.8 and 6.9, respectively. Since the theoretical coefficients do not depend on the airfoil section thickness, they will be compared with data from Abbott and Doenhoff (1949) for a NACA 2418 airfoil as well as for a NACA 2412 airfoil. For both airfoil sections, the maximum camber is 2% of the chord length and is located at $x = 0.4c$. The maximum thickness is 12% of chord for the NACA 2412 airfoil section and is 18% of the chord for the NACA 2418 airfoil section.

The correlation between the theoretical and the experimental values of lift is satisfactory for both airfoils (Fig. 6.8) until the angle of attack becomes so large that viscous phenomena significantly affect the flow field. The theoretical value for the zero lift angle of attack agrees very well with the measured values for the two airfoils. The theoretical value of $C_{l,\alpha}$ is 2π per radian. Based on the measured lift coefficients for angles of attack for 0° to 10°, the experimental value of $C_{l,\alpha}$ is approximately 6.0 per radian for the NACA 2412 airfoil and approximately 5.9 per radian for the NACA 2418 airfoil.

The experimental values of the moment coefficient referred to the aerodynamic center (approximately −0.045 for the NACA 2412 section and −0.050 for the NACA

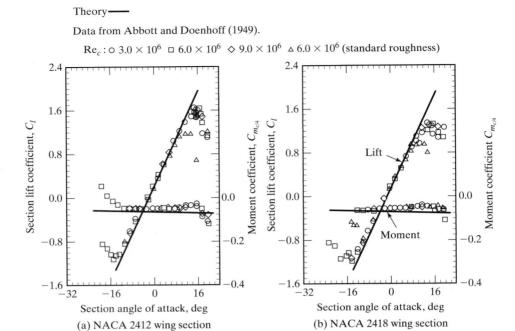

Theory ———

Data from Abbott and Doenhoff (1949).

$Re_c : \circ\ 3.0 \times 10^6$ $\square\ 6.0 \times 10^6$ $\diamond\ 9.0 \times 10^6$ $\triangle\ 6.0 \times 10^6$ (standard roughness)

(a) NACA 2412 wing section

(b) NACA 2418 wing section

Figure 6.8 Comparison of the aerodynamic coefficients calculated using thin airfoil theory for cambered airfoils: (a) NACA 2412 wing section; (b) NACA 2418 wing section. [Data From Abbott and von Doenhoff (1949).]

2418 airfoil) compare favorably with the theoretical value of −0.053 (Fig. 6.9). The correlation between the experimental values of the moment coefficient referred to the quarter chord, which vary with the angle of attack, and the theoretical value is not as good. Note also that the experimentally determined location of the aerodynamic center for these two airfoils is between $0.239c$ and $0.247c$. As noted previously, the location is normally between $0.23c$ and $0.27c$ for a real fluid flow, as compared with the value of $0.25c$ calculated using thin-airfoil theory.

Although the thickness ratio of the airfoil section does not enter into the theory, except as an implied limit to its applicability, the data of Figs. 6.8 and 6.9 show thickness-related variations. Note that the maximum value of the experimental lift coefficient is consistently greater for the NACA 2412 and that it occurs at a higher angle of attack. Also note that, as the angle of attack increases beyond the maximum lift value, the measured lift coefficients decrease more sharply for the NACA 2412. Thus, the thickness ratio influences the interaction between the adverse pressure gradient and the viscous boundary layer. The interaction, in turn, affects the aerodynamic coefficients. In Fig. 6.10, $C_{l,max}$ is presented as a function of the thickness ratio for the NACA 24XX series airfoils. The data of Abbott and von Doenhoff (1949) and the results of McCormick (1967) are presented. McCormick notes that below a thickness ratio of approximately 12%, $C_{l,max}$ decreases rapidly with decreasing thickness. Above a thickness ratio of 12%, the variation in $C_{l,max}$ is less pronounced.

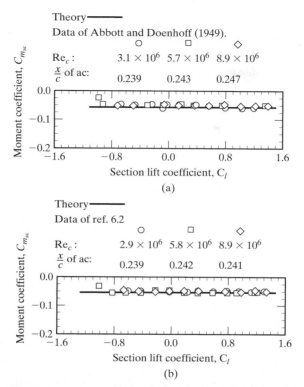

Figure 6.9 Comparison of the theoretical and the experimental section moment coefficient (about the aerodynamic center) for two cambered airfoils: (a) NACA 2412 wing section; (b) NACA 2418 wing section. [Data From Abbott and von Doenhoff (1949).]

The correlations presented in Figs. 6.6 and 6.8 indicate that, at low angles of attack, the theoretical lift coefficients based on thin-airfoil theory are in good agreement with the measured values from Abbott and von Doenhoff (1949). However, to compute the airfoil lift and pitching moment coefficients for various configurations exposed to a wide range of flow environments, especially if knowledge of the maximum section lift coefficient $C_{l,\max}$ is important, it is necessary to include the effects of the boundary layer and of the separated wake. Using repeated application of a panel method (see Chapter 3) to solve for the separated wake displacement surface, Henderson (1978) discussed the relative importance of separation effects. These effects are illustrated in Fig. 6.11. The lift coefficient calculated using potential flow analysis with no attempt to account for the effects either of the boundary layer or of separation is compared with wind-tunnel data in Fig. 6.11a. At low angles of attack where the boundary layer is thin and there is little, if any, separation, potential flow analysis of the surface alone is a fair approximation to the data; but as the angle of attack is increased, the correlation degrades.

Henderson (1978) notes, "Rarely will the boundary layers be thin enough that potential flow analysis of the bare geometry will be sufficiently accurate." By including the effect of the boundary layer but not the separated wake in the computational flow

Data of Abbott and von Doenhoff (1949):
Re_c : ○ 3×10^6 □ 6×10^6 ◇ 9×10^6
Fairings of McCormick (1967).

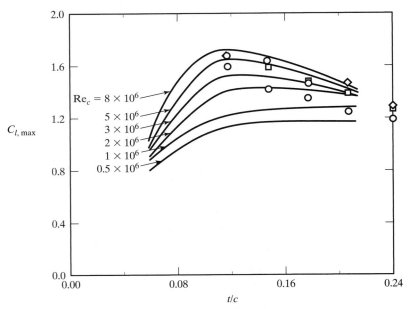

Figure 6.10 Effect of the thickness ratio on the maximum lift coefficient, NACA 24XX series airfoil sections.

model, the agreement between the theoretical lift coefficients and the wind-tunnel values is good at low angles of attack, as shown in Fig. 6.11b. When the angle of attack is increased and separation becomes important, the predicted and the measured lift coefficients begin to diverge.

Separation effects must be modeled in order to predict the maximum lift coefficient. As shown in Fig. 6.11c, when one accounts for the boundary layer and the separated wake, there is good agreement between theoretical values and experimental values through $C_{l,\max}$. This will be the case for any gradually separating section, such as the GAW-1, used in the example of Fig. 6.11.

6.6 LAMINAR-FLOW AIRFOILS

Airplane designers have long sought the drag reduction that would be attained if the boundary layer over an airfoil were largely laminar rather than turbulent (see Section 5.4.4 for details about boundary-layer transition and its impact on drag). Designers since the 1930s have developed airfoils that could reduce drag by maintaining laminar flow, culminating in the NACA developing laminar-flow airfoils for use on full-scale aircraft (Jacobs, 1939). Comparing equations (4.25) and (4.81) shows that there is a fairly significant reduction in

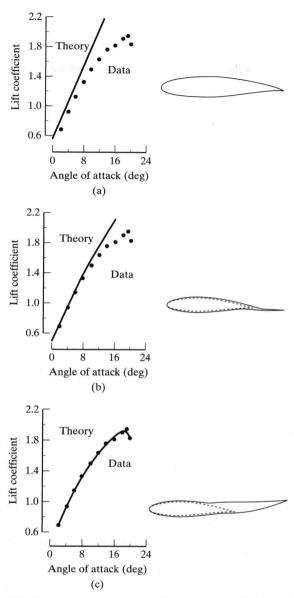

Figure 6.11 Relative importance of separation effects: (a) analysis of geometry alone; (b) analysis with boundary layer modeled; (c) analysis with boundary layer and separated wake modeled. [From Henderson (1978).]

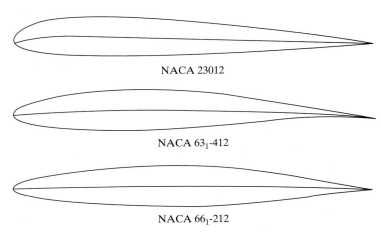

NACA 23012

NACA 63_1-412

NACA 66_1-212

Figure 6.12 Shapes of two NACA laminar-flow airfoil sections compared with the NACA 23012 airfoil section. [Loftin (1985).]

skin-friction drag (at reasonably high Reynolds numbers) if the boundary layers are laminar rather than turbulent.

$$\overline{C}_f = \frac{1.328}{\sqrt{\mathrm{Re}_L}} \qquad (4.25)$$

$$\overline{C}_f = \frac{0.074}{(\mathrm{Re}_L)^{0.2}} \qquad (4.81)$$

The early attempts at designing laminar-flow airfoils centered around modifications to the airfoil geometry that would maintain a favorable pressure gradient over a majority of the airfoil surface, as shown in Fig. 6.12. This was accomplished primarily by moving the maximum thickness location of the airfoil further aft, preferably to the mid-chord or beyond.

An entire series of these airfoils were designed and tested, and many of the resulting shapes can be found in *Theory of Wing Sections* by Abbott and von Doenhoff (1949) as the 6-digit airfoil series. These airfoil sections long have been used on general aviation aircraft, including airplanes like the Piper Archer. In the wind tunnel, these airfoils initially showed very promising drag reduction at cruise angles of attack, as shown in Fig. 6.13. The "bucket" in the drag curve for the laminar-flow airfoil occurs at angles of attack that normally might be required for cruise, and show a potential drag reduction of up to 25% over the conventional airfoil. Notice, however, that when the wind-tunnel model had typical surface roughness, the flow transitioned to turbulent, and the laminar flow benefits were greatly reduced, even eliminated.

The P-51 was the first production aircraft to utilize these laminar flow airfoils in an attempt to improve range by increasing the wing size and fuel volume for the same amount of drag as a turbulent-flow airfoil (see Fig. 6.14). Unfortunately, laminar-flow airfoils do not function properly if the boundary layer transitions to turbulent, which can happen easily if the wing surface is not smooth. Keeping an airfoil smooth is something

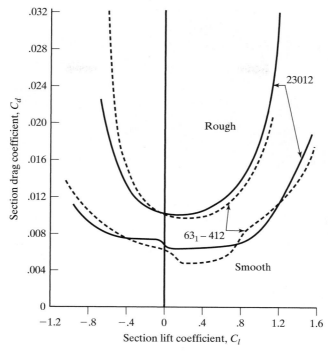

Figure 6.13 Drag characteristics of NACA laminar flow and conventional airfoils sections with both smooth and rough leading edges. [Loftin (1985).]

Figure 6.14 Restored NACA P-51 with laminar flow airfoil sections. [Courtesy of NASA Dryden Flight Research Center.]

that is relatively easy to achieve with wind-tunnel models but rarely takes place with production aircraft. "As a consequence, the use of NACA laminar-flow airfoil sections has never resulted in any significant reduction in drag as a result of the achievement of laminar flow" [Loftin (1985)]. This has led to a variety of flow-control devices being used to actively maintain laminar flow, but most of these devices require additional power sources (such as boundary-layer suction or blowing, as shown in Section 13.4.2), which usually does not make them practical as a drag-reduction concept.

Figure 6.15 Candidate laminar-flow airfoil aircraft: Black Widow Micro UAV (left) and Helios high-altitude solar-powered aircraft (right). (Black Widow from Grassmeyer and Keennon (2001) and Helios courtesy of NASA Dryden Flight Research Center.)

The relatively high Reynolds numbers of full-scale aircraft flying at common flight altitudes made the realization of lower drag using laminar flow impractical, but new applications have revived interest in laminar flow airfoils, including micro UAVs [Grassmeyer and Keennon (2001)] and high-altitude aircraft (see Fig. 6.15). These aircraft have vastly different configurations, ranging from very low aspect-ratio "flying discs" to very high aspect-ratio aircraft. The difference in design is dictated by the difference in application, with the micro UAV flying at such low Reynolds numbers that typical design thinking about aspect ratio no longer holds. Heavier aircraft often have higher aspect ratios, because induced drag is so much larger than skin-friction drag at higher Reynolds numbers. As the size of the vehicle decreases (and as weight and Reynolds number also decreases), aspect ratio no longer is the dominant factor in creating drag—skin-friction drag becomes more important, hence the low aspect-ratio design common for micro UAVs [Drela (2003)]. The high-altitude aircraft (such as Helios) also fly at low Reynolds numbers but require fairly heavy weights in order to carry the solar panels and batteries required for propulsion—the high aspect-ratio aircraft once again becomes more efficient.

The most important fluid dynamics characteristics for the design of laminar airfoils are laminar separation bubbles and transition. Laminar flow separates easier than turbulent flow and often leads to separation bubbles, as shown in Fig. 6.16. These separated flow regions reattach, but the boundary layer usually transitions to turbulent through the separation process, leading to higher drag due to the bubble and turbulent flow after the bubble. Methods to overcome the laminar separation bubble include the tailoring of the airfoil geometry ahead of the bubble formation or by using transition trips. While transition trips do increase the skin-friction drag, if used properly they also can lead to a net reduction in drag due to the elimination of the bubble [Gopalarathnam et al. (2001)].

The fact that the uses for laminar airfoils vary a great deal (ranging from UAVs, to gliders and to high-altitude aircraft) means that there is no single optimum airfoil: each application requires a different wing and airfoil design in order to optimize performance [Torres and Mueller (2004)]. This led many researchers to wind-tunnel testing of laminar-flow airfoils so that designers could choose optimum airfoil sections depending on their

Figure 6.16 Laminar separation bubble on an airfoil shown by surface oil flow where separation and reattachment are visible. [Selig (2003).]

requirements [Selig et al. (1989 and 2001)]. A good overview of the type of airfoils that work for various uses is presented by Selig (2003), including wind turbines, airfoils with low pitching moments, high-lift airfoils, and radio-controlled sailplanes (candidate airfoils are shown in Fig. 6.17). Another approach is to use various numerical prediction methods which have been developed over the years, including Eppler's code [Eppler et al. (1980 and 1990)], XFOIL [Drela (1989)], and inverse design methods such as PROFOIL [Selig and Maughmer (1992)] and its applications [Jepson and Gopalarathnam (2004)]. These codes partially rely on semi-empirical or theoretical methods for predicting laminar separation bubbles and transition and have been found to produce reasonable results.

6.7 HIGH-LIFT AIRFOIL SECTIONS

As noted by Smith (1975), "The problem of obtaining high lift is that of developing the lift in the presence of boundary layers—getting all the lift possible without causing separation. Provided that boundary-layer control is not used, our only means of obtaining

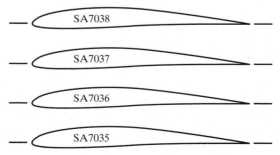

Figure 6.17 Candidate airfoils for radio controlled sailplanes. [Selig (2003)].

higher lift is to modify the geometry of the airfoil. For a one-piece airfoil, there are several possible means for improvement—changed leading-edge radius, a flap, changed camber, a nose flap, a variable-camber leading edge, and changes in detail shape of a pressure distribution."

Thus, if more lift is to be generated, the circulation around the airfoil section must be increased, or, equivalently, the velocity over the upper surface must be increased relative to the velocity over the lower surface. However, once the effect of the boundary layer is included, the Kutta condition at the trailing edge requires that the upper-surface and the lower-surface velocities assume a value slightly less than the free-stream velocity. Hence, when the higher velocities over the upper surface of the airfoil are produced in order to get more lift, larger adverse pressure gradients are required to decelerate the flow from the maximum velocity to the trailing-edge velocity. Again, referring to [Smith (1975)], "The process of deceleration is critical, for if it is too severe, separation develops. The science of developing high lift, therefore, has two components: (1) analysis of the boundary layer, prediction of separation, and determination of the kinds of flows that are most favorable with respect to separation; and (2) analysis of the inviscid flow about a given shape with the purpose of finding shapes that put the least stress on a boundary layer."

Stratford (1959) has developed a formula for predicting the point of separation in an arbitrary decelerating flow. The resultant Stratford pressure distribution, which recovers a given pressure distribution in the shortest distance, has been used in the work of Liebeck (1973). To develop a class of high-lift airfoil sections, Liebeck used a velocity distribution that satisfied "three criteria: (1) the boundary layer does not separate; (2) the corresponding airfoil shape is practical and realistic; and (3) maximum possible C_l is obtained." The optimized form of the airfoil velocity distribution is markedly different than that for a typical airfoil section (which is presented in Fig. 6.18). The velocity distribution is presented as a function of s, the distance along the airfoil surface, where s begins at the lower-surface trailing edge and proceeds clockwise around the airfoil surface to the upper- surface trailing edge. In the s-coordinate system, the velocities are negative on the lower surface and positive on the upper surface. The "optimum" velocity distribution, modified to obtain a realistic airfoil, is presented in Fig. 6.19. The lower-surface velocity is as low as possible in the interest of obtaining the maximum lift and increases continuously from the leading-edge stagnation point to the trailing-edge velocity. The upper-surface acceleration region is shaped to provide good off-design performance. A short boundary-layer transition ramp (the region where the flow decelerates, since an adverse pressure gradient promotes transition) is used to ease the boundary layer's introduction to the severe initial Stratford gradient.

Once the desired airfoil velocity distribution has been defined, there are two options available for calculating the potential flow. One method uses conformal mapping of the flow to a unit circle domain to generate the airfoil [e.g., Eppler and Somers (1980) and Liebeck (1976)]. A second method uses the panel method for the airfoil analysis [e.g., Stevens et al. (1971)]. Olson et al. (1978) note that, in the potential flow analysis, the airfoil section is represented by a closed polygon of planar panels connecting the input coordinate pairs. The boundary condition for the inviscid flow—that there be no flow through the airfoil surface—is applied at each of the panel centers. An additional equation, used to close the system, specifies that the upper- and lower-surface velocities have a common limit at

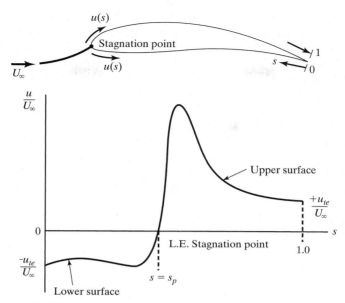

Figure 6.18 General form of the velocity distribution around a typical airfoil. [From Liebeck (1973).]

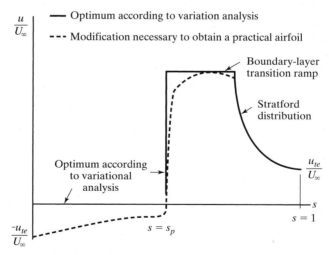

Figure 6.19 "Optimized" velocity distribution for a high-lift, single-element airfoil section. [From Liebeck (1973).]

the trailing edge (i.e., the Kutta condition). The effect of boundary-layer displacement is simulated by piecewise linear source distributions on the panels describing the airfoil contour. Thus, instead of modifying the airfoil geometry by an appropriate displacement thickness to account for the boundary layer, the boundary condition is modified by introducing surface transpiration. Miranda (1984) notes that "The latter approach is more satisfactory because the surface geometry and the computational grid are not affected by the boundary layer. This means that, for panel methods, the aerodynamic influence coefficients and, for finite difference methods, the computational grid do not have to be recomputed at each iteration." The boundary condition on the surface panels requires that the velocity normal to the surface equals the strength of the known source distribution.

Liebeck has developed airfoil sections which, although they "do not appear to be very useful" [the quotes are from Liebeck (1973)], develop an l/d of 600. The airfoil section, theoretical pressure distribution, the experimental lift curve and drag polar, and the experimental pressure distributions for a more practical, high-lift section are presented in Figs. 6.20 through 6.22. The pressure distributions indicate that the flow remained attached all the way to the trailing edge. The flow remained completely attached until the stalling angle was reached, at which point the entire recovery region separated instantaneously. Reducing the angle of attack less than $0.5°$ resulted in an instantaneous and complete reattachment, indicating a total lack of hysteresis effect on stall recovery.

Improvements of a less spectacular nature have been obtained for airfoil sections being developed by NASA for light airplanes. One such airfoil section is the General Aviation (Whitcomb) number 1 airfoil, GA(W)-1, which is 17% thick with a blunt nose and a cusped lower surface near the trailing edge. The geometry of the GA(W)-1 section is

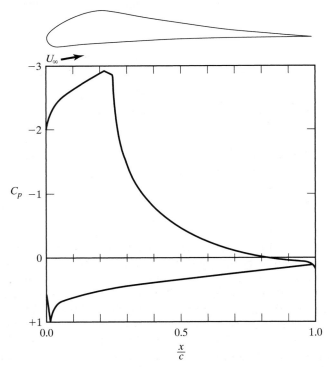

Figure 6.20 Theoretical pressure distribution for high-lift, single-element airfoil, $Re_c = 3 \times 10^6$, $t_{max} = 0.125c$, $C_l = 1.35$. [From Liebeck (1973).]

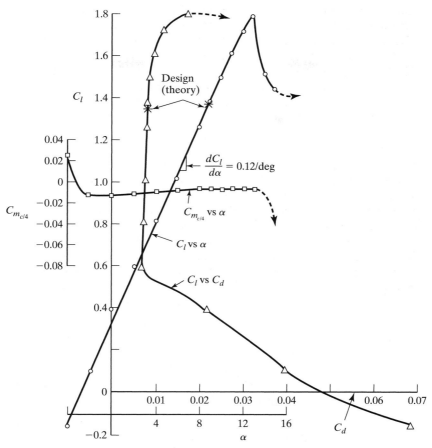

Figure 6.21 Experimental lift curve, drag polar, and pitching curve for a high-lift, single-element airfoil, $Re_c = 3 \times 10^6$. [From Liebeck (1973).]

similar to that of the supercritical airfoil, which is discussed in Chapter 9. Experimentally determined lift coefficients, drag coefficients, and pitching moment coefficients, which are taken from McGhee and Beasley (1973), are presented in Fig. 6.23. Included for comparison are the corresponding correlations for the NACA 65_2–415 and the NACA 65_3–418 airfoil sections. Both the GA(W)-1 and the NACA 65_3–418 airfoils have the same design lift coefficient (0.40), and both have roughly the same mean thickness distribution in the region of the structural box ($0.15c$ to $0.60c$). However, the experimental value of the maximum section lift coefficient for the GA(W)-1 was approximately 30% greater than for the NACA 65 series airfoil for a Reynolds number of 6×10^6. Since the section drag coefficient remains approximately constant to higher lift coefficients for the GA(W)-1, significant increases in the lift/drag ratio are obtained. At a lift coefficient of 0.90, the lift/drag ratio for the GA(W)-1 was approximately 70,

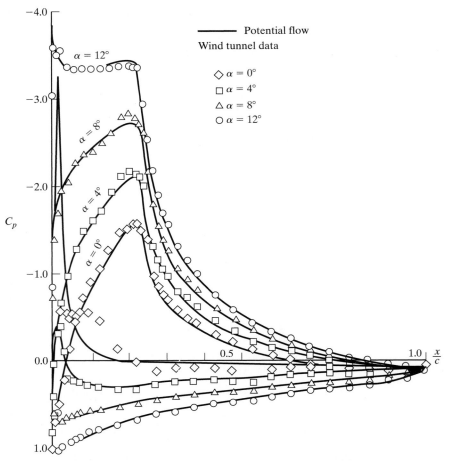

Figure 6.22 Comparison of the theoretical potential-flow and the experimental pressure distribution of a high-lift, single-element airfoil, $\text{Re}_c = 3 \times 10^6$. [From Liebeck (1973).]

which is 50% greater than that for the NACA 65_3–418 section. This is of particular importance from a safety standpoint for light general aviation airplanes, where large values of section lift/drag ratio at high lift coefficients result in improved climb performance.

6.8   MULTIELEMENT AIRFOIL SECTIONS FOR GENERATING HIGH LIFT

As noted by Meredith (1993), "High-lift systems are used on commercial jet transports to provide adequate low speed performance in terms of take-off and landing field lengths, approach speed, and community noise. The importance of the high-lift system is illustrated by the following trade factors derived for a generic large twin engine transport.

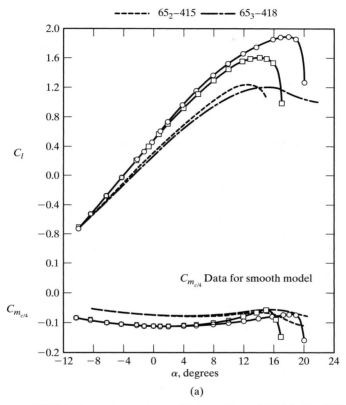

NASA GA(W)-1 airfoil

o   NASA standard roughness

□   NACA standard roughness

NACA airfoils, NACA standard roughness

- - - - - 65_2–415   — · — 65_3–418

Figure 6.23(a) Aerodynamic coefficients for a NASA GA(W)-1 airfoil, for a NACA 65_2–415 airfoil, and for a NACA 65_3–418 airfoil; $M_\infty = 0.20$, $\mathrm{Re}_c = 6 \times 10^6$: (a) lift coefficient and pitching moment coefficient curves; (b) drag polars. [Data from McGhee and Beasley (1973).]

1. A 0.10 increase in lift coefficient at constant angle of attack is equivalent to reducing the approach attitude by about one degree. For a given aft body-to-ground clearance angle, the landing gear may be shortened resulting in a weight savings of 1400 lb.

2. A 1.5% increase in the maximum lift coefficient is equivalent to a 6600 lb increase in payload at a fixed approach speed.

3. A 1% increase in take-off L/D is equivalent to a 2800 lb increase in payload or a 150 nm increase in range.

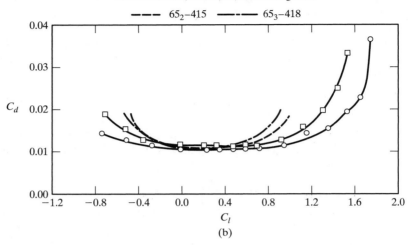

Figure 6.23(b)

While necessary, high-lift systems increase the airplane weight, cost, and complexity significantly. Therefore, the goal of the high-lift system designer is to design a high-lift system which minimizes these penalties while providing the required airplane take-off and landing performance."

Jasper et al. (1993) noted, "Traditionally (and for the foreseeable future) high-lift systems incorporate multi-element geometries in which a number of highly-loaded elements interact in close proximity to each other." Figure 6.24 shows a sketch depicting the cross section of a typical configuration incorporating four elements: a leading-edge slat,

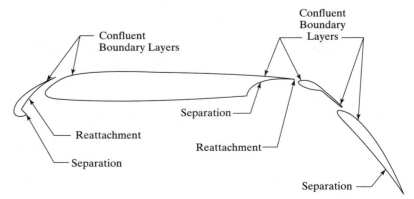

Figure 6.24 Sketch of the cross section of a typical high-lift multielement airfoil section. [Taken from Jasper et al. (1993).]

the main-element airfoil, a flap vane, and trailing-edge flap. Jasper et al. (1993) continued, "Such configurations generate very complex flowfields containing regions of separated flow, vortical flow, and confluent boundary layers. Laminar, turbulent, transitional, and relaminarizing boundary layers may exist. Although high lift systems are typically deployed at low freestream Mach numbers, they still exhibit compressibility effects due to the large pressure gradients generated .... It should be noted that many of the flowfield phenomena (e.g., separation, transition, turbulence, etc.) are areas of intense research in the computational community and are not yet fully amenable to computational analysis."

As noted by Yip et al. (1993), "Two-dimensional multi-element flow issues include the following:

1. compressibility effects including shock/boundary-layer interaction on the slat;
2. laminar separation-induced transition along the upper surfaces;
3. confluent turbulent boundary layer(s) — the merging and interacting of wakes from upstream elements with the boundary layers of downstream elements;
4. cove separation and reattachment; and
5. massive flow separation on the wing/flap upper surfaces."

The complex flowfields for high-lift multielement airfoils are very sensitive to Reynolds number–related phenomena and to Mach number–related phenomena. As noted in the previous paragraph, many of the relevant flow-field issues (e.g., separation, transition, and turbulence, etc.) are difficult to model numerically.

The airfoil configuration that is chosen based on cruise requirements determines a lot of important parameters for the high-lift devices, such as the chord and the thickness distribution. Only the type of the high-lift devices, the shape, the spanwise extensions, and the settings can be chosen by the designer of the high-lift system. Even then, the designer is limited by several constraints. As noted by Flaig and Hilbig (1993), "Usually the chordwise extension of the high-lift devices is limited by the location of the front spar and rear spar respectively, which can not be changed due to considerations of wing stiffness (twist, bending) and internal fuel volume." These constraints are depicted in the sketches of Fig. 6.25.

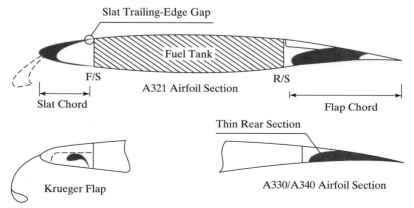

Figure 6.25 General constraints on the design of high-lift multi-element airfoil sections. [Taken from Flaig and Hilbig (1993).]

Flaig and Hilbig (1993) note further, "Especially the required fuel capacity for a long-range aircraft can be of particular significance in the wing sizing. Moreover, the inner wing flap chord of a typical low set wing aircraft is limited by the required storage space for the retracted main undercarriage.

After the chordwise extension of the leading edge and trailing edge devices has been fixed, the next design item is the optimization of their shapes.

The typical leading edge devices of today's transport aircraft are slats and Krueger flaps. In the case of a slat, the profile of upper and lower surface is defined by the cruise wing nose shape. Therefore only the shape of the slat inner side and the nose of the fixed-wing can be optimized.

A Krueger flap with a folded nose or flexible shape, as an example, generally offers greater design freedom to achieve an ideal upper surface shape, and thus gains a little in L/D and $C_{L,\max}$. But, trade-off studies carried out in the past for A320 and A340 have shown that this advantage for the Krueger flap is compromised by a more complex and heavier support structure than required for a slat."

The required maximum lift capability for the landing configuration determines the complexity of the high-lift system. In particular, the number of slots (or elements) of trailing-edge devices has a significant effect on $C_{L,\max}$. The degrading effect of wing sweep on the maximum lift coefficient necessitates an increase in the complexity of the high-lift system.

The general trend of the maximum lift efficiency is presented as a function of the system complexity in Fig. 6.26, which is taken from Flaig and Hilbig (1993). Note that the maximum value for the coefficient of lift for unpowered high-lift systems is approximately 3 (on an aircraft with typical 25-degree wing sweep). Powered high-lift systems with additional active boundary-layer control may achieve maximum values of the lift coefficient up to 7.

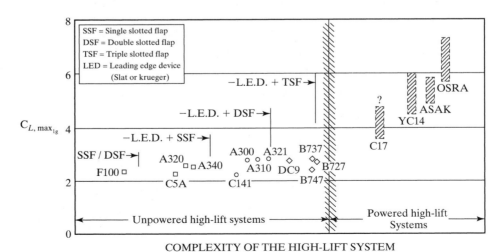

COMPLEXITY OF THE HIGH-LIFT SYSTEM

Figure 6.26 The maximum lift coefficient as a function of the complexity of the high-lift system. [Taken from Flaig and Hilbig (1993).]

The problem of computing the aerodynamic characteristics of multielement airfoils can be subdivided into the following broad topical areas, each requiring models for the computer program [Stevens, et al. (1971)]:

1. Geometry definition
2. Solution for the inviscid, potential flow
3. Solution for the conventional boundary layer
4. Solution for the viscous wakes and slot-flow characteristics
5. Combined inviscid/viscous solution

Stevens et al. (1971) note that the geometric modeling of the complete airfoil, including slots, slats, vanes, and flaps, requires a highly flexible indexing system to ensure that conventional arrangements of these components can be readily adapted to the code.

To compute the inviscid, potential flow, Stevens et al. (1971) and Olson et al. (1978) use distributed vortex singularities as the fundamental solution to the Laplace equation. Olson et al.(1978), note that viscous calculations can be separated into three types of flows: conventional boundary layers, turbulent wakes, and confluent boundary layers (i.e., wakes merging with conventional boundary layers). These are illustrated in Fig. 6.27. To obtain a complete viscous calculation, the conventional boundary layers on the upper and lower surfaces of the main airfoil are first analyzed. These calculations provide the initial conditions to start the turbulent-wake analysis at the trailing edge of the principal airfoil. The calculations proceed downstream until the wake merges with the outer edge of the boundary

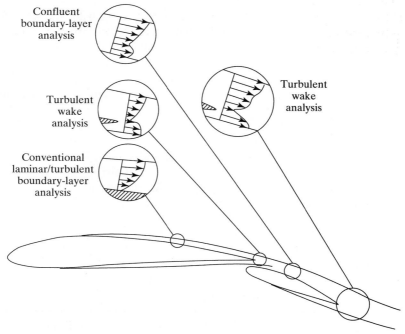

Figure 6.27 Theoretical flow models for the various viscous regions. [From Olson et al. (1978).]

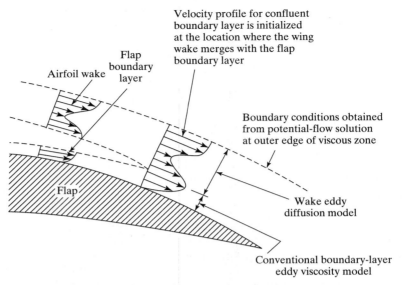

Figure 6.28 Flow model for merging of the wake from the principal airfoil with the boundary layer on the flap to form the confluent boundary layer on the upper surface of the flap. [From Olson et al. (1978).]

layer on the upper surface of the flap, as shown in Fig. 6.28. The wake from the principal airfoil and the boundary layer of the flap combine into a single viscous layer at this point, a so-called confluent boundary layer. The calculation procedure continues stepwise downstream to the flap trailing edge. At the flap trailing edge, this confluent boundary-layer solution merges with the boundary layer from the lower surface of the flap. The calculation then continues downstream into the wake along a potential-flow streamline.

Although the techniques used to calculate the viscous effects differ from those described in the preceding paragraph, the importance of including the viscous effects is illustrated in Fig. 6.29. Using repeated application of a panel method to solve for the separated wake displacement surface, Henderson (1978) found a significant effect on the pressure distributions both on the principal airfoil and the flap for the GA(W)-1 airfoil for which α was 12.5° and the flap angle was 40°. Although a separation wake occured for both models, the agreement between the calculated pressures and the experimental values was quite good.

As aircraft become more and more complex and as computational and experimental tools improve, the high-lift design process has matured a great deal. As was stated earlier, including viscous effects in high-lift design is important, but even with modern computer systems a high-lift design still may require a combination of viscous and inviscid numerical predictions. The Boeing 777 high-lift system was designed with various codes at different phases of the design process: a three-dimensional lifting surface code was used during preliminary design, two-dimensional viscous-inviscid coupled codes were used to design the multielement airfoil sections, and three-dimensional panel codes were used to evaluate flow interactions. Navier-Stokes and Euler codes were not used during

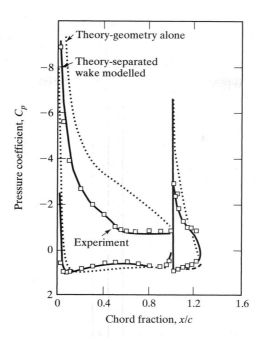

NASA GA(W)-1 airfoil, 30% Fowler flap,
Angle of attack = 12.5° , flap angle = 40°

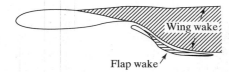

Figure 6.29 Comparison of experimental and calculated pressure distributions on a two-element airfoil with separation from both surfaces. [From Henderson (1978).]

the design process [Brune and McMasters (1990) and Nield (1995)]. This approach allowed for a reduction in wind-tunnel testing and resulted in a double-slotted flap that was more efficient than the triple-slotted flaps used on previous Boeing aircraft.

In fact, as the design process for high-lift systems has matured, the systems have become less complex, more affordable, more dependable, and more efficient. Figure 6.30 shows how high-lift airfoils designed by both Boeing/Douglas and Airbus have improved over the past twenty years. The improvement has been brought about by the increased use of numerical predictions, including the addition of Navier-Stokes and Euler methods to the predictions of high-lift airfoils [Rogers et al. (2001) and van Dam (2002)]. This evolution has been from triple-slotted flaps (such as on the Boeing 737), to double-slotted flaps (such as on the Boeing 777), and now to single-slotted flaps (such as on the Airbus A380 and Boeing 787). However, numerical predictions still require further improvement, including the addition of unsteady effects and improved

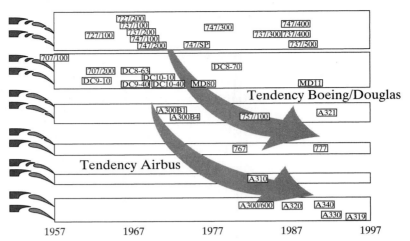

Figure 6.30 Design evolution of high-lift trailing edge systems. [From Reckzeh (2003).]

turbulence models, before high-lift design will be as evolved as one might hope [Rumsey and Ying (2002) and Cummings et al. (2004)].

6.9   HIGH-LIFT MILITARY AIRFOILS

As noted by Kern (1996), "There are two major geometric differences that distinguish modern high performance multi-role strike/fighter military airfoils from commercial configurations: (1) leading edge shape and (2) airfoil thickness. Integration of stealth requirements typically dictates sharp leading edges and transonic and supersonic efficiency dictates thin airfoils on the order of 5 to 8% chord .... The Navy also depends on low-speed high-lift aerodynamics, since it enables high performance multi-role strike aircraft to operate from a carrier deck." To obtain high lift at low speeds, the advanced fighter wing sections are configured with a plain leading-edge flap and a slotted trailing-edge flap. The schematic presented in Fig. 6.31 indicates some of the features of the complex flow field. The sharp leading edge causes the flow to separate, resulting in a shear layer that convects either above or below the airfoil surface. Depending on the angle of attack, the shear layer may or may not reattach to the surface of the airfoil. The flow field also contains cove flow, slot flow, merging shear layers, main element wake mixing, and trailing-edge flap separation.

Hobbs et al. (1996) presented the results of an experimental investigation using a 5.75% thick airfoil, which has a 14.07% chord plain leading-edge (L. E.) flap, a single slotted 30% chord trailing-edge (T. E.) flap, and a 8.78% chord shroud. Reproduced in Fig. 6.32 are the experimentally determined lift coefficients for the airfoil with δ_n (the leading-edge flap deflection angle) equal to 34°, with δ_f (the trailing-edge flap deflection angle) equal to 35°, and with δ_s (the shroud deflection angle) equal to 22.94°. This configuration provides the aircraft with the maximum lift required for the catapult and for the approach configurations. Note that, because of the leading-edge flow separation bubble, the lift curve displays

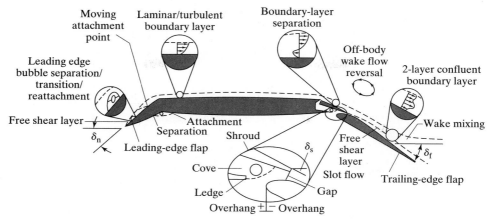

Figure 6.31 Sketch of the flow field for a military airfoil in a high-lift configuration. [A composite developed from information presented in Kern (1996) and Hobbs (1996).]

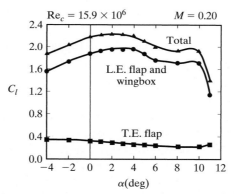

Figure 6.32 Total and component lift curves, $\delta_n = 34°$, $\delta_f = 35°$, $\delta_s = 22.94°$. [Taken from Hobbs (1996).]

no "linear" dependence on the angle of attack. The maximum lift coefficient of approximately 2.2 occurs at an angle of attack of 2°. The airfoil then gradually stalls, until total separation occurs at an angle of attack of 10°, with a rapid decrease in the section lift coefficient. As noted by Kern (1996), "This behavior seems less surprising when considering split-flap NACA 6% thick airfoils which all stall around $\alpha = 4°$."

Because the flow field includes trailing viscous wakes, confluent boundary layers, separated flows, and different transition regions, the Reynolds number is an important parameter in modeling the resultant flow field. The maximum values of the measured lift coefficients are presented as a function of the Reynolds number in Fig. 6.33. The data, which were obtained at a Mach number of 0.2, indicate that the maximum lift coefficient is essentially constant for Reynolds number beyond 9×10^6. These results (for

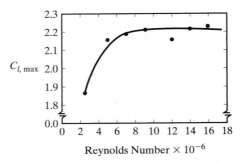

Figure 6.33 The effect of Reynolds number on $C_{l,\max}$, $\delta_n = 34°$, $\delta_f = 35°$, $\delta_s = 22.94°$. [Taken from Hobbs (1996).]

a two-dimensional flow) suggest that airfoils should be tested at a Reynolds number of 9×10^6, or more, in order to simulate maximum lift performance at full-scale flight conditions. Conversely, testing at a Reynolds number of 9×10^6 is sufficient to simulate full-scale maximum lift performance.

PROBLEMS

6.1. Using the identity given in equation (6.10), show that the vorticity distribution

$$\gamma(\theta) = 2\alpha U_\infty \frac{1 + \cos\theta}{\sin\theta}$$

satisfies the condition that flow is parallel to the surface [i.e., equation (6.8)]. Show that the Kutta condition is satisfied. Sketch the $2\gamma/U_\infty$ distribution as a function of x/c for a section lift coefficient of 0.5. What is the physical significance of $2\gamma/U_\infty$? What angle of attack is required for a symmetric airfoil to develop a section lift coefficient of 0.5?

Using the vorticity distribution, calculate the section pitching moment about a point 0.75 chord from the leading edge. Verify your answer, using the fact that the center of pressure (x_{cp}) is at the quarter chord for all angles of attack and the definition for lift.

6.2. Calculate C_l and $C_{m_{0.25c}}$ for a NACA 0009 airfoil that has a plain flap whose length is $0.2c$ and which is deflected $25°$. When the geometric angle of attack is $4°$, what is the section lift coefficient? Where is the center of pressure?

6.3. The mean camber line of an airfoil is formed by a segment of a circular arc (having a constant radius of curvature). The maximum mean camber (which occurs at midchord) is equal to kc, where k is a constant and c is a chord length. Develop an expression for the γ distribution in terms of the free-stream velocity U_∞ and the angle of attack α. Since kc is small, you can neglect the higher-order terms in kc in order to simplify the mathematics. What is the angle of attack for zero lift (α_{0l}) for this airfoil section? What is the section moment coefficient about the aerodynamic center ($C_{m_{ac}}$)?

6.4. The numbering system for wing sections of the NACA five-digit series is based on a combination of theoretical aerodynamic characteristics and geometric characteristics. The first integer indicates the amount of camber in terms of the relative magnitude of the design lift coefficient; the design lift coefficient in tenths is three halves of the first integer. The second and third integers together indicate the distance from the leading edge to the location of the maximum camber; this distance in percent of the chord is one-half the number represented by these integers. The last two integers indicate the section thickness in percent of the chord. The NACA 23012 wing section thus has a design lift coefficient of 0.3, has its maximum camber at 15% of the chord, and has a maximum thickness of $0.12c$. The equation for the mean camber line is

$$\frac{z}{c} = 2.6595\left[\left(\frac{x}{c}\right)^3 - 0.6075\left(\frac{x}{c}\right)^2 + 0.11471\left(\frac{x}{c}\right)\right]$$

for the region $0.0c \leq x \leq 0.2025c$ and

$$\frac{z}{c} = 0.022083\left(1 - \frac{x}{c}\right)$$

for the region $0.2025c \leq x \leq 1.000c$.

Calculate the A_0, A_1, and A_2 for this airfoil section. What is the section lift coefficient, C_l? What is the angle of attack for zero lift, α_{0l}? What angle of attack is required to develop the design lift coefficient of 0.3? Calculate the section moment coefficient about the theoretical aerodynamic center. Compare your theoretical values with the experimental values in Fig. P6.4 that are reproduced from the work of Abbott and von Doenhoff (1949). When the geometric angle of attack is 3°, what is the section lift coefficient? What is the x/c location of the center of pressure?

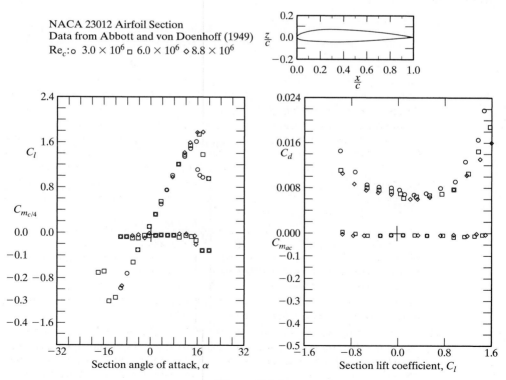

Figure P6.4

6.5. Look at the three airfoil geometries shown in Fig. 6.12. Discuss the geometric modifications to the laminar flow airfoils that make them distinct from the typical airfoil (NACA 23012). Include in your description airfoil geometric parameters such as camber, thickness, location of maximum camber, location of maximum thickness, leading edge radius, and trailing edge shape. Why were these modifications successful in creating a laminar flow airfoil?

6.6. What is a laminar separation bubble? What impact does it have on airfoil aerodynamics? What airfoil design features could be changed to eliminate (or largely reduce) the separation bubble?

6.7. What has enabled the evolution of commercial aircraft high-lift systems from triple-slotted to double- or even single-slotted geometries (see Fig. 6.30)? What are the advantages of these changes to aircraft design?

REFERENCES

Abbott IH, von Doenhoff AE. 1949. *Theory of Wing Sections*. New York: Dover

Brune GW, McMasters JH. 1990. Computational aerodynamics applied to high-lift systems. In *Computational Aerodynamics*, Ed. Henne PA. Washington, DC: AIAA

Carlson HW, Mack RJ. 1980. Studies of leading-edge thrust phenomena. *J. Aircraft* 17:890–897

Cummings RM, Morton SA, Forsythe JR. 2004. *Detached-eddy simulation of slat and flap aerodynamics for a high-lift wing*. Presented at AIAA Aerosp. Sci. Meet., 42nd, AIAA Pap. 2004–1233, Reno, NV

Drela M. 1989. XFOIL: An analysis and design system for low Reynolds number airfoils. In *Low Reynolds Number Aerodynamics*, Ed. Mueller TJ. New York: Springer-Verlag

Drela M, Protz JM, Epstein AH. 2003. *The role of size in the future of aeronautics*. Presented at Intl. Air and Space Symp., AIAA Pap. 2003–2902, Dayton, OH

Eppler R, Somers DM. 1980. A computer program for the design and analysis of low-speed airfoils. *NASA Tech. Mem. 80210*

Eppler R. 1990. *Airfoil Design and Data*. New York: Springer-Verlag

Flaig A, Hilbig R. 1993. High-lift design for large civil aircraft. In *High-Lift System Aerodynamics, AGARD CP 515*

Gopalarathnam A, Broughton BA, McGranahan BD, Selig MS. 2001. *Design of low Reynolds number airfoils with trips*. Presented at Appl. Aerodyn. Conf., 19th, AIAA Pap. 2001–2463, Anaheim, CA

Grassmeyer JM, Keennon MT. 2001. *Development of the Black Widow micro air vehicle*. Presented at AIAA Aerosp. Sci. Meet., 39th, AIAA Pap. 2001–0127, Reno, NV

Henderson ML. 1978. *A solution to the 2-D separated wake modeling problem and its use to predict C_{Lmax} of arbitrary airfoil sections*. Presented at AIAA Aerosp. Sci. Meet., 16th, AIAA Pap. 78–156, Huntsville, AL

Hobbs CR, Spaid FW, Ely WL, Goodman WL. 1996. *High lift research program for a fighter-type, multi-element airfoil at high Reynolds numbers*. Presented at AIAA Aerosp. Sci. Meet., 34th, AIAA Pap. 96–0057, Reno, NV

Jacobs EN. 1939. Preliminary report on laminar-flow airfoils and new methods adopted for airfoil and boundary-layer investigation. *NACA WR L–345*

Jasper DW, Agrawal S, Robinson BA. 1993. Navier-Stokes calculations on multi-element airfoils using a chimera-based solver. In *High-Lift System Aerodynamics, AGARD CP 515*

Jepson JK, Gopalarathnam A. 2004. *Inverse design of adapative airfoils with aircraft performance consideration.* Presented at AIAA Aerosp. Sci. Meet., 42nd, AIAA Pap. 2004–0028, Reno, NV

Kern S. 1996. *Evaluation of turbulence models for high lift military airfoil flowfields.* Presented at AIAA Aerosp. Sci. Meet., 34th, AIAA Pap. 96–0057, Reno, NV

Liebeck RH. 1973. A class of airfoils designed for high lift in incompressible flows. *J. Aircraft* 10:610–617

Liebeck RH. 1976. On *the design of subsonic airfoils for high lift.* Presented at Fluid and Plasma Dyn. Conf., 9th, AIAA Pap. 76–406, San Diego, CA

Loftin LK. 1985. Quest for performance: the evolution of modern aircraft. *NASA SP-468*

McCormick BW. 1967. *Aerodynamics of V/STOL Flight.* New York: Academic Press

McGhee RJ, Beasley WD. 1973. Low-speed aerodynamic characteristics of a 17-percent-thick section designed for general aviation applications. *NASA Tech. Note D-7428*

Meredith PT. 1993. Viscous phenomena affecting high-lift systems and suggestions for future CFD Ddevelopment. In *High-Lift System Aerodynamics, AGARD CP 515*

Miranda LR. 1984. Application of computational aerodynamics to airplane design. *J. Aircraft* 21: 355–369

Nield BN. 1995. An overview of the Boeing 777 high lift aerodynamic design. *The Aeronaut. J.* 99:361–371

Olson LE, James WD, McGowan PR. 1978. *Theoretical and experimental study of the drag of multielement airfoils.* Presented at Fluid and Plasma Dyn. Conf., 11th, AIAA Pap. 78–1223, Seattle, WA

Reckzeh D. 2003. Aerodynamic design of the high-lift wing for a megaliner aircraft. *Aerosp. Sci. Tech.* 7:107–119

Rogers SE, Roth K, Cao HV, Slotnick JP, Whitlock M, Nash SM, Baker D. 2001. Computation of viscous flow for a Boeing 777 aircraft in landing configuration. *J. Aircraft* 38: 1060–1068

Rumsey CL, Ying SX. 2002. Prediction of high lift: review of present CFD capability. *Progr. Aerosp. Sci.* 38:145–180

Selig MS, Donovan JF, Fraser DB. 1989. *Airfoils at Low Speeds.* Virginia Beach, VA: HA Stokely

Selig MS, Maughmer MD. 1992. A multi-point inverse airfoil design method based on conformal mapping. *AIAA J.* 30:1162–1170

Selig MS, Gopalaratham A, Giguére P, Lyon CA. 2001. Systematic airfoil design studies at low Reynolds number. In *Fixed and Flapping Wing Aerodynamics for Micro Air Vehicle Applications,* Ed. Mueller TJ. New York: AIAA, pp. 143–167

Selig MS. 2003. Low Reynolds number airfoil design. In *Low Reynolds Number Aerodynamics of Aircraft,* VKI Lecture Series

Smith AMO. 1975. High-lift aerodynamics. *J. Aircraft* 12:501–530

Stevens WA, Goradia SH, Braden JA. 1971. Mathematical model for two-dimensional multi-component airfoils in viscous flow. *NASA CR 1843*

Stratford BS. 1959. The prediction of separation of the turbulent boundary layer. *J. Fluid Mech.* 5:1–16

Torres GE, Mueller TJ. 2004. Low-aspect ratio wing aerodynamics at low Reynolds number. *AIAA J.* 42:865–873

Van Dam CP. 2002. The aerodynamic design of multi-element high-lift systems for transport airplanes. *Progr. Aerosp. Sci.* 38:101–144

Yip LP, Vijgen PMHW, Hardin JD, van Dam CP. 1993. In-flight pressure distributions and skin-friction measurements on a subsonic transport high-lift wing section. In *High-Lift System Aerodynamics, AGARD CP 515*

7 INCOMPRESSIBLE FLOW ABOUT WINGS OF FINITE SPAN

7.1 GENERAL COMMENTS

The aerodynamic characteristics for subsonic flow about an unswept *airfoil* have been discussed in Chapters 5 and 6. Since the span of an *airfoil* is infinite, the flow is identical for each spanwise station (i.e., the flow is two dimensional). The lift produced by the pressure differences between the lower surface and the upper surface of the airfoil section, and therefore the circulation (integrated along the chord length of the section), does not vary along the span. For a wing of finite span, the high-pressure air beneath the wing spills out around the wing tips toward the low-pressure regions above the wing. As a consequence of the tendency of the pressures acting on the top surface near the tip of the wing to equalize with those on the bottom surface, the lift force per unit span decreases toward the tips. A sketch of a representative aerodynamic load distribution is presented in Fig. 7.1. As indicated in Fig. 7.1a, there is a chordwise variation in the pressure differential between the lower surface and the upper surface. The resultant lift force acting on a section (i.e., a unit span) is obtained by integrating the pressure distribution over the chord length. A procedure that can be used to determine the sectional lift coefficient has been discussed in Chapter 6.

As indicated in the sketch of Fig. 7.1b, there is a spanwise variation in the lift force. As a result of the spanwise pressure variation, the air on the upper surface flows inboard toward the root. Similarly, on the lower surface, air will tend to flow outward toward the wing tips. The resultant flow around a wing of finite span is three dimensional, having both

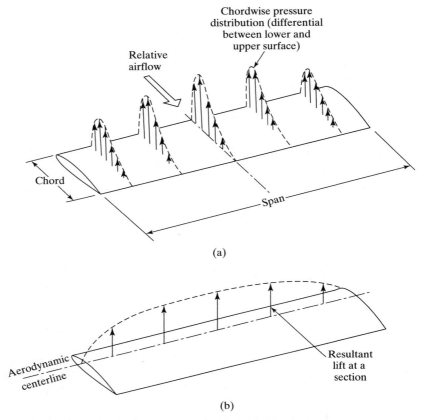

Figure 7.1 Aerodynamic load distribution for a rectangular wing in subsonic airstream: (a) differential pressure distribution along the chord for several spanwise stations; (b) spanwise lift distribution.

chordwise and spanwise velocity components. Where the flows from the upper surface and the lower surface join at the trailing edge, the difference in spanwise velocity components will cause the air to roll up into a number of streamwise vortices, distributed along the span. These small vortices roll up into two large vortices just inboard of the wing tips (see Fig. 7.2). The formation of a tip vortex is illustrated in the sketch of Fig. 7.2c and in the filaments of smoke in the photograph taken in the U.S. Air Force Academy's Smoke Tunnel (Fig. 7.2d). Very high velocities and low pressures exist at the core of the wing-tip vortices. In many instances, water vapor condenses as the air is drawn into the low-pressure flow field of the tip vortices. Condensation clearly defines the tip vortices (just inboard of the wing tips) of the Shuttle Orbiter *Columbia* on approach to a landing at Edwards Air Force Base (see Fig. 7.3).

At this point, it is customary to assume (1) that the vortex wake, which is of finite thickness, may be replaced by an infinitesimally thin surface of discontinuity, designated the trailing vortex sheet, and (2) that the trailing vortex sheet remains flat as it extends downstream from the wing. Spreiter and Sacks (1951) note that "it has been

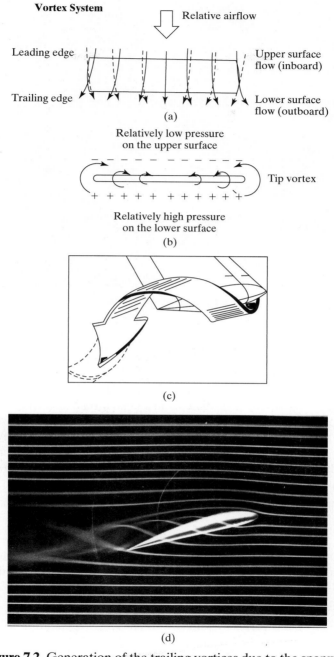

Figure 7.2 Generation of the trailing vortices due to the spanwise load distribution: (a) view from bottom; (b) view from trailing edge; (c) formation of the tip vortex. (d) smoke-flow pattern showing tip vortex. (Photograph courtesy U.S. Air Force Academy.)

Figure 7.3 Condensation marks the wing-tip vortices of the Space
Shuttle Orbiter *Columbia*. (Courtesy NASA.)

firmly established that these assumptions are sufficiently valid for the prediction of the
forces and moments on finite-span wings."

Thus, an important difference in the three-dimensional flow field around a wing
(as compared with the two-dimensional flow around an airfoil) is the spanwise variation
in lift. Since the lift force acting on the wing section at a given spanwise location is related
to the strength of the circulation, there is a corresponding spanwise variation in circulation, such that the circulation at the wing tip is zero. Procedures that can be used to
determine the vortex-strength distribution produced by the flow field around a three-
dimensional lifting wing are presented in this chapter.

7.2 VORTEX SYSTEM

A solution is sought for the vortex system which would impart to the surrounding air a
motion similar to that produced by a lifting wing. A suitable distribution of vortices
would represent the physical wing in every way except that of thickness. The vortex
system consists of

1. The bound vortex system
2. The trailing vortex system
3. The "starting" vortex

As stated in Chapter 6, the "starting" vortex is associated with a change in circulation
and would, therefore, relate to changes in lift that might occur at some time.

The representation of the wing by a bound vortex system is not to be interpreted as a rigorous flow model. However, the idea allows a relation to be established between

1. The physical load distribution for the wing (which depends on the wing geometry and on the aerodynamic characteristics of the wing sections)
2. The trailing vortex system

7.3 LIFTING-LINE THEORY FOR UNSWEPT WINGS

For this section, we are interested in developing a model that can be used to estimate the aerodynamic characteristics of a wing, which is unswept (or is only slightly swept) and which has an aspect ratio of 4.0, or greater. The spanwise variation in lift, $l(y)$, is similar to that depicted in Fig. 7.1b. Prandtl and Tietjens (1957) hypothesized that each airfoil section of the wing acts as though it is an isolated two-dimensional section, provided that the spanwise flow is not too great. Thus, each section of the finite-span wing generates a section lift equivalent to that acting on a similar section of an infinite-span wing having the same section circulation. We will assume that the lift acting on an incremental spanwise element of the wing is related to the local circulation through the Kutta-Joukowski theorem (see Section 3.15.2). That is,

$$l(y) = \rho_\infty U_\infty \Gamma(y) \tag{7.1}$$

Orloff (1980) showed that the spanwise lift distribution could be obtained from flow field velocity surveys made behind an airfoil section of the wing only and related to the circulation around a loop containing that airfoil section. The velocity surveys employed the integral form of the momentum equation in a manner similar to that used to estimate the drag in Problems 2.18 through 2.22.

Thus, we shall represent the spanwise lift distribution by a system of vortex filaments, the axis of which is normal to the plane of symmetry and which passes through the aerodynamic center of the lifting surface. Since the theoretical relations developed in Chapter 6 for inviscid flow past a thin airfoil showed that the aerodynamic center is at the quarter chord, we shall place the bound-vortex system at the quarter-chord line. The strength of the bound-vortex system at any spanwise location $\Gamma(y)$ is proportional to the local lift acting at that location $l(y)$. However, as discussed in Chapter 3, the vortex theorems of Helmholtz state that a vortex filament cannot end in a fluid. Therefore, we shall model the lifting character of the wing by a large number of vortex filaments (i.e., a large bundle of infinitesimal-strength filaments) that lie along the quarter chord of the wing. This is the bound-vortex system, which represents the spanwise loading distribution, as shown in Fig. 7.4(a). At any spanwise location y, the sum of the strengths of all of the vortex filaments in the bundle at that station is $\Gamma(y)$. When the lift changes at some spanwise location [i.e., $\Delta l(y)$], the total strength of the bound-vortex system changes proportionally [i.e., $\Delta\Gamma(y)$]. But vortex filaments cannot end in the fluid. Thus, the change $\Delta\Gamma(y)$ is represented in our model by having some of the filaments from our bundle of filaments turn 90° and continue in the streamwise direction (i.e., in the x direction). The strength of the trailing vortex at any y location is equal to the change in the strength of the bound-vortex system. The strength of the vortex filaments continuing

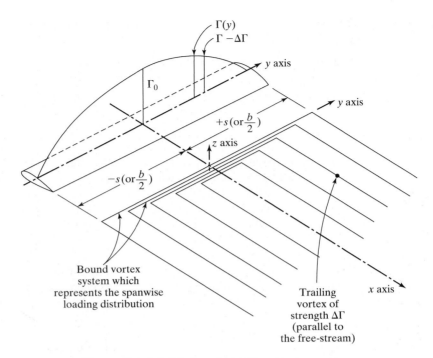

Figure 7.4 (a) Schematic trailing-vortex system.

in the bound-vortex system depends on the spanwise variation in lift and, therefore, depends upon parameters such as the wing planform, the airfoil sections that make up the wing, the geometric twist of the wing, etc.

Thus, as shown in Fig. 7.4a, if the strength of the vortex filaments in the bundle making up the bound-vortex system change by the amount $\Delta\Gamma$, a trailing vortex of strength $\Delta\Gamma$ must be shed in the x direction.

Thus, the vortex filaments that make up the bound-vortex system do not end in the fluid when the lift changes, but turn backward at each end to form a pair of vortices in the trailing-vortex system. For steady flight conditions, the *starting* vortex is left far behind, so that the trailing-vortex pair effectively stretches to infinity. The three-sided vortex, which is termed a *horseshoe vortex*, is presented in Fig. 7.4a. Thus, for practical purposes, the system consists of a the bound-vortex system and the related system of trailing vortices. Also included in Fig. 7.4a is a sketch of a symmetrical lift distribution, which the vortex system represents.

A number of vortices are made visible by the condensation of water vapor in the flow over an F/A-18 Hornet in the photograph of Fig. 7.4b. The two streamwise vortices associated with the flow around the edges of the strakes are easily seen on either side of the fuselage. The flow around wing/strake configurations will be discussed further in Section 7.8, "Leading-Edge Extensions." In addition, streamwise vorticity filaments originating in the wing-leading-edge region can be seen across the whole wing. The streamwise condensation pattern that appears across the wing in Fig. 7.4b is not normally

Figure 7.4 Concluded. (b) Streamwise vorticity from the F/A-18 Hornet wing leading edge during maneuver. (Copyrighted image reproduced courtesy of The Boeing Company.)

observed. It is believed that these streamwise vorticity filaments correspond to the trailing vortices shed by the spanwise variation in vorticity across the wing that is depicted in the schematic of Fig. 7.4a.

Conventional Prandtl lifting-line theory (PLLT) provides reasonable estimates of the lift and of the induced drag until boundary-layer effects become important. Thus, there will be reasonable agreement between the calculations and the experimental values for a single lifting surface having no sweep, no dihedral, and an aspect ratio of 4.0, or greater, operating at relatively low angles of attack. Of course, the skin friction component of drag will not be represented in the PLLT calculations at any angle of attack.

Because improvements continue to be made in calculation procedures [e.g., Rasmussen and Smith (1999) and Phillips and Snyder (2000)] and in ways of accounting for the nonlinear behavior of the aerodynamic coefficients [e.g., Anderson, Corda, and Van Wie (1980)], the lifting-line theory is still widely used today.

7.3.1   Trailing Vortices and Downwash

A consequence of the vortex theorems of Helmholtz is that a bound-vortex system does not change strength between two sections unless a vortex filament equal in strength to the change joins or leaves the vortex bundle. If $\Gamma(y)$ denotes the strength of the circulation along the y axis (the spanwise coordinate), a semiinfinite vortex of strength $\Delta\Gamma$ trails from the segment Δy, as shown in Fig. 7.5. The strength of the trailing vortex is given by

$$\Delta\Gamma = \frac{d\Gamma}{dy}\Delta y$$

It is assumed that each spanwise strip of the wing (Δy) behaves as if the flow were locally two dimensional. To calculate the influence of a trailing vortex filament located at y,

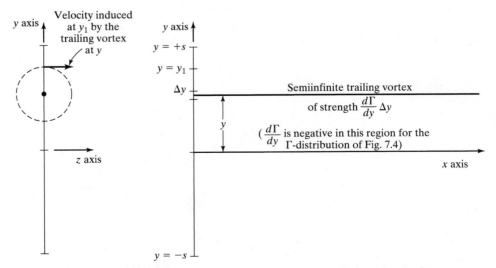

Figure 7.5 Geometry for the calculation of the induced velocity at $y = y_1$.

consider the semiinfinite vortex line, parallel to the x axis (which is parallel to the free-stream flow) and extending downstream to infinity from the line through the aerodynamic center of the wing (i.e., the y axis). The vortex at y induces a velocity at a general point y_1 on the aerodynamic centerline which is one-half the velocity that would be induced by an infinitely long vortex filament of the same strength:

$$\delta w_{y1} = \frac{1}{2}\left[+\frac{d\Gamma}{dy}\, dy\, \frac{1}{2\pi(y - y_1)} \right]$$

The positive sign results because, when both $(y - y_1)$ and $d\Gamma/dy$ are negative, the trailing vortex at y induces an upward component of velocity, as shown in Fig. 7.5, which is in the positive z direction.

To calculate the resultant induced velocity at any point y_1 due to the cumulative effect of all the trailing vortices, the preceding expression is integrated with respect to y from the left wing tip $(-s)$ to the right wing tip $(+s)$:

$$w_{y1} = +\frac{1}{4\pi} \int_{-s}^{+s} \frac{d\Gamma/dy}{y - y_1}\, dy \tag{7.2}$$

The resultant induced velocity at y_1 is, in general, in a downward direction (i.e., negative) and is called the *downwash*. As shown in the sketch of Fig. 7.6, the downwash angle is

$$\varepsilon = \tan^{-1}\left(-\frac{w_{y1}}{U_\infty} \right) \approx -\frac{w_{y1}}{U_\infty} \tag{7.3}$$

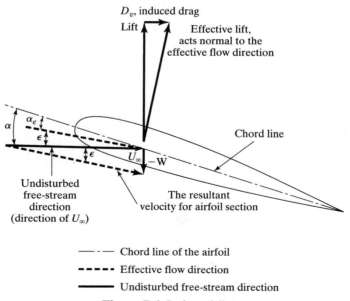

Figure 7.6 Induced flow.

The downwash has the effect of "tilting" the undisturbed air, so the effective angle of attack at the aerodynamic center (i.e., the quarter chord) is

$$\alpha_e = \alpha - \varepsilon \tag{7.4}$$

Note that, if the wing has a geometric twist, both the angle of attack (α) and the downwash angle (ε) would be a function of the spanwise position. Since the direction of the resultant velocity at the aerodynamic center is inclined downward relative to the direction of the undisturbed free-stream air, the effective lift of the section of interest is inclined aft by the same amount. Thus, the effective lift on the wing has a component of force parallel to the undisturbed free-stream air (refer to Fig. 7.6). This drag force is a consequence of the lift developed by a finite wing and is termed *vortex drag* (or the *induced drag* or the *drag-due-to-lift*). Thus, for subsonic flow past a finite-span wing, in addition to the skin friction drag and the form (or pressure) drag, there is a drag component due-to-lift. As a result of the induced downwash velocity, the lift generated by a finite-span wing composed of a given airfoil section, which is at the geometric angle of attack, α, is less than that for an infinite-span airfoil composed of the same airfoil section and which is at the same angle of attack α. See Fig. 5.25. Thus, at a given α, the three-dimensional flow over a finite-span wing generates less lift than the two-dimensional flow over an infinite-span airfoil.

Based on the Kutta-Joukowski theorem, the lift on an elemental airfoil section of the wing is

$$l(y) = \rho_\infty U_\infty \Gamma(y) \tag{7.1}$$

while the vortex drag is

$$d_v(y) = -\rho_\infty w(y)\Gamma(y) \tag{7.5}$$

The minus sign results because a downward (or negative) value of w produces a positive drag force. Integrating over the entire span of the wing, the total lift is given by

$$L = \int_{-s}^{+s} \rho_\infty U_\infty \Gamma(y)\, dy \tag{7.6}$$

and the total vortex drag is given by

$$D_v = -\int_{-s}^{+s} \rho_\infty w(y)\Gamma(y)\, dy \tag{7.7}$$

Note that for the two-dimensional airfoil (i.e., a wing of infinite span) the circulation strength Γ is constant across the span (i.e., it is independent of y) and the induced downwash velocity is zero at all points. Thus, $D_v = 0$, as discussed in Chapter 6. As a consequence of the trailing vortex system, the aerodynamic characteristics are modified significantly from those of a two-dimensional airfoil of the same section.

7.3.2   Case of Elliptic Spanwise Circulation Distribution

An especially simple circulation distribution, which also has significant practical implications, is given by the elliptic relation (see Fig. 7.7).

$$\Gamma(y) = \Gamma_0 \sqrt{1 - \left(\frac{y}{s}\right)^2} \tag{7.8}$$

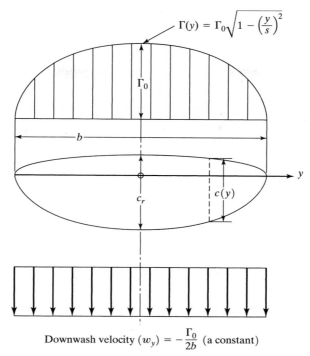

Downwash velocity $(w_y) = -\dfrac{\Gamma_0}{2b}$ (a constant)

Figure 7.7 Elliptic-circulation distribution and the resultant downwash velocity.

Since the lift is a function only of the free-stream density, the free-stream velocity, and the circulation, an elliptic distribution for the circulation would produce an elliptic distribution for the lift. However, to calculate the section lift coefficient, the section lift force is divided by the product of q_∞ and the local chord length at the section of interest. Hence, only when the wing has a rectangular planform (and c is, therefore, constant) is the spanwise section lift coefficient distribution (C_1) elliptic when the spanwise lift distribution is elliptic.

For the elliptic spanwise circulation distribution of equation (7.8), the induced downwash velocity is

$$w_{y1} = -\frac{\Gamma_0}{4\pi s} \int_{-s}^{+s} \frac{y}{\sqrt{s^2 - y^2}(y - y_1)} \, dy$$

which can be rewritten as

$$w_{y1} = -\frac{\Gamma_0}{4\pi s} \left[\int_{-s}^{+s} \frac{(y - y_1)\,dy}{\sqrt{s^2 - y^2}(y - y_1)} + \int_{-s}^{+s} \frac{y_1\,dy}{\sqrt{s^2 - y^2}(y - y_1)} \right]$$

Integration yields

$$w_{y1} = -\frac{\Gamma_0}{4\pi s}(\pi + y_1 I) \tag{7.9}$$

where

$$I = \int_{-s}^{+s} \frac{dy}{\sqrt{s^2 - y^2}(y - y_1)}$$

Since the elliptic loading is symmetric about the pitch plane of the vehicle (i.e., $y = 0$), the velocity induced at a point $y_1 = +a$ should be equal to the velocity at a point $y_1 = -a$. Referring to equation (7.9), this can be true if $I = 0$. Thus, for the elliptic load distribution

$$w_{y1} = w(y) = -\frac{\Gamma_0}{4s} \tag{7.10}$$

The induced velocity is independent of the spanwise coordinate.

The total lift for the wing is

$$L = \int_{-s}^{+s} \rho_\infty U_\infty \Gamma_0 \sqrt{1 - \left(\frac{y}{s}\right)^2} \, dy$$

Using the coordinate transformation

$$y = -s \cos \phi$$

the equation for lift becomes

$$L = \int_0^\pi \rho_\infty U_\infty \Gamma_0 \sqrt{1 - \cos^2 \phi} \, s \sin \phi \, d\phi$$

Thus,

$$L = \rho_\infty U_\infty \Gamma_0 s \frac{\pi}{2} = \frac{\pi}{4} b \rho_\infty U_\infty \Gamma_0 \tag{7.11}$$

The lift coefficient for the wing is

$$C_L = \frac{L}{\frac{1}{2}\rho_\infty U_\infty^2 S} = \frac{\pi b \Gamma_0}{2 U_\infty S} \tag{7.12}$$

Similarly, we can calculate the total vortex (or induced) drag for the wing.

$$D_v = \int_{-s}^{+s} \frac{\rho_\infty \Gamma_0}{4s} \Gamma_0 \sqrt{1 - \left(\frac{y}{s}\right)^2} \, dy$$

Introducing the coordinate transformation again, we have

$$D_v = \frac{\rho_\infty \Gamma_0^2}{4s} \int_0^\pi \sqrt{1 - \cos^2 \phi} \, s \sin \phi \, d\phi$$

$$= \frac{\pi}{8} \rho_\infty \Gamma_0^2 \tag{7.13}$$

and the drag coefficient for the induced component is

$$C_{Dv} = \frac{D_v}{\frac{1}{2}\rho_\infty U_\infty^2 S} = \frac{\pi \Gamma_0^2}{4 U_\infty^2 S} \tag{7.14}$$

Rearranging equation (7.12) to solve for Γ_0 gives

$$\Gamma_0 = \frac{2C_L U_\infty S}{\pi b} \tag{7.15}$$

Thus,

$$C_{Dv} = \frac{\pi}{4U_\infty^2 S}\left(\frac{2C_L U_\infty S}{\pi b}\right)^2$$

or

$$C_{Dv} = \frac{C_L^2}{\pi}\left(\frac{S}{b^2}\right)$$

Since the aspect ratio is defined as

$$AR = \frac{b^2}{S}$$

$$C_{Dv} = \frac{C_L^2}{\pi \cdot AR} \tag{7.16}$$

Note that once again we see that the induced drag is zero for a two-dimensional airfoil (i.e., a wing with an aspect ratio of infinity). Note also that the trailing vortex drag for an inviscid flow around a wing is not zero but is proportional to C_L^2.

The induced drag coefficient given by equation (7.16) and the measurements for a wing whose aspect ratio is 5 are compared in Fig. 7.8. The experimental values of the induced drag coefficient, which were presented by Schlichting and Truckenbrodt (1969), closely follow the theoretical values up to an angle of attack of 20°. The relatively constant difference between the measuredvalues and the theoretical values is due to the influence of skin friction, which was not included in the development of equation (7.16). Therefore, as noted in Chapter 5, the drag coefficient for an incompressible flow is typically written as

$$C_D = C_{D0} + kC_L^2 \tag{7.17}$$

where C_{D0} is the drag coefficient at zero lift and kC_L^2 is the lift-dependent drag coefficient. The lift-dependent drag coefficient includes that part of the viscous drag and of the form drag, which results as the angle of attack changes from α_{0l}.

These relations describing the influence of the aspect ratio on the lift and the drag have been verified experimentally by Prandtl and Betz. If one compares the drag polars for two wings which have aspect ratios of AR_1 and AR_2, respectively, then for a given value of the lift coefficient,

$$C_{D,2} = C_{D,1} + \frac{C_L^2}{\pi}\left(\frac{1}{AR_2} - \frac{1}{AR_1}\right) \tag{7.18}$$

where $C_{D0,1}$ has been assumed to be equal to $C_{D0,2}$. The data from Prandtl (1921) for a series of rectangular wings are reproduced in Fig. 7.9. The experimentally determined drag polars are presented in Fig. 7.9a. Equation (7.18) has been used to convert the drag polars for the different aspect ratio wings to the equivalent drag polar for a wing whose aspect ratio is 5 (i.e., $AR_2 = 5$). These converted drag polars, which

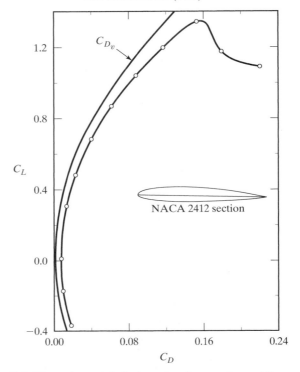

Figure 7.8 Experimental drag polar for a wing with an aspect ratio of 5 compared with the theoretical induced drag.

are presented in Fig. 7.9b, collapse quite well to a single curve. Thus, the correlation of the measurements confirms the validity of equation (7.18).

A similar analysis can be used to examine the effect of aspect ratio on the lift. Combining the definition for the downwash angle [equation (7.3)], the downwash velocity for the elliptic load distribution [equation (7.10)], and the correlation between the lift coefficient and Γ_0 [equation (7.15)], we obtain

$$\varepsilon = \frac{C_L}{\pi \cdot AR} \tag{7.19}$$

One can determine the effect of the aspect ratio on the correlation between the lift coefficient and the geometric angle of attack. To calculate the geometric angle of attack α_2 required to generate a particular lift coefficient for a wing of AR_2, if a wing with an aspect ratio of AR_1 generates the same lift coefficient at α_1, we use the equation

$$\alpha_2 = \alpha_1 + \frac{C_L}{\pi}\left(\frac{1}{AR_2} - \frac{1}{AR_1}\right) \tag{7.20}$$

Experimentally determined lift coefficients [from Prandtl (1921)] are presented in Fig. 7.10. The data presented in Fig. 7.10a are for the same rectangular wings of Fig. 7.9.

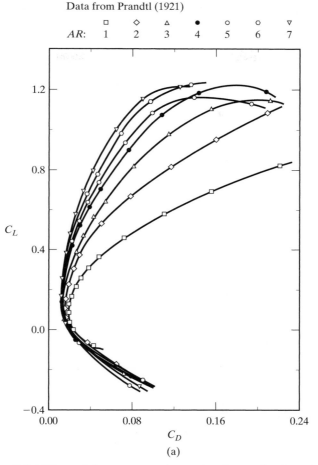

Data from Prandtl (1921)

Figure 7.9 Effect of the aspect ratio on the drag polar for rectangular wings (*AR* from 1 to 7): (a) measured drag polars.

The results of converting the coefficient-of-lift measurements using equation (7.20) in terms of a wing whose aspect ratio is 5 (i.e., $AR_2 = 5$) are presented in Fig. 7.10b. Again, the converted curves collapse into a single correlation. Therefore, the validity of equation (7.20) is experimentally verified.

7.3.3 Technique for General Spanwise Circulation Distribution

Consider a spanwise circulation distribution that can be represented by a Fourier sine series consisting of N terms:

$$\Gamma(\phi) = 4sU_\infty \sum_{1}^{N} A_n \sin n\phi \qquad \textbf{(7.21)}$$

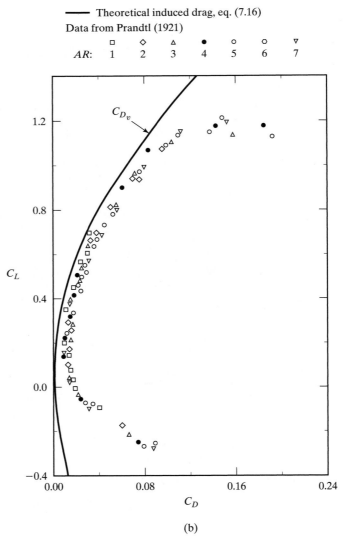

Figure 7.9 (*continued*) (b) drag polars converted to $AR = 5$.

As was done previously, the physical spanwise coordinate (y) has been replaced by the ϕ coordinate:

$$\frac{y}{s} = -\cos \phi$$

A sketch of one such Fourier series is presented in Fig. 7.11. Since the spanwise lift distribution represented by the circulation of Fig. 7.11 is symmetrical, only the odd terms remain.

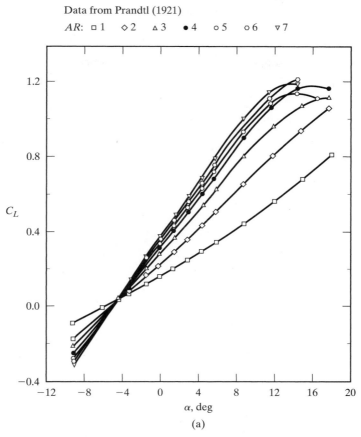

Data from Prandtl (1921)

AR: □ 1   ◇ 2   △ 3   ● 4   ○ 5   ○ 6   ▽ 7

Figure 7.10 Effect of aspect ratio on the lift coefficient for rectangular wings (*AR* from 1 to 7): (a) measured lift coefficients.

The section lift force [i.e., the lift acting on that spanwise section for which the circulation is $\Gamma(\phi)$] is given by

$$l(\phi) = \rho_\infty U_\infty \Gamma(\phi) = 4\rho_\infty U_\infty^2 s \sum_1^N A_n \sin n\phi \qquad (7.22)$$

To evaluate the coefficients $A_1, A_2, A_3, \ldots, A_N$, it is necessary to determine the circulation at N spanwise locations. Once this is done, the N-resultant linear equations can be solved for the A_n coefficients. Typically, the series is truncated to a finite series and the coefficients in the finite series are evaluated by requiring the lifting-line equation to be satisfied at a number of spanwise locations equal to the number of terms in the series. This method, known as the collocation method, will be developed in this section.

Recall that the section lift coefficient is defined as

$$C_l(\phi) = \frac{\text{lift per unit span}}{\frac{1}{2}\rho_\infty U_\infty^2 c}$$

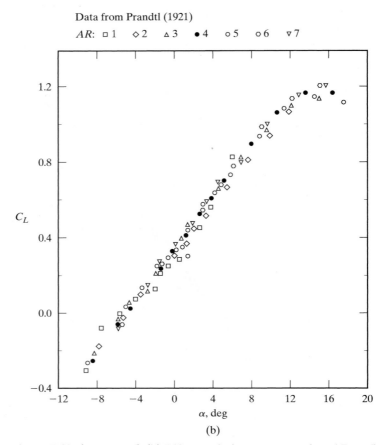

Figure 7.10 (*continued*) (b) Lift correlations converted to $AR = 5$.

Using the local circulation to determine the local lift per unit span, we obtain

$$C_l(\phi) = \frac{\rho_\infty U_\infty \Gamma(\phi)}{\frac{1}{2}\rho_\infty U_\infty^2 c} = \frac{2\Gamma(\phi)}{U_\infty c} \tag{7.23}$$

It is also possible to evaluate the section lift coefficient by using the linear correlation between the lift and the angle of attack for the equivalent two-dimensional flow. Thus, referring to Fig. 7.12 for the nomenclature, we have

$$C_l = \left(\frac{dC_l}{d\alpha}\right)_0 (\alpha_e - a_{0l}) \tag{7.24}$$

We now have two expressions for calculating the section lift coefficient at a particular spanwise location ϕ. We will set the expression in equation (7.23) equal to that in equation (7.24) to form equation (7.25).

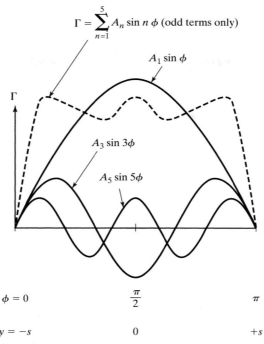

$$\Gamma = \sum_{n=1}^{5} A_n \sin n\phi \ (\text{odd terms only})$$

Figure 7.11 Symmetric spanwise lift distribution as represented by a sine series.

Let the equivalent lift-curve slope $(dC_l/d\alpha)_0$ be designated by the symbol a_0. Note that since $\alpha_e = \alpha - \varepsilon$, equations (7.23) and (7.24) can be combined to yield the relation

$$\frac{2\Gamma(\phi)}{c(\phi)a_0} = U_\infty[\alpha(\phi) - \alpha_{0l}(\phi)] - U_\infty \varepsilon(\phi) \tag{7.25}$$

For the present analysis, five parameters in equation (7.25) may depend on the spanwise location ϕ (or, equivalently, y) at which we will evaluate the terms. The five parameters are (1) Γ, the local circulation; (2) ε, the downwash angle, which depends on the circulation distribution; (3) c, the chord length, which varies with ϕ for a tapered wing planform; (4) α, the local geometric angle of attack, which varies with ϕ when the wing is twisted (i.e., geometric twist, which is illustrated in Fig. 5.7), and (5) α_{0l}, the zero lift angle of attack, which varies with ϕ when the airfoil section varies in the spanwise direction (which is known as *aerodynamic twist*). Note that

$$U_\infty \varepsilon = -w = -\frac{1}{4\pi} \int_{-s}^{+s} \frac{d\Gamma/dy}{y - y_1} dy$$

Using the Fourier series representation for Γ and the coordinate transformation, we obtain

$$-w = U_\infty \frac{\sum nA_n \sin n\phi}{\sin \phi}$$

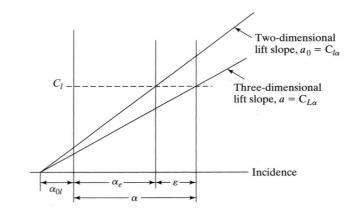

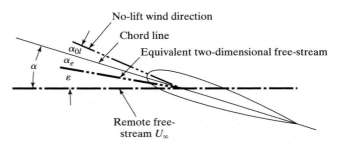

Figure 7.12 Nomenclature for wing/airfoil lift.

Equation (7.25) can be rewritten as

$$\frac{2\Gamma}{ca_0} = U_\infty(\alpha - \alpha_{0l}) - U_\infty\frac{\sum nA_n \sin n\phi}{\sin \phi}$$

Since $\Gamma = 4sU_\infty \sum A_n \sin n\phi$, the equation becomes

$$\frac{8s}{ca_0} \sum A_n \sin n\phi = (\alpha - \alpha_{0l}) - \frac{\sum nA_n \sin n\phi}{\sin \phi}$$

Defining $\mu = ca_0/8s$, the resultant governing equation is

$$\mu(\alpha - \alpha_{0l}) \sin \phi = \sum A_n \sin n\phi(\mu n + \sin \phi) \qquad \textbf{(7.26)}$$

which is known as the *monoplane equation.* If we consider only symmetrical loading distributions, only the odd terms of the series need be considered. That is, as shown in the sketch of Fig. 7.11,

$$\Gamma(\phi) = 4sU_\infty(A_1 \sin \phi + A_3 \sin 3\phi + A_5 \sin 5\phi + \dots)$$

7.3.4 Lift on the Wing

$$L = \int_{-s}^{+s} \rho_\infty U_\infty \Gamma(y)\, dy = \int_0^\pi \rho_\infty U_\infty s \Gamma(\phi) \sin\phi\, d\phi$$

Using the Fourier series for $\Gamma(\phi)$ gives us

$$L = 4\rho_\infty U_\infty^2 s^2 \int_0^\pi \sum A_n \sin n\phi \sin\phi\, d\phi$$

Noting that $\sin A \sin B = \frac{1}{2}\cos(A - B) - \frac{1}{2}\cos(A + B)$, the integration yields

$$L = 4\rho_\infty U_\infty^2 s^2 \left\{ A_1 \left[\frac{\phi}{2} + \frac{\sin 2\phi}{4}\right]\Bigg|_0^\pi + \sum_3^N \frac{1}{2} A_n \left[\frac{\sin(n - 1)\phi}{n - 1} - \frac{\sin(n + 1)\phi}{n + 1}\right]\Bigg|_0^\pi \right\}$$

The summation represented by the second term on the right-hand side of the equation is zero, since each of the terms is zero for $n \neq 1$. Thus, the integral expression for the lift becomes

$$L = (4s^2)\left(\tfrac{1}{2}\rho_\infty U_\infty^2\right) A_1 \pi = C_L\left(\tfrac{1}{2}\rho_\infty U_\infty^2\right)(S)$$

$$C_L = A_1 \pi \cdot \mathrm{AR} \qquad\qquad (7.27)$$

The lift depends only on the magnitude of the first coefficient, no matter how many terms may be present in the series describing the distribution.

7.3.5 Vortex-Induced Drag

$$D_v = -\int_{-s}^{+s} \rho_\infty w \Gamma\, dy$$

$$= \rho_\infty \int_0^\pi \underbrace{\frac{U_\infty \sum n A_n \sin n\phi}{\sin\phi}}_{-w}\ \underbrace{4 s U_\infty \sum A_n \sin n\phi s}_{\Gamma}\ \underbrace{\sin\phi\, d\phi}_{dy}$$

$$= 4\rho_\infty s^2 U_\infty^2 \int_0^\pi \sum n A_n \sin n\phi \sum A_n \sin n\phi\, d\phi$$

The integral

$$\int_0^\pi \sum n A_n \sin n\phi \sum A_n \sin n\phi\, d\phi = \frac{\pi}{2} \sum n A_n^2$$

Thus, the coefficient for the vortex-induced drag

$$C_{Dv} = \pi \cdot AR \sum n A_n^2 \qquad\qquad (7.28)$$

Since $A_1 = C_L/(\pi \cdot AR)$,

$$C_{Dv} = \frac{C_L^2}{\pi \cdot AR} \sum n \left(\frac{A_n}{A_1}\right)^2$$

where only the odd terms in the series are considered for the symmetric load distribution.

$$C_{Dv} = \frac{C_L^2}{\pi \cdot AR} \left[1 + \left(\frac{3A_3^2}{A_1^2} + \frac{5A_5^2}{A_1^2} + \frac{7A_7^2}{A_1^2} + \cdots\right)\right]$$

or

$$C_{Dv} = \frac{C_L^2}{\pi \cdot AR}(1 + \delta) \qquad (7.29)$$

where

$$\delta = \frac{3A_3^2}{A_1^2} + \frac{5A_5^2}{A_1^2} + \frac{7A_7^2}{A_1^2} + \cdots$$

Since $\delta \geq 0$, the drag is minimum when $\delta = 0$. In this case, the only term in the series representing the circulation distribution is the first term:

$$\Gamma(\phi) = 4sU_\infty A_1 \sin\phi$$

which is the elliptic distribution.

EXAMPLE 7.1:   Use the mooplane equation to compute the aerodynamic coefficients for a wing

The monoplane equation [i.e., equation (7.26)] will be used to compute the aerodynamic coefficients of a wing for which aerodynamic data are available. The geometry of the wing to be studied is illustrated in Fig. 7.13. The wing, which is unswept at the quarter chord, is composed of NACA 65–210 airfoil sections. Referring to the data of Abbott and von Doenhoff (1949), the zero-lift angle of attack (α_{0l}) is approximately $-1.2°$ across the span. Since the wing is untwisted, the geometric angle of attack is the same at all spanwise positions. The aspect ratio (AR) is 9.00. The taper ratio λ (i.e., c_t/c_r) is 0.40. Since the wing planform is trapezoidal,

$$S = 0.5(c_r + c_t)b = 0.5c_r(1 + \lambda)b$$

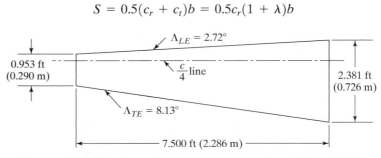

Figure 7.13 Planform for an unswept wing, AR = 9.00. $\lambda = 0.40$, airfoil section NACA 65–210.

and

$$AR = \frac{2b}{c_r + c_t}$$

Thus, the parameter μ in equation (7.26) becomes

$$\mu = \frac{ca_0}{4b} = \frac{ca_0}{2(AR) \cdot c_r(1 + \lambda)}$$

Solution: Since the terms are to be evaluated at spanwise stations for which $0 \leq \phi \leq \pi/2$ [i.e., $-s \leq y \leq 0$ (which corresponds to the port wing or left side of the wing)],

$$\mu = \frac{a_0}{2(1 + \lambda)AR}[1 + (\lambda - 1)\cos \phi]$$

$$= 0.24933(1 - 0.6\cos \phi) \qquad \textbf{(7.30)}$$

where the equivalent lift-curve slope (i.e., that for a two-dimensional flow over the airfoil section a_0) has been assumed to be equal to 2π. It might be noted that numerical solutions for lift and the vortex-drag coefficients were essentially the same for this geometry whether the series representing the spanwise circulation distribution included four terms or ten terms. Therefore, so that the reader can perform the required calculations with a pocket calculator, a four-term series will be used to represent the spanwise loading. Equation (7.26) is

$$\mu(\alpha - \alpha_{0l})\sin \phi = A_1 \sin \phi(\mu + \sin \phi) + A_3 \sin 3\phi(3\mu + \sin \phi)$$
$$+ A_5 \sin 5\phi(5\mu + \sin \phi) + A_7 \sin 7\phi(7\mu + \sin \phi) \quad \textbf{(7.31)}$$

Since there are four coefficients (i.e., A_1, A_3, A_5, and A_7) to be evaluated, equation (7.31) must be evaluated at four spanwise locations. The resultant values for the factors are summarized in Table 7.1. Note that, since we are considering the left side of the wing, the y coordinate is negative.

For a geometric angle of attack of $4°$, equation (7.31) becomes

$$0.00386 = 0.18897A_1 + 0.66154A_3 + 0.86686A_5 + 0.44411A_7$$

TABLE 7.1   Values of the Factor for Equation (7.31)

Station	ϕ	$-\dfrac{y}{s}$ $(= \cos \phi)$	$\sin \phi$	$\sin 3\phi$	$\sin 5\phi$	$\sin 7\phi$	μ
1	22.5°	0.92388	0.38268	0.92388	0.92388	0.38268	0.11112
2	45.0°	0.70711	0.70711	0.70711	−0.70711	−0.70711	0.14355
3	67.5°	0.38268	0.92388	−0.38268	−0.38268	0.92388	0.19208
4	90.0°	0.00000	1.00000	−1.00000	1.00000	−1.00000	0.24933

for $\phi = 22.5°$ (i.e., $y = -0.92388s$). For the other stations, the equation becomes

$$0.00921 = 0.60150A_1 + 0.80451A_3 - 1.00752A_5 - 1.21053A_7$$
$$0.01611 = 1.03101A_1 - 0.57407A_3 - 0.72109A_5 + 2.09577A_7$$
$$0.02263 = 1.24933A_1 - 1.74799A_3 + 2.24665A_5 - 2.74531A_7$$

The solution of this system of linear equations yields

$$A_1 = 1.6459 \times 10^{-2}$$
$$A_3 = 7.3218 \times 10^{-5}$$
$$A_5 = 8.5787 \times 10^{-4}$$
$$A_7 = -9.6964 \times 10^{-5}$$

Using equation (7.27), the lift coefficient for an angle of attack of 4° is

$$C_L = A_1 \pi \cdot AR = 0.4654$$

The theoretically determined lift coefficients are compared in Fig. 7.14 with data for this wing. In addition to the geometric characteristics already described, the wing had a dihedral angle of 3°. The measurements reported by Sivells (1947) were obtained at a Reynolds number of approximately

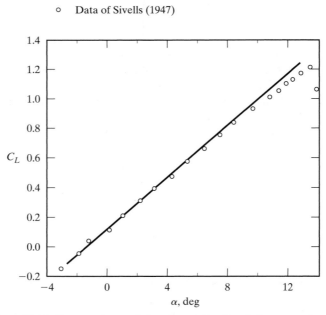

Figure 7.14 Comparison of the theoretical and the experimental lift coefficients for an unswept wing in a subsonic stream. (Wing is that of Fig. 7.13.)

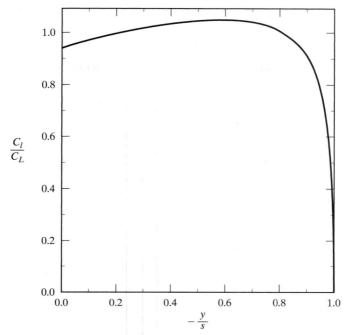

Figure 7.15 Spanwise distribution of the local lift coefficient, $AR = 9, \lambda = 0.4$, untwisted wing composed of NACA 65–210 airfoil sections.

4.4×10^6 and a Mach number of approximately 0.17. The agreement between the theoretical values and the experimental values is very good.

The spanwise distribution for the local lift coefficient of this wing is presented in Fig. 7.15. As noted by Sivells (1947), the variation of the section lift coefficient can be used to determine the spanwise position of initial stall. The local lift coefficient is given by

$$C_l = \frac{\rho_\infty U_\infty \Gamma}{0.5 \rho_\infty U_\infty^2 c}$$

which for the trapezoidal wing under consideration is

$$C_l = 2(AR)(1 + \lambda)\frac{c_r}{c} \sum A_{2n-1} \sin(2n - 1)\phi \qquad \textbf{(7.32)}$$

The theoretical value of the induced drag coefficient for an angle of attack of 4°, as determined using equation (7.29), is

$$C_{Dv} = \frac{C_L^2}{\pi \cdot AR}\left(1 + \frac{3A_3^2}{A_1^2} + \frac{5A_5^2}{A_1^2} + \frac{7A_7^2}{A_1^2}\right)$$

$$= 0.00766(1.0136) = 0.00776$$

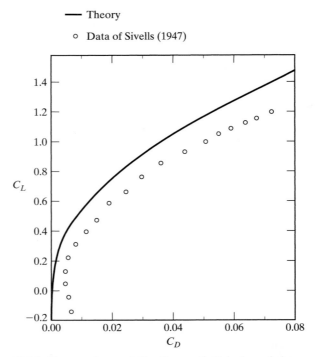

Figure 7.16 Comparison of the theoretical induced drag coefficients and the measured drag coefficients for an unswept wing in a subsonic stream. (Wing is that of Fig. 7.13.)

The theoretically determined induced drag coefficients are compared in Fig. 7.16 with the measured drag coefficients for this wing. As has been noted earlier, the theoretical relations developed in this chapter do not include the effects of skin friction. The relatively constant difference between the measured values and the theoretical values is due to the influence of skin friction.

The effect of the taper ratio on the spanwise variation of the lift coefficient is illustrated in Fig. 7.17. Theoretical solutions are presented for untwisted wings having taper ratios from 0 to 1. The wings, which were composed of NACA 2412 airfoil sections, all had an aspect ratio of 7.28. Again, the local lift coefficient has been divided by the overall lift coefficient for the wings. Thus,

$$\frac{C_l}{C_L} = \frac{2(1 + \lambda)}{\pi A_1} \frac{c_r}{c} \sum A_{2n-1} \sin(2n - 1)\phi$$

The values of the local (or section) lift coefficient near the tip of the highly tapered wings are significantly greater than the overall lift coefficient for that planform. As noted earlier, this result is important relative to the separation (or stall) of the boundary layer for a particular planform when it is operating at a relatively high angle of attack.

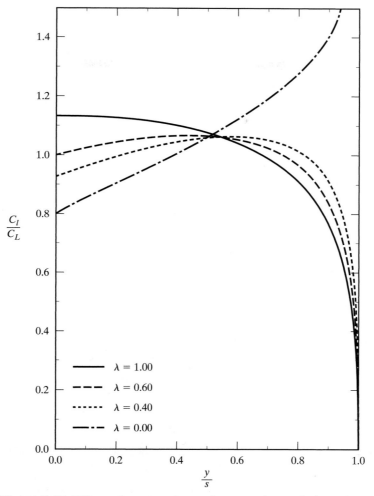

Figure 7.17 Effect of taper ratio on the spanwise variation of the lift coefficient for an untwisted wing.

Sketches of stall patterns are presented in Fig. 7.18. The desirable stall pattern for a wing is a stall which begins at the root sections so that the ailerons remain effective at high angles of attack.

The spanwise load distribution for a rectangular wing indicates stall will begin at the root and proceed outward. Thus, the stall pattern is favorable. The spanwise load distribution for a wing with a moderate taper ratio ($\lambda = 0.4$) approximates that of an elliptical wing (i.e., the local lift coefficient is roughly constant across the span). As a result, all sections will reach stall at essentially the same angle of attack.

Tapering of the wing reduces the wing-root bending moments, since the inboard portion of the wing carries more of the wing's lift than the tip. Furthermore, the longer wing-root chord makes it possible to increase the actual thickness of the wing while

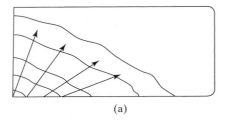

(a)

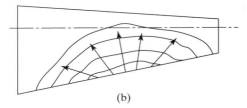

(b)

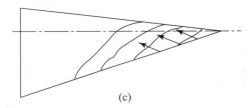

(c)

Figure 7.18 Typical stall patterns: (a) rectangular wing, $\lambda = 1.0$; (b) moderately tapered wing, $\lambda = 0.4$; (c) pointed wing, $\lambda = 0.0$.

maintaining a low thickness ratio, which is needed if the airplane is to operate at high speeds also. While taper reduces the actual loads carried outboard, the lift coefficients near the tip are higher than those near the root for a tapered wing. Therefore, there is a strong tendency to stall near (or at) the tip for the highly tapered (or pointed) wings.

In order to prevent the stall pattern from beginning in the region of the ailerons, the wing may be given a geometric twist, or washout, to decrease the local angles of attack at the tip (refer to Table 5.1). The addition of leading edge slots or slats toward the tip increases the stall angle of attack and is useful in avoiding tip stall and the loss of aileron effectiveness.

7.3.6   Some Final Comments on Lifting-Line Theory

With continuing improvements, lifting-line theory is still used to provide rapid estimates of the spanwise load distributions and certain aerodynamic coefficients for unswept, or slightly swept, wings. In the Fourier series analysis of Rasmussen and Smith (1999), the planform and the twist distributions for general wing configurations are represented explicitly. The spanwise circulation distribution $\Gamma(y)$ is obtained explicitly in terms of the Fourier coefficients for the chord distribution and the twist distribution.

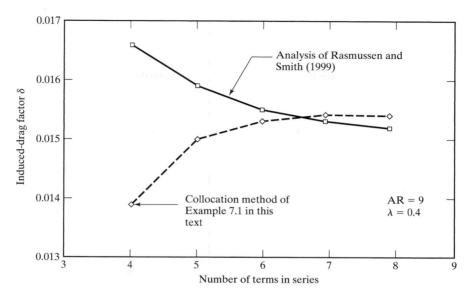

Figure 7.19 Convergence properties of induced-drag factor δ for tapered wing, $\lambda = 0.4$. [Taken from Rasmussen and Smith (1999).]

The method of Rasmussen and Smith (1999) was used to solve for the aerodynamic coefficients for the wing of Example 7.1. The induced-drag factor δ, as taken from Rasmussen and Smith (1999), is reproduced in Fig. 7.19. The values of the induced-drag factor are presented as a function of the number of terms in the Fourier series. The values of δ, which were computed using the method of Rasmussen and Smith (1999), are compared with the values computed using the collocation method of Example 7.1. The two methods produce values which are very close, when six, seven, or eight terms are used in the Fourier series. Rasmussen and Smith (1999) claim, "The method converges faster and is more accurate for the same level of truncation than collocation methods."

The *lifting-line theory* of Phillips and Snyder (2000) is in reality the vortex-lattice method applied using only a single lattice element in the chordwise direction for each spanwise subdivision of the wing. Thus, the method is very much like that used in Example 7.2 (see Fig. 7.31), except that many more panels are used to provide better resolution of the spanwise loading.

Incorporating empirical information into the modeling, one can extend the range of *applicability* of lifting-line theory. Anderson, Corda, and Van Wie (1980) noted that "certain leading-edge modifications can favorably tailor the high-lift characteristics of wings for light, single-engine general aviation airplanes so as to inhibit the onset of stall/spins. Since more than 30% of all general aviation accidents are caused by stall/spins, such modifications are clearly of practical importance. A modification of current interest is an abrupt extension and change in shape of the leading edge along a portion of the wing span—a so called 'drooped' leading edge. This is shown

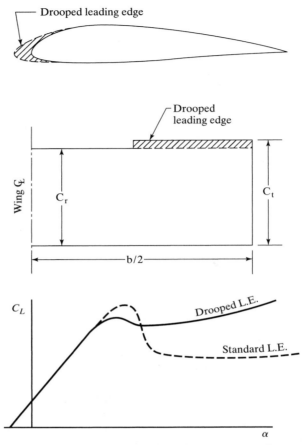

Figure 7.20 Drooped leading-edge characteristic. [Taken from Anderson, Corda, and Van Wie (1980).]

schematically in Fig. 7.20, where at a given spanwise location, the chord and/or the leading-edge shape of the wing change discontinuously. The net aerodynamic effect of this modification is a smoothing of the normally abrupt drop in lift coefficient C_L at stall, and the generation of a relatively large value of C_L at very high poststall angles of attack, as also shown in Fig. 7.20. As a result, an airplane with a properly designed drooped leading edge has increased resistance toward stalls/spins." As shown in Fig. 7.20, the chord is extended approximately 10% over a portion of the span.

The poststall behavior was modeled by introducing the experimentally determined values of the lift-curve slope in place of a_0 [Anderson, Corda, and Van Wie (1980)] The reader is referred to the discussion leading up to equations (7.25) and (7.26). The authors noted that the greatest compromise in using lifting-line theory into the stall angle-of-attack range and beyond is the use of data for the *two-dimensional flow around an airfoil*. The actual flow for this configuration is a complex, three-dimensional flow with separation.

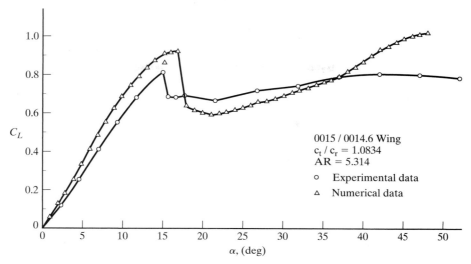

Figure 7.21 Lift coefficient versus angle of attack for a drooped leading-edge wing; comparison between experiment and numerical results. [Taken from Anderson, Corda, and Van Wie (1980).]

Nevertheless, with the use of experimental values for the lift-curve slope of the airfoil section, lifting-line theory generates reasonable estimates for C_L, when compared with experimental data, as shown in Fig. 7.21.

7.4   PANEL METHODS

Although lifting-line theory (i.e., the monoplane equation) provides a reasonable estimate of the lift and of the induced drag for an unswept, thin wing of relatively high aspect ratio in a subsonic stream, an improved flow model is needed to calculate the lifting flow field about a highly swept wing or a delta wing. A variety of methods have been developed to compute the flow about a thin wing which is operating at a small angle of attack so that the resultant flow may be assumed to be steady, inviscid, irrotational, and incompressible.

The basic concept of panel methods is illustrated in Fig. 7.22. The configuration is modeled by a large number of elementary quadrilateral panels lying either on the actual aircraft surface, or on some mean surface, or a combination thereof. To each elementary panel, there is attached one or more types of singularity distributions, such as sources, vortices, and doublets. These singularities are determined by specifying some functional variation across the panel (e.g., constant, linear, quadratic, etc.), whose actual value is set by corresponding strength parameters. These strength parameters are determined by solving the appropriate boundary condition equations. Once the singularity strengths have been determined, the velocity field and the pressure field can be computed.

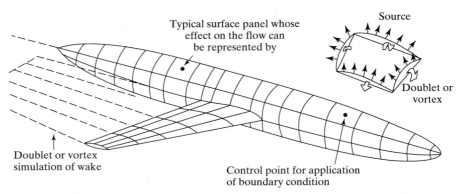

Figure 7.22 Representation of an airplane flowfield by panel (or singularity) methods.

7.4.1 Boundary Conditions

Johnson (1980) noted that, as a general rule, a boundary-value problem associated with LaPlace's equation, see equation (3.26), is well posed if either φ or $(\partial\varphi/\partial n)$ is specified at every point of the surface of the configuration which is being analyzed or designed. "Fluid flow boundary conditions associated with LaPlace's equation are generally of analysis or design type. Analysis conditions are employed on portions of the boundary where the geometry is considered fixed, and resultant pressures are desired. The permeability of the fixed geometry is known; hence, analysis conditions are of the Neumann type (specification of normal velocity). Design boundary conditions are used wherever a geometry perturbation is allowed for the purpose of achieving a specific pressure distribution. Here a perturbation to an existing tangential velocity vector field is made; hence, design conditions are fundamentally of the Dirichlet type (specification of potential). The design problem in addition involves such aspects as stream surface lofting (i.e., integration of streamlines passing through a given curve), and the relationship between a velocity field and its potential."

Neumann boundary conditions (specification of $\partial\varphi/\partial n$ at every point on the surface) arise naturally in the analysis of fixed configurations bounded by surfaces of known permeability. If the surface of the configuration is impermeable (as is the case for almost every application discussed in this text), the normal component of the resultant velocity must be zero at every point of the surface. Once a solution φ has been found to the boundary-value problem, the pressure coefficient at each point on the surface of the impermeable boundary can computed using

$$C_p = 1 - (U_t/U_\infty)^2 \tag{7.33}$$

The reader should note that the tangential velocity at the "surface" of a configuration in an inviscid flow is represented by the symbols U and U_t in equation (3.13) and equation (7.33), respectively.

Johnson (1980) noted further that Dirichlet boundary conditions (specification of φ) arise in connection with the inverse problem (i.e., that of solving for a specified pressure distribution on the surface of the configuration). The specification of φ guarantees a predetermined tangential velocity vector field and, therefore, a predetermined pressure coefficient distribution, as related through equation (7.33). However, the achievement of a desired pressure distribution on the surface is not physically significant without restrictions on the flux through the surface. To achieve both a specified pressure distribution and a normal flow distribution on the surface, the position of the surface must, in general, be perturbed, so that the surface will be a stream surface of the flow field. The total design problem is thus composed of two problems. The first is to find a perturbation potential for the surface that yields the desired distribution for the pressure coefficient and the second is to update the surface geometry so that it is a stream surface of the resultant flow. Johnson (1980) concluded, "The two problems are coupled and, in general, an iterative procedure is required for solution."

7.4.2 Methods

The first step in a panel method is to divide the boundary surface into a number of panels. A finite set of control points (equal in number to the number of singularity parameters) is selected at which the boundary conditions are imposed. The construction of each network requires developments in three areas: (1) the definition of the surface geometry; (2) the definition of the singularity strengths; and (3) the selection of the control points and the specification of the boundary conditions.

Numerous codes using panel-method techniques have been developed [e.g., Bristow and Grose (1978)] the variations depending mainly on the choice of type and form of singularity distribution, the geometric layout of the elementary panels, and the type of boundary condition imposed. The choice of combinations is not a trivial matter. Although many different combinations are in principle mathematically equivalent, their numerical implementation may yield significantly different results from the point of view of numerical stability, computational economy, accuracy, and overall code robustness. Bristow and Grose (1978) note that there is an important equivalence between surface doublet distributions and vorticity distributions. A surface doublet distribution can be replaced by an equivalent surface vortex distribution. The theoretical equivalency between vorticity distributions and doublet distributions does not imply equivalent simplicity in a numerical formulation.

For each control point (and there is one control point per panel), the velocities induced at that control point by the singularities associated with each of the panels of the configuration are summed, resulting in a set of linear algebraic equations that express the exact boundary condition of flow tangency on the surface. For many applications, the aerodynamic coefficients computed using panel methods are reasonably accurate. Bristow and Grose (1978) discuss some problems with the source panel class of methods when used on thin, highly loaded surfaces such as the aft portion of a supercritical airfoil. In such cases, source strengths with strong gradients can degrade the local velocity calculations.

Margason et al. (1985) compare computed aerodynamic coefficients using one vortex lattice method (VLM), one source panel method, two low-order surface potential distributions, and two high-order surface potential distributions. The computed values of C_L are presented as a function of α for a 45° swept-back and a 45° swept-forward wing in Fig. 7.23a and b, respectively. The five surface panel methods

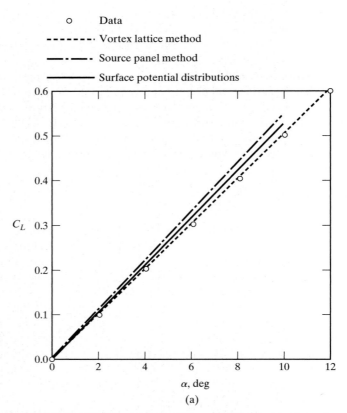

Figure 7.23 Comparison of the lift coefficient as a function of angle of attack: (a) $\Lambda_{c/4} = 45°$, NACA 64A010 section, $AR = 3.0$, $\lambda = 0.5$, aft swept wing. [From Margason et al. (1985).]

consistently overpredict the experimental data with little difference between the lift coefficients predicted by the various surface panel methods. As Margason et al. (1985) note, "The VLM predicts the experimental data very well, due to the fact that vortex lattice methods neglect both thickness and viscosity effects. For most cases, the effect of viscosity offsets the effect of thickness, fortuitously yielding good agreement between the VLM and experiment."

7.5   VORTEX LATTICE METHOD

The vortex lattice method is the simplest of the methods reviewed by Margason, et al. (1985). The VLM represents the wing as a surface on which a grid of horseshoe vortices is superimposed. The velocities induced by each horseshoe vortex at a specified control point are calculated using the law of Biot-Savart. A summation is performed for all control points on the wing to produce a set of linear algebraic equations for the horseshoe

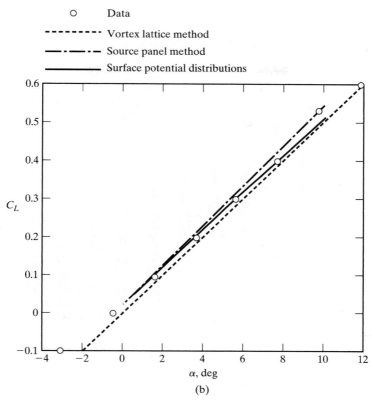

Figure 7.23 (*continued*) (b) $\Lambda_{c/4} = -45°$, NACA 64A112 section, $AR = 3.55, \lambda = 0.5$, forward swept wing. [From Margason et al. (1985).]

vortex strengths that satisfy the boundary condition of no flow through the wing. The vortex strengths are related to the wing circulation and the pressure differential between the upper and lower wing surfaces. The pressure differentials are integrated to yield the total forces and moments.

In our approach to solving the governing equation, the continuous distribution of bound vorticity over the wing surface is approximated by a finite number of discrete horseshoe vortices, as shown in Fig. 7.24. The individual horseshoe vortices are placed in trapezoidal panels (also called *finite elements* or *lattices*). This procedure for obtaining a numerical solution to the flow is termed the vortex lattice method.

The bound vortex coincides with the quarter-chord line of the panel (or element) and is, therefore, aligned with the local sweepback angle. In a rigorous theoretical analysis, the vortex lattice panels are located on the mean camber surface of the wing and, when the trailing vortices leave the wing, they follow a curved path. However, for many engineering applications, suitable accuracy can be obtained using linearized theory in which straight-line trailing vortices extend downstream to infinity. In the linearized approach, the trailing

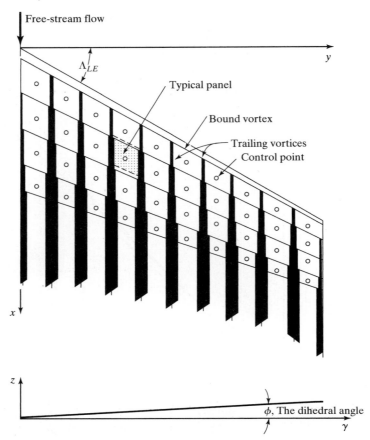

Figure 7.24 Coordinate system, elemental panels, and horseshoe vortices for a typical wing planform in the vortex lattice method.

vortices are aligned either parallel to the free stream or parallel to the vehicle axis. Both orientations provide similar accuracy within the assumptions of linearized theory. In this text we shall assume that the trailing vortices are parallel to the axis of the vehicle, as shown in Fig. 7.25. This orientation of the trailing vortices is chosen because the computation of the influences of the various vortices (which we will call the *influence coefficients*) is simpler. Furthermore, these geometric coefficients do not change as the angle of attack is changed. Application of the boundary condition that the flow is tangent to the wing surface at "the" control point of each of the $2N$ panels (i.e., there is no flow through the surface) provides a set of simultaneous equations in the unknown vortex circulation strengths. The control point of each panel is centered spanwise on the three-quarter-chord line midway between the trailing-vortex legs.

An indication of why the three-quarter-chord location is used as the control point may be seen by referring to Fig. 7.26. A vortex filament whose strength Γ represents the lifting character of the section is placed at the quarter-chord location. It induces a velocity,

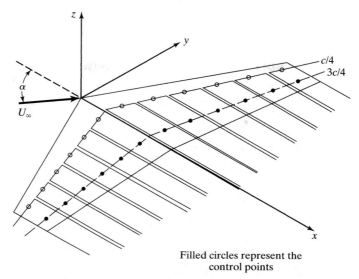

Filled circles represent the
control points

Figure 7.25 Distributed horseshoe vortices representing the
lifting flow field over a swept wing.

$$U = \frac{\Gamma}{2\pi r}$$

at the point c, the control point which is a distance r from the vortex filament. If the
flow is to be parallel to the surface at the control point, the incidence of the surface
relative to the free stream is given by

$$\alpha \approx \sin\alpha = \frac{U}{U_\infty} = \frac{\Gamma}{2\pi r U_\infty}$$

But, as was discussed in equations (6.11) and (6.12),

$$l = \tfrac{1}{2}\rho_\infty U_\infty^2 c 2\pi\alpha = \rho_\infty U_\infty \Gamma$$

Combining the preceding relations gives us

$$\pi\rho_\infty U_\infty^2 c \frac{\Gamma}{2\pi r U_\infty} = \rho_\infty U_\infty \Gamma$$

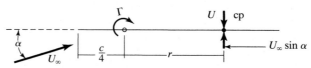

Figure 7.26 Planar airfoil section indicating location of control
point where flow is parallel to the surface.

Solving for r yields

$$r = \frac{c}{2}$$

Thus, we see that the control point is at the three-quarter-chord location for this two-dimensional geometry. The use of the chordwise slope at the 0.75-chord location to define the effective incidence of a panel in a finite-span wing has long been in use [e.g., Falkner (1943) and Kalman et al. (1971)].

Consider the flow over the swept wing that is shown in Fig. 7.25. Note that the bound-vortex filaments for the port (or left-hand) wing are not parallel to the bound-vortex filaments for the starboard (or right-hand) wing. Thus, for a lifting swept wing, the bound-vortex system on one side of the wing produces downwash on the other side of the wing. This downwash reduces the net lift and increases the total induced drag produced by the flow over the finite-span wing. The downwash resulting from the bound-vortex system is greatest near the center of the wing, while the downwash resulting from the trailing-vortex system is greatest near the wing tips. Thus, for a swept wing, the lift is reduced both near the center and near the tips of the wing. This will be evident in the spanwise lift distribution presented in Fig. 7.33 for the wing of Example 7.2. (See Fig. 7.31.)

7.5.1 Velocity Induced by a General Horseshoe Vortex

The velocity induced by a vortex filament of strength Γ_n and a length of dl is given by the *law of Biot and Savart* [see Robinson and Laurmann (1956)]:

$$\overrightarrow{dV} = \frac{\Gamma_n(\overrightarrow{dl} \times \vec{r})}{4\pi r^3} \tag{7.34}$$

Referring to the sketch of Fig. 7.27, the magnitude of the induced velocity is

$$dV = \frac{\Gamma_n \sin\theta \, dl}{4\pi r^2} \tag{7.35}$$

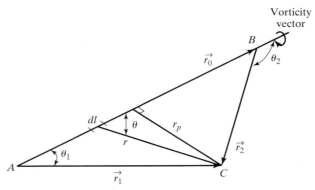

Figure 7.27 Nomenclature for calculating the velocity induced by a finite-length vortex segment.

Since we are interested in the flow field induced by a horseshoe vortex which consists of three straight segments, let us use equation (7.34) to calculate the effect of each segment separately. Let AB be such a segment, with the vorticity vector directed from A to B. Let C be a point in space whose normal distance from the line AB is r_p. We can integrate between A and B to find the magnitude of the induced velocity:

$$V = \frac{\Gamma_n}{4\pi r_p} \int_{\theta_1}^{\theta_2} \sin\theta \, d\theta = \frac{\Gamma_n}{4\pi r_p}(\cos\theta_1 - \cos\theta_2) \tag{7.36}$$

Note that, if the vortex filament extends to infinity in both directions, then $\theta_1 = 0$ and $\theta_2 = \pi$. In this case

$$V = \frac{\Gamma_n}{2\pi r_p}$$

which is the result used in Chapter 6 for the infinite-span airfoils. Let $\vec{r}_0, \vec{r}_1$, and $\vec{r}_2$ designate the vectors $\overrightarrow{AB}, \overrightarrow{AC}$, and $\overrightarrow{BC}$, respectively, as shown in Fig. 7.27. Then

$$r_p = \frac{|\vec{r}_1 \times \vec{r}_2|}{r_0} \qquad \cos\theta_1 = \frac{\vec{r}_0 \cdot \vec{r}_1}{r_0 r_1} \qquad \cos\theta_2 = \frac{\vec{r}_0 \cdot \vec{r}_2}{r_0 r_2}$$

In these equations, if a vector quantity (such as $\vec{r}_0$) is written without a superscript arrow, the symbol represents the magnitude of the parameter. Thus, r_0 is the magnitude of the vector $\vec{r}_0$. Also note that $|\vec{r}_1 \times \vec{r}_2|$ represents the magnitude of the vector cross product. Substituting these expressions into equation (7.36) and noting that the direction of the induced velocity is given by the unit vector

$$\frac{\vec{r}_1 \times \vec{r}_2}{|\vec{r}_1 \times \vec{r}_2|}$$

yields

$$\vec{V} = \frac{\Gamma_n}{4\pi} \frac{\vec{r}_1 \times \vec{r}_2}{|\vec{r}_1 \times \vec{r}_2|^2} \left[\vec{r}_0 \cdot \left(\frac{\vec{r}_1}{r_1} - \frac{\vec{r}_2}{r_2} \right) \right] \tag{7.37}$$

This is the basic expression for the calculation of the induced velocity by the horseshoe vortices in the VLM. It can be used regardless of the assumed orientation of the vortices.

We shall now use equation (7.37) to calculate the velocity that is induced at a general point in space (x, y, z) by the horseshoe vortex shown in Fig. 7.28. The horseshoe vortex may be assumed to represent that for a typical wing panel (e.g., the nth panel) in Fig. 7.24. Segment AB represents the bound vortex portion of the horseshoe system and coincides with the quarter-chord line of the panel element. The trailing vortices are parallel to the x axis. The resultant induced velocity vector will be calculated by considering the influence of each of the elements.

For the bound vortex, segment $\overrightarrow{AB}$,

$$\vec{r}_0 = \overrightarrow{AB} = (x_{2n} - x_{1n})\hat{i} + (y_{2n} - y_{1n})\hat{j} + (z_{2n} - z_{1n})\hat{k}$$

$$\vec{r}_1 = (x - x_{1n})\hat{i} + (y - y_{1n})\hat{j} + (z - z_{1n})\hat{k}$$

$$\vec{r}_2 = (x - x_{2n})\hat{i} + (y - y_{2n})\hat{j} + (z - z_{2n})\hat{k}$$

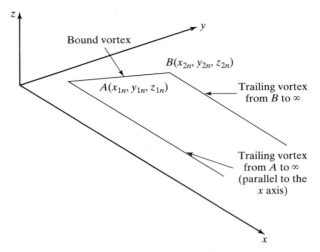

Figure 7.28 "Typical" horseshoe vortex.

Using equation (7.37) to calculate the velocity induced at some point $C(x, y, z)$ by the vortex filament AB (shown in Figs. 7.28 and 7.29),

$$\vec{V}_{AB} = \frac{\Gamma_n}{4\pi} \{\text{Fac1}_{AB}\}\{\text{Fac2}_{AB}\} \tag{7.38a}$$

where

$$\{\text{Fac1}_{AB}\} = \frac{\vec{r}_1 \times \vec{r}_2}{|\vec{r}_1 \times \vec{r}_2|^2}$$

$$= \{[(y - y_{1n})(z - z_{2n}) - (y - y_{2n})(z - z_{1n})]\hat{i}$$

$$- [(x - x_{1n})(z - z_{2n}) - (x - x_{2n})(z - z_{1n})]\hat{j}$$

$$+ [(x - x_{1n})(y - y_{2n}) - (x - x_{2n})(y - y_{1n})]\hat{k}\}/$$

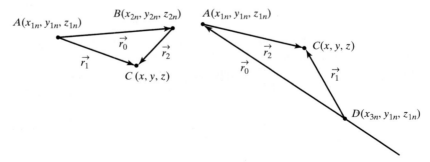

Figure 7.29 Vector elements for the calculation of the induced velocities.

$$\{[(y - y_{1n})(z - z_{2n}) - (y - y_{2n})(z - z_{1n})]^2$$
$$+ [(x - x_{1n})(z - z_{2n}) - (x - x_{2n})(z - z_{1n})]^2$$
$$+ [(x - x_{1n})(y - y_{2n}) - (x - x_{2n})(y - y_{1n})]^2\}$$

and

$$\{Fac2_{AB}\} = \left(\vec{r}_0 \cdot \frac{\vec{r}_1}{r_1} - \vec{r}_0 \cdot \frac{\vec{r}_2}{r_2}\right)$$

$$= \{[(x_{2n} - x_{1n})(x - x_{1n}) + (y_{2n} - y_{1n})(y - y_{1n}) + (z_{2n} - z_{1n})(z - z_{1n})]/$$
$$\sqrt{(x - x_{1n})^2 + (y - y_{1n})^2 + (z - z_{1n})^2}$$
$$- [(x_{2n} - x_{1n})(x - x_{2n}) + (y_{2n} - y_{1n})(y - y_{2n}) + (z_{2n} - z_{1n})(z - z_{2n})]/$$
$$\sqrt{(x - x_{2n})^2 + (y - y_{2n})^2 + (z - z_{2n})^2}\}$$

To calculate the velocity induced by the filament that extends from A to ∞, let us first calculate the velocity induced by the collinear, finite-length filament that extends from A to D. Since $\vec{r}_0$ is in the direction of the vorticity vector,

$$\vec{r}_0 = \overrightarrow{DA} = (x_{1n} - x_{3n})\hat{i}$$
$$\vec{r}_1 = (x - x_{3n})\hat{i} + (y - y_{1n})\hat{j} + (z - z_{1n})\hat{k}$$
$$\vec{r}_2 = (x - x_{1n})\hat{i} + (y - y_{1n})\hat{j} + (z - z_{1n})\hat{k}$$

as shown in Fig. 7.29. Thus, the induced velocity is

$$\overrightarrow{V_{AD}} = \frac{\Gamma_n}{4\pi}\{Fac1_{AD}\}\{Fac2_{AD}\}$$

where

$$\{Fac1_{AD}\} = \frac{(z - z_{1n})\hat{j} + (y_{1n} - y)\hat{k}}{[(z - z_{1n})^2 + (y_{1n} - y)^2](x_{3n} - x_{1n})}$$

and

$$\{Fac2_{AD}\} = (x_{3n} - x_{1n})\left\{\frac{x_{3n} - x}{\sqrt{(x - x_{3n})^2 + (y - y_{1n})^2 + (z - z_{1n})^2}}\right.$$

$$\left. + \frac{x - x_{1n}}{\sqrt{(x - x_{1n})^2 + (y - y_{1n})^2 + (z - z_{1n})^2}}\right\}$$

Letting x_3 go to ∞, the first term of $\{Fac2_{AD}\}$ goes to 1.0. Therefore, the velocity induced by the vortex filament which extends from A to ∞ in a positive direction parallel to the

x axis is given by

$$\overrightarrow{V_{A\infty}} = \frac{\Gamma_n}{4\pi} \left\{ \frac{(z - z_{1n})\hat{j} + (y_{1n} - y)\hat{k}}{[(z - z_{1n})^2 + (y_{1n} - y)^2]} \right\}$$

$$\left[1.0 + \frac{x - x_{1n}}{\sqrt{(x - x_{1n})^2 + (y - y_{1n})^2 + (z - z_{1n})^2}} \right] \qquad \textbf{(7.38b)}$$

Similarly, the velocity induced by the vortex filament that extends from B to ∞ in a positive direction parallel to the x axis is given by

$$\overrightarrow{V_{B\infty}} = -\frac{\Gamma_n}{4\pi} \left\{ \frac{(z - z_{2n})\hat{j} + (y_{2n} - y)\hat{k}}{[(z - z_{2n})^2 + (y_{2n} - y)^2]} \right\}$$

$$\left[1.0 + \frac{x - x_{2n}}{\sqrt{(x - x_{2n})^2 + (y - y_{2n})^2 + (z - z_{2n})^2}} \right] \qquad \textbf{(7.38c)}$$

The total velocity induced at some point (x, y, z) by the horseshoe vortex representing one of the surface elements (i.e., that for the nth panel) is the sum of the components given in equation (7.38). Let the point (x, y, z) be the control point of the mth panel, which we will designate by the coordinates (x_m, y_m, z_m). The velocity induced at the mth control point by the vortex representing the nth panel will be designated as $\overrightarrow{V_{m,n}}$. Examining equation (7.38), we see that

$$\overrightarrow{V_{m,n}} = \overrightarrow{C_{m,n}}\Gamma_n \qquad \textbf{(7.39)}$$

where the influence coefficient $\overrightarrow{C_{m,n}}$ depends on the geometry of the nth horseshoe vortex and its distance from the control point of the mth panel. Since the governing equation is linear, the velocities induced by the $2N$ vortices are added together to obtain an expression for the total induced velocity at the mth control point:

$$\overrightarrow{V_m} = \sum_{n=1}^{2N} \overrightarrow{C_{m,n}}\Gamma_n \qquad \textbf{(7.40)}$$

We have $2N$ of these equations, one for each of the control points.

7.5.2 Application of the Boundary Conditions

Thus, it is possible to determine the resultant induced velocity at any point in space, if the strengths of the $2N$ horseshoe vortices are known. However, their strengths are not known a priori. To compute the strengths of the vortices, Γ_n, which represent the lifting flow field of the wing, we use the boundary condition that the surface is a streamline. That is, the resultant flow is tangent to the wing at each and every control point (which is located at the midspan of the three-quarter-chord line of each elemental panel). If the flow is tangent to the wing, the component of the induced velocity normal to the wing at the control point balances the normal component of the free-stream velocity. To evaluate the induced velocity components, we must introduce at this point our convention that the trailing vortices are parallel to the

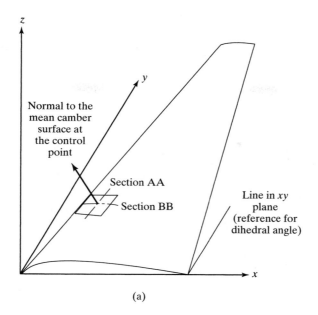

(a)

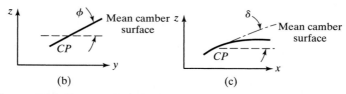

(b)   (c)

Figure 7.30 Nomenclature for the tangency requirement: (a) normal to element of the mean camber surface; (b) section AA; (c) section BB.

vehicle axis [i.e., the x axis for equation (7.38) is the vehicle axis]. Referring to Fig. 7.30, the tangency requirement yields the relation

$$-u_m \sin \delta \cos \phi - v_m \cos \delta \sin \phi + w_m \cos \phi \cos \delta$$

$$+ U_\infty \sin(\alpha - \delta) \cos \phi = 0 \qquad (7.41)$$

where ϕ is the dihedral angle, as shown in Fig. 7.24, and δ is the slope of the mean camber line at the control point. Thus,

$$\delta = \tan^{-1}\left(\frac{dz}{dx}\right)_m$$

For wings where the slope of the mean camber line is small and which are at small angles of attack, equation (7.41) can be replaced by the approximation

$$w_m - v_m \tan \phi + U_\infty\left[\alpha - \left(\frac{dz}{dx}\right)_m\right] = 0 \qquad (7.42)$$

This approximation is consistent with the assumptions of linearized theory. The unknown circulation strengths (Γ_n) required to satisfy these tangent flow boundary conditions are determined by solving the system of simultaneous equations represented by equation (7.40). The solution involves the inversion of a matrix.

7.5.3   Relations for a Planar Wing

Equations (7.38) through (7.42) are those for the VLM where the trailing vortices are parallel to the x axis. As such, they can be solved to determine the lifting flow for a twisted wing with dihedral. Let us apply these equations to a relatively simple geometry, a planar wing (i.e., one that lies in the xy plane), so that we can learn the significance of the various operations using a geometry which we can readily visualize. For a planar wing, $z_{1n} = z_{2n} = 0$ for all the bound vortices. Furthermore, $z_m = 0$ for all the control points. Thus, for our planar wing,

$$\overrightarrow{V}_{AB} = \frac{\Gamma_n}{4\pi} \frac{\hat{k}}{(x_m - x_{1n})(y_m - y_{2n}) - (x_m - x_{2n})(y_m - y_{1n})}$$

$$\left[\frac{(x_{2n} - x_{1n})(x_m - x_{1n}) + (y_{2n} - y_{1n})(y_m - y_{1n})}{\sqrt{(x_m - x_{1n})^2 + (y_m - y_{1n})^2}} \right.$$

$$\left. - \frac{(x_{2n} - x_{1n})(x_m - x_{2n}) + (y_{2n} - y_{1n})(y_m - y_{2n})}{\sqrt{(x_m - x_{2n})^2 + (y_m - y_{2n})^2}} \right] \quad \textbf{(7.43a)}$$

$$\overrightarrow{V}_{A\infty} = \frac{\Gamma_n}{4\pi} \frac{\hat{k}}{y_{1n} - y_m} \left[1.0 + \frac{x_m - x_{1n}}{\sqrt{(x_m - x_{1n})^2 + (y_m - y_{1n})^2}} \right] \quad \textbf{(7.43b)}$$

$$\overrightarrow{V}_{B\infty} = -\frac{\Gamma_n}{4\pi} \frac{\hat{k}}{y_{2n} - y_m} \left[1.0 + \frac{x_m - x_{2n}}{\sqrt{(x_m - x_{2n})^2 + (y_m - y_{2n})^2}} \right] \quad \textbf{(7.43c)}$$

Note that, for the planar wing, all three components of the vortex representing the nth panel induce a velocity at the control point of the mth panel which is in the z direction (i.e., a downwash). Therefore, we can simplify equation (7.43) by combining the components into one expression:

$$w_{m,n} = \frac{\Gamma_n}{4\pi} \left\{ \frac{1}{(x_m - x_{1n})(y_m - y_{2n}) - (x_m - x_{2n})(y_m - y_{1n})} \right.$$

$$\left[\frac{(x_{2n} - x_{1n})(x_m - x_{1n}) + (y_{2n} - y_{1n})(y_m - y_{1n})}{\sqrt{(x_m - x_{1n})^2 + (y_m - y_{1n})^2}} \right.$$

$$\left. - \frac{(x_{2n} - x_{1n})(x_m - x_{2n}) + (y_{2n} - y_{1n})(y_m - y_{2n})}{\sqrt{(x_m - x_{2n})^2 + (y_m - y_{2n})^2}} \right.$$

$$+ \frac{1.0}{y_{1n} - y_m} \left[1.0 + \frac{x_m - x_{1n}}{\sqrt{(x_m - x_{1n})^2 + (y_m - y_{1n})^2}} \right]$$

$$\left. - \frac{1.0}{y_{2n} - y_m} \left[1.0 + \frac{x_m - x_{2n}}{\sqrt{(x_m - x_{2n})^2 + (y_m - y_{2n})^2}} \right] \right\} \qquad \textbf{(7.44)}$$

Summing the contributions of all the vortices to the downwash at the control point of the mth panel,

$$w_m = \sum_{n=1}^{2N} w_{m,n} \qquad \textbf{(7.45)}$$

Let us now apply the tangency requirement defined by equations (7.41) and (7.42). Since we are considering a planar wing in this section, $(dz/dx)_m = 0$ everywhere and $\phi = 0$. The component of the free-stream velocity perpendicular to the wing is $U_\infty \sin \alpha$ at any point on the wing. Thus, the resultant flow will be tangent to the wing if the total vortex-induced downwash at the control point of the mth panel, which is calculated using equation (7.45) balances the normal component of the free-stream velocity:

$$w_m + U_\infty \sin \alpha = 0 \qquad \textbf{(7.46)}$$

For small angles of attack,

$$w_m = -U_\infty \alpha \qquad \textbf{(7.47)}$$

In Example 7.2, we will solve for the aerodynamic coefficients for a wing that has a relatively simple planform and an uncambered section. The vortex lattice method will be applied using only a single lattice element in the chordwise direction for each spanwise subdivision of the wing. Applying the boundary condition that there is no flow through the wing at only one point in the chordwise direction is reasonable for this flat-plate wing. However, it would not be adequate for a wing with cambered sections or a wing with deflected flaps.

EXAMPLE 7.2: **Use the vortex lattice method (VLM) to calculate the aerodynamic coefficients for a swept wing**

Let us use the relations developed in this section to calculate the lift coefficient for a swept wing. So that the calculation procedures can be easily followed, let us consider a wing that has a relatively simple geometry (i.e., that illustrated in Fig. 7.31). The wing has an aspect ratio of 5, a taper ratio of unity (i.e., $c_r = c_t$), and an uncambered section (i.e., it is a flat plate). Since the taper ratio is unity, the leading edge, the quarter-chord line, the three-quarter-chord line, and the trailing edge all have the same sweep, 45°. Since

$$AR = 5 = \frac{b^2}{S}$$

and since for a swept, untapered wing

$$S = bc$$

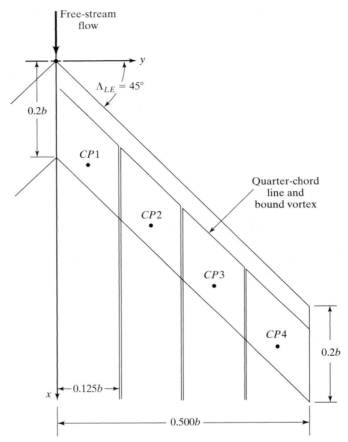

Figure 7.31 Four-panel representation of a swept planar wing, taper ratio of unity, AR = 5, Λ = 45°.

it is clear that $b = 5c$. Using this relation, it is possible to calculate all of the necessary coordinates in terms of the parameter b. Therefore, the solution does not require that we know the physical dimensions of the configuration.

Solution: The flow field under consideration is symmetric with respect to the $y = 0$ plane (xz plane); that is, there is no yaw. Thus, the lift force acting at a point on the starboard wing $(+y)$ is equal to that at the corresponding point on the port wing $(-y)$. Because of symmetry, we need only to solve for the strengths of the vortices of the starboard wing. Furthermore, we need to apply the tangency condition [i.e., equation (7.47)] only at the control points of the starboard wing. However, we must remember to include the contributions of the horseshoe vortices of the port wing to the velocities induced at these control points (of the starboard wing). Thus, for this planar symmetric flow, equation (7.45) becomes

$$w_m = \sum_{n=1}^{N} w_{m,ns} + \sum_{n=1}^{N} w_{m,np}$$

where the symbols s and p represent the starboard and port wings, respectively.

The planform of the starboard wing is divided into four panels, each panel extending from the leading edge to the trailing edge. By limiting ourselves to only four spanwise panels, we can calculate the strength of the horseshoe vortices using only a pocket electronic calculator. Thus, we can more easily see how the terms are to be evaluated. As before, the bound portion of each horseshoe vortex coincides with the quarter-chord line of its panel and the trailing vortices are in the plane of the wing, parallel to the x axis. The control points are designated by the solid symbols in Fig. 7.31. Recall that $(x_m, y_m, 0)$ are the coordinates of a given control point and that $(x_{1n}, y_{1n}, 0)$ and $(x_{2n}, y_{2n}, 0)$ are the coordinates of the "ends" of the bound-vortex filament AB. The coordinates for a 4×1 lattice (four spanwise divisions and one chordwise division) for the starboard (right) wing are summarized in Table 7.2.

Using equation (7.44) to calculate the downwash velocity at the CP of panel 1 (of the starboard wing) induced by the horseshoe vortex of panel 1 of the starboard wing,

$$
\begin{aligned}
w_{1,1s} = \frac{\Gamma_1}{4\pi} \Bigg\{ & \frac{1.0}{(0.1625b)(-0.0625b) - (0.0375b)(0.0625b)} \\
& \left[\frac{(0.1250b)(0.1625b) + (0.1250b)(0.0625b)}{\sqrt{(0.1625b)^2 + (0.0625b)^2}} \right. \\
& \left. - \frac{(0.1250b)(0.0375b) + (0.1250b)(-0.0625b)}{\sqrt{(0.0375b)^2 + (-0.0625b)^2}} \right] \\
& + \frac{1.0}{-0.0625b}\left[1.0 + \frac{0.1625b}{\sqrt{(0.1625b)^2 + (0.0625b)^2}} \right] \\
& - \frac{1.0}{0.0625b}\left[1.0 + \frac{0.0375b}{\sqrt{(0.0375b)^2 + (-0.0625b)^2}} \right] \Bigg\} \\
= & \frac{\Gamma_1}{4\pi b}(-16.3533 - 30.9335 - 24.2319)
\end{aligned}
$$

TABLE 7.2 Coordinates of the Bound Vortices and of the Control Points of the Starboard (Right) Wing

Panel	x_m	y_m	x_{1n}	y_{1n}	x_{2n}	y_{2n}
1	0.2125b	0.0625b	0.0500b	0.0000b	0.1750b	0.1250b
2	0.3375b	0.1875b	0.1750b	0.1250b	0.3000b	0.2500b
3	0.4625b	0.3125b	0.3000b	0.2500b	0.4250b	0.3750b
4	0.5875b	0.4375b	0.4250b	0.3750b	0.5500b	0.5000b

Note that, as one would expect, each of the vortex elements induces a negative (downward) component of velocity at the control point. The student should visualize the flow induced by each segment of the horseshoe vortex to verify that a negative value for each of the components is intuitively correct. In addition, the velocity induced by the vortex trailing from A to ∞ is greatest in magnitude. Adding the components together, we find

$$w_{1,1s} = \frac{\Gamma_1}{4\pi b}(-71.5187)$$

The downwash velocity at the CP of panel 1 (of the starboard wing) induced by the horseshoe vortex of panel 1 of the port wing is

$$w_{1,1p} = \frac{\Gamma_1}{4\pi}\left\{ \frac{1.0}{(0.0375b)(0.0625b) - (0.1625b)(0.1875b)} \right.$$

$$\left[\frac{(-0.1250b)(0.0375b) + (0.1250b)(0.1875b)}{\sqrt{(0.0375b)^2 + (0.1875b)^2}} \right.$$

$$\left. - \frac{(-0.1250b)(0.1625b) + (0.1250b)(0.0625b)}{\sqrt{(0.1625b)^2 + (0.0625b)^2}} \right]$$

$$+ \frac{1.0}{-0.1875b}\left[1.0 + \frac{0.0375b}{\sqrt{(0.0375b)^2 + (0.1875b)^2}} \right]$$

$$\left. - \frac{1.0}{-0.0625b}\left[1.0 + \frac{0.1625b}{\sqrt{(0.1625b)^2 + (0.0625b)^2}} \right] \right\}$$

$$= \frac{\Gamma_1}{4\pi b}[-6.0392 - 6.3793 + 30.9335]$$

$$= \frac{\Gamma_1}{4\pi b}(18.5150)$$

Similarly, using equation (7.44) to calculate the downwash velocity at the CP of panel 2 induced by the horseshoe vortex of panel 4 of the starboard wing, we obtain

$$w_{2,4s} = \frac{\Gamma_4}{4\pi}\left\{ \frac{1.0}{(-0.0875b)(-0.3125b) - (-0.2125b)(-0.1875b)} \right.$$

$$\left[\frac{(0.1250b)(-0.0875b) + (0.1250b)(-0.1875b)}{\sqrt{(-0.0875b)^2 + (-0.1875b)^2}} \right.$$

$$\left. - \frac{(0.1250b)(-0.2125b) + (0.1250b)(-0.3125b)}{\sqrt{(-0.2125b)^2 + (-0.3125b)^2}} \right]$$

$$+ \frac{1.0}{0.1875b}\left[1.0 + \frac{-0.0875b}{\sqrt{(-0.0875b)^2 + (-0.1875b)^2}} \right]$$

$$- \frac{1.0}{0.3125b}\left[1.0 + \frac{-0.2125b}{\sqrt{(-0.2125b)^2 + (-0.3125b)^2}} \right]\Bigg\}$$

$$= \frac{\Gamma_4}{4\pi b}[-0.60167 + 3.07795 - 1.40061]$$

$$= \frac{\Gamma_4}{4\pi b}(1.0757)$$

Again, the student should visualize the flow induced by each segment to verify that the signs and the relative magnitudes of the components are individually correct.

Evaluating all of the various components (or influence coefficients), we find that at control point 1

$$w_1 = \frac{1}{4\pi b}[(-71.5187\Gamma_1 + 11.2933\Gamma_2 + 1.0757\Gamma_3 + 0.3775\Gamma_4)_s$$

$$+ (+18.5150\Gamma_1 + 2.0504\Gamma_2 + 0.5887\Gamma_3 + 0.2659\Gamma_4)_p]$$

At CP 2,

$$w_2 = \frac{1}{4\pi b}[(+20.2174\Gamma_1 - 71.5187\Gamma_2 + 11.2933\Gamma_3 + 1.0757\Gamma_4)_s$$

$$+ (+3.6144\Gamma_1 + 1.1742\Gamma_2 + 0.4903\Gamma_3 + 0.2503\Gamma_4)_p]$$

At CP 3,

$$w_3 = \frac{1}{4\pi b}[(+3.8792\Gamma_1 + 20.2174\Gamma_2 - 71.5187\Gamma_3 + 11.2933\Gamma_4)_s$$

$$+ (+1.5480\Gamma_1 + 0.7227\Gamma_2 + 0.3776\Gamma_3 + 0.2179\Gamma_4)_p]$$

At CP 4,

$$w_4 = \frac{1}{4\pi b}[(+1.6334\Gamma_1 + 3.8792\Gamma_2 + 20.2174\Gamma_3 - 71.5187\Gamma_4)_s$$

$$+ (+0.8609\Gamma_1 + 0.4834\Gamma_2 + 0.2895\Gamma_3 + 0.1836\Gamma_4)_p]$$

Since it is a planar wing with no dihedral, the no-flow condition of equation (7.47) requires that

$$w_1 = w_2 = w_3 = w_4 = -U_\infty\alpha$$

Thus

$$-53.0037\Gamma_1 + 13.3437\Gamma_2 + 1.6644\Gamma_3 + 0.6434\Gamma_4 = -4\pi bU_\infty\alpha$$

$$+23.8318\Gamma_1 - 70.3445\Gamma_2 + 11.7836\Gamma_3 + 1.3260\Gamma_4 = -4\pi bU_\infty\alpha$$

$$+5.4272\Gamma_1 + 20.9401\Gamma_2 - 71.1411\Gamma_3 + 11.5112\Gamma_4 = -4\pi bU_\infty\alpha$$

$$+2.4943\Gamma_1 + 4.3626\Gamma_2 + 20.5069\Gamma_3 - 71.3351\Gamma_4 = -4\pi bU_\infty\alpha$$

Solving for $\Gamma_1, \Gamma_2, \Gamma_3$, and Γ_4, we find that

$$\Gamma_1 = +0.0273(4\pi bU_\infty\alpha) \qquad \textbf{(7.48a)}$$

$$\Gamma_2 = +0.0287(4\pi bU_\infty\alpha) \qquad \textbf{(7.48b)}$$

$$\Gamma_3 = +0.0286(4\pi bU_\infty\alpha) \qquad \textbf{(7.48c)}$$

$$\Gamma_4 = +0.0250(4\pi bU_\infty\alpha) \qquad \textbf{(7.48d)}$$

Having determined the strength of each of the vortices by satisfying the boundary conditions that the flow is tangent to the surface at each of the control points, the lift of the wing may be calculated. For wings that have no dihedral over any portion of the wing, all the lift is generated by the free-stream velocity crossing the spanwise vortex filament, since there are no side-wash or backwash velocities. Furthermore, since the panels extend from the leading edge to the trailing edge, the lift acting on the nth panel is

$$l_n = \rho_\infty U_\infty \Gamma_n \qquad \textbf{(7.49)}$$

which is also the lift per unit span. Since the flow is symmetric, the total lift for the wing is

$$L = 2 \int_0^{0.5b} \rho_\infty U_\infty \Gamma(y)\, dy \qquad \textbf{(7.50a)}$$

or, in terms of the finite-element panels,

$$L = 2\rho_\infty U_\infty \sum_{n=1}^{4} \Gamma_n \Delta y_n \qquad \textbf{(7.50b)}$$

Since $\Delta y_n = 0.1250b$ for each panel,

$$L = 2\rho_\infty U_\infty 4\pi bU_\infty\alpha(0.0273 + 0.0287 + 0.0286 + 0.0250)0.1250b$$

$$= \rho_\infty U_\infty^2 b^2 \pi\alpha(0.1096)$$

To calculate the lift coefficient, recall that $S = bc$ and $b = 5c$ for this wing. Therefore,

$$C_L = \frac{L}{q_\infty S} = 1.096\pi\alpha$$

Furthermore,

$$C_{L,\alpha} = \frac{dC_L}{d\alpha} = 3.443 \text{ per radian} = 0.0601 \text{ per degree}$$

Comparing this value $C_{L,\alpha}$ with that for an unswept wing (such as the results presented in Fig. 7.14), it is apparent that an effect of sweepback is the reduction in the lift-curve slope.

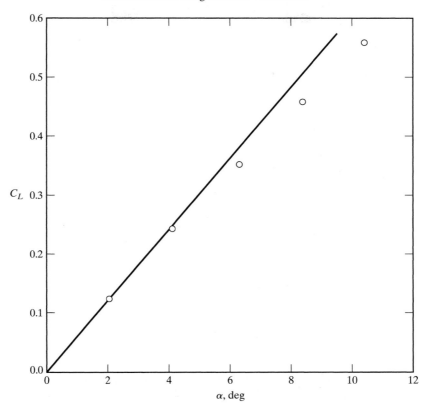

Figure 7.32 Comparison of the theoretical and the experimental lift coefficients for the swept wing of Fig. 7.31 in a subsonic stream.

The theoretical lift curve generated using the VLM is compared in Fig. 7.32 with experimental results reported by Weber and Brebner (1958). The experimentally determined values of the lift coefficient are for a wing of constant chord and of constant section, which was swept 45° and which had an aspect ratio of 5. The theoretical lift coefficients are in good agreement with the experimental values.

Since the lift per unit span is given by equation (7.49), the section lift coefficient for the nth panel is

$$C_{l(n\text{th})} = \frac{l_n}{\frac{1}{2}\rho_\infty U_\infty^2 c_{\text{av}}} = \frac{2\Gamma}{U_\infty c_{\text{av}}} \tag{7.51}$$

When the panels extend from the leading edge to the trailing edge, such as is the case for the 4×1 lattice shown in Fig. 7.31, the value of Γ given in equation (7.48) is used

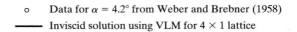

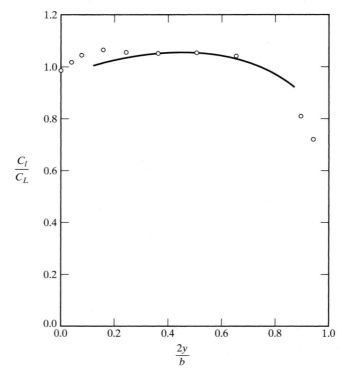

Figure 7.33 Comparison of the theoretical and the experimental spanwise lift distribution for the wing of Fig. 7.31.

in equation (7.51). When there are a number of discrete panels in the chordwise direction, such as the 10×4 lattice shown in Fig. 7.24, you should sum (from the leading edge to the trailing edge) the values of Γ for those bound-vortex filaments at the spanwise location (i.e., in the chordwise strip) of interest. For a chordwise row,

$$\frac{C_l c}{c_{av}} = \sum_{j=1}^{J_{max}} \left(\frac{l}{q_\infty c_{av}} \right)_j \tag{7.52}$$

where c_{av} is the average chord (and is equal to S/b), c is the local chord, and j is the index for an elemental panel in the chordwise row. The total lift coefficient is obtained by integrating the lift over the span

$$C_L = \int_0^1 \frac{C_l c}{c_{av}} d\left(\frac{2y}{b} \right) \tag{7.53}$$

The spanwise variation in the section lift coefficient is presented in Fig. 7.33. The theoretical distribution is compared with the experimentally determined spanwise load distribution for an angle of attack of 4.2°, which was presented by Weber and Brebner (1958).

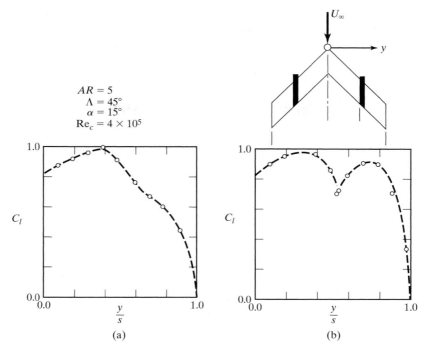

Figure 7.34 Effect of a boundary-layer fence on the spanwise distribution of the local lift coefficient: (a) without fence; (b) with fence. [Data from Schlichting (1960).]

The increased loading of the outer wing sections promotes premature boundary-layer separation there. This unfavorable behavior is amplified by the fact that the spanwise velocity component causes the already decelerated fluid particles in the boundary layer to move toward the wing tips. This transverse flow results in a large increase in the boundary-layer thickness near the wing tips. Thus, at large angles of attack, premature separation may occur on the suction side of the wing near the tip. If significant tip stall occurs on the swept wing, there is a loss of the effectiveness of the control surfaces and a forward shift in the wing center of pressure that creates an unstable, nose-up increase in the pitching moment.

Boundary-layer fences are often used to break up the spanwise flow on swept wings. The spanwise distribution of the local-lift coefficient [taken from Schlichting (1960)] without and with a boundary-layer fence is presented in Fig. 7.34. The essential effect of the boundary-layer fence does not so much consist in the prevention of the transverse flow but, much more important, in that the fence divides each wing into an inner and an outer portion. Both transverse flow and boundary-layer separation may be present, but to a reduced extent. Boundary-layer fences are evident on the swept wings of the Trident, shown in Fig. 7.34.

Once we have obtained the solution for the section lift coefficient (i.e., that for a chordwise strip of the wing), the induced-drag coefficient may be calculated using the relation given by Multhopp (1950):

$$C_{Dv} = \frac{1}{S} \int_{-0.5b}^{+0.5b} C_l c \alpha_i \, dy \tag{7.54}$$

Figure 7.35 Trident illustrating boundary-layer fences on the wing. (Courtesy of British Aerospace.)

where α_i, which is the induced incidence, given by

$$\alpha_i = -\frac{1}{8\pi} \int_{-0.5b}^{+0.5b} \frac{C_l c}{(y - \eta)^2} \, d\eta \tag{7.55}$$

For a symmetrical loading, equation (7.55) may be written

$$\alpha_i = -\frac{1}{8\pi} \int_{0}^{0.5b} \left[\frac{C_l c}{(y - \eta)^2} + \frac{C_l c}{(y + \eta)^2} \right] d\eta \tag{7.56}$$

Following the approach of Kalman, Giesing, and Rodden (1970), we consider the mth chordwise strip, which has a semiwidth of e_m and whose centerline is located at $\eta = y_m$. Let us approximate the spanwise lift distribution across the strip by a parabolic function:

$$\left(\frac{C_l c}{C_L \bar{c}} \right)_m = a_m \eta^2 + b_m \eta + c_m \tag{7.57}$$

To solve for the coefficients a_m, b_m, and c_m, note that

$$y_{m+1} = y_m + (e_m + e_{m+1})$$

$$y_{m-1} = y_m - (e_m + e_{m-1})$$

Thus,

$$c_m = \left(\frac{C_l c}{C_L \bar{c}} \right)_m - a_m \eta_m^2 - b_m \eta_m$$

$$a_m = \frac{1}{d_{mi} d_{mo}(d_{mi} + d_{mo})} \left\{ d_{mi} \left(\frac{C_l c}{C_L \bar{c}} \right)_{m+1} - (d_{mi} + d_{mo}) \left(\frac{C_l c}{C_L \bar{c}} \right)_m + d_{mo} \left(\frac{C_l c}{C_L \bar{c}} \right)_{m-1} \right\}$$

and

$$b_m = \frac{1}{d_{mi}d_{mo}(d_{mi} + d_{mo})} \left\{ d_{mo}(2\eta_m - d_{mo}) \left[\left(\frac{C_lc}{C_L\bar{c}}\right)_m - \left(\frac{C_lc}{C_L\bar{c}}\right)_{m-1} \right] \right.$$

$$\left. - d_{mi}(2\eta_m - d_{mi}) \left[\left(\frac{C_lc}{C_L\bar{c}}\right)_{m+1} - \left(\frac{C_lc}{C_L\bar{c}}\right)_m \right] \right\}$$

where

$$d_{mi} = e_m + e_{m-1}$$

and

$$d_{mo} = e_m + e_{m+1}$$

For a symmetric load distribution, we let

$$\left(\frac{C_lc}{C_L\bar{c}}\right)_{m-1} = \left(\frac{C_lc}{C_L\bar{c}}\right)_m$$

and

$$e_{m-1} = e_m$$

at the root. Similarly, we let

$$\left(\frac{C_lc}{C_L\bar{c}}\right)_{m+1} = 0$$

and

$$e_{m+1} = 0$$

at the tip. Substituting these expressions into equations (7.56) and (7.57), we then obtain the numerical form for the induced incidence:

$$\frac{\alpha_i(y)}{C_Lc} = -\frac{1}{4\pi} \sum_{m=1}^{N} \left\{ \frac{y^2(y_m + e_m)a_m + y^2b_m + (y_m + e_m)c_m}{y^2 - (y_m + e_m)^2} \right.$$

$$- \frac{y^2(y_m - e_m)a_m + y^2b_m + (y_m - e_m)c_m}{y^2 - (y_m - e_m)^2}$$

$$+ \frac{1}{2}ya_m \log\left[\frac{(y - e_m)^2 - y_m^2}{(y + e_m)^2 - y_m^2}\right]^2$$

$$\left. + \frac{1}{4}b_m \log\left[\frac{y^2 - (y_m + e_m)^2}{y^2 - (y_m - e_m)^2}\right]^2 + 2e_ma_m \right\} \qquad \textbf{(7.58)}$$

We then assume that the product $C_lc\alpha_i$ also has a parabolic variation across the strip.

$$\left[\left(\frac{C_lc}{C_L\bar{c}}\right)\left(\frac{\alpha_i}{C_L\bar{c}}\right)\right]_n = a_ny^2 + b_ny + c_n \qquad \textbf{(7.59)}$$

The coefficients a_n, b_n, and c_n can be obtained using an approach identical to that employed to find a_m, b_m, and c_m. The numerical form of equation (7.54) is then a generalization of Simpson's rule:

$$\frac{C_{Dv}}{C_L^2} = \frac{4}{AR} \sum_{n=1}^{N} e_n \left\{ \left[y_n^2 + \left(\frac{1}{3}\right) e_n^2 \right] a_n + y_n b_n + c_n \right\} \tag{7.60}$$

The lift developed along the chordwise bound vortices in a chordwise row of horseshoe vortices varies from the leading edge to the trailing edge of the wing because of the longitudinal variation of both the sidewash velocity and the local value of the vortex strength for planforms that have a nonzero dihedral angle. For techniques to compute the lift, the pitching moment, and the rolling moment for more general wings, the reader is referred to Margason and Lamar (1971).

Numerous investigators have studied methods for improving the convergence and the rigor of the VLM. Therefore, there is an ever-expanding body of relevant literature with which the reader should be familiar before attempting to analyze complex flow fields. Furthermore, as noted in the introduction to this section, VLM is only one (and, indeed, a relatively simple one) of the methods used to compute the aerodynamic coefficients.

7.6 FACTORS AFFECTING DRAG DUE-TO-LIFT AT SUBSONIC SPEEDS

The term representing the lift-dependent drag coefficient in equation (7.17) includes those parts of the viscous drag and of the form drag, which result when the angle of attack changes from α_{ol}. For the present analysis, the "effective leading-edge suction" (s) will be used to characterize the drag-due-to-lift. For 100% suction, the drag-due-to-lift coefficient is the potential flow induced vortex drag, which is represented by the symbol C_{Dv}. For an elliptically loaded wing, this value would be $C_L^2/(\pi AR)$ as given in equation (7.16). Zero percent suction corresponds to the condition where the resultant force vector is normal to the chord line, as a result of extensive separation at the wing leading edge.

Aircraft designed to fly efficiently at supersonic speeds often employ thin, highly swept wings. Values of the leading-edge suction reported by Henderson (1966) are reproduced in Fig. 7.36. The data presented in Fig. 7.36 were obtained at the lift coefficient for which (L/D) is a maximum. This lift coefficient is designated $C_{L,\text{opt}}$. Values of s are presented as a function of the free-stream Reynolds number based on the chord length for a wing with an aspect ratio of 1.4 and whose sharp leading edge is swept 67° in Fig. 7.36a. The values of s vary only slightly with the Reynolds number. Thus, the suction parameter can be presented as a function of the leading-edge sweep angle (independent of the Reynolds number). Values of s are presented in Fig. 7.36b as a function of the leading-edge sweep angle for several sharp-edge wings. Even for relatively low values for the sweep angle, suction values no higher than about 50% were obtained.

Several features that can increase the effective leading-edge suction can be incorporated into a wing design. The effect of two of the features (leading-edge flaps and wing warp) is illustrated in Fig. 7.37. Values of s are presented as a function of the Reynolds

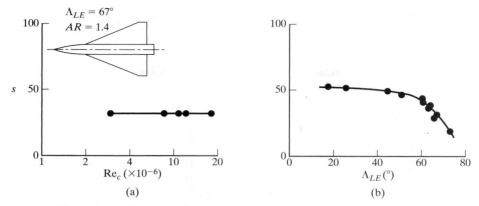

Figure 7.36 The variation of the effective leading-edge suction (s) as (a) a function of the Reynolds number and (b) of the leading-edge sweep angle for a wing with a sharp leading edge, $C_{L,\text{opt}}$, and $M < 0.30$. [Taken from Henderson (1966).]

number for a wing swept 74° for $C_{L,\text{opt}}$. The data [again taken from Henderson (1966)] show that both leading-edge flaps and wing warp significantly increase the values of the effective leading-edge suction relative to those for a symmetrical, sharp leading-edge wing.

A third feature that can be used to improve the effective leading-edge suction is the wing leading-edge radius. Data originally presented by Henderson (1966) are reproduced in Fig. 7.38. Values of s are presented as a function of the Reynolds number for a wing with an aspect ratio of 2.0 and whose leading edge is swept 67°. Again, the wing with a sharp leading edge generates relatively low values of s, which are essentially independent of the Reynolds number. The values of s for the wing with a rounded leading edge ($r_{\text{LE}} = 0.0058c$) exhibit large increases in s as the Reynolds number increases. As noted by Henderson (1966), the increase in the effective leading-edge suction due to the leading-edge radius accounts for 2/3 of the increase in $(L/D)_{\text{max}}$ from 8 to 12. A reduction in the skin friction drag accounts for the remaining 1/3 of the increase in $(L/D)_{\text{max}}$.

The total trim drag has two components: (1) the drag directly associated with the tail and (2) an increment in drag on the wing due to the change in the wing lift coefficient required to offset the tail load. McKinney and Dollyhigh (1971) state "One of the more important considerations in a study of trim drag is the efficiency with which the wing operates. Therefore, a typical variation of the subsonic drag-due-to-lift parameter (which is a measure of wing efficiency) is presented in Fig. 7.39. The full leading-edge suction line and the zero leading-edge suction line are shown for reference. The variation of leading-edge suction with Reynolds number and wing geometry is discussed in a paper by Henderson (1966). The cross-hatched area indicates a typical range of values of the drag-due-to-lift parameter for current aircraft including the use of both fixed camber and twist or wing flaps for maneuvering. At low lift coefficients ($C_L \approx 0.30$) typical of 1-g flight, drag-due-to-lift values approaching those corresponding to full leading-edge suction are generally obtained. At the high lift coefficient ($C_L \approx 1.0$) which

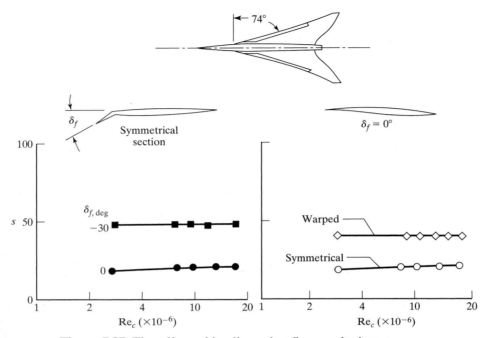

Figure 7.37 The effect of leading-edge flaps and wing warp on the effective leading-edge suction (s) for $C_{L,\mathrm{opt}}$, $M < 0.30$. [Taken from Henderson (1966).]

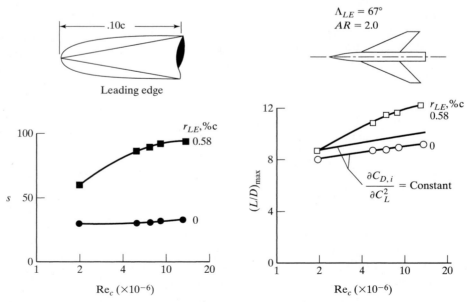

Figure 7.38 The effect of the leading-edge shape, $C_{L,\mathrm{opt}}$, $M < 0.30$. [Taken from Henderson (1966).]

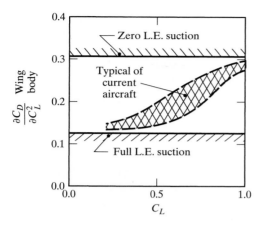

Figure 7.39 Typical variation of drag-due-to-lift parameter; aspect ratio = 2.5; subsonic speeds. [Taken from Henderson (1966).]

corresponds to the maneuvering case . . . , the drag-due-to-lift typically approaches the zero leading-edge suction line even when current maneuver flap concepts are considered. The need for improving drag-due-to-lift characteristics of wings at the high-lift coefficients by means such as wing warp, improved maneuver devices, and so forth, is recognized."

7.7   DELTA WINGS

As has been discussed, a major aerodynamic consideration in wing design is the prediction and the control of flow separation. However, as the sweep angle is increased and the section thickness is decreased in order to avoid undesirable compressibility effects, it becomes increasingly more difficult to prevent boundary-layer separation. Although many techniques have been discussed to alleviate these problems, it is often necessary to employ rather complicated variable-geometry devices in order to satisfy a wide range of conflicting design requirements which result due to the flow-field variations for the flight envelope of high-speed aircraft. Beginning with the delta-wing design of Alexander Lippisch in Germany during World War II, supersonic aircraft designs have often used thin, highly swept wings of low aspect ratio to minimize the wave drag at the supersonic cruise conditions. It is interesting to note that during the design of the world's first operational jet fighter, the Me 262, the outer panels of the wing were swept to resolve difficulties arising when increasingly heavier turbojets caused the center of gravity to move. Thus, the introduction of sweepback did not reflect an attempt to reduce the effects of compressibility [Voight (1976)]. This historical note is included here to remind the reader that many parameters enter into the design of an airplane; aerodynamics is only one of them. The final configuration will reflect design priorities and trade-offs.

At subsonic speeds, however, delta-wing planforms have aerodynamic characteristics which are substantially different from those of the relatively straight, high-aspect-ratio wings designed for subsonic flight. Because they operate at relatively high angles of attack, the boundary layer on the lower surface flows outward and separates as it goes over the leading edge, forming a free shear layer. The shear layer curves upward and inboard, eventually rolling up into a core of high vorticity, as shown in Fig. 7.40.

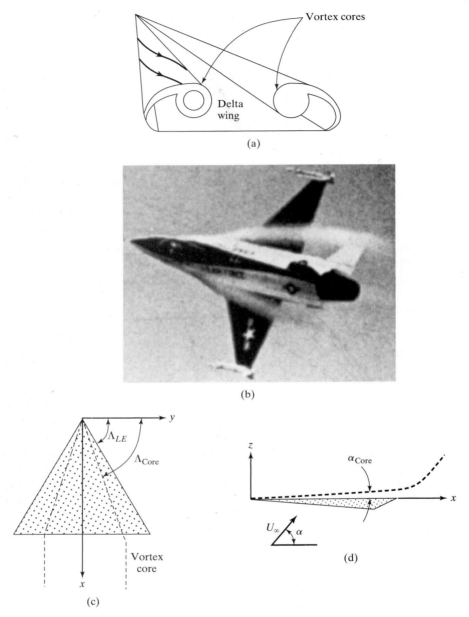

Figure 7.40 (a) Vortex core that develops for flow over a delta wing. (b) Water vapor condenses due to the pressure drop revealing the vortex core for a General Dynamic F-16. (Photograph courtesy of General Dynamics.) (c) Trajectory of the leading-edge vortex. [Taken from Visbal (1995).] (d) Inclination angle of the vortex trajectory. [Taken from Visbal (1995).]

There is an appreciable axial component of motion and the fluid spirals around and along the axis. A spanwise outflow is induced on the upper surface, beneath the coiled vortex sheet, and the flow separates again as it approaches the leading edge [Stanbrook and Squire (1964)]. The size and the strength of the coiled vortex sheets increase with increasing incidence and they become a dominant feature of the flow, which remains steady throughout the range of practical flight attitudes of the wing. The formation of these vortices is responsible for the nonlinear aerodynamic characteristics that exist over the angle-of-attack range [Hummel (2004)].

The leading-edge suction analogy developed by Polhamus (1971a and 1971b) can be used to calculate the lift and the drag-due-to-lift characteristics which arise when the separated flow around sharp-edge delta wings reattaches on the upper surface. Thus, the correlations apply to thin wings having neither camber nor twist. Furthermore, the method is applicable to wings for which the leading edges are of sufficient sharpness that separation is fixed at the leading edge. Since the vortex flow induces reattachment and since the Kutta condition must, therefore, still be satisfied at the trailing edge, the total lift coefficient consists of a potential-flow term and a vortex-lift term. The total lift coefficient can be represented by the sum

$$C_L = K_p \sin \alpha \cos^2 \alpha + K_v \sin^2 \alpha \cos \alpha \qquad (7.61)$$

The constant K_p is simply the normal-force slope calculated using the potential-flow lift-curve slope. The constant K_v can be estimated from the potential flow leading-edge suction calculations. Using the nomenclature for arrow-, delta-, and diamond-planform wings illustrated in Fig. 7.41, K_p and K_v are presented as a function of the planform parameters in Figs. 7.42 and 7.43, respectively.

Values of the lift coefficient calculated using equation (7.61) are compared in Fig. 7.44 with experimental values for uncambered delta wings that have sharp leading edges. Data are presented for wings of aspect ratio from 1.0 [Peckham (1958)] to 2.0 (Bartlett and Vidal, 1955). Since the analytical method is based on an analogy with potential-flow leading-edge suction, which requires that flow reattaches on the upper surface inboard of the vortex, the correlation between theory and data breaks down as flow reattachment fails to occur. The lift coefficients calculated using equation (7.61) for $AR = 1.0$ and 1.5 are in good agreement with the measured values up to angles of attack in excess of $20°$. However, for the delta wing, whose aspect ratio is 2.0, significant deviations between the calculated values and the experimental values exist for angles of attack above $15°$.

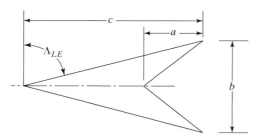

Figure 7.41 Wing geometry nomenclature.

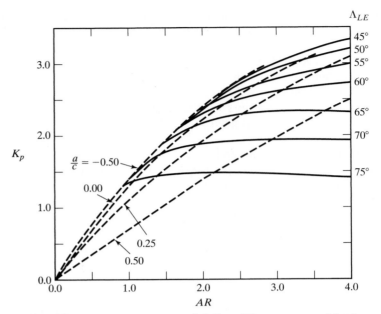

Figure 7.42 Variation of potential-flow lift constant with plan-form parameters. [Taken from Pollhamus (1971b).]

If the leading edges of the wing are rounded, separation occurs well inboard on the wing's upper surface. The result is that, outboard of the loci of the separation points, the flow pattern is essentially that of potential flow and the peak negative pressures at the section extremity are preserved. However, as noted by Peckham (1958), increasing the thickness causes a reduction in net lift. The combined effect of the thickness ratio and of the shape of the leading edges is illustrated in the experimental lift coefficients presented in Fig. 7.45, which are taken from Bartlett and Vidal (1955). The lift coefficients are less for the thicker wings, which also have rounded leading edges.

The lift coefficients for a series of delta wings are presented as a function of the angle of attack in Fig. 7.46. The lift-curve slope, $dC_L/d\alpha$, becomes progressively smaller as the aspect ratio decreases. However, for all but the smallest aspect ratio wing, the maximum value of the lift coefficient and the angle of attack at which it occurs increase as the aspect ratio decreases.

For a thin flat-plate model, the resultant force acts normal to the surface. Thus, the induced drag ΔC_D for a flat-plate wing would be

$$\Delta C_D = C_D - C_{D0} = C_L \tan \alpha \tag{7.62}$$

Using equation (7.61) to evaluate C_L,

$$C_D = C_{D0} + K_p \sin^2 \alpha \cos \alpha + K_v \sin^3 \alpha \tag{7.63}$$

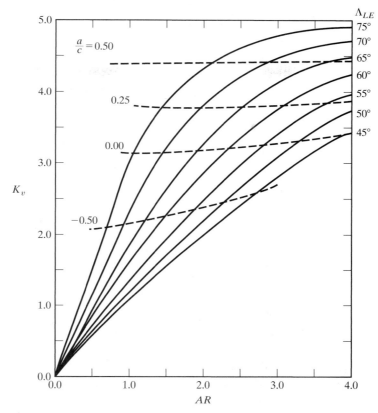

Figure 7.43 Variation of vortex-lift constant with planform parameters. [Taken from Pollhamus (1971b).]

Experimental values of the drag coefficient from Bartlett and Vidal (1955) are compared with the correlation of equation (7.62) in Fig. 7.47. The experimental drag coefficient increases with angle of attack (see Fig. 7.47). The correlation is best for the higher values of the lift coefficient.

The flow field over a delta wing is such that the resultant pressure distribution produces a large nose-down (negative) pitching moment about the apex, as illustrated by the experimental values from Bartlett and Vidal (1955) presented in Fig. 7.48. The magnitude of the negative pitching moment increases as the angle of attack is increased. The resultant aerodynamic coefficients present a problem relating to the low-speed performance of a delta-wing aircraft which is designed to cruise at supersonic speeds, since the location of the aerodynamic center for subsonic flows differs from that for supersonic flows. At low speeds (and, therefore, at relatively low values of the dynamic pressure), delta wings must be operated at relatively high angles of attack in order to generate sufficient lift, since

$$L = \tfrac{1}{2}\rho_\infty U_\infty^2 S C_L$$

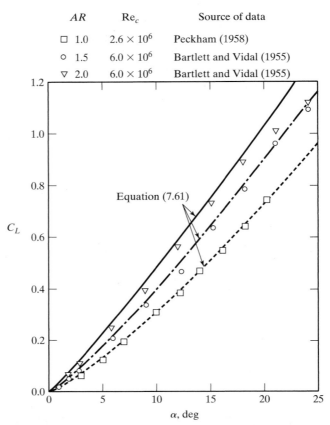

Figure 7.44 Comparison of the calculated and the experimental lift coefficients for thin, flat delta wings with sharp leading edges.

However, if the wing is at an angle of attack that is high enough to produce the desired C_L, a large nose-down pitching moment results. Thus, the basic delta configuration is often augmented by a lifting surface in the nose region (called a canard), which provides a nose-up trimming moment. The canards may be fixed, such as those shown in Fig. 7.49 on North American's XB-70 Valkyrie, or retractable, such as those on the Dassault Mirage. An alternative design approach, which is used on the Space Shuttle *Orbiter*, uses a reflexed wing trailing edge (i.e., negative camber) to provide the required trimming moment.

As noted by Gloss and Washburn (1978), "the proper use of canard surfaces on a maneuvering aircraft can offer several attractive features such as potentially higher trimmed-lift capability, improved pitching moment characteristics, and reduced trimmed drag. In addition, the geometric characteristics of close-coupled canard configurations offer a potential for improved longitudinal progression of cross-sectional area which could result in reduced wave drag at low supersonic speeds and placement of the horizontal control surfaces out of the high wing downwash and jet exhaust." These benefits are primarily associated with the additional lift developed by the canard and with the beneficial interaction between the canard flow field and that of the wing. These

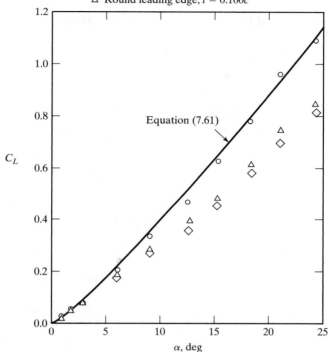

Figure 7.45 Effect of the leading-edge shape on the measured lift coefficient for thin, flat delta wings for which $AR = 1.5$, $Re_c = 6 \times 10^6$. [Data from Bartlett and Vidal (1955).]

benefits may be accompanied by a longitudinal instability (or pitch up) at the higher angles of attack because of the vortex lift developed on the forward canards.

As discussed by Visbal (1995), "For a highly swept wing, boundary layer separation occurs at the sharp leading-edge, beginning at small angles of incidence, and results in a 3-D shear layer which spirals into a pair of counterrotating primary vortices above the wing (Fig. 7.40a). These longitudinal vortices are the mechanisms for the downstream convection of the vorticity shed at the edge, and their induced suction on the wing upper surface results in vortex-induced lift which persists at relatively large angles of attack as compared to the 2-D wing situation. Over most of the wing extent, the trajectory of the leading-edge vortex is straight. For a fixed leading-edge sweep (Λ_{LE}), and over the range of interest in angle of attack, the vortex sweep angle Λ_{core} (see Fig. 7.40c) is nearly independent of α and this constant value is approximately proportional to Λ_{LE}. The inclination angle of the vortex trajectory relative to the wing surface (Fig. 7.40d) also increases linearly with angle of attack. As the vortex reaches the wing trailing edge it is deflected and in a short distance in the near wake aligns itself with the freestream direction."

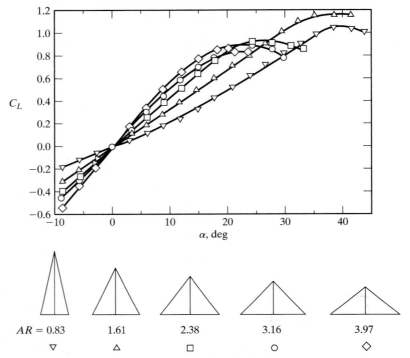

Figure 7.46 Lift coefficients for delta wings of various aspect ratios; $t = 0.12c$, $\mathrm{Re}_c \approx 7 \times 10^5$. [Data from Schlichting and Truckenbrodt (1969).]

Visbal (1995) also noted, "The leading-edge vortices above a delta wing at high angle of attack experience a dramatic form of flow disruption termed 'vortex breakdown' or 'vortex bursting'. This phenomenon is characterized by reverse axial flow and swelling of the vortex core, and is accompanied by marked flow fluctuations downstream. Vortex breakdown poses severe limitations on the performance of agile aircraft due to its sudden effects on the aerodynamic forces and moments and their impact on stability and control. For maneuvering delta wings, the onset of vortex development with its inherent long time scales results in dynamic hysteresis and lags in the vortex development and aerodynamic loads. In addition, the coherent fluctuations within the breakdown region can promote a structural response in aircraft surfaces immersed in the vortex path. An important example is that of 'tail buffet' in twin-tailed aircraft where the fluid/structure interaction may result in significant reduction of the service life of structural components."

In a related comment, Visbal (1995) notes, "As the leading-edge sweep and angle of attack increase, the interaction between the counterrotating leading-edge vortices increases and may lead to flow asymmetry. This phenomenon becomes more severe when vortex breakdown is present given the sudden onset and sensitivity and the accompanying swelling of the vortex core."

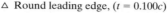

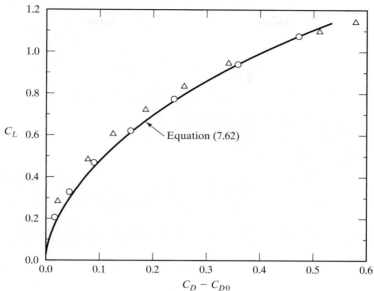

Figure 7.47 Drag correlation for thin, flat delta wings for which $AR = 1.5$, $\mathrm{Re}_c = 6 \times 10^6$. [Data from Bartlett and Vidal (1955).]

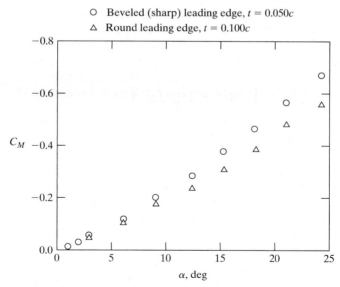

Figure 7.48 Moment coefficient (about the apex) for thin, flat delta wings for which $AR = 1.5$, $\mathrm{Re}_c = 6 \times 10^6$. [Data from Bartlett and Vidal (1955).]

Figure 7.49 North American XB-70 illustrating the use of canards. (Courtesy of NASA.)

As noted in the previous paragraphs and as depicted in Fig. 7.40, the flow over a delta wing with sharp leading edges is dominated by vortices inboard of the leading edge. The pressure on the wing's leeward surface beneath these vortices is very low and contributes significantly to the lift. When the aircraft is at moderate to high angles of attack, the axes of the vortices move away from the leeward surface of the wing. With the vortex core away from the surface, the vortex breaks down, or bursts, at some distance from the leading edge. Vortex breakdown, or bursting, is depicted in the water-tunnel flow of Fig. 7.52. Note that the diameter of the vortex increases suddenly when bursting, or breakdown, occurs. When vortex breakdown occurs, there are large reductions in the peak velocities both in the transverse and in the axial directions.

As the angle of attack increases, the vortex-burst location moves forward, ahead of the trailing edge toward the apex. The forward movement of the location of vortex breakdown and the attendant increase in the pressure on the leeward surface of the wing beneath the vortices are characteristic of the stall of a delta wing.

The location of the vortex breakdown during the unsteady pitch-up motion is downstream of its steady-state location, resulting in a lift overshoot. The magnitude of the lift overshoot depends on the leading-edge sweep angle.

Defining the positive attributes of a delta wing, Herbst (1980) noted, "There is a considerable basic aerodynamic potential in terms of lift to drag improvements in delta wings. Theoretical tools are available now to refine the wing planform, profile, and twist distribution. In particular, the leading edge suction can be improved considerably by means of proper leading edge profile, thus improving drag due to lift, the weakness of any highly swept wing .... With the help of an electronic digital control system, properly designed with modern aerodynamic tools and suitably equipped with a canard control surface, a delta wing could be designed to maintain its classically good supersonic performance without sacrificing cruise and subsonic performance compared to a more conventional trapezoidal wing."

7.8   LEADING-EDGE EXTENSIONS

The designs of lightweight fighters that can cruise supersonically and maneuver transonically employ additional, highly swept areas ahead of the main wing. These forward areas are called strakes, gloves, fillets, apex regions, and leading-edge extensions. The strake/wing combinations are also known by different names (e.g., ogee and double delta).

Lamar and Frink (1982) note that the mutual benefits derived from strake/wing configurations "include for the wing: (1) minimal interference at or below the cruise angle of attack, (2) energizing of the upper surface boundary layer with the resulting flow reattachment on the outer wing panel at moderate to high angle of attack due to the strake vortex, and (3) reduced area required for maneuver lift. Benefits for the strake are (1) upwash from the main wing strengthens the strake vortex and (2) the need for only a small area (hence wetted area and comparatively lightweight structure) to generate its significant contribution to the total lift, because the strake provides large amounts of vortex lift. It should be mentioned that these vortex-induced benefits are realized when the strake vortex is stable and maintaining a well-organized vortex system over the wing. Once the angle of attack becomes sufficiently large that strake-vortex breakdown progresses ahead of the wing trailing edge, these favorable effects deteriorate significantly."

A method for estimating the aerodynamic forces and moments for strake/wing/body configurations estimates the vortex flow effects with the suction analogy and the basic potential flow effects with the vortex lattice method [Lamar and Gloss (1975)]. As noted by Lamar and Frink (1982) and illustrated in Fig. 7.50, "at low angle of α, it was concluded that the configuration had attached flow away from the leading and side edges, and downstream from the component tips, vortical flow existed. However, at the higher values of α, the strake vortex becomes much larger and tends to displace the wing-vortex-flow system off the wing, so that this system can no longer cause flow reattachment to occur on

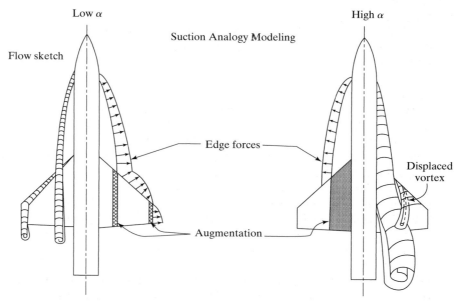

Figure 7.50 Theoretical vortex model for strake/wing configuration. [Taken from Lamar and Frink (1982).]

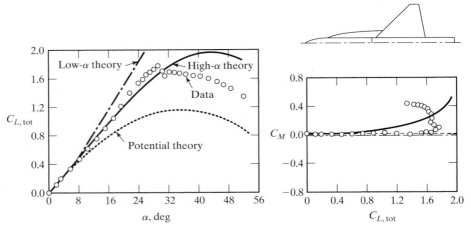

Figure 7.51 Predicted and measured $C_{L,\text{tot}}$ and C_M for strake/wing configuration at $M_\infty = 0.2$. [Taken from Lamar and Frink (1982).]

the wing. This lack of reattachment causes a large portion of theoretically available aerodynamic effects to be effectively lost to the configuration."

The two theoretical solutions along with a potential flow solution are compared with data [as taken from Lamar and Frink (1982)] in Fig. 7.51. The C_L data are better estimated by the high-α theory. However, there may be an angle-of-attack range below the α for $C_{L,\text{max}}$ for which the C_L data are underpredicted. This is most likely associated with the exclusion of any vortex lift from the wing in the high-α theory.

A flow visualization photograph of the vortices shed from the sharp leading edges of slender wings and wing leading-edge extensions (LEXs) or wing-body strakes at high angles of attack is presented in Fig. 7.52. The data were obtained in the Northrop Water Tunnel. Although the Reynolds number was relatively low, the separation point does not vary with Reynolds number, provided that flow separation occurs from a sharp leading edge. Thus, Erickson (1982) concludes that "the water tunnel provides a good representation of the wake shed from a wing."

Note that, in the caption of Fig. 7.52, it is stated that the Reynolds number is 10^4 for this test in a water tunnel. The comparisons of the vortex burst locations on the F/A-18 from experiments that spanned the range from 1/48-scale models in water tunnels to actual flight tests of the full-scale aircraft are discussed in Komerath et al. (1992). It was noted, "The correlation between model-scale tests and flight tests is surprisingly accurate. At large angles of attack, the flow separates quite close to the wing leading edge regardless of Reynolds number. Also, most high-α flight occurs at relatively low Mach numbers due to structural limitations. Thus, the effects of compressibility, although present, are not primary. Erickson, et al. (1989) show some discrepancies between data obtained at different Reynolds and Mach numbers, showing that the effects of surface roughness, freestream turbulence, tunnel wall effects, and model support interference must be considered in each study."

The flow field computed for an F-18C at an angle of attack of $32°$ is presented in Fig. 7.53 [Wurtzler and Tomaro (1999)]. The free-stream Mach number for this flow is 0.25 and the free-stream Reynolds number (based on the model length) is 15×10^6. The solution was generated using the turbulence model of Spalart and Allmaras (1992)

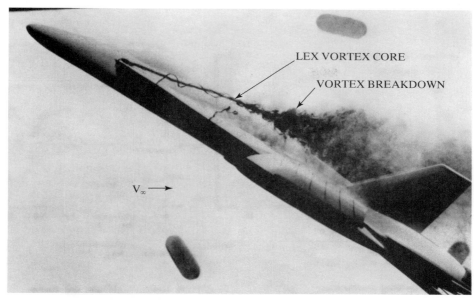

Figure 7.52 Flow visualization of a LEX vortex on a Subscale Advanced Fighter Model; $\alpha = 32°$, Reynolds number $= 10^4$ from the Northrop Water Tunnel. [Taken from Erickson (1982).]

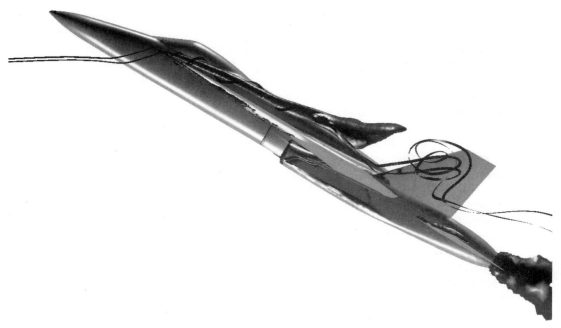

Figure 7.53 Flow-field computations for an F-18C at an angle of attack of $32°$, $M_\infty = 0.25$, $\mathrm{Re}_L = 15 \times 10^6$. [Provided by Wurtzler and Tomaro (1999).]

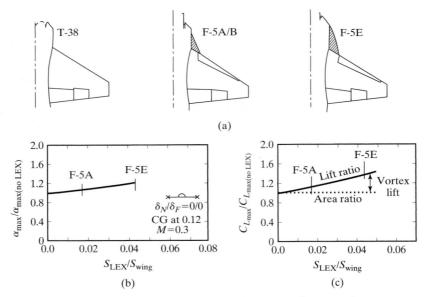

(a)

(b)   (c)

Figure 7.54 Effect of wing leading-edge extentions on the maximum angle of attack and the maximum lift for the F-5 family of aircraft: (a) configurations; (b) maximum angle of attack; (c) maximum lift. [From Stuart (1978).]

in the Cobalt$_{60}$ code [Strang, Tomaro, and Grismer (1999)]. Note that the computation of a turbulent separated flow is one of the greatest challenges to the computational fluid dynamicist at the time this fifth edition is written (which is during the summer of the year 2007). Nevertheless, the solution exhibits several interesting phenomena.

The surface marking a constant-entropy contour is highlighted in Fig. 7.53, so that the reader can visualize the vortex core. Note that vortex breakdown, or bursting, occurs at an x coordinate between the engine inlet and the leading edge of the vertical tail. Also included in Fig. 7.53 are two streamlines of the flow that are affected by the vortex. Note that, near the leading edge of the wing, the streamlines form spirals as they are entrained by the vortex. Once the vortex breaks down, the streamlines exhibit chaotic behavior. Note that despite the differences in the flow conditions and in the geometry, the flow fields of Figs. 7.52 and of 7.53 are qualitatively similar.

As noted in Stuart (1978), the original T-38 wing planform did not include a leading-edge extension (LEX) at the wing root. The effects of the LEX in terms of maximum trimmed angle of attack and trimmed $C_{L\,max}$ are illustrated in Fig. 7.54. The F-5E LEX with an area increase of 4.4% of the wing reference area provides a $C_{L\,max}$ increment of 38% of the no-LEX value. Northrop studies suggest a practical LEX upper size limit for any given configuration, and the F-5E LEX is close to this value. The higher trimmed angle of attack capability increases air combat effectiveness. The higher drag associated with the higher angle of attack is very useful in producing overshoots of the attacking aircraft. Although strakes (or leading-edge extensions) produce significant increases in the maximum angle of attack and in the maximum lift coefficient, the load distribution may concurrently produce large nose-up (positive) pitching moments at the high angles of attack.

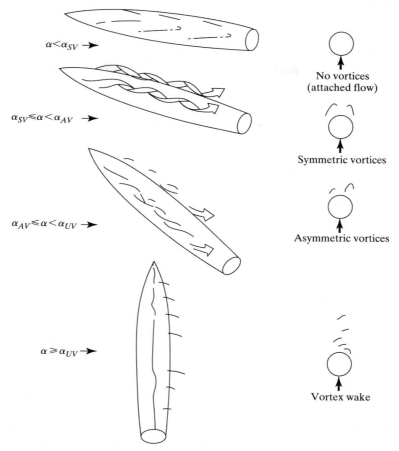

Figure 7.55 Effect of angle of attack on the leeside flowfield.
[From Ericsson and Reding (1986).]

7.9 ASYMMETRIC LOADS ON THE FUSELAGE AT HIGH ANGLES OF ATTACK

Ericsson and Reding (1986) noted that it has long been recognized that asymmetric vortex shedding can occur on bodies of revolution at high angles of attack. Large asymmetric loads can be induced on the body itself, even at zero sideslip. They note that experimental results have shown that the vortex-induced side force can exceed the normal force.

As a slender body of revolution is pitched through the angle-of-attack range from 0 to 90°, there are four distinct flow patterns that reflect the diminishing influence of the axial flow component. Sketches of these four flow patterns are presented in Fig. 7.55. At low angles of attack $(0° < \alpha < \alpha_{SV})$, the axial-flow component dominates and the flow is attached, although the cross-flow effects will generate a thick viscous layer on the leeside. For these vortex-free flows, the normal force is approximately linear with angle of attack, and the side force is zero. At intermediate angles of attack $(\alpha_{SV} \leq \alpha < \alpha_{AV})$ [i.e., the angle of attack is equal to or greater than the minimum for the formation of symmetric

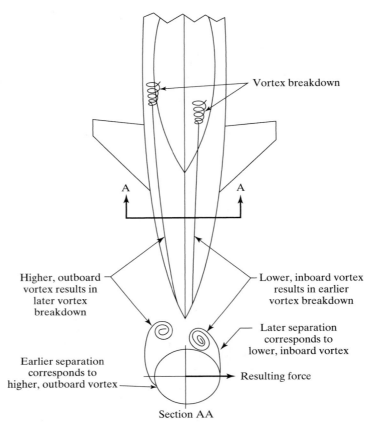

Figure 7.56 Sketch of asymmetric vortices shed from aircraft fore-body for $\alpha_{AV} \leq \alpha < \alpha_{UV}$. [Taken from Cobleigh (1994).]

vortices α_{SV} but less than that for the formation of asymmetric vortices (α_{AV})], the cross flow separates and generates a symmetric pair of vortices. The normal force increases nonlinearly in response to the added *vortex lift*, but there continues to be no side force.

7.9.1 Asymmetric Vortex Shedding

In the angle-of-attack range where the vortex shedding is asymmetric ($\alpha_{AV} \leq \alpha < \alpha_{UV}$), as shown in Fig. 7.55c, the axial flow component is still sufficient to produce steady vortices. However, the vortex pattern is asymmetric, producing a side force and a yawing moment, even at zero sideslip. As observed by Cobleigh (1994), "This side force is the result of surface pressure imbalances around the forebody of the aircraft caused by an asymmetric forebody boundary layer and vortex system. In this scenario, the boundary layer on each side of the forebody separates at different locations as shown in Fig. 7.56. At separation, corresponding vortex sheets are generated which roll up into an asymmetrically positioned vortex pair. The forces on the forebody are generated primarily by

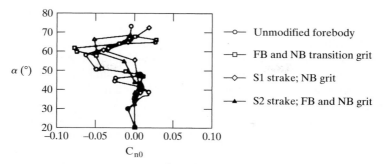

Figure 7.57 Variation of yawing moment asymmetry with angle of attack for an X-31 for a 1-g maneuver. [Taken from Cobleigh (1994).]

the boundary layer and to a lesser extent by the vortices, depending on their proximity to the forebody surface. Fig 7.56 shows a typical asymmetrical arrangement where the lower, more inboard vortex corresponds to a boundary layer which separates later and, conversely, the higher, more outboard vortex corresponds to the boundary layer which separated earlier. The suction generated by the more persistent boundary layer and the closer vortex combine to create a net force in their direction. Since the center of gravity of the aircraft is well aft of the forebody, a sizeable yawing moment asymmetry develops."

There are a variety of techniques that can be used to modify the asymmetric vortex pattern, including physical surfaces, such as strakes, and boundary-layer control, such as blowing or suction at the surface. The yawing moment asymmetry for an X-31 during slow decelerations (essentially 1g) to high AOA (angle of attack) conditions, as taken from Cobleigh (1994), is reproduced in Fig. 7.57. For the basic aircraft, the asymmetric forces started to build up, beginning at an angle of attack of 48° reaching a peak value of C_{n0} (the yawing moment coefficient at zero sideslip) of -0.063 at an angle of attack of approximately 57°. The asymmetric forces decreased rapidly with alpha, becoming relatively small by an angle of attack of 66°. In an attempt to reduce these asymmetries, transition grit strips were installed on both sides of the forebody (FB) and along the sides of the nose boom (NB). The data indicated that the side forces actually increased when grit strips were used. Although the largest asymmetry began to build at the same angle of attack as for the basic aircraft (an AOA of 48°), the peak asymmetry increased to C_{n0} of -0.078. For the aircraft with these transition strips, the largest asymmetry occurred in the alpha range from 58° to 61°. The use of a strake along with blunting of the nose tip delayed the initiation of the yawing moment asymmetry up to an AOA of 55° (refer to Fig. 7.57). The peak asymmetric yawing moment coefficient C_{n0} of -0.040 occurred at an angle of attack of 60°, after which the asymmetric yawing moment coefficient decreased. As evident in the data presented in Fig. 7.57, the use of both a strake and boundary-layer transition strips produced significant yawing moments which persisted over a broad angle-of-attack range.

In summary, Cobleigh (1994) reported, "Several aerodynamic modifications were made to the X-31 forebody with the goal of minimizing the asymmetry. A method for determining the yawing moment asymmetry from flight data was developed and an analysis of the various configuration changes completed. The baseline aircraft were found to have significant asymmetries above 45° -angle of attack with the largest asymmetry typically occurring around 60° -angle of attack. Applying symmetrical boundary-layer transition strips along the forebody sides increased the magnitude of the asymmetry and widened the angle-of-attack range over which the largest asymmetry acted. Installing longitudinal forebody strakes and rounding the sharp nose of the aircraft caused the yawing moment asymmetry magnitude to be reduced. The transition strips and strakes made the asymmetry characteristic of the aircraft more repeatable than the clean forebody configuration. Although no geometric differences between the aircraft were known, ship 2 consistently had larger yawing moment asymmetries than ship 1."

For axisymmetric bodies with wings or fins, the interaction with the asymmetric vortex pattern can produce rolling moments. Typically, the effectiveness of the vertical tail and rudder to control the yawing moment falls off as the angle of attack increases, because the vertical tail lies in the wake of the wing and of the fuselage.

Ericsson and Reding (1986) note that "the seriousness of the problem is illustrated by the wind-tunnel test results for the F-111, which show the vortex-induced yawing moment to exceed, by an order of magnitude, the available control capability through full rudder deflection. The problem can be cured by use of nose strakes or other types of body reshaping that cause the forebody vortices to be shed symmetrically."

An article by Dornheim (1994) included the paragraph, "The X-31's main problem is directional instability from asymmetric nose vortices around 60 deg. AOA, which caused an unintended departure in flight test. At that 165 knots condition, the yaw moment from the vortices was estimated to be 2.5 times more powerful than the thrust vectoring, and as powerful as the rudder itself would be at 10 deg. AOA. A subtle reshaping of the nose and adding vortex control strakes has tamed the vortices, and the X-31 should be able to fly at least 300 knots, where the 6g structural limit comes into play, before the remaining vortices can overpower the thrust vectoring at high AOA."

7.9.2 Wakelike Flows

Finally, at very high angles of attack ($\alpha_{UV} \leq \alpha < 90°$) the axial-flow component has less and less influence, so that the vortex shedding becomes unsteady, starting on the aft body and progressing toward the nose with increasing angle of attack. In this angle-of-attack range, the leeside flow resembles the wake of a two-dimensional cylinder normal to the flow. The mean side force decreases toward zero in this angle-of-attack range.

7.10 FLOW FIELDS FOR AIRCRAFT AT HIGH ANGLES OF ATTACK

As noted by Mendenhall and Perkins (1996), "Many modern high-performance fighter aircraft have mission requirements which necessitate rapid maneuvers at high angles of attack and large angular rotation rates. Under these flow conditions, the vehicle may operate in a flow regime in which the aerodynamic characteristics are dominated by unsteady nonlinear effects induced by flow separation, vortex shedding, and vortex lag effects (Fig. 7.58). During extreme multiple-axis maneuvering conditions, the dynamic and time-dependent effects of these nonlinear flow characteristics contribute significantly to the behavior and maneuvering capability of aircraft."

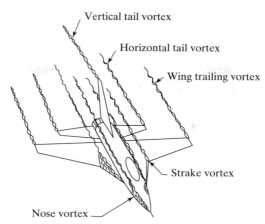

Figure 7.58 Vortex wakes near a maneuvering fighter. [Taken from Mendenhall and Perkins (1996).]

Mendenhall and Perkins continue, "The presence of the vortex wake introduces memory into the flow problem, and the nonlinear forces and moments on the vehicle depend on the time history of the motion and the wake development. For example, vorticity shed from the nose of the vehicle will pass downstream to influence the loads on the wing and tail surfaces. The vortex-induced loads depend on the motion of the vehicle and the vortex wake during the time it takes the vorticity to be transported from the nose to the tail."

The vortices associated with flow over highly swept wings (such as delta wings), with the flow over leading-edge extensions (LEXs) and with the flow over the nose region at high angles of attack, have been discussed individually in the previous sections of this chapter. Numerical predictions [e.g., Cummings et al. (1992)] and in-flight flow-field measurements [e.g., Del Frate and Zuniga (1990)] have been presented for a modified F-18 aircraft called the high-alpha research vehicle (HARV). The HARV was a single-place F-18 aircraft built by McDonnell Douglas and Northrop Aircraft and was powered by two General Electric F404-GE-400 turbofan engines with afterburners. The aircraft featured a midwing with leading- and trailing-edge flaps. Leading-edge extensions extend from the wing roots to just forward of the windscreen. A sketch illustrating the location of the forebody vortex core, of the LEX vortex, and of the LEX/forebody vortex interaction, as taken from Del Frate and Zuniga (1990), is reproduced in Fig. 7.59. The pattern is for an angle

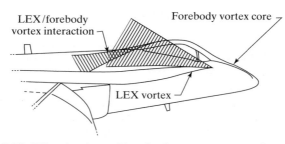

Figure 7.59 Wingtip view of forebody vortex system. [Taken from Del Frate and Zuniga (1990).]

of attack of 38.7° and for zero sideslip. As noted by Del Frate and Zuniga (1990), "These vortex cores are generated at moderate to high angles of attack by the shape of the fore-body and the sharp leading edge of each LEX. The LEX vortex cores are tightly wound and extend downstream until experiencing vortex core breakdown. Visible evidence of the vortex core breakdown is a stagnation of flow in the core with a sudden expansion in the core diameter. Similarly, the forebody vortex cores extend downstream until they interact with the LEX vortices. This interaction results in the forebody vortex cores being pulled beneath the LEX vortices and then redirected outboard."

"Lessons learned from comparisons between ground-based tests and flight mea-surements for the high-angle-angle-of-attack programs on the F-18 High Alpha Re-search Vehicle (HARV), the X-29 forward-swept wing aircraft, and the X-31 enhanced fighter maneuverability aircraft" were presented by Fisher et al. (1998). "On all three ve-hicles, Reynolds number effects were evident on the forebodies at high angles of attack. The correlation between flight and wind tunnel forebody pressure distribution for the F-18 HARV were improved by using twin longitudinal grit strips on the forebody of the wind-tunnel model. Pressure distributions obtained on the X-29 wind-tunnel model at flight Reynolds numbers showed excellent correlation with the flight data up to $\alpha = 50°$. Above $\alpha = 50°$ the pressure distributions for both flight and wind tunnel became asym-metric and showed poorer agreement, possibly because of the different surface finish of the model and aircraft. The detrimental effect of a very sharp nose apex was demon-strated on the X-31 aircraft. Grit strips on the forebody of the X-31 reduced the ran-domness but increased the magnitude of the asymmetry. Nose strakes were required to reduce the forebody yawing moment asymmetries and the grit strips on the flight test noseboom improved the aircraft handling qualities."

7.11 UNMANNED AIR VEHICLE WINGS

As was mentioned in Chapter 6, unmanned air vehicles (UAV) of various sizes have led to a revolution in airfoil and wing designs in recent years. The tried and true methods for designing wings often can fall short when applied to either very large or very small aircraft, as Reynolds number effects change the basic characteristics of the aerody-namics of these vehicles. In an attempt to classify these vehicles and their resulting aerodynamics, Wood (2002) devised four definitions for the size of unmanned air vehicles (see Table 7.3 and Fig. 7.60). These definitions are used to classify vehicles with similar aerodynamic and technical issues involved in their design.

TABLE 7.3 Classification of UAVs [Wood (2002)]

UAV Type	Weight (lbs)	Wing Span (ft)
Micro	<1	<2
Meso	1 to 2,000	2 to 30
Macro	2,000 to 10,000	30 to 150
Mega	>10,000	>150

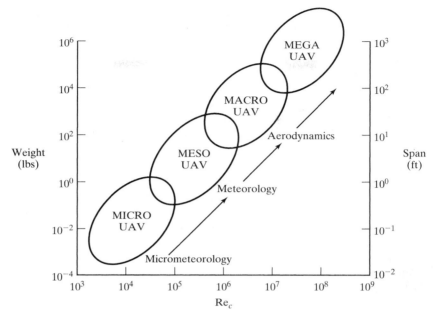

Figure 7.60 Classification of UAVs by weight, span, and Reynolds number. [From Wood (2002).]

As was discussed in Chapter 6 for laminar-flow airfoils, as the size of the vehicle decreases (and as weight and Reynolds number also decreases), aspect ratio no longer is the dominant factor in creating drag (see Fig. 7.61) — skin friction drag becomes more

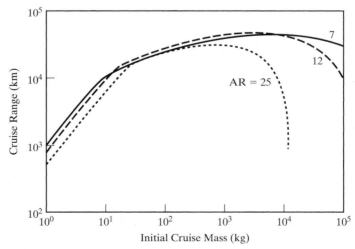

Figure 7.61 The influence of wing aspect ratio on aircraft range as a function of mass (all other factors kept constant). [From Drela et al. (2003).]

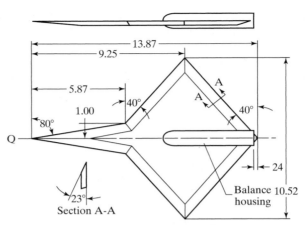

Figure 7.62 Unusual UAV configuration. [From Hammons and Thompson (2006).]

important, hence the low aspect-ratio designs common for micro UAVs [Drela et al. (2003).] The high-altitude aircraft also fly at low Reynolds numbers but require fairly high weights in order to carry the solar panels and batteries required for propulsion. The high-aspect ratio aircraft once again becomes more efficient. In fact, Drela et al. (2003) found that for each aspect ratio examined there was an aircraft size (weight) that would maximize the range: "this is because, as the aspect ratio increases, the wing is proportionately heavier for a large vehicle than for a smaller vehicle, while at small size, drag is more important so that the increasing friction drag with aspect ratio offsets the reduction in induced drag" [Drela et at. (2003)]. This is related to the well known square-cube law of aircraft design: The weight of the wing varies with the cube of the wing span while the area of the wing varies with the square of the wing span [see for example, McMasters (1998)]. This is one of the reasons these wings have such largely different design characteristics. Couple the square-cube law with the various amounts of laminar and turbulent flow the wing can experience, and the resulting wings can be quite unusual when compared with typical aircraft, as shown in Fig. 7.62.

7.12 SUMMARY

The shape of the wing is a dominant parameter in determining the performance and handling characteristics of an airplane. The planform of a wing is defined by the aspect ratio, sweep, and taper ratio. At subsonic speeds, the lift-dependent drag can be reduced by increasing the aspect ratio. A high-aspect-ratio wing produces a high-lift-curve slope, which is useful for takeoff and landing but has a weight penalty since, for a given wing area, more massive root sections are needed to handle the higher bending moments. Furthermore, a swept wing has a greater structural span than a straight wing of the same area and aspect ratio. Thus, the use of sweepback may also lead to a growth in wing weight. As a result, the choice of aspect ratio requires a trade study in which the designer may chose to emphasize one criterion at the expense of another (see the discussion relating to transport aircraft in Chapter 13).

Tapering the wing reduces the wing-root bending moments, since the inboard portion of the wing carries more of the wing's lift than the portion near the tip. While taper reduces the actual loads carried by the outboard sections, the wing-tip sections are subjected to higher lift coefficients and tend to stall first (see Fig. 7.18). Thus, progressively reducing the incidence of the local section through the use of geometric twist from the root to the tip (i.e., washout) may be used either to reduce local loading or to prevent tip stalling.

PROBLEMS

7.1. Consider an airplane that weighs 14,700 N and cruises in level flight at 300 km/h at an altitude of 3000 m. The wing has a surface area of 17.0 m^2 and an aspect ratio of 6.2. Assume that the lift coefficient is a linear function of the angle of attack and that $\alpha_{0l} = -1.2°$. If the load distribution is elliptic, calculate the value of the circulation in the plane of symmetry (Γ_0), the downwash velocity (w_{y1}), the induced-drag coefficient (C_{Dv}), the geometric angle of attack (α), and the effective angle of attack (α_e).

7.2. Consider the case where the spanwise circulation distribution for a wing is parabolic,

$$\Gamma(y) = \Gamma_0\left(1 - \frac{y^2}{s^2}\right)$$

If the total lift generated by the wing with the parabolic circulation distribution is to be equal to the total lift generated by a wing with an elliptic circulation distribution, what is the relation between the Γ_0 values for the two distributions? What is the relation between the induced downwash velocities at the plane of symmetry for the two configurations?

7.3. When a GA(W)-1 airfoil section (i.e., a wing of infinite span) is at an angle of attack of 4°, the lift coefficient is 1.0. Using equation (7.20), calculate the angle of attack at which a wing whose aspect ratio is 7.5 would have to operate to generate the same lift coefficient. What would the angle of attack have to be to generate this lift coefficient for a wing whose aspect ratio is 6.0?

7.4. Consider a planar wing (i.e., no geometric twist) which has a NACA 0012 section and an aspect ratio of 7.0. Following Example 7.1, use a four-term series to represent the load distribution. Compare the lift coefficient; the induced-drag coefficient, and the spanwise lift distribution for taper ratios of **(a)** 0.4, **(b)** 0.5, **(c)** 0.6, and **(d)** 1.0.

7.5. Consider an airplane that weighs 10,000 N and cruises in level flight at 185 km/h at an altitude of 3.0 km. The wing has a surface area of 16.3 m^2, an aspect ratio of 7.52, and a taper ratio of 0.69. Assume that the lift coefficient is a linear function of the angle of attack and the airfoil section is a NACA 2412 (see Chapter 6 for the characteristics of this section). The incidence of the root section is +1.5°; the incidence of the tip section is −1.5°. Thus, there is a geometric twist of −3° (washout). Following Example 7.1, use a four-term series to represent the load distribution and calculate
(a) The lift coefficient (C_L)
(b) The spanwise load distribution ($C_l(y)/C_L$)
(c) The induced drag coefficient (C_{Dv})
(d) The geometric angle of attack (a)

7.6. Use equation (7.37) to calculate the velocity induced at some point $C(x, y, z)$ by the vortex filament AB (shown in Fig. 7.29); that is, derive equation (7.38a).

7.7. Use equation (7.37) to calculate the velocity induced at some point $C(x, y, 0)$ by the vortex filament AB in a planar wing; that is, derive equation (7.43a).

7.8. Calculate the downwash velocity at the CP of panel 4 induced by the horseshoe vortex of panel 1 of the starboard wing for the flow configuration of Example 7.2.

7.9. Following the VLM approach used in Example 7.2, calculate the lift coefficient for a swept wing. The wing has an aspect ratio of 8, a taper ratio of unity (i.e., $c_r = c_t$), and an uncambered section (i.e., it is a flat plate). Since the taper ratio is unity, the leading edge, the quarter-chord line, the three-quarter chord line, and the trailing edge all have the same sweep, 45°. How does the lift coefficient for this aspect ratio (8) compare with that for an aspect ratio of 5 (i.e., that computed in Example 7.2)? Is this consistent with our knowledge of the effect of aspect ratio (e.g., Fig. 7.10)?

7.10. Following the VLM approach used in Example 7.2, calculate the lift coefficient for a swept wing. The wing has an aspect ratio of 5, a taper ratio of 0.5 (i.e., $c_t = 0.5 c_r$), an uncambered section, and the quarter chord is swept 45°. Since the taper ratio is not unity, the leading edge, the quarter-chord line, the three-quarter-chord line, and the trailing edge have different sweep angles. This should be taken into account when defining the coordinates of the horseshoe vortices and the control points.

7.11. Following the VLM approach used in Example 7.2, calculate the lift coefficient for the forward swept wing of Fig. 7.23b. The quarter chord is swept forward 45°, the aspect ratio is 3.55, and the taper ratio 0.5. The airfoil section (perpendicular to the quarter chord) is a NACA 64A112. For this airfoil section $\alpha_{o2} = -0.94°$ and $C_{l,\alpha} = 6.09$ per radian. For purposes of applying the no-flow boundary condition at the control points, assume that the wing is planar. Prepare a graph of the lift coefficient. How does this compare with that of Fig. 7.23?

7.12. Following the VLM approach used in Example 7.2, calculate the lift coefficient for a delta wing whose aspect ratio is 1.5. What is the sweep angle of the leading edge? The fact that the quarter-chord and the three-quarter-chord lines have different sweeps should be taken into account when defining the coordinates of the horseshoe vortices and the control points. How do the calculated values for the lift coefficient compare with the experimental values presented in Fig. 7.44?

7.13. Use equation (7.61) to calculate the lift coefficient as a function of the angle of attack for a flat delta wing with sharp leading edges. The delta wing has an aspect ratio of 1.5. Compare the solution with the data of Fig. 7.44.

7.14. Use equation (7.62) to calculate the induced drag for a flat delta wing with sharp leading edges. The delta wing has an aspect ratio of 1.5. Compare the solution with the data of Fig. 7.47.

7.15. Assume that the wing area of an airplane is proportional to the square of the wing span and the volume, and thus the weight, is proportional to the cube of the wing span (this is the square-cube law). Find the wing loading of the aircraft as a function of wing span, b. Using these relationships explain why very large aircraft (like the Boeing 747 or the Airbus 380) have to fly with very large wing loadings. If you wanted to re-design an existing aircraft that weighed 92,000 lbs with a wing span of 75 ft, what wing span would you need to add an additional 10% to the weight?

REFERENCES

Abbott IH, von Doenhoff AE. 1949. *Theory of Wing Sections*. New York: Dover

Anderson JD, Corda S, Van Wie DM. 1980. Numerical lifting line theory applied to drooped leading-edge wings below and above stall. *J. Aircraft* 17:898–904

Bartlett GE, Vidal RJ. 1955. Experimental investigations of influence of edge shape on the aerodynamic characteristics of low aspect ratio wings at low speeds. *J. Aeron. Sci.* 22:517–533, 588

Bristow DR, Grose GG. 1978. Modification of the Douglas Neumann program to improve the efficient of predicting component and high lift charactersitics. *NASA CR 3020*

Cobleigh BR. 1994. High-angle-of-attack yawing moment asymmetry of the X-31 aircraft from flight test. Presented at Appl. Aerodyn. Conf., 12th, AIAA Pap. 94–1803, Colorado Springs, CO

Cummings RM, Rizk YM, Schiff LB, Chaderjian NM. 1992. Navier-Stokes prediction for the F-18 wing and fuselage at large incidence. *J. Aircraft* 29: 565–574

Del Frate JH, Zuniga FA. 1990. In-flight flow field analysis on the NASA F-18 high alpha research vehicle with comparison to the ground facility data. Presented at AIAA Aerosp. Sci. Meet., 28th, AIAA Pap. 90–0231, Reno, NV

Dornheim MA. 1994. X-31, F-16 MATV, F/A-18 HARV explore diverse missions. *Aviat. Week Space Tech.* 140:46–47

Drela M, Protz JM, Epstein AH. 2003. The role of size in the future of aeronautics. Presented at Intl. Air Space Symp., AIAA Pap. 2003–2902, Dayton, OH

Ericsson LE, Reding JP. 1986. Asymmetric vortex shedding from bodies of revolution. In *Tactical Missile Aerodynamics*, Ed. Hemsch MH, Nielsen JN. Washington, DC: AIAA

Erickson GE. 1982. Vortex flow correlation. Presented at Congr. Intl. Coun. Aeron. Sci., 13th, ICAS Pap. 82–6.61, Seattle, WA

Erickson GE, Hall RM, Banks DW, DelFrate JH, Schreiner JA, Hanley RJ, Pulley CT. 1989. Experimental investigation of the F/A-18 vortex flows at subsonic through transonic speeds. Presented at Appl. Aerodyn. Conf., 7th, AIAA Pap. 89–2222, Seattle, WA

Falkner VM. 1943. The calculation of aerodynamic loading on surfaces of any shape. *ARC R&M 1910*

Fisher DF, Cobleigh BR, Banks, DW, Hall RM, Wahls RA. 1998. Reynolds number effects at high angles of attack. Presented at Adv. Meas. Ground Test. Tech. Conf., 20th, AIAA Pap. 98–2879, Albuquerque, NM

Gloss BB, Washburn KE. 1978. Load distribution on a close-coupled wing canard at transonic speeds. *J. Aircraft* 15:234–239

Hammons C, Thompson DS. 2006. A numerical investigation of novel planforms for micro UAVs. Presented at AIAA Aerosp. Sci. Meet., 44th, AIAA Pap. 2006–1265, Reno, NV

Henderson WP. 1966. Studies of various factors affecting drag due to lift at subsonic speeds. *NASA Tech. Note D-3584*

Herbst WB. 1980. Future fighter technologies. *J. Aircraft* 17:561–566

Hummel D. 2004. Effects of boundary layer formation on the vortical flow above a delta wing. Presented at NATO RTO Appl. Veh. Tech. Spec. Meet., RTO-MP-AVT-111, Paper 30-1, Prague, Czech Republic.

Johnson FT. 1980. A general panel method for the analysis and design of arbitrary configurations in incompressible flows. *NASA CR 3079*

Kalman TP, Giesing JP, Rodden WP. 1970. Spanwise distribution of induced drag in subsonic flow by the vortex lattice method. *J. Aircraft* 7:574–576

Kalman TP, Rodden WP, Giesing JP. 1971. Application of the doublet-lattice method to nonplanar configurations in subsonic flow. *J. Aircraft* 8:406–413

Komerath NMS, Liou SG, Schwartz RJ, Kim JM. 1992. Flow over a twin-tailed aircraft at angle of attack; part I: spatial characteristic. *J. Aircraft* 29:413–420

Lamar JE, Gloss BB. 1975. Subsonic aerodynamic characteristics of interacting lifting surfaces with separated flow around sharp edges predicted by vortex-lattice method. *NASA Tech. Note D-7921*

Lamar JE, Frink NT. 1982. Aerodynamic features of designed strake-wing configurations. *J. Aircraft* 19:639–646

Margason RJ, Lamar JE. 1971. Vortex-lattice Fortran program for estimating subsonic aerodynamic charactersitics of complex planforms. *NASA Tech. Note D-6142*

Margason RJ, Kjelgaard SO, Sellers WL, Morris CEK, Walkey KB, Shields EW. 1985. Subsonic panel methods—a comparison of several production codes. Presented at AIAA Aerosp. Sci. Meet., 23rd, AIAA Pap. 85–0280, Reno, NV

McKinney LW, Dollyhigh SM. 1971. Some trim drag considerations for maneuvering aircraft. *J. Aircraft* 8:623–629

McMasters JH.1998. Advanced configurations for very large transport airplanes. *Aircraft Design* 1: 217–242

Mendenhall MR, Perkins SC. 1996. Predicted high-α aerodynamic characteristics of maneuvering aircraft. Presented at Appl. Aerodyn. Conf., 14th, AIAA Pap. 96–2433, New Orleans, LA

Multhopp H. 1950. Method for calculating the lift distribution of wings (subsonic lifting surface theory). *ARC T&M 2884*

Orloff KL. 1980. Spanwise lift distribution on a wing from flowfield velocity surveys. *J. Aircraft* 17:875–882

Peckham DH. 1958. Low-speed wind tunnel tests on a series of uncambered slender pointed wings with sharp edges. *ARC R&M 3186*

Phillips WF, Snyder DO. 2000. Application of lifting-line theory to systems of lifting surfaces. Presented at AIAA Aerosp. Sci. Meet., 38th, AIAA Pap. 2000–0653, Reno, NV

Polhamus ED. 1971a. Predictions of vortex-lift characteristics by a leading-edge suction analogy. *J. Aircraft* 8:193–199

Polhamus ED. 1971b. Charts for predicting the subsonic vortex-lift characteristics of arrow, delta, and diamond wings. *NASA Tech. Note D-6243*

Prandtl L. 1921. Application of modern hydrodynamics to aeronautics. *NACA Report 116*

Prandtl L, Tietjens OG. 1957. *Applied Hydro- and Aeromechanics*. New York: Dover

Rasmussen ML, Smith DE. 1999. Lifting-line theory for arbitrarily shaped wings. *J. Aircraft* 36:340–348

Robinson A, Laurmann JA. 1956. *Wing Theory*. Cambridge: Cambridge University Press

Schlichting H. 1960. Some developments in boundary layer research in the past thirty years. *J. Roy. Aeron. Soc.* 64:64–79

Schlichting H, Truckenbrodt E. 1969. *Aerodynamik des Flugzeuges*. Berlin: Spring-Verlag

Sivells JC. 1947. Experimental and calculated characteristics of three wings of NACA 64-210 and 65-210 airfoil sections with and without washout. *NACA Tech. Note 1422*

Spalart PR, Allmaras R. 1992. A one-equation turbulence model for aerodynamics flows. Presented at AIAA Aerosp. Sci. Meet., 30th, AIAA Pap. 92–0439, Reno, NV

Spreiter JR, Sacks, AH. 1951. The rolling up of the vortex sheet and its effect on the downwash behinds wings. *J. Aeron. Sci.* 18:21–32

Stanbrook A, Squire LC. 1964. Possible types of flow at swept leading edges. *The Aeron. Quart.* 15:72–82

Strang WZ, Tomaro RF, Grismer MJ. 1999. The defining methods of Cobalt$_{60}$: a parallel, implicit, unstructured Euler/Navier-Stokes flow solver. Presented at AIAA Aerosp. Sci. Meet., 37th, AIAA Pap. 99–0786, Reno, NV

Stuart WG. 1978. *Northrop F-5 Case Study in Aircraft Design.* Washington, DC: AIAA

Visbal MR. 1995. Computational and physical aspects of vortex breakdown on delta wings. Presented at AIAA Aerosp. Sci. Meet., 33rd, AIAA Pap. 95–0585, Reno, NV

Voight W. 1976. Gestation of the swallow. *Air Intern.* 10:135–139, 153

Weber J, Brebner GG. 1958. Low-speed tests on 45-deg swept-back wings, part I: pressure measurements on wings of aspect ratio 5. *ARC R&M 2882*

Wood RM. 2002. A discussion of aerodynamic control effectors for unmanned air vehicles. Presented at Tech. Conf. Workshop Unmanned Aerosp. Veh., 1st, AIAA Pap. 2002–3494, Portsmouth, VA

Wurtzler KE, Tomaro RF. 1999. Unsteady aerodynamics of aircraft maneuvering at high angle of attack. Presented at DoD High Perf. Comp. User's Group Meet., Monterey, CA

8 DYNAMICS OF A COMPRESSIBLE FLOW FIELD

Thus far, we have studied the aerodynamic forces for incompressible (constant density) flows past the vehicle. At low flight Mach numbers (e.g., below a free-stream Mach number M_∞ of approximately 0.3 to 0.5), Bernoulli's equation [equation (3.10)] provides the relation between the pressure distribution about an aircraft and the local velocity changes of the air as it flows around the various components of the vehicle. However, as the flight Mach number increases, changes in the local air density also affect the magnitude of the local static pressure. This leads to discrepancies between the actual aerodynamic forces and those predicted by incompressible flow theory. For our purposes, the Mach number is the parameter that determines the extent to which compressibility effects are important. The purpose of this chapter is to introduce those aspects of compressible flows (i.e., flows in which the density is not constant) that have applications to aerodynamics

Throughout this chapter, air will be assumed to behave as a thermally perfect gas (i.e., the gas obeys the equation of state):

$$p = \rho R T \tag{1.10}$$

We will assume that the gas is also calorically perfect; that is, the specific heats, c_p and c_v, of the gas are constant. These specific heats are discussed further in the next section. The term *perfect gas* will be used to describe a gas that is both thermally and calorically perfect.

Even though we turn our attention in this and in the subsequent chapters to compressible flows, we may still divide the flow around the vehicle into (1) the boundary layer near the surface, where the effects of viscosity and heat conduction are important, and (2) the external flow, where the effects of viscosity and heat conduction can be neglected. As has been true in previous chapters, the inviscid flow is conservative.

8.1 THERMODYNAMIC CONCEPTS

Having reviewed the fundamentals of thermodynamics in Chapter 2 and having derived the energy equation therein, let us turn our attention to the related aerodynamic concepts.

8.1.1 Specific Heats

For simplicity (and without loss of generality), let us consider a system in which there are no changes in the kinetic and the potential energies. Equation (2.23) becomes

$$\delta q - p \, dv = du_e \tag{8.1}$$

As an extension of our discussion of fluid properties in Chapter 1, we note that, for any simple substance, the specific internal energy is a function of any other two independent fluid properties. Thus, consider $u_e = u_e(v, T)$. Then, by the chain rule of differentiation,

$$du_e = \left(\frac{\partial u_e}{\partial T} \right)_v dT + \left(\frac{\partial u_e}{\partial v} \right)_T dv \tag{8.2}$$

where the subscript is used to designate which variable is constant during the differentiation process.

From the principles of thermodynamics, it may be shown that for a thermally perfect gas, that is, one obeying equation (1.10),

$$\left(\frac{\partial u_e}{\partial v} \right)_T = 0$$

which is equivalent to saying that the internal energy of a perfect gas does not depend on the specific volume or, equivalently, the density, and hence depends on the temperature alone. Thus, equation (8.2) becomes

$$du_e = \left(\frac{\partial u_e}{\partial T} \right)_v dT = c_v \, dT \tag{8.3}$$

In equation (8.3) we have introduced the definition that

$$c_v \equiv \left(\frac{\partial u_e}{\partial T} \right)_v$$

which is the *specific heat at constant volume*. It follows that

$$\Delta u_e = u_{e2} - u_{e1} = \int_1^2 c_v \, dT \tag{8.4}$$

Experimental evidence indicates that for most gases, c_v is constant over a wide range of conditions. For air below a temperature of approximately 850 K and over a wide range of pressure, c_v can be treated as a constant. The value for air is

$$c_v = 717.6 \frac{\text{N} \cdot \text{m}}{\text{kg} \cdot \text{K}} \left(\text{or} \frac{\text{J}}{\text{kg} \cdot \text{K}} \right) = 0.1717 \frac{\text{Btu}}{\text{lbm} \cdot {}^\circ\text{R}}$$

The assumption that c_v is constant is contained within the more general assumption that the gas is a perfect gas. Thus, for a perfect gas,

$$\Delta u_e = c_v \, \Delta T \tag{8.5}$$

Since c_v and T are both properties of the fluid and since the change in a property between any two given states is independent of the process used in going between the two states, equation (8.5) is valid even if the process is not one of constant volume. Thus, equation (8.5) is valid for any simple substance undergoing any process where c_v can be treated as a constant.

Substituting equation (8.3) into equation (8.1), one sees that c_v is only directly related to the heat transfer if the process is one in which the volume remains constant (i.e., $dv = 0$). Thus, the name "specific heat" can be misleading. Physically, c_v is the proportionality constant between the amount of heat transferred to a substance and the temperature rise in the substance held at constant volume.

In analyzing many flow problems, the terms u_e and pv appear as a sum, and so it is convenient to define a symbol for this sum:

$$h \equiv u_e + pv \equiv u_e + \frac{p}{\rho} \tag{8.6}$$

where h is called the *specific enthalpy*. Substituting the differential form of equation (8.6) into equation (8.1) and collecting terms yields

$$\delta q + v \, dp = dh \tag{8.7}$$

which is the first law of thermodynamics expressed in terms of the enthalpy rather than the internal energy.

Since any property of a simple substance can be written as a function of any two other properties, we can write

$$h = h(p, T) \tag{8.8}$$

Thus,

$$dh = \left(\frac{\partial h}{\partial p} \right)_T dp + \left(\frac{\partial h}{\partial T} \right)_p dT \tag{8.9}$$

From the definition of the enthalpy, it follows that h is also a function of temperature only for a thermally perfect gas, since both u_e and p/ρ are functions of the temperature only. Thus,

$$\left(\frac{\partial h}{\partial p} \right)_T = 0$$

and equation (8.9) becomes

$$dh = \left(\frac{\partial h}{\partial T}\right)_p dT = c_p\, dT \tag{8.10}$$

We have introduced the definition

$$c_p \equiv \left(\frac{\partial h}{\partial T}\right)_p \tag{8.11}$$

which is the *specific heat at constant pressure*. In general, c_p depends on the composition of the substance and its pressure and temperature. It follows that

$$\Delta h = h_2 - h_1 = \int_1^2 c_p\, dT \tag{8.12}$$

Experimental evidence indicates that for most gases, c_p is essentially independent of temperature and of pressure over a wide range of conditions. Again, we conclude that, provided that the temperature extremes in a flow field are not too widely separated, c_p can be treated as a constant so that

$$\Delta h = c_p\, \Delta T \tag{8.13}$$

For air below a temperature of approximately 850 K, the value of c_p is 1004.7 N $\cdot$ m/kg $\cdot$ K (or J/kg $\cdot$ K), which is equal to 0.2404 Btu/lbm $\cdot$ °R.

An argument parallel to that used for c_v shows that equation (8.13) is valid for any simple substance undergoing any process where c_p can be treated as a constant. Again, we note that the term *specific heat* is somewhat misleading, since c_p is only directly related to the heat transfer if the process is isobaric.

8.1.2 Additional Relations

A gas which is thermally and calorically perfect is one that obeys equations (1.10), (8.5), and (8.13). In such a case, there is a simple relation between c_p, c_v, and R. From the definitions of c_p and h, the perfect-gas law, and the knowledge that h depends upon T alone, we write

$$c_p \equiv \left(\frac{\partial h}{\partial T}\right)_p = \frac{dh}{dT} = \frac{du_e}{dT} + \frac{d}{dT}\left(\frac{p}{\rho}\right) = c_v + R \tag{8.14}$$

Let us introduce another definition, one for the ratio of specific heats:

$$\gamma \equiv \frac{c_p}{c_v} \tag{8.15}$$

For the most simple molecular model, the kinetic theory of gases shows that

$$\gamma = \frac{n + 2}{n}$$

where n is the number of degrees of freedom for the molecule. Thus, for a monatomic gas, such as helium, $n = 3$ and $\gamma = 1.667$. For a diatomic gas, such as nitrogen, oxygen, or air, $n = 5$ and $\gamma = 1.400$. Extremely complex molecules, such as Freon or tetrafluoromethane, have large values of n and values of γ which approach unity. In many treatments of air at high temperature and high pressure, approximate values of γ (e.g., 1.1 to 1.2) are used to approximate "real-gas" effects.

Combining equations (8.14) and (8.15), we can write

$$c_p = \frac{\gamma R}{\gamma - 1} \quad \text{and} \quad c_v = \frac{R}{\gamma - 1} \qquad \textbf{(8.16)}$$

8.1.3   Second Law of Thermodynamics and Reversibility

The first law of thermodynamics does not place any constraints regarding what types of processes are physically possible and what types are not, providing that equation (2.22) is satisfied. However, we know from experience that not all processes permitted by the first law actually occur in nature. For instance, when one rubs sandpaper across a table, both the sandpaper and the table experience a rise in temperature. The first law is satisfied because the work done on the system by the sander's arm, which is part of the surroundings and which is, therefore, negative work for equation (2.23), is manifested as an increase in the internal energy of the system, which consists of the sandpaper and the table. Thus, the temperatures of the sandpaper and of the table increase. However, we do not expect that we can extract all the work back from the system and have the internal energy (and thus the temperature) decrease back to its original value, even though the first law would be satisfied. If this could occur, we would say the process was reversible, because the system and its surroundings would be restored to their original states.

The possibility of devising a reversible process, such as the type outlined previously, cannot be settled by a theoretical proof. Experience shows that a truly reversible process has never been devised. This empirical observation is embodied in the *second law of thermodynamics*. For our purposes, an irreversible process is one that involves viscous (friction) effects, shock waves, or heat transfer through a finite temperature gradient. Thus, in regions outside boundary layers, viscous wakes, and planar shock waves, one can treat the flow as reversible. Note that the flow behind a curved shock wave can be treated as reversible only along a streamline.

The second law of thermodynamics provides a way to quantitatively determine the degree of reversibility (or irreversibility). Since the effects of irreversibility are dissipative and represent a loss of available energy (e.g., the kinetic energy of an aircraft wake, which is converted to internal energy by viscous stresses, is directly related to the aircraft's drag), the reversible process provides an ideal standard for comparison to real processes. Thus, the second law is a valuable tool available to the engineer.

There are several logically equivalent statements of the second law. In the remainder of this text we will usually be considering adiabatic processes, processes in which there is no heat transfer. This is not a restrictive assumption, since heat transfer in aerodynamic problems usually occurs only in the boundary layer and has a negligible effect on the flow in the inviscid region. The most convenient statement of the second law, for our purposes, is

$$ds \geq 0 \qquad \textbf{(8.17)}$$

for an adiabatic process. Thus, when a system is isolated from all heat exchange with its surroundings, s, the entropy of the system either remains the same (if the process is reversible) or increases (if it is irreversible). It is not possible for a process to occur if the entropy of the system and its surroundings decreases. Thus, just as the first law led to the definition of internal energy as a property, the second law leads to the definition of entropy as a property.

The entropy change for a reversible process can be written as

$$\delta q = T\,ds$$

Thus, for a reversible process in which the only work done is that done at the moving boundary of the system,

$$T\,ds = du_e + p\,dv \tag{8.18}$$

However, once we have written this equation, we see that it involves only changes in properties and does not involve any path-dependent functions. We conclude, therefore, that this equation is valid for all processes, both reversible and irreversible, and that it applies to the substance undergoing a change of state as the result of flow across the boundary of an open system (i.e., a control volume) as well as to the substance comprising a closed system (i.e., a control mass).

For a perfect gas, we can rewrite equation (8.18) as

$$ds = c_v \frac{dT}{T} + R\frac{dv}{v}$$

This equation can be integrated to give

$$s_2 - s_1 = c_v \ln\left\{\left[\left(\frac{v_2}{v_1}\right)^{\gamma-1}\right]\frac{T_2}{T_1}\right\} \tag{8.19a}$$

Applying the equation of state for a perfect gas to the two states,

$$\frac{v_2}{v_1} = \frac{\rho_1}{\rho_2} = \frac{p_1}{p_2}\frac{T_2}{T_1}$$

equation (8.19a) can be written

$$s_2 - s_t = R \ln\left\{\left[\left(\frac{T_2}{T_1}\right)^{\gamma/(\gamma-1)}\right]\frac{p_1}{p_2}\right\} \tag{8.19b}$$

Equivalently,

$$s_2 - s_1 = c_v \ln\left\{\left[\left(\frac{\rho_1}{\rho_2}\right)^{\gamma}\right]\frac{p_2}{p_1}\right\} \tag{8.19c}$$

Using the various forms of equation (8.19), one can calculate the entropy change in terms of the properties of the end states.

In many compressible flow problems, the flow external to the boundary layer undergoes processes that are isentropic (i.e., adiabatic and reversible). If the entropy is constant at each step of the process, it follows from equation (8.19) that p, ρ, and T are interrelated. The following equations describe these relations for isentropic flow:

$$\frac{p}{\rho^{\gamma}} = \text{constant} \tag{8.20a}$$

$$\frac{T^{\gamma/(\gamma-1)}}{p} = \text{constant} \tag{8.20b}$$

and

$$Tv^{(\gamma-1)} = \text{constant} \tag{8.20c}$$

8.1.4 Speed of Sound

From experience, we know that the speed of sound in air is finite. To be specific, the *speed of sound* is defined as the rate at which infinitesimal disturbances are propagated from their source into an undisturbed medium. These disturbances can be thought of as small pressure pulses generated at a point and propagated in all directions. We shall learn later that finite disturbances such as shock waves propagate at a greater speed than that of sound waves.

Consider a motionless point source of disturbance in quiescent, homogeneous air (Fig. 8.1a). Small disturbances generated at the point move outward from the point in a spherically symmetric pattern. The distance between wave fronts is determined by the

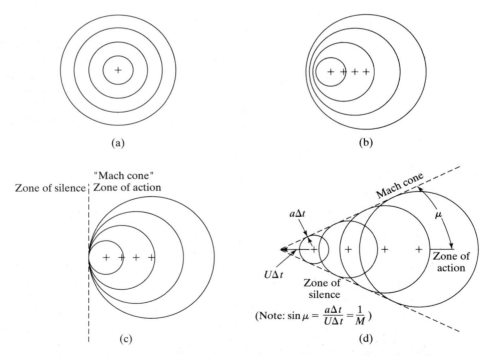

Figure 8.1 Wave pattern generated by pulsating disturbance of infinitesimal strength; (a) disturbance is stationary ($U = 0$); (b) disturbance moves to the left at subsonic speed ($U < a$); (c) disturbance moves to the left at sonic speed ($U = a$); (d) disturbance moves to the left at supersonic speed ($U > a$).

frequency of the disturbance. Since the disturbances are small, they leave the air behind them in essentially the same state it was before they arrived. The radius of a given wave front is given by

$$r = at \qquad (8.21)$$

where a is the speed of propagation (speed of sound) of the wave front and t is the time since the particular disturbance was generated.

Now, suppose that the point source begins moving (Fig. 8.1b) from right to left at a constant speed U which is less than the speed of sound a. The wave-front pattern will now appear as shown in Fig. 8.1b. A stationary observer ahead of the source will detect an increase in frequency of the sound while one behind it will note a decrease. Still, however, each wave front is separate from its neighbors.

If the speed of the source reaches the speed of sound in the undisturbed medium, the situation will appear as shown in Fig. 8.1c. We note that the individual wave fronts are still separate, except at the point where they coalesce.

A further increase in source speed, such that $U > a$, leads to the situation depicted in Fig. 8.1d. The wave fronts now form a conical envelope, which is known as the *Mach cone*, within which the disturbances can be detected. Outside of this "zone of action" is the "zone of silence," where the pulses have not arrived.

We see that there is a fundamental difference between subsonic $(U < a)$ and supersonic $(U > a)$ flow. In subsonic flow, the effect of a disturbance propagates upstream of its location, and, thus, the upstream flow is "warned" of the approach of the disturbance. In supersonic flow, however, no such "warning" is possible. Stating it another way, disturbances cannot propagate upstream in a supersonic flow relative to a source-fixed observer. This fundamental difference between the two types of flow has significant consequences on the flow field about a vehicle.

We note that the half-angle of the Mach cone dividing the zone of silence from the zone of action is given by

$$\sin \mu = \frac{1}{M} \qquad (8.22)$$

where

$$M \equiv \frac{U}{a}$$

is the Mach number. At $M = 1$ (i.e., when $U = a$), $\mu = \mu/2$, and as $M \to \infty$, $\mu \to 0$.

To determine the speed of sound a, consider the wave front in Fig. 8.1a propagating into still air. A small portion of curved wave front can be treated as planar. To an observer attached to the wave, the situation appears as shown in Fig. 8.2. A control volume is also shown attached to the wave. The boundaries of the volume are selected so that the flow is normal to faces parallel to the wave and tangent to the other faces. We make the key assumption (borne out by experiment) that, since the strength of the disturbance is infinitesimal, a fluid particle passing through the wave undergoes a process that is both reversible and adiabatic (i.e., isentropic).

The integral forms of the continuity and the momentum equations for a one-dimensional, steady, inviscid flow give

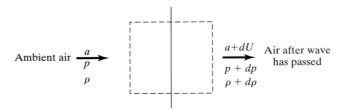

Figure 8.2 Control volume used to determine the speed of sound. (A velocity of equal magnitude and of opposite direction has been superimposed on a wave of Fig. 8.1a so that the sound wave is stationary in this figure.)

$$\rho a\, dA = (\rho + d\rho)(a + dU)\, dA \tag{8.23}$$

$$p\, dA - (p + dp)\, dA = [(a + dU) - a]\rho a\, dA \tag{8.24}$$

Simplifying equation (8.24), dividing equations (8.23) and (8.24) by dA, and combining the two relations gives

$$dp = a^2\, d\rho \tag{8.25}$$

However, since the process is isentropic,

$$a^2 = \left(\frac{\partial p}{\partial \rho}\right)_s \tag{8.26}$$

where we have indicated that the derivative is taken with entropy fixed, as originally assumed.

For a perfect gas undergoing an isentropic process, equation (8.20a) gives

$$p = c\rho^\gamma$$

where c is a constant. Thus,

$$a^2 = \left(\frac{\partial p}{\partial \rho}\right)_s = \gamma c\rho^{\gamma-1} = \frac{\gamma p}{\rho} \tag{8.27}$$

Using the equation of state for a perfect gas, the speed of sound is

$$a = \sqrt{\gamma R T} \tag{8.28}$$

8.2 ADIABATIC FLOW IN A VARIABLE-AREA STREAMTUBE

For purposes of derivation, consider the one-dimensional flow of a perfect gas through a variable-area streamtube (see Fig. 8.3). Let us apply the integral form of the energy equation [i.e., equation (2.35)] for steady, one-dimensional flow. Let us assume that there is no heat transfer through the surface of the control volume (i.e., $\dot{Q} = 0$) and that only flow work (pressure-volume work) is done. Work is done on the system by the pressure forces acting at station 1 and is, therefore, negative. Work is done by the system at station 2. Thus,

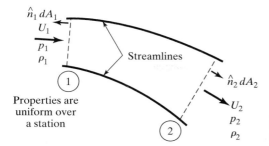

Figure 8.3 One-dimensional flow in a streamtube.

$$+ \; p_1 U_1 A_1 - p_2 U_2 A_2 = -\left(u_{e1} \rho_1 U_1 A_1 + \frac{U_1^2}{2} \rho_1 U_1 A_1 \right)$$

$$+ \left(u_{e2} \rho_2 U_2 A_2 + \frac{U_2^2}{2} \rho_2 U_2 A_2 \right)$$

Rearranging, noting that $\rho_1 U_1 A_1 = \rho_2 U_2 A_2$ by continuity, and using the definition for the enthalpy, we obtain

$$H_t = h_1 + \frac{U_1^2}{2} = h_2 + \frac{U_2^2}{2} \tag{8.29}$$

where, as is usually the case in aerodynamics problems, changes in potential energy have been neglected. The assumption of one-dimensional flow is valid provided that the streamtube cross-sectional area varies smoothly and gradually in comparison to the axial distance along the streamtube.

For a perfect gas, equation (8.29) becomes

$$c_p T_1 + \frac{U_1^2}{2} = c_p T_2 + \frac{U_2^2}{2} \tag{8.30}$$

or

$$T_1 + \frac{U_1^2}{2c_p} = T_2 + \frac{U_2^2}{2c_p} \tag{8.31}$$

By definition, the *stagnation temperature* T_t is the temperature reached when the fluid is brought to rest adiabatically.

$$T_t = T + \frac{U^2}{2c_p} \tag{8.32}$$

Since the locations of stations 1 and 2 are arbitrary,

$$T_{t1} = T_{t2} \tag{8.33}$$

That is, the stagnation temperature is a constant for the adiabatic flow of a perfect gas and will be designated simply as T_t. Note that, whereas it is generally true that the stagnation enthalpy is constant for an adiabatic flow, as indicated by equation (8.29), the stagnation temperature is constant only if the additional condition that c_p is constant (i.e., the gas behaves as a perfect gas) is satisfied.

For any one-dimensional adiabatic flow,

$$\frac{T_t}{T} = 1 + \frac{U^2}{2[\gamma R/(\gamma - 1)]T} = 1 + \frac{\gamma - 1}{2}M^2 \qquad \textbf{(8.34)}$$

Note that we have used the perfect-gas relations that $c_p = \gamma R/(\gamma - 1)$ and that $a^2 = \gamma RT$.

It is interesting to note that, when the flow is isentropic, Euler's equation for one-dimensional, steady flow (i.e., the inviscid-flow momentum equation) gives the same result as the energy equation for an adiabatic flow. To see this, let us write equation (3.2) for steady, one-dimensional, inviscid flow:

$$\rho u \frac{du}{ds} = -\frac{dp}{ds} \qquad \textbf{(8.35a)}$$

$$\int \frac{dp}{\rho} + \int u \, du = 0 \qquad \textbf{(8.35b)}$$

Note that for a one-dimensional flow $u = U$. For an isentropic process (which is of course adiabatic),

$$p = c\rho^{\gamma}$$

Differentiating, substituting the result into equation (8.35b), and integrating, we obtain

$$\frac{\gamma}{\gamma - 1}c\rho^{\gamma - 1} + \frac{U^2}{2} = \text{constant}$$

Thus,

$$\frac{1}{\gamma - 1}\frac{\gamma p}{\rho} + \frac{U^2}{2} = \text{constant}$$

Using the perfect-gas equation of state, we obtain

$$h + \frac{U^2}{2} = \text{constant}$$

which is equation (8.29).

EXAMPLE 8.1:   An indraft, supersonic wind tunnel

We are to design a supersonic wind tunnel using a large vacuum pump to draw air from the ambient atmosphere into our tunnel, as shown in Fig. 8.4. The air is continuously accelerated as it flows through a convergent/divergent nozzle so that flow in the test section is supersonic. If the ambient air is at the standard sea-level conditions, what is the maximum velocity that we can attain in the test section?

Solution: To calculate this maximum velocity, all we need is the energy equation for a steady, one-dimensional, adiabatic flow. Using equation (8.29), we have

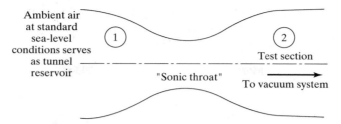

Figure 8.4 Indraft, supersonic wind tunnel.

$$h_1 + \tfrac{1}{2}U_1^2 = h_2 + \tfrac{1}{2}U_2^2$$

Since the ambient air (i.e., that which serves as the tunnel's "stagnation chamber" or "reservoir") is at rest, $U_1 = 0$. The maximum velocity in the test section occurs when $h_2 = 0$ (i.e., when the flow expands until the static temperature in the test section is zero).

$$(1004.7)288.15 \text{ J/kg} = \frac{U_2^2}{2}$$

$$U_2 = 760.9 \text{ m/s}$$

Of course, we realize that this limit is physically unattainable, since the gas would liquefy first. However, it does represent an upper limit for conceptual considerations.

EXAMPLE 8.2:   A simple model for the Shuttle Orbiter flow field

During a nominal reentry trajectory, the Space Shuttle Orbiter flies at 3964 ft/s at an altitude of 100,000 ft. The corresponding conditions at the stagnation point (point 2 in Fig. 8.5) are $p_2 = 490.2$ lbf/ft^2 and $T_2 = 1716.0°$R. The static pressures for two nearby locations (points 3 and 4 of Fig. 8.5) are $p_3 = 259.0$ lbf/ft^2 and $p_4 = 147.1$ lbf/ft^2. All three points are outside the boundary layer. What are the local static temperature, the local velocity, and the local Mach number at points 3 and 4?

Solution: At these conditions, the air can be assumed to behave approximately as a perfect gas. Furthermore, since all three points are outside the boundary layer and downstream of the bow shock wave, we will assume that the flow expands isentropically from point 2 to point 3 and then to point 4. (Note that because the shock wave is curved, the entropy will vary through the shock layer. For a further discussion of the rotational flow downstream of a curved shock wave, refer to Chapter 12. Thus, validity of the assumption that the expansion is isentropic should be verified for a given application.)

For an isentropic process, we can use equation (8.20b) to relate the temperature ratio to the pressure ratio between two points. Thus,

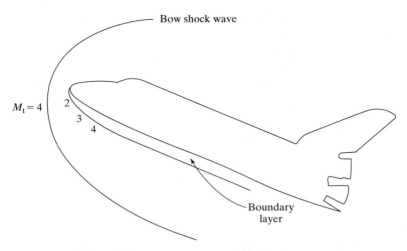

Figure 8.5 Shuttle flow field for Example 8.2.

$$T_3 = T_2\left(\frac{p_3}{p_2}\right)^{(\gamma-1)/\gamma} = (1716.0)(0.83337) = 1430.1°\text{R}$$

Similarly,

$$T_4 = (1716.0)(0.70899) = 1216.6°\text{R}$$

Using the energy equation for the adiabatic flow of a perfect gas [i.e., equation (8.31)], and noting that $U_2 = 0$, since point 2 is a stagnation point,

$$U_3 = [2c_p(T_2 - T_3)]^{0.5}$$

$$= \left[2\left(0.2404\frac{\text{Btu}}{\text{lbm}\cdot°\text{R}}\right)\left(32.174\frac{\text{ft}\cdot\text{lbm}}{\text{lbf}\cdot\text{s}^2}\right)\left(778.2\frac{\text{ft}\cdot\text{lbf}}{\text{Btu}}\right)(285.9°\text{R})\right]^{0.5}$$

$$= 1855.2\ \text{ft/s}$$

Similarly,

$$U_4 = 2451.9\ \text{ft/s}$$

Using equation (1.14b) for the speed of sound in English units,

$$M_3 = \frac{U_3}{a_3} = \frac{1855.2}{49.02(1430.1)^{0.5}} = 1.001$$

$$M_4 = \frac{U_4}{a_4} = \frac{2451.9}{49.02(1216.6)^{0.5}} = 1.434$$

Thus, the flow accelerates from the stagnation conditions at point 2 to sonic conditions at point 3, becoming supersonic at point 4.

8.3 ISENTROPIC FLOW IN A VARIABLE-AREA STREAMTUBE

It is particularly useful to study the isentropic flow of a perfect gas in a variable-area streamtube, since it reveals many of the general characteristics of compressible flow. In addition, the assumption of constant entropy is not too restrictive, since the flow outside the boundary layer is essentially isentropic except while crossing linear shock waves or downstream of curved shock waves.

Using equations (8.20) and (8.34), we can write

$$\frac{p_{t1}}{p} = \left(1 + \frac{\gamma - 1}{2}M^2\right)^{\gamma/(\gamma-1)} \tag{8.36}$$

$$\frac{\rho_{t1}}{\rho} = \left(1 + \frac{\gamma - 1}{2}M^2\right)^{1/(\gamma-1)} \tag{8.37}$$

where p_{t1} and ρ_{t1} are the *stagnation pressure* and the *stagnation density*, respectively. Applying these equations between two streamwise stations shows that if T_t is constant and the flow is isentropic, the stagnation pressure p_{t1} is a constant. The equation of state requires that ρ_{t1} be constant also.

To get a feeling for the deviation between the pressure values calculated assuming incompressible flow and those calculated using the compressible flow relations, let us expand equation (8.36) in powers of M^2:

$$\frac{p_{t1}}{p} = 1 + \frac{\gamma}{2}M^2 + O(M^4) + \cdots \tag{8.38}$$

Since the flow is essentially incompressible when the Mach number is relatively low, we can neglect higher order terms. Retaining only terms of order M^2 yields

$$\frac{p_{t1}}{p} = 1 + \frac{\gamma}{2}M^2 \tag{8.39}$$

which for a perfect gas becomes

$$\frac{p_{t1}}{p} = 1 + \frac{\gamma}{2}\frac{U^2}{\gamma RT} = 1 + \frac{U^2}{2p/\rho} \tag{8.40}$$

Rearranging,

$$p_{t1} = p + \frac{\rho U^2}{2} \tag{8.41}$$

Thus, for low Mach numbers, the general relation, equation (8.36), reverts to Bernoulli's equation for incompressible flow. The static pressures predicted by equation (8.36) are compared with those of equation (8.41) in terms of percent error as a function of Mach number in Fig. 8.6. An error of less than 1% results when Bernoulli's equation is used if the local Mach number is less than or equal to 0.5 in air.

In deriving equations (8.34), (8.36), and (8.37), the respective stagnation properties have been used as references to nondimensionalize the static properties. Since the continuity equation for the one-dimensional steady flow requires that ρUA be a constant, the area becomes infinitely large as the velocity goes to zero. Let us choose the area

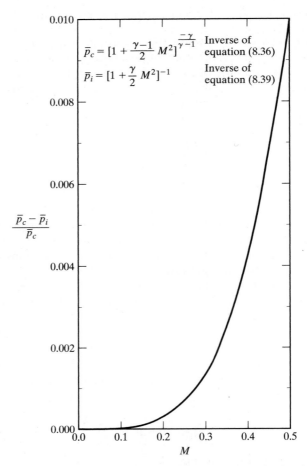

Figure 8.6 Effect of compressibility on the theoretical value for the pressure ratio.

where the flow is sonic (i.e., $M = 1$) as the reference area to relate to the streamtube area at a given station. Designating the sonic conditions by a (*) superscript, the continuity equation yields

$$\frac{A^*}{A} = \frac{\rho U}{\rho^* U^*} = \frac{\rho_{t1}(\rho/\rho_{t1})\sqrt{\gamma R T_t}\sqrt{T/T_t}M}{\rho_{t1}(\rho^*/\rho_{t1})\sqrt{\gamma R T_t}\sqrt{T^*/T_t}} \quad (8.42)$$

since $M^* = 1$. Noting that ρ^*/ρ_{t1} and T^*/T_t are to be evaluated at $M = M^* = 1$,

$$\frac{A^*}{A} = M\left[\frac{2}{\gamma + 1}\left(1 + \frac{\gamma - 1}{2}M^2\right)\right]^{-(\gamma+1)/2(\gamma-1)} \quad (8.43)$$

Given the area, A, and the Mach number, M, at any station, one could compute an A^* for that station from equation (8.43). A^* is the area the streamtube would have to be if the flow were accelerated or decelerated to $M = 1$ isentropically. Equation (8.43) is especially useful in streamtube flows that are isentropic, and therefore where A^* is a constant.

In order to aid in the solution of isentropic flow problems, the temperature ratio [equation (8.34)], the pressure ratio [equation (8.36)], the density ratio [equation (8.37)], and the area ratio [equation (8.43)] are presented as a function of the Mach number in Table 8.1. A more complete tabulation of these data is given in Ames Research Center Staff (1953). The results of Table 8.1 are summarized in Fig. 8.7.

In order to determine the mass-flow rate in a streamtube,

$$\dot{m} = \rho U A = \rho_{t1}\left(\frac{\rho}{\rho_{t1}}\right) M \sqrt{\gamma R T_t} \sqrt{\frac{T}{T_t}} A$$

$$\frac{\dot{m}}{A} = \sqrt{\frac{\gamma}{R}} \frac{p_{t1}}{\sqrt{T_t}} \frac{M}{\{1 + [(\gamma - 1)/2]M^2\}^{(\gamma+1)/2(\gamma-1)}} \tag{8.44}$$

Thus, the mass-flow rate is proportional to the stagnation pressure and inversely proportional to the square root of the stagnation temperature. To find the condition of maximum flow per unit area, one could compute the derivative of $(\dot{m}/A)$ as given by equation (8.44) with respect to Mach number and set the derivative equal to zero. At this condition, one would find that $M = 1$. Thus, setting $M = 1$ in equation (8.44) yields

$$\left(\frac{\dot{m}}{A}\right)_{max} = \frac{\dot{m}}{A^*} = \sqrt{\frac{\gamma}{R}\left(\frac{2}{\gamma + 1}\right)^{(\gamma+1)/(\gamma-1)}} \frac{p_{t1}}{\sqrt{T_t}} \tag{8.45}$$

The maximum flow rate per unit area occurs where the cross-section area is a minimum (designated A^*), where the Mach number is one. Thus, the maximum flow rate per unit area occurs at the sonic throat when the flow is choked.

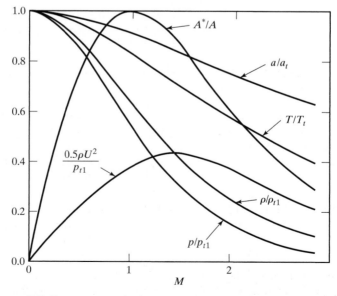

Figure 8.7 Property variations as functions of Mach number for isentropic flow for $\gamma = 1.4$.

TABLE 8.1   Correlations for a One-Dimensional, Isentropic Flow of Perfect AIR ($\gamma = 1.4$)

M	$\dfrac{A}{A^*}$	$\dfrac{p}{p_{t1}}$	$\dfrac{\rho}{\rho_{t1}}$	$\dfrac{T}{T_t}$	$\dfrac{A}{A^*}\dfrac{p}{p_{t1}}$
0	∞	1.00000	1.00000	1.00000	∞
0.05	11.592	0.99825	0.99875	0.99950	11.571
0.10	5.8218	0.99303	0.99502	0.99800	5.7812
0.15	3.9103	0.98441	0.98884	0.99552	3.8493
0.20	2.9635	0.97250	0.98027	0.99206	2.8820
0.25	2.4027	0.95745	0.96942	0.98765	2.3005
0.30	2.0351	0.93947	0.95638	0.98232	1.9119
0.35	1.7780	0.91877	0.94128	0.97608	1.6336
0.40	1.5901	0.89562	0.92428	0.96899	1.4241
0.45	1.4487	0.87027	0.90552	0.96108	1.2607
0.50	1.3398	0.84302	0.88517	0.95238	1.12951
0.55	1.2550	0.81416	0.86342	0.94295	1.02174
0.60	1.1882	0.78400	0.84045	0.93284	0.93155
0.65	1.1356	0.75283	0.81644	0.92208	0.85493
0.70	1.09437	0.72092	0.79158	0.91075	0.78896
0.75	1.06242	0.68857	0.76603	0.89888	0.73155
0.80	1.03823	0.65602	0.74000	0.88652	0.68110
0.85	1.02067	0.62351	0.71361	0.87374	0.63640
0.90	1.00886	0.59126	0.68704	0.86058	0.59650
0.95	1.00214	0.55946	0.66044	0.84710	0.56066
1.00	1.00000	0.52828	0.63394	0.83333	0.52828
1.05	1.00202	0.49787	0.60765	0.81933	0.49888
1.10	1.00793	0.46835	0.58169	0.80515	0.47206
1.15	1.01746	0.43983	0.55616	0.79083	0.44751
1.20	1.03044	0.41238	0.53114	0.77640	0.42493
1.25	1.04676	0.38606	0.50670	0.76190	0.40411
1.30	1.06631	0.36092	0.48291	0.74738	0.38484
1.35	1.08904	0.33697	0.45980	0.73287	0.36697
1.40	1.1149	0.31424	0.43742	0.71839	0.35036
1.45	1.1440	0.29272	0.41581	0.70397	0.33486
1.50	1.1762	0.27240	0.39498	0.68965	0.32039
1.55	1.2115	0.25326	0.37496	0.67545	0.30685
1.60	1.2502	0.23527	0.35573	0.66138	0.29414
1.65	1.2922	0.21839	0.33731	0.64746	0.28221
1.70	1.3376	0.20259	0.31969	0.63372	0.27099
1.75	1.3865	0.18782	0.30287	0.62016	0.26042
1.80	1.4390	0.17404	0.28682	0.60680	0.25044
1.85	1.4952	0.16120	0.27153	0.59365	0.24102
1.90	1.5555	0.14924	0.25699	0.58072	0.23211
1.95	1.6193	0.13813	0.24317	0.56802	0.22367

(*continued on next page*)

TABLE 8.1 continued

M	$\dfrac{A}{A^*}$	$\dfrac{p}{p_{t1}}$	$\dfrac{\rho}{\rho_{t1}}$	$\dfrac{T}{T_t}$	$\dfrac{A}{A^*}\dfrac{p}{p_{t1}}$
2.00	1.6875	0.12780	0.23005	0.55556	0.21567
2.05	1.7600	0.11823	0.21760	0.54333	0.20808
2.10	1.8369	0.10935	0.20580	0.53135	0.20087
2.15	1.9185	0.10113	0.19463	0.51962	0.19403
2.20	2.0050	0.09352	0.18405	0.50813	0.18751
2.25	2.0964	0.08648	0.17404	0.49689	0.18130
2.30	2.1931	0.07997	0.16458	0.48591	0.17539
2.35	2.2953	0.07396	0.15564	0.47517	0.16975
2.40	2.4031	0.06840	0.14720	0.46468	0.16437
2.45	2.5168	0.06327	0.13922	0.45444	0.15923
2.50	2.6367	0.05853	0.13169	0.44444	0.15432
2.55	2.7630	0.05415	0.12458	0.43469	0.14963
2.60	2.8960	0.05012	0.11787	0.42517	0.14513
2.65	3.0359	0.04639	0.11154	0.41589	0.14083
2.70	3.1830	0.04295	0.10557	0.40684	0.13671
2.75	3.3376	0.03977	0.09994	0.39801	0.13276
2.80	3.5001	0.03685	0.09462	0.38941	0.12897
2.85	3.6707	0.03415	0.08962	0.38102	0.12534
2.90	3.8498	0.03165	0.08489	0.37286	0.12185
2.95	4.0376	0.02935	0.08043	0.36490	0.11850
3.00	4.2346	0.02722	0.07623	0.35714	0.11527
3.50	6.7896	0.01311	0.04523	0.28986	0.08902
4.00	10.719	0.00658	0.02766	0.23810	0.07059
4.50	16.562	0.00346	0.01745	0.19802	0.05723
5.00	25.000	$189(10)^{-5}$	0.01134	0.16667	0.04725
6.00	53.189	$633(10)^{-6}$	0.00519	0.12195	0.03368
7.00	104.143	$242(10)^{-6}$	0.00261	0.09259	0.02516
8.00	190.109	$102(10)^{-6}$	0.00141	0.07246	0.01947
9.00	327.189	$474(10)^{-7}$	0.000815	0.05814	0.01550
10.00	535.938	$236(10)^{-7}$	0.000495	0.04762	0.01263
∞	∞	0	0	0	0

Figure 8.7 shows that, for each value of A^*/A, there are two values of M: one subsonic, the other supersonic. Thus, from Fig. 8.7 we see that, while all static properties of the fluid monotonically decrease with Mach number, the area ratio does not. We conclude that to accelerate a subsonic flow to supersonic speeds, the streamtube must first converge in an isentropic process until sonic conditions are reached. The flow then accelerates in a diverging streamtube to achieve supersonic Mach numbers. Note that just because a streamtube is convergent/divergent, it does not necessarily follow that the flow is sonic at

the narrowest cross section (or throat). The actual flow depends on the pressure distribution as well as the geometry of the streamtube. However, if the Mach number is to be unity anywhere in the streamtube, it must be so at the throat.

For certain calculations (e.g., finding the true airspeed from Mach number and stagnation pressure) the ratio $0.5\,\rho U^2/p_{t1}$ is useful.

$$\frac{0.5\rho U^2}{p_{t1}} = \frac{\frac{1}{2}(p/RT)(\gamma/\gamma)U^2}{p_{t1}} = \frac{1}{2}\frac{\gamma p}{p_{t1}}\frac{U^2}{\gamma RT} = \frac{\gamma M^2}{2}\frac{p}{p_{t1}} \tag{8.46}$$

Thus,

$$\frac{0.5\rho U^2}{p_{t1}} = \frac{\gamma M^2}{2}\left(1 + \frac{\gamma - 1}{2}M^2\right)^{-\gamma/(\gamma-1)} \tag{8.47}$$

The ratio of the local speed of sound to the speed of sound at the stagnation conditions is

$$\frac{a}{a_t} = \sqrt{\frac{\gamma RT}{\gamma RT_t}} = \left(\frac{T}{T_t}\right)^{0.5} = \left(1 + \frac{\gamma - 1}{2}M^2\right)^{-0.5} \tag{8.48}$$

Note: The nomenclature used herein anticipates the discussion of shock waves. Since the stagnation pressure varies across a shock wave, the subscript $t1$ has been used to designate stagnation properties evaluated upstream of a shock wave (which correspond to the free-stream values in a flow with a single shock wave). Since the stagnation temperature is constant across a shock wave (for perfect-gas flows), it is designated by the simple subscript t.

8.4 CHARACTERISTIC EQUATIONS AND PRANDTL-MEYER FLOWS

Consider a two-dimensional flow around a slender airfoil shape. The deflection of the streamlines as flow encounters the airfoil are sufficiently small that shock waves are not generated. Thus, the flow may be considered as isentropic (excluding the boundary layer, of course). For the development of the equations for a more general flow, refer to Hayes and Probstein (1966). For the present type of flow, the equations of motion in a natural (or streamline) coordinate system, as shown in Fig. 8.8, are as follows: the continuity equation,

$$\frac{\partial(\rho U)}{\partial s} + \rho U\frac{\partial\theta}{\partial n} = 0 \tag{8.49}$$

the *s*-momentum equation,

$$\rho U\frac{\partial U}{\partial s} + \frac{\partial p}{\partial s} = 0 \tag{8.50}$$

and the *n*-momentum equation,

$$\rho U^2\frac{\partial\theta}{\partial s} + \frac{\partial p}{\partial n} = 0 \tag{8.51}$$

Since the flow is isentropic, the energy equation provides no unique information and is, therefore, not used. However, since the flow is isentropic, the change in pressure with re-

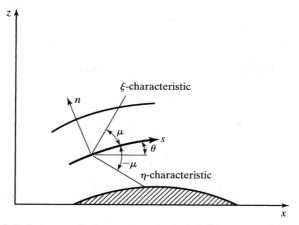

Figure 8.8 Supersonic flow around an airfoil in natural (streamline) coordinates.

spect to the change in density is equal to the square of the speed of sound. Thus,

$$\frac{\partial p}{\partial \rho} = a^2 \tag{8.52}$$

and the continuity equation becomes

$$\frac{\partial p}{\partial s}\frac{M^2 - 1}{\rho U^2} + \frac{\partial \theta}{\partial n} = 0 \tag{8.53}$$

Combining equations (8.51) and (8.53) and introducing the concept of the directional derivative [e.g., Wayland (1957)], we obtain

$$\frac{\partial p}{\partial \xi} + \frac{\rho U^2}{\sqrt{M^2 - 1}}\frac{\partial \theta}{\partial \xi} = 0 \tag{8.54a}$$

along the line having the direction

$$\frac{dn}{ds} = \tan \mu = \frac{1}{\sqrt{M^2 - 1}}$$

(i.e., the left-running characteristic ξ of Fig. 8.8). A *characteristic* is a line which exists only in supersonic flows. Characteristics should not be confused with finite-strength waves, such as shock waves. The ξ characteristic is inclined to the local streamline by the angle μ, which is the Mach angle,

$$\mu = \sin^{-1}\left(\frac{1}{M}\right)$$

The ξ characteristics correspond to the left-running Mach waves, which are so called because to an observer looking downstream, the Mach wave appears to be going downstream in a leftward direction. Equivalently,

TABLE 8.2   Mach Number and Mach Angle as a Function of Prandt-Meyer Angle

ν (deg)	M	μ (deg)	ν (deg)	M	μ (deg)
0.0	1.000	90.000	25.0	1.950	30.847
0.5	1.051	72.099	25.5	1.968	30.536
1.0	1.082	67.574	26.0	1.986	30.229
1.5	1.108	64.451	26.5	2.004	29.928
2.0	1.133	61.997	27.0	2.023	29.632
2.5	1.155	59.950	27.5	2.041	29.340
3.0	1.177	58.180	28.0	2.059	29.052
3.5	1.198	56.614	28.5	2.078	28.769
4.0	1.218	55.205	29.0	2.096	28.491
4.5	1.237	53.920	29.5	2.115	28.216
5.0	1.256	52.738	30.0	2.134	27.945
5.5	1.275	51.642	30.5	2.153	27.678
6.0	1.294	50.619	31.0	2.172	27.415
6.5	1.312	49.658	31.5	2.191	27.155
7.0	1.330	48.753	32.0	2.210	26.899
7.5	1.348	47.896	32.5	2.230	26.646
8.0	1.366	47.082	33.0	2.249	26.397
8.5	1.383	46.306	33.5	2.269	26.151
9.0	1.400	45.566	34.0	2.289	25.908
9.5	1.418	44.857	34.5	2.309	25.668
10.0	1.435	44.177	35.0	2.329	25.430
10.5	1.452	43.523	35.5	2.349	25.196
11.0	1.469	42.894	36.0	2.369	24.965
11.5	1.486	42.287	36.5	2.390	24.736
12.0	1.503	41.701	37.0	2.410	24.510
12.5	1.520	41.134	37.5	2.431	24.287
13.0	1.537	40.585	38.0	2.452	24.066
13.5	1.554	40.053	38.5	2.473	23.847
14.0	1.571	39.537	39.0	2.495	23.631
14.5	1.588	39.035	39.5	2.516	23.418
15.0	1.605	38.547	40.0	2.538	23.206
15.5	1.622	38.073	40.5	2.560	22.997
16.0	1.639	37.611	41.0	2.582	22.790
16.5	1.655	37.160	41.5	2.604	22.585
17.0	1.672	36.721	42.0	2.626	22.382
17.5	1.689	36.293	42.5	2.649	22.182
18.0	1.706	35.874	43.0	2.671	21.983
18.5	1.724	35.465	43.5	2.694	21.786
19.0	1.741	35.065	44.0	2.718	21.591
19.5	1.758	34.673	44.5	2.741	21.398
20.0	1.775	34.290	45.0	2.764	21.207
20.5	1.792	33.915	45.5	2.788	21.017
21.0	1.810	33.548	46.0	2.812	20.830
21.5	1.827	33.188	46.5	2.836	20.644
22.0	1.844	32.834	47.0	2.861	20.459
22.5	1.862	32.488	47.5	2.886	20.277
23.0	1.879	32.148	48.0	2.910	20.096
23.5	1.897	31.814	48.5	2.936	19.916
24.0	1.915	31.486	49.0	2.961	19.738
24.5	1.932	31.164	49.5	2.987	15.561

(continued on next page)

TABLE 8.2 (*continued*)

ν (deg)	M	μ (deg)	ν (deg)	M	μ (deg)
50.0	3.013	19.386	76.0	4.903	11.768
50.5	3.039	19.213	76.5	4.955	11.642
51.0	3.065	19.041	77.0	5.009	11.517
51.5	3.092	18.870	77.5	5.063	11.392
52.0	3.119	18.701	78.0	5.118	11.268
52.5	3.146	18.532	78.5	5.175	11.145
53.0	3.174	18.366	79.0	5.231	11.022
53.5	3.202	18.200	79.5	5.289	10.899
54.0	3.230	18.036	80.0	5.348	10.777
54.5	3.258	17.873	80.5	5.408	10.656
55.0	3.287	17.711	81.0	5.470	10.535
55.5	3.316	17.551	81.5	5.532	10.414
56.0	3.346	17.391	82.0	5.596	10.294
56.5	3.375	17.233	82.5	5.661	10.175
57.0	3.406	17.076	83.0	5.727	10.056
57.5	3.436	16.920	83.5	5.795	9.937
58.0	3.467	16.765	84.0	5.864	9.819
58.5	3.498	16.611	84.5	5.935	9.701
59.0	3.530	16.458	85.0	6.006	9.584
59.5	3.562	16.306	85.5	6.080	9.467
60.0	3.594	16.155	86.0	6.155	9.350
60.5	3.627	16.006	86.5	6.232	9.234
61.0	3.660	15.856	87.0	6.310	9.119
61.5	3.694	15.708	87.5	6.390	9.003
62.0	3.728	15.561	88.0	6.472	8.888
62.5	3.762	15.415	88.5	6.556	8.774
63.0	3.797	15.270	89.0	6.642	8.660
63.5	3.832	15.126	89.5	6.729	8.546
64.0	3.868	14.983	90.0	6.819	8.433
64.5	3.904	14.840	90.5	6.911	8.320
65.0	3.941	14.698	91.0	7.005	8.207
65.5	3.979	14.557	91.5	7.102	8.095
66.0	4.016	14.417	92.0	7.201	7.983
66.5	4.055	14.278	92.5	7.302	7.871
67.0	4.094	14.140	93.0	7.406	7.760
67.5	4.133	14.002	93.5	7.513	7.649
68.0	4.173	13.865	94.0	7.623	7.538
68.5	4.214	13.729	94.5	7.735	7.428
69.0	4.255	13.593	95.0	7.851	7.318
69.5	4.297	13.459	95.5	7.970	7.208
70.0	4.339	13.325	96.0	8.092	7.099
70.5	4.382	13.191	96.5	8.218	6.989
71.0	4.426	13.059	97.0	8.347	6.881
71.5	4.470	12.927	97.5	8.480	6.772
72.0	4.515	12.795	98.0	8.618	6.664
72.5	4.561	12.665	98.5	8.759	6.556
73.0	4.608	12.535	99.0	8.905	6.448
73.5	4.655	12.406	99.5	9.055	6.340
74.0	4.703	12.277	100.0	9.210	6.233
74.5	4.752	12.149	100.5	9.371	6.126
75.0	4.801	12.021	101.0	9.536	6.019
75.5	4.852	11.894	101.5	9.708	5.913
			102.0	9.885	5.806

$$\frac{\partial p}{\partial \eta} - \frac{\rho U^2}{\sqrt{M^2 - 1}} \frac{\partial \theta}{\partial \eta} = 0 \tag{8.54b}$$

along the line whose direction is

$$\frac{dn}{ds} = \tan(-\mu)$$

(i.e., the right-running characteristic η of Fig. 8.8). Equations (8.54a) and (8.54b) provide a relation between the local static pressure and the local flow inclination.

     Euler's equation for a steady, inviscid flow, which can be derived by neglecting the viscous terms and the body forces in the momentum equation (3.1), states that

$$dp = -\rho U \, dU$$

Thus, for the left-running characteristic, equation (8.54a) becomes

$$\frac{dU}{U} = \frac{d\theta}{\sqrt{M^2 - 1}} \tag{8.55}$$

But from the adiabatic-flow relations for a perfect gas,

$$\left(\frac{U}{a_t}\right)^2 = M^2\left(1 + \frac{\gamma - 1}{2}M^2\right)^{-1} \tag{8.56}$$

where a_t is the speed of sound at the stagnation conditions. Differentiating equation (8.56) and substituting the result into equation (8.55) yields

$$d\theta = \frac{\sqrt{M^2 - 1} \, dM^2}{2M^2\{1 + [(\gamma - 1)/2]M^2\}} \tag{8.57}$$

Integration of equation (8.57) yields the relation, which is valid for a left-running characteristic:

$$\theta = \nu + \text{constant of integration}$$

where ν is a function of the Mach number as given by

$$\nu = \sqrt{\frac{\gamma + 1}{\gamma - 1}} \arctan \sqrt{\frac{\gamma - 1}{\gamma + 1}(M^2 - 1)} - \arctan\sqrt{M^2 - 1} \tag{8.58}$$

Tabulations of the *Prandtl-Meyer angle* (ν), the corresponding Mach number, and the corresponding Mach angle (μ) are presented in Table 8.2.

     Thus, along a left-running characteristic,

$$\nu - \theta = R \tag{8.59a}$$

a constant. Similarly, along a right-running characteristic,

$$\nu + \theta = Q \tag{8.59b}$$

which is another constant. The use of equations (8.59a) and (8.59b) is simplified if the slope of the vehicle surface is such that one only needs to consider the waves of a single family (i.e., all the waves are either left-running waves or right-running waves). To make the application clear, let us work through a sample problem.

EXAMPLE 8.3: Use Prandtl-Meyer relations to calculate the aerodynamic coefficients for a thin airfoil

Consider the infinitesimally thin airfoil which has the shape of a parabola:

$$x^2 = -\frac{c^2}{z_{max}}(z - z_{max})$$

where $z_{max} = 0.10c$, moving through the air at $M_\infty = 2.059$. The leading edge of the airfoil is parallel to the free stream. The thin airfoil will be represented by five linear segments, as shown in Fig. 8.9. For each segment Δx will be $0.2c$. Thus, the slopes of these segments are as follows:

Segment	a	b	c	d	e
θ	$-1.145°$	$-3.607°$	$-5.740°$	$-8.048°$	$-10.370°$

Solution: For the free-stream flow,

$$\nu_\infty = 28.000° \qquad \frac{p_\infty}{p_{t1}} = 0.11653 \qquad \theta_\infty = 0°$$

Since the turning angles are small, it will be assumed that both the acceleration of the flow over the upper surface and the deceleration of the flow

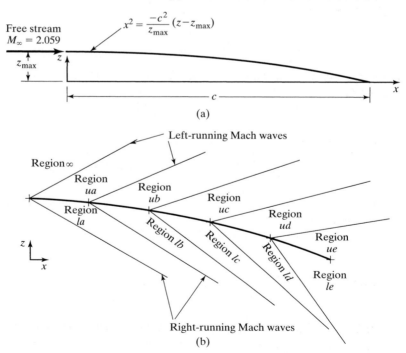

Figure 8.9 Mach waves for supersonic flow past a thin airfoil: (a) airfoil section; (b) wave pattern.

over the lower surface are isentropic processes. Note that the expansion waves on the upper surface diverge as the flow accelerates, but the compression waves of the lower surface coalesce. Since the flow is isentropic, we can use equations (8.59a) and (8.59b). Furthermore, the stagnation pressure is constant throughout the flow field and equal to p_{t1} (which is the value for the free-stream flow).

In going from the free-stream (region ∞ in Fig. 8.9) to the first segment on the upper surface (region ua), we move along a right-running characteristic to cross the left-running Mach wave shown in the figure. Thus,

$$\nu + \theta = Q$$

or

$$d\nu = -d\theta$$

so

$$\nu_{ua} = \nu_\infty - (\theta_{ua} - \theta_\infty)$$
$$= 28.000° - (-1.145°) = 29.145°$$

and

$$M_{ua} = 2.1018$$

Using Table 8.1, $p_{ua}/p_{t1} = 0.1091$.

Similarly, in going from the free stream to the first segment on the lower surface (region la) we move along a left-running characteristic to cross the right-running Mach wave shown in the figure. Thus,

$$\nu - \theta = R$$

or

$$d\nu = d\theta$$

so

$$\nu_{la} = \nu_\infty + (\theta_{la} - \theta_\infty)$$
$$= 28.000° + (-1.145°) = 26.855°$$

Summarizing,

Segment	Upper surface			Lower surface		
	ν_u	M_u	$\dfrac{p_u}{p_{t1}}$	ν_l	M_l	$\dfrac{p_l}{p_{t1}}$
a	29.145°	2.1018	0.1091	26.855°	2.0173	0.1244
b	31.607°	2.1952	0.0942	24.393°	1.9286	0.1428
c	33.740°	2.2784	0.0827	22.260°	1.8534	0.1604
d	36.048°	2.3713	0.0715	19.952°	1.7733	0.1813
e	38.370°	2.4679	0.0615	17.630°	1.6940	0.2045

Let us now calculate the lift coefficient and the drag coefficient for the airfoil. Note that we have not been given the free-stream pressure (or, equivalently, the altitude at which the airfoil is flying) or the chord length of the airfoil. But that is not critical, since we seek the force coefficients.

$$C_l = \frac{l}{\frac{1}{2}\rho_\infty U_\infty^2 c}$$

Referring to equation (8.46), it is evident that, for a perfect gas,

$$q_\infty = \frac{1}{2}\rho_\infty U_\infty^2 = \frac{\gamma}{2}p_\infty M_\infty^2 \qquad (8.60)$$

Thus,

$$C_l = \frac{l}{(\gamma/2)p_\infty M_\infty^2 c} \qquad (8.61)$$

Referring again to Fig. 8.9, the incremental lift force acting on any segment (i.e., the ith segment) is

$$dl_i = (p_{li} - p_{ui})\, ds_i \cos \theta_i = (p_{li} - p_{ui})\, dx_i$$

Similarly, the incremental drag force for any segment is

$$dd_i = (p_{li} - p_{ui})\, ds_i \sin \theta_i = (p_{li} - p_{ui})\, dx_i \tan \theta_i$$

Segment	$\dfrac{p_{li}}{p_\infty}$	$\dfrac{p_{ui}}{p_\infty}$	$\dfrac{dl_i}{p_\infty}$	$\dfrac{dd_i}{p_\infty}$
a	1.070	0.939	0.0262c	0.000524c
b	1.226	0.810	0.0832c	0.004992c
c	1.380	0.710	0.1340c	0.01340c
d	1.559	0.614	0.1890c	0.0264c
e	1.759	0.529	0.2460c	0.0443c
Sum			0.6784c	0.0896c

Finally,

$$C_l = \frac{\Sigma\, dl_i}{(\gamma/2)p_\infty M_\infty^2 c} = \frac{0.6784c}{0.7(4.24)c} = 0.2286$$

$$C_d = \frac{\Sigma\, dd_i}{(\gamma/2)p_\infty M_\infty^2 c} = \frac{0.0896c}{0.7(4.24)c} = 0.0302$$

$$\frac{l}{d} = \frac{C_l}{C_d} = 7.57$$

8.5   SHOCK WAVES

The formation of a shock wave occurs when a supersonic flow decelerates in response to a sharp increase in pressure or when a supersonic flow encounters a sudden, compressive change in direction. For flow conditions where the gas is a continuum, the shock wave is a narrow region (on the order of several molecular mean free paths thick, $\sim 6 \times 10^{-6}$ cm) across which there is an almost instantaneous change in the values of the flow parameters. Because of the large streamwise variations in velocity, pressure, and temperature, viscous and heat-conduction effects are important within the shock wave. The difference between a shock wave and a Mach wave should be kept in mind. A *Mach wave* represents a surface across which some derivatives of the flow variables (such as the thermodynamic properties of the fluid and the flow velocity) may be discontinuous while the variables themselves are continuous. A *shock wave* represents a surface across which the thermodynamic properties and the flow velocity are essentially discontinuous. Thus, the characteristic curves, or Mach lines, are patching lines for continuous flows, whereas shock waves are patching lines for discontinuous flows.

Consider the curved shock wave illustrated in Fig. 8.10. The flow upstream of the shock wave, which is stationary in the body-fixed coordinate system, is supersonic. At the plane of symmetry, the shock wave is normal (or perpendicular) to the free-stream flow, and the flow downstream of the shock wave is subsonic. Away from the plane of symmetry, the shock wave is oblique and the downstream flow is often supersonic. The velocity and the thermodynamic properties upstream of the shock wave are designated by the subscript 1. Note that whereas the subscript 1 designates the free-stream (∞) properties for flows such as those in Fig. 8.10, it designates the local flow properties just upstream of the shock wave when it occurs in the midchord region of a transonic airfoil (see Chapter 9). The downstream values are designated by the subscript 2. We will analyze

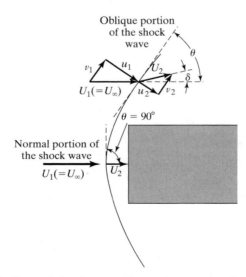

Figure 8.10 Curved shock wave illustrating nomenclature for normal shock wave and for oblique shock wave.

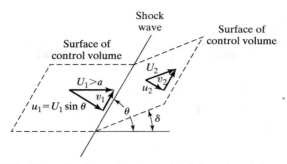

Figure 8.11 Control volume for analysis of flow through an oblique shock wave.

oblique shock waves by writing the continuity, the momentum, and the energy equations for the flow through the control volume, shown in Fig. 8.11. For a steady flow, the integral equations of motion yield the following relations for the flow across an oblique segment of the shock wave:

1. Continuity:

$$\rho_1 u_1 = \rho_2 u_2 \tag{8.62}$$

2. Normal component of momentum:

$$p_1 + \rho_1 u_1^2 = p_2 + \rho_2 u_2^2 \tag{8.63}$$

3. Tangential component of momentum:

$$\rho_1 u_1 v_1 = \rho_2 u_2 v_2 \tag{8.64}$$

4. Energy:

$$h_1 + \tfrac{1}{2}(u_1^2 + v_1^2) = h_2 + \tfrac{1}{2}(u_2^2 + v_2^2) \tag{8.65}$$

In addition to describing the flow across an oblique shock wave such as shown in Fig. 8.11, these relations can be used to describe the flow across a normal shock wave, or that portion of a curved shock wave which is perpendicular to the free stream, by letting $v_1 = v_2 = 0$.

Equation (8.63) can be used to calculate the maximum value of the pressure coefficient in the hypersonic limit as $M_1 \to \infty$. In this case, the flow is essentially stagnated ($u_2 \approx 0$) behind a normal shock wave. Thus, equation (8.63) becomes

$$p_2 - p_1 \approx \rho_1 u_1^2 \tag{8.66}$$

As a result,

$$C_{p\,max} = \frac{p_2 - p_1}{\tfrac{1}{2}\rho_1 U_1^2} \approx 2 \tag{8.67}$$

Note that, at the stagnation point of a vehicle in a hypersonic stream, C_p approaches 2.0. The value of C_p at the stagnation point of a vehicle in a supersonic stream is a function of the free-stream Mach number and is greater than 1.0. Recall that it is 1.0 for a low-speed stream independent of the velocity, provided that the flow is incompressible.

Comparing equation (8.62) with (8.64), one finds that for the oblique shock wave,

$$v_1 = v_2 \tag{8.68}$$

That is, the tangential component of the velocity is constant across the shock wave and we need not consider equation (8.64) further. Thus, the energy equation becomes

$$h_1 + \tfrac{1}{2}u_1^2 = h_2 + \tfrac{1}{2}u_2^2 \tag{8.69}$$

There are four unknowns (p_2, ρ_2, u_2, h_2) in the three equations (8.62), (8.63), and (8.69). Thus, we need to introduce an equation of state as the fourth equation. For hypervelocity flows where the shock waves are strong enough to cause dissociation or ionization, one can solve these equations numerically using the equation of state in tabular or in graphical form [e.g., Moeckel and Weston (1958)]. However, for a perfect-gas flow,

$$p = \rho RT$$

and

$$h = c_p T$$

A comparison of the properties downstream of a normal shock wave as computed using the charts for air in thermodynamic equilibrium with those computed using the perfect-gas model for air is presented in Chapter 12.

Note that equations (8.62), (8.63), and (8.69) involve only the component of velocity normal to the shock wave:

$$u_1 = U_1 \sin \theta \tag{8.70}$$

Hence, the property changes across an oblique shock wave are the same as those across a normal shock wave when they are written in terms of the upstream Mach number component perpendicular to the shock. The tangential component of the velocity is unchanged. This is the *sweepback principle*, that the oblique flow is reduced to the normal flow by a uniform translation of the axes (i.e., a Galilean transformation). Note that the tangential component of the Mach number does change, since the temperature (and therefore the speed of sound) changes across the shock wave.

Since the flow through the shock wave is adiabatic, the entropy must increase as the flow passes through the shock wave. Thus, the flow must decelerate (i.e., the pressure must increase) as it passes through the shock wave. One obtains the relation between the shock-wave (θ) and the deflection angle (δ),

$$\cot \delta = \tan \theta \left[\frac{(\gamma + 1)M_1^2}{2(M_1^2 \sin^2 \theta - 1)} - 1 \right] \tag{8.71}$$

From equation (8.71) it can be seen that the deflection angle is zero for two "shock"-wave angles. (1) The flow is not deflected when $\theta = \mu$, since the Mach wave results from an infinitesimal disturbance (i.e., a zero-strength shock wave). (2) The flow is not deflected when it passes through a normal shock wave (i.e., when $\theta = 90°$).

Solutions to equation (8.71) are presented in graphical form in Fig. 8.12a. Note that for a given deflection angle δ, there are two possible values for the shock-wave angle θ. The larger of the two values of θ is called the *strong* shock wave, while the

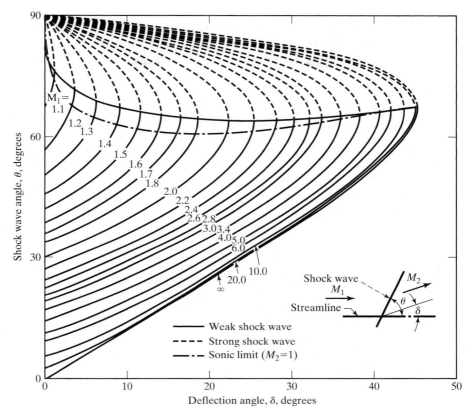

Figure 8.12 Variation of shock-wave parameters with wedge flow-deflection angle for various upstream Mach numbers, $\gamma = 1.4$: (a) shock-wave angle.

smaller value is called the *weak* shock wave. In practice, the weak shock wave typically occurs in external aerodynamic flows. However, the strong shock wave occurs if the downstream pressure is sufficiently high. The high downstream pressure associated with the strong shock wave may occur in flows in wind tunnels, in engine inlets, or in other ducts.

If the deflection angle exceeds the maximum value for which it is possible that a weak shock can be generated, a strong, detached shock wave will occur. For instance, a flat-plate airfoil can be inclined 34° to a Mach 3.0 stream and still generate a weak shock wave. This is the maximum deflection angle for a weak shock wave to occur. If the airfoil were to be inclined at 35° to the Mach 3.0 stream, a strong curved shock wave would occur with a complex subsonic/supersonic flow downstream of the shock wave.

Once the shock-wave angle θ has been found for the given values of M_1 and δ, the other downstream properties can be found using the following relations:

$$\frac{p_2}{p_1} = \frac{2\gamma M_1^2 \sin^2 \theta - (\gamma - 1)}{\gamma + 1} \tag{8.72}$$

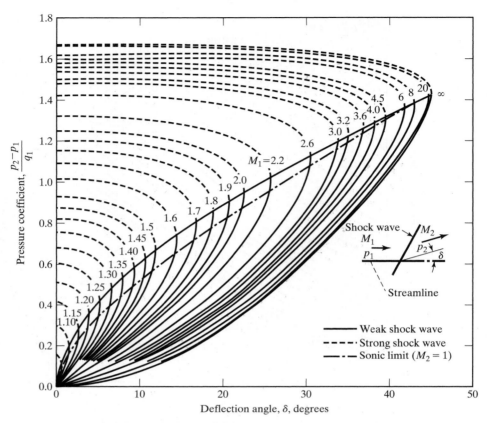

Figure 8.12 (continued) (b) pressure coefficient.

$$\frac{\rho_2}{\rho_1} = \frac{(\gamma + 1)M_1^2 \sin^2 \theta}{(\gamma - 1)M_1^2 \sin^2 \theta + 2} \tag{8.73}$$

$$\frac{T_2}{T_1} = \frac{[2\gamma M_1^2 \sin^2 \theta - (\gamma - 1)][(\gamma - 1)M_1^2 \sin^2 \theta + 2]}{(\gamma + 1)^2 M_1^2 \sin^2 \theta} \tag{8.74}$$

$$M_2^2 = \frac{(\gamma - 1)M_1^2 \sin^2 \theta + 2}{[2\gamma M_1^2 \sin^2 \theta - (\gamma - 1)] \sin^2 (\theta - \delta)}$$

$$\frac{p_{t2}}{p_{t1}} = e^{-\Delta s/R} \tag{8.75}$$

$$= \left[\frac{(\gamma + 1)M_1^2 \sin^2 \theta}{(\gamma - 1)M_1^2 \sin^2 \theta + 2} \right]^{\gamma/(\gamma-1)} \left[\frac{\gamma + 1}{2\gamma M_1^2 \sin^2 \theta - (\gamma - 1)} \right]^{1/(\gamma-1)} \tag{8.76}$$

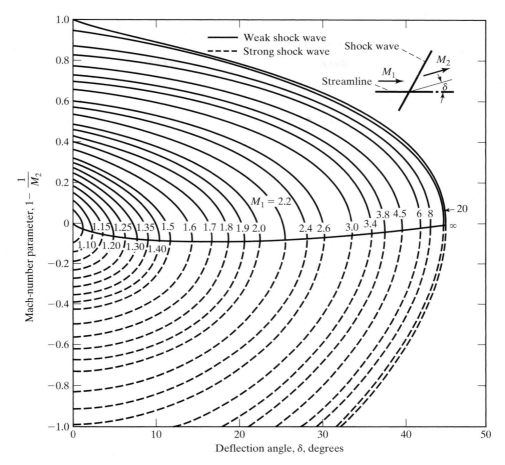

Figure 8.12 (continued) (c) downstream Mach number.

and

$$C_p = \frac{p_2 - p_1}{q_1} = \frac{4(M_1^2 \sin^2 \theta - 1)}{(\gamma + 1)M_1^2} \tag{8.77}$$

The pressure coefficient is presented in Fig. 8.12b as a function of δ and M_1. Equation (8.77) is consistent with equation (8.67) for a normal shock since $\gamma \rightarrow 1$ as $M_1 \rightarrow \infty$ due to the dissociation of molecules in the air at high Mach numbers. The values for many of these ratios are presented for a normal shock wave in Table 8.3 and in Fig. 8.13. The values for the pressure ratios, the density ratios, and the temperature ratios for an oblique shock wave can be read from Table 8.3 provided that $M_1 \sin \theta$ is used instead of M_1 in the first column. Note that since it is the tangential component of the velocity which is unchanged and not the tangential component of the Mach number, we cannot use Table 8.3 to calculate the downstream Mach number. The downstream Mach number is presented in Fig. 8.12c as a func-

TABLE 8.3 Correlation of Flow Properties Across a Normal Shock Wave as a Function of the Upstream Mach Number for air, $\gamma = 1.4$

M_1	M_2	$\dfrac{p_2}{p_1}$	$\dfrac{\rho_2}{\rho_1}$	$\dfrac{T_2}{T_1}$	$\dfrac{p_{t2}}{p_{t1}}$
1.00	1.00000	1.00000	1.00000	1.00000	1.00000
1.05	0.95312	1.1196	1.08398	1.03284	0.99987
1.10	0.91177	1.2450	1.1691	1.06494	0.99892
1.15	0.87502	1.3762	1.2550	1.09657	0.99669
1.20	0.84217	1.5133	1.3416	1.1280	0.99280
1.25	0.81264	1.6562	1.4286	1.1594	0.98706
1.30	0.78596	1.8050	1.5157	1.1909	0.97935
1.35	0.76175	1.9596	1.6027	1.2226	0.96972
1.40	0.73971	2.1200	1.6896	1.2547	0.95819
1.45	0.71956	2.2862	1.7761	1.2872	0.94483
1.50	0.70109	2.4583	1.8621	1.3202	0.92978
1.55	0.68410	2.6363	1.9473	1.3538	0.91319
1.60	0.66844	2.8201	2.0317	1.3880	0.89520
1.65	0.65396	3.0096	2.1152	1.4228	0.87598
1.70	0.64055	3.2050	2.1977	1.4583	0.85573
1.75	0.62809	3.4062	2.2781	1.4946	0.83456
1.80	0.61650	3.6133	2.3592	1.5316	0.81268
1.85	0.60570	3.8262	2.4381	1.5694	0.79021
1.90	0.59562	4.0450	2.5157	1.6079	0.76735
1.95	0.58618	4.2696	2.5919	1.6473	0.74418
2.00	0.57735	4.5000	2.6666	1.6875	0.72088
2.05	0.56907	4.7363	2.7400	1.7286	0.69752
2.10	0.56128	4.9784	2.8119	1.7704	0.67422
2.15	0.55395	5.2262	2.8823	1.8132	0.65105
2.20	0.54706	5.4800	2.9512	1.8569	0.62812
2.25	0.54055	5.7396	3.0186	1.9014	0.60554
2.30	0.53441	6.0050	3.0846	1.9468	0.58331
2.35	0.52861	6.2762	3.1490	1.9931	0.56148
2.40	0.52312	6.5533	3.2119	2.0403	0.54015
2.45	0.51792	6.8362	3.2733	2.0885	0.51932
2.50	0.51299	7.1250	3.3333	2.1375	0.49902
2.55	0.50831	7.4196	3.3918	2.1875	0.47927
2.60	0.50387	7.7200	3.4489	2.2383	0.46012
2.65	0.49965	8.0262	3.5047	2.2901	0.44155
2.70	0.49563	8.3383	3.5590	2.3429	0.42359
2.75	0.49181	8.6562	3.6119	2.3966	0.40622
2.80	0.48817	8.9800	3.6635	2.4512	0.38946
2.85	0.48470	9.3096	3.7139	2.5067	0.37330
2.90	0.48138	9.6450	3.7629	2.5632	0.35773
2.95	0.47821	9.986	3.8106	2.6206	0.34275
3.00	0.47519	10.333	3.8571	2.6790	0.32834
3.50	0.45115	14.125	4.2608	3.3150	0.21295
4.00	0.43496	18.500	4.5714	4.0469	0.13876
4.50	0.42355	23.458	4.8119	4.8761	0.09170
5.00	0.41523	29.000	5.0000	5.8000	0.06172
6.00	0.40416	41.833	5.2683	7.941	0.02965
7.00	0.39736	57.000	5.4444	10.469	0.01535
8.00	0.39289	74.500	5.5652	13.387	0.00849
9.00	0.38980	94.333	5.6512	16.693	0.00496
10.00	0.38757	116.50	5.7413	20.388	0.00304
∞	0.37796	∞	6.000	∞	0

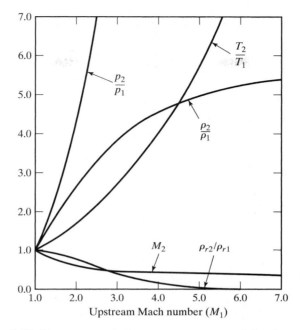

Figure 8.13 Property variations across a normal shock wave.

tion of the deflection angle and of the upstream Mach number. An alternative procedure to calculate the Mach number behind the shock wave would be to convert the value of M_2 in Table 8.3 (which is the normal component of the Mach number) to the normal component of velocity using T_2 to calculate the local speed of sound. Then we can calculate the total velocity downstream of the shock wave:

$$U_2 = \sqrt{u_2^2 + v_2^2}$$

from which we can calculate the downstream Mach number.

For supersonic flow past a cone at zero angle of attack, the shock-wave angle θ_c depends on the upstream Mach number M_1 and the cone half-angle δ_c. Whereas all properties are constant downstream of the weak, oblique shock wave generated when supersonic flow encounters a wedge, this is not the case for the conical shock wave. In this case, properties are constant along rays (identified by angle ω) emanating from the vertex of the cone, as shown in the sketch of Fig. 8.14. Thus, the static pressure varies with distance back from the shock along a line parallel to the cone axis. The shock-wave angle, the pressure coefficient

$$C_p = \frac{p_c - p_1}{q_1}$$

(where p_c is the static pressure along the surface of the cone), and the Mach number of the inviscid flow at the surface of the cone M_c are presented in Fig. 8.15 as a function of the cone semivertex angle δ_c and the free-stream Mach number M_1.

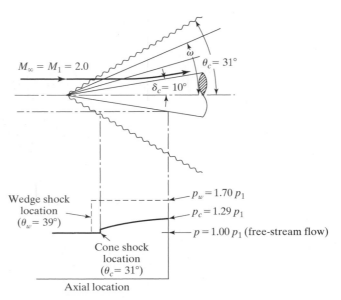

Figure 8.14 Supersonic flow past a sharp cone at zero angle of attack.

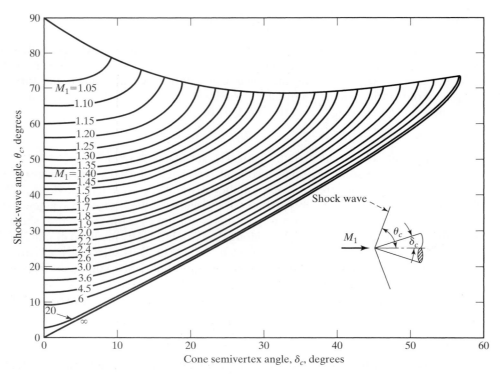

Figure 8.15 Variations of shock-wave parameters with cone semivertex angle for various upstream Mach numbers, $\gamma = 1.4$: (a) shock-wave angle.

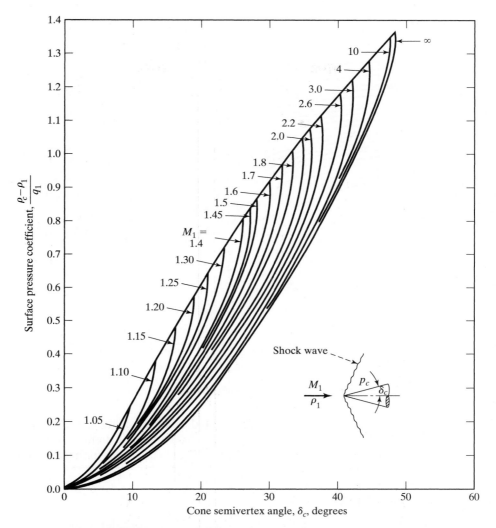

Figure 8.15 (continued) (b) pressure coefficient.

EXAMPLE 8.4: **Supersonic flow past a sharp cone at zero angle-of-attack**

Consider the cone whose semivertex angle is 10° exposed to a Mach 2 stream, as shown in Fig. 8.14. Using Fig. 8.15a, the shock-wave angle is 31°. The pressure just downstream of the shock wave is given by equation (8.72) as

$$\frac{p_2}{p_1} = \frac{2(1.4)(4)(0.5150)^2 - 0.4}{2.4} = 1.07$$

The pressure increases across the shock layer (i.e., moving parallel to the cone axis), reaching a value of $1.29p_1$ (or $1.29p_\infty$) at the surface of the cone, as can be calculated using Fig. 8.15b. Included for comparison is the pressure downstream of the weak, oblique shock for a wedge with the same turning angle. Note

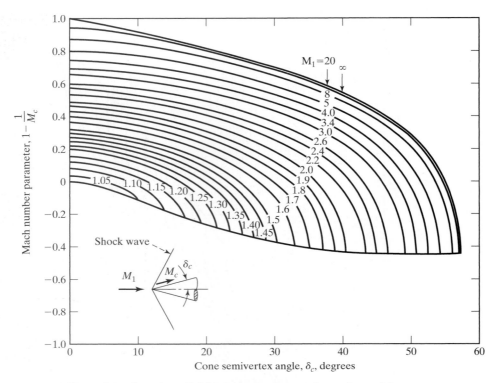

Figure 8.15 (continued) (c) Mach number on the surface of the cone.

that the shock-wave angle θ_w and the pressure in the shock layer p_w are greater for the wedge. The difference is due to axisymmetric effects, which allow the flow to spread around the cone and which do not occur in the case of the wedge.

8.6 VISCOUS BOUNDARY LAYER

In our analysis of boundary layers in Chapter 4, we considered mostly flows for which the density is constant. The correlations for the skin friction coefficient which were developed for these low-speed flows were a function of the Reynolds number only. However, as the free-stream Mach number approaches the transonic regime, shock waves occur at various positions on the configuration, as will be discussed in Chapter 9. The presence of the boundary layer and the resultant shock-wave/boundary-layer interaction can radically alter the flow field. Furthermore, when the free-stream Mach number exceeds two, the work of compression and of viscous energy dissipation produces considerable increases in the static temperature in the boundary layer. Because the temperature-dependent properties, such as the density and the viscosity, are no longer constant, our solution technique must include the energy equation as well as the continuity equation and the momentum equation.

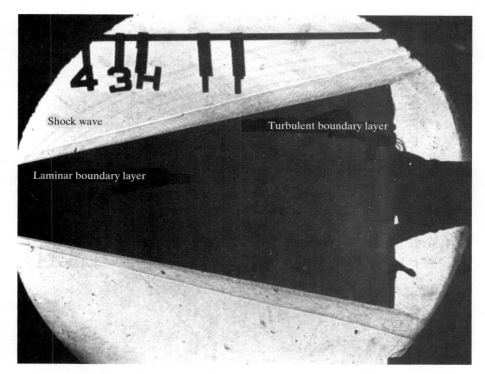

Figure 8.16 Hypersonic flow past a slender cone: $M_\infty = 11.5$; $\mathrm{Re}_{\infty,L} = 4.28 \times 10^6$; $\delta_c = 12°$. (Courtesy of Vought Corp.)

Because the density gradients in the compressible boundary layer affect a parallel beam of light passing through a wind-tunnel test section, we can photographically record the boundary layer. A shadowgraph of the flow field for a cone ($\delta_c = 12°$) in a supersonic, Mach 11.5 stream is presented in Fig. 8.16. Boundary-layer transition occurs approximately one-quarter of the way along the portion of the conical generator which appears in the photograph. Upstream (nearer the apex), the laminar boundary layer is thin and "smooth." Downstream (toward the base of the cone), the vortical character of the turbulent boundary layer is evident. The reader might also note that the shock-wave angle is approximately 14.6°. The theoretical value, as calculated using Fig. 8.15a, is 14.3°. That a slightly larger angle is observed experimentally is due in part to the displacement effect of the boundary layer.

In addition to the calculations of the aerodynamic forces and moments, the fluid dynamicist must address the problem of heat transfer. In this section, we will discuss briefly

1. The effects of compressibility

2. Shock-wave/boundary-layer interactions

8.6.1 Effects of Compressibility

As noted previously, considerable variations in the static temperature occur in the supersonic flow field around a body. We can calculate the maximum temperature that occurs in the flow of a perfect gas by using the energy equation for an adiabatic flow [i.e., equation (8.34)]. This maximum temperature, which is the *stagnation temperature*, is

$$T_t = T_\infty\left(1 + \frac{\gamma - 1}{2}M_\infty^2\right)$$

For example, let us calculate the stagnation temperature for a flow past a vehicle flying at a Mach number of 4.84 and at an altitude of 20 km. Referring to Table 1.2, T_∞ is 216.65 K. Also, since a_∞ is 295.069 m/s, U_∞ is 1428 m/s (or 5141 km/h):

$$T_t = 216.65[1 + 0.2(4.84)^2] = 1231.7\,\text{K} = T_{te}$$

which is the stagnation temperature of the air outside of the boundary layer (where the effects of heat transfer are negligible). Thus, for this flow, we see that the temperature of the air at the stagnation point is sufficiently high that we could not use an aluminum structure. We have calculated the stagnation temperature, which exists only where the flow is at rest relative to the vehicle and where there is no heat transferred from the fluid. However, because heat is transferred from the boundary layer of these flows, the static temperature does not reach this value. A numerical solution for the static temperature distribution across a laminar boundary layer on a flat-plate wing exposed to this flow is presented in Fig. 8.17. Although the maximum value of the static temperature is well below the stagnation temperature, it is greater than either the temperature at the wall or at the edge of the boundary layer. Thus, the designer of vehicles that fly at supersonic speeds must consider problems related to convective heat transfer (i.e., the heat transfer due to fluid motion).

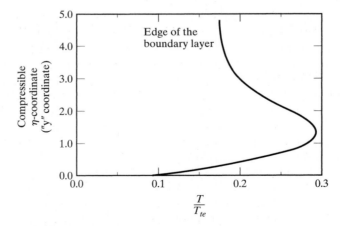

Figure 8.17 Static temperature distribution across a compressible, laminar boundary layer; $M_e = 4.84\ T_w = 0.095\,T_{te}$.

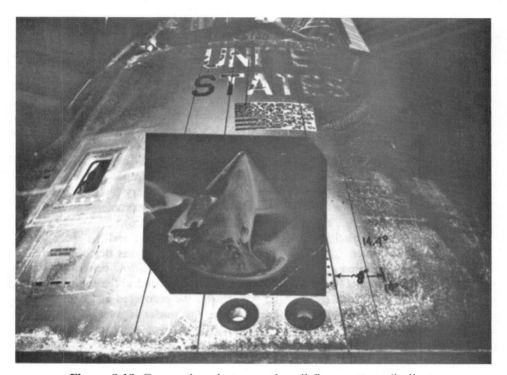

Figure 8.18 Comparison between the oil flow pattern (indicating skin friction) obtained in the wind tunnel and the char patterns on a recovered *Apollo* spacecraft. (Courtesy of NASA.)

The correlation between convective heat transfer and the shear forces acting at the wall, which is known as *Reynolds analogy* and was discussed in Chapter 4, is clearly illustrated in Fig. 8.18. The streaks in the oil-flow pattern obtained in a wind tunnel show that the regions of high shear correspond to the regions of high heating, which is indicated by the char patterns on the recovered spacecraft. For further information about convective heat transfer, the reader is referred to Chapter 4 and to texts such as such as Kays (1966) and Chapman (1974).

For high-speed flow past a flat plate, the skin friction coefficient depends on the local Reynolds number, the Mach number of the inviscid flow, and the temperature ratio, T_w/T_e. Thus,

$$C_f = C_f\left(\text{Re}_x, M_e, \frac{T_w}{T_e}\right) \tag{8.78}$$

Spalding and Chi (1964) developed a calculation procedure based on the assumption that there is a unique relation between $F_c C_f$ and $\text{Re}_x F_{\text{Re}}$, where C_f is the skin friction coefficient, Re_x is the local Reynolds number, and F_c and F_{Re} are correlation parameters which depend only on the Mach number and on the temperature ratio. F_c and F_{Re} are presented as functions of M_e and of T_w/T_e in Tables 8.4 and 8.5, respectively. Thus, given

TABLE 8.4   Values of F_c as a Function of M_e and T_w/T_e

$\dfrac{T_w}{T_e}$	M_e						
	0.0	1.0	2.0	3.0	4.0	5.0	6.0
0.05	0.3743	0.4036	0.4884	0.6222	0.7999	1.0184	1.2759
0.10	0.4331	0.4625	0.5477	0.6829	0.8628	1.0842	1.3451
0.20	0.5236	0.5530	0.6388	0.7756	0.9584	1.1836	1.4491
0.30	0.5989	0.6283	0.7145	0.8523	1.0370	1.2649	1.5337
0.40	0.6662	0.6957	0.7821	0.9208	1.1069	1.3370	1.6083
0.50	0.7286	0.7580	0.8446	0.9839	1.1713	1.4031	1.6767
0.60	0.7873	0.8168	0.9036	1.0434	1.2318	1.4651	1.7405
0.80	0.8972	0.9267	1.0137	1.1544	1.3445	1.5802	1.8589
1.00	1.0000	1.0295	1.1167	1.2581	1.4494	1.6871	1.9684
2.00	1.4571	1.4867	1.5744	1.7176	1.9130	2.1572	2.4472
3.00	1.8660	1.8956	1.9836	2.1278	2.3254	2.5733	2.8687
4.00	2.2500	2.2796	2.3678	2.5126	2.7117	2.9621	3.2611
5.00	2.6180	2.6477	2.7359	2.8812	3.0813	3.3336	3.6355
6.00	2.9747	3.0044	3.0927	3.2384	3.4393	3.6930	3.9971
8.00	3.6642	3.6938	3.7823	3.9284	4.1305	4.3863	4.6937
10.00	4.3311	4.3608	4.4493	4.5958	4.7986	5.0559	5.3657

Source: Spalding and Chi (1964).

TABLE 8.5   Values of F_{Re} as a Function of M_e and T_w/T_e

$\dfrac{T_w}{T_e}$	M_e						
	0.0	1.0	2.0	3.0	4.0	5.0	6.0
0.05	221.0540	232.6437	256.5708	278.2309	292.7413	300.8139	304.3061
0.10	68.9263	73.0824	82.3611	91.2557	97.6992	101.7158	103.9093
0.20	20.4777	22.0029	25.4203	28.9242	31.6618	33.5409	34.7210
0.30	9.8486	10.6532	12.5022	14.4793	16.0970	17.2649	18.0465
0.40	5.7938	6.2960	7.4742	8.7703	9.8686	10.6889	11.2618
0.50	3.8127	4.1588	4.9812	5.9072	6.7120	7.3304	7.7745
0.60	2.6969	2.9499	3.5588	4.2576	4.8783	5.3658	5.7246
0.80	1.5487	1.7015	2.0759	2.5183	2.9247	3.2556	3.5076
1.00	1.0000	1.1023	1.3562	1.6631	1.9526	2.1946	2.3840
2.00	0.2471	0.2748	0.3463	0.4385	0.5326	0.6178	0.6903
3.00	0.1061	0.1185	0.1512	0.1947	0.2410	0.2849	0.3239
4.00	0.0576	0.0645	0.0829	0.1079	0.1352	0.1620	0.1865
5.00	0.0356	0.0400	0.0516	0.0677	0.0856	0.1036	0.1204
6.00	0.0240	0.0269	0.0349	0.0460	0.0586	0.0715	0.0834
8.00	0.0127	0.0143	0.0187	0.0248	0.0320	0.0394	0.0466
10.00	0.0078	0.0087	0.0114	0.0153	0.0198	0.0246	0.0294

TABLE 8.6 Values of $F_c C_f$ as a Function of $F_{Re} Re_x$

$F_c C_f$	$F_{Re} Re_x$	$F_c C_f$	$F_{Re} Re_x$
0.0010	5.758×10^{10}	0.0055	8.697×10^4
0.0015	4.610×10^8	0.0060	5.679×10^4
0.0020	4.651×10^7	0.0065	3.901×10^4
0.0025	9.340×10^6	0.0070	2.796×10^4
0.0030	2.778×10^6	0.0075	2.078×10^4
0.0035	1.062×10^6	0.0080	1.592×10^4
0.0040	4.828×10^5	0.0085	1.251×10^4
0.0045	2.492×10^5	0.0090	1.006×10^4
0.0050	1.417×10^5		

Source: Spalding and Chi (1964).

the flow conditions and the surface temperature, we can calculate F_c and F_{Re}. Then we can calculate the product $F_{Re} Re_x$ and find the corresponding value of $F_c C_f$ in Table 8.6. Since F_c is known, we can solve for C_f.

The ratio of the experimental skin friction coefficient to the incompressible value at the same Reynolds number [as taken from Stalmach (1958)] is presented in Fig. 8.19 as a function of the Mach number. The experimental skin friction coefficients are for adiabatic flows; that is, the surface temperature was such that there was no heat transferred from the fluid to the wall. The experimental values are compared with those given by the Spalding-Chi correlation.

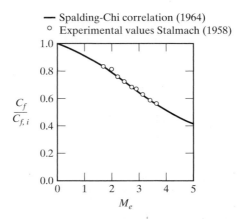

Figure 8.19 Ratio of the compressible, turbulent experimental skin friction coefficient to the incompressible value at the same Reynolds number as a function of Mach number.

EXAMPLE 8.5:   Skin-friction coefficient for a supersonic, turbulent boundary layer

What is the skin-friction coefficient for a turbulent boundary layer on a flat plate, when $M_e = 2.5$, $\mathrm{Re}_x = 6.142 \times 10^6$, and $T_w = 3.0 T_e$?

Solution: For these calculations,

$$F_c = 2.056 \ (\text{see Table 8.4})$$

and

$$F_{\mathrm{Re}} = 0.1729 \ (\text{see Table 8.5})$$

Thus,

$$\mathrm{Re}_x F_{\mathrm{Re}} = 1.062 \times 10^6$$

Using Table 8.6, we obtain

$$F_c C_f = 0.0035$$

so that

$$C_f = 1.70 \times 10^{-3}$$

8.7   THE ROLE OF EXPERIMENTS FOR GENERATING INFORMATION DEFINING THE FLOW FIELD

The tools that are available to the designers of aerospace vehicles include analytical methods, numerical methods, experimental programs using ground-testing facilities, and experimental programs using flight-test programs. Most of the material defining the role of experiments in generating information defining the flow field has been taken from an article written by Bertin and Cummings for the 2006 *Annual Review of Fluid Mechanics*.

8.7.1   Ground-Based Tests

Despite the remarkable advances in hardware and software for computational fluid dynamics (CFD) tools, when two critical "return-to-flight" (RTF) concerns were identified in the aerothermodynamic environment of the Space Shuttle, very extensive wind-tunnel programs were conducted. One RTF concern dealt with the heating in the bipod region during launch. The second concern was related to the aerodynamics forces and moments during reentry. Wind-tunnel tests provided the majority of the aerothermal information required by the "return-to-flight" advisory teams. Computational fluid dynamics (CFD) solutions were used primarily to generate numerical values for use in comparisons with the experimental measurements.

Since there is no single ground-based facility capable of duplicating the high-speed flight environment, different facilities are used to address various aspects of the design problems associated with supersonic flight. Most of the measurements that are used during the design process to define the aerothermodynamic environment (i. e., the aerodynamic forces and moments, the surface pressure distribution, and the heat-transfer distribution) are obtained in the following:

1. Conventional wind-tunnels
2. Shock-heated wind-tunnels
3. Shock tubes
4. Arc-heated test facilities
5. Ballistic, free-flight ranges

The parameters that can be simulated in ground-based test facilities include:

1. Free-stream Mach number
2. Free-stream Reynolds number (and its influence on the character of the boundary layer)
3. Free-stream velocity
4. Pressure altitude
5. Total enthalpy of the flow
6. Density ratio across the shock wave
7. Test gas
8. Wall-to-total temperature ratio
9. Thermochemistry of the flow field

Note that some of the parameters are interrelated (e. g., the free-stream velocity, the total enthalpy of the flow, the free-stream Mach number, and the wall-to-total temperature ratio). The critical heating to the bipod region occurs in the Mach number range from 3.5 to 4.0—just prior to staging at approximately 30,480 m (100,000 ft). Because the total temperature for Mach 4 flow at 30,480 m (100,000 ft) is approximately 950 K (1710°R), one can simultaneously simulate the free-stream velocity, the total enthalpy of the flow, the free-stream Mach number, and the free-stream Reynolds number in conventional wind-tunnel facilities. Two facilities in which one can match these four parameters are the Supersonic Aerothermal Tunnel C at the Arnold Engineering Development Center (AEDC) [Anderson and Matthews (1993)] and the LENS II facility at the Calspan-University at Buffalo Research Center (CUBRC) [Holden, et al. (1995)].

Note also that, during a specific run of a wind-tunnel test program, the model temperature usually starts out at room temperature. The temperature measurements that are used to determine the heat-transfer rates to the model usually are made early during the test run. Thus, the temperature of model surface remains relatively cool during the data-recording portion of the run. On the other hand, the relatively long exposure to the high-temperature flight environment causes the surface of the vehicle to become very hot. As a result, the wall-to-total-temperature ratio in ground-based tests is usually well below the flight value. To match the flight value of the wall-to-total-temperature ratio, the total temperature of the tunnel flow is reduced below the flight value. Thus, even for this relatively benign flow, one does not match the flight values of all of the flow-field parameters.

In addition to the nine flow-field related parameters that were identified earlier in this section, additional factors must be considered when developing a test plan. The additional factors include:

1. Model scale
2. Test time
3. Types of data available
4. Flow quality (including uniformity, noise, cleanliness, and steadiness)

In a statement attributed to J. Leith Potter, Trimmer, et al. (1986) noted that "Aerodynamic modeling is the art of partial simulation." Thus, the test engineer must decide which parameters are critical to accomplishing the objectives of the test program. In fact, during the development of a particular vehicle, the designers most likely will utilize many different facilities with the run schedule, the model, the instrumentation, and the test conditions for each program tailored to answer specific questions. As stated by Matthews, et al. (1984), "A precisely defined test objective coupled with comprehensive pretest planning are essential for a successful test program."

There are many reasons for conducting ground-based test programs. Some objectives include the following:

Objective 1. Obtain data to define the aerodynamic forces and moments and/or the heat-transfer distributions for complete configurations whose complex flow fields resist computational modeling.

Objective 2. Use partial configurations to obtain data defining local-flow phenomena, such as the inlet flow field for hypersonic air-breathing engines or the shock/boundary-layer interactions associated with deflected control surfaces (a body flap).

Objective 3. Obtain detailed flow-field data to be used in developing numerical models for use in a computational algorithm (code validation).

Objective 4. Obtain measurements of parameters, such as the heat transfer and the drag, to be used in comparison with computed flow-field solutions over a range of configuration geometries and of flow conditions (code calibration).

Objective 5. Obtain data that can be used to develop empirical correlations for phenomena that resist analytical and/or numerical modeling, such as boundary-layer transition and turbulence modeling.

Objective 6. Certify the performance of air-breathing engines.

Even today, extensive ground-based test-programs are conducted during the design process in order to define the aerothermodynamic environment for the entire vehicle. When the access to space study that was conducted by NASA in the early 1990s recommended the development of a fully reusable launch vehicle (RLV) [Bekey et al. (2001)], NASA joined an industry-led technology-development effort for the X-33/RLV.

As part of the industry/government partnership, personnel and facilities at the Langley Research Center (NASA) were assigned the task of providing information regarding the aerodynamic forces and moments, the surface heating, and the criteria for boundary-layer transition to Lockheed Martin in support of X-33 development and design.

A special section in the September–October 2001 issue of the *Journal of Spacecraft and Rockets* presented five archival journal articles, documenting the results from this cooperative effort. Two articles presented information relating to objective (1) "Obtain data to define the aerodynamic forces and moments and/or the heat-transfer distributions for complete configurations whose complex flow fields resist computational modeling." They are Horvath, et al. (2001) and Murphy, et al. (2001). Two articles presented information relating to objective (4) "Obtain measurements of parameters, such as the heat transfer and the drag, to be used in comparison with computed flow-field solutions over a range of configuration geometries and of flow conditions (code calibration)." They are Hollis, et al. (2001a) and Hollis, et al. (2001b). One article [Berry, et al. (2001a)] related to objective (5) "Obtain data that can be used to develop empirical correlations for phenomena that resist analytical and/or numerical modeling, such as boundary-layer transition and turbulence modeling."

"In 1996 NASA initiated the Hyper-X program as part of an initiative to mature the technologies associated with hypersonic air-breathing propulsion." [Engelund (2001)]. "The primary goals of the Hyper-X program are to demonstrate and validate the technologies, the experimental techniques, and the computational methods and tools required to design and develop hypersonic aircraft with airframe-integrated, dual-mode scramjet propulsion systems." Although hypersonic air-breathing propulsion systems have been studied in the laboratory environment for over forty years, a complete airframe-integrated vehicle configuration had never been flight tested. Again, personnel and facilities at the Langley Research Center (NASA) were used to define the aerodynamic and surface heating environments, including information relating to the boundary-layer transition criteria, as part of the design, the development, the construction, and the flight-test program, for the X-43.

A special section in the November–December 2001 issue of the *Journal of Spacecraft and Rockets* presented seven archival journal articles summarizing the results from this program. Two articles presented information relating to objective (1) "Obtain data to define the aerodynamic forces and moments and/or the heat-transfer distributions for complete configurations whose complex flow fields resist computational modeling." They are Engelund, et al. (2001) and Holland, et al. (2001). One article presented information relating to objective (4) "Obtain measurements of parameters, such as the heat transfer and the drag, to be used in comparison with computed flow-field solutions over a range of configuration geometries and of flow conditions (code calibration)." That was by Cockrell, et al. (2001). One article [Berry, et al. (2001b)] related to objective (5) "Obtain data that can be used to develop empirical correlations for phenomena that resist analytical and/or numerical modeling, such as boundary-layer transition and turbulence modeling."

As described by Woods, et al. (2001), the Hyper-X Research Vehicle also called (HXRV or free flyer) required a booster to deliver the vehicle to the engine test points. The test conditions for the first flight included $M_\infty = 7$ and $q_\infty = 47,879$ N/m^2 (1000 psf) at an altitude of approximately 28,956 m (95,000 ft). The Hyper-X Launch Vehicle

(HXLV) stack was initially carried aloft under the wing of a B-52. The HXLV was dropped, the Pegasus ignited, and the stack accelerated to the desired test Mach number. When the stack reached the desired test conditions and attitude, a stage-separation sequence of events separated the free flyer from the booster. The free-flying research vehicle then followed a preprogrammed trajectory. Although occurring in less than 500 ms, stage separation was critical to reaching the engine test point and, hence, was critical to mission success.

Buning, et al. (2001) reported that: "Even following the AEDC test, several aerodynamic issues remained in fully understanding the dynamics of the stage separation maneuver. This understanding was complicated by the unsteady nature of the event, the number of degrees of freedom associated with the booster, research vehicle, and control surfaces; and limits in the amount of wind-tunnel data available. These issues were in three basic areas: unsteady effects, aerodynamic database extrapolation, and differences between wind-tunnel and flight conditions." Viscous and inviscid computational fluid dynamics (CFD) techniques were used to quantify unsteady effects, to examine the cause and the extent of interference between the booster and the research vehicle, and to identify differences between the wind-tunnel and the flight environments.

8.7.2 Flight Tests

Flight tests are very expensive. Flight tests take a long time to plan and to execute successfully. Furthermore, it is difficult to obtain quality data at well-defined test conditions. Flight tests will never replace ground-based tests or CFD in the design process. Nevertheless, flight tests are critical to our understanding of the hypersonic aerothermodynamic environment, since they provide data which cannot be obtained elsewhere.

Neumann (1988) suggests that there are a variety of reasons for conducting flight tests. Four reasons were suggested by Buck et al. (1963):

1. To demonstrate interactive technologies and to identify unanticipated problems.
2. To form a catalyst (or a focus) for technology.
3. To gain knowledge not only from the flights but also from the process of development.
4. To demonstrate technology in flight so that it is credible for larger-scale applications.

Neumann added three reasons of his own:

5. To verify ground-test data and/or to understand the bridge between ground-test simulations and actual flight.
6. To validate the overall performance of the system.
7. To generate information not available on the ground.

Williamson (1992) noted: "Due to the large investment in flight testing, it is then desirable to make as many measurements as possible during the flight to help verify predictions or explain any modeling inaccuracies. The instrumentation should measure as directly as possible the things that have been predicted. This is often not possible and it is often necessary to infer predictions from related but not direct measurements."

Neumann (1989) stated: "Heat transfer is a quantity which cannot be directly measured: It is interpreted within the context of a thermal model rather than measured." Thus, one must be able to develop a numerical model that describes the relation between the measured temperature and the sensor's design, in order to obtain a reasonably accurate value of the experimentally-determined heat transfer. However, one also must develop a numerical model depicting how the heat-transfer sensor responds from its location in the vehicle. Serious problems can occur when the heat sensor is not properly integrated into the flight structure such that minimal thermal distortion is produced.

Williamson (1992) notes that "Measurements fall into three groups. These include atmospheric properties measurements, offboard vehicle related sensor measurements, and onboard vehicle related measurements telemetered to the ground or stored on tape and retrieved post-flight."

There are two types of flight-test programs: (1) Research and Development (R and D) programs and (2) flights of prototype or operational vehicles. R and D programs are focused on technical issues which drive the design of the vehicle and its flight operations. Iliff and Shafer (1992) noted that: "In the 1960's, several programs successfully generated aerothermodynamic flight data to improve the understanding and interpretation of theoretical and ground test results. The ASSET and PRIME programs were flown in the early 1960's and provided aerothermodynamic flight data for ablative and metallic thermal protection (TPS) concepts." Project Fire provided calorimeter heating measurements on a large-scale blunt body entering the earth's atmosphere at an initial velocity 11.35 km/s (37.24 kft/s). As discussed by Cornette (1966), the forebody of the "Apollo-like" reentry capsule was constructed of three beryllium calorimeter shields, which were alternated with phenolic-asbestos heat shields. This multiple-layer arrangement provided three distinct time periods when measurements defining the aerothermodynamic environment could be obtained. The FIRE program provided flight data that was used to understand radiative heat-transfer that become important for flight through the earth's atmosphere and when returning from lunar and from planetary missions [Ried, et al. (1972) and Sutton (1985)].

Flight-test data have been obtained on prototype vehicles, [e. g., the flights of the unmanned Apollo Command Modules (017 and 020) as discussed in Lee and Goodrich (1972)] and on operational vehicles [e. g., the Space Shuttle Orbiter, NASA (1995)]. Throckmorton (1993) noted that the concept of using the Shuttle Orbiter as a flight research vehicle as an adjunct to its normal operational mission was a topic of discussion within the research community throughout the 1970s. Aerothermodynamic parameters based on flight-test data obtained from thermocouples embedded in the Space Shuttle Thermal Protection System (TPS) were used to expand the flight envelope for the Orbiter. As noted by Hodge and Audley (1983), "Requirements for the technique include an analytical model for the simulation of the heat transfer to a point on the TPS, flight test maneuvers which cause the thermocouples imbedded near the surface of the TPS to respond sufficiently above the noise levels, and a parameter estimation program to reduce flight thermocouple data. ... The data reduction program correlates heating with the variables such as angle of attack, sideslip, control surface deflection, and Reynolds number. This technique could also be used for wind tunnel data reduction."

The interactions between impinging shock waves and the bow shock wave can produce locally severe heat-transfer rates. Edney (1968) identified six different shock/shock interactions patterns. The positive deflection of the Orbiter body flap creates a shock wave which interacts with the boundary layer on the Orbiter. The adverse pressure gradient produced by the shock wave causes the boundary layer to thicken and (in many cases) to separate. The upstream extent of the interaction-produced perturbations depends on the size of the subsonic portion of the approach boundary layer and on the strength of the shock wave produced by the turning of the flow. Thus, as noted by Bertin (1994), the parameters that influence the extent of an interaction are (1) whether the approach boundary layer is laminar or turbulent, (2) the Mach number of the approach flow, (3) the Reynolds number of the approach flow, (4) the surface temperature, (5) the deflection angle of the ramp, and (6) the chemical state of the gas. Bertin et al. (1996) and Fujii et al. (2001) investigated the perturbed flow fields due to deflected control surfaces using the experimental values of the heat transfer measured during hypersonic flight. The HYFLEX vehicle [Fujii et al. (2001)] was a small-scale R and D vehicle whose length was 4.40 m (14.44 ft). A two-stage launcher provided an initial altitude of 107 km (351 kft) at a velocity of 3.88 km/s (12.73 kft/s).

Despite the oft-mentioned problems with developing boundary-layer transition correlations using flight-test measurements, there are numerous flight-test programs where boundary-layer transition was a major focus of the data gathering efforts. Weston and Fitzkee (1963) noted that observed boundary-layer transition Reynolds numbers on the Mercury capsule "agree well with Reynolds numbers obtained for wake transition behind spheres flown in a hypervelocity gun facility." Also flown in the 1960's was the Reentry F vehicle which provided hypersonic boundary-layer transition and is still used today. The Reentry F flight provided boundary-layer transition data at Mach numbers up to 20 and altitudes down to 24.38 km (80.00 kft) [Wright and Zoby (1977)].

EXAMPLE 8.6:   Wind-tunnel simulation of supersonic missile flow fields

Assume that you are given the task of determining the aerodynamic forces and moments acting on a slender missile which flies at a Mach number of 3.5 at an altitude of 27,432 m (90,000 ft). Aerodynamic coefficients for the missile, which is 20.0 cm (7.874 in) in diameter and 10 diameters long, are required for angles of attack from 0° to 55°. The decision is made to obtain experimental values of the required coefficients in the Vought High-Speed Wind Tunnel (in Dallas, Texas). Upstream of the model shock system, the flow in the wind tunnel is isentropic and air, at these conditions, behaves as a perfect gas. Thus, the relations developed in this section can be used to calculate the wind-tunnel test conditions.

Solution:

1. *Flight conditions.*   Using the properties for a standard atmosphere (such as were presented in Chapter 1), the relevant parameters for the flight condition include

$$U_\infty = 1050 \text{ m/s}$$

$$p_\infty = 1.7379 \times 10^{-2} p_{\text{SL}} = 1760.9 \text{ N/m}^2$$

$$T_\infty = 224 \text{ K}$$

$$\text{Re}_{\infty,d} = \frac{\rho_\infty U_\infty d}{\mu_\infty} = 3.936 \times 10^5$$

$$M_\infty = 3.5$$

2. *Wind-tunnel conditions.*   Information about the operational characteristics of the Vought High-Speed Wind Tunnel is contained in the tunnel hand-book [Arnold (1968)]. To ensure that the model is not so large that its presence alters the flow in the tunnel (i.e., the model dimensions are within the allowable blockage area), the diameter of the wind-tunnel model, d_{wt}, will be 4.183 cm (1.6468 in.).

Based on the discussion in Chapter 2, the Mach number and the Reynolds number are two parameters which we should try to simulate in the wind tunnel. The free-stream unit Reynolds number (U_∞/ν_∞) is presented as a function of the free-stream Mach number and of the stagnation pressure for a stagnation temperature of 311 K (100°F) in Fig. 8.20 (which has been taken from Arnold (1968) and has been left in English units). The student can use the equations of this section to verify the value for the unit Reynolds number given the conditions in the stagnation chamber and the free-stream Mach number. In order to match the flight Reynolds number of 3.936×10^5 in the wind tunnel,

$$\left(\frac{U_\infty}{\nu_\infty} \right)_{\text{wt}} = \frac{3.936 \times 10^5}{d_{\text{wt}}} = 2.868 \times 10^6/\text{ft}$$

But as indicated in Fig. 8.20, the lowest unit Reynolds number possible in this tunnel at $M_\infty = 3.5$ is approximately $9.0 \times 10^6/\text{ft}$. Thus, if the model is 4.183 cm in diameter, the lowest possible tunnel value of $\text{Re}_{\infty,d}$ is 1.235×10^6, which is greater than the flight value. This is much different than the typical subsonic flow, where (as discussed in Chapter 5) the maximum wind-tunnel Reynolds number is usually much less than the flight value. To obtain the appropriate Reynolds number for the current supersonic flow, we can choose to use a smaller model. Using a model that is 1.333 cm in diameter would yield a Reynolds number of 3.936×10^5 (as desired). If this model is too small, we cannot establish a tunnel condition which matches both the flight Mach number and the Reynolds number. In this case, the authors would choose to simulate the Mach number exactly rather than seek a compromise Mach-number/Reynolds-number test condition, since the pressure coefficients and the shock interaction phenomena on the control surfaces are Mach-number-dependent in this range of free-stream Mach number.

The conditions in the stagnation chamber of the tunnel are $T_t = 311$ K (560°R) and $p_{t1} = 5.516 \times 10^5$ N/m² (80 psia). Thus, using either equations (8.34) and (8.36) or the values presented in Table 8.1, one finds that $T_\infty = 90.18$ K (162.32°R) and $p_\infty = 7.231 \times 10^3$ N/m² (1.049 psia). The cold

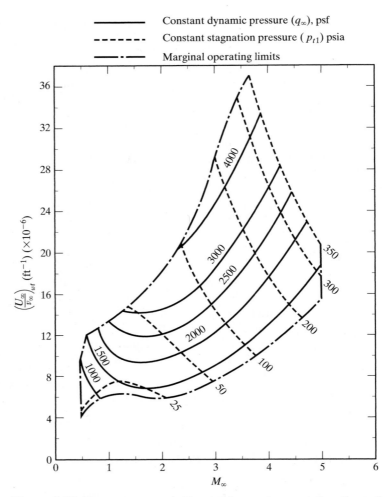

Figure 8.20 Free-stream unit Reynolds number as a function of the free-stream Mach number for $T_t = 100°F$ in the Vought High-Speed Wind Tunnel. [From Arnold (1968).]

free-stream temperature is typical of supersonic tunnels (e.g., most high-speed wind tunnels operate at temperatures near liquefaction of oxygen). Thus, the free-stream speed of sound is relatively low and, even though the free-stream Mach number is 3.5, the velocity in the test section U_∞ is only 665 m/s (2185 ft/s). In summary, the relevant parameters for the wind-tunnel conditions include

$$U_\infty = 665 \text{ m/s}$$

$$p_\infty = 7.231 \times 10^3 \text{ N/m}^2$$

$$T_\infty = 90.18 \text{ K}$$

$$\mathrm{Re}_{\infty,d} = 3.936 \times 10^5 \text{ if } d = 1.333 \text{ cm or } 1.235 \times 10^6 \text{ if } d = 4.183 \text{ cm}$$

$$M_\infty = 3.5$$

Because of the significant differences in the dimensional values of the flow parameters (such as U_∞, p_∞, and T_∞), we must again nondimensionalize the parameters so that correlations of wind-tunnel measurements can be related to the theoretical solutions or to the design flight conditions.

8.8 COMMENTS ABOUT THE SCALING/CORRECTION PROCESS(ES) FOR RELATIVELY CLEAN CRUISE CONFIGURATIONS

In an article for the *Annual Review of Fluid Mechanics*, Bushnell (2006) noted that "a survey of 12 commercial transport aircraft, constructed by three major American manufacturers over a 20-year period showed that you are just as likely to estimate too high a drag as too low a drag. Six of the twelve (scaling) 'predictions' were low, four predictions were high, and two were right on. The drag predictions were as much as 22% low and 10% high." Citing the work of other researchers, Bushnell (2006) noted that similar values/discrepancies were obtained for airbreathing missiles. "As the experience from the various applications discussed herein indicates, 'surprise' occurs far too often and flow complexity can seriously degrade the 'scaled predictions'. Typical correction levels .... include 6% wall interference, −5% Reynolds number, +2% roughness, and −4% sting and aeroelastic distortion effects. Total correction(s) from just these issues/concerns/effects are the order of +12%. On the other hand, the Viking aeroshell drag observed during Martian entry agreed within some 2% of the scaled ground results, admittedly for a very bluff body with a fairly well characterized flow field."

8.9 SHOCK-WAVE/BOUNDARY-LAYER INTERACTIONS

Severe problems of locally high heating or premature boundary-layer separation may result due to viscous/inviscid interactions which occur during flight at supersonic Mach numbers. The shock wave generated by a deflected flap will interact with the upstream boundary layer. The interaction will generally cause the upstream boundary layer to separate with locally high heating rates occurring when the flow reattaches. The extent of the separation, which can cause a loss of control effectiveness, depends on the character of the upstream boundary-layer. Other viscous interaction problems can occur when the shock waves generated by the forebody and other external components impinge on downstream surfaces of the vehicle. Again, locally severe heating rates or boundary-layer separation may occur.

The basic features of the interaction between a shock wave and a laminar boundary layer for a two-dimensional flow are shown in Fig. 8.21a. The pressure rise induced by the shock wave is propagated upstream through the subsonic portion of the boundary layer. Recall that pressure disturbances can affect the upstream flow only if the flow is subsonic. As a result, the boundary-layer thickness increases and the momentum decreases. The thickening boundary layer deflects the external stream and creates a series of compression

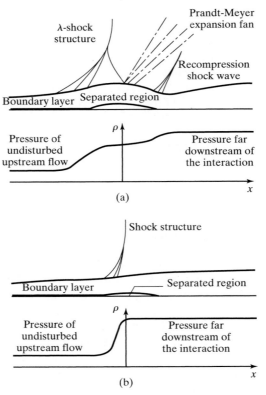

Figure 8.21 Flow field for shock-wave boundary-layer interaction: (a) laminar boundary-layer; (b) turbulent boundary-layer.

waves to form a λ-like shock structure. If the shock-induced adverse-pressure-gradient is great enough, the skin friction will be reduced to zero and the boundary layer will separate. The subsequent behavior of the flow is a strong function of the geometry. For a flat plate, the flow reattaches at some distance downstream. In the case of a convex body, such as an airfoil, the flow may or may not reattach, depending upon the body geometry, the characteristics of the boundary layer, and the strength of the shock wave.

   If the flow reattaches, a Prandtl-Meyer expansion fan results as the flow turns back toward the surface. As the flow reattaches and turns parallel to the plate, a second shock wave (termed the *reattachment shock*) is formed. Immediately downstream of reattachment, the boundary-layer thickness reaches a minimum. It is in this region that the maximum heating rates occur.

   In the case of a shock interaction with a turbulent boundary-layer (Fig. 8.21b), the length of the interaction is considerably shorter than the interaction length for a laminar boundary layer. This results because the air particles near the wall of a turbulent boundary-layer have greater momentum than do those near the wall in a laminar boundary-layer and can therefore overcome a greater adverse pressure gradient. Furthermore, since the subsonic portion of a turbulent boundary layer is relatively thin, the region through which the shock-induced pressure rise can propagate upstream is limited. As a result, a much greater pressure rise is required to cause a turbulent boundary layer to separate.

Pressure distributions typical of the shock-wave/boundary-layer interactions are presented in Fig. 8.21 for a laminar boundary layer and for a turbulent one. As one would expect from the preceding description of the shock interaction, the pressure rise is spread over a much longer distance when the boundary layer is laminar.

PROBLEMS

8.1. The test section of the wind tunnel of Example 8.6 has a square cross section that is 1.22 m × 1.22 m.
   (a) What is the mass-flow rate of air through the test section for the flow conditions of the example (i.e., $M_\infty = 3.5$, $p_\infty = 7.231 \times 10^3 \, \text{N/m}^2$, and $T_\infty = 90.18 \, \text{K}$)?
   (b) What is the volume flow rate of air through the test section?

8.2. A practical limit for the stagnation temperature of a continuously operating supersonic wind tunnel is 1000 K. With this value for the stagnation (or total) temperature, prepare a graph of the static free-stream temperature as a function of the test-section Mach number. The fluid in the free stream undergoes an isentropic expansion. The static temperature should not be allowed to drop below 50 K, since at low pressure oxygen begins to liquefy at this temperature. With this as a lower bound for temperature, what is the maximum Mach number for the facility? Assume that the air behaves as a perfect gas with $\gamma = 1.4$.

8.3. Tunnel B at the Arnold Enginering Development Center (AEDC) in Tennessee is often used for determining the flow field and/or the heating rate distributions. The Mach number in the test section is 8. If the stagnation temperature is 725 K and the stagnation pressure can be varied from $6.90 \times 10^5 \, \text{N/m}^2$ to $5.90 \times 10^6 \, \text{N/m}^2$, what is the range of Reynolds number that can be obtained in this facility? Assume that the characteristic dimension is 0.75 m, which could be the length of a hypersonic waverider model.

8.4. Given the flow conditions discussed in Problem 8.3,
   (a) What is the (range of the) static temperature in the test section? If there is only one value of the static temperature, state why. To what range of altitude (as given in Table 1.2) if any, do these temperatures correspond?
   (b) What is the (range of the) velocity in the test section? Does it depend on the pressure?
   (c) What is the range of static pressure in the test section? To what range of altitude (as given in Table 1.2) do these pressures correspond?

8.5. A convergent-only nozzle, which exhausts into a large tank, is used as a transonic wind tunnel (Fig. P8.5). Assuming that the air behaves as a perfect gas, answer the following.

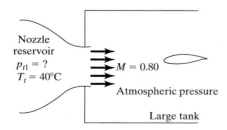

Figure P8.5

(a) If the pressure in the tank is atmospheric (i.e., 1.01325×10^5 N/m²), what should the stagnation pressure in the nozzle reservoir be so that the Mach number of the exhaust flow is 0.80?

(b) If the stagnation temperature is 40°C, what is the static temperature in the test stream?

(c) A transonic airfoil with a 30-cm chord is located in the test stream. What is Re_c for the airfoil?

$$Re_c = \frac{\rho_\infty U_\infty c}{\mu_\infty}$$

(d) What is the pressure coefficient C_p at the stagnation point of the airfoil?

8.6. A small hole exists in the body of an airplane and serves as a convergent nozzle, as shown in Fig. P8.6. The air in the cabin is at 0.50×10^5 N/m². Assume that the cabin volume is sufficiently large that the cabin serves essentially as a stagnation chamber and that the conditions in the cabin remain constant for a period of time, independent of the conditions outside the airplane. Furthermore, assume that the flow in the nozzle is isentropic. Sketch the pressure distribution and calculate the static pressure in the nozzle exit plane when the airplane is at the following altitudes: (a) 6 km, (b) 8 km, (c) 10 km, and (d) 12 km.

Since the air expands isentropically from constant reservoir conditions, the back pressure (i.e., the pressure outside the airplane), p_b, is an important parameter. For the standard atmosphere:

(a) $p_b = 0.47218 \times 10^5$ N/m² $= 0.94435 p_{t1}$

(b) $p_b = 0.35652 \times 10^5$ N/m² $= 0.71303 p_{t1}$

(c) $p_b = 0.26500 \times 10^5$ N/m² $= 0.53000 p_{t1}$

(d) $p_b = 0.19399 \times 10^5$ N/m² $= 0.38799 p_{t1}$

Where $p_{t1} = p_c = 0.5 \times 10^5$ N/m².

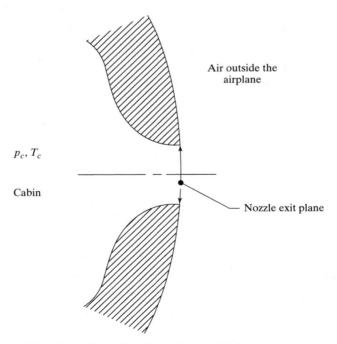

Air outside the
airplane

p_c, T_c

Cabin

Nozzle exit plane

Figure P8.6 Sketch for Problems 8.6 and 8.7

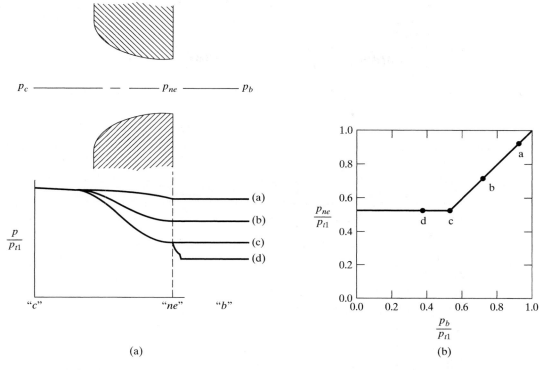

Figure P8.6 (continued) Solution for Problem 8.6 (i) pressure distribution: (ii) pressure ratios.

8.7. If the temperature in the cabin is 22°C and the exit diameter is 0.75 cm, what is the mass flow rate through the hole of Problem 8.6 when the altitude is 6 km? 12 km?

8.8. Consider the flow of air through the convergent-divergent nozzle shown in Fig. P8.8. The conditions in the stagnation chamber are $p_{t1} = 100$ psia and $T_t = 200°F$. The cross-sectional area of the test section is 2.035 times the throat area. The pressure in the test section can be varied by controlling the valve to the vacuum tank. Assuming isentropic flow in the nozzle, calculate the static pressure, static temperature, Mach number, and velocity in the test section for the following back pressures.

(a) $p_b = 100.000$ psia
(b) $p_b = 97.250$ psia
(c) $p_b = 93.947$ psia
(d) $p_b = 9.117$ psia

Let us first convert the stagnation temperature to °R. $T_t = 659.6°R$. Since the back pressure is equal to the stagnation pressure for condition (a), there is no flow. Thus, for (a).

$$p_1 = 100.000 \text{ psia}, \quad T_1 = 659.6°R, \quad M_1 = 0, \quad U_1 = 0$$

8.9. Consider the flow of air through the convergent-divergent nozzle shown in Fig. P8.9. The conditions in the stagnation chamber are $p_{t1} = 100$ psia and $T_t = 200°F$. The cross-sectional area of the test section is 2.035 times the throat area. Thus far we have repeated the conditions of Problem 8.8. Calculate the static pressure, static temperature, Mach number, and velocity in the test section for the following back pressures:

(a) $p_b = 51.38$ psia
(b) $p_b = 75.86$ psia

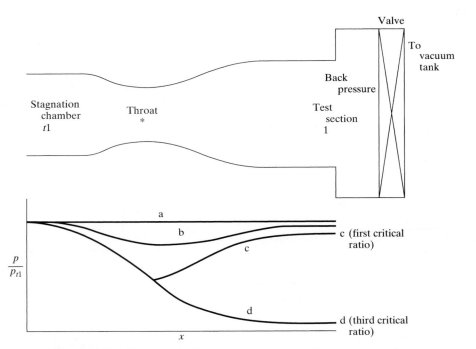

Figure P8.8 Isentropic flow in a convergent-divergent nozzle.

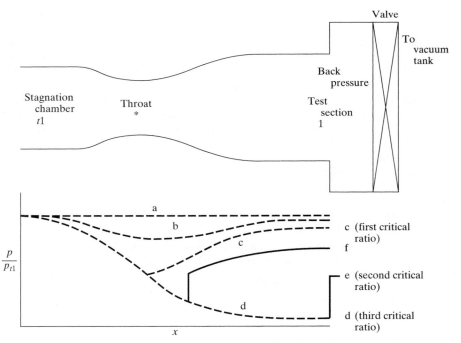

Figure P8.9 Flow in a convergent-divergent nozzle with a shock wave in the divergent section (dashed lines are from Figure P8.7)

8.10. Air flows through the insulated variable-area streamtube such that it may be considered one dimensional and steady. At one end of the streamtube, $A_1 = 5.0\,\text{ft}^2$ and $M_1 = 3.0$. At the other end of the streamtube, $p_{t2} = 2116\,\text{psfa}$, $p_2 = 2101\,\text{psfa}$, $T_{t2} = 500°\text{R}$, and $A_2 = 5.0\,\text{ft}^2$. What is the flow direction; that is, is the flow from (1) to (2) or from (2) to (1)?

8.11. A pitot tube in a supersonic stream produces a curved shock wave standing in front of the nose part, as shown in Fig. P8.11. Assume that the probe is at zero angle of attack and that the shock wave is normal in the vicinity of the nose. The probe is designed to sense the stagnation pressure behind a normal shock (p_{t2}) and the static pressure behind a normal shock (p_2). Derive an expression for M_∞ in terms of the stagnation (p_{t2}) and static (p_2) pressures sensed by the probe.

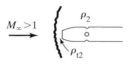

Figure P8.11

8.12. Consider the flow in a streamtube as it crosses a normal shock wave (Fig. P8.12).
(a) Determine the ratio A_2^*/A_1^*.
(b) What are the limits of A_2^*/A_1^* as $M_1 \to 1$ and as $M_1 \to \infty$?
(c) What is the significance of A_1^*? of A_2^*??

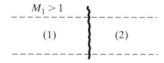

Figure P8.12

8.13. You are to measure the surface pressure on simple models in a supersonic wind tunnel. The air flows from right to left. To evaluate the experimental accuracy, it is necessary to obtain theoretical pressures for comparison with the data. If a 30° wedge is to be placed in a Mach 3.5 stream (Fig. P8.13), calculate
(a) The surface pressure in N/m^2
(b) The pressure difference (in cm Hg) between the columns of mercury in U-tube manometer between the pressure experienced by the surface orifice and the wall orifice (which is used to measure the static pressure in the test section)
(c) The dynamic pressure of the free-stream flow
Other measurements are that the pressure in the reservoir is $6.0 \times 10^5\,\text{N/m}^2$ and the barometric pressure is 75.2 cm Hg.

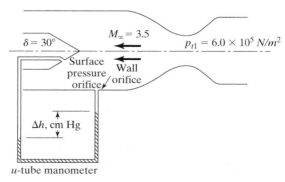

Figure P8.13

8.14. It is desired to turn a uniform stream of air compressively by $10°$. The upstream Mach number is 3.
  (a) Determine the final Mach number and net change in entropy, if the turning is accomplished by (1) a single $10°$ sharp turn, (2) two successive $5°$ sharp turns, or (3) an infinite number of infinitesimal turns.
  (b) What do you conclude from the results in part (a)?
  (c) In light of the results in part (b), can we make any conclusions as to whether it is better to make *expansive* turns gradually or abruptly, when the flow is supersonic?

8.15. A flat-plate airfoil, whose length is c, is in a Mach 2.0 stream at an angle of attack of $10°$ (Fig. P8.15).
  (a) Use the oblique shock-wave relations to calculate the static pressure in region (2) in terms of the free-stream value p_1.
  (b) Use the Prandtl-Meyer relations to calculate the static pressure in region (3) in terms of p_1.
  (c) Calculate C_l, C_d, and $C_{m0.5c}$ (the pitching moment about the midchord). Do these coefficients depend on the free-stream pressure (i.e., the altitude)?

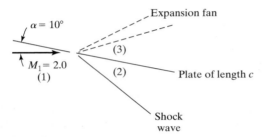

Figure P8.15

8.16. A conical spike whose half-angle (δ_c) is $8°$ is located in the inlet of a turbojet engine (Fig. P8.16). The engine is operating in a $M_\infty = 2.80$ stream such that the angle of attack of the spike is zero. If the radius of the engine inlet is 1.0 m, determine l, the length of the spike extension, such that the conical shock just grazes the lip of the nacelle.

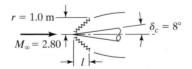

Figure P8.16

8.17. Consider the two-dimensional inlet for a turbojet engine. The upper surface deflects the flow $10°$, the lower surface deflects the flow $5°$ (Fig. P8.17). The free-stream Mach number is 2.5 (i.e., $M_1 = 2.5$) and the pressure p_1 is $5.0 \times 10^3 \, \text{N/m}^2$. Calculate the static pressure,

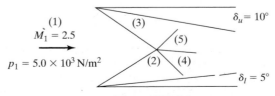

Figure P8.17

the Mach number, and the flow direction in regions (2), (3), (4), and (5). Note that since regions (4) and (5) are divided by a fluid/fluid interface,

$$p_4 = p_5 \quad \text{and} \quad \theta_4 = \theta_5$$

That is, the static pressure in region (4) is equal to that in region (5), and the flow direction in region (4) is equal to that in region (5).

8.18. A single wedge airfoil is located on the centerline of the test section of a Mach 2.0 wind tunnel (Fig. P8.18). The airfoil, which has a half-angle δ of 5°, is at zero angle of attack. When the weak, oblique shock wave generated at the leading edge of the airfoil encounters the wall, it is reflected so that the flow in region (3) is parallel to the tunnel wall. If the test section is 30.0 cm high, what is the maximum chord length (c) of the airfoil so that it is not struck by the reflected shock wave? Neglect the effects of shock-wave/boundary-layer interactions at the wall.

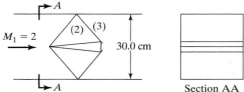

Section AA   **Figure P8.18**

8.19. An airplane flies 600 mi/h at an altitude of 35,000 ft where the temperature is −75° and the ambient pressure is 474 psfa. What is the temperature and the pressure of the air (outside the boundary layer) at the nose (stagnation point) of the airplane? What is the Mach number for the airplane?

8.20. A three-dimensional bow shock wave is generated by the Shuttle *Orbiter* during entry. However, there is a region where the shock wave is essentially normal to the free-stream flow, as illustrated in Fig. 8.10. The velocity of the Shuttle is 7.62 km/s at an altitude of 75.0 km. The free-stream temperature and the static pressure at this altitude are 200.15 K and 2.50 N/m², respectively. Use the normal shock relations to calculate the values of the following parameters downstream of the normal shock wave:

p_2: static pressure

p_{t2}: stagnation pressure

T_2: Static temperature

T_{t2}: stagnation temperature

M_2: Mach number

What is the pressure coefficient at the stagnation point; that is,

$$C_{p,t2} = \frac{p_{t2} - p_\infty}{q_\infty} = ?$$

Use the perfect-gas relations and assume that $\gamma = 1.4$. The perfect-gas assumption is valid for the free-stream flow. However, it is not a good assumption for calculating the flow across the shock wave. Therefore, many of the perfect-gas theoretical values for the shock-flow properties will not even be close to the actual values. See Chapter 12 for further discussions of this problem.

8.21. Consider the hypersonic flow past the cone shown in Fig. 8.16. The cone semivertex angle (δ_c) is 12°. The free-stream flow is defined by

$$M_\infty = 11.5$$
$$T_t = 1970 \text{ K}$$
$$p_\infty = 1070 \text{ N/m}^2$$

Assume that the flow is inviscid, except for the shock wave itself (i.e., neglect the boundary layer on the cone). Assume further that the gas obeys the perfect-gas laws with $\gamma = 1.4$. Using Fig. 8.15b, calculate the static pressure at the surface of the cone (p_c). Using Fig. 8.15c, calculate the Mach number at the surface of the cone (M_c), which is, in practice, the Mach number at the edge of the boundary layer. Calculate the stagnation pressure of the flow downstream of the shock wave. Calculate the Reynolds number at a point 10.0 cm from the apex of the cone.

8.22. Repeat Problem 8.21 for a planar symmetric wedge that deflects the flow 12° (i.e., $\delta = 12°$ in Fig. 8.12).

8.23. An explosion generates a shock wave that moves through the atmosphere at 1000 m/s. The atmospheric conditions ahead of the shock wave are those of the standard sea-level atmosphere. What are the static pressure, static temperature, and velocity of the air behind the shock wave?

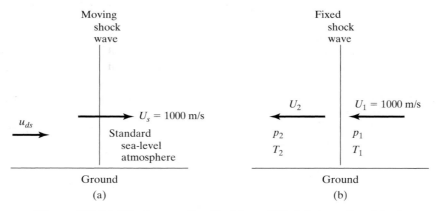

Figure P8.23 Blast wave for Problem 8.23 (a) traveling blast wave; (b) transformed steady flow.

REFERENCES

Ames Research Center Staff. 1953. Equations, tables, and charts for compressible flow. *NACA Report 1135*

Anderson A, Matthews RK. 1993. Aerodynamic and Aerothermal Facilities, I Hypersonic Wind Tunnels. In *Methodology of Hypersonic Testing*, VKI/AEDC Special Course, Rhode-Saint-Genèse, Belgium

Arnold JW. 1968. High Speed Wind Tunnel Handbook. Vought Aeronautics Division AER-EIR-13552-B

Bekey I, Powell R, Austin R. 2001. NASA access to space. *Aerospace America* 32(5):38–43

Berry SA, Horvath TJ, Hollis BR, Thompson RA, Hamilton HH. 2001a. X-33 hypersonic boundary-layer transition. *J. Spacecr. Rockets* 38:646–657

Berry SA, Auslender H., Dilley AD, Calleja JF. 2001b. Hypersonic boundary-layer trip development for Hyper-X. *J. Spacecr. Rockets* 38:853–864

Bertin JJ. 1994. *Hypersonic Aerothermodynamics*. Washington, DC: AIAA

Bertin JJ, Bouslog SA, Wang KC, Campbell CH. 1996. Recent aerothermodynamic flight measurements during Shuttle Orbiter re-entry. *J. Spacecr. Rockets* 33:457–462

Bertin JJ, Cummings RM. 2006. Critical hypersonic aerothermodynamic phenomena. *Annu. Rev. Fluid Mech.* 38:129–157

Buck ML, Benson BR, Sieron TR, Neumann RD. 1963. Aerodynamic and performance analyses of a superorbital re-entry vehicle. In *Dynamics of Manned Lifting Planetary Entry*, Ed. Scala SM, Harrison AC, Rogers M. New York: John Wiley

Buning PG, Wong TC, Dilley AD, Pao JL. 2001. Computational fluid dynamics prediction of Hyper-X stage separation aerodynamics. *J. Spacecr. Rockets* 38:820–827

Burcham FW, Nugent J. 1970. Local flow field around a pylon-mounted dummy ramjet engine on the X-15-2 airplane for Mach numbers from 2.0 to 6.7. *NASA Tech. Note D-5638*

Bushnell DM. 2006. Scaling: wind tunnel to flight. *Annu. Rev. Fluid Mech.* 36: 111–128

Chapman AJ. 1974. *Heat Transfer*. New York: Macmillan

Cockrell CE, Engelund WC, Bittner RD, Jentink TN, Dilley AD, Frendi, A. 2001. Integrated aeropropulsive computational fluid dynamics methodology for the Hyper-X flight experiment. *J. Spacecr. Rockets* 38:836–843

Cornette ES. 1966. Forebody temperatures and calorimeter heating rates measured during project Fire II reentry at 11.35 kilometers per second. *NASA TMX-1035*

Edney BE. 1968. Effects of shock impingement on the heat transfer around blunt bodies. *AIAA J.* 6:15–21.

Engelund WC. 2001. Hyper-X aerodynamics: The X-43A airframe-integrated scramjet propulsion flight-test experiments. *J. Spacecr. Rockets* 38:801–802

Engelund WC, Holland SD, Cockrell CE, Bittner RD. 2001. Aerodynamic database development for the Hyper-X airframe-integrated scramjet propulsion experiments. *J. Spacecr. Rockets* 38:803–810

Fujii K, Watanabe S, Kurotaki T, Shirozu M. 2001. Aerodynamic heating measurements on nose and elevon of hypersonic flight experiment vehicle. *J. Spacecr. Rockets* 38:8–14

Hayes WD, Probstein RF. 1966. *Hypersonic Flow Theory, Vol. I, Inviscid Flows*. New York: Academic Press

Hodge JK, Audley SR. 1983. Aerothermodynamic parameter estimation from shuttle thermocouple data during flight test maneuvers. *J. Spacecr. Rockets* 20:453–460

Holden M, Chadwick K, Kolly J. 1995. *Hypervelocity studies in the LENS facility*. Presented at Intl. Aerosp. Planes Hypers. Tech. Conf., 6th, AIAA Pap. 95–6040, Chattanooga, TN

Holland SD, Woods WC, Engelund WC. 2001. Hyper-X research vehicle experimental aerodynamics Ttest program overview. *J. Spacecr. Rockets* 38:828–835

Hollis BR, Horvath TJ, Berry SA, Hamilton HH, Thompson RA, Alter SJ. 2001a. X-33 computational aeroheating predictions and comparisons with experimental data. *J. Spacecr. Rockets* 38:658–669

Hollis BR, Thompson RA, Murphy KJ, Nowak RJ, Riley CJ, Wood WA, Alter SJ, Prabhu R. 2001b. X-33 aerodynamic computations and comparisons with wind-tunnel data. *J. Spacecr. Rockets* 38:684–691

Horvath TJ, Berry SA, Hollis BR, Liechty DS, Hamilton HH, Merski NR. 2001. X-33 experimental aeroheating at Mach 6 using phosphor thermography. *J. Spacecr. Rockets* 38:634–645

Iliff KW, Shafer MF. 1992. Space Shuttle hypersonic flight research and the comparison to ground test results. Presented at Aerosp. Ground Test. Conf., 17th, AIAA Pap. 92–3988, Nashville, TN,

Kays WM. 1966. *Convective Heat and Mass Transfer*. New York: McGraw-Hill

Lee DB, Goodrich WD 1972. The aerothermodynamic environment of the Apollo Command Module during suborbital entry. *NASA Tech. Note D-6792*

Matthews RK, Nutt KW, Wannenwetsch GD, Kidd CT, Boudreau AH. 1985. Developments in aerothermal test techniques at the AEDC supersonic-hypersonic wind tunnels. Presented at Thermophysics Conf., 20th, AIAA Pap. 85–1003, Williamsburg, VA

Moeckel WE, Weston KC. 1958. Composition and thermodynamic properties of air in chemical equilibrium. *NACA Tech. Note 4265*

Murphy KJ, Nowak RJ, Thompson RA, Hollis BR, Prabhu R. 2001. X-33 hypersonic aerodynamic characteristics. *J. Spacecr. Rockets* 38:670–683

Neumann RD. 1988. Missions and requirements. AGARD Report 761, *Special Course on Aerothermodynamics of Hypersonic Vehicles*, Neuilly sur Seine, France

Neumann RD. 1989. Defining the aerothermodynamic methodology. In *Hypersonics, Volume I Defining the Hypersonic Environment*, Ed. Bertin JJ, Glowinski R, Periaux J, Boston: Birkhaeuser Boston

Ried RC, Rochelle WC, Milhoan JD. 1972. Radiative heating to the Apollo Command Module: engineering prediction and flight measurements. *NASA TMX-58091*

Spalding DB, Chi SW. 1964. The drag of a compressible turbulent boundary layer on a smooth plate with and without heat transfer. *J. Fluid Mech.* 18:117–143

Stalmach CJ. 1958. Experimental investigation of the surface impact pressure probe method of measuring local skin friction at supersonic speeds. Defense Research Laboratory, The University of Texas at Austin, DRL-410

Sutton K. 1985. Air radiation revisited. In *Progress in Astronautics and Aeronautics Vol. 96: Thermal Design of Aeroassisted Orbiital Transfer Vehicles*, Ed. Nelson HF, pp. 419–441

Throckmorton DA. 1993. Shuttle entry aerothermodynamic flight research: the orbiter experiments program. *J. Spacecr. Rockets* 30:449–465.

Throckmorton DA, Ed. 1995. Orbiter Experiments (OEX) Aerothermodynamics Symposium. *NASA CP 3248*

Trimmer LL, Cary AM, Voisinet RL. 1986. The optimum hypersonic wind tunnel. Presented at Aerodyn. Test. Conf., 14th, AIAA Pap. 86–0739, West Palm Beach, FL

Wayland H. 1957. *Differential Equations Applied in Science and Engineering,* Princeton, NJ: Van Nostrand

Weston KC, Fitzkee AL. 1963. Afterbody heat transfer measurements obtained during reentry of the spacecraft of the Mercury-Atlas 5 mission. *NASA TM X-564*

Williamson WE. 1992. *Hypersonic flight testing*. Presented at Aerosp. Ground Test. Conf., 17th, AIAA Pap. 92–3989, Nashville, TN

Woods WC, Holland SD, DiFulvio M. 2001. Hyper-X stage separation wind-tunnel test program. *J. Spacecr. Rockets* 38:881–819

Wright RL, Zoby EV. 1977. *Flight boundary layer transition measurements on a slender cone at Mach 20*. Presented at Fluid and Plasmadynamics Conf., 10th, AIAA Pap. 77–0719, Albuquerque, NM

9 COMPRESSIBLE, SUBSONIC FLOWS AND TRANSONIC FLOWS

In Chapters 3 through 7, flow-field solutions were generated for a variety of configurations using the assumption that the density was constant throughout the flow field. As noted when discussing Fig. 8.6, an error of less than 1% results when the incompressible-flow Bernoulli equation is used to calculate the local pressure provided that the local Mach number is less than or equal to 0.5 in air. Thus, if the flight speed is small compared with the speed of sound, the changes in pressure which are generated by the vehicle motion are small relative to the free-stream static pressure, and the influence of compressibility can be neglected. As shown in Fig. 9.1, the streamlines converge as the incompressible flow accelerates past the midsection of the airfoil. The widening of the streamtubes near the nose and the contraction of the streamtubes

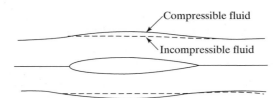

Figure 9.1 Comparison of streamlines for an incompressible flow past an airfoil with those for a subsonic, compressible flow.

in the regions of increased velocity lead to a progressive reduction in the curvature of the streamlines. As a result, there is a rapid attenuation of the flow disturbance with distance from the airfoil.

As the flight speed is increased, the flow may no longer be considered as incompressible. Even though the flow is everywhere subsonic, the density decreases as the pressure decreases (or, equivalently, as the velocity increases). The variable-density flow requires a relatively high velocity and diverging streamlines in order to get the mass flow past the midsection of the airfoil. The expansion of the minimum cross section of the streamtubes forces the streamlines outward so that they conform more nearly to the curvature of the airfoil surface, as shown in Fig. 9.1. Thus, the disturbance caused by the airfoil extends vertically to a greater distance.

Increasing the flight speed further, we reach the *critical Mach number*, the name given to the lowest (subsonic) free-stream Mach number for which the maximum value of the local velocity first becomes sonic. Above the critical Mach number, the flow field contains regions of locally subsonic and locally supersonic velocities in juxtaposition. Such mixed subsonic/supersonic flow fields are termed *transonic flows*.

9.1 COMPRESSIBLE, SUBSONIC FLOW

For completely subsonic flows, a compressible flow retains a basic similarity to an incompressible flow. In particular, the compressible flow can be considered to be an irrotational potential motion in many cases. In addition to the absence of significant viscous forces, the existence of potential motion for a compressible flow depends on the existence of a unique relation between the pressure and the density. In an inviscid flow, the fluid elements are accelerated entirely by the action of the pressure gradient. Thus, if the density is a function of the pressure only, the direction of the pressure gradient will coincide with that of the density gradient at all points. The force on each element will then be aligned with the center of gravity of the fluid element and the pressure forces will introduce no rotation.

9.1.1 Linearized Theory for Compressible Subsonic Flow About a Thin Wing at Relatively Small Angles of Attack

The continuity equation for steady, three-dimensional flow is

$$\frac{\partial(\rho u)}{\partial x} + \frac{\partial(\rho v)}{\partial y} + \frac{\partial(\rho w)}{\partial z} = 0 \tag{9.1}$$

In the inviscid region of the flow field (i.e., outside the thin boundary layer), the components of the momentum equation may be written as

$$u\frac{\partial u}{\partial x} + v\frac{\partial u}{\partial y} + w\frac{\partial u}{\partial z} = -\frac{1}{\rho}\frac{\partial p}{\partial x} \tag{9.2a}$$

$$u\frac{\partial v}{\partial x} + v\frac{\partial v}{\partial y} + w\frac{\partial v}{\partial z} = -\frac{1}{\rho}\frac{\partial p}{\partial y} \tag{9.2b}$$

$$u\frac{\partial w}{\partial x} + v\frac{\partial w}{\partial y} + w\frac{\partial w}{\partial z} = -\frac{1}{\rho}\frac{\partial p}{\partial z} \tag{9.2c}$$

The speed of sound is defined as the change in pressure with respect to the change in density for an isentropic process. Thus,

$$a^2 = \left(\frac{\partial p}{\partial \rho}\right)_s$$

However, since the flow we are studying actually is isentropic, this may be written in terms of the actual pressure and density changes which result due to the fluid motion. Thus,

$$\left(\frac{\partial p}{\partial \rho}\right) = a^2 \tag{9.3}$$

Combining equations (9.1) through (9.3) and noting that the flow is irrotational, one obtains

$$\left(1 - \frac{u^2}{a^2}\right)\frac{\partial u}{\partial x} + \left(1 - \frac{v^2}{a^2}\right)\frac{\partial v}{\partial y} + \left(1 - \frac{w^2}{a^2}\right)\frac{\partial w}{\partial z} - 2\frac{uv}{a^2}\frac{\partial u}{\partial y}$$

$$-2\frac{vw}{a^2}\frac{\partial v}{\partial z} - 2\frac{wu}{a^2}\frac{\partial w}{\partial x} = 0 \tag{9.4}$$

A useful simplification of equation (9.4) can be made for the case of a slender body moving in the x direction at the velocity U_∞. As shown in Fig. 9.2, the magnitude and the direction of the local velocity are changed only slightly from the free-stream velocity. Thus, the resultant velocity at any point can be represented as the vector sum of the free-stream velocity (a constant) together with the perturbation velocities, u', v', and w'. Thus,

$$u = U_\infty + u'$$
$$v = v'$$
$$w = w'$$

Since the perturbation velocities are considered to be small in magnitude when compared with the free-stream velocity, equation (9.4) becomes

$$\left(1 - \frac{u^2}{a^2}\right)\frac{\partial u}{\partial x} + \frac{\partial v}{\partial y} + \frac{\partial w}{\partial z} = 0 \tag{9.5}$$

where u/a is essentially equal to the local Mach number. Equation (9.5) can be simplified further if one recalls that the local speed of sound can be determined using the energy equation for adiabatic flow:

$$\frac{a^2}{a_\infty^2} = 1 - \frac{\gamma - 1}{2}\left(\frac{u^2 + v^2 + w^2}{U_\infty^2} - 1\right)M_\infty^2 \tag{9.6}$$

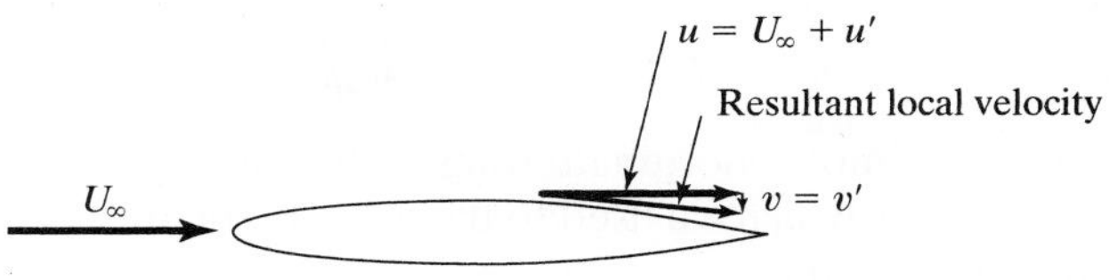

Figure 9.2 Velocity components for subsonic, compressible flow past a thin airfoil at a small angle of attack.

Since only small perturbations are considered, the binomial theorem can be used to generate the relation

$$\frac{a_\infty^2}{a^2} = 1 + \frac{\gamma - 1}{2} M_\infty^2 \left(2 \frac{u'}{U_\infty} + \frac{u'^2 + v'^2 + w'^2}{U_\infty^2} \right) \tag{9.7}$$

To simplify equation (9.5), note that

$$1 - \frac{u^2}{a^2} = 1 - \frac{(U_\infty + u')^2}{a^2} \frac{U_\infty^2}{U_\infty^2} \frac{a_\infty^2}{a_\infty^2}$$

$$= 1 - \frac{U_\infty^2 + 2u'U_\infty + u'^2}{U_\infty^2} M_\infty^2 \frac{a_\infty^2}{a^2} \tag{9.8}$$

Substituting equation (9.7) into equation (9.8) and neglecting the higher-order terms yields the expression

$$1 - \frac{u^2}{a^2} = 1 - M_\infty^2 \left[1 + \frac{2u'}{U_\infty} \left(1 + \frac{\gamma - 1}{2} M_\infty^2 \right) \right] \tag{9.9}$$

Using equation (9.9), equation (9.5) can be rewritten as

$$(1 - M_\infty^2) \frac{\partial u}{\partial x} + \frac{\partial v}{\partial y} + \frac{\partial w}{\partial z} = M_\infty^2 \left(1 + \frac{\gamma - 1}{2} M_\infty^2 \right) \frac{2u'}{U_\infty} \frac{\partial u}{\partial x}$$

This equation can be rewritten in terms of the perturbation velocities as

$$(1 - M_\infty^2) \frac{\partial u'}{\partial x} + \frac{\partial v'}{\partial y} + \frac{\partial w'}{\partial z} = \frac{2}{U_\infty} \left(1 + \frac{\gamma - 1}{2} M_\infty^2 \right) M_\infty^2 u' \frac{\partial u'}{\partial x} \tag{9.10}$$

Furthermore, the term on the right-hand side often can be neglected, as it is of second order in the perturbation velocity components. As a result, one obtains the linearized equation

$$(1 - M_\infty^2) \frac{\partial u'}{\partial x} + \frac{\partial v'}{\partial y} + \frac{\partial w'}{\partial z} = 0 \tag{9.11}$$

Since the flow is everywhere isentropic, it is irrotational. The condition of irrotationality allows us to introduce a velocity potential ϕ, which is a point function with continuous derivatives. Let ϕ be the potential function for the perturbation velocity:

$$u' = \frac{\partial \phi}{\partial x} \qquad v' = \frac{\partial \phi}{\partial y} \qquad w' = \frac{\partial \phi}{\partial z} \tag{9.12}$$

The resultant expression, which applies to a completely subsonic, compressible flow, is the linearized potential equation

$$(1 - M_\infty^2)\phi_{xx} + \phi_{yy} + \phi_{zz} = 0 \tag{9.13}$$

By using a simple coordinate transformation, equation (9.13) can be reduced to Laplace's equation, which we used to describe incompressible, irrotational flows. If the "affine" transformation

$$x' = \frac{x}{\sqrt{1 - M_\infty^2}} \tag{9.14a}$$

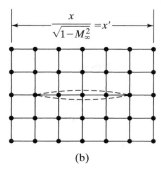

(a)   (b)

Figure 9.3 Distribution of points having equal values of ϕ in the linearized transformation for subsonic, compressible flow: (a) compressible flow; (b) corresponding incompressible flow.

$$y' = y \qquad\qquad (9.14b)$$
$$z' = z \qquad\qquad (9.14c)$$

is introduced, equation (9.13) becomes

$$\phi_{x'x'} + \phi_{y'y'} + \phi_{z'z'} = 0 \qquad\qquad (9.15)$$

Thus, if the potential field for incompressible flow past a given configuration is known, a corresponding solution for the linearized compressible flow can be readily obtained. The potential distribution for an incompressible flow and the corresponding "foreshortened" distribution (which satisfies the compressible flow equation) are compared in Fig. 9.3 at points having the same value of ϕ. Although the calculation of a compressible flow field from a known incompressible flow is relatively straightforward, care must be taken in the determination of the boundary conditions satisfied by the compressible flow field.

Referring to equation (9.14), we see that the transformation in effect changes the ratio of the x dimension to the y and z dimensions. Although the spanwise dimensions are unaltered, the transformed chordwise dimension is. Thus, although the airfoil section of the corresponding wings remain geometrically similar, the aspect ratios for the wings differ. The compressible flow over a wing of aspect ratio AR at the Mach number M_∞ is related to the incompressible flow over a wing of aspect ratio $AR \sqrt{1 - M_\infty^2}$. This is illustrated in the sketch of Fig. 9.4. A study of the changes in a completely subsonic flow field around a given wing as the Mach number is increased corresponds to an investigation of the incompressible flow around a series of wings of progressively reduced aspect ratio.

Using the linearized approximation, the pressure coefficient for the compressible flow is given by

$$C_p = -\frac{2u'}{U_\infty} = -\frac{2}{U_\infty}\frac{\partial \phi}{\partial x} \qquad\qquad (9.16a)$$

which is related to the pressure coefficient for the corresponding incompressible flow (C'_p) through the correlation

$$C_p = \frac{C'_p}{\sqrt{1 - M_\infty^2}} \qquad\qquad (9.16b)$$

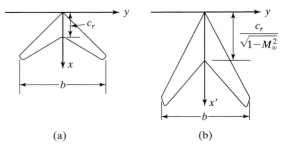

(a)   (b)

Figure 9.4 Wings for flows related by the linearized transformation: (a) wing for compressible flow; (b) corresponding wing for incompressible flow.

The effect of compressibility on the flow past an airfoil system is to increase the horizontal perturbation velocities over the airfoil surface by the factor $1/\sqrt{1 - M_\infty^2}$. The correlation is known as the *Prandtl-Glauert formula*.

We can calculate the lift by integrating the pressure distribution over the airfoil surface from the leading edge to the trailing edge. Based on equation (9.16b), we find that the section lift coefficient for a compressible subsonic flow (C_l) also exceeds the corresponding value for an incompressible flow (C_l') by a factor of $1/[(1 - M_\infty^2)^{0.5}]$. Thus,

$$C_l = \frac{C_l'}{\sqrt{1 - M_\infty^2}}$$

The resultant variation for the lift-curve slope with Mach number is presented for a two-dimensional unswept airfoil in Fig. 9.5.

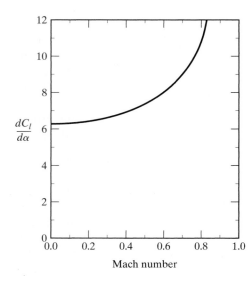

Mach number

Figure 9.5 Variation of lift-curve slope with Mach number using Prandtl-Glauert formula.

Furthermore, based on the pressure correlation of equation (9.16b), the position of the resultant aerodynamic force for a compressible, subsonic flow is the same as that for an incompressible flow. This, of course, is true provided that there are no shock waves. Thus, in the absence of shock waves, the drag acting on an airfoil in an inviscid, compressible, subsonic flow is the same as the drag acting on an airfoil in an inviscid, incompressible flow. That is, the section drag is zero.

Although the Prandtl-Glauert relation provides a simple method for calculating the flow around an airfoil, Jones and Cohen (1960) warn that the method generally underestimates the effect of compressibility on the magnitude of disturbances for airfoils of finite thickness. As the free-stream Mach number approaches a value of unity, the quantity $1/\sqrt{1 - M_\infty^2}$ approaches infinity, causing various perturbation parameters to approach infinity. Hence, the Prandtl-Glauert formula begins to show increasing departures from reality as the Mach number approaches a value of unity. The relative inaccuracy at a particular Mach number depends on parameters such as the section thickness and the angle of attack.

Foss and Blay (1987) described a test flight of the YP-38 in September 1940. Major Signa Gilkey peeled off into a steep dive starting at 35,000 ft. "As he headed down and the airspeed built up, the airplane began to shake, mildly at first and then more violently. Even worse, the airplane wanted to nose down further and increase the severity of the dive. The control column was shaking and the control forces had become heavy, and it was impossible to pull the column back to counteract the nose-down tendency. The airplane was out of control!" As noted by Foss and Blay, Major Gilkey elected to stay with the plane and recovered at 7000 ft. "Major Gilkey had become the first pilot to encounter a new, uncharted high-speed flight regime. He had attained air speeds where new aerodynamic phenomena known as compressibility effects were created."

In seeking ways to control the aircraft experiencing these compressibility effects, the designers turned to the references of the time. Foss and Blay (1987) note, "In hindsight, it is interesting to note that no one was anticipating the nose down pitching moment tendencies, which were hinted at in the section on pitching moment data. And of course, there were no data to suggest that hinge moments would become unmanageable."

Using a P-38 model in the high-speed wind tunnel of the Ames Research Center, model drag-rise characteristics were measured for the first time. The wind-tunnel correlations, as taken from Foss and Blay (1987), are reproduced in Fig. 9.6. Pressure data and flow visualization clearly revealed the shock-stall conditions and helped confirm the flight-test findings.

9.2 TRANSONIC FLOW PAST UNSWEPT AIRFOILS

The section lift coefficient measurements presented in Farren (1956) as a function of Mach number are reproduced in Fig. 9.7. The data indicate that the flow is essentially unchanged up to approximately one-third the speed of sound. The variations in the section lift coefficient with Mach number indicate complex changes in the flow field through the transonic speed range. Attention is called to the section lift coefficient at five particular Mach numbers (identified by the letters a through e). Significant differences exist between the flow fields at these five Mach numbers. To illustrate the essential changes in the flow, line drawings made from schlieren photographs are reproduced in Fig. 9.8.

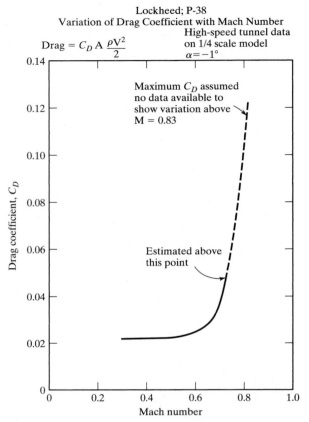

Figure 9.6 High-speed drag variation derived from wind-tunnel tests. [From Foss and Blay (1987).]

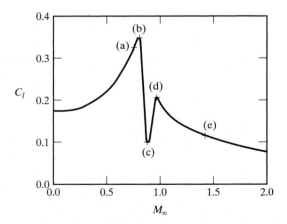

Figure 9.7 Section lift coefficient as a function of Mach number to illustrate the effect of compressibility. [From Farren (1956).] Refer to Fig. 9.8 for the flow fields corresponding to the lettered points on this graph.

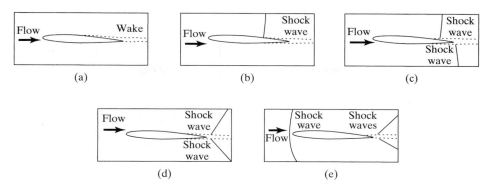

Figure 9.8 Flow field around an airfoil in transonic streams based on schlieren photographs: (a) Mach number $M_\infty = 0.75$; (b) Mach number $M_\infty = 0.81$; (c) Mach number $M_\infty = 0.89$; (d) Mach number $M_\infty = 0.98$; (e) Mach number $M_\infty = 1.4$. [From Farren (1956).]

(a) When the free-stream Mach number is 0.75, the flow past the upper surface decelerates from local flow velocities which are supersonic without a shock wave. The section lift coefficient is approximately 60% greater than the low-speed values at the same angles of attack.

(b) At $M_\infty = 0.81$, the section lift coefficient reaches its maximum value, which is approximately twice the low-speed value. As indicated in Figs. 9.8b and 9.9a, flow is supersonic over the first 70% of the surface, terminating in a shock wave. The flow on the lower surface is subsonic everywhere. Because the viscous flow separates at the foot of the shock wave, the wake is appreciably wider than for (a).

(c) At $M_\infty = 0.89$, flow is supersonic over nearly the entire lower surface and deceleration to subsonic speed occurs through a shock wave at the trailing edge. As a result, the lower surface pressures are lower at $M_\infty = 0.89$ than at $M_\infty = 0.81$. Flow on the upper surface is not greatly different than that for Fig. 9.8b. As a result, the lift is drastically reduced. Separation at the foot of the upper surface shock wave is more conspicuous and the turbulent wake is wide. The shock wave at the trailing edge of the lower surface effectively isolates the upper surface from the lower surface. As a result, the pressure on the upper surface near the trailing edge is greater than that on the lower surface. The corresponding pressure and local Mach number distributions are presented in Fig. 9.9b.

(d) When the free-stream Mach number is 0.98, the shock waves both for the upper surface and for the lower surface have reached the trailing edge. The local Mach number is supersonic for most of the airfoil (both upper and lower surfaces).

(e) When the free-stream flow is supersonic, a bow shock wave (i.e., the detached shock wave in front of the leading edge) is generated. The flow around the airfoil is supersonic everywhere except very near the rounded nose. The shock waves at the trailing edge remain, but they have become weaker.

The data presented in Figs. 9.7 through 9.9 illustrate the effect of Mach number for a given airfoil section at a particular angle of attack. Parameters such as thickness ratio,

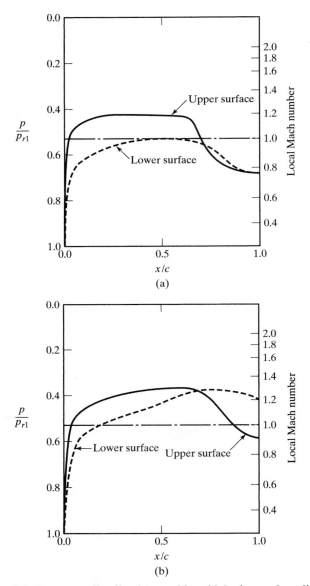

Figure 9.9 Pressure distribution and local Mach number distribution for transonic flows around an airfoil: (a) Flow at the trailing edge is subsonic, $M_\infty = 0.81$; (b) flow at the lower surface trailing edge is supersonic, $M_\infty = 0.89$. [From Farren (1956).]

camber, and nose radius also influence the magnitude of the compressibility effects. Transonic flows are very sensitive to the contour of the body surface, since changes in the surface slope affect the location of the shock wave and, therefore, the inviscid flow field as well as the downstream boundary-layer.

Furthermore, as was discussed in Chapter 8, the shock-wave/boundary-layer interaction and the possible development of separation downstream of the shock wave are sensitive to the character of the boundary-layer, to its thickness, and to its velocity profile at the interaction location. Since a turbulent boundary-layer can negotiate higher adverse pressure gradients than can a laminar one, the shock-wave/boundary-layer interaction is smaller for a turbulent boundary-layer. Thus, it is important to consider the Reynolds number when simulating a flow in the wind tunnel. A large Reynolds number difference between the desired flow and its simulation may produce significant differences in shock-wave location and the resultant flow field. The use of artificial trips to force transition to occur at a specified point (as discussed in Chapter 5) may be unsatisfactory for transonic flows, since the shock-wave location as well as the extent of flow separation can become a function of the artificial tripping.

The Cobalt$_{60}$ code [Strang, et al. (1999)] has been used to compute the flow field around a NACA 0012 airfoil section at an angle of attack of 3°. The turbulence model of Spalart and Allmaras (1992) was used to represent the viscous boundary-layer near the surface of the airfoil. Contours of constant Mach number, as computed by Forsythe and Blake (2000) for a free-stream Mach number of 0.8 at a free-stream Reynolds number (based on the chord length) of 3.0×10^6, are presented in Fig. 9.10. The Mach 1 contours for the upper surface and for the lower surface are highlighted in white. Note that the relatively large area of supersonic flow on the upper surface is terminated by a shock

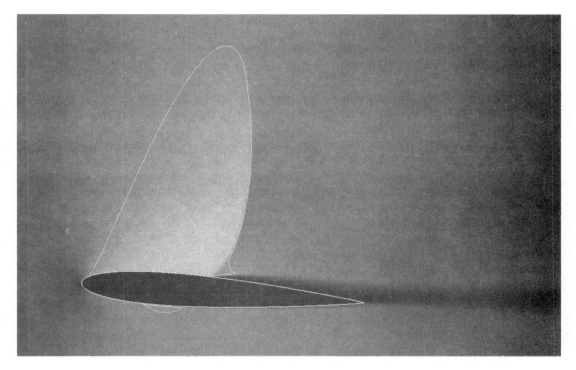

Figure 9.10 Constant-density contours for a NACA 0012 airfoil in a Mach 0.8 stream at $\alpha = 3°$. [As provided by Forsythe and Blake (2000).]

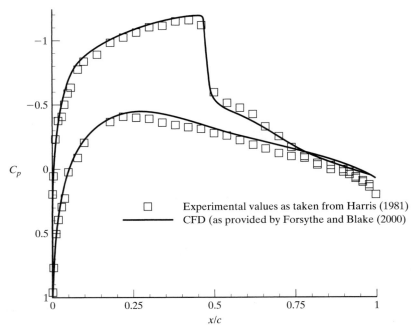

Figure 9.11 Pressure distributions for a NACA 0012 in a Mach 0.8 stream at $\alpha = 3°$.

wave at the downstream (or right-hand) side of the region of supersonic flow. The shock wave is evident in the sudden increase in pressure evident both in the experimental and in the computed pressure distributions for this airfoil, which are taken from Forsythe, et al. (2000) and reproduced in Fig. 9.11. The flow conditions for the experimental data, which are taken from Harris (1981), include a free-stream Mach number of 0.799, a Reynolds number of 9.0×10^6, and an angle of attack of 2.86°. The sudden increase in pressure due to the shock wave occurs on the upper surface at midchord. The adverse pressure gradient associated with the shock-wave/boundary-layer interaction causes the boundary layer to separate from the aft-end surface of the airfoil. This is evident in the (darkened region) Mach number contours presented in Fig. 9.10.

The section lift coefficient for this two-dimensional flow [Forsythe and Blake (2000)] is presented as a function of the free-stream Mach number in Fig. 9.12. Note the similarity between the Mach-number dependence of the computed section lift coefficient, as presented in Fig. 9.12, and that for the measurements presented by Farren (1956), which are reproduced in Fig. 9.7.

The computed flow field for transonic flow over a NACA 0012 airfoil that was presented in Fig. 9.10 depicts regions (both above and below the airfoil) where the flow has accelerated to the point where the flow is locally supersonic. Since the pressure is decreasing, the acceleration of the flow is often termed an expansion. During the isentropic expansion of the flow, there is a corresponding decrease in the temperature.

If the airplane is flying through an airstream where the relative humidity is high, the local temperature of the expanding flow can decrease below the dew point. Water

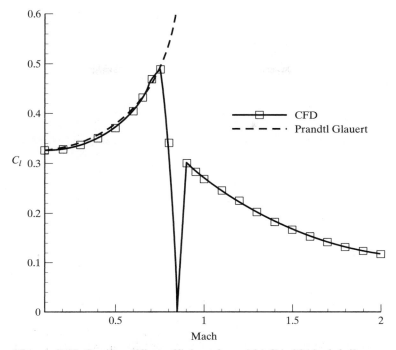

Figure 9.12 Section lift coefficient for a NACA 0012 airfoil as a function of the free-stream Mach number. [As provided by Forsythe and Blake (2000).]

vapor will condense and become visible. Further downstream, as the flow decelerates (perhaps as it passes through a shock wave that terminates the locally supersonic flow), the pressure and the temperature increase. As the local temperature rises above the dew point, the condensed particles evaporate (or vaporize) and are no longer visible. Thus, when an airplane is flying at transonic speeds in a relatively humid atmosphere, the relatively large accelerations (expansions) and decelerations (compressions) yield a condensation pattern, such as that for the F-15I, which is presented in Fig. 9.13a.

The schlieren technique, which allows someone to see shock waves in a wind-tunnel test, such as depicted on the cover and in Fig. 9.8, is based on density gradients in the flow field that bend light rays passing through the flow. Under certain circumstances, these density gradients allow us to "see" the shock waves for an airplane in flight, as shown for the transonic flight of an F-111 in Fig. 9.13b.

The experimentally determined lift coefficients for an untwisted rectangular wing whose aspect ratio is 2.75 and whose airfoil section is a NACA 65A005 (a symmetric profile for which $t = 0.05c$) are presented in Fig. 9.14 as a function of angle of attack. The corresponding drag polars are presented in Fig. 9.15. These data from [Stahl and Mackrodt (1965)] were obtained at Reynolds numbers between 1.0×10^6 and 1.8×10^6. The lift-curve slope is seen to be a function of the free-stream Mach number. Furthermore, the linear relation between the lift coefficient and the angle of attack remains valid to higher angles of attack for supersonic flows.

(a)

(b)

Figure 9.13 (a) Condensation and evaporation of the water vapor delineates regions where the flow has expanded for an F-15I in transonic flight. (Copyrighted image reproduced courtesy of The Boeing Company.) (b) F-111 flying at transonic speeds with shock waves clearly visible. (Courtesy of R. C. Maydew and S. McAlees of the Sandia National Laboratories.)

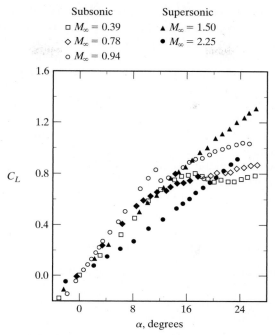

Figure 9.14 Effect of Mach number on the lift-coefficient/angle-of-attack correlation for a rectangular wing, $AR = 2.75$. [Data from Stahl and Mackrodt (1965).]

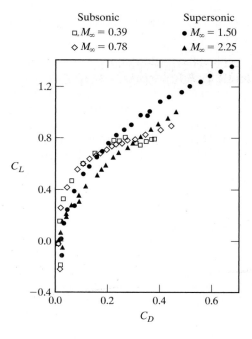

Figure 9.15 Effect of the Mach number on drag polars for a rectangular wing, $AR = 2.75$. [Data from Stahl and Mackrodt (1965).]

Bushnell (2004) noted: "shock waves are usually detrimental, requiring mitigation. The volume and lift-engendered drag associated with shock waves is additive to the usual friction and vortex drag-due-to-lift, comprising one-third of total aircraft drag in the supersonic cruise case, even for the well-designed configuration. The reason for the decades-long Mach .8ish cruise-speed plateau associated with conventional long-haul transport aircraft (requiring over ten hours of flight time transpacific) is avoidance of the major drag increases due to strong shock formation on the upper wing surfaces at higher speeds. Shock wave drag is also a major reason there are no economically viable supersonic transports yet extant. Due to wave drag, the aerodynamic efficiency of a supersonic transport (SST) is the order of one-half or less of a conventional (subsonic) transport. Usual SST designs require a fuel fraction approaching 60%, including reserves. A 1% overall drag reduction for these designs translates to a 5 to 10% increase in payload."

9.3 WAVE DRAG REDUCTION BY DESIGN

9.3.1 Airfoil Contour Wave Drag Approaches

Three features which contribute to the location of the shock wave and, therefore, to the pressure distribution [Holder (1964)] are

1. The flow in the supersonic region ahead of the shock wave
2. The pressure rise across the shock wave itself (which involves considerations of the interaction with the boundary layer)
3. The subsonic flow downstream of the shock wave (which involves considerations of the boundary-layer development between the shock wave and the trailing edge and of the flow in the near wake)

If the trailing-edge pressure changes as a result of flow separation from the upper surface, the lower surface flow must adjust itself to produce a similar change in pressure (since the pressure near the trailing edge must be approximately equal for the two surfaces, unless the flow is locally supersonic at the trailing edge). The conditions for divergence of the trailing-edge pressure correspond to those for a rapid drop in the lift coefficient and to the onset of certain unsteady flow phenomena, such as buffeting.

9.3.2 Supercritical Airfoil Sections

The Mach-number/lift-coefficient flight envelopes of modern jet aircraft operating at transonic speeds are limited by the compressibility drag rise and by the buffeting phenomenon. Airfoil section designs which alleviate or delay the onset of the drag rise and buffeting can contribute to higher maximum speeds (transport applications) or better lift performance (fighter applications). Using intuitive reasoning and substantiating experiment Whitcomb and Clark (1965) noted that R.T. Whitcomb and his coworkers have developed a "supercritical" airfoil shape which delays the subsonic drag rise. The

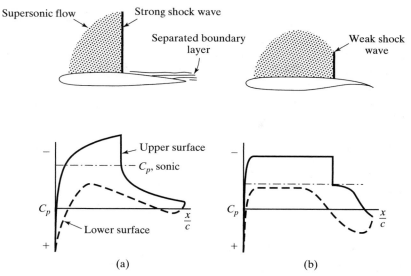

Figure 9.16 Comparison of transonic flow over a NACA 64A series airfoil with that over a "supercritical" airfoil section: (a) NACA 64A series, $M = 0.72$; (b) supercritical airfoil, $M = 0.80$. [From Ayers (1972).]

principal differences between the transonic flow field for a conventional airfoil and that for a supercritical airfoil are illustrated by the data presented in Fig. 9.16. At supercritical Mach numbers, a broad region of locally supersonic flow extends vertically from both airfoils, as indicated by the pressure coefficients above the sonic value and by the shaded areas of the flow fields in Fig. 9.16. The region of locally supersonic flow usually terminates in a shock wave, which results in increased drag. The much flatter shape of the upper surface of the supercritical airfoil causes the shock wave to occur downstream and therefore reduces its strength. Thus, the pressure rise across the shock wave is reduced with a corresponding reduction in drag. However, the diminished curvature of the upper surface also results in a reduction of the lift carried by the midchord region of the airfoil section. To compensate for the loss in lift from the midchord region, additional lift must be generated from the region of the airfoil behind the shock wave, particularly on the lower surface. The increase of lift in this area is achieved by substantial positive camber and incidence of the aft region of the airfoil, especially of the lower surface. The increased lift generated by the concave surface near the trailing edge of the lower surface is evident in the experimental pressure distributions presented in Fig. 9.17. The midchord region of the lower surface should be designed to maintain subcritical flow over the range of operating conditions. If not, when the pressure rise associated with a shock wave is superimposed on the pressure rise caused by the cusp, separation of the lower-surface boundary-layer would occur. To minimize the surface curvatures (and, therefore, the induced velocities) in the midchord regions both for the upper and the lower surfaces, the leading-edge radius is relatively large.

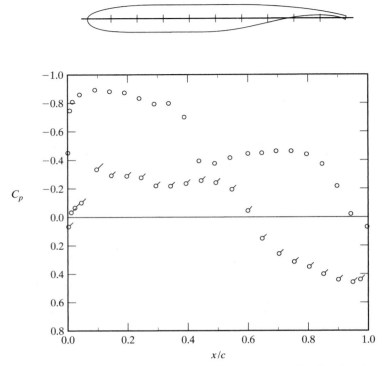

Figure 9.17 Experimentally determined pressure distribution for a supercritical airfoil section, $M_\infty = 0.80$, $C_l = 0.54$, $Re_c = 3.0 \times 10^6$ (flagged symbols are for the lower surface). [Data from Hurley et al. (1975).]

9.4   SWEPT WINGS AT TRANSONIC SPEEDS

In the late 1930s, two aerodynamicists who had been taught by Prandtl, Adolf Busemann and Albert Betz, discovered that drag at transonic and supersonic speeds could be reduced by sweeping back the wings. [The interested reader is referred to Miller and Sawers (1970)]. At the Volta conference in Rome in 1935, in his paper on high-speed flight, Busemann showed that sweepback would reduce drag at supersonic speeds. Betz, in 1939, was the first person to draw attention to the significant reduction in transonic drag which comes when the wing is swept back enough to avoid the formation of shock waves which occur when the flow over the wing is locally supersonic. The basic principle is that the component of the main flow parallel to the wing leading edge is not perturbed by the wing, so the critical conditions are reached only when the component of the free-stream velocity normal to the leading edge has been locally accelerated at some point on the wing to the local sonic speed. This simple principle is obviously only true (if at all) on an infinite span wing of constant section. Nevertheless, the initial suggestion of Betz led to wind-tunnel tests which substantiated the essence of the theory. The

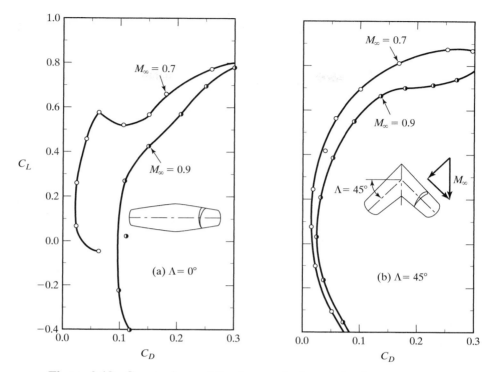

Figure 9.18 Comparison of the transonic drag polar for an unswept wing with that for a swept wing. [Data from Schlichting (1960).]

results of the wind-tunnel measurements performed at Göttingen in 1939 by H. Ludwieg [as presented in Schlichting (1960)] are reproduced in Fig. 9.18. These data show that the effect of the shock waves that occur on the wing at high subsonic speeds are delayed to higher Mach numbers by sweepback.

The experimentally-determined lift coefficients and drag polars for a delta wing whose aspect ratio is 2.31 and whose airfoil section is NACA 65A005 (a symmetric profile for which $t = 0.05c$) are presented in Fig. 9.19. These data from Stahl and Mackrodt (1965) were obtained at Reynolds numbers between 1.0×10^6 and 1.8×10^6. At subsonic speeds, the lift-coefficient/angle-of-attack correlation for the delta wing is markedly different than that for a rectangular wing. (In addition to the data presented in Figs. 9.14 and 9.19, the reader is referred to Chapter 7.) Even in subsonic streams, the lift is a linear function of the angle of attack up to relatively high inclinations. However, over the angle-of-attack range for which the lift coefficient is a linear function of the angle of attack, the lift-curve slope $(dC_L/d\alpha)$ is greater for the rectangular wing for all the free-stream Mach numbers considered.

One can use a variable-geometry (or swing-wing) design to obtain a suitable combination of low-speed and high-speed characteristics. In the highly swept, low-aspect-ratio configuration, the variable-geometry wing provides low wave drag and eliminates the need for wing fold in the case of naval aircraft. At the opposite end of the sweep range,

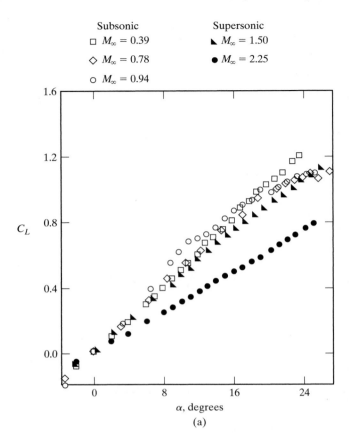

Figure 9.19 Effect of Mach number on the aerodynamic characteristics of a delta wing; $AR = 2.31$, $\Lambda_{LE} = 60°$: (a) lift-coefficient/angle-of-attack correlation. [Data from Stahl and Mackrodt (1965).]

one obtains efficient subsonic cruise and loiter and good maneuverability at the lower speeds. Negative factors in a swing-wing design are complexity, a loss of internal fuel capacity, and the considerable weight of the hinge/pivot structure. A variable-geometry design, the Rockwell International B1, is presented in Fig. 9.20.

9.4.1 Wing–Body Interactions and the "Area Rule"

The chordwise pressure distribution on a plane, finite-span, swept-back wing varies across the span such that the maximum local velocity is reached much farther aft at the root and farther forward at the tip compared to the basic infinite-wing distribution [Rogers and Hall (1960)]. These distortions at the root and at the tip determine the flow pattern. On a swept wing with a square-cut tip, a shock wave first occurs near the tip. This "initial tip shock" is relatively short in extent. As the Mach number is increased, a rear shock is developed which dominates the flow. Because of the variation of the component of velocity normal to the wing leading edge, streamlines over the wing surface will tend to be curved. However, the flow at the wing root is constrained to follow the fuselage. Therefore, for a straight fuselage, a set of compression waves originate in the wing-root

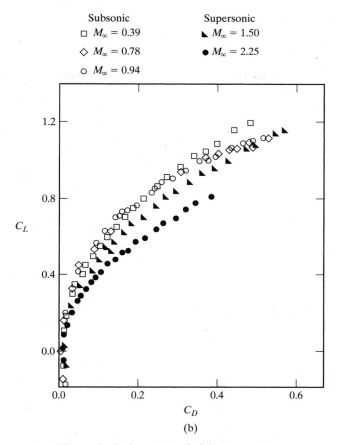

Figure 9.19 (*continued*) (b) drag polar.

region to turn the flow parallel to the fuselage. As shown in Fig. 9.21, the compression waves propagate across the wing span and ultimately may coalesce near the tip to form the rear shock.

Thus, it is clear that the interaction between the central fuselage and the swept wing has a significant effect on the transonic flow over the wing. In the early 1950s, Küchemann (1957) recognized that a properly shaped waistlike indentation on the fuselage could be designed to produce a velocity distribution on the wing near the root similar to that on the corresponding infinite-span wing. Küchemann noted that "the critical Mach number can be raised and the drag reduced in the transonic and the supersonic type of flow, the wing behaving in some respects like a thinner wing." Whitcomb (1954) found that the zero-lift drag rise is due primarily to shock waves. Furthermore, the shock-wave formations about relatively complex swept-wing/body combinations at zero lift near the speed of sound are similar to those which occur for a body of revolution with the same axial development of cross-sectional area normal to the airstream. Whitcomb concluded that "near the speed of sound, the zero-lift drag

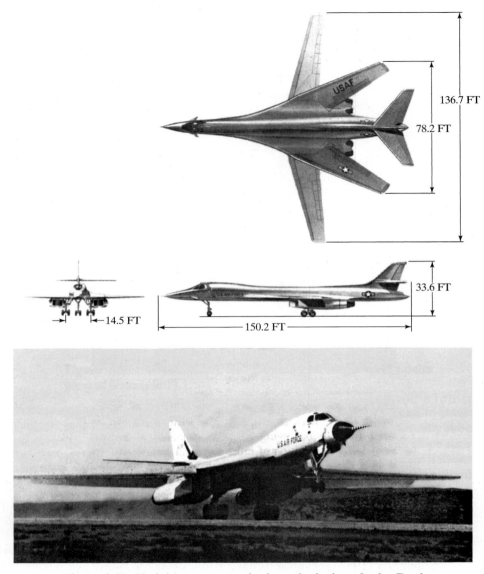

Figure 9.20 Variable geometry (swing-wing) aircraft, the Rockwell International B-1: (a) three-view sketches illustrating the variable-geometry wing; (b) low-speed configuration. (Courtesy of Rockwell International.)

rise of a low-aspect-ratio thin-wing/body combination is primarily dependent on the axial development of the cross-sectional areas normal to the air stream." Therefore, the drag-rise increments near the speed of sound are less for fuselage/wing configurations which have a more gradual change in the cross-sectional area (including the

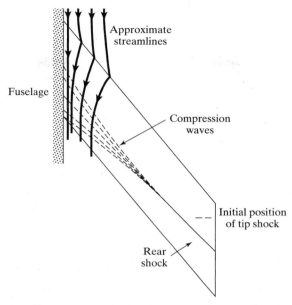

Figure 9.21 Formation of rear shock from compression waves associated with flow near the root. [From Rogers and Hall (1960).]

fuselage and the wing) with axial position (as well as a reduction in the relative magnitude of the maximum area). Whitcomb noted that it would be expected that indenting the body of a wing-body combination, so that the combination has nearly the same axial distribution of cross-sectional area as the original body alone, would result in a large reduction in the transonic drag rise. Design applications of this "theory," which is often known as *Whitcomb's area rule* for transonic configurations, are illustrated in Fig. 9.22. Early flight tests revealed that the prototypes of the YF-102 had serious deficiencies in performance. A major redesign effort included the first application of Whitcomb's area rule to reduce the transonic drag. The modified YF-102A achieved supersonic flight satisfactorily, providing the U.S. Air Force with its first operational delta-wing design. For additional information, the reader is referred to Taylor (1969).

The relation between the distribution of the cross-sectional area and the transonic drag rise had a significant effect on the design of the Convair B-58, the Air Force's first supersonic bomber. Free-flight, rocket-powered models of the original Convair design, designated MX-1626, were tested at the Wallops Island Flight Test Range (NACA). The peak drag coefficient at Mach 1.02 was almost twice as high as that predicted and the model did not achieve supersonic speeds. An evaluation of the longitudinal cross-sectional area distribution in accordance with the area rule of R. T. Whitcomb along with data obtained in the Helium Gun Facility at Wallops on a body of revolution having the same longitudinal distribution of the cross-sectional area as the MX-1626 provided an explanation of the unexpectedly high drag. Presented in Fig. 9.23, which is taken from Shortal (1978), are a sketch of the planform of the MX-1626, the longitudinal distribution of the cross-sectional area, and the drag coefficient for the

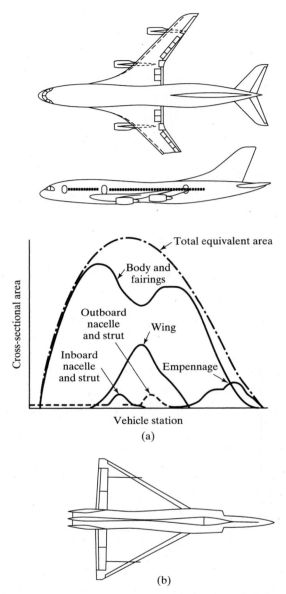

Figure 9.22 Application of the area rule to the axial distribution of the cross-sectional area: (a) area rule applied to the design of a near-sonic transport; (b) sketch of the Convair F 102 A. [From Goodmanson and Gratzer (1973).]

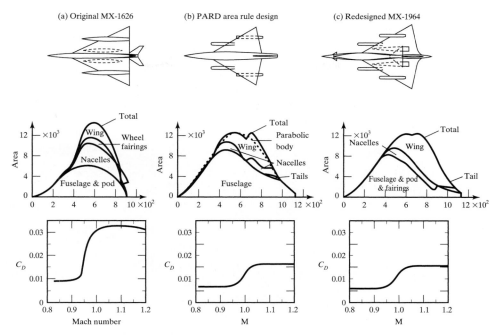

Figure 9.23 Planform sketches, longitudinal distributions of cross-sectional areas, and drag coefficients for equivalent bodies of revolution for supersonic bomber models. [Taken from Shortal (1978).]

equivalent axisymmetric configuration as a function of the Mach number. Note how the wing, the wheel fairings, and the nacelles all add area to the fuselage and the weapons/fuel pod, creating a large total cross-sectional area. Furthermore, the placement of these components results in large longitudinal variations in the total cross-sectional area.

To reflect changes in the design requirements due to the unavailability of the originally planned large jet engines and to changes in the function of the pod, the Air Force changed the Project Number from MX-1626 to MX-1964. R. N. Hopko, R. O. Piland, and J. R. Hall set out to design a vehicle which met the general requirements of the new MX-1964 yet reduced the transonic drag rise. The design, designated PARD Area-Rule Design in Fig. 9.23, included a 3% thick, 60-degree delta wing, modified by a 10-degree forward swept trailing edge. The diamond-shaped wing provides a more gradual variation in the cross-sectional area distribution. To this wing, they added four separate nacelles, staggered chordwise, and made the fuselage a body of revolution. The resultant planform for the PARD Area-Rule Design, the longitudinal cross-sectional area distribution, and the measured drag coefficient for the equivalent body of revolution as a function of the Mach number are reproduced in Fig. 9.23b.

Based on these data, Convair evaluated four alternative designs. Data for the configuration which had the lowest drag rise of the four Convair configurations are presented in Fig. 9.23c. Note that the configuration designated the redesigned MX-1964

Figure 9.24 Photograph of B-58. (From the collection of Tim Valdez.)

had both a smooth progression of cross-sectional area and the minimum value of the maximum cross-sectional area.

While touring the Air Force Museum at the Wright-Patterson Air Force Base, H. Hillaker (1994) noted, "The area-rule requirements of the B-58 required that the inboard and outboard nacelles be staggered—the inlet face of the outboard nacelle began on the same plane as the exhaust face of the inboard nacelle. We previously had siamese nacelles (engines side-by-side as in a B-52 nacelle) which was lighter in weight but the higher drag more than offset the lower weight." This is illustrated in the photograph of Fig. 9.24.

Applications of the area rule in the design of the F-5 are illustrated in Figs. 9.25 through 9.27. The benefit due to area ruling, considering a straight-sided fuselage as a base, is a large decrease in drag at transonic speeds, as shown in Fig. 9.25. As a result, there is a corresponding increase in transonic acceleration, which is very important to this fighter's design, since [According to Stuart (1978) and Bradley (1982)] the major and decisive portions of air-to-air combat take place at altitudes below 30,000 ft at speeds from Mach 0.5 to 1.0.

Area ruling was also applied to the tip-tank design for the F-5A/B aircraft. Initial prototype flight and wind-tunnel tests indicated that transonic buffet and longitudinal instabilities existed with the originally proposed wingtip tank. Continued development tests in the transonic wind tunnel proved that an area ruling concept (as shown in Fig. 9.26) could be used to essentially eliminate the pitch instabilities in the Mach 0.90 to 0.95 region. A cruise drag benefit also existed because of the improvement in wing-tip airflow characteristics. A slight instability which remained in the wind-tunnel data and was thought to be Reynolds number dependent was never found during the flight tests. The photograph of an F5 (Fig. 9.27) illustrates these applications of the area rule.

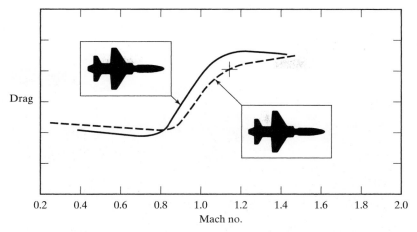

Figure 9.25 Reduction in the transonic drag rise due to the application of area rule to the F-5 fuselage. [From Stuart (1978).]

By carefully designing the wing and the tail configurations, one can obtain a smooth variation of the cross-section area even without adjusting the fuselage cross section. Such is the case for the swept-wing cruise missile which is presented in Fig. 9.28. As shown in Fig. 9.29, the cross-section area distribution is relatively smooth and symmetric about the midbody station. This result is due, in part, to the high sweep angles for the wing and for the tail. As might be expected, this shape is conducive to low transonic drag and the drag-rise Mach number is relatively high (i.e., approximately 0.95). Furthermore, the configuration could be trimmed with zero control deflection near the lift coefficient required for the maximum lift-to-drag ratio at a Mach number of 0.95.

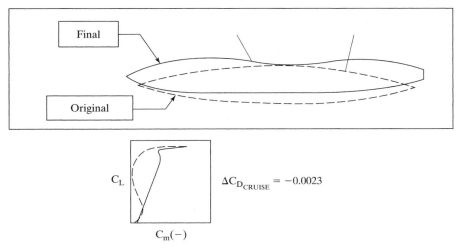

Figure 9.26 Application of area rule to the design of the tip tank for the F-5A/B aircraft. [From Stuart (1978).]

Figure 9.27 Illustration of the application of the area rule to the tip tanks of the F-5. (From the author's collection.)

9.4.2   Second-Order Area-Rule Considerations

Optimization of the wing/body/nacelle configuration plays a critical role in the design of business jets. Unlike previous business jets developed for cruising speeds near a Mach number of 0.7, new aircraft are designed to cruise at speeds near Mach 0.8. Added to the requirement of more speed is the desire for significant increases in range.

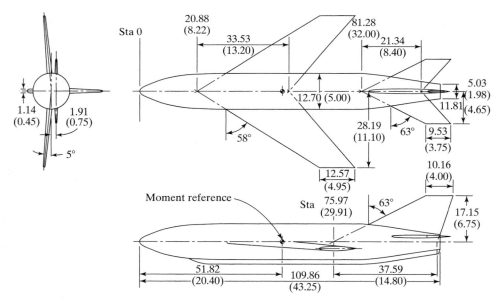

Figure 9.28 Details of a model of a swept-wing cruise missile. Linear dimensions are in centimeters (inches). [From Spearman and Collins (1972).]

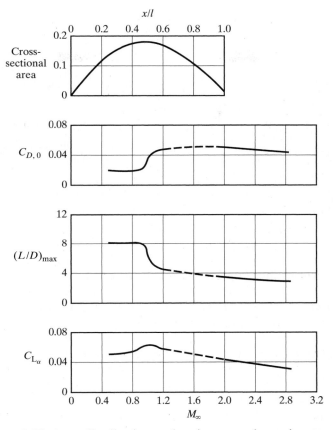

Figure 9.29 Area distribution and various aerodynamic parameters for a swept-wing cruise missile. [From Spearman and Collins (1972).]

To meet these requirements, the designs employ wings with more sweep and carefully designed cross sections that minimize the onset of the compressibility drag. Furthermore, these high-speed wings must be compatible with typical business jet configurations that have engines mounted on the fuselage near the wing trailing edge. As noted by Gallman, et al. (1996). "These requirements create an aerodynamic design problem that benefits from modern computational fluid dynamics (CFD) and aerodynamic shape optimization .... Full potential analysis of the wing-body-nacelle configuration provided an accurate assessment of the influence of the fuselage mounted engines on wing pressures." Gallman, et al. (1996) used codes similar to those discussed (briefly) in Chapter 14.

The area rule is essentially a linear-theory concept for zero lift. Whitcomb (1976) noted that "to achieve the most satisfactory drag characteristics at lifting conditions the fuselage shape had to be modified from that defined by the simple application of the area rule as previously described to account for the nonlinearity

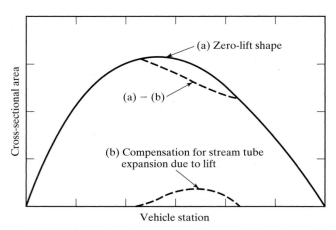

Figure 9.30 Second-order area rule considerations. [From Whitcomb (1976).]

of the flow at such conditions. For lifting conditions at near sonic speeds there is a substantial local region of supercritical flow above the wing surface which results in local expansions of the streamtube areas. In the basic considerations of the area rule concept, this expansion is equivalent to an increase in the physical thickness of the wing. To compensate for this effect the fuselage indentation required to eliminate the far-field effects of the wing must be increased." The additional correction to the cross-sectional areas required for a transonic transport, as taken from Whitcomb (1976) is illustrated in Fig. 9.30. "The fuselage indentation based on this corrected cross-sectional area distribution resulted in a significant (0.02) delay in the drag rise Mach number compared with that for the indentation based on the zero lift distribution."

However, as noted by Ayers (1972), the fuselage cross-sectional area needed for storing landing gear and aircraft subsystems and for accommodating passenger seating and headroom, conflicts with the area-rule requirements in some cases.

Carlsen (1995) noted: "That the sonic area rule is a far field method of predicting and understanding the wave drag due to shock losses [Whitcomb (1956)]. It is based on the idea of perfect pressure disturbance communication between the wing, or other external features, and the fuselage. It proposes that a body of revolution with the same axial development of cross-sectional area will have a wave drag that is similar to the original configuration [Whitcomb (1956)].

Carlsen (1995) noted that the communication between the flow on the wing and the flow on the fuselage changes in the transonic regime because of the mixed supersonic and subsonic flow. "In transonic flow, dissipation of disturbances occurs in the subsonic regions and the stream-tube areas are no longer invariant. As a result of this dissipation, it is erroneous to subtract the total wing volume from the fuselage. Part of the pressure changes created by the indentations in the fuselage is dissipated before it reaches the wing by passing through embedded subsonic regions. To account for the majority of this dissipation, the new transonic method applies a weighting function to the sonic area rule that modifies the volume removed from the fuselage. Only the volume that will relieve the flow on the wing should be removed. This minimizes the volume removed from the fuselage and still maintains the drag rise delay."

Carlsen (1995) recommended the use of weighting functions that would be applied "during the integration of the area of the wing that is intersected by a given Mach 1 plane. Area at the wingtip is given less value than the area near the wing root. The weighted wing area is then subtracted from the fuselage area that is intersected by this same Mach plane. The procedure is repeated at a series of locations along the fuselage axis, resulting in a net volume removal from the fuselage."

Carlsen (1995) concludes that the proposed use of a weighting function adjusts for the effects of mixed flows, i.e., the communication between the flow over the fuselage and that over its external parts that accounts for the dissipation due to embedded subsonic regions. The dissipating effect of the subsonic regions in the transonic regime significantly influences the approach to area ruling. Drag-rise delays that match the traditional sonic-area-rule can be obtained by modifying the aircraft to only 60% of what traditional sonic-area-rule prescribes.

9.4.3   Forward Swept Wing

The aerodynamic effects (e.g., increased critical Mach number, decreased lift curve slope, etc.) resulting from wing sweepback are also present for forward swept wings (FSW) as well. However, there are some comparative advantages to a FSW in relation to an aft swept wing (ASW), which will be discussed in this section.

Historically, the use of FSW in aircraft designs was, with minor exceptions, precluded by the relatively low (in comparison with ASW) Mach numbers at which they experience aeroelastic flutter and divergence when they are constructed of conventional aircraft materials such as metals. The reasons for this will be discussed later, but, as an example, Diederich and Budiansky (1948) show that uniform FSWs with sweep angles from −20 to −60° exhibit decreases as high as 78% in aeroelastic divergence dynamic pressures in comparison to straight wings. To eliminate this problem using conventional materials requires a prohibitive structural weight penalty. Thus, until recently, the FSW, along with some of its desirable characteristics, has been excluded from the aircraft designer's stable of design options. However, recent advances in the development of composite materials for aircraft structures have permitted the design and fabrication of aeroelastically tailored wings which substantially eliminate the divergence problem.

A swept wing constructed of conventional materials twists under load. An ASW twists down at the tips as the lift increases while a FSW twists up. In the former case, the tips tend to unload, while in the latter case they tend to load the structure further. Thus, the FSW constructed of conventional materials has a much lower speed for structural divergence which occurs when the elastic restoring forces can no longer overcome the aerodynamic forces causing the deformation. The twisting behavior occurs because the axis of section centers of pressure along the span of an ASW is behind the structural axis of the wing in typical designs, while just the opposite is true for the FSW [Uhuad, et al. (1983)]. The structural axis can be thought of as the locus of points along the wing where a load produces only wing bending with no twist.

The solution to this problem for the FSW is to design an aeroelastically tailored wing using composite materials arranged in unidirectional layers or plies which lie at selected angles relative to one another. By careful design and fabrication, the wing can be made to deform under loading so that the divergence problem just described can be delayed to a much higher Mach number. This permits the designer to capitalize on some of the advantages of FSWs compared to ASWs without paying an exhorbitant weight penalty. Incidentally, aeroelastic tailoring is not limited to FSWs, but can be applied to straight and aft swept wings as well [Brown et al. (1980)].

The relative advantages of the FSW over the ASW are, of course, a function of the aircraft mission and fall into aerodynamic and other categories. Excellent overall discussions of these FSW characteristics are given in Whitford (1987) and in Krone (1980). We restrict our discussion here to aerodynamic advantages of the FSW. Key aerodynamic advantages of the FSW fall into two categories: reduced drag and enhanced maneuverability at transonic Mach numbers and high angles-of-attack.

Modern aircraft designs rely on wing sweep and supercritical airfoil sections to delay adverse compressibility effects and reduce their severity when they do occur. Good design practices force the shock on the upper surface of the wing to occur as close to the trailing edge as possible and to be as weak as possible. Typical shock locations for a supercritical wing occur in the vicinity of the 70% chord line along the span. As noted by Whitford (1987), this results in a more highly swept shock for an FSW compared to an ASW with the same leading-edge sweep angle, wing area, and taper ratio. The more highly swept shock, of course, results in a lower wave drag penalty. Further, for designs with the same shock sweep angle, the FSW requires less leading-edge sweep, which results in a higher lift curve slope than the comparable ASW and a reduction in induced drag at subsonic conditions. And although the lower leading-edge sweep will result in higher wave drag at supersonic conditions, drag due to shock-induced separation has been found to be less. Thus, as noted earlier, careful attention to the aircraft mission will determine the design solution to be adopted in each case. Whitford (1987) also points out that an FSW with the same planform area, shock sweep, and span as an ASW has its aerodynamic center closer to the wing root and hence experiences a lower wing root bending moment. Conversely, an FSW can be designed with a higher aspect ratio than a comparable ASW and produce the same bending moment at the wing root. The higher aspect ratio in turn produces lower induced drag.

Flow over an ASW has a component along the span and toward the tips which results in a thicker boundary layer in the tip region than would otherwise exist. This, coupled with the more highly loaded tips of the ASW, can lead to tip stall, with the stall region moving inward toward the root as the stall progresses. This phenomenon can result in loss of aileron control effectiveness at high angles-of-attack with a consequent loss of maneuverability. An additional and undesirable offshoot of this phenomenon is an overall pitch-up tendency since the loss of lift typically occurs aft of the aircraft center of gravity. For the FSW the spanwise flow component is toward the wing root leading to root stall first. Thus, aileron effectiveness is preserved up to higher angles-of-attack with a consequent improvement in maneuverability. However, the FSW with root stall still suffers the pitch-up tendency since the loss of lift is once again typically aft of the aircraft center of gravity.

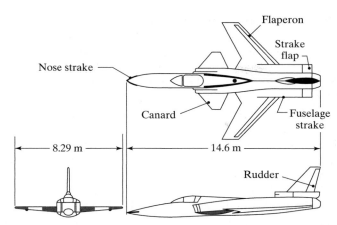

Figure 9.31 Three-view sketch of full-scale X-29 configuration.

To investigate further and to validate by flight test these and other potential advantages of the FSW relative to the ASW, the Defense Advanced Research Projects Agency (DARPA), NASA, and the Air Force have funded the design and fabrication of the X-29. This aircraft, produced by Grumman, is an experimental flight-test vehicle designed to explore a variety of advanced technologies, including FSW and close-coupled canards. The primary objective of the X-29 program is to give designers confidence in the FSW approach and to validate the advantages of FSW and other technologies for use in future aircraft designs [Putnam (1984)].

The X-29 configuration, illustrated in the three view sketches of Fig. 9.31, has several distinctive design features. These include a three-surface longitudinal control design, consisting of an all-movable close-coupled canard with large deflection capability, symmetric deflection of wing flaperons, and deflectable aft-fuselage strake flaps. Roll control is provided by differential flaperon deflection (or, equivalently, aileron deflection). A conventional rudder provides directional control. The X-29 has a long slender forebody, similar to the F-5 design, which produces strong vortical flow in the stall/poststall region.

Using data obtained in free-flight wind-tunnel tests at the Langley Research Center (NASA), Croom et al. (1988) studied the low-speed, high-angle-of-attack flight dynamics of the X-29 configuration. As noted by Croom et al. (1988), "The most dominant characteristic of the configuration is the extreme level of inherent static pitch instability (-0.35 static margin). Clearly, the basic airframe is unflyable without stability augmentation. However, the all-movable canard and strake flap control surfaces provide significant control authority throughout the trim angle-of-attack range .... Moreover, the large control moments in conjunction with the relatively low pitch inertia, associated with the relatively lightweight design indicate the potential for a high level of pitch agility, which is very desirable for fighter aircraft."

A major benefit arising from the use of forward-swept wings at high angles-of-attack is the favorable progression of the stall pattern. Flow separation progresses from the wing-root region outboard, resulting in the retention of aileron control power beyond the stall angle-of-attack. Thus, significant roll control moments persist to very high angles of attack and lateral control should be comparatively good throughout the trim angle-of-attack range.

As noted by Croom et al. (1988) "The data show good levels of static directional stability below 20-deg. angle of attack. However, as the angle of attack is increased above 25-deg., the directional stability degrades significantly, resulting in unstable to neutrally stable values for trim canard incidences. The reduction in stability is due to the vertical tail becoming immersed in the low energy wake of the wing. However, at poststall angles of attack, the directional stability is re-established. Past studies have shown that under sideslip conditions, the flat elliptical cross-section forebody generates an asymmetric vortex system that produces a suction force on the windward side of the forebody resulting in yawing moments into the sideslip (i.e., stabilizing)."

9.5 TRANSONIC AIRCRAFT

Ayers (1972) also notes that "wind-tunnel studies have indicated that combining the supercritical airfoil, the area rule, and wing sweep can push the cruising speed of subsonic aircraft very near Mach 1.0." Using computer-oriented numerical techniques, numerous investigators have solved the nonlinear equation for the disturbance velocity potential in transonic flow, which can be written [refer to Newman and Allison (1971)]

$$(1 - M_\infty^2)\phi_{xx} + \phi_{yy} + \phi_{zz} = K\phi_x\phi_{xx} \tag{9.17}$$

Comparing equation (9.17) with equation (9.10), one would find that, for the particular assumptions made during the development of equation (9.10),

$$K = \frac{2}{U_\infty}\left(1 + \frac{\gamma - 1}{2}M_\infty^2\right)M_\infty^2$$

In the transonic Mach number range, this term cannot be neglected as it was at lower speeds; see equation (9.10).

Bailey and Ballhaus (1975) used a relaxation procedure to solve the transonic small-disturbance equation:

$$(1 - M_\infty^2)\phi_{xx} + \phi_{yy} + \phi_{zz} = \left(\frac{\gamma + 1}{2}M_\infty^n\phi_x^2\right)_x \tag{9.18}$$

The parameter n reflects the nonuniqueness of the equation and, quoting Bailey and Ballhaus (1975), "can be adjusted to better approximate the exact sonic pressure coefficient." A finite-difference equation is derived by applying the divergence theorem to the integral of equation (9.18) over an elemental, rectangular, computation volume (or cell). The boundary conditions for the wing include the Kutta condition, which requires that the pressure, which is proportional to ϕ_x, be continuous at the trailing edge. This fixes the section circulation, which is equal to the difference in

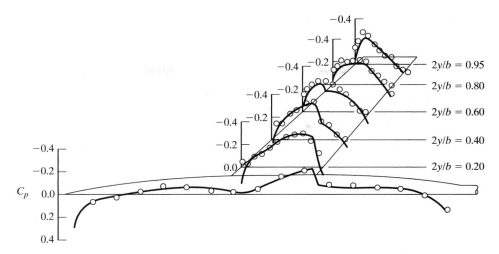

Figure 9.32 Comparison of computed and experimental pressure coefficients C_p for swept-wing/fuselage configuration, $M_\infty = 0.93$; $\alpha = 0°$, $\Lambda_{c/4} = 45°$, AR $= 4$, $\lambda = 0.6$, NACA 65A006 streamwise section. [From Bailey and Ballhaus (1975).]

potential at the section trailing edge linearly extrapolated from points above and below. The solution of the difference equation is obtained by a relaxation scheme with the iterations viewed as steps in pseudo-time. The combination of new and old values in the difference operators is chosen so that the related time-dependent equation represents a properly posed problem whose steady-state solution approaches that of the steady-state equation. The calculated and the experimental pressure distributions for a swept-wing/fuselage configuration at $M_\infty = 0.93$ and $\alpha = 0°$ are compared in Fig. 9.32. The wing has an aspect ratio of 4 and a taper ratio of 0.6. The quarter chord of the wing is swept 45° and the streamwise airfoil section is a NACA 65A006. The computed results were obtained using a Cartesian grid (x, y, z) of $91 \times 59 \times 27$ for the wing as well as for the fuselage. The experimental data were obtained at $\text{Re}_c = 2.0 \times 10^6$. As was noted earlier in this chapter, the maximum velocity is reached farther aft for the stations near the root and moves toward the leading edge at stations nearer the tip. The agreement with experiment on the fuselage centerline and the two inboard panels is good. In the computed results, the wing-root shock propagates laterally to $2y/b = 0.60$, but the experimental shock dissipates before reaching that point. Thus, there is a discrepancy between the experimental values and the theoretical predictions. This deficiency results for wings with moderate-to-large sweep angles. Ballhaus, Bailey, and Frick (1976) note that a modified small-disturbance equation, containing cross-flow terms which have been previously neglected, would be a suitable approximation to the full potential equation over a wide range of sweep angles. Thus, a small-disturbance differential equation which could be used to describe the resultant three-dimensional flow is

$$(1 - M_\infty^2)\phi_{xx} + \phi_{yy} + \phi_{zz} - (\gamma + 1)M_\infty^n \phi_x \phi_{xx}$$

$$= 2M_\infty^2 \phi_y \phi_{xy} + (\gamma - 1)M_\infty^2 \phi_x \phi_{yy} + \frac{\gamma + 1}{2} M_\infty^2 \phi_x^2 \phi_{xx} \quad \textbf{(9.19)}$$

Note that the terms on the left-hand side of this equation are those of equation (9.18) and those on the right-hand side are the additional terms of the modified small-disturbance formulation. Boppe (1978) notes that the cross-flow terms, $\phi_y \phi_{xy}$ and $\phi_x \phi_{yy}$, provide the ability to define shock waves which are swept considerably relative to the free-stream flow. Boppe recommends that the higher-order term, $\phi_x^2 \phi_{xx}$, be included to provide an improved approximation to the full potential equation at the critical velocity. Ballhaus et al. (1976) note that the use of an improved form of the governing equation alone does not guarantee that the shock waves will be properly represented. The finite-difference scheme must also adequately describe the physics of the problem.

Flow-field solutions for complex airplane configurations flying at transonic speeds can be computed using numerical programs that solve the Euler equations. The Euler equations for three-dimensional flows are presented in Appendix A. Using a structured, surface-fitted grid, Kusunose, et al. (1987) generated solutions for the flow over advanced turboprop (with an ultra-high-bypass propulsion system) airplane configurations. A structured surface-fitted grid on the configuration surface, as shown in Fig. 9.33, was

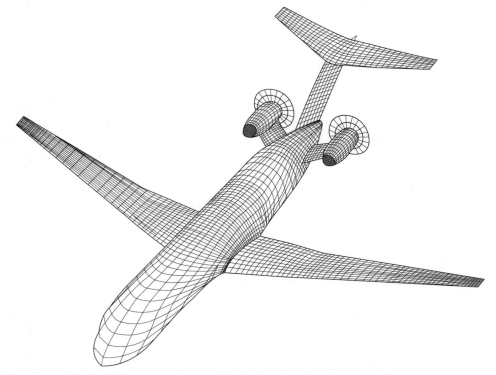

Figure 9.33 Structured surface-fitted grid for an advanced prop-fan airplane. [From Kusunose et al. (1987).]

generated first and then the flow-field grids were obtained using an elliptic grid generation scheme. Kusunose et al. (1987) chose this scheme because it offered "two major advantages: (1) the structural grid requires minimum storage for the flow solution since the grid points are well ordered and (2) the surface-fitted grid allows easier and more accurate implementation of the boundary conditions." The presence of trailing vortex wakes produced by the wing, the strut, and the horizontal tail as well as the slip-stream and the rotational flow (with different total pressure and total temperature) produced by the propellor led Kusunose et al. (1987) to use the Euler equations in formulating the solution algorithm. "When these are solved, trailing vortex wakes come out automatically as part of the solution, and the rotational flows are captured naturally through the Euler equation formulation." To account for the viscous effects on the wing, the Euler solver is coupled with a three-dimensional boundary-layer method McLean et al. (1978). For the approach of Kusunose et al. (1987) the viscous effects are simulated by using the boundary-layer displacement thickness to create an effective "displacement body." Miranda (1984) indicates that modifying the boundary condition (i.e., the surface transpiration concept) is preferable to the displacement body approach because the surface geometry and the computational grid are not affected by the boundary layer.

Miranda (1984) notes that the problem of grid generation is a crucial one for the practical application of numerical flow simulation. Both methods of solution, singularity and field, require the construction of complex computational grids: mesh-surface grids for singularity methods and spatial grids for field methods.

9.6 SUMMARY

Transonic flows are inherently nonlinear and highly three dimensional for wing-body combinations. The nonlinear inviscid flow and the resultant shock waves present a considerable challenge to the analyst. Strong shock-wave/boundary-layer interactions cause rapid increases in drag and can lead to buffeting. Shock-induced boundary-layer separation can also limit the lift coefficient. However, experimental investigations have provided considerable insights into desirable design practices for transonic configurations. Furthermore, three-dimensional transonic flow computational techniques are developing at a rapid pace.

PROBLEMS

9.1. Starting with the energy equation for steady, one-dimensional, adiabatic flow of a perfect gas [i.e., equation (8.29)], derive equation (9.7).

9.2. A rectangular wing having an aspect ratio of 3.5 is flying at $M_\infty = 0.85$ at 12 km. A NACA 0006 airfoil section is used at all spanwise stations. What is the airfoil section and the aspect ratio for the equivalent wing in an incompressible flow?

9.3. When discussing Whitcomb's area rule, it was noted that, for lifting conditions, there is a substantial local region of supercritical flow above the wing surface which results in local expansions of the streamtube areas. Consider flow past a two-dimensional airfoil in a stream where $M_\infty = 0.85$. Calculate the distance between two streamlines, dy_e, at a point where the local inviscid Mach number, M_e, is 1.20 in terms of the distance between these two streamlines in the undisturbed flow, dy_∞. Use the integral form of the continuity equation [equation (2.5)],

$$dy_e = \frac{\rho_\infty U_\infty}{\rho_e U_e}\, dy_\infty$$

Assume an isentropic expansion of a perfect gas.

REFERENCES

Ayers TG. 1972. Supercritical aerodynamics: worthwhile over a range of speeds. *Astronaut. Aeronaut.* 10(8):32–36

Bailey FR, Ballhaus WF. 1975. Comparisons of computed and experimental pressures for transonic flows about isolated wings and wing-fuselage configurations. *NASA Spec. Pub. 347*

Ballhaus WF, Bailey FR, Frick J. 1976. Improved computational treatment of transonic flow about swept wings. *NASA Conf. Proc. 2001*

Boppe CW. 1978. *Computational transonic flow about realistic aircraft configurations*. Presented at AIAA Aerosp. Sci. Meet., 16th, AIAA Pap. 78–104, Huntsville, AL

Bradley RG. 1982. Practical aerodynamic problems—military aircraft. In *Transonic Aerodynamics*. Ed Nixon D, Washington, DC: AIAA

Brown LE, Price MA, Gringrich PB. 1980. *Aeroelastically tailored wing design*. Presented at The Evolution of Wing Design Symp., AIAA Pap. 80–3046, Dayton, OH

Carlsen WD. 1995. Development of transonic area-rule methodology. *J. Aircraft* 32:1056–1061

Croom MA, Whipple RD, Murri DG, Grafton SB, Fratello DJ. 1988. *High-alpha flight dynamics research on the X-29 configuration using dynamic model test techniques*. Proceedings of the Aerosp. Tech. Conf., SAE Pap. 881420, Anaheim, CA

Diederich FW, Budiansky B. 1948. Divergence of swept wings. *NACA Tech. Note 1680*

Farren WS. 1956. The aerodynamic art. *J. Roy. Aero. Soc.* 60:431–449

Forsythe JR, Blake DC. 2000. Private Transmittal

Forsythe JR, Strang WZ, Hoffmann KA. 2000. *Validation of several Reynolds averaged turbulence models in a 3-D unstructured grid code*. Presented at AIAA Fluids 2000 Conf., AIAA Pap. 2000–2552, Denver, CO

Foss RL, Blay R. 1987. From propellers to jets in fighter aircraft design. *Lockheed Horizons* 23:2–17

Gallman JW, Reuther JJ, Pfeiffer NJ, Forrest WC, Bernstorf DJ. 1996. *Business jet wing design using aerodynamic shape optimization*. Presented at AIAA Aerosp. Sci. Meet., 34th, AIAA Pap. 96–0554, Reno, NV

Goodmanson LT, Gratzer LB. 1973. Recent advances in aerodynamics for transport aircraft. *Aeronaut. Astronaut.* 11(12):30–45

Harris CD. 1981. Two-dimensional aerodynamic characteristics of the NACA 0012 airfoil in the Langley 8-foot transonic pressure tunnel. *NASA Tech. Mem. 81927*

Hillaker H. 1994. Private Conversation

Holder DW. 1964. The transonic flow past two-dimensional aerofoils. *J. Roy. Aero. Soc.* 68:501–516

Hurley FX, Spaid FW, Roos FW, Stivers LS, Bandettini A. 1975. Detailed transonic flow field measurements about a supercritical airfoil section. *NASA Tech. Mem. X-3244*

Jones RT, Cohen D. 1960. *High Speed Wing Theory*. Princeton NJ: Princeton University Press

Krone NJ. 1980. *Forward swept wing design*. Presented at The Evolution of Wing Design Symp., AIAA Pap. 80–3047, Dayton, OH

Küchemann D. 1957. Methods of reducing the transonic drag of swept-back wings at zero lift. *J. Roy. Aero. Soc.* 61:37–42

Kusunose K, Mascum DL, Chen HC, Yu NJ. 1987. *Transonic analysis for complex airplane configurations*. Presented at AIAA Fluid Dyn., Plasma Dyn., and Lasers Conf., 19th, AIAA Pap. 87–1196, Honolulu, HI

McLean JD, Randall JL. 1978. Computer program to calculate three-dimensional boundary-layer flows over wings with wall mass transfer. *NASA Contr. Rep. 3123*

Miller R, Sawers D. 1970. *The Technical Development of Modern Aviation*. New York: Praeger Publishers

Miranda LR. 1984. Application of computational aerodynamics to airplane design. *J. Aircraft* 21:355–370

Newman PA, Allison DO. 1971. An annotated bibliography on transonic flow theory. *NASA Tech. Mem. X-2353*

Putnam TW. 1984. The X-29 flight research program. *AIAA Student J.* 22:2–12, 39

Rogers EWE, Hall IM. 1960. An introduction to the flow about plane swept-back wings at transonic speeds. *J. Roy. Aero. Soc.* 64:449–464

Schlichting H. 1960. Some developments in boundary layer research in the past thirty years. *J. Roy. Aero. Soc.* 64:64–79

Shortal JA. 1978. A new dimension, Wallops Island flight test range: the first fifteen years. *NASA RP-1028*

Spalart PR, Allmaras SR. 1992. *A one-equation turbulence model for aerodynamic flows*. Presented at AIAA Aerosp. Sci, Meet., 30th, AIAA Pap. 92–0439, Reno, NV

Spearman ML, Collins IK. 1972. Aerodynamic characteristics of a swept-wing cruise missile at Mach numbers from 0.50 to 2.86. *NASA Tech. Note D-7069*

Stahl W, Mackrodt PA. 1965. Dreikomponentenmessungen bis zu grossen anstellwinkeln an fuenf tragfluegeln mit verschieden umrissformen in unterschall und ueberschallstroemung. *Z. Flugwissensch.* 13:447–453

Strang WZ, Tomaro RF, Grismer MJ. 1999. *The defining methods of Cobalt$_{60}$: a parallel, implicit, unstructured Euler/Navier-Stokes flow solver*. Presented at AIAA Aerosp. Sci. Meet., 37th, AIAA Pap. 99–0786, Reno, NV, Jan. 1999

Stuart WG. 1978. *Northrop F-5 Case Study in Aircraft Design*. Washington, DC: AIAA

Taylor JWR. (Ed) 1969. *Combat Aircraft of the World*. New York: G. P. Putnam Sons

Uhuad GC, Weeks TM, Large R. 1983. Wind tunnel investigation of the transonic aerodynamic characteristics of forward swept wings. *J. Aircraft* 20:195–202

Whitcomb RT. 1956. A study of the zero-lift drag-rise characteristics of wing-body combinations near the speed of sound. *NACA Rep. 1273*

Whitcomb RT. 1976. "Advanced Transonic Aerodynamic Technology", Presented in NASA CP 2001, *Adv. Eng. Sci. 4*, Nov. 1976

Whitcomb RT, Clark LR. 1965. An airfoil shape for efficient flight at supercritical Mach numbers. *NASA Tech. Mem. X-1109*

Whitford R. 1987. *Design for Air Combat*. London: Jane's Publishing

10 TWO-DIMENSIONAL, SUPERSONIC FLOWS AROUND THIN AIRFOILS

The equations that describe inviscid supersonic flows around thin airfoils at low angles of attack will be developed in this chapter. The airfoil is assumed to extend to infinity in both directions from the plane of symmetry (i.e., it is a wing of infinite aspect ratio). Thus, the flow field is the same for any cross section perpendicular to the wing and the flow is two dimensional. However, even though the relations developed in this chapter neglect the effects of viscosity, there will be a significant drag force on a two-dimensional airfoil in a supersonic stream. This drag component is known as *wave drag*. Wave drag can exist without shock waves.

In Chapter 8 we derived the Prandtl-Meyer relations to describe the isentropic flow which results when a supersonic flow undergoes an expansive or a compressive change in direction which is sufficiently small that shock waves do not occur. In Example 8.3, the Prandtl-Meyer relations were used to calculate the aerodynamic coefficients for supersonic flow past a thin airfoil. When relatively large compressive changes in the flow direction occur, it is necessary to use the relations describing the nonisentropic flow through an oblique shock wave. (As will be shown, when the compressive changes in direction are only a few degrees, the pressure increase calculated using the Prandtl-Meyer relations is essentially equal to that calculated using the oblique shock-wave relations.) Provided that the assumptions made in the derivations of these techniques are valid, they can be combined to solve for the two-dimensional flow about an airfoil if the shock wave(s) at the leading edge is (are) attached and planar. When the leading-edge shock wave(s) is (are) planar, the flow downstream of the shock wave(s)

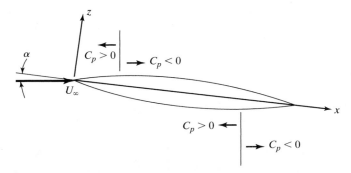

Figure 10.1 General features for linearized supersonic flow past a thin airfoil.

is isentropic. Thus, the isentropic relations developed in Chapter 8 can be used to describe the subsequent acceleration of the flow around the airfoil.

Experience has shown that the leading edge and the trailing edge of supersonic airfoils should be sharp (or only slightly rounded) and the section relatively thin. If the leading edge is not sharp (or only slightly rounded), the leading-edge shock wave will be detached and relatively strong, causing relatively large wave drag. We shall consider, therefore, profiles of the general cross section shown in Fig. 10.1. For these thin airfoils at relatively small angles of attack, we can apply the method of small perturbations to obtain theoretical approximations to the aerodynamic characteristics of the two-dimensional airfoils.

By "thin" airfoil, we mean that the thickness, camber, and angle of attack of the section are such that the local flow direction at the airfoil surface deviates only slightly from the free-stream direction. First we treat the Ackeret, or linearized, theory for thin airfoils and then higher-order theories. The coefficients calculated using the linearized and higher-order theories will be compared with the values calculated using the techniques of Chapter 8.

10.1 LINEAR THEORY

The basic assumption of linear theory is that pressure waves generated by thin sections are sufficiently weak that they can be treated as Mach waves. Under this assumption, the flow is isentropic everywhere. The pressure and the velocity changes for a small expansive change in flow direction (an acceleration) have already been derived in Chapter 8 [i.e., equation (8.55)]. Let us define the free-stream flow direction to be given by $\theta_\infty = 0$. For small changes in θ, we can use equation (8.55) to calculate the change in pressure:

$$p - p_\infty = -\rho_\infty U_\infty (U - U_\infty) \tag{10.1a}$$

$$\frac{U_\infty - U}{U_\infty} = \frac{\theta}{\sqrt{M_\infty^2 - 1}} \tag{10.1b}$$

We will define the angle θ so that we obtain the correct sign for the pressure coefficient both for left-running characteristics and for right-running characteristics. Combining these relations yields

$$C_p = +\frac{2\theta}{\sqrt{M_\infty^2 - 1}} \qquad \textbf{(10.1c)}$$

which can be used to calculate the pressure on the airfoil surface, since θ is known at every point on the airfoil surface.

A positive pressure coefficient is associated with a compressive change in flow direction relative to the free-stream flow. If the flow is turned toward the upstream Mach waves, the local pressure coefficient is positive and is greatest where the local inclination is greatest. Thus, for the double-convex-arc airfoil section shown in Fig. 10.1, the pressure is greatest at the leading edge, being greater on the lower surface when the airfoil is at a positive angle of attack. Flow accelerates continuously from the leading edge to the trailing edge for both the lower surface and the upper surface. The pressure coefficient is zero (i.e., the local static pressure is equal to the free-stream value) at those points where the local surface is parallel to the free stream. Downstream, the pressure coefficient is negative, which corresponds to an expansive change in flow direction.

The pressure coefficients calculated using the linearized approximation and Busemann's second-order approximation (to be discussed in the next section) are compared in Fig. 10.2 with the exact values of Prandtl-Meyer theory for expansive turns and of oblique shock-wave theory for compressive turns. For small deflections, linear theory provides suitable values for engineering calculations.

Since the slope of the surface of the airfoil section measured with respect to the free-stream direction is small, we can set it equal to its tangent. Referring to Fig. 10.3, we can write

$$\theta_u = \frac{dz_u}{dx} - \alpha \qquad \textbf{(10.2a)}$$

$$\theta_l = -\frac{dz_l}{dx} + \alpha \qquad \textbf{(10.2b)}$$

The lift, the drag, and the moment coefficients for the section can be determined using equations (10.1c) and (10.2).

10.1.1 Lift

Referring to Fig. 10.4, we see that the incremental lift force (per unit span) acting on the chordwise segment $ABCD$ of the airfoil section is

$$dl = p_l\,ds_l\cos\theta_l - p_u\,ds_u\cos\theta_u \qquad \textbf{(10.3)}$$

Employing the usual thin-airfoil assumptions, equation (10.3) can be written as

$$dl \approx (p_l - p_u)\,dx \qquad \textbf{(10.4)}$$

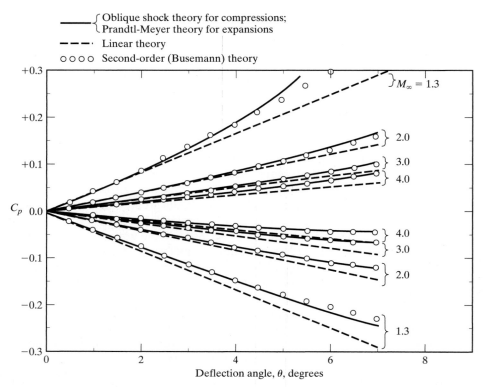

Figure 10.2 Theoretical pressure coefficients as a function of the deflection angle (relative to the stream) for various two-dimensional theories.

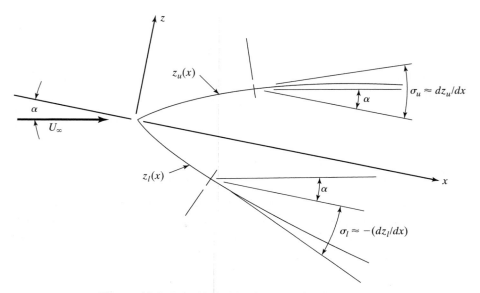

Figure 10.3 Detailed sketch of an airfoil section.

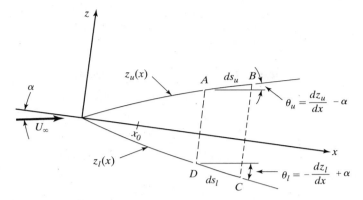

Figure 10.4 Thin-airfoil geometry for determining C_l, C_d, and $C_{m_{x0}}$.

In coefficient form, we have

$$dC_l \approx (C_{pl} - C_{pu})d\left(\frac{x}{c}\right) \tag{10.5}$$

Using equations (10.1c) and (10.2), equation (10.5) becomes

$$dC_l = \frac{2}{\sqrt{M_\infty^2 - 1}}\left(2\alpha - \frac{dz_l}{dx} - \frac{dz_u}{dx}\right)d\left(\frac{x}{c}\right) \tag{10.6}$$

where, without loss of generality, we have assumed that positive values both for θ_u and θ_l represent compressive changes in the flow direction from the free-stream flow.

We can calculate the total lift of the section by integrating equation (10.6) from $x/c = 0$ to $x/c = 1$. Note that since the $z_u = z_l = 0$ at both the leading edge and the trailing edge,

$$\int_0^1 \frac{dz_l}{dx}d\left(\frac{x}{c}\right) = 0 \tag{10.7a}$$

and

$$\int_0^1 \frac{dz_u}{dx}d\left(\frac{x}{c}\right) = 0 \tag{10.7b}$$

Thus,

$$C_l = \frac{4\alpha}{\sqrt{M_\infty^2 - 1}} \tag{10.8}$$

We see that, in the linear approximation for supersonic flow past a thin airfoil, the lift coefficient is independent of the camber and of the thickness distribution. Furthermore, the angle of attack for zero lift is zero. The lift-curve slope is seen to be only a function of the free-stream Mach number, since

$$\frac{dC_l}{d\alpha} = \frac{4}{\sqrt{M_\infty^2 - 1}} \tag{10.9}$$

Examining equation (10.9), we see that for $M_\infty \gtrsim 1.185$, the lift-curve slope is less than the theoretical value for incompressible flow past a thin airfoil, which is 2π per radian.

10.1.2   Drag

The incremental drag force due to the inviscid flow acting on the arbitrary chordwise element $ABCD$ of Fig. 10.4 is

$$dd = p_l \, ds_l \sin \theta_l + p_u \, ds_u \sin \theta_u \qquad (10.10)$$

Again, using the assumptions common to small deflection angles, equation (10.10) becomes

$$dd = p_l \theta_l \, dx + p_u \theta_u \, dx \qquad (10.11)$$

In coefficient form, we have

$$dC_d = (C_{pl}\theta_l + C_{pu}\theta_u)d\!\left(\frac{x}{c}\right) + \frac{2}{\gamma M_\infty^2}(\theta_l + \theta_u)d\!\left(\frac{x}{c}\right) \qquad (10.12)$$

Using equation (10.1c) for compressive turns and equation (10.2) to approximate the angles, equation (10.12) yields

$$dC_d = \frac{2}{\sqrt{M_\infty^2 - 1}}\left[2\alpha^2 + \left(\frac{dz_u}{dx}\right)^2 + \left(\frac{dz_l}{dx}\right)^2\right]d\!\left(\frac{x}{c}\right)$$

$$+ \left[\frac{-4\alpha}{\sqrt{M_\infty^2 - 1}}\left(\frac{dz_l}{dx} + \frac{dz_u}{dx}\right) + \frac{2}{\gamma M_\infty^2}\left(\frac{dz_u}{dx} - \frac{dz_l}{dx}\right)\right]d\!\left(\frac{x}{c}\right) \qquad (10.13)$$

Using equation (10.7), we find that the integration of equation (10.13) yields

$$C_d = \frac{4\alpha^2}{\sqrt{M_\infty^2 - 1}} + \frac{2}{\sqrt{M_\infty^2 - 1}}\int_0^1\left[\left(\frac{dz_u}{dx}\right)^2 + \left(\frac{dz_l}{dx}\right)^2\right]d\!\left(\frac{x}{c}\right) \qquad (10.14)$$

Note that for the small-angle assumptions commonly used in analyzing flow past a thin airfoil,

$$\frac{dz_u}{dx} = \tan \sigma_u \approx \sigma_u$$

Thus,

$$\frac{1}{c}\int_0^c \left(\frac{dz_u}{dx}\right)^2 dx = \overline{\sigma_u^2} \qquad (10.15a)$$

Similarly, we can write

$$\frac{1}{c}\int_0^c \left(\frac{dz_l}{dx}\right)^2 dx = \overline{\sigma_l^2} \qquad (10.15b)$$

We can use these relations to replace the integrals of equation (10.14) by the average values that they represent. Thus, the section-drag coefficient for this frictionless flow model is

$$C_d = \frac{d}{q_\infty c} = \frac{4\alpha^2}{\sqrt{M_\infty^2 - 1}} + \frac{2}{\sqrt{M_\infty^2 - 1}}(\overline{\sigma_u^2} + \overline{\sigma_l^2}) \tag{10.16}$$

Note that the drag is not zero even though the airfoil is of infinite span and the viscous forces have been neglected. This drag component, which is not present in subsonic flows, is known as *wave drag*. Note also that, as this small perturbation solution shows, it is not necessary that shock waves be present for wave drag to exist. Such was also the case in Example 8.3, which examined the shock-free flow past an infinitesimally thin, parabolic arc airfoil.

Let us examine the character of the terms in equation (10.16). Since the lift is directly proportional to the angle of attack and is independent of the section thickness, the first term is called the *wave drag due to lift* or the *induced wave drag* and is independent of the shape of the airfoil section. The second term is often referred to as the *wave drag due to thickness* and depends only on the shape of the section. Equation (10.16) also indicates that, for a given configuration, the wave-drag coefficient decreases with increasing Mach number. If we were to account for the effects of viscosity, we could write

$$C_d = C_{d,\text{due to lift}} + C_{d,\text{thickness}} + C_{d,\text{friction}} \tag{10.17a}$$

where

$$C_{d,\text{due to lift}} = \frac{4\alpha^2}{\sqrt{M_\infty^2 - 1}} = \alpha C_l \tag{10.17b}$$

and

$$C_{d,\text{thickness}} = \frac{2}{\sqrt{M_\infty^2 - 1}}(\overline{\sigma_u^2} + \overline{\sigma_l^2}) \tag{10.17c}$$

Note also that $C_{d,\text{thickness}}$ is the C_{d0} of previous chapters.

10.1.3 Pitching Moment

Let us now use linear theory to obtain an expression for the pitching moment coefficient. Referring to Fig. 10.4, the incremental moment (taken as positive for nose up relative to the free stream) about the arbitrary point x_0 on the chord is

$$dm_{x0} = (p_u - p_l)(x - x_0)dx \tag{10.18}$$

where we have incorporated the usual small-angle assumptions and have neglected the contributions of the chordwise components of p_u and p_l to the pitching moment.

In coefficient form, we have

$$dC_{m_{x0}} = (C_{pu} - C_{pl})\frac{x - x_0}{c}d\left(\frac{x}{c}\right) \tag{10.19}$$

Substituting equation (10.1c) into equation (10.19) yields

$$dC_{m_{x0}} = \frac{2}{\sqrt{M_\infty^2 - 1}}(\theta_u - \theta_l)\frac{x - x_0}{c}d\left(\frac{x}{c}\right) \tag{10.20}$$

Substituting equation (10.2) into equation (10.20) and integrating along the chord gives

$$C_{m_{x0}} = \frac{-4\alpha}{\sqrt{M_\infty^2 - 1}}\left(\frac{1}{2} - \frac{x_0}{c}\right) + \frac{2}{\sqrt{M_\infty^2 - 1}}\int_0^1 \left(\frac{dz_u}{dx} + \frac{dz_l}{dx}\right)\frac{x - x_0}{c}d\left(\frac{x}{c}\right) \quad (10.21)$$

Note that the average of the upper surface coordinate z_u and the lower surface coordinate z_l defines the mean camber coordinate z_c,

$$\tfrac{1}{2}(z_u + z_l) = z_c$$

We can then write equation (10.21) as

$$C_{m_{x0}} = \frac{-4\alpha}{\sqrt{M_\infty^2 - 1}}\left(\frac{1}{2} - \frac{x_0}{c}\right) + \frac{4}{\sqrt{M_\infty^2 - 1}}\int_0^1 \frac{dz_c}{dx}\frac{x - x_0}{c}d\left(\frac{x}{c}\right) \quad (10.22)$$

where we have assumed that $z_u = z_l$ both at the leading edge and at the trailing edge.

As discussed in Section 5.4.2, the aerodynamic center is that point about which the pitching moment coefficient is independent of the angle of attack. It may also be considered to be that point along the chord at which all changes in lift effectively take place. Thus, equation (10.22) shows that the aerodynamic center is at midchord for a thin airfoil in a supersonic flow. This is in contrast to the thin airfoil in an incompressible flow where the aerodynamic center is at the quarter chord.

EXAMPLE 10.1: **Use linear theory to calculate the lift coefficient, the wave-drag coefficient, and the pitching-moment coefficient**

Let us use the linear theory to calculate the lift coefficient, the wave-drag coefficient, and the pitching-moment coefficient for the airfoil section whose geometry is illustrated in Fig 10.5. For purposes of discussion, the flow field has been divided into numbered regions, which correspond to each of the facets of the double-wedge airfoil, as shown. In each region the flow properties are such that the static pressure and the Mach number are constant, although they differ from region to region. We seek the lift coefficient, the drag coefficient, and the pitching-moment coefficient per unit span of the airfoil given the free-stream flow conditions, the angle of attack, and the geometry of the airfoil neglecting the effect of the viscous boundary layer. The only forces acting on the airfoil are the pressure forces. Therefore, once we have determined the static pressure in each region, we can then integrate to find the resultant forces and moments.

Solution: Let us now evaluate the various geometric parameters required for the linearized theory:

$$z_u(x) = \begin{cases} x\tan 10° & \text{for } 0 \le x \le \frac{c}{2} \\ (c - x)\tan 10° & \text{for } \frac{c}{2} \le x \le c \end{cases}$$

$$z_l(x) = \begin{cases} -x\tan 10° & \text{for } 0 \le x \le \frac{c}{2} \\ -(c - x)\tan 10° & \text{for } \frac{c}{2} \le x \le c \end{cases}$$

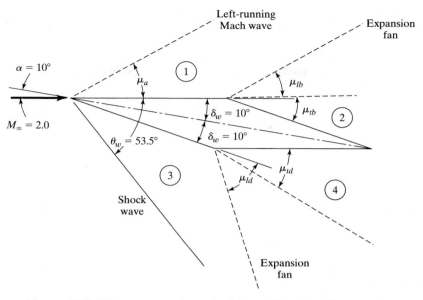

Figure 10.5 Wave pattern for a double-wedge airfoil in a Mach 2 stream.

Furthermore,

$$\overline{\sigma_l^2} = \int_0^1 \sigma_l^2 d\left(\frac{x}{c}\right) = \delta_w^2$$

and

$$\overline{\sigma_u^2} = \int_0^1 \sigma_u^2 d\left(\frac{x}{c}\right) = \delta_w^2$$

We can use equation (10.8) to calculate the section lift coefficient for a 10° angle of attack at $M_\infty = 2.0$:

$$C_l = \frac{4(10\pi/180)}{\sqrt{2^2 - 1}} = 0.4031$$

Similarly, the drag coefficient can be calculated using equation (10.16):

$$C_d = \frac{4(10\pi/180)^2}{\sqrt{2^2 - 1}} + \frac{2}{\sqrt{2^2 - 1}}\left[\left(\frac{10}{57.296}\right)^2 + \left(\frac{10}{57.296}\right)^2\right]$$

Therefore,

$$C_d = 0.1407$$

The lift/drag ratio is

$$\frac{l}{d} = \frac{C_l}{C_d} = \frac{0.4031}{0.1407} = 2.865$$

Note that in order to clearly illustrate the calculation procedures, this airfoil section is much thicker (i.e., $t = 0.176c$) than typical supersonic airfoil sections for which $t \approx 0.05c$ (see Table 5.1). The result is a relatively low lift/drag ratio.

Similarly, equation (10.22) for the pitching moment coefficient gives

$$C_{m_{x0}} = \frac{-4\alpha}{\sqrt{M_\infty^2 - 1}}\left(\frac{1}{2} - \frac{x_0}{c}\right)$$

since the mean-camber coordinate x_c is everywhere zero. At midchord, we have

$$C_{m_{0.5c}} = 0$$

This is not a surprising result, since equation (10.22) indicates that the moment about the aerodynamic center of a symmetric (zero camber) thin airfoil in supersonic flow vanishes.

10.2 SECOND-ORDER THEORY (BUSEMANN'S THEORY)

Equation (10.1c) is actually the first term in a Taylor series expansion of Δp in powers of θ. Busemann showed that a more accurate expression for the pressure change resulted if the θ^2 term were retained in the expansion. His result [given in Edmondson, et al. (1945)] in terms of the pressure coefficient is

$$C_p = \frac{2\theta}{\sqrt{M_\infty^2 - 1}} + \left[\frac{(\gamma + 1)M_\infty^4 - 4M_\infty^2 + 4}{2(M_\infty^2 - 1)^2}\right]\theta^2 \qquad \textbf{(10.23a)}$$

or

$$C_p = C_1\theta + C_2\theta^2 \qquad \textbf{(10.23b)}$$

Again, θ is positive for a compression turn and negative for an expansion turn. We note that the θ^2 term in equation (10.23) is always a positive contribution. Table 10.1 gives C_1 and C_2 for various Mach numbers in air.

It is important to note that since the pressure waves are treated as Mach waves, the turning angles must be small. These assumptions imply that the flow is isentropic everywhere. Thus, equations (10.5), (10.12), and (10.19), along with equation (10.23), can still be used to find C_l, C_d, and $C_{m_{x0}}$. Note that Fig. 10.2 shows that Busemann's theory agrees even more closely with the results obtained from oblique shock and Prandtl-Meyer expansion theory than does linear theory.

EXAMPLE 10.2:   **Use Busemann's theory to calculate the lift coefficient, the wave-drag coefficient, and the pitching-moment coefficient**

Let us calculate the pressure coefficient on each panel of the airfoil in Example 10.1 using Busemann's theory. We will use equation (10.23).

TABLE 10.1   Coefficients C_1 and C_2 for the Busemann Theory for Perfect Air, $\gamma = 1.4$

M_∞	C_1	C_2
1.10	4.364	30.316
1.12	3.965	21.313
1.14	3.654	15.904
1.16	3.402	12.404
1.18	3.193	10.013
1.20	3.015	8.307
1.22	2.862	7.050
1.24	2.728	6.096
1.26	2.609	5.356
1.28	2.503	4.771
1.30	2.408	4.300
1.32	2.321	3.916
1.34	2.242	3.599
1.36	2.170	3.333
1.38	2.103	3.109
1.40	2.041	2.919
1.42	1.984	2.755
1.44	1.930	2.614
1.46	1.880	2.491
1.48	1.833	2.383
1.50	1.789	2.288
1.52	1.747	2.204
1.54	1.708	2.129
1.56	1.670	2.063
1.58	1.635	2.003
1.60	1.601	1.949
1.70	1.455	1.748
1.80	1.336	1.618
1.90	1.238	1.529
2.00	1.155	1.467
2.50	0.873	1.320
3.00	0.707	1.269
3.50	0.596	1.248
4.00	0.516	1.232
5.00	0.408	1.219
10.0	0.201	1.204
∞	0	1.200

Solution: Since panel 1 is parallel to the free stream, $C_{p1} = 0$ as before. For panel 2,

$$C_{p2} = \frac{2(-20\pi/180)}{\sqrt{2^2 - 1}} + \frac{(1.4 + 1)(2)^4 - 4(2)^2 + 4}{2(2^2 - 1)^2}\left(\frac{20\pi}{180}\right)^2$$

$$= -0.4031 + 0.1787$$

$$= -0.2244$$

As noted, the airfoil in this sample problem is relatively thick, and therefore the turning angles are quite large. As a result, the differences between linear theory and higher-order approximations are significant but not unexpected. For panel 3,

$$C_{p3} = 0.4031 + 0.1787$$
$$= 0.5818$$

For panel 4,

$$C_{p4} = 0$$

since the flow along surface 4 is parallel to the free stream.

Having determined the pressures acting on the individual facets of the double-wedge airfoil, let us now determine the section lift coefficient:

$$C_l = \frac{\Sigma p \cos \theta (0.5c/\cos \delta_w)}{(\gamma/2)p_\infty M_\infty^2 c} \tag{10.24a}$$

where δ_w is the half-angle of the double-wedge configuration. We can use the fact that the net force in any direction due to a constant pressure acting on a closed surface is zero to get

$$C_l = \frac{1}{2\cos \delta_w}\Sigma C_p \cos \theta \tag{10.24b}$$

where the signs assigned to the C_p terms account for the direction of the force. Thus,

$$C_l = \frac{1}{2\cos 10°}(-C_{p2}\cos 20° + C_{p3}\cos 20°)$$
$$= 0.3846$$

Similarly, we can calculate the section wave-drag coefficient:

$$C_d = \frac{\Sigma p \sin \theta (0.5c/\cos \delta_w)}{(\gamma/2)p_\infty M_\infty^2 c} \tag{10.25a}$$

or

$$C_d = \frac{1}{2\cos \delta_w}\Sigma C_p \sin \theta \tag{10.25b}$$

Applying this relation to the airfoil section of Fig 10.5, the section wave-drag coefficient for $\alpha = 10°$ is

$$C_d = \frac{1}{2\cos 10°}(C_{p3}\sin 20° - C_{p2}\sin 20°)$$
$$= 0.1400$$

Let us now calculate the moment coefficient with respect to the midchord of the airfoil section (i.e., relative to $x = 0.5c$). As we have seen, the theoretical solutions for linearized flow show that the midchord point is the aerodynamic center for a thin airfoil in a supersonic flow. Since the pressure is constant on each of the facets of the double-wedge airfoil of Fig. 10.5 (i.e., in each numbered region), the force acting on a given facet will be normal to the surface and will act at the midpoint of the panel. Thus,

$$C_{m_{0.5c}} = \frac{m_{0.5c}}{\frac{1}{2}\rho_\infty U_\infty^2 c^2} = (-p_1 + p_2 + p_3 - p_4)\frac{c^2/8}{\frac{1}{2}\rho_\infty U_\infty^2 c^2}$$

$$+ (p_1 - p_2 - p_3 + p_4)\frac{(c^2/8)\tan^2\delta_w}{\frac{1}{2}\rho_\infty U_\infty^2 c^2} \quad \textbf{(10.26)}$$

Note that, as usual, a nose-up pitching moment is considered positive. Also note that we have accounted for terms proportional to $\tan^2\delta_w$. Since the pitching moment due to a uniform pressure acting on any closed surface is zero, equation (10.26) can be written as

$$C_{m_{0.5c}} = (-C_{p1} + C_{p2} + C_{p3} - C_{p4})\frac{1}{8}$$

$$+ (C_{p1} - C_{p2} - C_{p3} + C_{p4})\frac{\tan^2\delta_w}{8} \quad \textbf{(10.27)}$$

The reader is referred to equations (5.11) through (5.15) for a review of the technique. Thus,

$$C_{m_{0.5c}} = 0.04329$$

10.3 SHOCK-EXPANSION TECHNIQUE

The techniques discussed thus far assume that compressive changes in flow direction are sufficiently small that the inviscid flow is everywhere isentropic. In reality, a shock wave is formed as the supersonic flow encounters the two-dimensional double-wedge airfoil of the previous example problems. Since the shock wave is attached to the leading edge and is planar, the downstream flow is isentropic. Thus, the isentropic Prandtl-Meyer relations developed in Chapter 8 can be used to describe the acceleration of the flow around the airfoil. Let us use this shock-expansion technique to calculate the flow field around the airfoil shown in Fig. 10.5.

EXAMPLE 10.3: **Use the shock-expansion theory to calculate the lift coefficient, the wave-drag coefficient, and the pitching-moment coefficient.**

For purposes of discussion, the flow field has been divided into numbered regions that correspond to each of the facets of the double-wedge airfoil, as shown in Fig. 10.5. As was true for the approximate theories, the flow properties in each region, such as the static pressure and the Mach number, are constant, although they differ from region to region. We will calculate the section lift coefficient, the section drag coefficient, and the section pitching moment coefficient for the inviscid flow.

Solution: Since the surface of region 1 is parallel to the free stream, the flow does not turn in going from the free-stream conditions (∞) to region 1. Thus, the properties in region 1 are the same as in the free stream. The pressure coefficient on the airfoil surface bounding region 1 is zero. Thus,

$$M_1 = 2.0 \qquad \nu_1 = 26.380° \qquad \theta_1 = 0° \qquad C_{p1} = 0.0$$

Furthermore, since the flow is not decelerated in going to region 1, a Mach wave (and not a shock wave) is shown as generated at the leading edge of the upper surface. Since the Mach wave is of infinitesimal strength, it has no effect on the flow. However, for completeness, let us calculate the angle between the Mach wave and the free-stream direction. The Mach angle is

$$\mu_a = \sin^{-1} \frac{1}{M_\infty} = 30°$$

Since the surface of the airfoil in region 2 "turns away" from the flow in region 1, the flow accelerates isentropically in going from region 1 to region 2. To cross the left-running Mach waves dividing region 1 from region 2, we move along right-running characteristics. Therefore,

$$d\nu = -d\theta$$

Since the flow direction in region 2 is

$$\theta_2 = -20°$$

ν_2 is

$$\nu_2 = \nu_1 - \Delta\theta = 46.380°$$

Therefore,

$$M_2 = 2.83$$

To calculate the pressure coefficient for region 2,

$$C_{p2} = \frac{p_2 - p_\infty}{\frac{1}{2}\rho_\infty U_\infty^2} = \frac{p_2 - p_\infty}{(\gamma/2)p_\infty M_\infty^2}$$

Thus,

$$C_{p2} = \left(\frac{p_2}{p_\infty} - 1\right)\frac{2}{\gamma M_\infty^2}$$

Since the flow over the upper surface of the airfoil is isentropic,

$$p_{t\infty} = p_{t1} = p_{t2}$$

Therefore,

$$C_{p2} = \left(\frac{p_2}{p_{t2}} \frac{p_{t\infty}}{p_\infty} - 1\right)\frac{2}{\gamma M_\infty^2}$$

Using Table 8.1. or equation (8.36), to calculate the pressure ratios given the values for M_∞ and for M_2,

$$C_{p2} = \left(\frac{0.0352}{0.1278} - 1\right)\frac{2}{1.4(4)} = -0.2588$$

The fluid particles passing from the free stream to region 3 are turned by the shock wave through an angle of $20°$. The shock wave decelerates the flow and the pressure in region 3 is relatively high. To calculate the pressure coefficient for region 3, we must determine the pressure increase across the shock wave. Since we know that $M_\infty = 2.0$ and $\theta = 20°$, we can use Fig. 8.12b to find the value of C_{p3} directly; it is 0.66. As an alternative procedure, we can use Fig. 8.12 to find the shock-wave angle, θ_w, which is equal to $\theta_w = 53.5°$. Therefore, as discussed in Section 8.5, we can use $M_\infty \sin\theta$ (instead of M_∞), which is 1.608, and the correlations of Table 8.3 to calculate the pressure increase across the oblique shock wave:

$$\frac{p_3}{p_\infty} = 2.848$$

Thus,

$$C_{p3} = \left(\frac{p_3}{p_\infty} - 1\right)\frac{2}{\gamma M_\infty^2} = 0.66$$

Using Fig. 8.12c, we find that $M_3 = 1.20$.
    Having determined the flow in region 3,

$$M_3 = 1.20 \qquad \nu_3 = 3.558° \qquad \theta_3 = -20°$$

one can determine the flow in region 4 using the Prandtl-Meyer relations. One crosses the right-running Mach waves dividing regions 3 and 4 on left-running characteristics. Thus,

$$d\nu = d\theta$$

Since $\theta_4 = 0°, d\theta = +20°$ and

$$\nu_4 = 23.558°$$

so

$$M_4 = 1.90$$

Note that because of the dissipative effect of the shock wave, the Mach number in region 4 (whose surface is parallel to the free stream) is less than the free-stream Mach number.
    Whereas the flow from region 3 to region 4 is isentropic, and

$$p_{t3} = p_{t4}$$

the presence of the shock wave causes

$$p_{t3} < p_{t\infty}$$

To calculate

$$C_{p4} = \left(\frac{p_4}{p_\infty} - 1\right)\frac{2}{\gamma M_\infty^2}$$

$$= \left(\frac{p_4}{p_3}\frac{p_3}{p_\infty} - 1\right)\frac{2}{\gamma M_\infty^2}$$

the ratio

$$\frac{p_4}{p_3} = \frac{p_4}{p_{t4}}\frac{p_{t3}}{p_3}$$

can be determined since both M_3 and M_4 are known. The ratio p_3/p_∞ has already been found to be 2.848. Thus,

$$C_{p4} = \left[\frac{0.1492}{0.4124}(2.848) - 1.0\right]\frac{2}{1.4(4.0)} = 0.0108$$

We can calculate the section lift coefficient using equation (10.24).

$$C_l = \frac{1}{2\cos 10°}(-C_{p1} - C_{p2}\cos 20° + C_{p3}\cos 20° + C_{p4})$$

$$= 0.4438$$

Similarly, using equation (10.25) to calculate the section wave-drag coefficient,

$$C_d = \frac{1}{2\cos 10°}(C_{p3}\sin 20° - C_{p2}\sin 20°)$$

Thus,

$$C_d = 0.1595$$

The lift/drag ratio in our example is

$$\frac{l}{d} = 2.782$$

for this airfoil section.

    In some cases, it is of interest to locate the leading and trailing Mach waves of the Prandtl-Meyer expansion fans at b and d. Thus, using the subscripts l and t to indicate leading and trailing Mach waves, respectively, we have

$$\mu_{lb} = \sin^{-1}\frac{1}{M_1} = 30°$$

$$\mu_{tb} = \sin^{-1}\frac{1}{M_2} = 20.7°$$

$$\mu_{ld} = \sin^{-1}\frac{1}{M_3} = 56.4°$$

$$\mu_{td} = \sin^{-1}\frac{1}{M_4} = 31.8°$$

Each Mach angle is shown in Fig. 10.5.

To calculate the pitching moment about the midchord point, we substitute the values we have found for the pressure coefficients into equation (10.27) and get

$$C_{m_{0.5c}} = 0.04728$$

Example 10.3 illustrates how to calculate the aerodynamic coefficients using the *shock-expansion technique*. This approach is exact provided the relevant assumptions are satisfied. A disadvantage of the technique is that it is essentially a numerical method which does not give a closed-form solution for evaluating airfoil performance parameters, such as the section lift and drag coefficients. However, if the results obtained by the method applied to a variety of airfoils are studied, one observes that the most efficient airfoil sections for supersonic flow are thin with little camber and have sharp leading edges. (The reader is referred to Table 5.1 to see that these features are used on the high-speed aircraft.) Otherwise, wave drag becomes prohibitive.

Thus, we have used a variety of techniques to calculate the inviscid flow field and the section aerodynamic coefficients for the double-wedge airfoil at an angle of attack of $10°$ in an airstream where $M_\infty = 2.0$. The theoretical values are compared in Table 10.2. Although the airfoil section considered in these sample problems is much thicker than those actually used on supersonic airplanes, there is reasonable agreement between the aerodynamic coefficients calculated using the various techniques. Thus, the errors in the local pressure coefficients tend to compensate for each other when the aerodynamic coefficients are calculated.

The theoretical values of the section aerodynamic coefficients as calculated using these three techniques are compared in Fig. 10.6, with experimental values taken from Pope (1958). The airfoil is reasonably thin and the theoretical values for the section lift coefficient and for the section wave-drag coefficient are in reasonable agreement with the data. The experimental values of the section moment coefficient exhibit the angle-of-attack dependence of the shock-expansion theory, but they differ in magnitude. Note that, for the airfoil shown in Fig. 10.6, C_l is negative at zero angle of attack. This is markedly different from the subsonic result, where the section lift coefficient is positive for a cambered airfoil at zero angle of attack. This is another example illustrating that the student should not apply intuitive ideas for subsonic flow to supersonic flows.

TABLE 10.2 Comparison of the Aerodynamic Parameters for the Two-Dimensional Airfoil Section of Fig. 10.5, $M_\infty = 2.0$, $\alpha = 10°$

	Linearized (Ackeret) theory	Second-order (Busemann) theory	Shock-expansion technique
C_{p1}	0.0000	0.0000	0.0000
C_{p2}	−0.4031	−0.2244	−0.2588
C_{p3}	+0.4031	+0.5818	+0.660
C_{p4}	0.0000	0.0000	+0.0108
C_l	0.4031	0.3846	0.4438
C_d	0.1407	0.1400	0.1595
$C_{m_{0.5c}}$	0.0000	0.04329	0.04728

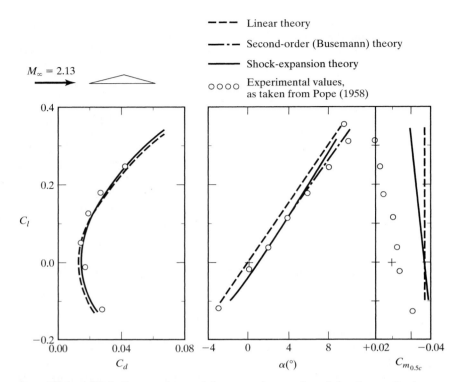

Figure 10.6 Comparison of the experimental and the theoretical values of C_l, C_d, and $C_{m_{0.5c}}$, for supersonic flow past an airfoil.

PROBLEMS

10.1. Consider supersonic flow past the thin airfoil shown in Fig. P10.1. The airfoil is symmetric about the chord line. Use linearized theory to develop expressions for the lift coefficient, the drag coefficient, and the pitching-moment coefficient about the midchord. The resultant expressions should include the free-stream Mach number, the constants a_1 and a_2, and the thickness ratio $t/c(\equiv \tau)$. Show that, for a fixed thickness ratio, the wave drag due to thickness is a minimum when $a_1 = a_2 = 0.5$.

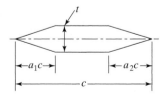

Figure P10.1

10.2. Consider the infinitesimally thin airfoil which has the shape of a parabola:

$$x^2 = -\frac{c^2}{z_{max}}(z - z_{max})$$

The leading edge of the profile is tangent to the direction of the oncoming airstream. This is the airfoil of Example 8.3. Use linearized theory for the following.

(a) Find the expressions for the lift coefficient, the drag coefficient, the lift/drag ratio, and the pitching-moment coefficient about the leading edge. The resultant expressions should include the free-stream Mach number and the parameter, z_{max}/c.

(b) Graph the pressure coefficient distribution as a function of x/c for the upper surface and for the lower surface. The calculations are for $M_\infty = 2.059$ and $z_{max} = 0.1c$. Compare the pressure distributions for linearized theory with those of Example 8.3.

(c) Compare the lift coefficient and the drag coefficient calculated using linearized theory for $M_\infty = 2.059$ and for $z_{max} = 0.1c$ with those calculated in Example 8.3.

10.3. Consider the double-wedge profile airfoil shown in Fig. P10.3. If the thickness ratio is 0.04 and the free-stream Mach number is 2.0 at an altitude of 12 km, use linearized theory to compute the lift coefficient, the drag coefficient, the lift/drag ratio, and the moment coefficient about the leading edge. Also, compute the static pressure and the Mach number in each region of the sketch. Make these calculations for angles of attack of 2.29° and 5.0°.

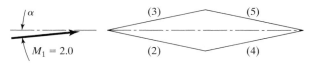

Figure P10.3

10.4. Repeat Problem 10.3 using second-order (Busemann) theory.

10.5. Repeat Problem 10.3 using the shock-expansion technique. What is the maximum angle of attack at which this airfoil can be placed and still generate a weak shock wave?

10.6. Calculate the lift coefficient, the drag coefficient, and the coefficient for the moment about the leading edge for the airfoil of Problem 10.3 and for the same angles of attack, if the flow were incompressible subsonic.

10.7. Verify the theoretical correlations presented in Fig. 10.6. Note that for this airfoil section,

$$\tau = \frac{t}{c} = 0.063$$

Furthermore, the free-stream Mach number is 2.13.

10.8. For linearized theory, it was shown [i.e., equation (10.17c)] that the section drag coefficient due to thickness is

$$C_{d,\,thickness} = \frac{2}{\sqrt{M_1^2 - 1}}(\overline{\sigma_u^2} + \overline{\sigma_l^2})$$

If τ is the thickness ratio, show that

$$C_{d,\,thickness} = \frac{4\tau^2}{\sqrt{M_1^2 - 1}}$$

for a symmetric, double-wedge airfoil section and that

$$C_{d,\,thickness} = \frac{5.33\tau^2}{\sqrt{M_1^2 - 1}}$$

for a biconvex airfoil section. In doing this problem, we are verifying the values for K_1 given in Table 11.1.

10.9. Consider the symmetric, diamond-shaped airfoil section (Fig. P10.9; as shown in the sketch, all four facets have the same value of δ_w) exposed to a Mach 2.20 stream in a wind tunnel. For the wind tunnel, $p_{t1} = 125$ psia $T_t = 600°$R. The airfoil is such that the maximum thickness t equals to $0.07c$. The angle of attack is $6°$.

   (a) Using the shock-wave relations where applicable and the isentropic expansion relations (Prandtl-Meyer) where applicable, calculate the pressures in regions 2 through 5.

   (b) Using the linear-theory relations, calculate the pressures in regions 2 through 5.

   (c) Calculate C_A, C_N, C_d, and $C_{m_{0.5c}}$ for this configuration. C_A is the axial force coefficient for the force along the axis (i.e., parallel to the chordline of the airfoil) and C_N is normal force coefficient (i.e., normal to the chordline of the airfoil).

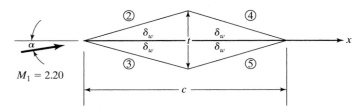

Figure P10.9

10.10. Consider the "cambered," diamond-shaped airfoil exposed to a Mach 2.00 stream in a wind tunnel (Fig. P10.10). For the wind tunnel, $p_{t1} = 125$ psia $T_t = 650°$R. The airfoil is such that $\delta_2 = 8°$, $\delta_3 = 2°$; the maximum thickness t equals to $0.07c$ and is located at $x = 0.40c$. The angle of attack is $10°$.

   (a) Using the shock-wave relations where applicable and the isentropic expansion relations (Prandtl-Meyer) where applicable, calculate the pressures in regions 2 through 5.

   (b) Using the linear-theory relations, calculate the pressures in regions 2 through 5.

   (c) Calculate C_A, C_N, C_l, C_d, and $C_{m_{0.5c}}$ for this configuration. C_A is the axial force coefficient for the force along the axis (i.e., parallel to the chordline of the airfoil) and C_N is the normal force coefficient (i.e., normal to the chordline of the airfoil).

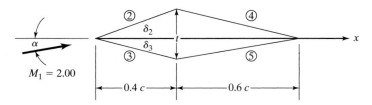

Figure P10.10

10.11. Consider the two-dimensional airfoil having a section which is a biconvex profile, i.e.,

$$R(x) = 2t\frac{x}{L}\left(1 - \frac{x}{L}\right)$$

The model is placed in a supersonic wind-tunnel, where the free-stream Mach number in the tunnel test section is 2.2; $p_{t_1} = 125$ lbf/in^2 and $T_t = 600°$R. The airfoil is such that the maximum thickness (t) equals $0.07L$ and it occurs at the mid chord. The angle of attack is $6°$.

Using the shock/Prandtl-Meyer expansion technique to model the flowfield for the biconvex airfoil section, calculate the pressure distributions both for the windward (bottom) and for the leeward (top) sides. Present the results in a single graph that compares the pressure distribution for the windward side with that for the leeward side.

10.12. Consider the two-dimensional airfoil having a section which is a biconvex profile, i.e.,

$$R(x) = 2t\frac{x}{L}\left(1 - \frac{x}{L}\right)$$

The model is placed in a supersonic wind-tunnel, where the free-stream Mach number in the tunnel test section is 2.2; $p_{t_1} = 125$ lbf/in^2 and $T_t = 600°$R. The airfoil is such that the maximum thickness (t) equals $0.07L$ and it occurs at the mid chord. The angle of attack is 6°.

Using the linearized theory relations to model the flowfield for the biconvex airfoil section, calculate the pressure distributions both for the windward (bottom) and for the leeward (top) sides. Present the results in a single graph that compares the pressure distribution for the windward side with that for the leeward side.

10.13. Consider the two-dimensional airfoil having a section which is a biconvex profile, i.e.,

$$R(x) = 2t\frac{x}{L}\left(1 - \frac{x}{L}\right)$$

The model is placed in a supersonic wind-tunnel, where the free-stream Mach number in the tunnel test section is 2.2; $p_{t_1} = 125$ lbf/in^2 and $T_t = 600°$R. The airfoil is such that the maximum thickness (t) equals $0.07L$ and it occurs at the mid chord.

Calculate C_d (the section drag coefficient) for the bi-convex airfoil section over an angle of attack range from 0° to 10°. Use the linearized theory relations to determine the pressure distribution that is required to calculate the form drag. Also, calculate C_d for the symmetric diamond-shaped airfoil section (see Problem 10.9).

Prepare a graph comparing the section drag coefficient (C_d) for the bi-convex airfoil for the angle of attack range from 0° to 10° with that for the symmetric, diamond-shaped airfoil section. The shock/Prandtl-Meyer technique is to be used to calculate the pressure distribution for both airfoil sections. What is the effect of airfoil cross-section on the form drag?

REFERENCES

Edmondson N, Murnaghan FD, Snow RM. 1945. The theory and practice of two-dimensional supersonic pressure calculations. *Johns Hopkins Univ Appl Physics Lab Bumblebee Report 26*, JHU/APL, December 1945.

Pope A. 1958. *Aerodynamics of Supersonic Flight*. 2nd Ed. New York: Putnam Publ. Corp.

11 SUPERSONIC FLOWS OVER WINGS AND AIRPLANE CONFIGURATIONS

The density variations associated with the flow over an aircraft in supersonic flight significantly affect the aerodynamic design considerations relative to those for subsonic flight. As noted in Chapter 10, the inviscid pressure distribution results in a drag component, known as wave drag, even if one assumes the flow to be isentropic (i.e., neglects the effects of shock waves). Wave drag represents a significant fraction of the total drag in supersonic flows and is related to the bluntness of the configuration. Thus, as illustrated in Table 5.1, the wing sections for supersonic aircraft tend to have a relatively small thickness ratio. Furthermore, the wing planform is such that the aspect ratio is relatively small.

Techniques by which the aerodynamic forces and moments can be computed can be classed either as panel (or singularity) methods or field methods. For panel methods, the configuration is modeled by a large number of quadrilateral panels, approximating the aircraft surface. One or more types of singularity distributions (e.g., sources) are assigned to each elementary panel to simulate the effect that panel has on the flow field. A variety of panel methods have been developed, e.g., that described in Cenko (1981), which can be used to generate solutions for compressible flow. Panel methods yield good results for slender bodies at small angles of attack. Furthermore, linearized-theory concepts also allow one to separate the drag-due-to-lift into two fundamental components: (1) the vortex drag associated with the spanwise distribution of the lifting force and the resultant downwash behind the wing and (2) the wave-drag-due-to-lift, which arises only for supersonic flow, associated with the longitudinal distribution of the lift and the resultant disturbance waves propagating into the surrounding air.

However, as the design lift coefficient is increased, methods based on linear-theory approximations no longer provide an adequate simulation of the complex flow fields. At high-lift conditions, the flow becomes extremely nonlinear, requiring the use of higher-order prediction techniques which take into account the various nonlinear flow phenomena in determining the configuration aerodynamics. The high-lift flow regime can be modeled using a fully three-dimensional, inviscid, attached-flow computational approach which employs the full-potential equation.

Field methods include a wide variety of assumptions in the flow models. Walkley and Smith (1987) describe a technique which uses a finite-difference formulation based on the characteristics theory of signal propagation to solve the conservative form of the full potential equation, including flows with embedded shock waves and subsonic flows. Additional rigor is obtained (at the cost of additional computational time and expense) if one solves the Euler equation [e.g., Allen and Townsend (1986)]. Solution techniques employing the Euler equations allow one to incorporate entropy terms that are neglected in the full potential model. For relatively slender vehicles at small angles of attack (so that the flow is attached and the boundary layer is thin), one should expect reasonable accuracy for the pressure distributions and force coefficients if one solves the flow field using the Euler equations. To obtain suitable accuracy when computing the flow fields for higher angles of attack or for configurations where viscous/inviscid interactions are important, it is necessary to model the viscous effects and their interdependence on the inviscid flow. For such applications, Navier-Stokes formulations are needed [e.g., Schiff and Steger (1979)].

It is beyond the scope of this text to treat the various methods available by which one can calculate the forces and moments acting on supersonic vehicles. Furthermore, even computer codes that are based on the Navier-Stokes equations employ simplifying assumptions in the solution algorithm. Thus, in reviewing techniques for predicting wing leading-edge vortex flows at supersonic speeds, Wood and Miller (1985) note that, based on their comparisons between computed flows and experimental data, (at least) one Euler code was not well suited for the analysis of wings with separated flow at high lift and low supersonic speeds. Instead, a code based on a linearized-theory method that was modified to account for both nonlinear attached-flow effects (lower surface) and nonlinear separated flow (upper surface) [Carlson and Mack (1980)] provided the best correlation with the experimentally measured vortex strength, vortex position, and total lifting characteristics. Thus, it is important to develop an understanding of the general features of supersonic flows and their analysis.

In this chapter we consider steady, supersonic, irrotational flow about wing of finite aspect ratio. The objective is to determine the influence of geometric parameters on the lift, the drag, and the pitching moment for supersonic flows past finite wings. In Chapter 10 we evaluated the effect of the section geometry for flows that could be treated as two dimensional (i.e., wings of infinite aspect ratio). In this chapter three-dimensional effects will be included since the wings are of finite span. However, we will continue to concentrate on configurations that can be handled by small disturbance (linear) theories.

After a discussion of the general characteristics of flow about supersonic wings, we proceed to a development of the governing equation and boundary conditions for the supersonic wing problem. We then outline the consequences (particularly as they pertain to determining drag) of linearity on the part of the equation and the boundary conditions, and proceed to discuss two solution methods: the conical-flow and the singularity-distribution methods. Example problems are worked using the latter method. We close the chapter with discussions of aerodynamic interaction effects among aircraft components in supersonic flight and of some design considerations for supersonic aircraft.

11.1   GENERAL REMARKS ABOUT LIFT AND DRAG

A typical lift/drag polar for a supersonic airplane is presented in Fig. 11.1. At supersonic speeds, aircraft drag is composed of

1. Skin-friction drag
2. Wave-drag-due-to-thickness (or volume), also known as the zero-lift wave drag
3. Drag-due-to-lift

Thus,

$$C_D = C_{D,\text{friction}} + C_{D,\text{thickness}} + C_{D,\text{due-to-lift}} \tag{11.1}$$

As noted earlier in this chapter, the drag-due-to-lift is itself composed of the vortex drag and of the wave-drag-due-to-lift. Experimental evidence indicates that equation (11.1) can be written using the approximation introduced in Chapter 5, that is, equation (5.46),

$$C_D = C_{D0} + kC_L^2 \tag{11.2}$$

where C_{D0}, the zero-lift drag coefficient, is composed of the sum of $C_{D,\text{friction}}$ and $C_{D,\text{thickness}}$. For supersonic flows, k (the drag-due-to-lift factor) is a strong function of the Mach number. For values illustrating the relationship between k and the Mach number, the reader is referred to Example 5.7 and to Problems 5.15 and 5.16.

The skin friction drag results from the presence of the viscous boundary layer near the surface of the vehicle (see Chapters 4 and 8). Reference temperature methods can be used in the calculation of skin friction coefficients for compressible flows (see Chapter 12). As noted in Middleton and Lundry (1980a), the zero-lift wave drag can be calculated using either far-field methods (i.e., the supersonic area rule) or near-field methods (i.e., the integration of the surface pressure distribution). The far-field method offers advantages for fuselage optimization according to area-rule concepts. The near-field method is used as an analysis tool for applications where the detailed pressure distributions are of interest. The drag-due-to-lift (which includes the trim drag) is computed from lifting-analysis programs. As noted in Middleton et al. (1980b), the wing-design and the lift-analysis programs are separate lifting-surface methods which solve the direct or the inverse problem of

1. Design—to define the wing-camber-surface shape required to produce a selected lifting-pressure distribution. The wing-design program includes methods for defining an optimum pressure distribution.

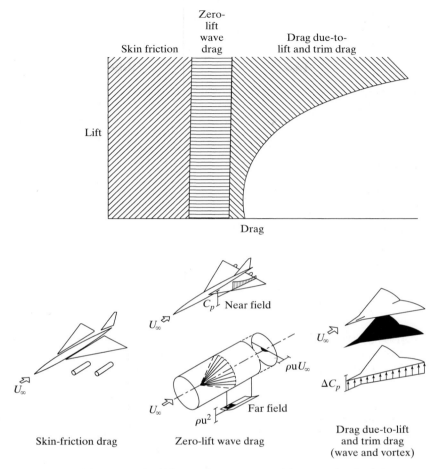

Figure 11.1 Drag buildup using superposition method of drag analysis. [From Middleton and Lundry (1980a).]

2. Lift analysis—to define the lifting pressure distribution acting on a given wing-camber-surface shape and to calculate the associated force coefficients.

For efficient flight at a lift coefficient which maximizes the lift-to-drag ratio, the drag-due-to-lift is about one half of the total, as noted in Carlson and Mann (1992).

11.2 GENERAL REMARKS ABOUT SUPERSONIC WINGS

The unique characteristics of supersonic flow lead to some interesting conclusions about wings in a supersonic stream. Consider the rectangular wing of Fig. 11.2. The pressure at a given point $P(x, y)$ on the wing is influenced only by pressure disturbances generated at points within the upstream Mach cone (determined by $\mu = \sin^{-1} 1/M_\infty$)

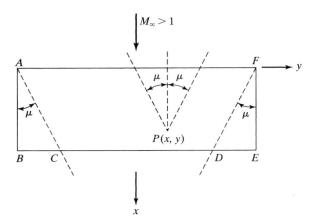

Figure 11.2 Rectangular wing in a supersonic stream.

emanating from *P*. As a result, the wing tips affect the flow only in the regions *BAC* and *DEF*. The remainder of the wing (*ACDF*) is not influenced by the tips and can be treated using the two-dimensional theory developed in Chapter 10.

In the case of an arbitrary planform (see Fig. 11.3), we have the following definitions:

1. A supersonic (subsonic) leading edge is that portion of the wing leading edge where the component of the free-stream velocity normal to the edge is supersonic (subsonic).

2. A supersonic (subsonic) trailing edge is that portion of the wing trailing edge where the component of the free-stream velocity normal to the edge is supersonic (subsonic).

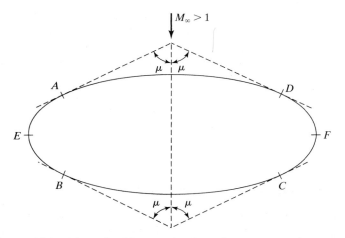

Figure 11.3 Wing of arbitrary planform in a supersonic stream.

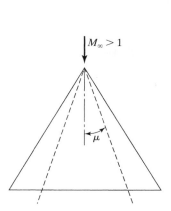

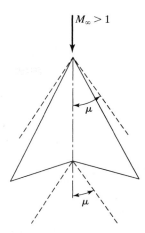

Figure 11.4 Delta wing with supersonic leading and trailing edges.

Figure 11.5 Arrow wing with a subsonic leading edge and a supersonic trailing edge.

In Fig. 11.3, AD and BC are supersonic leading and trailing edges, respectively. AE and DF are subsonic leading edges, and EB and FC are subsonic trailing edges. Note that the points, A, D, B, and C are the points of tangency of the free-stream Mach cone with the leading and trailing edges.

The delta wing of Fig. 11.4 has supersonic leading and trailing edges, while the arrow wing of Fig. 11.5 has a subsonic leading edge and a supersonic trailing edge.

Points on the upper surface within two-dimensional regions that are bounded by supersonic edges have flows that are independent of lower surface flow and vice versa, for points on the lower surface. Thus, in many cases, portions of supersonic wings can be treated by the two-dimensional theory of Chapter 10.

The conclusion drawn in Chapter 10—that good aerodynamic efficiency in supersonic flight depends on thin-airfoil sections with sharp leading and trailing edges—carries over to finite aspect ratio wings. We also find that the benefits of sweepback (discussed in Chapter 9) in decreasing wave drag are also present in the supersonic regime.

An experimental investigation was conducted to determine the aerodynamic characteristics of a potential high-speed civil transport [Hernandez et al. (1993)]. As noted therein, "The inboard wing panel has a leading-edge sweep of 79°, which produces a subsonic normal Mach number at the Mach 3.0 cruise condition. Because of the subsonic leading-edge normal Mach number, relatively blunt leading-edges were possible without a substantial zero-lift wave drag penalty."

11.3 GOVERNING EQUATION AND BOUNDARY CONDITIONS

In Chapter 9 we derived the small perturbation (or linear potential) equation (9.13). Although the derivation was for the subsonic case, the assumptions made in that derivation are satisfied by thin wings in supersonic flow as well.

In mathematical form, the assumptions made in deriving equation (9.13) are

$$M_\infty^2 \left(\frac{u'}{U_\infty}\right)^2 \ll 1 \qquad M_\infty^2 \left(\frac{v'}{U_\infty}\right)^2 \ll 1 \qquad M_\infty^2 \left(\frac{w'}{U_\infty}\right)^2 \ll 1$$

$$M_\infty^2 \left(\frac{u'v'}{U_\infty^2}\right) \ll 1 \qquad M_\infty^2 \left(\frac{u'w'}{U_\infty^2}\right) \ll 1 \qquad M_\infty^2 \left(\frac{v'w'}{U_\infty^2}\right) \ll 1$$

$$\frac{M_\infty^2}{|1 - M_\infty^2|}\left(\frac{u'}{U_\infty}\right) \ll 1 \qquad M_\infty^2 \left(\frac{v'}{U_\infty}\right) \ll 1 \qquad M_\infty^2 \left(\frac{w'}{U_\infty}\right) \ll 1 \qquad \textbf{(11.3)}$$

One observes from equation (11.3) that the assumptions are satisfied for thin wings in supersonic flow provided the free-stream Mach number (M_∞) is not too close to one (transonic regime), nor too great (hypersonic regime). Practically speaking, this restricts supersonic linear theory to the range $1.2 \le M_\infty \le 5$.

Rewriting equation (9.13) in standard form (to have a positive factor for the ϕ_{xx} term) yields

$$(M_\infty^2 - 1)\phi_{xx} - \phi_{yy} - \phi_{zz} = 0 \qquad \textbf{(11.4)}$$

where ϕ is the perturbation potential. This is a linear, second-order partial differential equation of the hyperbolic type, whereas equation (9.13) is of the elliptic type (when $M_\infty < 1$). This fundamental mathematical difference between the equations governing subsonic and supersonic small perturbation flow has already been discussed in Chapter 8. There, we saw that a small disturbance in a subsonic stream affects the flow upstream and downstream of the disturbance, whereas in a supersonic flow the influence of the disturbance is present only in the "zone of action" defined by the Mach cone emanating in the downstream direction from the disturbance. These behaviors are characteristic of the solutions to elliptic and hyperbolic equations, respectively.

A boundary condition imposed on the flow is that it must be tangent to the surface at all points on the wing. Mathematically, we have

$$\left(\frac{w'}{U_\infty + u'}\right)_{\text{surface}} = \frac{dz_s}{dx} \qquad \textbf{(11.5)}$$

which is the same as the condition imposed on the flow in the subsonic case. See the equation immediately following equation (7.41). Note that the equation of the surface is given by $z_s = z_s(x, y)$.

Consistent with the definition of the perturbation potential, equation (11.5) becomes

$$\left(\frac{\phi_z}{U_\infty + \phi_x}\right)_{\text{surface}} = \frac{dz_s}{dx} \qquad \textbf{(11.6)}$$

Applying the assumption that the flow perturbations are small, we have

$$(\phi_z)_{z=0} = U_\infty \frac{dz_s}{dx} \qquad \textbf{(11.7)}$$

as the flow tangency boundary condition since $U_\infty + u' \approx U_\infty$ and the surface corresponds to $z_s \approx 0$. An additional condition that must be applied at a subsonic trailing edge is the *Kutta condition*. This condition is

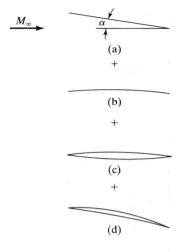

Figure 11.6 Effects of angle of attack, camber, and thickness are additive in linear theory: (a) angle of attack; (b) camber distribution; (c) thickness distribution; (d) resultant wing.

$$C_{pu_{te}} = C_{pl_{te}} \tag{11.8}$$

where the subscripts u_{te} and l_{te} stand for upper and lower wing surface at the trailing edge, respectively. Physically, the condition represented by equation (11.8) means that the local lift at a subsonic trailing edge is zero.

11.4 CONSEQUENCES OF LINEARITY

In Chapter 10 it was shown that the effects of angle of attack, camber, and thickness distribution were additive; refer to equations (10.8), (10.14), and (10.22). In general, a wing can, for the purpose of analysis, be replaced by three components (see Fig. 11.6): (a) a flat plate of the same planform at the same angle of attack; (b) a thin plate with the same planform and camber at zero angle of attack; and (c) a wing of the same planform and thickness distribution but with zero camber and zero angle of attack. The perturbation potential for each of these components can be determined separately and added together to get the combined potential describing the flow about the actual wing.

Thus, the linear nature of the governing equation and the boundary conditions allows us to break the general wing problem into parts, solve each part by methods appropriate to it, and linearly combine the results to arrive at the final flow description. The ability to treat thin wings in this manner greatly simplifies what would otherwise be very difficult problems.

11.5 SOLUTION METHODS

We learned in Chapter 9 that subsonic, compressible flows could be treated by applying an affine transformation [equation (9.14)] to [equation (9.13)] and the boundary conditions. This resulted in an equivalent incompressible flow problem [equation (9.15)], which could be handled by the methods of Chapters 6 and 7 for two- and three-dimensional flows, respectively. However, no affine transformation exists which can be used to transform [equation (11.4)] into [equation (9.15)].

We will discuss two solution methods for equation (11.4). The first, the conical-flow method, was first proposed by Busemann (1947) and was used extensively before the advent of high-speed digital computers. The second, the singularity-distribution method, has been known for some time [Lomax et al. (1951), Shapiro (1954), and Ferri (1949)] but was not generally exploited until the high-speed digital computer was commonly available. The latter method is particularly suited to such computers and is more easily applied to general configurations than the former. Thus, it is more widely used today. However, the solutions generated using the conical-flow method (where applicable) serve as comparison checks for solutions using the computerized singularity-distribution method.

11.6 CONICAL-FLOW METHOD

A conical flow exists when flow properties, such as the velocity, the static pressure, and the static temperature, are invariant along rays (e.g., *PA* in Fig. 11.7) emanating from a point (e.g., point *P* at the wing tip). If equation (11.4) is transformed from the *x, y, z* coordinate system to a conical coordinate system [as was done in Snow (1948)], the resulting equation has only two independent variables, since properties are invariant along rays from the apex of the cone. A further transformation [Shapiro (1954)] results in Laplace's equation in two independent variables, which is amenable to solution by well-known methods (complex variable theory, Fourier series, etc.).

Since the conical-flow technique is not generally applicable and is not as adaptable to computers, we will not go through the mathematical details of its development. However, we will present some results that are applicable to simple wing shapes. The interested reader is referred to Shapiro (1954), Ferri (1949), Carafoli (1956) and Jones and Cohen (1957), for in-depth presentations of conical flow theory and its applications.

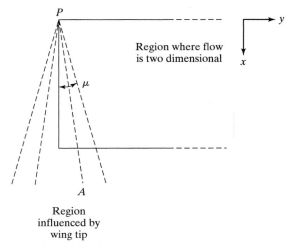

Figure 11.7 In a conical flow, properties are invariant along rays emanating from a point.

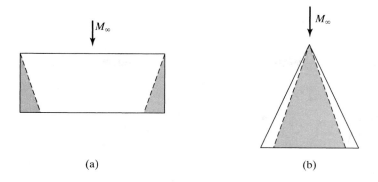

Shaded regions: can be analyzed with conical theory
Unshaded regions: can be analyzed with two-dimensional theory

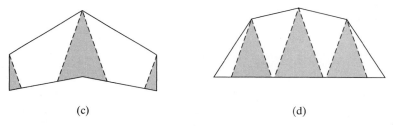

Figure 11.8 Examples of regions that can be treated with conical theory: (a) rectangular wing; (b) delta wing; (c) swept swing; (d) "double" delta wing.

Regions of various wings that can be treated using conical-flow theory are illustrated in Fig. 11.8.

11.6.1 Rectangular Wings

Bonney (1947) has shown that the lift inside the Mach cone at the tip of a rectangular wing is one-half the lift of a two-dimensional flow region of equal area. This is illustrated in the pressure distribution for an isolated rectangular wing tip, which is presented in Fig. 11.9. The curve of Fig. 11.9 represents the equation

$$\frac{\Delta p}{\Delta p_{2-d}} = \frac{2}{\pi} \sin^{-1} \sqrt{\frac{\tan \mu'}{\tan \mu}}$$

The analysis can be extended to the interaction of the two tip flows when their respective Mach cones intersect (or overlap) on the wing surface. The case where the entire trailing edge of the wing is in the overlap region of the tip Mach cones is illustrated in

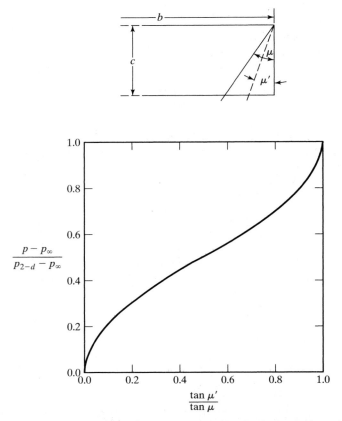

Figure 11.9 Pressure distribution at the tip of a rectangular wing (p = actual pressure due to tip loss, p_{2-d} = corresponding two-dimensional pressure). [From Bonney (1947).]

Fig. 11.10. The pressure distribution in the region of overlap (when $1 \leq \beta \cdot AR \leq 2$) is determined by adding the pressures due to each tip and subtracting from them the two-dimensional pressure field determined by Busemann's second-order equation (10.23). Note that $\beta^2 = M_\infty^2 - 1$.

The extent of the overlap region is determined by the parameter $\beta \cdot AR$. Three cases are shown in Fig. 11.11. Conical-flow theory is not applicable in the regions indicated in the figure. Such cases will occur when $\beta \cdot AR < 1$.

The key assumptions used in this development are as follows:

1. Secondary tip effects originating at the point of maximum thickness of a double-wedge airfoil are neglected.
2. Tip effects extend to the limits of the Mach cone defined by the free-stream Mach number M_∞ and not to the Mach cone defined by the local Mach number.
3. Flow separation does not occur.
4. Linear theory applies.

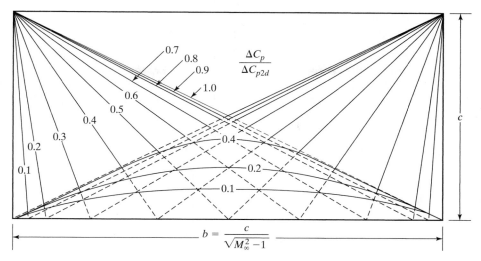

Figure 11.10 Effect of a subsonic wing tip on the pressure distribution for a rectangular wing for which $\beta \cdot AR = 1$.

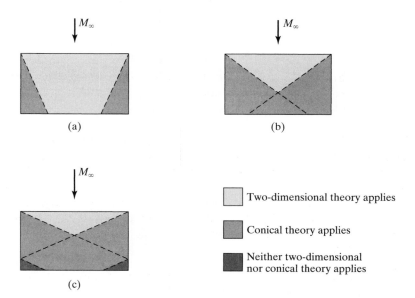

Figure 11.11 Regions where conical flow applies for rectangular wings: (a) $\beta \cdot AR > 2$: no overlap, (b) $1 \leq \beta \cdot AR \leq 2$: overlap along the trailing edge; (c) $\beta \cdot AR < 1$: overlap extends beyond the wing tips.

A summary of the results from Bonney (1947) for the case of nonoverlapping tip effects is given in Table 11.1. Conclusions to be drawn from this analysis are as follows:

1. A decrease in the aspect ratio for a given supersonic Mach number and airfoil section results in decreases in the coefficients for the drag-due-to-lift, the lift, and the pitching moment. Note that the behavior of the drag-due-to-lift here is in direct contrast to its behavior in subsonic flow. The center of pressure will move forward with a decrease in aspect ratio.

2. When the thickness ratio is increased, the coefficients of lift and of drag-due-to-lift for finite-span wings increase slightly, but the moment coefficient about the leading edge decreases. Further, the center of pressure moves forward both for airfoils and for wings as the thickness ratio is increased.

3. Airfoils having the same cross-sectional area will have the same center of pressure location.

4. The thickness drag will vary with the square of the thickness ratio for a given cross-sectional shape.

5. Airfoils of symmetrical cross section with a maximum thickness at the midchord point will have the least drag for a given thickness ratio.

A comparison of conical flow predictions with data obtained from Nielsen et al. (1948) is given in Fig. 11.12 for a double-wedge airfoil for the conditions shown in the figure.

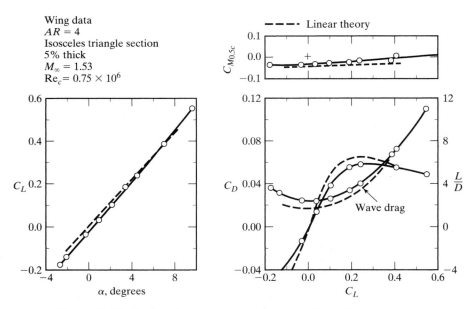

Figure 11.12 Comparison of linear theory results with experimental data for a rectangular wing. [Data from Nielsen et al. (1948).]

TABLE 11.1 Conical-Flow Results for Rectangular Wings Using Busemann Second-Order Approximation Theory

Type	Flat-Plate Airfoil	Flat-Plate Wing	Finite-Thickness Airfoil	Finite-Thickness Wing
C_L	$\dfrac{4\alpha}{\beta}$	$\dfrac{4\alpha}{\beta}\left(1 - \dfrac{1}{2AR\cdot\beta}\right)$	$\dfrac{4\alpha}{\beta}$	$\dfrac{4\alpha}{\beta}\left[1 - \dfrac{1}{2AR\cdot\beta}(1 - C_3A')\right]$
C_D	$\dfrac{4\alpha^2}{\beta} + C_{D,\text{friction}}$	$\dfrac{4\alpha^2}{\beta}\left(1 - \dfrac{1}{2AR\cdot\beta}\right) + C_{D,\text{friction}}$	$\dfrac{K_1\tau^2}{\beta} + \dfrac{4\alpha^2}{\beta} + C_{D,\text{friction}}$	$\dfrac{K_1\tau^2}{\beta} + C_{D,\text{friction}} + \dfrac{4\alpha^2}{\beta}\left[1 - \dfrac{1}{2AR\cdot\beta}(1 - C_3A')\right]$
C_{M_0}	$\dfrac{2\alpha}{\beta}$	$\dfrac{2\alpha}{AR\cdot\beta^2}\left(AR\cdot\beta - \dfrac{2}{3}\right)$	$\dfrac{2\alpha}{\beta}(1 - C_3A')$	$\dfrac{2\alpha}{AR\cdot\beta^2}\left[AR\cdot\beta - \dfrac{2}{3} - C_3A'(AR\cdot\beta - 1)\right]$
x_{cp}	$\dfrac{c}{2}$	$\left[\dfrac{AR\cdot\beta - \frac{2}{3}}{2AR\cdot\beta - 1}\right]c$	$(1 - C_3A')\dfrac{c}{2}$	$\left[\dfrac{AR\cdot\beta - \frac{2}{3} - C_3A'(AR\cdot\beta - 1)}{2AR\cdot\beta - 1 + C_3A'}\right]c$

$$C_3 = \frac{\gamma M_\infty^4 + (M_\infty^2 - 2)^2}{2(M_\infty^2 - 1)^{3/2}}; \quad A' = \frac{\text{airfoil cross-sectional area}}{\text{chord squared}}; \quad \alpha \text{ in radians.}$$

Values of A', K_1:

Type of airfoil	A'	K_1
Double wedge	$\frac{1}{2}\tau$	4
Modified double wedge	$\frac{2}{3}\tau$	6
Biconvex	$\frac{2}{3}\tau$	5.33

Source: Bonney (1947).

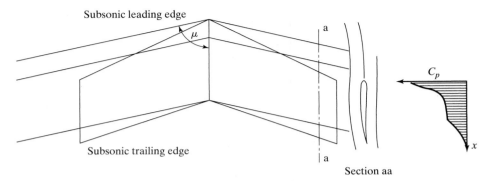

Subsonic leading edge

μ

a

C_p

Subsonic trailing edge

a

x

Section aa

Figure 11.13 Pressure distribution over the wing chord for a section of a swept wing with subsonic leading and trailing edges.

11.6.2 Swept Wings

If the wing leading edge is swept aft of the Mach cone originating at the apex of the wing as shown in Fig. 11.13, the disturbances propagate along the Mach lines "warning" the approaching flow of the presence of the wing. As noted earlier, a leading edge that is swept within the Mach cone is referred to as subsonic leading edge and the flow approaching the wing is similar to subsonic flow, even though the flight speed is supersonic. Thus, when the leading edge is subsonic, the wing can be treated by the methods of Chapters 7 and 9. Furthermore, for sufficiently swept leading edges, they can be rounded similar to those used for subsonic speeds.

There are penalties associated with sweepback (some of which show up at subsonic speeds) including reduction of the lift-curve slope, increased drag-due-to-lift, tip stalling, and reduced effectiveness of high-lift devices. There are structural considerations also. A swept wing has a greater structural span than a straight wing having the same area and aspect ratio. Sweepback introduces additional (possibly severe) torsion because the applied loads on the wing act aft of the wing root.

Consider the flow depicted in Fig. 11.14, where the leading and trailing edges of the wing are supersonic. The tips and center portion of a swept wing can be treated with

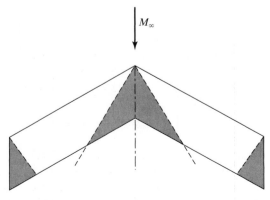

M_∞

Shaded regions: conical flow theory applies
Unshaded regions: two-dimensional flow theory applies

Figure 11.14 Regions of conical flow and of two-dimensional flow for a swept wing.

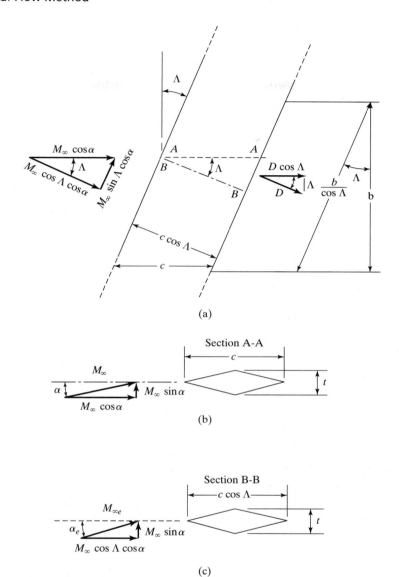

Figure 11.15 Nomenclature for flow around a sweptback wing of infinite aspect ratio: (a) view in plane of wing; (b) view in plane parallel to direction of flight; (c) view in plane normal to the leading edge. [From Shapiro (1954).]

conical-flow theory while the remaining portion of the wing can be analyzed by the two-dimensional techniques of Chapter 10, if an appropriate coordinate transformation is made. Refer to Fig. 11.15, in which a segment of an infinitely long sweptback wing with sweepback angle Λ and angle of attack α are presented. The free-stream Mach number can be broken into the three components, as shown in figure. The component tangent to the leading edge is unaffected by the presence of the wing (if we neglect viscous effects). Thus, we may consider the equivalent free-stream Mach number M_{∞_e} normal to the leading edge. This will be the flow as seen by an observer moving spanwise at the tangential Mach number $M_{\infty} \sin \Lambda \cos \alpha$. Note that the airfoil section exposed to M_{∞_e} will be that taken in a plane normal to the leading edge. The flow at M_{∞_e} about this section can be treated with the two-dimensional theory of Chapter 10.

Referring to the geometry presented in Fig. 11.15, we can see that

$$M_{\infty_e} = [(M_{\infty} \sin \alpha)^2 + (M_{\infty} \cos \alpha \cos \Lambda)^2]^{0.5}$$

or

$$M_{\infty_e} = M_{\infty}(1 - \sin^2 \Lambda \cos^2 \alpha)^{0.5} \tag{11.9}$$

Also,

$$\alpha_e = \tan^{-1} \frac{M_{\infty} \sin \alpha}{M_{\infty} \cos \alpha \cos \Lambda} = \tan^{-1} \frac{\tan \alpha}{\cos \Lambda} \tag{11.10}$$

$$\tau_e = \frac{t}{c \cos \Lambda} = \frac{\tau}{\cos \Lambda} \tag{11.11}$$

where $\tau \equiv t/c$ is the thickness ratio.

Now, the total lift per unit span is not changed by the spanwise motion of the observer; and only M_{∞_e} generates pressure forces necessary to create lift. Thus,

$$C_l = \frac{l}{(\gamma/2) p_{\infty} M_{\infty}^2 c}$$

and

$$C_{le} = \frac{l}{(\gamma/2) p_{\infty} M_{\infty_e}^2 c \cos \Lambda (1/\cos \Lambda)} \tag{11.12}$$

Similarly, ignoring viscous effects and noting that the wave drag is normal to the leading edge,

$$C_{de} = \frac{d}{(\gamma/2) p_{\infty} M_{\infty_e}^2 c \cos \Lambda (1/\cos \Lambda)} \tag{11.13}$$

while

$$C_d = \frac{d \cos \Lambda}{(\gamma/2) p_{\infty} M_{\infty}^2 c} \tag{11.14}$$

where $d \cos \Lambda$ is the drag component in the free-stream direction.

Combining equations (11.9) and (11.12), we get

$$C_l = C_{le} \left(\frac{M_{\infty e}}{M_{\infty}} \right)^2 = C_{le}(1 - \sin^2 \Lambda \cos^2 \alpha) \tag{11.15}$$

For the drag, we combine equations (11.9), (11.13), and (11.14) to get

$$C_d = C_{de} \cos \Lambda \left(\frac{M_{\infty e}}{M_\infty} \right)^2 = C_{de} \cos \Lambda (1 - \sin^2 \Lambda \cos^2 \alpha) \tag{11.16}$$

The results just derived are true in general. If we restrict ourselves to the assumptions of the linear (Ackeret) theory, which was discussed in Chapter 10, then

$$C_{le} = \frac{4\alpha_e}{\sqrt{M_{\infty e}^2 - 1}} \tag{11.17}$$

$$C_{de} = \frac{4}{\sqrt{M_{\infty e}^2 - 1}} \left(\alpha_e^2 + \frac{\overline{\sigma_{ue}^2} + \overline{\sigma_{le}^2}}{2} \right) \tag{11.18}$$

These results are identical to those obtained in Chapter 10 for infinite aspect ratio wings with leading edges normal to the free-stream flow direction. See equations (10.8) and (10.16).

The results of Ivey and Bowen (1947) for flow about sweptback airfoils with double-wedge profiles are presented in Fig. 11.16. Note that significant improvement in performance can be realized with sweep back. The results in Fig. 11.16 are based on the exact relations [i.e., equations (11.15) and (11.16)], the shock-expansion theory (not linear theory), and the assumption that the skin friction drag coefficient per unit span is 0.006.

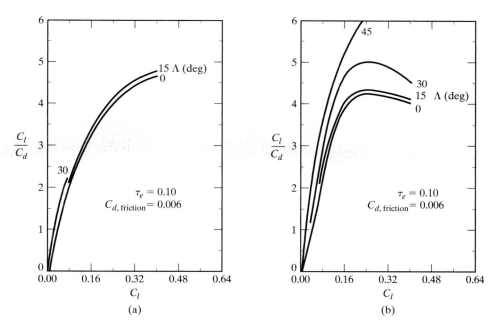

Figure 11.16 Theoretical effect of sweepback for double-wedge section airfoils with supersonic leading edges: (a) lift-to-drag ratio for $M_\infty = 1.5$; (b) lift-to-drag ratio for $M_\infty = 2.0$;

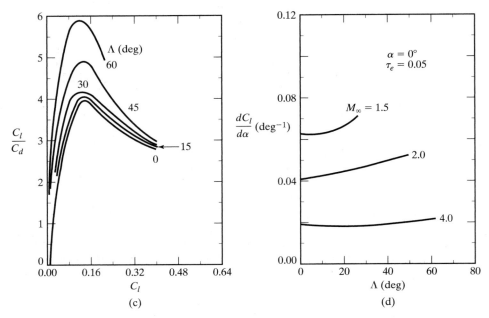

Figure 11.16 (*Continued*) (c) lift-to-drag ratio for $M_\infty = 4.0$; (d) lift-curve slope. [From Ivey and Bowen (1947)]

11.6.3 Delta and Arrow Wings

Puckett and Stewart (1947) used a combination source-distribution and conical-flow theory to investigate the flow about delta- and arrow-shaped planforms (see Fig. 11.17). Cases studied included subsonic and/or supersonic leading and trailing edges with double-wedge airfoil sections. Stewart (1946) and Puckett (1946) used conical-flow theory to investigate the flow about simple delta planforms.

Two significant conclusions about delta and arrow planforms that can be drawn from these studies are as follows:

1. For wings where the sweepback of both leading and trailing edges is relatively small, the strong drag peak at Mach 1 (characteristic of two-dimensional wings) is replaced by a weaker peak at a higher Mach number, corresponding to coincidence of the Mach wave with the leading edge.
2. Delta and arrow wings with subsonic leading edges can have lift curve slopes $(dC_L/d\alpha)$ approaching the two-dimensional value $(4/\beta)$ with much lower values of C_D/τ^2 than those characteristic of two-dimensional wings of the same thickness.

A theoretical comparison of a rectangular, delta, and arrow wing is given in Table 11.2. As noted in Wright et al. (1978), "One of the prominent advantages of the arrow wing is in

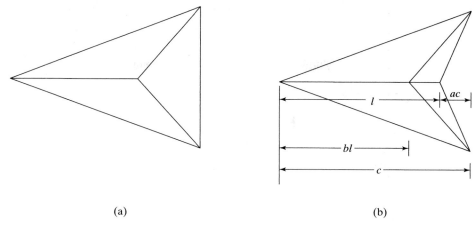

(a)   (b)

Figure 11.17 Delta- and arrow-wing planforms with double-wedge sections: (a) delta planform; (b) arrow planform.

TABLE 11.2 Comparison of Aerodynamic Coefficients for Rectangular, Delta, and Arrow Wing Planforms for $M_\infty = 1.50$

Wing Planform	Rectangular	Delta [a]	Arrow[a]
	$\Lambda = 70°$	$\Lambda = 70°$ $b = 0.2$ $\beta \cot \Lambda = 0.4$	$\Lambda = 70°$ $b = 0.2$ $\beta \cot \Lambda = 0.4$
	$AR = 1$	$a = 0$	$a = 0.25$
$\dfrac{\beta}{4} \dfrac{dC_L}{d\alpha}$	0.554	0.544	0.591
$\dfrac{dC_L}{d\alpha}$	1.98	1.94	2.11
Relative area, S	1.00	1.018	0.938
Relative root chord, l	1.00	1.69	1.26
Root thickness ratio, τ	0.10	0.059	0.080
$C_{D,\,thickness}$	0.0119	0.0048	0.0070
$C_{D,\,friction}$	0.0060	0.0060	0.0060
C_{D0}	0.0179	0.0108	0.0130
$\left(\dfrac{C_L}{C_D}\right)_{max}$	5.25	8.6	9.3

[a] See Fig. 11.17 for a definition of a and b.
Source: Puckett and Stewart (1947).

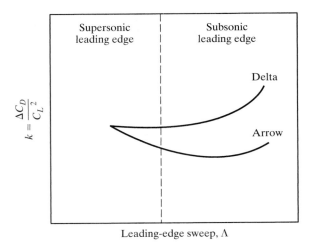

Figure 11.18 Comparison of induced drag for delta- and arrow-wing planforms. *Note*: $\Delta C_D = \Delta C_{D0}$. [From Wright et al. (1978).]

the area of induced drag. This is illustrated qualitatively in Fig. 11.18, where the planform with a trailing edge cut out or notch ratio is shown to have lower induced drag. The second advantage of the arrow wing is its ability to retain a subsonic round leading edge at an aspect ratio that is of the same level as that of the lesser swept supersonic leading edge delta. The advantages of the subsonic leading edge are a lower wave drag at cruise and a high L/D for subsonic flight operations due to increased leading edge suction."

11.7 SINGULARITY-DISTRIBUTION METHOD

The second method that can be used to solve equation (11.4) is the singularity-distribution method. Detailed treatment of the mathematical aspects of the theory as well as applications of it to various wing planforms are presented in Lomax et al. (1951), Shapiro (1954), Jones and Cohen (1957) and Carlson and Miller (1974). For simple planforms, the singularity-distribution method can provide exact analytical closed-form solutions to three-dimensional wing problems [see Shapiro (1954), Jones and Cohen (1957), and Puckett and Stewart (1946)]. However, the method is quite adaptable for use with computers to solve for flow about complex shapes, and this is where it is most extensively applied today.

In Chapter 3 we learned that the governing equation [i.e., equation (3.26)] for incompressible, irrotational flow is linear even without the assumption of small disturbances. This allowed us to combine elementary solutions (i.e., source, sink, doublet, vortex, etc.) of the governing equation to generate solutions for incompressible flows about shapes of aerodynamic interest. In supersonic flow, where the small disturbance assumption is necessary to linearize the governing equation [e.g., equation (11.4)], there are analogs to the simple solutions for the incompressible case. Owing to a mathematical similarity to their incompressible counterparts, the supersonic solutions are quite naturally referred to as supersonic sources, sinks, doublets, and vortices. However, the physical relationship to their subsonic counterparts is not quite so direct and will not be pursued here.

The supersonic source (recall that a sink is simply a negative source), the doublet, and the horseshoe vortex potentials given by Lomax (1951) are as follows:

$$\text{Source: } \phi_s = -\frac{Q}{r_c} \tag{11.19a}$$

$$\text{Doublet: } \phi_d = +\frac{Qz\beta^2}{r_c^3} \tag{11.19b}$$

where the axis of the doublet is in the positive z direction.

$$\text{Vortex: } \phi_v = -\frac{Qzv_c}{r_c} \tag{11.19c}$$

In equation (11.19), Q is the strength of the singularity and

$$r_c = \{(x - x_1)^2 - \beta^2[(y - y_1)^2 + z^2]\}^{0.5}$$
$$\beta^2 = M_\infty^2 - 1$$
$$v_c = \frac{x - x_1}{(y - y_1)^2 + z^2}$$

One can verify by direct substitution that equations (11.19a) through (11.19c) satisfy equation (11.4). Note that the point (x_1, y_1, z_1) is the location of the singularity. Since the wing is in the $z = 0$ plane, $z_1 = 0$ for every singularity for this approximation. This is because, in the singularity distribution method, the wing is replaced by a distribution of singularities in the plane of the wing. The hyperbolic radius, r_c, is seen to be imaginary outside the Mach cone extending downstream from the location of the singularity in each instance. Thus, the influence of the singularity is only present in the zone of action downstream of the point $(x_1, y_1, 0)$.

Using high-speed digital computers, modern numerical techniques consist of modeling a wing or body by replacing it with singularities at discrete points. These singularities are combined linearly to create a flow pattern similar to that about the actual body. The strengths of the singularities are determined so that the boundary condition requiring that the flow be tangent to the body's surface is satisfied at selected points. For configurations with sharp trailing edges it is also necessary to satisfy the Kutta condition at those sharp trailing edges which are subsonic.

Once the singularity distribution is determined, the potential at a given point is obtained by summing the contributions of all the singularities to the potential at that point. Velocity and pressure distributions follow from equations (9.12) and (9.16a).

Four types of problems that can be treated by the singularity-distribution method [see Lomax et al. (1951)] are as follows:

Two nonlifting cases:

1. Given the thickness distribution and the planform shape, find the pressure distribution on the wing.
2. Given the pressure distribution on a wing of symmetrical section, find the wing shape (i.e., find the thickness distribution and the planform).

Two lifting cases:

3. Given the pressure distribution on a lifting surface (zero thickness), find the slope at each point of the surface.

4. Given a lifting surface, find the pressure distribution on it. Here, it is necessary to impose the Kutta condition for subsonic trailing edges when they are present.

Cases 1 and 3 are called "direct" problems because they involve integrations with known integrands. Cases 2 and 4 are "indirect" or "inverse" problems, since the unknown to be found appears inside the integral sign. Thus, the solution of inverse problems involves the inversion of an integral equation.

One might expect that cases 1 and 4 would be the only ones of practical interest. However, this is not the case. Many times, a designer wishes to specify a given loading distribution (e.g., either for structural or for stability analyses) and solve for the wing shape which will give that prescribed loading distribution. Thus, the engineer may encounter any one of the four cases in aircraft design work. A variation of cases 2 and 3 is to specify the potential on a surface instead of the pressure distribution.

Cases 1 and 2 are most conveniently solved using source or doublet distributions, while cases 3 and 4 are most often treated using vortex distributions.

11.7.1 Find the Pressure Distribution Given the Configuration

Consider a distribution of supersonic sources in the xy plane. The contribution to the potential at any point $P(x, y, z)$ due to an infinitesimal source at $P'(x_1, y_1, 0)$ in the plane is, from equation (11.19a),

$$d\phi(x, y, z) = -\frac{C(x_1, y_1)\, dx_1 dy_1}{\sqrt{(x - x_1)^2 - \beta^2[(y - y_1)^2 + z^2]}} \tag{11.20}$$

where $C(x_1, y_1)$ is the source strength per unit area. Consistent with the linearity assumption, the flow tangency condition [equation (11.7)] gives the vertical (z direction) velocity component in the xy plane as

$$w'(x, y, 0) = \left[\frac{\partial\phi(x, y, z)}{\partial z}\right]_{z=0} = U_\infty \frac{dz_s(x, y)}{dx} \tag{11.21}$$

Taking the derivative with respect to z of equation (11.20) gives

$$\frac{\partial[d\phi(x, y, z)]}{\partial z} = dw'(x, y, z) = -\frac{C(x_1, y_1)\beta^2 z\, dx_1\, dy_1}{\{(x - x_1)^2 - \beta^2[(y - y_1)^2 + z^2]\}^{1.5}} \tag{11.22}$$

Note that x_1 and y_1 are treated as constants. Taking the limit of equation (11.22) as $z \to 0$, we get $dw'(x, y, 0) = 0$, except very near the point (x_1, y_1) where the limit is indeterminant (i.e., of the form 0/0). We conclude that the vertical velocity at a point in the xy plane is due only to the source at that point and to no other sources. In other words, a source induces a vertical velocity at its location and nowhere else. A source does, however, contribute to u' and v' (i.e., the x and y components of the perturbation velocity) at other locations. We still must determine the contribution of the source at the point (x_1, y_1) to the vertical velocity at $P'(x_1, y_1, 0)$. Puckett (1946) shows that the latter contribution is

$$dw'(x_1, y_1, 0) = \pi C(x_1, y_1) \qquad \textbf{(11.23a)}$$

However, since this is the entire contribution,

$$w'(x_1, y_1, 0) = dw'(x_1, y_1, 0) = \pi C(x_1, y_1) \qquad \textbf{(11.23b)}$$

Using equation (11.21), we see that

$$C(x_1, y_1) = \frac{U_\infty}{\pi} \left[\frac{dz_s(x_1, y_1)}{dx_1} \right] \equiv \frac{U_\infty \lambda(x_1, y_1)}{\pi} \qquad \textbf{(11.23c)}$$

where λ is the local slope of the wing section. Thus, we obtain the important result that the source distribution strength at a point is proportional to the local surface slope at the point.

Substituting equation (11.23c) into equation (11.20) and integrating gives

$$\phi(x, y, 0) = -\frac{U_\infty}{\pi} \int \int_S \frac{\lambda(x_1, y_1) \, dx_1 \, dy_1}{[(x - x_1)^2 - \beta^2(y - y_1)^2]^{0.5}} \qquad \textbf{(11.24)}$$

where S is the region in the xy plane within the upstream Mach cone with apex at $P(x, y)$. Once the potential is known, the pressure distribution follows from equation (9.16); that is,

$$C_p = -\frac{2(\partial\phi/\partial x)}{U_\infty} \qquad \textbf{(9.16a)}$$

Since source distributions are used where the airfoil section is symmetric (see the preceding cases 1 and 2), the pressure distribution determined by equations (11.24) and (9.16) is the same on the upper and lower surfaces. Because linear theory requires that the deflection angles are small, the wave-drag coefficient at zero lift is

$$C_{Dw} = 2 \int \int_{S_u} C_{pu}(x, y) \lambda_u(x, y) \, dx \, dy \qquad \textbf{(11.25)}$$

where S is the surface of the wing and the subscript u indicates the upper surface. The formula contains the factor 2 to include the contribution of the lower surface since the section is symmetric.

EXAMPLE 11.1: Determine the pressure distribution for the single-wedge delta wing

Let us determine the pressure distribution for the simple wing shape shown in Fig. 11.19, which is a zero angle of attack. This is a single-wedge delta with subsonic leading edges. The leading edge is swept by the angle Λ_{LE} Granted, this wing is not practical because of its blunt trailing edge. However, neglecting the effects of the presence of the boundary layer, the effects of the trailing

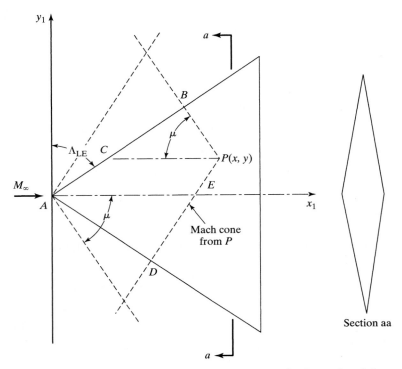

Figure 11.19 Nomenclature and geometry for single-wedge delta wing of Example 11.1.

edge are not propagated upstream in the supersonic flow. Thus, to obtain the flow about a wing with a sharp trailing edge, we can add (actually subtract) the solutions for the flow around two delta wings of constant slope such that the desired airfoil section can be obtained. This additive process is illustrated in Fig. 11.20. Thus, the simple case considered here can be used as a building block to construct more complex flow fields about wings of more practical shape [see Shapiro (1954) and Stewart (1946)].

Solution: Consider the point $P(x, y)$ in Fig. 11.19. The flow conditions at P are a result of the combined influences of all the sources within the upstream Mach cone from P. From symmetry, the vertical velocity perturbation vanishes ahead of the wing, and the source distribution simulating the wing extends only to the leading edges. Thus, the source distribution which affects P is contained entirely in the region $ABPD$.

The potential at P is given by equation (11.24):

$$\phi(x, y, 0) = -\frac{\lambda}{\pi}U_\infty \iint\limits_{ABPD} \frac{dx_1\, dy_1}{\left[(x - x_1)^2 - \beta^2(y - y_1)^2\right]^{0.5}}$$

where we have moved λ outside the integral sign, since it is a constant in this example.

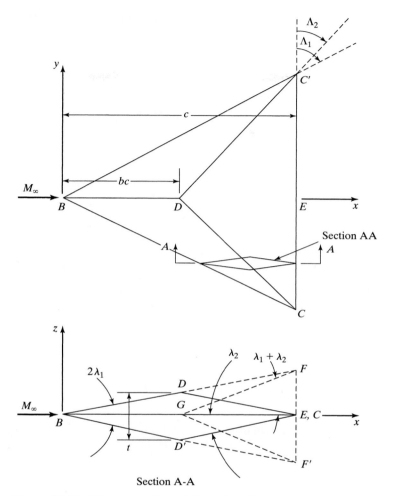

Figure 11.20 Geometry for a delta-wing planform with a double-wedge section. [From Shapiro (1954).]

To carry out the integration, it is convenient to break $ABPD$ into three separate areas, thus:

$$\phi(x, y, 0) = -\frac{\lambda}{\pi} U_\infty \left\{ \iint\limits_{ADE} \frac{dx_1 \, dy_1}{[(x - x_1)^2 - \beta^2(y - y_1)^2]^{0.5}} \right.$$

$$+ \iint\limits_{AEPC} \frac{dx_1 \, dy_1}{[(x - x_1)^2 - \beta^2(y - y_1)^2]^{0.5}}$$

$$\left. + \iint\limits_{CPB} \frac{dx_1 \, dy_1}{[(x - x_1)^2 - \beta^2(y - y_1)^2]^{0.5}} \right\}$$

To define the limits of integration, we note the following relationships from geometry:

$$\text{Along } AD: \quad x_1 = -y_1 \tan \Lambda_{LE}$$
$$\text{Along } ACB: \quad x_1 = +y_1 \tan \Lambda_{LE}$$
$$\text{Along } BP: \quad x_1 = x - \beta(y_1 - y)$$
$$\text{Along } DEP: \quad x_1 = x + \beta(y_1 - y)$$

Finally, the coordinates of point B and D are

$$B\left[\frac{(x + \beta y)\tan \Lambda_{LE}}{\tan \Lambda_{LE} + \beta}, \frac{x + \beta y}{\tan \Lambda_{LE} + \beta}\right]$$

$$D\left[\frac{(x - \beta y)\tan \Lambda_{LE}}{\tan \Lambda_{LE} + \beta}, \frac{-(x - \beta y)}{\tan \Lambda_{LE} + \beta}\right]$$

Thus, we have

$$\phi(x, y, 0) = -\frac{\lambda}{\pi}U_\infty\left[\int_{\frac{-(x-\beta y)}{\tan \Lambda_{LE}+\beta}}^{0} dy_1 \int_{-y_1 \tan \Lambda_{LE}}^{x+\beta(y_1-y)} \frac{dx_1}{[(x-x_1)^2 - \beta^2(y-y_1)^2]^{0.5}}\right.$$

$$+ \int_0^y dy_1 \int_{y_1 \tan \Lambda_{LE}}^{x+\beta(y_1-y)} \frac{dx_1}{[(x-x_1)^2 - \beta^2(y-y_1)^2]^{0.5}}$$

$$\left.+ \int_y^{\frac{(x+\beta y)}{\tan \Lambda_{LE}+\beta}} dy_1 \int_{y_1 \tan \Lambda_{LE}}^{x-\beta(y_1-y)} \frac{dx_1}{[(x-x_1)^2 - \beta^2(y-y_1)^2]^{0.5}}\right]$$

One can use standard integral tables and relationships involving inverse hyperbolic functions [i.e., see Hodgman (1977)] to show that the result of this integration and subsequent differentiation with respect to x is [see Puckett (1946)]

$$u'(x, y, 0) = +\frac{\partial \phi}{\partial x}$$

$$= -\frac{\lambda}{\pi}U_\infty \frac{2}{\beta[(\tan^2 \Lambda_{LE}/\beta^2) - 1]^{0.5}} \cosh^{-1}\left\{\left(\frac{\tan \Lambda_{LE}}{\beta}\right)\left[\frac{1 - (\beta y/x)^2}{1 - (y^2 \tan^2 \Lambda_{LE})/x^2}\right]^{0.5}\right\}$$

By equation (9.16a) the pressure distribution on the wing is

$$C_p(x, y, 0) = -\frac{2u'}{U_\infty}$$

Notice that C_p is invariant along rays ($y/x =$ constant) from the apex of the wing. Thus, as we might suspect from the geometry, this is a conical flow.

The wave-drag coefficient can be determined using equation (11.25). Figure 11.21 presents pressure distributions and wave drag for various configurations of single- and double-wedge delta wings (see Figs. 11.19 and 11.20).

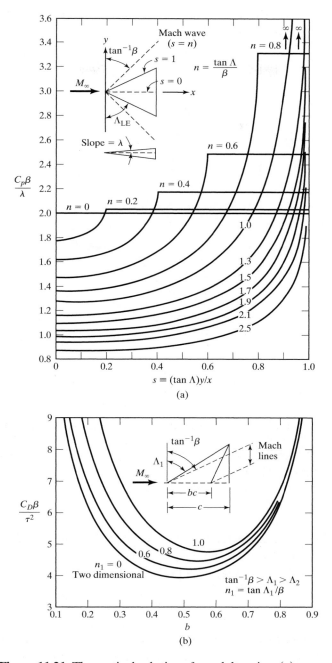

Figure 11.21 Theoretical solutions for a delta wing. (a) pressure distribution for a single-wedge delta wing at $\alpha = 0$. (b) Thickness drag of a double-wedge delta wing with a supersonic leading edge and a supersonic line of maximum thickness. [From Puckett (1946).]

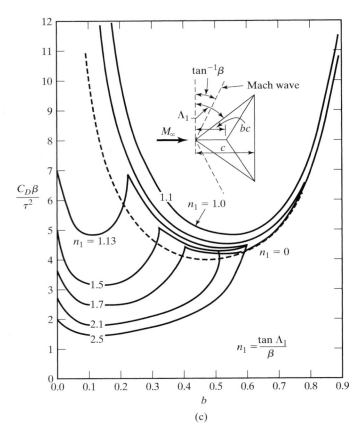

Figure 11.21 (*continued*) (c)Thickness drag of a double-wedge delta wing with a subsonic line of maximum thickness.

EXAMPLE 11.2:   Prepare graphs of the pressure distribution for a single-wedge delta wing

Consider a single-wedge delta wing with a leading-edge sweep angle of 60°, flying at a Mach number of 2.2. The wing has a thickness-to-chord ratio in the plane of symmetry of 0.04. Thus, referring to Fig. 11.21a, the surface slope λ is 0.02. Prepare graphs of C_p as a function $(x - x_{LE})/c(y)$ for the four planes shown in the sketch of Fig. 11.22 (i.e., at stations $y = 0.125b$; $y = 0.250b$; $y = 0.375b$; and $y = 0.450b$).

Solution: Let us first locate the Mach wave originating at the apex of the delta wing.

$$\mu = \sin^{-1}\frac{1}{M_\infty} = \frac{1}{2.2} = 27.04°$$

As shown in the sketch of Fig. 11.23, the Mach wave is downstream of the wing leading edge. Thus, wing leading edge is a supersonic leading edge. In

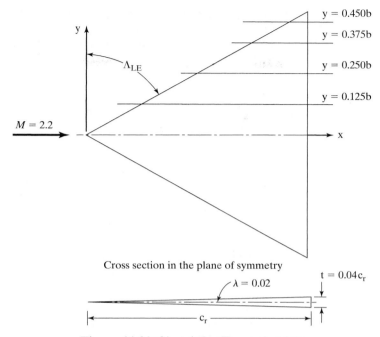

Figure 11.22 Sketch for Example 11.2.

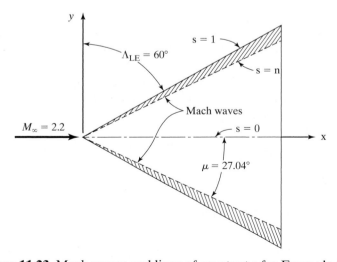

Figure 11.23 Mach waves and lines of constant s for Example 11.2.

the shaded region representing that portion of the wing between the leading edge and the Mach wave, the pressure must be constant and the same as that for a two-dimensional oblique airfoil. In this region,

$$C_p = -\frac{2u'}{U_\infty} = \frac{2\lambda}{\beta\sqrt{1 - n^2}} = \frac{2\lambda}{\sqrt{\beta^2 - \tan^2 \Lambda_{LE}}} \qquad \textbf{(11.26)}$$

The nomenclature used in equation (11.26) is consistent with that used in Figs. 11.21a and 11.23. Thus,

$$\beta = \sqrt{M_\infty^2 - 1} = 1.9596$$

and

$$n = \frac{\tan \Lambda_{LE}}{\beta} = 0.88388$$

The expression given in equation (11.26) can be rearranged to the format used in Fig. 11.21a, so that

$$\frac{C_p \beta}{\lambda} = \frac{2}{\sqrt{1 - n^2}} = 4.276$$

The reader should compare this numerical value with the graphical information presented in Fig. 11.21a. In the region between the wing leading edge and the Mach wave, the pressure coefficient itself is equal to

$$C_p = 0.0436$$

Let us use the expression developed by Puckett (1946) to determine the pressure at points on the wing that lie between the Mach wave and the plane of symmetry, when the wing leading edge is supersonic. The points are located in terms of the coordinate s, where

$$s = (\tan \Lambda_{LE}) \frac{y}{x}$$

Thus, $s = 0$ corresponds to the plane of symmetry, $s = 1$ corresponds to the wing leading edge, and $s = n$ corresponds to the Mach wave. The pressure coefficient in this region is given by

$$C_p = \frac{4\lambda}{\beta\pi\sqrt{1 - n^2}} \text{Real} \left\{ \frac{\pi}{2} - \sin^{-1}\sqrt{\frac{n^2 - s^2}{1 - s^2}} \right\} \qquad \textbf{(11.27)}$$

Graphs of the pressure distributions for the four stations of interest are presented in Fig. 11.24. Note that the Mach wave intersects the wing trailing edge at $y = 0.442b$. As a result, the plane $y = 0.450b$ lies entirely within the shaded region. Thus, the pressure coefficient in the $y = 0.450b$ plane is independent of x and equal to 0.0436.

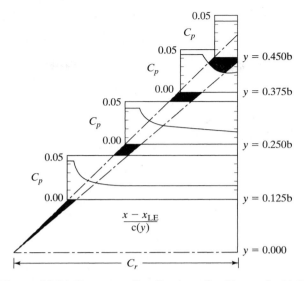

Figure 11.24 Pressure distributions for Example 11.2.

11.7.2 Numerical Method for Calculating the Pressure Distribution Given the Configuration

Carlson and Miller (1974) presented a numerical application of the vortex distribution method which is applicable to thin wings of arbitrary profile and of arbitrary planform. In accordance with the concepts of linearized theory, the wing is assumed to have negligible thickness and is assumed to lie approximately in the $z = 0$ plane. We will discuss the application of their method to determine the lifting pressure coefficient at the field point (x, y).

The equation governing the differential pressure coefficient $(\Delta C_p = C_{pl} - C_{pu})$ is

$$\Delta C_p(x, y) = -\frac{4}{\beta} \frac{\partial z_c(x, y)}{\partial x} + \frac{1}{\pi} \oiint_{s} R(x - x_1, y - y_1) \Delta C_p(x_1, y_1)\, d\beta y_1\, dx_1 \qquad \textbf{(11.28)}$$

where

$$R(x - x_1, y - y_1) \equiv \frac{x - x_1}{\beta^2(y - y_1)^2[(x - x_1)^2 - \beta^2(y - y_1)^2]^{0.5}}$$

and $z_c(x, y)$ is the z coordinate of the camber line. The function R may be thought of as an influence function relating the local loading at the point (x_1, y_1) to its influence on the flow field.

The integral in equation (11.28) represents the influence of a continuous distribution of horseshoe vortices originating from wing elements with vanishingly small chords and spans. The region of integration, S, originating at the field point (x, y), is shown in

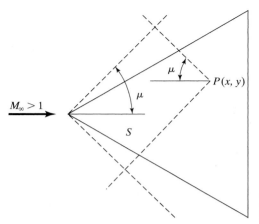

Figure 11.25 S is the region of integration for the supersonic vortex lattice method.

Fig. 11.25. The integral gives the appearance of being improper and divergent because of the singularity at $y_1 = y$ within the region of integration. However, the integral can be treated according to the concept of the generalization of the Cauchy principal value [see Lomax et al. (1951)], as indicated by the double-dash marks on the integral signs.

In order to replace the indicated integration in equation (11.28) with an algebraic summation, it is first necessary to replace the Cartesian coordinate system shown in Fig. 11.26a with the grid system shown in Fig. 11.26b. The region of integration, originally bound by the wing leading edge and the Mach lines, now consists of a set of grid elements approximating that region shown by the shaded area of Fig. 11.26b. Inclusion of partial as well as full grid elements provides a better definition of the wing leading edge and tends to reduce any irregularities that may arise in local surface slopes for elements in the vicinity of the leading edge.

Thus, we will determine the lifting pressure distribution numerically using a system of grid elements similar to those shown in Fig. 11.26b. Note that, in practice, many more elements would be used. The numbers L and N identify the spaces in the grid which replace the integration element $dx_1 \, d\beta y_1$. L^* and N^* identify the element associated with, and immediately in front of, the field point $(x, \beta y)$. Note that $L^* = x$ and $N^* = \beta y$, and x and βy take on only integer values. The region of integration, originally bounded by the leading edge and the forecone Mach lines from $(x, \beta y)$, is now approximated by the grid elements within the Mach forecone emanating from $(x, \beta y)$. Note that in the $(x, \beta y)$ coordinate system, the Mach cone half-angle is always $45°$.

The summation approximation to equation (11.28) then becomes

$$\Delta C_p(L^*, N^*) = -\frac{4}{\beta} \frac{\partial z_c(L^*, N^*)}{\partial x}$$

$$+\frac{1}{\pi} \sum_{N_{\min}}^{N_{\max}} \sum_{L_{LE}}^{L^*-|N^*-N|} \bar{R}(L^*-L, N^*-N) A(L,N) B(L,N) C(L,N) \Delta C_p(L,N) \quad \textbf{(11.29)}$$

where $\bar{R}$ is the average value of R within an element and is given by

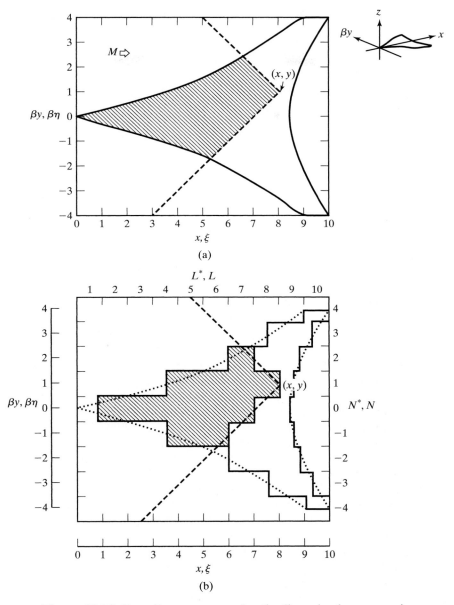

Figure 11.26 Coordinate systems for the linearized, supersonic vortex lattice method: (a) Cartesian coordinate system; (b) grid system used in numerical solution.

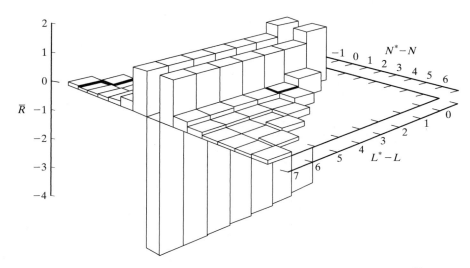

Figure 11.27 Numerical representation of the influence factor $\bar{R}$ (the $\bar{R}$ function).

$$\bar{R}(L^* - L, N^* - N) = \frac{[(L^* - L + 0.5)^2 - (N^* - N - 0.5)^2]^{0.5}}{(L^* - L + 0.5)(N^* - N - 0.5)}$$

$$- \frac{[(L^* - L + 0.5)^2 - (N^* - N + 0.5)^2]^{0.5}}{(L^* - L + 0.5)(N^* - N + 0.5)} \quad \textbf{(11.30)}$$

A graphical representation of this factor is presented in Fig. 11.27. Note the relatively small variations of the factor in the x (or L) direction contrasted with the larger variations in the y (or N) direction. For a given $L^* - L$ set of elements, the spanwise summation of the $\bar{R}$ values is found to be zero, due to the single negative value at $N^* - N = 0$ balancing all the others. At $L^* - L = 0$, where there is only one element in the spanwise summation, the $\bar{R}$ value of that element is zero. This fact ensures that an element will have no influence on itself. Furthermore, the fact that the spanwise summation of the $\bar{R}$ values is found to be zero ensures that the complete wing will produce a flow field which consists of equal amounts of upwash and of downwash and, therefore, introduces no net vertical displacements of the medium through which it is moving.

The limits on L in the summation of equation (11.29) are those of the wing leading edge (i.e., $L_{LE} = 1 + [x_{LE}]$, where $[x_{LE}]$ designates the whole-number part of the quantity) and of the Mach forecone at the selected N value. The vertical lines are used in $|N^* - N|$ to designate the absolute value of the enclosed quantity.

The factor $A(L, N)$ is a weighting factor which allows consideration of partial elements in the summation process and permits a better definition of the wing leading-edge shape. The factor $A(L, N)$ takes on values from 0 to 1, as given by

$$A(L, N) = 0 \qquad L - x_{LE} \leq 0$$
$$A(L, N) = L - x_{LE} \qquad 0 < L - x_{LE} < 1$$
$$A(L, N) = 1 \qquad L - x_{LE} \geq 1$$

The factor $B(L, N)$ is a weighting factor for the wing trailing-edge shape, which also takes on values from 0 to 1, as given by

$$B(L, N) = 0 \qquad\qquad L - x_{TE} \geq 1$$
$$B(L, N) = 1 - (L - x_{TE}) \qquad 0 < L - x_{TE} < 1$$
$$B(L, N) = 1 \qquad\qquad L - x_{TE} \leq 0$$

The factor $C(L, N)$ is a weighting factor for elements at the wing tip, which takes on values either of 0.5 or of 1.0, as given by

$$C(L, N) = 0.5 \qquad N = N_{max}$$
$$C(L, N) = 1 \qquad N \neq N_{max}$$

The differential, or lifting, pressure coefficient at a field point $\Delta C_p(L^*, N^*)$ can be determined for a wing of arbitrary surface shape provided the calculations are carried out in the proper sequence. The order of calculating $\Delta C_p(L^*, N^*)$ is from the apex rearward (i.e., increasing values of L^*). If this sequence is followed, at no time will there be an unknown $\Delta C_p(L, N)$ in the summation in equation (11.29), since the pressure coefficients for all of the points within the Mach fore cone originating at the field point (L^*, N^*) will have been already computed. Note that an element has no influence on itself since $\bar{R}(0, 0) = 0$ from equation (11.30). Thus, $\Delta C_p(L = L^*, N = N^*)$ is not required in the summation term of equation (11.29). Also, note that equation (11.29) is the summation to approximate what was originally an integral equation [see Lomax et al. (1951)] and thus already accounts for the flow tangency boundary condition, equation (11.21). This condition is satisfied exactly only at control points located at the midspan of the trailing edge of each grid element (see Fig. 11.28b).

The $\Delta C_p(L^*, N^*)$ given by equation (11.29) is defined at the trailing edge of the (L^*, N^*) element. To eliminate large oscillations in the pressure coefficient which can occur with this numerical technique, a smoothing operation is required. The procedure, taken directly from Carlson and Miller (1974), is as follows:

1. Calculate and retain, temporarily, the preliminary ΔC_p values for a given row, with $L^* =$ constant. Designate this as $\Delta C_{p,a}(L^*, N^*)$.
2. Calculate and retain, temporarily, ΔC_p values for the following row with $L^* =$ constant $+ 1$, by using the $\Delta C_{p,a}$ values obtained in the previous step for contributions from the row with $L^* =$ constant. Designate this as $\Delta C_{p,b}(L^*, N^*)$.
3. Calculate a final ΔC_p value from a fairing of integrated preliminary ΔC_p results.

For leading-edge elements, defined as $L^* - x_{LE}(N^*) \leq 1$,

$$\Delta C_p(L^*, N^*) = \frac{1}{2}\left[1 + \frac{A(L^*, N^*)}{1 + A(L^*, N^*)}\right]\Delta C_{p,a}(L^*, N^*)$$
$$+ \frac{1}{2}\left[\frac{A(L^*, N^*)}{1 + A(L^*, N^*)}\right]\Delta C_{p,b}(L^*, N^*) \qquad \textbf{(11.31)}$$

For all other elements, defined as $L^* - x_{LE}(N^*) > 1$,

$$\Delta C_p(L^*, N^*) = \tfrac{3}{4}\Delta C_{p,a}(L^*, N^*) + \tfrac{1}{4}\Delta C_{p,b}(L^*, N^*) \qquad \textbf{(11.32)}$$

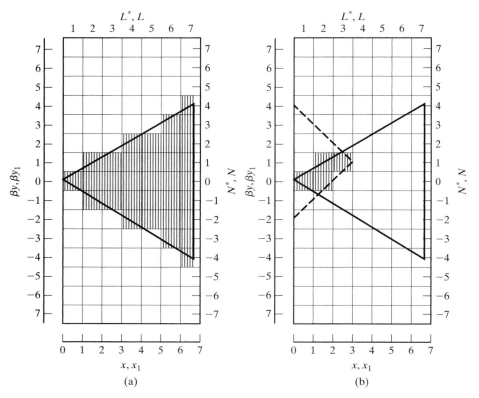

Figure 11.28 Grid element geometry for the supersonic vortex latice method: (a) general pattern; (b) region of integration for element $(3, 1)$ of Example 11.3.

where

$$A(L^*, N^*) = A(L, N)$$

The nomenclature $A(L^*, N^*)$ is used here to be consistent with that of Carlson and Miller (1974). Note that there is an error in Carlson and Miller (1974) for equation (11.31). The formulation given here is correct [see Carlson and Mack (1978)].

EXAMPLE 11.3:   **Calculate the pressure distribution for a flat-plate delta planform**

To illustrate the method, let us show how to calculate manually the pressure distribution at $M_\infty = 1.5$ for the flat-plate (zero-camber) delta planform of Fig. 11.28a for the subsonic leading edge case where $\beta \cot \Lambda_{\mathrm{LE}} = 0.6$. We will use the grid element set shown in Fig. 11.28a in order to keep the number of manual calculations within bounds. In an actual application, of course, one would use a much larger number of elements. Note that in Fig. 11.28a we use some partial elements along the leading and the trailing edges. Since ΔC_p is treated as a constant over a given element, partial elements affect only the reference area used when integrating the pressures to calculate the lift coefficient, the drag-due-to-lift coefficient, etc.

Solution: For this case, the wing is assumed to have negligible thickness and is assumed to lie approximately in the $z = 0$ plane. The wing streamwise slope $[\partial z_c(x, y)/\partial x]$ is a constant for the flat plate at incidence and is equal to the negative of the tangent of the wing angle of attack

$$\frac{\partial z_c(x, y)}{\partial x} = \frac{dz_c}{dx} = -\tan \alpha \approx -\alpha$$

since, even here, we are still restricting ourselves to linear theory, which implies that all changes in flow direction about the free-stream direction are small.

For this case, where the free-stream Mach number is 1.5:

$$\beta \equiv \sqrt{M_\infty^2 - 1} = 1.118$$

Therefore, the leading-edge sweepback angle corresponding to $\beta \cot \Lambda_{LE}$ = 0.6 is 61.78°. Thus, when the transformation is made from the (x, y) to the $(x, \beta y)$ plane, the angle that the leading edge makes with the βy axis is 59.04°. This is shown in Fig. 11.28a along with the grid system.

The equation of the wing leading and trailing edges in the $(x, \beta y)$ system is

$$x_{LE} = \frac{\beta |y|}{0.6} \qquad x_{TE} = \frac{N_{max}}{\beta \cot \Lambda_{LE}} = 6.6667$$

Note that the value of x_{TE} is defined by selection of a maximum N value, N_{max}. For the purposes of this example, we arbitrarily select $N_{max} = 4$. The wing is then scaled to give a semispan of N_{max}. This ensures that the weighting factor $C(L, N) = 0.5$ for $N = N_{max}$ is appropriate. This is important for streamwise tips of nonzero chord. Using this and the relationships defining $A(L, N)$, $B(L, N)$, and $C(L, N)$, we have the following:

A(L, N)

N \ L	1	2	3	4	5	6	7
0	1.0000	1.0000	1.0000	1.0000	1.0000	1.0000	1.0000
±1	0.0000	0.3333	1.0000	1.0000	1.0000	1.0000	1.0000
±2	0.0000	0.0000	0.0000	0.6667	1.0000	1.0000	1.0000
±3	0.0000	0.0000	0.0000	0.0000	0.0000	1.0000	1.0000
±4	0.0000	0.0000	0.0000	0.0000	0.0000	0.0000	0.3333

$B(L, N)$:

$$B(7, N) = 0.6667$$
$$B(L, N) = 1 \qquad \text{for all other grid elements}$$

$C(L, N)$:

$$C(7, \pm 4) = 0.5$$
$$C(L, N) = 1 \qquad \text{for all other grid elements}$$

We now proceed to calculate values of ΔC_p for several grid elements to show the technique. We begin at the apex with element $(1,0)$.

Element $(1,0)$.

For this element, $L^* = 1$, $N^* = 0$, and there are no other elements which contribute to the differential pressure coefficient at this element. Thus,

$$\Delta C_{p,a}(1,0) = \frac{4\alpha}{\beta}$$

To determine $\Delta C_p(1,0)$, we must calculate a preliminary value $\Delta C_{p,b}(1,0)$. This involves consideration of the element at $(2,0)$. The only element contributing to the pressure differential at $(2,0)$. is the one at $(1,0)$. Thus,

$$\Delta C_{p,b}(1,0) = \frac{4\alpha}{\beta} + \frac{1}{\pi}\left\{ \frac{[(2-1+0.5)^2 - (0-0-0.5)^2]^{0.5}}{(2-1+0.5)(0-0-0.5)} \right.$$
$$\left. - \frac{[(2-1+0.5)^2 - (0-0+0.5)^2]^{0.5}}{(2-1+0.5)(0-0+0.5)} \right\} \frac{4\alpha}{\beta}$$

or

$$\Delta C_{p,b}(1,0) = \frac{4\alpha}{\beta}(1 - 1.2004) = -0.8016\frac{\alpha}{\beta}$$

Therefore,

$$\Delta C_p(1,0) = \frac{1}{2}\left(1 + \frac{1}{1+1}\right)\frac{4\alpha}{\beta} + \frac{1}{2}\left(\frac{1}{1+1}\right)(-0.8016)\frac{\alpha}{\beta}$$

or

$$\Delta C_p(1,0) = 2.7996\frac{\alpha}{\beta}$$

Element $(2,0)$.

For this element, $L^* = 2$, $N^* = 0$, and the only element that contributes to the differential pressure coefficient at this element is the one at $(1,0)$. Thus,

$$\Delta C_{p,a}(2,0) = 4\frac{\alpha}{\beta} + \frac{1}{\pi}\left\{ \frac{[(2-1+0.5)^2 - (0-0-0.5)^2]^{0.5}}{(2-1+0.5)(0-0-0.5)} \right.$$
$$\left. - \frac{[(2-1+0.5)^2 - (0-0+0.5)^2]^{0.5}}{(2-1+0.5)(0-0+0.5)} \right\} 2.7996\frac{\alpha}{\beta}$$

or

$$\Delta C_{p,a}(2,0) = 0.6393\frac{\alpha}{\beta}$$

To determine $\Delta C_p(2,0)$, we must calculate a preliminary value $\Delta C_{p,b}(2,0)$. This involves consideration of the element at $(3,0)$. But we note that elements at $(1,0), (2,0), (2,1),$ and $(2,-1)$ contribute to the differential pressure at $(3,0)$. Thus, we must go ahead and find $\Delta C_{p,a}(2,1)$ and $\Delta C_{p,a}(2,-1)$. These, of course, are equal from symmetry.

The only element influencing the element at $(2,1)$ is the one at $(1,0)$. Thus,

$$\Delta C_{p,a}(2,1) = \frac{4\alpha}{\beta} + \frac{1}{\pi}\left\{ \frac{[(2-1+0.5)^2 - (1-0-0.5)^2]^{0.5}}{(2-1+0.5)(1-0-0.5)} \right.$$
$$\left. - \frac{[(2-1+0.5)^2 - (1-0+0.5)^2]^{0.5}}{(2-1+0.5)(1-0+0.5)} \right\}2.7996\frac{\alpha}{\beta}$$

or

$$\Delta C_{p,a}(2,1) = 5.6804\frac{\alpha}{\beta} = \Delta C_{p,a}(2,-1)$$

Therefore,

$$\Delta C_{p,b}(2,0) = \frac{4\alpha}{\beta} + \frac{1}{\pi}\left\{ \frac{[(3-1+0.5)^2 - (0-0-0.5)^2]^{0.5}}{(3-1+0.5)(0-0-0.5)} \right.$$
$$\left. - \frac{[(3-1+0.5)^2 - (0-0+0.5)^2]^{0.5}}{(3-1+0.5)(0-0+0.5)} \right\}2.7996\frac{\alpha}{\beta}$$
$$+ \frac{1}{\pi}\left\{ \frac{[(3-2+0.5)^2 - (0-0-0.5)^2]^{0.5}}{(3-2+0.5)(0-0-0.5)} \right.$$
$$\left. - \frac{[(3-2+0.5)^2 - (0-0+0.5)^2]^{0.5}}{(3-2+0.5)(0-0+0.5)} \right\}0.6393\frac{\alpha}{\beta}$$
$$+ \frac{2}{\pi}\left\{ \frac{[(3-2+0.5)^2 - (0-1-0.5)^2]^{0.5}}{(3-2+0.5)(0-1-0.5)} \right.$$
$$\left. - \frac{[(3-2+0.5)^2 - (0-1+0.5)^2]^{0.5}}{(3-2+0.5)(0-1+0.5)} \right\}(0.3333)(5.6804)\frac{\alpha}{\beta}$$

where the factor of 2 in front of the last bracketed term accounts for the fact that elements $(2,1)$ and $(2,-1)$ have an equal influence on element $(3,0)$.

Thus,

$$\Delta C_{p,b}(2,0) = 2.0128\frac{\alpha}{\beta}$$

Finally,

$$\Delta C_p(2,0) = \frac{3}{4}(0.6393)\frac{\alpha}{\beta} + \frac{1}{4}(2.0128)\frac{\alpha}{\beta}$$

or

$$\Delta C_p(2,0) = 0.9827\frac{\alpha}{\beta}$$

Element (2, 1).

For this element, $L^* = 2$, $N^* = 1$, and the only element that contributes to the differential pressure coefficient at this element is the one at $(1, 0)$. Thus, as shown previously,

$$\Delta C_{p,a}(2,1) = \frac{4\alpha}{\beta} + \frac{1}{\pi}\left\{ \frac{[(2 - 1 + 0.5)^2 - (1 - 0 - 0.5)^2]^{0.5}}{(2 - 1 + 0.5)(1 - 0 - 0.5)} \right.$$
$$\left. - \frac{[(2 - 1 + 0.5)^2 - (1 - 0 + 0.5)^2]^{0.5}}{(2 - 1 + 0.5)(1 - 0 + 0.5)} \right\} 2.7996 \frac{\alpha}{\beta}$$

or

$$\Delta C_{p,a}(2,1) = 5.6804 \frac{\alpha}{\beta}$$

To determine $\Delta C_p(2, 1)$, we must calculate a preliminary value $\Delta C_{p,b}$ $(2, 1)$. This involves consideration of the element at $(3, 1)$. We note, however, that elements at $(1, 0), (2, 0)$, and $(2, 1)$ contribute to the differential pressure at $(3, 1)$. Thus,

$$\Delta C_{p,b}(2,1) = \frac{4\alpha}{\beta} + \frac{1}{\pi}\left\{ \frac{[(3 - 1 + 0.5)^2 - (1 - 0 - 0.5)^2]^{0.5}}{(3 - 1 + 0.5)(1 - 0 - 0.5)} \right.$$
$$\left. - \frac{[(3 - 1 + 0.5)^2 - (1 - 0 + 0.5)^2]^{0.5}}{(3 - 1 + 0.5)(1 - 0 + 0.5)} \right\} 2.7996 \frac{\alpha}{\beta}$$
$$+ \frac{1}{\pi}\left\{ \frac{[(3 - 2 + 0.5)^2 - (1 - 0 - 0.5)^2]^{0.5}}{(3 - 2 + 0.5)(1 - 0 - 0.5)} \right.$$
$$\left. - \frac{[(3 - 2 + 0.5)^2 - (1 - 0 + 0.5)^2]^{0.5}}{(3 - 2 + 0.5)(1 - 0 + 0.5)} \right\} 0.6393 \frac{\alpha}{\beta}$$
$$+ \frac{1}{\pi}\left\{ \frac{[(3 - 2 + 0.5)^2 - (1 - 1 - 0.5)^2]^{0.5}}{(3 - 2 + 0.5)(1 - 1 - 0.5)} \right.$$
$$\left. - \frac{[(3 - 2 + 0.5)^2 - (1 - 1 + 0.5)^2]^{0.5}}{(3 - 2 + 0.5)(1 - 1 + 0.5)} \right\} (0.3333)(5.6804) \frac{\alpha}{\beta}$$

or

$$\Delta C_{p,b}(2,1) = 3.3820 \frac{\alpha}{\beta}$$

Note that we use preliminary (i.e., $\Delta C_{p,a}$) values for the ΔC_p's of influencing elements in the same "L" row as the field-point element under consideration. Thus, we have used $\Delta C_{p,a}(2, 0) = 0.6393$ instead of $\Delta C_p(2, 0) = 0.9827$ as the factor of the second major bracketed term in the preceding equation. As an alternative procedure we could have used ΔC_p instead of $\Delta C_{p,a}$, but we have chosen to be consistent with the method of Carlson and Miller (1974).

For influencing elements not in the same "L" row as the field-point element, we use the final (i.e., averaged ΔC_p values) as given by equations (11.31) and (11.32).

Finally,

$$\Delta C_p(2,1) = \frac{1}{2}\left(1 + \frac{0.3333}{1 + 0.3333}\right)5.6804\frac{\alpha}{\beta}$$
$$+ \frac{1}{2}\left(\frac{0.3333}{1 + 0.3333}\right)3.3820\frac{\alpha}{\beta}$$

or

$$\Delta C_p(2,1) = 3.9730\frac{\alpha}{\beta} = \Delta C_p(2,-1)$$

from symmetry.

We have now calculated the differential pressure coefficients for elements $(1,0),(2,0)$, and $(2,\pm1)$. One can continue in this manner and determine ΔC_p for the remaining elements.

Typical numerical results for this case at three chordwise stations are presented in Fig. 11.29. The number of grid elements used in Fig. 11.29 is approximately 2000.

Lifting-pressure distributions for a flat-plate delta wing ($\beta \cot \Lambda_{\text{LE}} = 1.20$) as computed by Middleton and Carlson (1965) are reproduced in Fig. 11.30. Agreement between the numerical and analytical solutions is quite good, although the numerical solution does not exhibit the sharp break at the Mach line that is characteristic of the analytical solution. The unevenness of the pressure coefficients observed in the subsonic leading-edge case does not occur because of the reduced effect of the averaging techniques with the less severe variations in the pressure near the leading edge. Figure 11.31 shows a comparison of the numerical results with those of exact linear theory for flat delta planforms such as the one used in Example 11.3.

Note that the numerical method is also applicable to wings of arbitrary camber and planform (e.g., supersonic or subsonic leading and/or trailing edges, etc.). To accommodate nonzero camber, one only need specify $\partial z_c(x,y)/\partial x$ in equation (11.29). A companion numerical method, also described in Carlson and Miller (1974), provides a means for the design of a camber surface corresponding to a specified loading distribution or to an optimum combination of loading distributions.

Once the lifting pressure coefficients, ΔC_p, have been determined for all the elements, it is possible to compute the aerodynamic coefficients for the wing. The lift coefficient for symmetric loads may be obtained from the following summation over all the elements:

$$C_L = \frac{2}{\beta S} \sum_{N^*=0}^{N^*=N_{\max}} \sum_{L^*=L_{\text{LE}}}^{L^*=L_{\text{TE}}} \left[\frac{3}{4}\Delta C_p(L^*,N^*)\right.$$

$$\left.+ \frac{1}{4}\Delta C_p(L^*+1,N^*)\right]A(L^*,N^*)B(L^*,N^*)C(L^*,N^*) \quad \textbf{(11.33)}$$

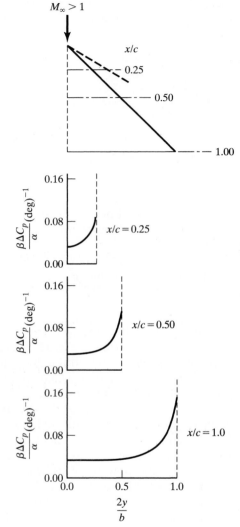

Figure 11.29 Numerical results for a flat-plate delta planform ($\beta \cot \Lambda_{\mathrm{LE}} = 0.6$). [From Carlson and Miller (1974).]

The pitching-moment coefficient about $x = 0$ is

$$
C_M = \frac{2}{\beta S \bar{c}} \sum_{N^*=0}^{N^*=N_{\max}} \sum_{L^*=L_{\mathrm{LE}}}^{L^*=L_{\mathrm{TE}}} (L^*) \left[\frac{3}{4} \Delta C_p(L^*, N^*) \right.
$$

$$
\left. + \frac{1}{4} \Delta C_p(L^* + 1, N^*) \right] A(L^*, N^*) B(L^*, N^*) C(L^*, N^*) \tag{11.34}
$$

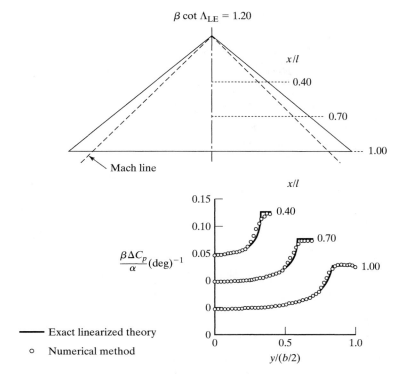

$\beta \cot \Lambda_{LE} = 1.20$

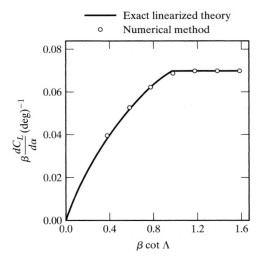

Figure 11.30 Lifting-pressure distributions for a flat-plate delta wing with a supersonic leading edge. $N_{max} = 50$. [From Middleton and Carlson (1965).]

Figure 11.31 Comparison of linear theory with vortex lattice results for a delta wing. [From Carlson and Miller (1974).]

The drag coefficient may be expressed as

$$C_D = -\frac{2}{\beta S} \sum_{N^*=0}^{N^*=N_{\max}} \sum_{L^*=L_{LE}}^{L^*=L_{TE}} \left[\frac{3}{4} \Delta C_p(L^*, N^*) + \frac{1}{4} \Delta C_p(L^* + 1, N^*)\right]$$

$$\left[\frac{3}{4} \frac{\partial z_c(L^*, N^*)}{\partial x} + \frac{1}{4} \frac{\partial z_c(L^* + 1, N^*)}{\partial x}\right] A(L^*, N^*)B(L^*, N^*)C(L^*, N^*) \quad \textbf{(11.35)}$$

This relationship does not consider any contribution of the theoretical leading-edge-suction force or of any separated flow effects associated with its exclusion and accounts only for the inclination of the normal force to the relative wind.

The element weighting factors are defined as follows. The leading-edge field-point-element weighting factor takes on values from 0 to 1.5.

$$A(L^*, N^*) = 0 \qquad (L^* - x_{LE} \leq 0)$$
$$A(L^*, N^*) = L^* - x_{LE} + 0.5 \qquad (0 < L^* - x_{LE} < 1)$$
$$A(L^*, N^*) = 1 \qquad (L^* - x_{LE} \geq 1)$$

The trailing-edge field-point-element weighting factor also takes on values from 0 to 1.5.

$$B(L^*, N^*) = 0 \qquad (L^* - x_{TE} \geq 0)$$
$$B(L^*, N^*) = 0.5 - (L^* - x_{TE}) \qquad (0 > L^* - x_{TE} > -1)$$
$$B(L^*, N^*) = 1 \qquad (L^* - x_{TE} \leq -1)$$

The centerline or wingtip grid element weighting factor is defined as

$$C(L^*, N^*) = 0.5 \qquad (N^* = 0)$$
$$C(L^*, N^*) = 1 \qquad (0 < N^* < N_{\max})$$
$$C(L^*, N^*) = 0.5 \qquad (N^* = N_{\max})$$

The wing area used in the expressions for the aerodynamic coefficients may be computed using the summation

$$S = \frac{2}{\beta} \sum_{N^*=0}^{N^*=N_{\max}} \sum_{L^*=1+[x_{LE}]}^{L^*=1+[x_{TE}]} A(L^*, N^*)B(L^*, N^*)C(L^*, N^*) \quad \textbf{(11.36)}$$

11.7.3 Numerical Method for the Determination of Camber Distribution

Equation (11.28) can be rearranged

$$\frac{\partial z_c(x, y)}{\partial x} = -\frac{\beta}{4} \Delta C_p(x, y)$$

$$+ \frac{\beta}{4\pi} \oiint \bar{R}(x - x_1, y - y_1) \Delta C_p(x_1, y_1) \, d\beta y_1 \, dx_1 \quad \textbf{(11.37)}$$

Equation (11.37) can be used to determine the wing surface shape necessary to support a specified lift distribution.

When discussing the numerical representation of the influence factor $\bar{R}$, as presented in Fig. 11.27, Middleton and Lundry (1980a) commented, "The physical significance of this $\bar{R}$ variation is that (for positive lift), all elements directly ahead of the field point element contribute only downwash and all other elements contribute upwash. An element at the leading edge near the wing tip of a subsonic leading-edge wing, therefore, sees a concentrated upwash field. It is this upwash field that makes the subsonic leading edge twisted and cambered wing attractive from the standpoint of drag-due-to-lift, since a local element may be inclined forward to produce both lift and thrust."

Thus, significantly better wing performance can be achieved through the generation of leading-edge thrust. This thrust occurs when the leading edge is subsonic (i.e., when the product $\beta \cot \Lambda_{\mathrm{LE}}$ is less than one) and is caused by the flow around the leading edge from a stagnation point on the windward, lower surface. In the case of a cambered wing, the thrust contributes both to the axial force and to the normal force.

Carlson and Mann (1992) recommend a suction parameter for rating the aerodynamic performance of various wing designs. The suction parameter is defined by the parameter S_s:

$$S_s = \frac{C_L \tan(C_L/C_{L\alpha}) - (C_D - C_{D0})}{C_L \tan(C_L/C_{L\alpha}) - C_L^2/\pi AR} \tag{11.38}$$

The term in the numerator $(C_D - C_{D0})$ equals ΔC_D, the drag coefficient due-to-lift. The factor $C_{L\alpha}$ in equation (11.38) represents the theoretical lift-curve slope at $\alpha = 0°$. Limits for the suction parameter can be evaluated using the sketches of the drag polar presented in Fig. 11.32. The lower bound for the suction parameter (i.e., $S_s = 0$)

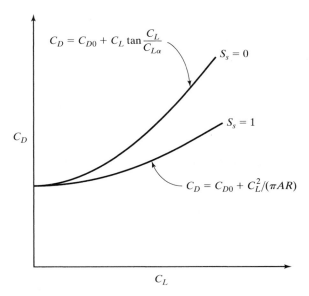

Figure 11.32 The relation between the suction parameter (S_s) and the drag polar. [From Carlson and Mann (1992).]

corresponds to the upper drag polar, where $C_D = C_{D0} + C_L \tan(C_L/C_{L\alpha})$. The drag coefficient for this limit is that for a flat wing with no leading-edge thrust and no vortex forces. The upper bound for the suction parameter (i.e., $S_s = 1$) corresponds to the lower drag polar, where $C_D = C_{D0} + C_L^2/(\pi AR)$. The drag coefficient for this limit is the drag for a wing with an elliptical span-load distribution (a uniform downwash) and the full amount of any leading-edge thrust that might be required. This limit is a carryover from subsonic flows, where the limit is reasonably achievable. There is no contribution representing the supersonic wave-drag-due-to-lift. In practice, the presence of wave-drag-due-to-lift at supersonic speeds prevents a close approach to this value. However, the simplicity of the expression and its repeatability make it a logical choice for use in the suction parameter.

Experience has shown that the maximum suction parameters actually achieved were lower than those predicted by linearized theory. Furthermore, the required surface for given design conditions was less severe (smaller departures from a flat surface) than that given by the linearized-theory design methods. Carlson and Mann (1992) suggest a design method, which employs two empirical factors, K_D (a design lift-coefficient factor) and K_s (a suction parameter factor). The design factor K_D, which is presented in Fig. 11.33a, provides a design lift coefficient for use in the theoretical wing design to replace the actual operational, or cruise, lift coefficient. Thus, for the selected design Mach number, the design lift-coefficient factor is taken from Fig. 11.33a and used to define the design lift coefficient:

$$C_{L,\text{des}} = K_D C_{L,\text{cruise}} \qquad (11.39)$$

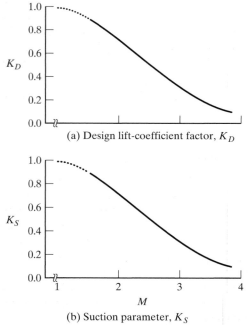

(a) Design lift-coefficient factor, K_D

(b) Suction parameter, K_S

Figure 11.33 Empirical method factors used to select the optimum design lift coefficients and to predict achievable suction parameters. [Taken from Carlson and Mann (1992).]

The design lift coefficient is used in the computer code definition of the lifting surface and of the theoretical performance, including $(S_{s,\text{max}})_{\text{th}}$. Next, the appropriate value of K_s is taken from Fig. 11.33b, so that one can estimate the magnitude of the suction parameter that can actually be achieved:

$$S_{s,\text{cruise}} = K_s(S_{s,\text{max}})_{\text{th}} \tag{11.40}$$

As noted by Carlson and Mann (1992), the code value of $(S_{s,\text{max}})_{\text{th}}$ for $C_{L,\text{des}}$ is used in this expression even though $C_{L,\text{des}}$ differs from $C_{L,\text{cruise}}$. This approximation is acceptable, since for a wing-design family of various $C_{L,\text{des}}$ values, there is very little change in the value of $S_{s,\text{max}}$.

The drag coefficient at the cruise lift coefficient can be estimated using the expression

$$C_{D,\text{cruise}} = C_{D0} + \frac{(C_{L,\text{cruise}})^2}{\pi AR}$$
$$+ (1 - S_{s,\text{cruise}})\left[C_{L,\text{cruise}} \tan\left(\frac{C_{L,\text{cruise}}}{C_{L\alpha}}\right) - \frac{(C_{L,\text{cruise}})^2}{\pi AR}\right] \tag{11.41}$$

Mann and Carlson (1994) note, "The use of the curves in Fig. 11.33 for designing a wing for given cruise conditions and estimating its performance may be summarized as follows. First select the cruise lift coefficient and Mach number. Next Fig. 11.33a is used to obtain the K_D factor and equation (11.39) is used to calculate $C_{L,\text{des}}$. A linearized theory computer code is then used to compute the optimum performance surface at $C_{L,\text{des}}$ and to obtain the variation of theoretical suction parameter with C_L. This is the wing surface that will give maximum performance at the cruise conditions. To estimate its performance, Fig. 11.33b is used to obtain a value for the suction parameter factor K_s. Note that $S_{s,\text{max,theory}}$ is essentially the maximum suction parameter for a wing designed for a lift coefficient equal to $C_{L,\text{cruise}}$ (not designed for $C_{L,\text{des}}$). However, a family of wings designed for different lift coefficients will have approximately the same maximum suction parameter. It is, therefore, sufficient to assume that $S_{s,\text{max,theory}}$ is equal to the maximum suction parameter for the wing designed at $C_{L,\text{des}}$. Equation (11.40) can then be used to obtain an estimate of $S_{s,\text{cruise}}$." The drag coefficient at the cruise lift coefficient can be estimated by use of equation (11.41).

11.8 DESIGN CONSIDERATIONS FOR SUPERSONIC AIRCRAFT

We have seen that theory predicts that highly swept wings (such that the leading edge is subsonic) have the potential for very favorable drag characteristics at supersonic flight speeds. However, experiment has shown that many of the theoretical aerodynamic benefits of leading-edge sweepback are not attained in practice because of separation of the flow over the upper surface of the wing. Elimination of separated flow can only be achieved in design by a careful blending of the effects of leading-and trailing-edge sweepback angles, leading-edge nose radius, camber, twist, body shape, wing/body junction aerodynamics, and wing planform shape and thickness distribution. Theory does not give all the answers here, and various empirical design criteria have been developed to aid in accounting for all of these design variables.

Squire and Stanbrook (1964) and Kulfan and Sigalla (1978) have studied the types of flow around highly swept edges. Sketches of the main types of flow on highly swept wings, as taken from their work, are presented in Fig. 11.34. Based on these investigations, we could classify the types of separation as

1. Leading-edge separation due to high suction pressures
2. Separation due to spanwise flow

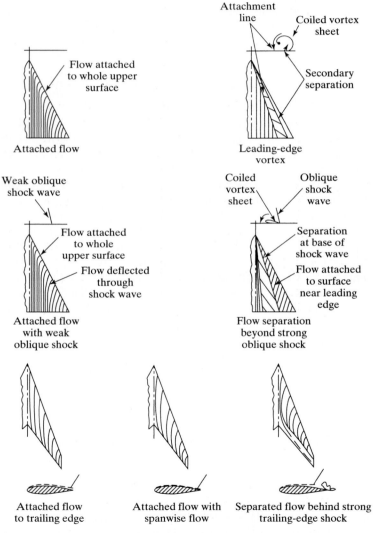

Figure 11.34 Main types of flow on highly swept wings. [From Squire and Stanbrook (1964) and Kulfan and Sigalla (1978).]

3. Inboard shock separation

4. Trailing-edge shock separation

In our analysis of swept wings, we found that the effective angle of attack of the wing was given by equation (11.10) as

$$\alpha_e = \tan^{-1}\frac{\tan \alpha}{\cos \Lambda}$$

for small angles. Thus, for highly swept wings with subsonic leading edges, separation can occur quite readily even for small values of wing angle of attack. This is particularly true if the leading edge is sharp. The observed flow from the leading edge is very similar to that for the delta wing in subsonic flow (see Chapter 7). The coiled vortices that form at the leading edges affect the flow below the wing as well, since the leading edge is subsonic, and therefore the top and bottom surfaces can "communicate."

The phenomenon of spanwise flow on a swept wing has also been discussed in Chapter 7. Spanwise flow results in a thickening of the boundary layer near the wing tips, and a thick boundary layer will separate more readily than a thin one exposed to the same adverse pressure gradient.

Inboard shock separation is primarily a function of the geometry of the wing/body junction near the leading edge of the wing. The upper surface flow in this region is toward the body, and therefore a shock is formed to turn the stream tangent to the body. The strength of the shock is, of course, dependent on the turning angle. If the shock is strong enough, the resulting pressure rise will separate the boundary layer.

Trailing-edge shock separation can occur if the wing trailing edge is supersonic (as is the case for delta wings, for example). A shock wave near the trailing edge is required to adjust the upper surface pressure back to the free-stream value. Again, if the required shock strength is too great, it will induce separation of the boundary layer.

The design criteria have been developed to eliminate separation are given in Kulfan and Sigalla (1978) and can be summarized as follows:

1. *Leading-edge separation.* Reject designs where the theoretical suction pressure exceeds 70% of vacuum pressure.

2. *Separation due to spanwise flow.* Use thin wing tips. (*Note*: An additional technique is "washout" or a lower section incidence at the wing tips relative to the incidence angles of inboard sections.)

3. *Inboard shock separation.* Use body contouring to keep the pressure rise across the inboard shock wave to less than 50%.

4. *Trailing-edge shock separation.* Keep the pressure ratio across the trailing-edge shock below $1 + 0.3M_1^2$, where M_1 is the local Mach number ahead of the shock. For swept trailing edges, use the local normal Mach number ($M_{N1} = M_1 \cos \Lambda_{\text{TE}}$) in the preceding equation.

Application of these criteria will not guarantee that the flow will not separate, however, and wind-tunnel tests of proposed designs must be undertaken. This is particularly true for military or sport aircraft designed for maneuvering at high load factors.

Carlson and Mack (1980) note that there are compensatory errors in linearized theory and the failure to account for nonlinearities may introduce little error in prediction of lift and drag. However, significant errors in the prediction of the pitching moment are common, especially for wings that depart from a delta planform. Additionally, for wings with twist and camber, appreciable errors in the prediction of drag due to the surface distortion (camber drag) often occur. In particular, linearized-theory methods do not indicate the proper camber surface for drag minimization (a function of the design lift coefficient). The method of Carlson and Mack (1980), which includes the estimation of nonlinearities associated with leading-edge thrust and the detached leading-edge vortex flow, provides improved prediction over linearized theory, as illustrated by the correlations reproduced in Fig. 11.35.

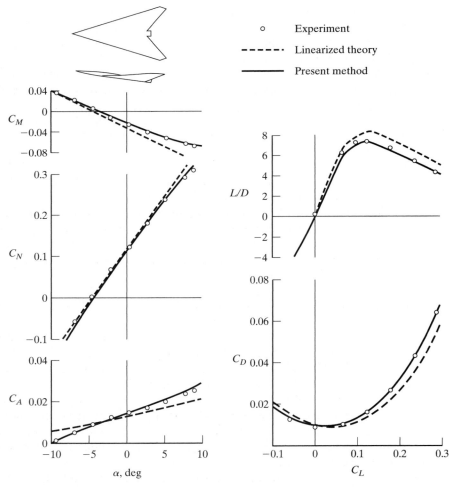

Figure 11.35 Comparison of the predicted and the measured forces and moments for 72.65° swept leading-edge arrow wing; $M_\infty = 2.6$; design value of $C_L = 0.12$. [Taken from Carlson and Mack (1980).]

11.9 SOME COMMENTS ABOUT THE DESIGN OF THE SST AND OF THE HSCT

11.9.1 The Supersonic Transport (SST), the Concorde

In an excellent review of the aerodynamic design of the Supersonic Transport, Wilde and Cormery (1970) reviewed the general design features of the Concorde. "From the beginning, Concorde was conceived to achieve the following overall objectives:

(a) the required level of performance on the long range North Atlantic mission, which requires the achievement of good lift/drag ratios at subsonic and supersonic conditions,

(b) good takeoff and landing characteristics, making full use of the very powerful ground effect,

(c) good flying qualities throughout the speed range by purely aerodynamic means, using artificial stability augmentation only to improve the crew workload, and

(d) operation without stall or buffet up to lift coefficients above those required for airworthiness demonstration purposes."

To achieve these objectives, it was understood "that:

(i) the wing, fuselage, and nacelles are aerodynamically interdependent,

(ii) the aerodynamic design of the wing must be integrated with the requirements of the structures, systems and production engineers,

(iii) no single speed can be taken as the design speed; supersonic design is very important but subsonic and transonic requirements must also be given proper attention, and

(iv) design for good performance cannot be separated from design for good handling qualities."

Area ruling of the fuselage was seriously considered in the project stages but was rejected on the basis of increased production costs.

Wilde and Cormery (1970) noted that "the performance in supersonic cruise is very important, of course, but about 40% of all fuel uplifted for a trans-Atlantic flight will either be used in subsonic flight conditions or is provisioned in reserves for possible extension of subsonic flight in diversion or holding."

Finally, considerable effort was expended reviewing the consequences of introducing changes to an existing design. For instance, "reducing the leading edge droop over the whole span would improve the supersonic drag, worsen the subsonic and transonic drag and increase the lift curve slope. It is to be noted that increase in lift curve slope is favourable, particularly at low speed since it permits either reduced takeoff and landing speeds at fixed attitude or reduced attitudes at given speeds."

"Inevitably the nacelles have forward facing surfaces which produce supersonic drag but this drag is offset to some extent by the induced interference pressures on the undersurface of the wing which produce lift."

11.9.2 The High-Speed Civil Transport (HSCT)

As noted in a Boeing study [Staff of Boeing Commercia (Airplanes (1989)], the aerodynamic tasks in the design of the High-Speed Civil Transport "included:

(a) Aerodynamic design integration of the study configurations.

(b) Integration of compatible high-lift system for each concept.

(c) Evaluation of the aerodynamic characteristics of all concepts to provide necessary resizing data for the airplane performance calculations. These included both flaps-up cruise configuration analyses as well as flaps-down takeoff and landing evaluation.

The aerodynamic design of the study configurations included optimized camber/twist distributions and area-ruled fuselages. The wing spanwise thickness distributions and airfoil shapes were constrained by structural depth requirements."

As the aircraft flight speed increases, careful integration of the propulsion system into the airframe design becomes increasingly important. Thus, as noted in NASA Contractor Report 4233 and in NASA Contractor Report 4234, "Nacelle shape, size, location, and operating conditions all influence the nacelle interference with other configuration components. The dominant interference effect is between the nacelles and the wing. To arrive at an acceptable nacelle design requires a balance between the isolated drag of the nacelle and the interference drag induced by the nacelle. The nacelles of each of the study configurations were placed below the wing and aft to (1) provide the inlet with a uniform flow field throughout the angle of attack range, (2) take advantage of the precompression caused by the wing shock, and (3) achieve favorable aerodynamic interference."

As noted in the NASA Contractor Reports "Environmental acceptability is a key element of any HSCT program. If not properly accounted for in the HSCT design, environmental limitations could substantially reduce use of the vehicle and, in the most extreme of circumstance, prohibit vehicle operations altogether. The primary area of environmental impact identified in the study were engine emissions effects, community noise, and sonic boom. . . . A viable HSCT must be designed so that the engine emissions have no significant impact on the Earth's ozone layer. . . . Operation out of conventional airports was determined to be a requirement for achieving adequate HSCT utilization. Accordingly, a viable HSCT must produce noise levels no higher than its subsonic competition. . . . The airplanes under study have been evaluated with subsonic flight profiles over land, which results in adverse economic and market impact. Thus, there is an impetus to explore low-boom designs that allow some form of overland supersonic operations."

11.9.3 Reducing the Sonic Boom

As noted in the previous section, a viable High Speed Civil Transport (HSCT) must produce noise levels no higher than its subsonic competition. "The problem is not building an aircraft capable of Mach 1+ speeds, but designing one that can fly over land with generating a sonic boom—and then getting the FAA and other international aviation regulatory agencies to certify it for such flights", as noted by Wilson (2007).

Gulfstream Aerospace moved closer to resolving that design issue by testing a design concept that involved attaching a telescoping spike to the nose of an F15B fighter jet. Note that the ultimate goal is to break up the typical shock-wave that creates a sonic boom into a series of smaller waves that do not.

Quiet Spike was another step in the ability to use technology to make a quiet supersonic business jet. The application of a telescoping spike to reduce the sonic boom would require that we demonstrate the ability to make a quiet supersonic business jet, to show that it could be incorporated into a telescoping structure, to put it on the front of an airplane, and show that the telescoping spike would reduce the noise.

Wilson (2007) further notes: "Without a spike you could design an aircraft that is quieter than what is flying today, but not quiet enough to . . . change the regulations [to make] supersonic flight over land unrestricted."

As noted by Croft (2004), "The noise problem with today's supersonic aircraft is caused by an N-wave pressure change that occurs as shock waves move past the observer. When an aircraft like the Concorde flies at supersonic speeds, conical shocks form at the front and rear of the aircraft and at every discontinuity in the flow along the way (for example, at the engine inlets, wing leading edges, and antennas) because the plane is moving faster than the air can move out of the way.".

"At a certain distance from the aircraft, the individual shocks coalesce into two stable conical shocks, one at the front of the plane and one at the tail. The effect on the listener is a double boom as each wave passes."

"According to [Peter] Coen [of NASA Langley], shock waves with greater than ambient pressure tend to move faster and collect at the nose, while those with lower than ambient pressure lag to the back. On the ground, the pressure distribution as the waves pass looks like an 'N': As the nose shock passes, the pressure spikes above the ambient pressure, ramps down below ambient, then snaps back to ambient as the tail shock passes."

The Defense Advanced Research Projects Agency (DARPA) funded a Shaped Sonic Boom Demonstration (SSBD) Program to determine if sonic booms could be substantially reduced by incorporating specialized aircraft shaping techniques. Pawlowski et al. (2005) wrote: "Although mitigation of the sonic boom via specialized shaping techniques was theorized decades ago, until now, this theory had never been tested with a flight vehicle subjected to flight conditions in a real atmosphere."

As shown in Figure 11.36, that was presented by Pawlowski et al. (2005): "Pressure measurements obtained on the ground and in the air confirmed that the specific modifications made to a Northrop Grumman F-5E aircraft not only changed the shape of the shock wave signature emanating from the aircraft, but also produced a 'flat-top' signature whose shape persisted, as predicted, as the pressure waves propagated through the atmosphere to the ground." Kandil et al. (2005) noted: "The modification was made primarily to the nose of the aircraft forward of the cockpit and engine inlets and was designed to propagate a 'flat top' pressure signature (sonic boom) to the ground. This signature, as the name indicates, has a flat pressure distribution downstream of the initial shock (see the sketch in Figure 11.36). Signatures of this type are desired since they yield weaker initial shocks relative to the N-wave signature normally produced by current supersonic aircraft."

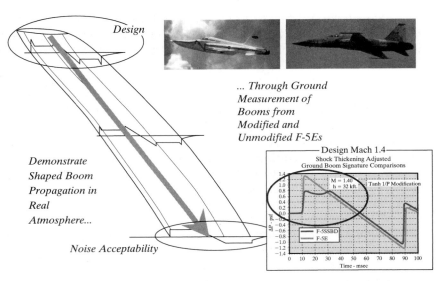

Figure 11.36 Impact of forebody shaping on N-wave production for the F-5E [From Pawlowski et al. (2005).]

11.9.4 Classifying High-Speed Aircraft Designs

One approach to classifying aircraft designs in relation to their speed regime is using the ratio of semispan ($b/2$) to length of the aircraft (l) as a measure of *configuration slenderness*. The reader is referred to Küchemann (1978) for more about the reasoning behind the use of this parameter. Harris (1992) presented this parameter for *configuration slenderness* as a function of the maximum cruise Mach number. "Subsonic aircraft, exemplified by the Boeing 747, McDonnell-Douglas MD-11, and Airbus A300, tend to optimize at high values of semispan-to-length ratio because of their high-aspect-ratio wings. The distinction between supersonic and hypersonic designs is taken as the dividing line where the semispan-to-length ratio is equal to the tangent of the Mach angle." From the definition of the Mach angle in equation (8.22), the reader can verify that the expression for the tangent of the Mach angle is

$$\tan \mu = \left[\frac{1}{\sqrt{M_\infty^2 - 1}} \right]$$

Harris (1992) continued, "Therefore, for the supersonic designs that may have subsonic leading edges, we can maintain relatively weak shock waves. . . . Long-range subsonic aircraft are essentially designed to achieve the highest cruise Mach number that can be attained and still avoid the adverse effects of shock waves.

"Supersonic aircraft, in order to achieve long-range capability, can be designed to minimize shock losses by keeping the configuration slender, that is, a low value of semispan-to-length ratio. As shown in Fig. 11.36, the short-range supersonic U.S. fighters, which are not designed for efficient cruise performance, fall in the semispan-to-length ratio range of about 0.25–0.35 in contrast to the lower values approaching 0.2 for the Concorde, Tu-144, XB-70, and SR-71 long-range supersonic cruise designs."

Flow-field solutions were generated using the Cobalt$_{60}$ code for an SR-71 configuration at an angle of attack of 5° [Tomaro and Wurtzler (1999)]. Interestingly enough, when the vehicle is at 5° angle of attack, the axis of the cone in each of the engine inlets is aligned with the free-stream velocity vector. Thus, the shock wave generated by the cone is symmetric with the engine flow path. Shock-wave patterns for these computed flow fields are presented in Fig. 11.38 for free-stream Mach numbers of 1.5, 3.2, and 3.8. According to the correlations presented in Fig. 11.37, a Mach number of 3.8 is approximately the highest cruise Mach number for the SR-71. Referring to Fig. 11.38c, the reader can see that, at this Mach number, the limiting portion of the bow-shock wave as it merges with the shock wave generated by the engine structure is tangent to the wing tips. Thus, the flow field computed using the full Navier-Stokes equations gives results that are consistent with the correlation presented in Fig. 11.38. That is, the tangent of the bow-shock wave angle is approximately equal to the semispan-to-length ratio.

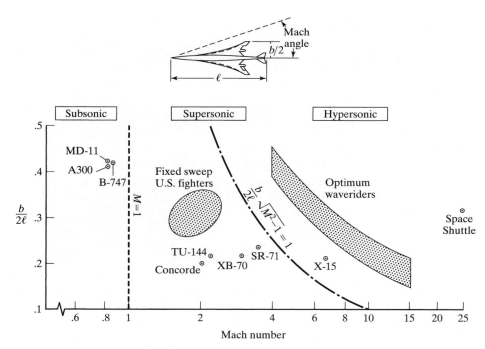

Figure 11.37 Aircraft classification by speed regime. [Taken from Harris (1992).]

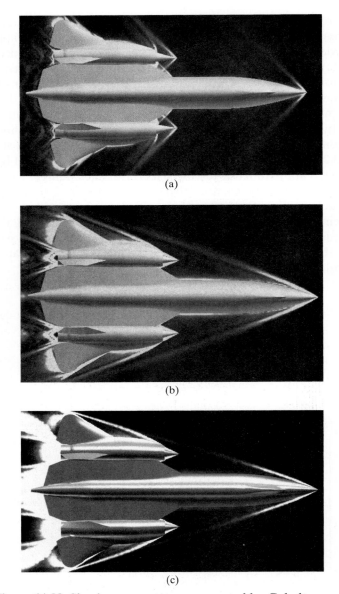

(a)

(b)

(c)

Figure 11.38 Shock-wave patterns generated by Cobalt$_{60}$ computations for an SR-71 at 5° angle of attack with (a) $M_\infty = 1.5$, (b) $M_\infty = 3.2$, and (c) $M_\infty = 3.8$. [From Tomaro and Wurtzler (1999).]

11.10 SLENDER BODY THEORY

Significant effort has been made in previous sections of this chapter to describe the aerodynamics of wings in supersonic flow. While wings are certainly the most important supersonic lifting devices, a considerable portion of supersonic lift also can come from the fuselage. Significant work has been done over the years to develop supersonic body theories [see Liepmann and Roshko (1957) for a summary]; and much of the development involves the method of small perturbations [van Dyke (1975)]. A brief review of the assumptions and resulting equations are presented here.

The starting point for slender body theory is the linearized potential equation (9.13)

$$(1 - M_\infty^2)\phi_{xx} + \phi_{yy} + \phi_{zz} = 0$$

While the formal derivation of the slender body theory concept is quite elaborate [see Ward (1955) for details], the concept is fairly straightforward and follows the concepts for supersonic linear-airfoil theory. For a long, slender body of revolution (see Fig. 11.39), the variation of velocity in the x direction can be assumed to be small compared with the velocity variations in the y or z directions.

$$\phi_{xx} << \phi_{yy}, \phi_{zz} \tag{11.42}$$

This simplification applied to equation (9.13) leads to a rather amazing result

$$\phi_{yy} + \phi_{zz} = 0 \tag{11.43}$$

which is Laplace's equation in the crossflow plane of the slender body (as shown in Fig. 11.40).

So the solution for slender bodies reduces to solving Laplace's equation in the y–z (crossflow) plane successively for each crossflow plane from the front to the rear of the body. This usually is accomplished by placing a distribution of sources or doublets along the axis of revolution to form the body shape and then performing inner and outer expansions of the flow-field to find the velocity and pressure fields around the body [Karamcheti (1980)]. The resulting pressure distribution on the body surface at zero angle of attack is formulated by Liepmann and Roshko (1957) as

$$C_p(x) = \frac{S''(x)}{\pi}\ln\left(\frac{2}{R(x)\sqrt{1 - M_\infty^2}}\right) + \frac{1}{\pi}\frac{d}{dx}\int_0^x S''(\xi)\ln(x - \xi)d\xi - \left(\frac{dR(x)}{dx}\right)^2 \tag{11.44}$$

where $R(x)$ is the cross-sectional radius distribution of the body and $S(x)$ is the cross-sectional area distribution of the body. The pressure distribution can be integrated along the length of the body, L, in order to determine the drag coefficient (excluding the drag on the base of the configuration) as

$$C_D = \frac{1}{S(L)}\int_0^L C_p\frac{dS(x)}{dx}dx \tag{11.45}$$

to obtain the zero-lift wave-drag coefficient ($C_D \equiv D/q_\infty S(L)$):

$$C_{D_0} = -\frac{1}{2\pi S(L)}\int_0^L\int_0^L S''(\xi)S''(x)\ln|x - \xi|d\xi\,dx \tag{11.46}$$

Figure 11.39 Long, slender body of revolution

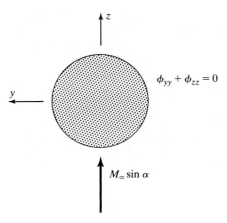

Figure 11.40 Crossflow plane for a slender body of revolution.

If the original formulation had retained the angle of attack of the slender body the lift-and-drag-coefficient would have been found to be

$$C_D = C_{D_0} + \alpha^2 \qquad C_L = 2\alpha \qquad \text{(11.47)}$$

which are very straightforward relationships that show a linear variation of lift with angle-of-attack and a squared relationship for the drag variation. As with subsonic flow, drag increase faster than lift with angle-of-attack, even though the drag is being produced by wave drag in this case. An improved formulation for the normal force coefficient of a slender body has been obtained as [Pitts et al. (1959)]

$$C_N = \sin(2\alpha)\cos(\alpha/2) + 2L/2R(L)\sin^2 \alpha \qquad \text{(11.48)}$$

which gives a nonlinear variation for the normal force as a function of angle-of-attack. For body shapes given by a simple power series, $R(X) = \varepsilon x^a$, a variety of body shapes can be obtained by choosing the factor ε and varying the coefficient a (in this case $\varepsilon = 0.1$ and $a = 0.7, 1.0,$ and 1.5) to obtain pressure-coefficient variations (Fig. 11.41) and wave-drag coefficients (Fig. 11.42).

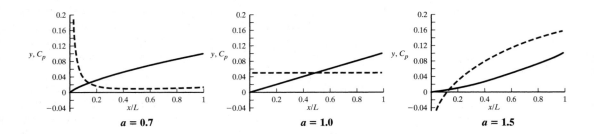

Figure 11.41 Pressure coefficients for various power series body shapes (body shape in solid line, pressure coefficient in dashed line).

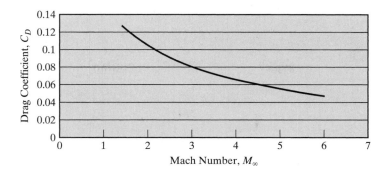

Figure 11.42 Mach variation for power series body ($\varepsilon = 0.1$ and $a = 1.5$).

11.11 AERODYNAMIC INTERACTION

Owing to the complexity of flow fields about most flight vehicles, aerodynamicists often develop theories for the flows about components of such vehicles. However, the designer is faced with the fact that aerodynamic loads on an entire aircraft are not simply the sum of the loads on the individual components; the difference is commonly referred to as *aerodynamic interaction*. Wind-tunnel experiments, flight tests, and physical reasoning have shown that interaction loads can be a significant contribution to the total loading on an aircraft. It was not until the high-speed digital computer was generally available that systematic treatment of the loads on rather arbitrary shapes was even feasible. Theories are still being developed to account for the effects of three-dimensional separated flows about arbitrary shapes. Thus, this is a rapidly changing area of aerodynamic theory. Significant progress has been made, however, and our purpose here is to give a brief account of the physical reasoning associated with the effects of aerodynamic interaction in supersonic flight, and to describe methods of determining these effects.

Consider Fig. 11.43, where a simple rectangular wing is mounted on an ogive cylinder with circular cross section. Two interaction effects are present: the effect of the wing on the body and the effect of the body on the wing. In supersonic flight, the suction pressure on the upper surface of the wing and the relatively higher pressure on the lower surface will be confined to the regions bounded by the Mach waves from the leading and trailing edges of the wing, as shown. This pressure differential will carry over onto the body and generate a net lift and wave-drag force on it. Thus, we have an example of the effect of the wing on the body.

Conversely, the effect of the body at angle of attack will be to produce an upwash about the sides of the body (see Fig. 11.44) which will increase the effective angle of attack of the wing. Provided that the resulting angle of attack will not cause the flow to separate on the upper wing surface (i.e., provided that the wing does not stall), the result will be greater wing lift in accordance with equation (10.8).

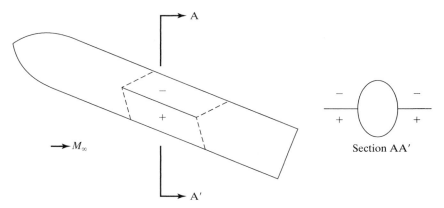

Figure 11.43 Wing-on-body interaction showing regions of positive and negative C_p. [From Hilton (1951).]

Hilton (1951) notes that the combined effects of the wing and body for the case shown in Fig. 11.45 are such that the wings produce the full lift to be expected from two-dimensional theory without tip effects. Taking into account tip effects and ignoring interactions, one would expect a wing lift of only three-fourths of this value. Thus, by introducing the body interaction effect, a full 25% increase in wing lift is experienced.

Other interaction effects can also be present (e.g., wing/tail, body/tail, etc.). Some simplifications arise in supersonic flow due to the fact that disturbances cannot propagate upstream. Thus, while one may want to consider the effects of the wings on the tail, the effects of the tail on the wings can usually be neglected unless they are very close to one another.

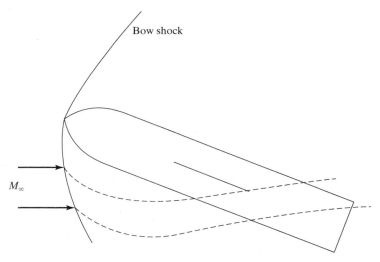

Figure 11.44 Body-on-wing interaction showing upwash-over-wing effect due to the body at angle of attack. [From Hilton (1951).]

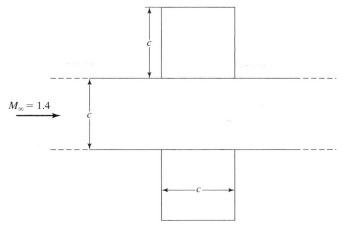

Figure 11.45 Configuration giving increased wing lift. [From Hilton (1951).]

Many treatments of interacting flows rely on the small disturbance theory governed by equation (11.4). The flow tangency boundary condition at surfaces and the Kutta condition at sharp trailing edges are involved in determining solutions. Problems are solved by distributing a series of singularities (e.g., sources, sinks, doublets, vortices) to simulate the vehicle in a uniform supersonic stream. The strength of these singularities is determined so as to satisfy the boundary conditions at discrete selected locations on the vehicle. In the method of Carmichael and Woodward (1966), wing camber and incidence are simulated by vortex distributions, while thickness is simulated by source and sink distributions. Thickness, camber, and incidence of the body are represented by line sources and doublets along the axis of the body. The effect of the interaction of the wing on the body is represented by a distribution of vortices on a cylinder whose radius is related to the average radius of the body, while the body-on-wing interaction is represented by a distribution of vortices on the wing camber surface [see Woodward (1968)].

An integrated supersonic design and analysis system is presented by Middleton and Lundry (1980). The analysis program can be used to predict the lifting pressure calculations for a wing camber surface at a selected angle of attack. It can also be used to calculate lifting pressure distributions and force coefficients for complete configurations over a range of angles of attack. The program described by Middleton and Lundry (1980a) carries two solutions along: one for the configuration at its input angle of attack, the other an incremental solution per degree angle of attack (called the flat-wing solution). The interference terms associated with the two solutions acting on the other surface shape are also calculated. The summation of these effects into the drag polar and the other force coefficients is performed by superposition. Calculation of the complete configuration lifting pressure solution can involve any or all of the following tasks:

1. Determine the isolated fuselage upwash field.
2. Determine the nacelle pressure field acting on the wing.
3. Compute the pressure field due to asymmetrical fuselage volume.

4. Generate a wing/canard solution in the presence of the fuselage upwash field.

5. Calculate the effects of the wing pressure field acting on the nacelles.

6. Determine the fuselage lift distribution in the presence of the wing downwash field.

7. Obtain the horizontal tail solution in the presence of the fuselage and wing flow fields.

8. Generate a solution for the complete configuration by superposition of the elemental solutions.

For the reader interested in pursuing the subject of aerodynamic interactions, there often appear surveys of the relevant phenomena and solution techniques [e.g., Tomaro and Wurtzler (1999)].

11.12 AERODYNAMIC ANALYSIS FOR COMPLETE CONFIGURATIONS IN A SUPERSONIC STREAM

There is a heirarchy of prediction/computation techniques that can be used to estimate the aerodynamic forces and moments for suitably complex configurations in a supersonic stream. Techniques based on linear theory have been described at length in this chapter. Techniques based on impact methods, on solutions of the Euler equations, and on solutions of the Navier-Stokes equations are discussed in Chapters 12 and 14.

Three-view oil-flow photographs of the Space Shuttle Orbiter are presented in Fig. 11.46 for

$$M_\infty = 1.25$$
$$\alpha = 10°$$

$\delta_{E,L} = -18.8°$ (the left elevon is deflected upward and the oil-flow pattern indicates a shock wave at the lower right of Fig. 11.46c, just below the OMS pod)

$\delta_{E,R} = +14.4°$ [the right elevon is deflected downward and the surface oil flows through the space between the OMS pod and the right elevon (Fig. 11.46a)]

$\delta_{SB} = -87.2°$ and $\delta_R = -25°$ [the speed brake is open 87.2° and deflected 25° to serve as a rudder, creating a relatively strong shock wave with an attendant viscous/inviscid interaction on one side of the vertical tail (Fig. 11.46a) but not on the other (Fig. 11.46c)]

$\delta_{BF} = 23.7°$ (the body flap is deflected 23.7°)

Also evident in these flow visualization photographs are free-shear layer separations and feather patterns associated with reattaching vortical flows. Because the complex nature of the flow field made it impossible to develop realistic flow models for numerical solutions, a total of 493 wind-tunnel tests using 52,993 hours were used in phase C/D of the Space Shuttle design development program in order to develop an aerodynamic data base [Whitnah and Hillje (1984)]. However, because wind-tunnel flows are also simulations (limited in model scale, high-temperature effects, etc.), wind-tunnel measurements should be correlated with the corresponding computed values based on solution techniques employing adequate flow models before extrapolating to flight.

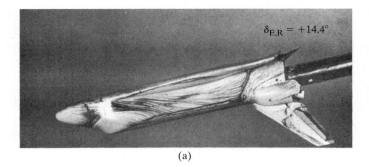

(a)

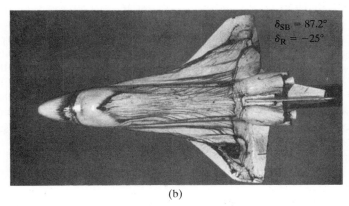

(b)

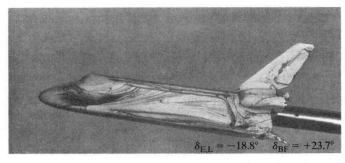

(c)

Figure 11.46 Oil flow patterns for Space Shuttle *Orbiter* at $M_\infty = 1.25$ and $\alpha = 10°$: (a) view of right side; (b) top view; (c) view of left side. (Courtesy of Johnson Space Center, NASA.)

PROBLEMS

11.1. Consider a flat-plate, rectangular wing. Derive the expressions for the lift coefficient, for the drag coefficient, for the coefficient of the moment about the leading edge, and for the location of the center of pressure given in Table 11.1. Assume that $\beta \cdot AR > 2$ so that the Mach cones emanating from the tip do not overlap.

11.2. Verify the statement given in Table 11.1 that A', which is defined as airfoil cross-sectional area divided by the square of chord, is equal to $\tau/2$ for a double-wedge airfoil section.

11.3. Consider a wing with a rectangular planform, whose aspect ratio is 4.0 and whose section is that shown in Fig. 10.5. Use Bonney's results in Table 11.1 to determine C_L and C_D for this wing for the flow conditions shown in Fig. 10.5.

11.4. **(a)** Using the relations given in Table 11.1, develop expressions for the lift coefficient as a function of a [i.e., $C_L(\alpha)$] and for the drag coefficient [i.e., $C_D(C_L)$] for the wing of Figs. 9.14 and 9.15. The wing has a rectangular planform with an aspect ratio of 2.75. Develop the relations for $M_\infty = 1.50$. Assume that the airfoil section is biconvex with a maximum thickness ratio of 0.05.

  (b) Compare the theoretical values with the experimental values presented in Figs. 9.14 and 9.15. What value of $C_{D,\text{friction}}$ (at $M_\infty = 1.50$) will cause your theoretical results to agree most closely with the data in the figures?

11.5. Consider the wing of the Northrop F-5E (see Table 5.1). If the airplane is flying at a Mach number of 1.23, will the quarter-chord line of the wing be in a supersonic or a subsonic condition relative to the free-stream flow? What must M_∞ be for the quarter-chord line to be in a sonic condition?

11.6. Derive the equation of the leading-edge sweep angle Λ_{LE} as a function of M_∞ for a sonic leading edge. Prepare a graph of the results. Assume small-angle approximations for α.

11.7. Show that the section lift coefficient for a swept airfoil with a supersonic leading edge is given by

$$C_l = \frac{4\cos\Lambda}{\sqrt{M_\infty^2 \cos\Lambda - 1}}\alpha$$

The thickness ratio and the angle of attack of the airfoil are sufficiently small that the small-angle approximations may be used.

11.8. Discuss the limits of validity of the result derived in Problem 11.7.

11.9. Using a Taylor's series expansion about $z = 0$, derive equation (11.7) from equation (11.6).

11.10. Consider the flat-plate rectangular wing of Problem 11.1. Assume that there is a plain flap along the entire trailing edge with hinge line at $x_f = fx$, where $0 \le f \le 1$. Derive a formula for C_L as a function of the flap deflection angle δ_f (see Fig. P11.10). Assume that $\beta \cdot \text{AR} > 2$.

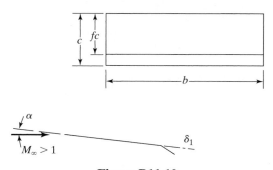

Figure P11.10

11.11. Derive the relation between the aspect ratio of a delta wing and the free-stream Mach number M_∞ if the leading edge is to be sonic.

11.12. Show that equation (11.30) follows from

$$R(x - x_1, y - y_1) \equiv \frac{x - x_1}{\beta^2 (y - y_1)^2 [(x - x_1)^2 - \beta^2 (y - y_1)^2]^{0.5}}$$

The variation of R with x may be assumed to be small.

11.13. Show that subtracting the single wedge FGF' from the single wedge BFF' yields the relations in the double wedge $BDED'$ (see Fig. 11.20). Thus, show that the source strength in region DCD' will be

$$C(x, y) = -\lambda_2 \frac{U_\infty}{\pi}$$

where $C = C_1 + C_2$ and C_1 is the source strength due to BFF' and C_2 is the source (sink) strength due to FGF'.

11.14. Determine $\Delta C_p(3, 0)$ for the flow described in Example 11.3.

11.15. Determine $\Delta C_p(3, \pm 1)$ for the flow described in Example 11.3.

11.16. Consider a flat-plate (zero-camber) delta planform in a $M_\infty = 1.5$ stream. The leading-edge sweepback angle (Λ_{LE}) is 42.974°. Is the leading edge subsonic or supersonic? What is the value of $\beta \cot \Lambda_{LE}$ for this flow?

11.17. For the configuration described in Problem 11.16 and with $N_{max} = 8$, determine $\Delta C_p(3, 0)$.

11.18. For the configuration described in Problem 11.16 and with $N_{max} = 8$, determine $\Delta C_p(3, \pm 1)$.

11.19. Develop the expression for dR/dx as a function of x/L for the slender, axially-symmetric body-of-revolution, which has a profile section of

$$R(x) = 2t \frac{x}{L} \left(1 - \frac{x}{L} \right)$$

Develop the expression for the pressure coefficient as a function of x/L.

11.20. Develop the expression for the pressure coefficient for a two-dimensional airfoil section that has the same bi-convex prfile as was specified in Problem 11.19, i.e.,

$$R(x) = 2t \frac{x}{L} \left(1 - \frac{x}{L} \right)$$

Prepare a graph comparing the pressure distribution for the axially-symmetric body-of-revolution with the pressure distribution for the two-dimensional airfoil section that has the same bi-convex profile.

11.21. Test are to be conducted in an indraft wind tunnel. An indraft wind tunnel takes air from the room in which the tunnel is located. The air then accelerates through a convergent-divergent nozzle to the free-stream conditions in the test section. The free-stream Mach number in the test section is 5.0. The "atmospheric" conditions in the room form the stagnation conditions for the tunnel: $p_{t_1} = 15$ lbf/in² and $T_t = 540°R$. Calculate the L/D ration for the two-dimensional airfoil secrtion as a function of the angle of attack from 0° to 10°.

REFERENCES

Allen JM, Townsend JC. 1986. Application of a supersonic Euler code. *J. Aircraft* 23:513–519

Ashley H, Rodden WP. 1972. Wing-body aerodynamic interaction. *Ann. Rev. Fluid Mech.* 4:431–472

Bonney EA. 1947. Characteristics of rectangular wings at supersonic speeds. *J. Aeron. Sci.* 14:110–116

Busemann A. 1947. Infinitesimal conical supersonic flow. *NACA Tech. Mem. 1100*

Carafoli E. 1956. *High Speed Aerodynamics*. Elmsford, NY: Pergamon Press

Carlson HW, Miller DS. 1974. Numerical analysis of wings at supersonic speeds. *NASA Tech. Note D-7713*

Carlson HW, Mack RJ. 1978. Estimation of leading edge thrust for supersonic wings of arbitrary planform. *NASA TP 1270*

Carlson HW, Mack RJ. 1980. Estimation of wing nonlinear aerodynamic characteristics at supersonic speeds. *NASA TP 1718*

Carlson HW, Mann MJ. 1992. Survey and analysis of research on supersonic drag-due-to-lift minimization with recommendations for wing design. *NASA TP 3365*

Carmichael RS, Woodward FA. 1966. Integrated approach to the analysis and design of wings and wing-body combination in supersonic flow. *NASA Tech. Note D-3685*

Cenko A, Tinoco EN, Dyer RD, DeJongh J. 1981. PAN AIR applications to weapons carriage and separation. *J. Aircraft* 18:128–134

Croft J. 2004. Engineering through the sound barrier. *Aerospace Am.* 42(9):24–26, 29–31

Ferri A. 1949. *Elements of Aerodynamics of Supersonic Flow*. New York: Macmillan Publishing

Harris RV. 1992. On the threshold—the outlook for supersonic and hypersonic aircraft. *J. Aircraft* 29:10–19

Hernandez G, Covell PF, McGraw ME. 1993. An experimental investigation of a Mach 2.0 high-speed civil transport at supersonic speeds. *NASA TP 3365*

Hilton WF. 1951. *High Speed Aerodynamics*. New York: Longmans, Green & Co.

Hodgman CD, ed. 1977. *Standard Mathematical Tables*, 12th ed. Cleveland: Chemical Rubber Co.

Ivey HR, Bowen EH. 1947. Theoretical supersonic lift and drag characteristics of symmetrical wedge-shape-airfoil sections as affected by sweepback outside the Mach cone. *NACA Tech. Note 1226*

Jones RT, Cohen D. 1957. Aerodynamics of wings at high speeds. In *Aerodynamic Components of Aircraft at High Speeds*, Princeton: Princeton University Press

Kandil OA, Ozcer IA, Zheng X, Bobbitt PJ. 2005. *Comparison of full-potential propagation-code computations with the F-5E "shaped sonic boom comparison" program*. Presented at Aerosp. Sci. Meet., 43rd, AIAA Pap. 2005–0013, Reno, NV

Karamcheti K. 1980. *Principles of Ideal-Fluid Aerodynamics*. 2nd ed. Melbourne, FL: Krieger

Küchemann D. 1978. *Aerodynamic Design of Aircraft*. Oxford: Pergamon Press

Kulfan RM, Sigalla A. 1978. *Real flow limitations in supersonic airplane design*. Presented at Aerosp. Sci. Meet., 16th, AIAA Pap. 78–0147, Huntsville, AL

Liepmann HW, Roshko A. 1957. *Elements of Gasdynamics*. New York: John Wiley

Lomax H, Heaslet MA, Fuller FB. 1951. Integrals and integral equations in linearized wing theory. *NACA Report 1054*

Mann MJ, Carlson HW. 1994. Aerodynamic design of supersonic cruise wings with a calibrated linearized theory. *J. Aircraft*. 31:25–41

Middleton WD, Carlson HW. 1965. A numerical method for calculating the flat-plate pressure distribution on supersonic wings of arbitrary planform. *NASA Tech. Note D–2570*

Middleton WD, Lundry JL. 1980a. A system for aerodynamic design and analysis of supersonic aircraft. *NASA Contr. Rep. 3351*

Middleton WD, Lundry JL, Coleman RG. 1980b. A system for aerodynamic design and analysis of supersonic aircraft. *NASA Contr. Rep. 3352*

Nielsen JN, Matteson FH, Vincenti WG. 1948. Investigation of wing characteristics at a Mach number of 1.53, III: unswept wings of differing aspect ratio and taper ratio. NACA Res. Mem. A8E06

Pawlowski JW, Graham DH, Boccadoro CH, Coen PG, Maglieri DJ. 2005. *Origins and overview of the shaped sonic boom demonstration program.* Presented at Aerosp. Sci. Meet., 43rd, AIAA Pap. 2005–0005, Reno, NV

Pitts WC, Nielsen JN, Kaatari GE. 1959. Lift and center of pressure of wing-body-tail combinations at subsonic, transonic, and supersonic speeds. *NACA Rep. 1307*

Puckett AE. 1946. Supersonic wave drag of thin airfoils. *J. Aeron. Sci.* 13:475–484

Puckett AE, Stewart HJ. 1947. Aerodynamic performance of delta wings at supersonic speeds. *J. Aeron. Sci.* 14:567–578

Schiff LB, Steger JL. 1979. *Numerical simulation of steady supersonic viscous flow.* Presented at Aerosp. Sci. Meet., 17th, AIAA Pap. 79–0130, New Orleans, LA

Shapiro AH. 1954. *The Dynamics and Thermodynamics of Compressible Fluid Flow.* New York: The Ronald Press

Snow RM. 1948. Aerodynamics of thin quadrilateral wings at supersonic speeds. *Q. Appl. Math.* 5:417–428

Squire LC, Stanbrook A. 1964. Possible types of flow at swept leading edges. *Aeronaut. Quart.* 15:72–82

Staff of Boeing Commerical Airplanes. 1989. High-speed civil transport study. *NASA Contr. Rep. 4233*

Staff of Boeing Commerical Airplanes. 1989. High-speed civil transport study, summary. *NASA Contr. Rep. 4234*

Stewart HJ. 1946. "The lift of a delta wing at supersonic speeds, " Quarterly of Applied Math., Vol. 4, No. 3, pp. 246–254.

Tomaro RF, Wurtzler KE. 1999. *High speed configuration aerodynamics: SR-71 to SMV.* Presented at Appl. Aerodyn. Conf., 17th, AIAA Pap. 99–3204, Norfolk, VA

Van Dyke M. 1975. *Perturbation Methods in Fluid Mechanics.* Stanford: Parabolic Press

Walkley KB, Smith GE. 1987. Wave drag analysis of realistic fighter aircraft using a full-potential method. *J. Aircraft.* 24:623–628

Ward GN. 1955. *Linearized Theory of Steady High Speed Flow.* Cambridge: Cambridge University Press

Whitnah AM, Hillje ER. 1984. Space Shuttle wind tunnel testing program summary. *NASA RP 1125*

Wilde MG, Cormery G. 1970. The aerodynamic derivation of the concorde wing. *Can. Aeronaut. Space J.* 16:175–184

Wilson JR. 2007. "Quiet spike, softening the sonic boom," Aerospace America, Oct. 2007, pp. 38–42

Wood RM, Miller DS. 1985. Assessment of preliminary prediction techniques for wing leading-edge vortex flows at supersonic speeds. *J. Aircraft.* 22:473–478

Woodward FA. 1968. Analysis and design of wing-body combinations at subsonic and supersonic speeds. *J. Aircraft.* 5:528–534

Wright BR, Bruckman F, Radovich NA. 1978. *Arrow wings for supersonic cruise aircraft.* Presented at Aerosp. Sci. Meet., 18th, AIAA Pap. 78–0151, Huntsville, AL

12 HYPERSONIC FLOWS

A vehicle flying through the atmosphere at hypersonic speeds generates a shock layer, that is, the region between the bow shock wave and the vehicle surface, in which the pressure, the temperature, and/or the density may change by two orders of magnitude, and more. Because the kinetic energy associated with hypersonic flight is converted into high temperatures within the shock layer, the flow-field trade studies conducted during the vehicle design include consideration both of the heat-transfer environment and of the aerodynamic forces and moments. Thus, one often speaks of the aerothermodynamic environment of a hypersonic vehicle.

By definition,

$$M_\infty = \frac{U_\infty}{a_\infty} \gg 1 \tag{12.1}$$

is the basic assumption for all hypersonic flow theories. Thus, the internal thermodynamic energy of the free-stream fluid particles is small compared with the kinetic energy of the free stream for hypersonic flows. In flight applications, this results because the velocity of the fluid particles is relatively large. The limiting case, where M_∞ approaches infinity because the free-stream velocity approaches infinity while the free-stream thermodynamic state remains fixed, produces extremely high temperatures in the shock layer. The high temperatures associated with hypersonic flight cannot be accommodated in ground test facilities. Therefore, in wind-tunnel applications, hypersonic Mach numbers are achieved through relatively low speeds of sound; that is, for the wind tunnel, M_∞

approaches infinity because the speed of sound (the temperature) goes to "zero" while the free-stream velocity is held fixed. As a result, the fluid temperatures remain below the levels that would damage the wind tunnel and the model.

Another assumption common to hypersonic flow is that

$$\varepsilon = \frac{\rho_\infty}{\rho_2} \ll 1 \tag{12.2}$$

which is known as the small-density-ratio assumption. Thus, this assumption relates primarily to the properties of the gas downstream of the shock wave. Recall that, for a perfect gas,

$$\varepsilon = \frac{\gamma - 1}{\gamma + 1} \tag{12.3}$$

for a normal shock wave as $M_\infty \to \infty$. Thus, ε is $\frac{1}{6}$ for perfect air. Note that typical hypersonic wind tunnels operate at conditions where the test gas can be approximated by the perfect-gas relations. However, during the reentry of the Apollo Command Module, the density ratio approached $\frac{1}{20}$.

For slender configurations, such as sharp cones and wedges, the strong shock assumption is

$$M_\infty \sin \theta_b \gg 1 \tag{12.4}$$

which is mixed in nature, since it relates both to the flow and to the configuration. The concept termed the "Mach number independence principle" depends on this assumption.

12.1 NEWTONIAN FLOW MODEL

As the free-stream Mach number approaches infinity and the hypersonic independence principle applies to the flow, the shock layer becomes very thin. As a result, one can assume that the speed and direction of the gas particles in the free stream remain unchanged until they strike the solid surface exposed to the flow. For this flow model (termed Newtonian flow since it is similar in character to one described by Newton in the seventeenth century), the normal component of momentum of the impinging fluid particle is wiped out, while the tangential component of momentum is conserved. Thus, using the nomenclature of Fig. 12.1 and writing the integral form of the momentum equation for a constant-area streamtube normal to the surface,

$$p_\infty + \rho_\infty (U_{\infty,n})^2 = p_\infty + \rho_\infty (U_\infty \sin \theta_b)^2 = p_s \tag{12.5}$$

Rearranging so that the local pressure is written in terms of the pressure coefficient gives

$$C_p = \frac{p_s - p_\infty}{\frac{1}{2}\rho_\infty U_\infty^2} = 2 \sin^2 \theta_b = 2 \cos^2 \phi \tag{12.6}$$

The pressure coefficient, as defined in equation (12.6), is known as the Newtonian value, where the 2 represents the pressure coefficient at the stagnation point (which is designated $C_{p,t2}$), since $\theta_b = 90°$ at the Newtonian stagnation point.

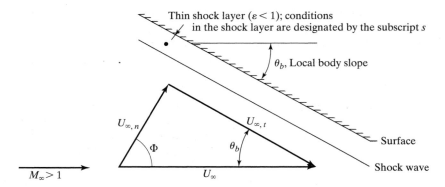

Figure 12.1 Nomenclature for Newtonian flow model.

The Newtonian flow model and the various theories for thin shock layers related to the Newtonian approximation are based on the small-density-ratio assumption. The small-density-ratio requirement for Newtonian theory also places implicit restrictions on the body shape in order that the shock layer be thin. The range of applicability for Newtonian theory, as defined by Marconi et al. (1976), is reproduced in Fig. 12.2.

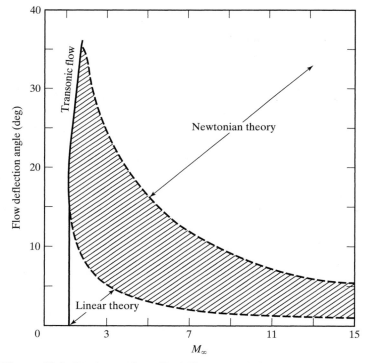

Figure 12.2 Regions of applicability of inviscid flow theories for the surface pressure on a sharp cone. [From Marconi, et al. (1976).]

Small perturbation theory yields accurate results only for the flow over slender bodies at small angles of attack in a low supersonic Mach number stream. However, Newtonian theory provides useful results when the Mach number is large and/or the flow deflection angle is large. This is equivalent to the strong shock assumption

$$M_\infty \sin \theta_b \gg 1$$

Consider the pressure coefficients presented in Fig. 12.3. Theoretical values of C_p (taken from Chapter 8) are presented as a function of M_∞ for a cone whose semi-vertex angle is $15°$ and for a $15°$ wedge. Note that, as $M_\infty \to \infty$, the pressure coefficients become independent of the Mach number (the Mach number independence principle). The shock layer is thinner for the cone, and the limiting value of C_p is approached at a lower Mach number. Also presented is the Newtonian pressure coefficient (which is independent of the Mach number). As $M_\infty \to \infty$, all three techniques yield roughly the same value for C_p. Therefore, it can be seen that, at hypersonic speeds, the pressure coefficient for these simple shapes depends primarily on the flow deflection angle.

The Mach number independence principle was derived for inviscid flow. Since pressure forces are much larger than the viscous forces for blunt bodies or for slender bodies at relatively large angles of attack when the Reynolds number exceeds 10^5, one would expect the Mach number independence principle to hold at these conditions.

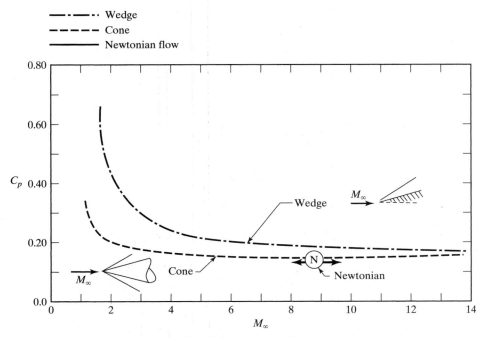

Figure 12.3 Pressure coefficient for the flow of perfect air past a wedge and a cone; deflection angle is $15°$.

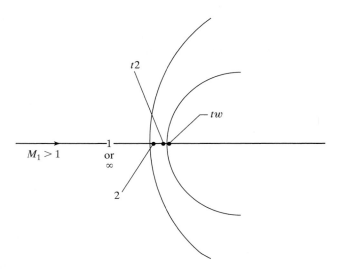

Figure 12.4 Nomenclature for the stagnation region. 1 or ∞, free-stream conditions; 2, conditions immediately downstream of the shock wave; $t2$, conditions at the stagnation point (downstream of the normal portion of the shock wave) but outside the boundary layer; tw, conditions at the wall at the downstream stagnation point.

12.2   STAGNATION REGION FLOW-FIELD PROPERTIES

The nomenclature for the flow near the stagnation point of a vehicle in a hypersonic stream is illustrated in Fig. 12.4. The free-stream flow (designated by the subscript ∞ or 1) passes through the normal portion of the shock wave reaching state 2 and then decelerates isentropically to state $t2$, which constitutes the edge condition for the thermal boundary layer at the stagnation point. The streamline from the shock wave to the stagnation point may be curved for nonaxisymmetric flow fields. The pressure and the heating rate at the stagnation point are useful reference values for characterizing hypersonic flows.

The relations for steady, one-dimensional, inviscid, adiabatic flow in a constant-area streamtube were used to compute the conditions across a normal shock wave:

$$\rho_1 U_1 = \rho_2 U_2 \tag{12.7}$$

$$p_1 + \rho_1 U_1^2 = p_2 + \rho_2 U_2^2 \tag{12.8}$$

$$h_1 + \tfrac{1}{2}U_1^2 = h_2 + \tfrac{1}{2}U_2^2 = H_t \tag{12.9}$$

where H_t is the total (or stagnation) enthalpy of the flow.

If one assumes that the gas is thermally perfect,

$$p = \rho R T \tag{12.10}$$

and calorically perfect,

$$h = c_p T \tag{12.11}$$

the ratio of the values of flow properties across the shock wave can be written as a unique function of M_1 (or M_∞, the free-stream Mach number) and γ (the ratio of specific heats). Referring to Chapter 8, the relations are

$$\frac{p_2}{p_1} = \frac{2\gamma M_1^2 - (\gamma - 1)}{\gamma + 1} \tag{12.12}$$

$$\frac{\rho_2}{\rho_1} = \frac{U_1}{U_2} = \frac{(\gamma + 1)M_1^2}{(\gamma - 1)M_1^2 + 2} \tag{12.13}$$

$$\frac{T_2}{T_1} = \frac{[2\gamma M_1^2 - (\gamma - 1)][(\gamma - 1)M_1^2 + 2]}{(\gamma + 1)^2 M_1^2} \tag{12.14}$$

If one assumes that the flow decelerates isentropically from the conditions at point 2 (immediately downstream of the normal portion of the shock wave) to the stagnation point outside of the thermal boundary layer (point *t*2),

$$\frac{p_{t2}}{p_1} = \left[\frac{(\gamma + 1)M_1^2}{2}\right]^{\gamma/(\gamma-1)} \left[\frac{\gamma + 1}{2\gamma M_1^2 - (\gamma - 1)}\right]^{1/(\gamma-1)} \tag{12.15}$$

$$\frac{T_{t2}}{T_1} = \frac{T_{t1}}{T_1} = 1 + \frac{\gamma - 1}{2}M_1^2 \tag{12.16}$$

Note that, whereas it is generally true that the stagnation enthalpy is constant across a normal shock wave for an adiabatic flow [see equation (12.9)], the stagnation temperature is constant across a normal shock wave only for the adiabatic flow of a perfect gas [see equation (12.16)]. Note also that for a perfect gas (i.e., one that is thermally perfect and calorically perfect), the ratio of the downstream value to the free-stream value for the flow properties (across a normal shock wave) can be written as a function of γ and $M_1(= M_\infty)$ only. Thus, the perfect-gas values for the ratios defined by equations (12.12) through (12.16) do not depend specifically on the altitude.

In reality, for hypersonic flight, the gas molecules that pass through the bow shock wave are excited to higher vibrational and chemical energy modes. This lowers the specific-heat ratio of the gas below the free-stream value if it is assumed that equilibrium exists and that dissociation is not driven to completion. A large amount of the energy that would have gone into increasing the static temperature behind the bow shock wave for a perfect gas is used instead to excite the vibrational energy levels or to dissociate the gas molecules. As additional energy is absorbed by the gas molecules entering the shock layer, the conservation laws and the thermophysics dictate certain changes in the forebody flow. The static temperature, the speed of sound, and the velocity in the shock layer are less for the equilibrium, real-gas flow than for a perfect-gas flow. The real-gas value of the static pressure is slightly larger than the perfect-gas value. The density is increased considerably, and, as a result, the shock layer thickness is reduced.

Equations (12.7) through (12.9) are not restricted to the perfect-gas assumption and can be applied to a high-temperature, hypersonic flow. Let us use the U.S. Standard Atmosphere (1976) to define the free-stream properties (i.e., p_1, ρ_1, and h_1, at 150,000 ft).

Since there are four unknowns in equations (12.7) through (12.9) (i.e., p_2, ρ_2, h_2, and U_2), but only three equations, additional relations are needed to obtain a solution. The graphs of Moeckel and Weston (1958) (see Fig. 12.5) were used in tabular form to define

$$\rho(p, h)$$

$$s(p, h)$$

$$T(h, s)$$

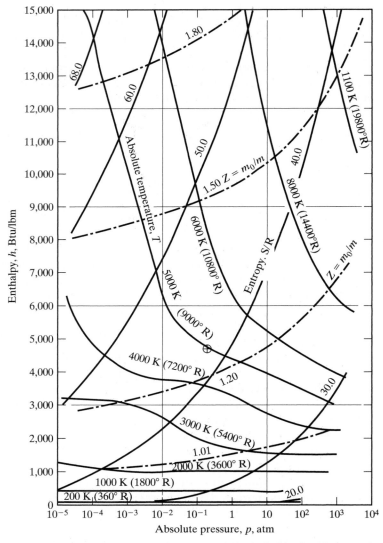

Figure 12.5 Thermodynamic properties of air in chemical equilibrium. [From Moeckel and Weston (1958).]

Let us calculate the flow downstream of a normal shock wave when $M_1 = 14$ at an altitude of 150,000 ft. Using the more generally applicable expression equation (12.9),

$$H_{t1} = h_1 + \tfrac{1}{2}U_1^2 = H_{t2} = H_t = 4636.77 \text{ Btu/lbm}$$

Assuming that the air remains in thermodynamic equilibrium as it crosses the shock wave, p_{t2} is 0.3386 atm (716.57 lb/ft^2) and T_{t2} is 8969.6°R, as represented by the $\oplus$ in Fig. 12.5. Using the perfect-gas relations of equations (12.12) through (12.16), p_{t2} is 0.3256 atm (689.12 lb/ft^2) and T_{t2} is 19,325°R.

The perfect-gas values and the equilibrium air values for T_{t2}/T_1, p_{t2}/p_1, and $C_{p,t2}$ (the pressure coefficient at the stagnation point) are presented as a function of the free-stream Mach number for an altitude of 150,000 ft in Figs. 12.6, 12.7, and 12.8, respectively. The comments made earlier regarding the qualitative differences between the perfect-gas values and those for equilibrium air (e.g., "The real-gas value of the static pressure is slightly larger than the perfect-gas value") are illustrated in Figs. 12.7 and 12.8.

The ratio of T_{t2}/T_1 is presented in Fig. 12.6. As noted earlier, the energy absorbed by the dissociation process causes the real-gas equilibrium temperature to

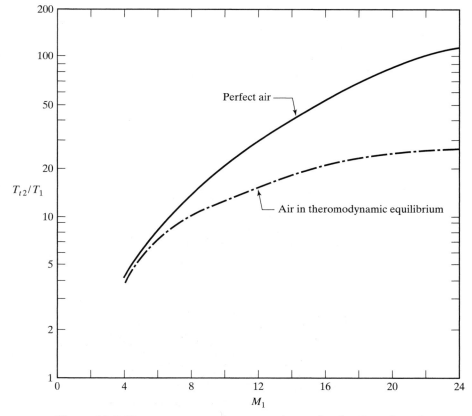

Figure 12.6 Temperature at the stagnation point (at the edge of the boundary layer) of a sphere ($R_N = 0.3048$ m) at an altitude of 45,721 m.

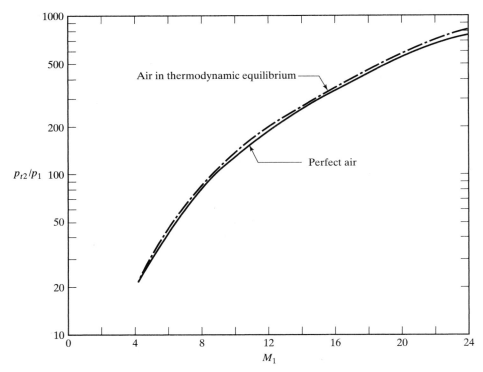

Figure 12.7 Pressure at the stagnation point of a sphere ($R_N = 0.3048$ m) at an altitude of 45,721 m.

be markedly lower than the perfect-gas value. The specific heat correlations that were presented by Hansen (1957) as a function of pressure and of temperature can be used to identify conditions where the dissociation of oxygen and of nitrogen affect the properties. Hansen notes that, at all pressures, the dissociation of oxygen is essentially complete before the dissociation of nitrogen begins. Based on Hansen's correlations, the oxygen dissociation reaction begins near Mach 7 and the nitrogen reaction begins near Mach 18 for the equilibrium air model at 150,000 ft.

The p_{t2}/p_1 ratio and the stagnation-point pressure coefficient are presented in Figs. 12.7 and 12.8, respectively. As noted earlier, the real-gas value of the static pressure is slightly greater than the perfect-gas value. Note that, at $M_1 = 4$, the stagnation pressure coefficient $C_{p,t2}$ is approximately 1.8 for both perfect air and for air in thermodynamic equilibrium. At $M_1 = 24$, $C_{p,t2}$ is 1.932 for the equilibrium air model as compared with 1.838 for perfect air and 2 for Newtonian flow.

From the preceding discussion, it should be clear that the changes in a fluid property across a normal shock wave are a function of M_1 and γ only for a perfect gas [see equation (12.12)], that is,

$$\frac{p_2}{p_1} = f(M_1, \gamma)$$

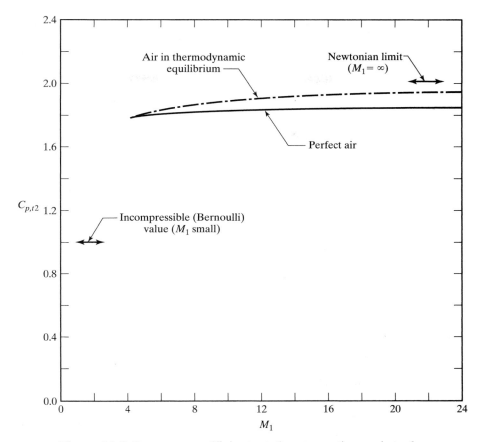

Figure 12.8 Pressure coefficient at the stagnation point of a sphere (R_N = 0.3048 m) at an altitude of 45,721 m.

However, for a reacting gas in chemical equilibrium, three free-stream parameters are necessary to obtain the ratios of properties across a normal shock wave (the free-stream velocity and two thermodynamic properties),

$$\frac{p_2}{p_1} = f(U_1, p_1, T_1)$$

12.3 MODIFIED NEWTONIAN FLOW

It is evident in the computed values for the pressure coefficient at the stagnation point presented in Fig. 12.8 that, even when $M_1 = 24$, $C_{p,t2}$ is 1.838 for perfect air and is 1.932 for air in thermodynamic equilibrium. Thus, as noted by Lees (1955), it would be more appropriate to compare the ratio $C_p/C_{p,\max}$ with $\sin^2 \theta_b$ (or, equivalently, $\cos^2 \phi$). Such a comparison is presented in Fig. 12.9 using data for hemispherically capped cylinders. Even though the data of Fig. 12.9 are for free-stream Mach numbers from 1.97 to 4.76, the $\sin^2 \theta_b$

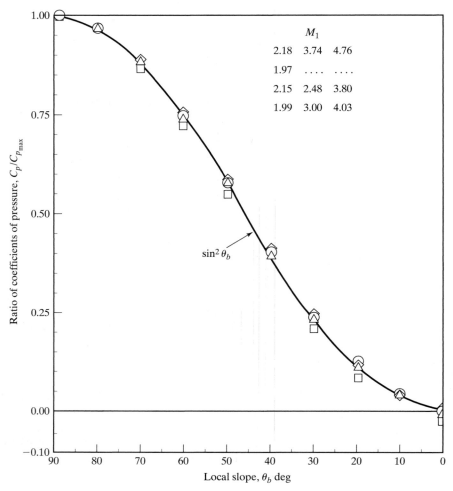

Figure 12.9 Correlation between $C_p/C_{p,\max}$ ratio and local body slope. [From Isaacson and Jones (1968).]

relation represents the data quite adequately for this blunt configuration. Therefore, an alternative representation of the pressure coefficient for hypersonic flow is

$$C_p = C_{p,t2} \sin^2 \theta_b = C_{p,t2} \cos^2 \phi \qquad \textbf{(12.17)}$$

which will be termed modified Newtonian flow.

EXAMPLE 12.1:   Derive an expression for the drag coefficient of a sphere

Neglecting the effects of skin friction and using the modified Newtonian flow model to describe the pressure distribution, derive an expression for the drag coefficient for a sphere.

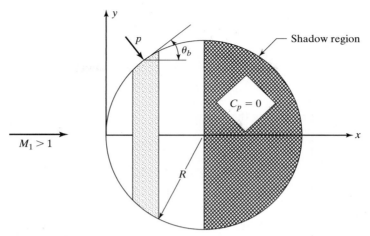

Figure 12.10 Sketch for calculating the modified Newtonian pressure drag on a sphere, Example 12.1.

Solution: Because the sphere is a blunt configuration, the pressure forces are the principal component of the drag force at high Reynolds numbers. Thus, using the coordinate system shown in Fig. 12.10, the drag on the sphere due to the pressure is

$$D = \int p[2\pi y (ds)] \sin \theta_b = \int p 2\pi y \, dy$$

where the incremental surface area on which the pressure acts is $[2\pi y (ds)]$; refer to the shaded region on the forebody in Fig. 12.10. As discussed in Chapter 5, the net resultant force in any direction due to a constant pressure acting over a closed surface is zero. Thus, the pressure coefficient can be used in our expression for the drag:

$$D = q_\infty \int C_p 2\pi y \, dy \tag{12.18}$$

Using modified Newtonian theory to define the pressure, as given by equation (12.17), with the coordinates shown in Fig. 12.10,

$$C_p = C_{p,t2} \sin^2 \theta_b = C_{p,t2} \left(\frac{dy}{ds} \right)^2 = C_{p,t2} \frac{(y')^2}{1 + (y')^2} \tag{12.19}$$

where $y' = dy/dx$. Combining equations (12.18) and (12.19), the expression for the drag becomes

$$D = q_\infty C_{p,t2} \int_0^R \frac{(y')^2}{1 + (y')^2} 2\pi y \, y' \, dx \tag{12.20}$$

The limits of the integration are $0 \le x \le R$, since, according to the Newtonian flow model, the pressure coefficient on the leeward side crosshatched region of the sphere in Fig. 12.10) is zero.

For the sphere,

$$y = (2xR - x^2)^{0.5}$$

and

$$\frac{dy}{dx} = \frac{R - x}{(2xR - x^2)^{0.5}}$$

Substituting these expressions into equation (12.20),

$$D = 2\pi q_\infty C_{p,t2} \int_0^R \frac{\left[\dfrac{R - x}{(2xR - x^2)^{0.5}}\right]^3}{1 + \dfrac{(R - x)^2}{2xR - x^2}}(2xR - x^2)^{0.5}\, dx$$

Simplifying and integrating yields

$$D = \frac{2\pi q_\infty C_{p,t2}}{R^2}\left[R^3 x - \frac{3}{2}x^2 R^2 + x^3 R - \frac{x^4}{4} \right]_0^R$$

$$= \frac{C_{p,t2}}{2} q_\infty \pi R^2$$

Thus,

$$C_D = \frac{D}{q_\infty \pi R^2} = \frac{C_{p,t2}}{2} \tag{12.21}$$

Let us consider the case where the modified Newtonian flow model is applied in a similar manner to calculate the pressure drag on a right circular cylinder whose axis is perpendicular to the free-stream flow. Although the pressure distribution is that given by Equation (12.19) and the cross section is a circle, as shown in Fig. 12.10, the flow around the cylinder is two dimensional, whereas the flow around a sphere is axisymmetric. As a result, the section pressure drag coefficient (per unit span) for the cylinder is

$$C_{d,p} = \tfrac{2}{3}C_{p,t2} \tag{12.22}$$

The values of section pressure drag coefficient calculated using equation (12.22) are compared in Fig. 12.11 with data that were presented in Koppenwallner (1969). Despite the simplifications inherent in this technique to calculate the theoretical drag coefficient, the agreement between the data and the theoretical values is outstanding. Note also that the pressure drag for the blunt, right circular cylinder reaches its hypersonic limiting value by $M_\infty = 4$.

Data presented by Koppenwallner (1969) indicate a significant increase in the total drag coefficient for a right circular cylinder occurs due to the friction drag when the Knudsen number (which is the ratio of the length of the molecular mean free path to a characteristic dimension of the flow field) is greater than 0.01. Data ‸nted by Koppenwallner are reproduced in Fig. 12.12. Using the Reynolds

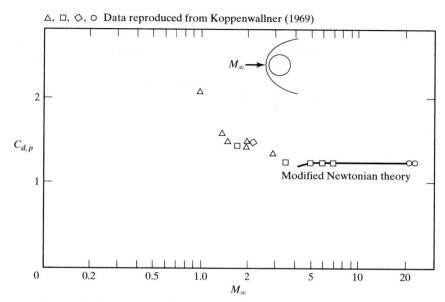

Figure 12.11 Pressure drag of a right circular cylinder as a function of Mach number. [From Koppenwallner (1969).]

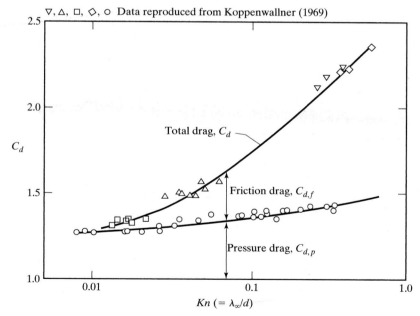

Figure 12.12 Contribution of the total drag due to friction drag and to pressure drag. [From Koppenwallner (1969).]

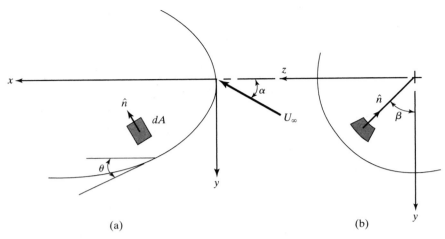

Figure 12.13 Coordinate system nomenclature for axisymmetric configurations; (a) side view; (b) front view.

number based on the flow conditions behind a normal shock wave as the characteristic parameter,

$$\text{Re}_2 = \frac{\rho_2 U_2 d}{\mu_2}$$

the friction drag for $\text{Re}_2 > 10$ is given by

$$C_{d,f} = \frac{5.3}{(\text{Re}_2)^{1.18}} \tag{12.23}$$

The data presented in Figs. 12.11 and 12.12 illustrate the significance of high-altitude effects on the aerodynamic coefficients.

The modified Newtonian flow model can be used to obtain a quick, engineering estimate of the pressure distribution. Consider the axisymmetric configuration shown in Fig. 12.13, where $\hat{n}$ is a unit vector that is normal to the surface element dA and is positive in the inward direction, θ is the local surface inclination, and β is the angular position of a point on the surface of the body. The angle η, the angle between the velocity vector $\vec{V}_\infty$ and the inward normal $\hat{n}$, is given by

$$\cos \eta = \frac{\vec{V}_\infty \cdot \hat{n}}{|\vec{V}_\infty||\hat{n}|} \tag{12.24}$$

where

$$\vec{V}_\infty = U_\infty \cos \alpha \hat{i} - U_\infty \sin \alpha \hat{j} \tag{12.25}$$

and

$$\hat{n} = \hat{i} \sin \theta - \hat{j} \cos \theta \cos \beta - \hat{k} \cos \theta \sin \beta \tag{12.26}$$

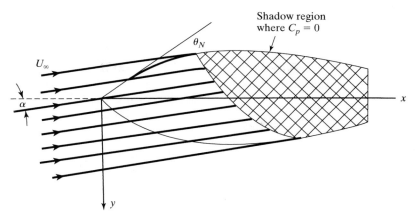

Figure 12.14 Region where $C_p = 0$, that is, that portion of the body which lies in the "shadow of the free stream."

Thus,

$$\cos \eta = \cos \alpha \sin \theta + \sin \alpha \cos \theta \cos \beta \quad \text{(12.27)}$$

so that the pressure coefficient is

$$C_p = C_{p,t2} \cos^2 \eta \quad \text{(12.28)}$$

In the Newtonian (or the modified Newtonian) model, the free-stream flow does not impinge on those portions of the body surface which are inclined away from the free-stream direction and which may, therefore, be thought of as lying in the "shadow of the free stream." This is illustrated in Fig. 12.14. For the modified Newtonian flow model, $C_p = 0$ in the shaded region of Fig. 12.14.

EXAMPLE 12.2: Determine the aerodynamic coefficients for a sharp cone

Consider hypersonic flow past a sharp cone where $-\theta_c \le \alpha \le \theta_c$, as shown in Fig. 12.15. Neglecting the effects of skin friction and using the modified Newtonian flow model to describe the pressure distribution, derive expressions for the lift coefficient, the drag coefficient, and the pitching moment coefficient.

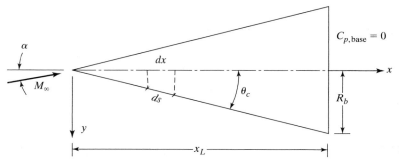

Figure 12.15 Nomenclature for hypersonic flow past a sharp cone.

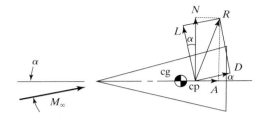

Figure 12.16 Resolving the forces acting on a vehicle.

Solution: For simplicity, let us first calculate the forces in a body-fixed coordinate system. As shown in Fig. 12.16, the forces in the body-fixed coordinate system are A, the force along the axis of the body, and N, the force normal to the body axis. Once A and N have been calculated, the lift and the drag can be calculated since

$$L = N \cos \alpha - A \sin \alpha \tag{12.29}$$

and

$$D = N \sin \alpha + A \cos \alpha \tag{12.30}$$

Applying equations (12.24) through (12.28) specifically for the sharp cone flow depicted in Fig. 12.15,

$$C_p = C_{p,t2}(\cos \alpha \sin \theta_c + \sin \alpha \cos \theta_c \cos \beta)^2 \tag{12.31}$$

since the deflection angle is the cone semivertex angle, θ_c, which is a constant.

The axial force coefficient is found by integrating the pressure force over the entire (closed) surface of the cone:

$$C_A = \frac{1}{\left(\frac{1}{2}\rho_\infty U_\infty^2\right)(\pi R_b^2)} \oiint_S (p - p_\infty)(\hat{n} \, dS) \cdot \hat{\imath} \tag{12.32}$$

Recall that, since we are integrating over a closed surface, the net force in any direction due to a constant pressure acting on that closed surface is zero. Hence, subtracting p_∞, as is done in equation (12.32), does not affect the value of the axial force. A differential element for the surface of the cone (dS) is

$$dS = r \, d\beta \, ds = (x \tan \theta_c) \, d\beta \frac{dx}{\cos \theta_c} \tag{12.33}$$

Combining equations (12.31) through (12.33) with equation (12.26) and noting that, since $C_{p,\text{base}} = 0$, the limits of integration are $0 \leq \beta \leq \pi$ (if we multiply by 2) and $0 \leq x \leq x_L$, the axial force coefficient is

$$C_A = \frac{2C_{p,t2}}{\pi R_b^2} \int_0^{x_L} \left[\int_0^{\pi} (\cos^2 \alpha \sin^2 \theta_c + 2\sin \alpha \cos \alpha \sin \theta_c \cos \theta_c \cos \beta \right.$$

$$\left. + \sin^2 \alpha \cos^2 \theta_c \cos^2 \beta)(x \tan \theta_c) \, d\beta \frac{dx}{\cos \theta_c} \sin \theta_c \right]$$

Integrating first over β yields

$$C_A = \frac{2C_{p,t2}}{\pi R_b^2} \int_0^{x_L} \left(\cos^2 \alpha \sin^2 \theta_c \pi + \sin^2 \alpha \cos^2 \theta_c \frac{\pi}{2} \right) \tan^2 \theta_c x \, dx$$

Integrating with respect to x gives us

$$C_A = \frac{2C_{p,t2}}{R_b^2} \left(\cos^2 \alpha \sin^2 \theta_c + \frac{1}{2} \sin^2 \alpha \cos^2 \theta_c \right) \tan^2 \theta_c \frac{x_L^2}{2}$$

Since $R_b = x_L \tan \theta_c$,

$$C_A = C_{p,t2} \left[\sin^2 \theta_c + \tfrac{1}{2} \sin^2 \alpha (1 - 3 \sin^2 \theta_c) \right] \tag{12.34}$$

Let us calculate the normal force coefficient due to the pressures acting over the closed surface of the cone. Since the normal force is positive in the negative y direction,

$$C_N = \frac{1}{\left(\frac{1}{2} \rho_\infty U_\infty^2 \right)(\pi R_b^2)} \oiint_S (p - p_\infty)(\hat{n} \, dS) \cdot (-\hat{j}) \tag{12.35}$$

Combining equations (12.26), (12.31), (12.33), and (12.35) and using the limits of integration, $0 \le \beta \le \pi$ and $0 \le x \le x_L$ (as discussed previously), we have

$$C_N = \frac{2C_{p,t2}}{\pi R_b^2} \int_0^{x_L} \left[\int_0^\pi (\cos^2 \alpha \sin^2 \theta_c + 2\sin \alpha \cos \alpha \sin \theta_c \cos \theta_c \cos \beta \right.$$

$$\left. + \sin^2 \alpha \cos^2 \theta_c \cos^2 \beta)(x \tan \theta_c) \, d\beta \, \frac{dx}{\cos \theta_c} (\cos \theta_c \cos \beta) \right]$$

Integrating first with respect to β gives us

$$C_N = \frac{2C_{p,t2}}{\pi R_b^2} \int_0^{x_L} \left(2 \sin \alpha \cos \alpha \sin \theta_c \cos \theta_c \frac{\pi}{2} \right) \tan \theta_c x \, dx$$

Integrating with respect to x yields

$$C_N = \frac{2C_{p,t2}}{R_b^2} (\sin \alpha \cos \alpha \sin \theta_c \cos \theta_c) \tan \theta_c \frac{x_L^2}{2}$$

so that

$$C_N = \frac{C_{p,t2}}{2} \sin 2\alpha \cos^2 \theta_c \tag{12.36}$$

Referring to equation (12.29), the lift coefficient is

$$C_L = C_N \cos \alpha - C_A \sin \alpha$$

Using the coefficients presented in equations (12.34) and (12.36), we have

$$C_L = C_{p,t2} \sin \alpha \left[\cos^2 \alpha \cos^2 \theta_c - \sin^2 \theta_c - \tfrac{1}{2} \sin^2 \alpha (1 - 3 \sin^2 \theta_c) \right] \tag{12.37}$$

Correspondingly,

$$C_D = C_{p,t2} \cos \alpha \left[\sin^2 \alpha \cos^2 \theta_c + \sin^2 \theta_c + \tfrac{1}{2} \sin^2 \alpha (1 - 3 \sin^2 \theta_c) \right] \quad \textbf{(12.38)}$$

To calculate the pitching moment, note that the moment due to the incremental pressure force acting at a radial distance $\vec{r}$ from the origin is

$$\overrightarrow{dM} = \vec{r} \times \overrightarrow{dF} = \vec{r} \times p\hat{n} \, dS$$

Note that the incremental moment $\overrightarrow{dM}$ is a vector, which can be written in terms of its components:

$$\overrightarrow{dM} = dL\,\hat{i} + dN\,\hat{j} + dM\,\hat{k} \quad \textbf{(12.39)}$$

In equation (12.39), L is the rolling moment (which is positive when causing the right wing to move down), N is the yawing moment (which is positive when causing the nose to move to the right), and M is the pitching moment (which is positive when causing a nose-up motion).

If we want to take the moments about the apex of the cone, the moment arm is

$$\vec{r} = x\hat{i} + x \tan \theta_c \cos \beta \hat{j} + x \tan \theta_c \sin \beta \hat{k} \quad \textbf{(12.40)}$$

Let us examine the $\vec{r} \times \hat{n}$ term of the moment expression:

$$\vec{r} \times \hat{n} = \begin{vmatrix} \hat{i} & \hat{j} & \hat{k} \\ x & x \tan \theta_c \cos \beta & x \tan \theta_c \sin \beta \\ \sin \theta_c & -\cos \theta_c \cos \beta & -\cos \theta_c \sin \beta \end{vmatrix}$$

Thus,

$$\vec{r} \times \hat{n} = \hat{i}[0] + \hat{j}[x \sin \beta (\tan \theta_c \sin \theta_c + \cos \theta_c)]$$
$$- \hat{k}[x \cos \beta (\tan \theta_c \sin \theta_c + \cos \theta_c)] \quad \textbf{(12.41)}$$

The first term (i.e., the contribution to the rolling moment) is always zero, since, in the absence of viscous forces, the only forces are the pressure forces that act normal to the surface and therefore act through the axis of symmetry for a body of revolution. The negative sign for the $\hat{k}$ term (i.e., the pitching moment) results because a pressure force acting in the first and fourth quadrants ($\pi/2 \geq \beta \geq -\pi/2$) produces a nose-down (negative) pitching moment. For this example we are interested in the pitching moment about the apex, which is considered positive when nose up,

$$M_0 = \oiint_S \vec{r} \times (p - p_\infty)\hat{n} \, dS \cdot \hat{k} \quad \textbf{(12.42)}$$

Taking the resultant cross product of the moment arm from the apex ($\vec{r}$) and the inward-directed unit normal to the surface area $\hat{n}$ with the dot product with $\hat{k}$, we obtain

$$(\vec{r} \times \hat{n}) \cdot \hat{k} = -x \cos \theta_c \cos \beta - x \sin \theta_c \tan \theta_c \cos \beta$$

Thus, the pitching moment coefficient

$$C_{M_0} = \frac{-2}{\pi R_b^2 R_b} \int_0^{x_L} \left[\int_0^{\pi} C_p x \tan \theta_c \, d\beta \, \frac{dx}{\cos \theta_c} (x \cos \theta_c \cos \beta \right.$$

$$\left. + x \sin \theta_c \tan \theta_c \cos \beta) \right] \quad \textbf{(12.43)}$$

We can simplify this expression if we focus on the terms containing the cone-half-angle (θ_c), so that

$$\tan \theta_c \frac{1}{\cos \theta_c} (\cos \theta_c + \sin \theta_c \tan \theta_c) = \tan \theta_c (1 + \tan^2 \theta_c) = \frac{\tan \theta_c}{\cos^2 \theta_c}$$

Substituting the modified-Newtonian-flow expression for the pressure coefficient and integrating with respect to β yields

$$C_{M_0} = -\frac{2C_{p,t2}}{\pi R_b^2 R_b} \frac{\tan \theta_c}{\cos^2 \theta_c} \int_0^{x_L} x^2 (\sin \alpha \cos \alpha \sin \theta_c \cos \theta_c \pi) \, dx$$

Integrating with respect to x and using the fact that $R_b = x_L \tan \theta_c$, we obtain

$$C_{M_0} = -\frac{C_{p,t2} \sin 2\alpha}{3 \tan \theta_c} \quad \textbf{(12.44)}$$

Although the resultant moment is produced by the integration of the distributed aerodynamic forces over the vehicle surface, it can be represented as composed of two components, one effectively due to the normal-force component and the other due to the axial-force component. Thus, we can isolate the two terms of equation (12.43):

$$C_{M_0} = -\frac{2}{\pi R_b^2 R_b} \int_0^{x_L} \left(\int_0^{\pi} C_p x \tan \theta_c \, d\beta \, \frac{dx}{\cos \theta_c} x \cos \theta_c \cos \beta \right)$$

$$- \frac{2}{\pi R_b^2 R_b} \int_0^{x_L} \left(\int_0^{\pi} C_p x \tan \theta_c \, d\beta \, \frac{dx}{\cos \theta_c} x \sin \theta_c \tan \theta_c \cos \beta \right)$$

Writing the total pitching moment as the sum of the two components, we have

$$C_{M_0} = C_{M_{0,N}} + C_{M_{0,A}} \quad \textbf{(12.45)}$$

where

$$C_{M_{0,N}} = -\frac{2 \tan \theta_c}{\pi R_b^2 R_b} \int_0^{x_L} \left(\int_0^{\pi} C_p \cos \beta \, d\beta \right) x^2 \, dx \quad \textbf{(12.46)}$$

and

$$C_{M_{0,A}} = -\frac{2 \tan^3 \theta_c}{\pi R_b^2 R_b} \int_0^{x_L} \left(\int_0^{\pi} C_p \cos \beta \, d\beta \right) x^2 \, dx \quad \textbf{(12.47)}$$

Substituting the modified Newtonian-flow expression for the pressure coefficient yields

$$
C_{M_{0,N}} = -\frac{2 \tan \theta_c C_{p,t2}}{\pi R_b^2 R_b} \int_0^{x_L} \left[\int_0^{\pi} (\cos^2 \alpha \sin^2 \theta_c \cos \beta \right.
$$

$$
\left. + 2 \sin \alpha \cos \alpha \sin \theta_c \cos \theta_c \cos^2 \beta + \sin^2 \alpha \cos^2 \theta_c \cos^3 \beta) \, d\beta \right] x^2 \, dx
$$

Integrating gives us

$$
C_{M_{0,N}} = -\frac{C_{p,t2} \sin 2\alpha}{R_b^2 R_b} \tan \theta_c \sin \theta_c \cos \theta_c \frac{x_L^3}{3}
$$

$$
C_{M_{0,N}} = -\frac{C_{p,t2} \sin 2\alpha \cos^2 \theta_c}{3} \frac{x_L}{R_b}
$$

Using equation (12.36) for the expression for C_N, we can rewrite

$$
C_{M_{0,N}} = C_N \left(-\frac{2}{3} \frac{x_L}{R_b} \right) \tag{12.48}
$$

Similarly,

$$
C_{M_{0,A}} = \frac{-2 \tan^3 \theta_c C_{p,t2}}{\pi R_b^2 R_b} \int_0^{x_L} \left[\int_0^{\pi} (\cos^2 \alpha \sin^2 \theta_c \cos \beta \right.
$$

$$
\left. + 2 \sin \alpha \cos \alpha \sin \theta_c \cos \theta_c \cos^2 \beta + \sin^2 \alpha \cos^2 \theta_c \cos^3 \beta) \, d\beta \right] x^2 \, dx
$$

Integrating yields

$$
C_{M_{0,A}} = -\frac{C_{p,t2} \sin 2\alpha \sin^2 \theta_c}{3} \frac{x_L}{R_b}
$$

Using equation (12.34) for the expression for C_A, we can rewrite

$$
C_{M_{0,A}} = C_A \frac{-\sin 2\alpha \sin^2 \theta_c}{3[\sin^2 \theta_c + 0.5 \sin^2 \alpha (1 - 3 \sin^2 \theta_c)]} \frac{x_L}{R_b} \tag{12.49}
$$

Following this division of the pitching moment into components, we can represent it as an effective net force acting at a "center of pressure," as a shown in Fig. 12.17. The pitching moment may be represented by

$$
M_0 = M_{0,N} + M_{0,A} = -x_{cp} N - y_{cp} A \tag{12.50}
$$

The two terms represent the effects of the normal force and of the axial force, respectively. In terms of the coefficients,

$$
C_{M_0} = \frac{M_0}{q_\infty S R_b} = -C_N \frac{x_{cp}}{R_b} - C_A \frac{y_{cp}}{R_b} \tag{12.51}
$$

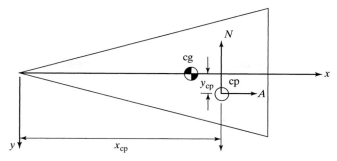

Figure 12.17 Aerodynamic forces acting at the center of pressure as effectively producing the aerodynamic pitching moment.

Comparing the formulations presented in equations (12.48), (12.49) and (12.51), we find that

$$x_{cp} = \frac{2}{3} x_L \tag{12.52}$$

and

$$y_{cp} = \frac{\sin 2\alpha \sin^2 \theta_c x_L}{3[\sin^2 \theta_c + 0.5 \sin^2 \alpha(1 - 3 \sin^2 \theta_c)]} \tag{12.53}$$

   If the vehicle is to be statically stable, the center of pressure should be located such that the aerodynamic forces produce a restoring moment when the configuration is perturbed from its "stable orientation." Thus, if the vehicle pitches up (i.e., α increases), the net pitching moment should be negative (causing a nose-down moment), which decreases α.

   The parameter

$$\text{S.M.} = \frac{x_{cp} - x_{cg}}{x_L} \tag{12.54}$$

is the static margin. The static margin must be positive for uncontrolled vehicles. For high-performance, hypersonic vehicles, the static margin is usually 3 to 5%. Note, however, as illustrated in Fig. 12.17, both the axial force and the normal force contribute to the pitching moment. Thus, it is possible that the vehicle is statically stable when $x_{cp} = x_{cg}$, if y_{cp} is below y_{cg} (as is the case in the sketch of Fig. 12.17). In this case the axial force will produce the required restoring moment.

EXAMPLE 12.3:   What is the total drag acting on a sharp cone?

We neglected the viscous forces and the base pressure in estimating the forces acting on the sharp cone of Example 12.2. Consider a sharp cone ($\theta_c = 10°$) exposed to the Mach 8 flow of Tunnel B at the Arnold Engineering Development

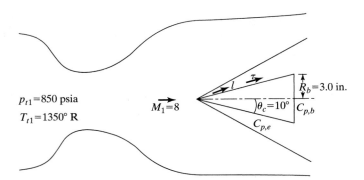

p_{t1}=850 psia

T_{t1}=1350° R

$\overrightarrow{M_1}$=8

θ_c=10°

R_b=3.0 in.

$C_{p,b}$

$C_{p,e}$

Figure 12.18 Sharp cone mounted in Tunnel B of AEDC.

Center (AEDC), as shown in Fig. 12.18. For this problem, the subscript $t1$ designates the flow conditions in the nozzle reservoir, the subscript 1 designates those in the test section, the subscript e designates those at the edge of the boundary layer of the sharp cone (and which are constant along the entire length of the cone), and the subscript w designates conditions at the wall of the cone. If $T_{t1} = 1350°R$, $p_{t1} = 850$ psia, $T_w = 600°R$, and $\alpha = 0°$, develop expressions for the forebody pressure coefficient ($C_{p,e}$), the base pressure coefficient ($C_{p,b}$), and the skin friction. What is the total drag acting on the vehicle?

Solution: Let us use Figs. 8.15b and c to obtain values for the pressure coefficient and for the Mach number of the inviscid flow at the edge of the boundary layer for a 10°-half-angle cone in a Mach 8 stream. Inherent in this use of these figures is the assumption that the boundary layer is thin and that there are no significant viscous/inviscid interactions which perturb the pressure field. Thus, $M_e = 6$ and $C_{p,e} = 0.07$. Since $q_1 = \frac{1}{2}\rho_1 U_1^2 = (\gamma/2)p_1 M_1^2$,

$$p_e = p_1\left(1 + \frac{\gamma}{2}M_1^2 C_{p,e}\right)$$

We can use Table 8.1 to calculate the free-stream static pressure (p_1) in the Mach 8 flow of the wind-tunnel test section:

$$p_1 = \frac{p_1}{p_{t1}}p_{t1} = (102 \times 10^{-6})(850) = 0.867 \text{ psia}$$

Thus,

$$p_e = 0.0867[1.0 + (0.7)(64)(0.07)] = 0.3586 \text{ psia}$$

Since $M_e = 6$, we can use Table 8.1 and the fact that the edge flow is that for an adiabatic flow of a perfect gas, so that $T_{t1} = T_{te}$. Thus,

$$T_e = \frac{T_e}{T_{te}}T_{te} = (0.12195)(1350) = 164.63°R$$

To aid in our decision as to whether the boundary layer is laminar or turbulent, let us calculate the Reynolds number for the flow at the boundary-layer edge:

$$\text{Re}_l = \frac{\rho_e U_e l}{\mu_e} = \left[\frac{p_e}{RT_e}\right]\left[M_e\sqrt{\gamma RT_e}\right]\left[\frac{T_e + 198.6}{2.27 \times 10^{-8}T_e^{1.5}}\right]l$$

where, as shown in Fig. 12.18, l is the wetted distance along a conical generator. Thus,

$$\text{Re}_l = \left(0.0001828\frac{\text{lbf}\cdot\text{s}^2}{\text{ft}^4}\right)\left(3773.80\frac{\text{ft}}{\text{s}}\right)\left(7.575 \times 10^6\frac{\text{ft}^2}{\text{lbf}\cdot\text{s}}\right)l$$

$$= 5.226 \times 10^6\,l$$

Based on the wetted length of a conical ray ($l_c = 17.276\,\text{in.}$),

$$\text{Re}_{lc} = 7.523 \times 10^6$$

It should be noted that, although hypersonic boundary-layers are very stable, boundary-layer transition should occur for this flow, probably within the first one-half of the cone length. Thus, we can use the correlation presented in Fig. 12.19 to determine the base pressure:

$$\frac{p_b}{p_e} = 0.02$$

so that $p_b = 0.00717$ psi.

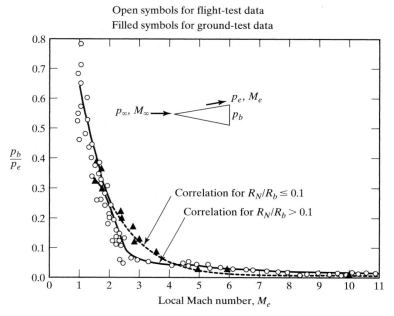

Figure 12.19 Base pressure ratio as a function of the local Mach number for turbulent flow. [From Cassanto (1973).]

Note that since $p_b = 0.02p_e$,

$$C_{p,b} = \frac{p_b - p_1}{q_1} = \left(\frac{p_b}{p_1} - 1\right)\frac{2}{\gamma M_1^2}$$

$$= \left(\frac{p_b}{p_e}\frac{p_e}{p_1} - 1\right)\frac{2}{\gamma M_1^2} = -0.0205$$

Recall that the assumption for Newtonian flow is $C_{p,b} = 0.0$.

To calculate the skin friction contribution, let us use Reynolds analogy (see Chapter 4). For supersonic flow past a sharp cone, the heat-transfer rate (or, equivalently, the Stanton number) can be calculated using the relatively simple approximation obtained through the Eckert reference temperature approach [Eckert (1955)]. In this approach, the heat-transfer rates are calculated using the incompressible relations [e.g., equation (4.103)] with the temperature-related parameters evaluated at Eckert's reference temperature (T^*). According to Eckert (1955),

$$T^* = 0.5(T_e + T_w) + 0.22r(T_{te} - T_e)$$

where r is the recovery factor. For turbulent flow, the recovery factor is $\sqrt[3]{Pr}$, which is equal to 0.888. Thus,

$$T^* = 0.5(164.63 + 600) + 0.22(0.888)(1350 - 164.63)$$

$$= 613.89°R$$

Thus,

$$\text{Re}_l^* = \left[\frac{p_e}{RT^*}\right][U_e]\left[\frac{T^* + 198.6}{2.27 \times 10^{-8}T^{*1.5}}\right]l$$

$$= \left(0.00004901\frac{\text{lbf} \cdot \text{s}^2}{\text{ft}^4}\right)\left(3773.80\frac{\text{ft}}{\text{s}}\right)\left(2.353 \times 10^6\frac{\text{ft}^2}{\text{lbf} \cdot \text{s}}\right)l$$

$$= 4.353 \times 10^5 l$$

Thus, using Reynolds' analogy, we have

$$C_f = \frac{0.0583}{(\text{Re}_l^*)^{0.2}} = \frac{0.004344}{l^{0.2}}$$

Thus,

$$\tau = C_f\left(\frac{\gamma}{2}p_eM_e^2\right) = \frac{0.03926}{l^{0.2}}\frac{\text{lbf}}{\text{in.}^2}$$

To calculate the total drag,

$$D = \int p_e 2\pi r \, dl \sin\theta_c + \int \tau_w 2\pi r \, dl \cos\theta_c - p_b\pi R_b^2$$

Noting that p_e is constant along the length of the conical generator, $dl \sin\theta_c = dr$, and $r = l \sin\theta_c$,

$$D = p_e \pi R_b^2 + 0.03926(2\pi)(\cos \theta_c \sin \theta_c) \int_0^{l_c} l^{0.8} \, dl - p_b \pi R_b^2$$

$$= (0.3586)(9\pi) + (0.04218)\frac{l_c^{1.8}}{1.8} - (0.00717)(9\pi)$$

$$= 10.139 + 3.956 - 0.203 = 13.892 \, \text{lbf}$$

Various researchers have developed numerical techniques using the two layer approach, that is, one in which an inviscid solution is first computed providing the flow conditions (e.g., the static pressure and the entropy) at the edge of the boundary layer. Having defined the pressure and the entropy distribution, the remaining flow properties can be calculated and the inviscid streamlines generated. Brandon and DeJarnette (1977) and Riley et al. (1990) have generated numerical solutions of Euler's equation complemented by streamline tracing techniques to define the inviscid properties and the inviscid streamlines for various blunt configurations. Once the boundary conditions are known, solutions of the boundary layer will provide the temperature, the velocity, and the gas-component distributions adjacent to the surface from which one can determine the convective heat-transfer rate and the skin friction. Techniques of various rigor can be used to generate distributions for the convective heat-transfer rate and the skin friction. Zoby et al. (1981) have computed the laminar heat-transfer distributions for the Shuttle *Orbiter* using an incompressible Blasius relation (a similarity transformation) with compressibility effects accounted for by Eckert's reference-enthalpy method. Bertin and Cline (1980) developed a nonsimilarity boundary-layer code to generate solutions for laminar, transitional, and turbulent boundary layers. Although these two techniques are very different, they produced very similar computations of the Shuttle's reentry environment.

12.4 HIGH *L/D* HYPERSONIC CONFIGURATIONS—WAVERIDERS

Nonweiler (1959) concluded that "The most effective method of alleviating the deceleration is by the introduction of lift, which also serves to decrease the peak heating rate, though the total heat absorbed is usually greater." To achieve high *L/D* (lift/drag), Nonweiler proposed all-wing designs of a delta planform, such that the shock wave(s) is (are) attached to the leading edges at the design Mach number and at the design angle of attack. Thus, the wing appears to ride on its shock wave, hence the name "waverider." Shapes of the wing undersurface of a delta amenable to treatment by exact shock-wave theory, as proposed by Nonweiler are reproduced in Fig. 12.20.

Experimental studies of Nonweiler's waverider configurations indicated lower aerodynamic performance (i.e., lower lift-to-drag ratio) than was predicted. The experimentally observed deficiencies in the aerodynamic performance of the caret waveriders have been attributed to their large wetted area and, hence, relatively high skin friction drag.

Squire (1976b) studied the conditions for which the flow on the upper surface of a wing was independent of that on the lower surface of the wing. If the Mach number normal to the leading edge, which is defined in equation (11.9), is sufficiently high, and

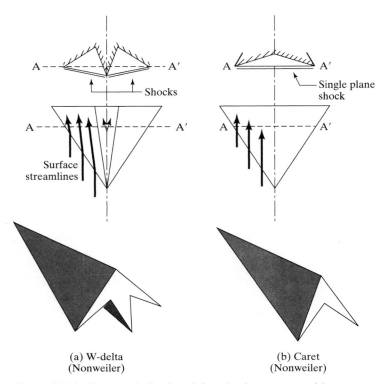

Figure 12.20 Shapes of all-wing delta planforms amenable to treatment by exact shock-wave theory, as proposed by Nonweiler (1959).

the incidence normal to the leading edge, which is defined in equation (11.10), is not too great, the flow near the leading edge is similar to that over an infinitely swept wing. That is, the shock wave below the wing is attached to the wing leading edges and there is a Prandtl-Meyer expansion around the edges, producing a region of uniform suction on the upper surface. This is designated as region C in Fig. 12.21, which is taken from Squire (1976a). As the Mach number normal to the leading edge decreases or as the angle of attack normal to the leading edge increases, the shock wave detaches from the windward surface. This is designated region B in Fig. 12.21. Once the shock wave detaches from the compression surface of a thin wing, flows on the two surfaces are no longer independent. For these conditions, the pressure near the leading edge of the compression surface is relatively high, so that there is a strong outflow around the leading edge on to the upper surface. However, since the detached shock wave (of region B) is close to the leading edge, the crossflow reaches sonic speed as it accelerates around the leading edge. Thus, the shape of the upper surface does not influence the flow on the lower surface. For region A, the conditions are such that the shock wave is well detached from the windward surface. For such conditions, the flow on the upper and on the lower surfaces are no longer independent. The flow on the suction surface is sensitive to changes in the shape of the windward (lower) surface.

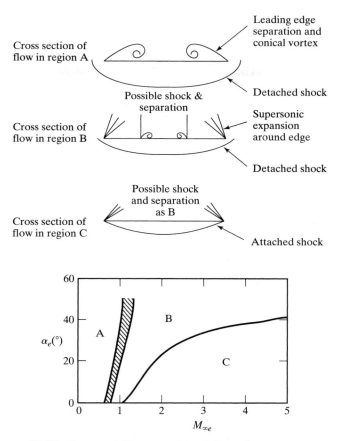

Figure 12.21 Types of flow on thin delta wings at supersonic speeds. [As taken from Squire (1976a).]

Anderson, et al. (1990) note, "A waverider is a supersonic or hypersonic vehicle which, at the design point, has an *attached* shock wave all along its leading edge, as sketched in Fig. 12.22a. Because of this, the vehicle appears to be riding on top of its shock wave, hence the term waverider. This is in contrast to a more conventional hypersonic vehicle, where the shock wave is usually detached from the leading edge, as sketched in Fig. 12.22b. The aerodynamic advantage of the waverider in Fig. 12.22a is that the high pressure behind the shock wave under the vehicle does not "leak" around the leading edge to the top surface; the flowfield over the bottom surface is contained, and the high pressure is preserved. In contrast, for the vehicle shown in Fig. 12.22b, there is communication between the flows over the bottom and top surfaces; the pressure tends to leak around the edge, and the general integrated pressure level on the bottom surface is reduced, resulting in less lift. Because of this, the generic vehicle in Fig. 12.22b must fly at a larger angle of attack, α, to produce the same lift as the waverider in Fig. 12.22a. This is illustrated in Fig. 12.23, where the lift curves (*L* versus α) are sketched for the two vehicles in Fig. 12.22. At the same lift, points

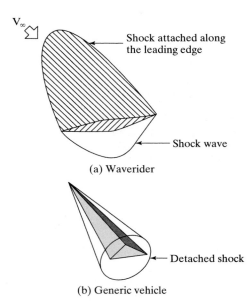

(a) Waverider

(b) Generic vehicle

Figure 12.22 Comparison of a waverider with a generic hypersonic configuration. [Taken from Anderson et al. (1990).]

1*a* and 1*b* in Fig. 12.23 represent the waverider and generic vehicles respectively. Also shown in Fig. 12.23 are typical variations of *L/D* versus α, which for slender hypersonic vehicles are not too different for shapes in Fig. 12.22a and 12.22b. However, note that because the waverider generates the same lift at the smaller α (point 1*a* in Fig. 12.23), the *L/D* for the waverider is considerably higher (point 1*aa*) than that for the generic shape (point 1*bb*)."

Anderson et al. (1990) note further that "when waveriders are optimized for maximum *L/D*, the previous studies have demonstrated that the driving parameter that alters the *L/D* ratio is the skin friction drag."

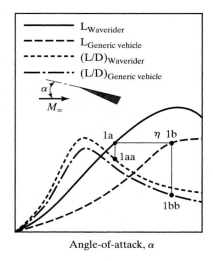

Angle-of-attack, α

Figure 12.23 Curves of lift and (L/D) versus angle of attack: Comparison between a waverider and a generic vehicle. [From Anderson et al. (1990).]

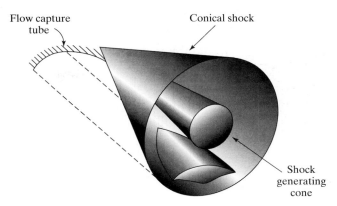

Figure 12.24 Derivation of waverider configurations from a conical flow field.

The design of a hypersonic waverider configuration employs a variety of computational methods. One basic design methodology, which was first developed for conical flow fields, is illustrated in Fig. 12.24. A capture flow tube (whose streamlines are parallel to the axis of the shock-generating cone) intersects the conical shock wave. The leading edge of the waverider is the intersection of the capture flow tube and the conical shock wave. The shape of the waverider leading edge is dependent on the shape of the capture flow tube, the shock angle, and the distance from the capture flow tube to the conical shock-wave centerline. The lower surface is defined by tracing streamlines from the leading edge to the desired base location of the waverider. Eggers and Radespiel (1993) state, "An obvious choice for the upper surface of hypersonic waveriders is the undisturbed Flow Capture Tube used to define the leading edge of the configuration. This results in zero pressure coefficient along the upper surface and the aerodynamic behavior is completely governed by the lower surface. However, with a more careful consideration of typical waverider missions, a designer would like to have more options at his disposal for several reasons. Firstly, variations of the flow state along the upper surface will result in a redistribution of the vehicle's volume which is useful to assist integration of fuels, systems, and payload. A careful use of flow expansion along the upper surface may also enhance *L/D* at the design point. Finally, flow expansion in the rear part of a waverider will decrease the base drag of vehicles in the subsonic, transonic, and supersonic part of the vehicle's trajectory." The process described in this paragraph can be used to derive a waverider from any axisymmetric flow field.

To achieve realistic estimates of the *L/D* ratio, the effect of skin friction drag, which represents a relatively large fraction of the total drag, must be included in deriving the waverider configuration. Bowcutt et al. (1987) have developed a numerical procedure for optimizing cone-derived waveriders, which includes the effect of skin friction.

The results of an experimental investigation which was conducted to determine the aerodynamic characteristics of waverider configurations are reported in Bauer et al. (1990). The Mach 4 waverider configuration, which was determined using the conical flow field process described previously (including both the pressure drag and the friction

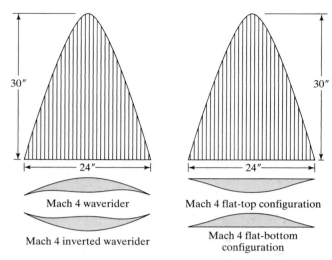

Figure 12.25 Waverider configurations tested in the Langley Unitary Plan Wind Tunnel (UPWT). [Taken from Bowcutt et al. (1987).]

drag in the optimization procedure), is depicted in Fig. 12.25. A second wind-tunnel model was built with a flat top, but having the same planform and cross-sectional area distribution. Data were also obtained with the two configurations inverted. The data were obtained in the Unitary Plan Wind Tunnel (UPWT) at the Langley Research Center (NASA) at a Mach number of 4 and at a unit Reynolds number of 2×10^6 per foot. To insure that the boundary layer on the model was fully turbulent, grit particles (with a particle diameter of 0.0215 in.) were applied 4.0 in. aft of the leading edge to promote the onset of transition at the desired location on the model.

Bauer et al. (1990) note that "The Mach 4 waverider is seen to have the most lifting potential followed by the flat-bottom configurations, which exhibit a 7% reduction in $C_{L\alpha}$. The Mach 4 waverider has 15% higher $C_{L\alpha}$ than when inverted."

The experimentally-determined drag coefficients, as determined for $(L/D)_{\max}$ and presented in Bauer et al. (1990), are reproduced in Fig. 12.26. The grit drag was found to be only a very small part of the total drag. Since the base drag was found to be 25 to 30% of the total drag, a significant effort was made to measure the base drag accurately. The estimates of the skin friction drag, $C_{D,f}$, were made using the reference temperature method. The skin friction drag, which is roughly 25% of the total forebody drag, was found to be slightly lower for the flat-top/flat-bottom configurations than for the waverider, because the flat-top configuration has a smaller wetted area. The wetted surface area for the flat-top configuration was 897.6 in.^{2}, while that for the waverider was 915.2 in.^{2} The zero-lift pressure drag $(C_{D0})_p$ was determined by subtracting the estimated value of $C_{D,f}$ from the measured value of C_{D0}. The value of $(C_{D0})_p$, thus determined for the flat-top/flat-bottomed configurations, was 13% lower than that for the waverider. This is the main reason that $(L/D)_{\max}$ is higher for the flat-top/flat-bottom configurations than for the waverider, as will be discussed for Fig. 12.27.

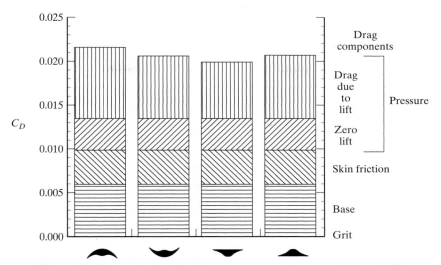

Figure 12.26 Breakdown of drag components at maximum *L/D* conditions, $M_\infty = 4$, $Re_{\infty/ft} = 2 \times 10^6$. [From Bauer et al. (1990).]

The aerodynamic characteristics at $(L/D)_{max}$, as experimentally determined for the four configurations and as computed using the design code, are reproduced in Fig. 12.27. Base drag and grit drag have been removed from the values presented in Fig. 12.27. The values both of C_L and of C_D when (L/D) is a maximum were found to be higher than the design values for all four configurations. However, none of the four configurations achieved a value of $(L/D)_{max}$ equal to that predicted using the design code. The experimental value of $(L/D)_{max}$ for the Mach 4 waverider was 13% lower than that predicted using the design code. $(L/D)_{max}$ was 5% higher for the flat-top configuration than for the waverider.

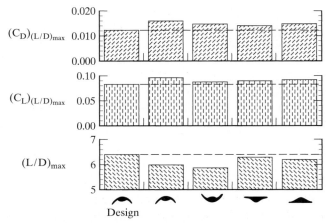

Figure 12.27 Aerodynamic characteristics at maximum *L/D* conditions, $M_\infty = 4$, $Re_{\infty/ft} = 2 \times 10^6$. [From Bauer et al. (1990).]

However, Bauer et al. (1990) noted that "the waverider has higher L/D values than the flat-top configuration for $C_L > 0.16$." Furthermore, Bauer et al. conclude that "These measured performance deficiencies may be attributed to the slight shock detachment that was observed at the design Mach number and angle of attack."

Not only is the aerodynamic performance of a waverider sensitive to viscous effects, but so is the heat transfer. Waveriders, by design, have sharp leading edges so that the bow shock wave is attached. However, for flight Mach numbers above 5, heat transfer to the leading edge can result in surface temperatures exceeding the limits of most structural materials. As will be discussed in the next section, "Aerodynamic Heating," the heat flux to the leading edge can be reduced by increasing the radius of the leading edge. However, while increasing the leading-edge radius can alleviate heating concerns, there is a corresponding increase in drag and, hence, a reduction in L/D. Eggers and Radespiel (1993) note that "Defining the thickness of the leading edge to be about 10/5000 [of the vehicle's length] due to model manufacture requirements one obtains significant detachment of the bow shock and corresponding wave drag. Moreover, the shock is no longer confined to the lower surface which reduces the overall lift. Consequently, the inviscid L/D reduces by about 25% for this large amount of bluntness." Thus, the specification of the leading-edge radius for an aerodynamically efficient waverider that is to operate at Mach numbers in excess of 5 leads to the need for compromise. The leading-edge radius should be sufficiently large to limit the heat flux, yet be as small as possible to minimize the leading-edge drag.

As noted by Haney and Beaulieu (1994), "However, a waverider by itself doesn't necessarily make a good hypersonic aircraft. To do that attention needs to be paid to volumetric efficiency, stability and control, and airframe-engine integration." Cervisi and Grantz (1994) note, "The forebody compression surface sets up the flow environment for the remainder of the vehicle. It constitutes a large fraction of the vehicle drag and provides airflow to the engine inlet. However, once a particular forebody is selected, the available design space for the rest of the vehicle becomes very constrained due to the highly integrated nature of hypersonic flight." Thus, the designer must consider the relation between the forebody compression surface and the boundary-layer transition front, fuselage lift and drag, the inlet capture area, etc.

The sketch presented in Fig. 12.28 illustrates how the inlet capture area can be quickly estimated from a conically derived forebody. For two-dimensional or axisymmetric inlets, the bow-shock-wave location defines the mass flow as a function of inlet width. For shock-on-lip designs (i.e., designs for which the bow shock wave intersects the cowl lip), this condition corresponds to the design point. Since increasing the capture area also increases the engine size and weight, finding the proper balance between the capture area and the nonflowpath drag is a goal for the forebody configuration.

12.5 AERODYNAMIC HEATING

As noted in the introduction to this chapter, the kinetic energy associated with hypersonic flight is converted to high temperatures within the shock layer. As a result, heat transfer is a very important factor in the design of hypersonic vehicles. Sharp leading edges would experience such extremely large heating rates in hypersonic flight that they

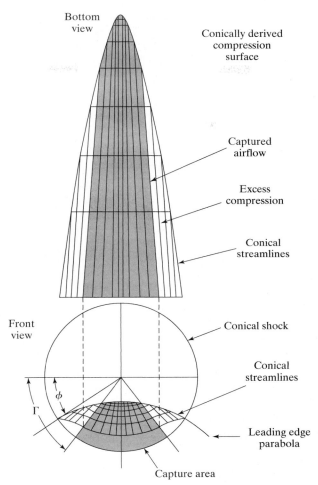

Figure 12.28 Compression surface derived from a conical flow field. [Taken from Cervisi and Grantz (1994).]

would quickly melt or ablate (depending on the material used). Therefore, if a low-drag configuration is desired, the design would probably involve a slender cone with a spherically blunted nose. For a manned reentry craft, where the time of flight can be long and the dissipation of kinetic energy at relatively high altitude is desirable, the resultant high-drag configurations may be an airplane-like Space Shuttle that flies at high angles of attack or a blunt, spherical segment such as the Apollo Command Module.

The expression for the modified-Newtonian-flow pressure coefficient [equation (12.17)] can be rearranged to give

$$\frac{p_s}{p_{t2}} = \sin^2\theta_b + \frac{p_\infty}{p_{t2}}\cos^2\theta_b = \cos^2\phi + \frac{p_\infty}{p_{t2}}\sin^2\phi \tag{12.55}$$

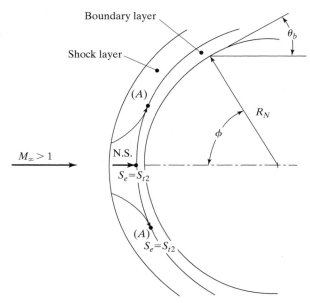

Figure 12.29 Sketch of the normal shock/isentropic expansion flow model for hypersonic flow over a blunt body. The air at the edge of the boundary layer at point A has passed through the (nearly) normal part of the shock wave. Thus, the entropy of the air particles at the edge of the boundary layer (S_e) is essentially equal to S_{t2} at all stations. In essence, the air has passed through the normal part of the shock wave (NS) and has undergone an isentropic expansion (IE) to the local pressures.

Let us apply equation (12.55) to the relatively simple flow depicted in Fig. 12.29. As the boundary layer grows in the streamwise direction, air is entrained from the inviscid portion of the shock layer. Thus, when determining the fluid properties at the edge of the boundary layer, one must determine the entropy of the streamline at the boundary-layer edge. Note that the streamlines near the axis of symmetry of a blunt-body flow, such as those depicted in Fig. 12.29, have passed through that portion of the bow shock wave which is nearly perpendicular to the free-stream flow. As a result, it can be assumed that all the air particles at the edge of the boundary layer have essentially the same entropy. Therefore, the entropy at the edge of the boundary layer and, as a result, p_{t2} are the same at all streamwise stations. The local flow properties are the same as if the air had passed through a normal shock wave and had undergone an isentropic expansion to the local pressure (designated an NS/IE process). For such an isentropic expansion (IE), the ratio of p_s/p_{t2} can then be used to define the remaining flow conditions (for an equilibrium flow). Note that, if the flow expands isentropically to a point where the local static pressure (p_s) is approximately $0.5p_{t2}$, the flow is essentially sonic for all values of γ. Solving equation (12.55), we find that $p_s \approx 0.5p_{t2}$ when $\theta_b = 45°$ (i.e., the sonic points occur when the local body slope is 45°).

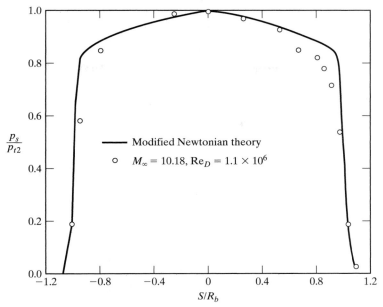

Figure 12.30 Comparison of the modified Newtonian pressures and the experimental pressures for the Apollo Command Module at $\alpha = 0°$. [Taken from Bertin (1966).]

The modified Newtonian pressures for the zero-angle-of-attack Apollo Command Module are compared in Fig. 12.30 with data obtained in Tunnel C at AEDC that are presented by Bertin (1966). The experimental pressures measured in Tunnel C at a nominal free-stream Mach number of 10 and at a Reynolds number $Re_{\infty, D}$ of 1.1×10^6 have been divided by the calculated value of the stagnation pressure behind a normal shock wave (p_{t2}). For reference, a sketch of the Apollo Command Module is presented in Fig. 12.31. Note that an S/R_b ratio of 0.965 defines the tangency point of the spherical heat shield and the toroidal surface, while an S/R_b ratio of 1.082 corresponds to the maximum body radius. Because the windward heat shield of the Apollo Command Module is a truncated spherical cap, the actual sonic points, which occur near the tangency point of the spherical heat shield and the toroidal surface, are inboard of the locations that they would occupy for a full spherical cap. As a result, the entire flow field in the subsonic portion of the shock layer is modified, and the streamwise velocity gradients are relatively large in order to produce sonic flow at the "corners" of the Command Module. Thus, significant differences exist between the modified Newtonian pressures and the measured values as one approaches the edge of the spherical heat shield. Because the velocity gradient at the stagnation point of a hemispherical segment is increased above the value for a full hemisphere, the stagnation point heating rate will also be increased. Investigations of the stagnation region velocity gradients as a function of R_b/R_N have been reported by Stoney (1958) and by Inouye et al. (1968).

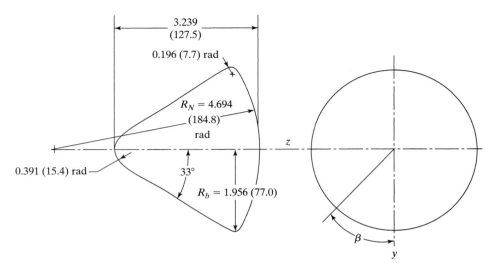

Figure 12.31 Clean (no protuberances) Apollo Command Module. Dimensions in meters (inches).

12.5.1 Similarity Solutions for Heat Transfer

The magnitude of the heat transfer from a compressible boundary layer composed of dissociating gases can be approximated by

$$|\dot{q}_w| = \left(k \frac{\partial T}{\partial y} \right)_w + \left(\rho \Sigma D_i h_i \frac{\partial C_i}{\partial y} \right)_w \tag{12.56}$$

where the heat is transported by conduction and by diffusion. [See Dorrance (1962).] Assume that the flow is such that the Lewis number is approximately equal to unity, the hot gas layer is in chemical equilibrium, and the surface temperature is much less than the external stream temperature. For these assumptions, the magnitude of the heat transferred to the wall is

$$|\dot{q}_w| = \left(\frac{k}{c_p} \frac{\partial h}{\partial y} \right)_w \tag{12.57}$$

There are many situations where a coordinate transformation can be used to reduce the governing partial differential equations for a laminar boundary layer to ordinary differential equations. [See Dorrance (1962).] Fay and Riddell (1958) note, "As is usual in boundary-layer problems, one first seeks solutions of restricted form which permit reducing exactly the partial differential equations to ordinary differential form. An easily recognizable case is that of the stagnation point flow, where, because of symmetry, all the dependent variables are chosen to be functions of y alone, except u which must be taken proportional to x times a function of y. This also appears to be the *only* case for which the exact ordinary differential equations may be obtained regardless of the recombination rate."

Using the similarity transformations suggested by Fay and Riddell (1958), we obtain

$$\eta(x_1, y_1) = \frac{\rho_w u_e r^k}{\sqrt{2S}} \int_0^{y_1} \frac{\rho}{\rho_w} dy \qquad (12.58)$$

and

$$S(x_1) = \int_0^{x_1} \rho_w \mu_w u_e r^{2k} \, dx \qquad (12.59)$$

where r is the cross-section radius of the body of revolution and k denotes whether the flow is axisymmetric ($k = 1$) (for example, a sphere) or two dimensional ($k = 0$) (for example, a cylinder whose axis is perpendicular to the free stream). Limiting approximations are used to describe the flow in the vicinity of the stagnation point, for example,

$$u_e \approx \left(\frac{du_e}{dx}\right)_{t2} x; \qquad r \approx x; \qquad \text{and } \rho_w = \text{constant}$$

Thus, at the stagnation point,

$$S \approx \rho_w \mu_w \left(\frac{du_e}{dx}\right)_{t2} \int xx^{2k} \, dx = \rho_w \mu_w \left(\frac{du_e}{dx}\right)_{t2} \frac{x^{2(k+1)}}{2(k+1)} \qquad (12.60)$$

and

$$\eta \approx \frac{(du_e/dx)_{t2} x x^k}{\sqrt{2S}} \int \rho \, dy = \left[\frac{(du_e/dx)_{t2}(k+1)}{\rho_w \mu_w}\right]^{0.5} \int \rho \, dy \qquad (12.61)$$

Note that using equation (12.61) as the expression for the transformed y coordinate, together with the definition for the heat transfer as given by equation (12.57), we find that, given the same flow condition,

$$(\dot{q}_{t,\text{ref}})_{\text{axisym}} = \sqrt{2}(\dot{q}_{t,\text{ref}})_{2\text{-dim}} \qquad (12.62)$$

(i.e., the stagnation-point heat-transfer rate for a sphere is the $\sqrt{2}$ times that for a cylinder).

The stagnation-point heat-transfer rate for a laminar boundary layer of a spherical cap may be written as

$$\dot{q}_{t,\text{ref}} = \frac{\text{Nu}_x}{\sqrt{\text{Re}_x}} \sqrt{\rho_{w,t} \mu_{w,t} \left(\frac{du_e}{dx}\right)_{t2}} \frac{H_{t2} - h_{w,t}}{Pr} \qquad (12.63)$$

For a Lewis number of 1, Fay and Riddell (1958) found that

$$\frac{\text{Nu}_x}{\sqrt{\text{Re}_x}} = 0.67 \left(\frac{\rho_{t2} \mu_{t2}}{\rho_{w,t} \mu_{w,t}}\right)^{0.4} \qquad (12.64)$$

for velocities between 5800 ft/s (1768 m/s) and 22,800 ft/s (6959 m/s) and at altitudes between 25,000 ft (7620 m) and 120,000 ft (36,576 m).

Let us use Euler's equation to evaluate the velocity gradient at the stagnation point:

$$\left(\frac{dp_e}{dx}\right)_{t2} = -\rho_e u_e \left(\frac{du_e}{dx}\right)_{t2} \approx -\rho_{t2}\left(\frac{du_e}{dx}\right)_{t2}^2 x \tag{12.65}$$

Using the modified Newtonian flow pressure distribution [i.e., equation (12.55)] to evaluate the pressure gradient at the stagnation point,

$$\left(\frac{dp_e}{dx}\right)_{t2} = -2p_{t2}\cos\phi\sin\phi\frac{d\phi}{dx} + 2p_\infty\cos\phi\sin\phi\frac{d\phi}{dx} \tag{12.66}$$

Since $\sin\phi \approx \phi = x/R_N$, $\cos\phi \approx 1$, and $d\phi/dx = 1/R_N$, we find that

$$\left(\frac{du_e}{dx}\right)_{t2} = \frac{1}{R_N}\sqrt{\frac{2(p_{t2} - p_\infty)}{\rho_{t2}}} \tag{12.67}$$

Combining the expressions given by equations (12.63) through (12.67), we obtain the equilibrium correlation of Fay and Riddell (1958):

$$\dot{q}_{t,\text{ref}} = \frac{0.67}{\Pr_{t2}}(\rho_{t2}\mu_{t2})^{0.4}(\rho_{w,t}\mu_{w,t})^{0.1}(H_{t2} - h_{w,t})\left[\frac{1}{R_N}\left(\frac{2(p_{t2} - p_\infty)}{\rho_{t2}}\right)^{0.5}\right]^{0.5} \tag{12.68}$$

This relation can be used either for wind-tunnel flows or for flight conditions. Detra, et al. (1957) have developed a correlation:

$$\dot{q}_{t,\text{ref}} = \frac{17,600}{(R_N)^{0.5}}\left(\frac{\rho_\infty}{\rho_{\text{S.L.}}}\right)^{0.5}\left(\frac{U_\infty}{U_{\text{c.o.}}}\right)^{3.15} \tag{12.69}$$

In equation (12.69), R_N is the nose radius in feet, $\rho_{\text{S.L.}}$ is the sea-level density, $U_{\text{c.o.}}$ is the velocity for a circular orbit, and $\dot{q}_{t,\text{ref}}$ is the stagnation point heat-transfer rate in $\text{Btu/ft}^2 \cdot \text{s}$.

The heat-transfer rate at the stagnation point of a 1-ft (0.3048-m) sphere flying at 150,000 ft (45,721 m) has been calculated using equation (12.68) for both the perfect-air model and the equilibrium-air model and using equation (12.69). The computed values are presented in Fig. 12.32 as a function of M_∞. Above $M_\infty = 14$, the perfect-gas model no longer provides realistic values of $\dot{q}_{t,\text{ref}}$.

12.6 A HYPERSONIC CRUISER FOR THE TWENTY-FIRST CENTURY?

Harris (1992) considered the flight-block time for global-range fraction for several cruise Mach numbers. A vehicle cruising at Mach 10 would have global range in three hours flight-block time. Thus, he concluded that Mach numbers greater than about 10 to 15 provide an insignificant improvement in flight-block time both for civil and for military aircraft.

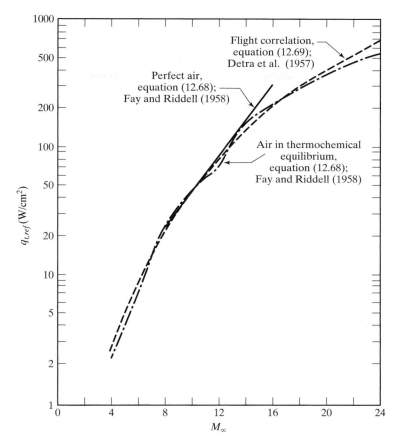

Figure 12.32 Heat-transfer rate at the stagnation point of a sphere ($R_N = 0.3048$ m) at an attitude of 45,721 m.

Hunt and Rausch (1998) noted that, for hypersonic airplanes carrying a given payload at a given cruise Mach number, range is a good figure of merit. This figure of merit is impacted by the fuel selection. Calculations indicate that Mach 8 is approximately the cruise speed limit for which a dual mode ramjet/scramjet engine can be cooled with state-of-the-art cooling techniques, when endothermic hydrocarbon fuels are used. On the other hand, liquid hydrogen has a much greater cooling capacity. Furthermore, hydrogen-fueled vehicles have considerably more range than hydrocarbon-fueled vehicles flying at the same Mach number. In addition, the range of hydrogen-fueled vehicles maximizes at about a Mach number of 10, which is beyond the maximum flight Mach number associated with the cooling limits for endothermic hydrocarbon fuels.

Thus, for hypersonic airplanes, the constraints on cooling the engine limit endothermic-hydrocarbon-fueled vehicles to Mach numbers of approximately 8 or less. Airplanes that cruise at Mach numbers of 8 or greater will be hydrogen fueled. However, hydrogen-fueled systems can be designed for vehicles that cruise at Mach numbers below 8. The shape of the vehicle and the systems that constitute it will be very different for endothermic-hydrocarbon-fueled vehicles than for the hydrogen-fueled vehicles. Because of the dramatic differences in the fuel density and in the planform required to generate the required loading, the fuel type will greatly impact the design of the vehicle.

Other than fuel, the greatest impact on the system architectures will come from issues relating to airframe/propulsion-system integration. The entire lower surface of the vehicle is considered to be a boundary for the flowpath for airbreathing-propulsion system. The forebody serves as an external precompression surface for the engine inlet and the aftbody serves as a high expansion-ratio nozzle. Since the propulsion system must take the vehicle from takeoff roll to cruise Mach number and back, the engine flowpath may contain a single duct, two ducts, or other concepts. The propulsion system may include auxiliary rocket components.

Configuration development studies for a global-reach, Mach 10 dual-fuel design have considered a variety of concepts. Bogar et al. (1996a) considered two classes of vehicles: a waverider and a lifting body. "When pressure forces alone are considered, the waverider enjoys a considerable advantage. However, as addition effects are added, this advantage erodes. Due to its larger wetted area, the waverider has higher viscous drag. And while the lifting body derives a benefit from trim effects, the waverider suffers a penalty because of the much larger forebody pitching moments which need to be balanced by the control surfaces." Bogar et al. (1996a) concluded that: "At Mach 10 cruise conditions, lifting bodies and waveriders provide comparable performance. Reshaping the vehicle to improve volumetric efficiency and provide moderate increases in fineness improves mission performance."

Bogar et al. (1996b) explored concepts for a Mach 10 vehicle capable of global-reach missions, which could reach a target 8500 nautical miles (nmi) away in less than 90 minutes after takeoff. A sketch of the concept, which is approximately 200 feet in length and has a reference area of approximately 10,000 square feet, is presented in Fig. 12.33. The vehicle has a takeoff gross weight of approximately 500,000 pounds. A dual-fuel concept (i.e., one that uses both hydrogen and endothermic hydrocarbons) was preferred for the return mission due to its capability for in-flight refueling on the return leg. The all-hydrogen vehicle was superior on the one-way mission due to its higher specific impulse (I_{sp}). The buildup of the components for the minimum drag for the baseline lifting body is presented in Fig. 12.34. The major components are pressure/wave drag, viscous drag, leading-edge drag, and base drag. At cruise speed, the pressure drag is responsible for 78% of the total drag, while the viscous drag accounts for 15% of the total drag. Note the relatively high transonic-drag rise. Thus, although the vehicle is designed to cruise at Mach 10, its shape is significantly affected by the transonic portion of the flight.

To help develop the technology base for the airbreathing-propulsion systems for hypersonic vehicles, NASA has initiated the Hyper-X Program. As discussed by McClinton et al. (1998), "The goal of the Hyper-X Program is to demonstrate and validate the technology, the experimental techniques, and computational methods and tools for design and performance prediction of hypersonic aircraft with airframe-integrated hydrogen-fueled, dual-mode combustion scramjet propulsion systems. Accomplishing this goal requires flight demonstration of and data from a hydrogen-fueled scramjet powered hypersonic aircraft."

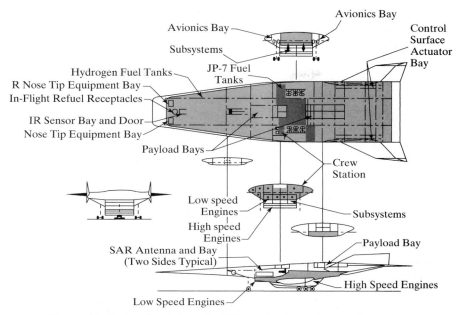

Figure 12.33 Mach 10 dual-fuel lifting-body cruise configuration. [Taken from Bogar et al. (1996b).]

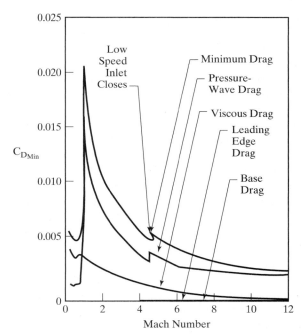

Figure 12.34 Minimum drag buildup for baseline vehicle. [Taken from Bogar et al. (1996b).]

McClinton et al. (1998) further state, "The Hyper-X Program concentrates on three main objectives required to significantly advance the Mach 5 to 10 scramjet technology leading to practical hypersonic flight:

1. vehicle design and flight test risk reduction—i.e., preflight analytical and experimental verification of the predicted aerodynamic, propulsive, structural, and integrated air-vehicle system performance and operability of the Hyper-X Research Vehicle (HXRV),

2. flight validation of design methods, and

3. methods enhancements—i.e., continued development of the advanced tools required to refine scramjet-powered vehicle designs.

"These objectives include experimental, analytical and numerical (CFD) activities applied to design the research vehicle and scramjet engine; wind tunnel verification of the vehicle aerodynamic, propulsion, and propulsion-airframe integration, performance and operability, vehicle aerodynamic database and thermal loads development; thermal-structural design; boundary layer transition analysis and control; flight control law development; and flight simulation model development. Included in the above is HXLV boost, stage separation and all other critical flight phases."

"Flight data will be utilized to verify the design methods, the wind tunnel methods and the overall utility of a scramjet powered vehicle."

12.7 IMPORTANCE OF INTERRELATING CFD, GROUND-TEST DATA, AND FLIGHT-TEST DATA

Woods et al. (1983) note that preflight predictions based on the aerodynamics in the Aerodynamics Design Data Book (ADDB) indicated that a 7.5° deflection of the body flap would be required to trim the Space Shuttle *Orbiter* for the center of gravity and for the vehicle configuration of STS-1. In reality, the body flap had to deflect to much larger values ($\delta_{BF} \approx 16°$) to maintain trim at the proper angle of attack ($\alpha = 40°$). Comparisons of equilibrium-air calculations and perfect-gas calculations indicate that at least part of this so-called "hypersonic anomaly" is due to real-gas effects at very high Mach numbers. At Mach 8, the flight data and the ADDB values agreed; as reported by Woods et al. (1983).

Consider the flow depicted in the sketch of Fig. 12.35. For perfect air ($\gamma = 1.4$), $\rho_2 = 6\rho_1$ across the normal portion of the shock wave, whereas $\rho_2 = 15\rho_1$ for air in thermodynamic equilibrium ($\gamma = 1.14$). Thus, for the equilibrium-air model the shock layer is thinner and the inclination of the bow shock wave relative to the free stream is less than that for the perfect-air model. Tangent-cone theory applied to the afterbody region shows a decrease in pressure with decreasing gamma. Using this simplistic flow model, one would expect the equilibrium-air pressures on the aft end of the vehicle to be less than those for perfect air. Computations of the flow field over simplified Orbiter geometries as reported by Woods et al. (1983) and by Maus et al. (1984) indicate that this is the case. The calculations of Maus et al. are reproduced in Fig. 12.36.

Maus et al. (1984) note further that the stagnation pressure increases with decreasing gamma. Thus, as presented in Fig. 12.7, the equilibrium-air value for the stagnation pressure

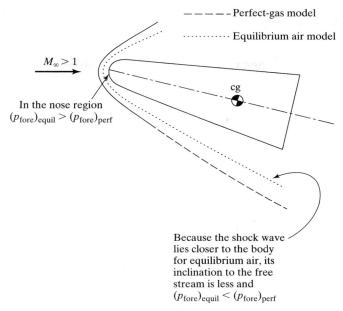

Figure 12.35 Hypersonic flow past an inclined spherically blunted cone comparing perfect-gas and equilibrium-air pressures.

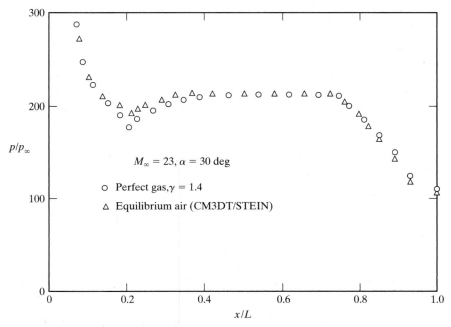

Figure 12.36 Comparison of perfect-gas and equilibrium-air calculations of the windward pitch-plane pressure distribution for the Space Shuttle *Orbiter* at $\alpha = 30°$. [From Maus et al. (1984).]

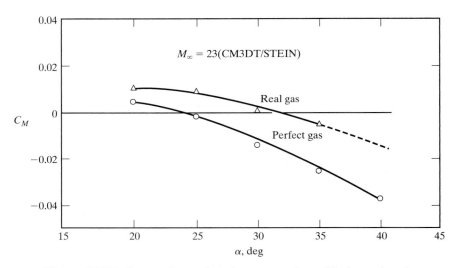

Figure 12.37 Comparison of perfect-gas and equilibrium-air calculations of the pitching moment for Space Shuttle *Orbiter* at $M_\infty = 23$. [From Maus et al. (1984).]

is greater than that for perfect air. This, too, is reflected in the nose region pressures presented in the more rigorous solutions of Woods et al. (1983) and Maus et al. (1984).

The differences between the equilibrium-air pressure distribution and the perfect-air pressure distribution may appear to be relatively small. Indeed, there is little difference in the normal force coefficients for the equilibrium-air model and for the perfect-air model.

However, because the equilibrium-air (real-gas) values are higher at the nose and lower at the tail, the real-gas effects tend to drive C_M more positive. The pitching moments for the Space Shuttle *Orbiter* at $M_\infty = 23$ presented by Maus et al. (1984) are reproduced in Fig. 12.37. Thus, detailed studies incorporating wind-tunnel data, flight-test data, and CFD solutions, as reported by Woods et al. (1983) and by Maus et al. (1984), provide insight into understanding a sophisticated aerodynamic problem.

Interest in hypersonic technology regularly waxes and wanes. Two books have been written to preserve this technology, one for the aerothermodynamic environment Bertin (1994)] and one for airbreathing propulsion systems [Heiser and Pratt (1994)] of vehicles which fly at hypersonic speeds.

12.8 BOUNDARY-LAYER-TRANSITION METHODOLOGY

Laminar-to-turbulent boundary-layer transition in high-speed boundary layers is critical to the prediction and control of heat transfer, skin friction, and other boundary-layer properties. Bertin (1994) noted that the mechanisms leading to transition are still poorly understood. The Defense Science Board (1992) found that boundary-layer transition

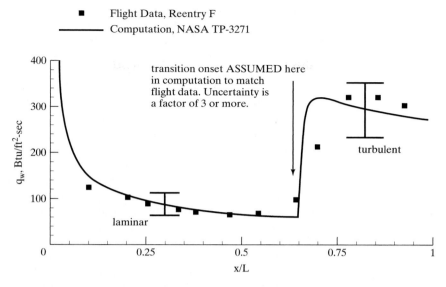

Figure 12.38 Heating-rate distribution along cone for reentry F.

was one of the two technical areas which needed further development before a demonstrator version of the National Aerospace Plane (NASP) could be justified.

Schneider (2004) notes that there are: "two major ways in which transition appears to be relevant for capsule flows. The first is transition on the blunt face, which can have a significant effect on heating, depending on ballistic coefficient, angle of attack, geometry, roughness, and so on. The second is the effect of transition on the shear layer that separates from the rim of the blunt face. This shear layer may be important, if it may reattach to the afterbody, or otherwise affect the aerodynamic stability or the aerothermodynamic heating. Sinha et al. (2004) shows that transition in the wake can have a significant effect on base heating. Transition may or may not occur in the shear layer in a significant way, again depending on the configuration and trajectory."

Schneider (2004) continued: "For the designer, one issue is whether transition has a significant effect on the thermal-protection system mass via increased heating. The other issue is the effect on lift-to-drag ratio through changes in the afterbody flow that affect moments and the trim angle of attack."

The high-heating rates of a turbulent boundary layer are illustrated in Figure 12.38, which presents computations and measurements of the surface heat transfer during the Reentry F flight test. Reda (1979) noted that the ballistic RV of the Reentry F was a 3.96-m (13-ft) beryllium cone that reentered at a peak Mach number of about 20 and a total enthalpy of about 18 MJ/kg (7753 Btu/lbm). Schneider (1999) summarized that the cone half-angle was 5°; the angle-of-attack was near zero; and the graphite nosetip had an initial radius of 0.25 cm (0.1 inch). The individual measurements, which are represented by the symbols, are compared with the heat-transfer rates that were computed using a variable-entropy boundary-layer code that included equilibrium chemistry. To provide the best agreement between the experimental heat-transfer rates and the numerical values, the computed boundary layer was assumed to transition instantaneously at an x/L 0.625, where x is the axial distance from the apex of the cone and L is the length of the cone. With the transition location for the computed boundary layer positioned to match the location based on the experimentally determined heat-transfer distribution, agreement was good both for the laminar and for the turbulent regions. Hamilton, who performed the computations, said that typical accuracies are 20 to 25% for the turbulent boundary layer and 15 to 20% for the laminar boundary layer. Schneider (2004) noted: "Present empirical correlations for the onset and the extent of transition are uncertain by a factor of three or more. Thus, our computational capabilities for laminar and turbulent heating in attached flows are fairly good, the uncertainty in prediction of the overall heating is now often dominated by the uncertainty in predicting the location of transition."

Schneider (2004) reported that: "A 1988 review by the Defense Science Board found that... estimates [of the point of transition] range from 20 to 80% along the body... The assumption made for the point of transition can affect the vehicle gross takeoff weight by a factor of two or more." This point is illustrated in Figure 12.39, which is taken from Whitehead (1989). A fully laminar boundary layer reduces the drag by approximately 6% for the baseline design of a single-stage-to-orbit National Aerospace Plane (NASP), while a fully turbulent boundary layer increases the drag by about 8%. For the baseline configuration, with substantial laminar flow on the forebody, the reduced drag and lower heating can result in a 60 to 70% increase in payload over the all-turbulent boundary-layer condition, as noted by Whitehead (1989).

Consider transition of the boundary layer of a sharp cone at zero angle-of-attack (AOA) in a supersonic/hypersonic stream. Despite the relative simplicity of the flow field, Schneider (2004) notes that the experimentally determined transition locations "in air at perfect-gas conditions are affected by cone half-angle, Mach number, tunnel size and noise, stagnation temperature, surface temperature distribution, surface roughness, and any blowing or ablation, as well as measurement technique. Sharp-cone, smooth-wall instability growth does not scale with Mach number, Reynolds number, and T_w/T_t, even under perfect-gas conditions [Kimmel and Poggie (2000)]. The mean boundary-layer profiles and their instability and transition depend on the absolute temperature; this is because the viscosity and heat-transfer coefficients depend on absolute temperature and do not scale. AOA effects are difficult to rule out, except by systematic azimuthal comparisons that are all too rare."

Bertin and Cummings (2003) note that: "The difficulty in developing criteria for predicting boundary-layer transition is complicated by the fact that the location of the onset of boundary-layer transition is very sensitive to the measurement technique used. The experimentally-determined heat-transfer rates increase above the laminar values at the upstream end of the boundary-layer transition process, i.e., at the onset of transitional

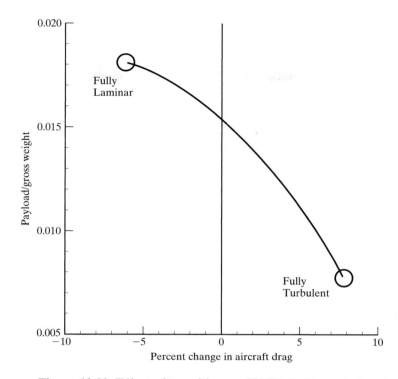

Figure 12.39 Effect of transition on NASP design.

flow. A schlieren photograph of the hypersonic flow field reveals vortices in the boundary layer associated with the various steps in the breakdown, i.e., in the transition of the boundary layer. However, for a flow as simple as a hypersonic flow over a slender, sharp cone, the boundary-layer transition location determined using the heat-transfer distribution along a conical generator is very different than that determined using a schlieren photograph."

Even if every researcher were to define and to measure transition in the same way, the scatter of transition data as a function of the Mach number still would be considerable. However, researchers employ a wide range of techniques in a wide variety of test simulations, which results in considerable scatter in transition correlations. This is illustrated in Figure 12.40, which was originally presented by Beckwith and Bertram (1972). This presentation of data makes it difficult to imagine a single relationship that could correlate all of the data. Making matters worse is the fact that transition rarely occurs along a line. Turbulent zones spread and merge in the longitudinal direction, which further complicates the development of correlations for boundary-layer transition.

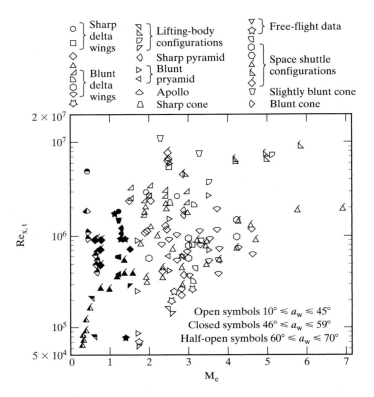

Figure 12.40 Transition Reynolds number as a function of local Mach number.

PROBLEMS

12.1. Consider a hypersonic vehicle flying through the earth's atmosphere at 10,000 ft/s at an altitude of 200,000 ft. At 200,000 ft, $p_\infty = 0.4715\,\text{lbf/ft}^2$, $\rho_\infty = 6.119 \times 10^{-7}\,\text{slug/ft}^3$, $T_\infty = 449°\text{R}$, and $h_\infty = 2.703 \times 10^6\,\text{ft}\cdot\text{lbf/slug}$. Using the thermodynamic properties of equilibrium air presented in Fig. 12.5, what are the static pressure (p_2), the static temperature (T_2), the enthalpy (h_2), and the static density (ρ_2) downstream of a normal shock wave? For comparison, use the normal-shock relations for perfect air, as given in Chapter 8, to calculate p_2, T_2, h_2, and ρ_2.

12.2. Consider a hypersonic vehicle reentering the earth's atmosphere at 20,000 ft/s at an altitude of 200,000 ft. Repeat Problem 12.1, that is, calculate p_2, ρ_2, T_2, and h_2 assuming that the air **(a)** is in thermodynamic equilibrium and **(b)** behaves as a perfect gas.

12.3. Consider a hypersonic vehicle reentering the earth's atmosphere at 26,400 ft/s at an altitude of 200,000 ft. Repeat Problem 12.1, that is, calculate p_2, ρ_2, T_2, and h_2 assuming that the air **(a)** is in thermodynamic equilibrium and **(b)** behaves as a perfect gas.

12.4. If the free-stream Mach number is 6, calculate the pressure coefficient on the surface of a wedge, whose deflection angle is 30°, using the correlations of Fig. 8.13b and using Newtonian theory. How does the altitude influence the resultant values of C_p?

12.5. If the free-stream Mach number is 6, calculate the pressure coefficient on the surface of a sharp cone, whose deflection angle is 30°, using the correlations of Fig. 8.16b and using Newtonian theory.

12.6. In equation (12.22), it was stated without proof that the section drag coefficient (per unit span) for modified Newtonian pressures acting on a right circular cylinder is: $C_{d,p} = \frac{2}{3} C_{p,t2}$. Prove that this equation is correct.

12.7. Consider the sharp cone in a hypersonic stream, which was the subject of Example 12.2. If $M_\infty = 10$ and $\theta_c = 10°$, prepare a graph of C_D as a function of α for $-10° \leq \alpha \leq 10°$.

12.8. Consider the sharp cone in a hypersonic stream, which was the subject of Example 12.2. If $M_\infty = 10$ and $\theta_c = 10°$, prepare a graph of C_L as a function of α for $-10° \leq \alpha \leq 10°$.

12.9. Consider the sharp cone in a hypersonic stream, which was the subject of Example 12.2. If $M_\infty = 10$ and $\theta_c = 10°$, prepare a graph of L/D as a function of α for $-10° \leq \alpha \leq 10°$.

12.10. Consider the sharp cone in a hypersonic stream, which was the subject of Example 12.2. If $M_\infty = 10$ and $\theta_c = 10°$, prepare a graph of C_M as a function of α for $-10° \leq \alpha \leq 10°$. Is the configuration statically stable if $x_{cg} = 0.6 x_L$?

12.11. A configuration that would generate lift at low angles of attack is a half-cone model, such as shown in Fig. P12.11. Neglecting the effects of skin friction and assuming the pressure is that of the Newtonian flow model, develop an expression for the lift-to-drag ratio as a function of θ_c for $0 \leq \alpha \leq \theta_c$.

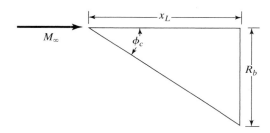

Figure P12.11

12.12. Develop expressions for y_{cp} and x_{cp} for the flow described in Problem 12.11.

12.13. Using the Newtonian flow model to describe the pressure field and neglecting the effects of viscosity, what is the drag coefficient on the $\theta_c = 10°$ sharp cone of Example 12.3? What would the drag coefficient be if you used the modified Newtonian flow model?

12.14. Following the procedure developed in Example 12.3, develop expressions for the forebody pressure coefficient ($C_{p,e}$), the base pressure coefficient ($C_{p,b}$), and the skin friction, if $\theta_c = 20°$ for the sharp cone exposed to the Mach 8 flow of Tunnel B.

12.15. Let us model the flow on the windward surface of a reentry vehicle and its body flap by the double wedge configuration shown in Fig. P12.15. The lower surface of the fuselage is inclined $10°(\alpha = 10°)$ to the free-stream flow, producing the supersonic flow in region 2. A body flap is deflected $10°$ (so that $\alpha + \delta_{BF} = 20°$).
  (a) Calculate the pressure coefficient for regions 2 and 3 using the Newtonian flow model for $M_1 = 2, 4, 6, 8, 10$, and 20.
  (b) Using the correlations of Fig. 8.13, calculate the pressure coefficients for regions 2 and 3 for $M_1 = 2, 4, 6, 8, 10$, and 20. The flow remains supersonic throughout the flow field

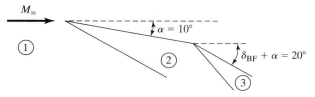

Figure P12.15

for these deflection angles. Note that to go from region 2 to 3 using the charts of Fig. 8.13, treat region 2 as an equivalent free-stream flow ("1"); then region 3 corresponds to region 2. Note, however, that $C_{p3} = (p_3 - p_1)/q_1$, that is, the reference conditions are those of the free stream.

REFERENCES

Anderson JD, Lewis MJ, Corda S. 1990. *Several families of viscous optimized waveriders – a review of waverider research at the University of Maryland.* Presented at International Hypersonic Waverider Symposium, 1st, College Park, MD

Bauer SXS, Covell PF, Forrest DK, McGrath BE. 1990. *Preliminary assessment of a Mach 4 and a Mach 6 waverider.* Presented at International Hypersonic Waverider Symposium, 1st, College Park, MD

Beckwith IE, Bertram MH. 1972. A survey of NASA Langley studies on high-speed transition and the Quiet Tunnel. *NASA Tech. Mem. X-2566*

Bertin JJ. 1966. The effect of protuberances, cavities, and angle of attack on the wind-tunnel pressure and heat-transfer distributions for the Apollo Command Module. *NASA TMX-11243*

Bertin JJ. 1994. *Hypersonic Aerothermodynamics.* Washington, DC: AIAA

Bertin JJ, Cline DD. 1980. Variable-grid-size transformation for solving nonsimilar laminar and turbulent boundary-layers. Ed. M. Gersten and P. R. Choudhury. *Proceedings of the 1980 Heat Transfer and Fluid Mechanics Institute,* Stanford, CA: Stanford University Press

Bertin JJ, Cummings RM. 2003. Fifty years of hypersonic, where we've been, where we're going. *Progr. Aerosp. Sci.* 39:511–536

Bogar TJ, Eiswirth EA, Couch LM, Hunt JL, McClinton CR. 1996a. *Conceptual design of a Mach 10 global reach reconnaisance aircraft.* Presented at Joint Propulsion Conference, 32nd, AIAA Pap. 96–2894, Lake Buena Vista, FL

Bogar TJ, Alberico JF, Johnson DB, Espinosa AM, Lockwood MK. 1996b. *Dual-fuel lifting body configuration development.* Presented at Intern. Space Planes and Hypersonic Systems and Techn. Conf., AIAA Pap. 96–4592, Norfolk, VA

Bowcutt KG, Anderson JD, Capriotti D. 1987. *Viscous optimized hypersonic waveriders.* Presented at Aerosp. Sci. Meet., 25th, AIAA Pap. 87–0272, Reno, NV

Brandon HJ, DeJarnette FR. 1977. *Three-dimensional turbulent heating on an ogive at angle of attack including effects of entropy-layer swallowing.* Presented at Thermophysics Conf., 12th, AIAA Pap. 77–754, Albuquerque, NM

Cassanto JM. 1973. A base pressure experiment for determining the atmospheric profile of the planets. *J. Spacecr. Rockets* 10:253–261

Cervisi RT, Grantz AC. 1994. *Efficient hypersonic accelerators derived from analytically defined flowfields.* Presented at Aerosp. Sci. Meet., 32nd, AIAA Pap. 94–0726, Reno, NV

Defense Science Board. 1988. *Report of the Defense Science Board Task Force on the National Aerospace Plane (NASP).* Washington, DC: Dept. of Defense

Defense Science Board. 1992. *Report of the Defense Science Board Task Force on the National Aerospace Plane.* Washington, DC: Dept. of Defense

Detra RW, Kemp NH, Riddell FR. 1957. Addendum to heat transfer to satellite vehicles reentering the atmosphere. *Jet Propulsion* 27:1256–1257

Dorrance WH. 1962. *Viscous Hypersonic Flow*. New York: McGraw-Hill Book Co.

Eckert ERG. 1955. Engineering relations for friction and heat transfer to surfaces in high velocity flow. *J. Aeronaut. Sci.* 22:585–587.

Eggers T, Radespiel R. 1993. *Design of waveriders*. Presented at Space Course on Low Earth Orbit Transportation, 2nd, Munich, Germany

Fay JA, Riddell FR. 1958. Theory of stagnation point heat transfer in dissociated air. *J. Aeronaut. Sci.* 25:73–85, 121

Haney JW, Beaulieu WD. 1994. *Waverider inlet integration issues*. Presented at Aerosp. Sci. Meet., 32nd, AIAA Pap. 94–0383, Reno, NV

Hansen CF. 1957. Approximations for the thermodynamic and transport properties of high-temperature air. *NACA Tech. Rep. R-50*

Harris RV. 1992. On the threshold – the outlook for supersonic and hypersonic aircraft. *J. Aircraft* 29:10–19

Heiser WH, Pratt D. 1994. *Hypersonic Airbreathing Propulsion*. Washington, DC: AIAA

Hunt JL, Rausch VL. 1998. *Airbreathing hypersonic Ssystems focus at NASA Langley Research Center*. Presented at International Spaceplanes and Hypersonic Systems and Tech. Conf., 8th, AIAA Pap. 98–1641, Norfolk, VA

Inouye M, Marvin JG, Sinclair AR. 1968. Comparison of experimental and theoretical shock shapes and pressure distributions on flat-faced cylinders at Mach 10.5. *NASA Tech. Note D-4397*

Isaacson LK, Jones JW. 1968. Prediction techniques for pressure and heat-transfer distributions over bodies of revolution in high subsonic to low-supersonic flight. *Naval Weapons Center TP 4570*

Kimmel RL, Poggie J. 2000. Effect of total temperature on boundary-layer stability at Mach 6. *AIAA J.* 38:1754–1755

Koppenwallner G. 1969. Experimentelle untersuchung der druckverteilung und des Wwiderstands von querangestroemten kreiszylindern bei hypersonischen Machzahlen in bereich von Kkontinuums-bis freier Mmolekularstroemung. *Z. Flugwissenschaften* 17:321–332.

Lees L. 1955. Hypersonic flow. *Proceedings of the 5th International Aeronautical Conference*, Los Angeles, CA, pp. 241–275

Marconi F, Salas M, Yeager L. 1976. Development of a computer code for calculating the steady supersonic/hypersonic inviscid flow around real configurations. *NASA Contr. Rep. 675*

Maus JR, Griiffith BJ, Szema KY, Best JT. 1984. Hypersonic Mach number and real gas effects on Space Shuttle Orbiter aerodynamics. *J. Spacecr. Rockets* 21:136–131

McClinton CR, Voland RT, Holland SC., Engelund WC, White JT, Pahle JW. 1998. *Wind tunnel testing, flight scaling, and flight validation with Hyper-X*. Presented at Advanced Measurement and Ground Test. Conf., 20th, AIAA Pap. 98–2866, Albuquerque, NM

Moeckel WE, Weston KC. 1958. Composition and thermodynamic properties of air in chemical equilibrium. *NACA Tech. Note 4265*

Nonweiler TRF. 1959. Aerodynamic problems of manned space vehicles. *J. Roy. Aeronaut. Soc.* 63:521–528

Reda WC. 1979. Boundary-layer-transition experiments on sharp, slender cones in supersonic free flight. *AIAA J.* 17:803–810

Riley CJ, DeJarnette FR, Zoby EV. 1990. Surface pressure and streamline effects on laminar calculations. *J. Spacecr. Rockets* 27:9–14

Schneider SP. 1999. Flight data for boundary-layer transition at hypersonic and supersonic speeds. *J. Spacecr. Rockets* 36:8–20

Schneider SP. 2004. Hypersonic laminar-turbulent transition on circular cones and scramjet forebodies. *Progr. Aerosp. Sci.* 40:1–50

Sinha K, Barnhardt M, Candler GV. 2004. Detached eddy simulation of hypersonic base flows with application to Fire *II experiments*. Presented at Fluid Dynamics Conf., 34th, AIAA Pap. 2004–2633, Portland, OR

Squire LC. 1976a. Flow regimes over delta wings at supersonic and hypersonic speeds. *Aeronaut. Quart.* 27:1–14

Squire LC. 1976b. The independence of upper and lower wing flows at supersonic speeds. *Aeronaut. J.* 80:452–456

Stoney WE. 1958. Aerodynamic heating of blunt-nose shapes at Mach numbers up to 14. *NACA 58E05a*

U. S. Standard Atmosphere 1976. Washington, DC: U.S. Government Printing Office.

Whitehead A. 1989. *NASP aerodynamics*. Presented at National Aerosp. Plane Conf., AIAA Pap. 89–5013, Dayton, OH

Woods WC, Arrington JP, Hamilton HH. 1983. A review of preflight estimates of real-gas effects on Space Shuttle aerodynamic characteristics. In *Shuttle Performance: Lessons Learned*, NASA 2283, Part I

Zoby EV, Moss JN, Sutton K. 1981. Approximate convective-heating equations for hypersonic flows. *J. Spacecr. Rockets* 18:64–70.

13 AERODYNAMIC DESIGN CONSIDERATIONS

In the previous chapters we have discussed techniques for obtaining flow-field solutions when the free-stream Mach number is either low subsonic, high subsonic, transonic, supersonic, or hypersonic. Many airplanes must perform satisfactorily over a wide speed range. Therefore, the thin, low-aspect ratio wings designed to minimize drag during supersonic cruise must deliver sufficient lift at low speeds to avoid unacceptably high landing speeds and/or landing field length. When these moderate aspect ratio, thin, swept wings operate at high angles of attack during high subsonic Mach number maneuvers, their performance is significantly degraded because of shock-induced boundary-layer separation and, at high angles of attack, because of leading-edge separation and wing stall. Furthermore, because of possible fuel shortages and sharp fuel price increases, the wings of a high-speed transport may be optimized for minimum fuel consumption instead of for maximum productivity. In this chapter we consider design parameters that improve the aircraft's performance over a wide range of speed.

13.1 HIGH-LIFT CONFIGURATIONS

Consider the case where the aerodynamic lifting forces acting on an airplane are equal to its weight:

$$W = L = \tfrac{1}{2}\rho_\infty U_\infty^2 S C_L \tag{13.1}$$

To support the weight of the airplane at relatively low speeds, we could either increase the surface area over which the lift forces act or increase the lift coefficient of the lifting surface.

13.1.1 Increasing the Area

During the early years of aviation, the relatively crude state of the art in structural analysis limited the surface area one could obtain with a single wing. Thus, as discussed in Cowley and Levy (1918). "In the attempt to increase the wing area in order to obtain the greatest lift out of an airfoil it was found that there was a point beyond which it was not advantageous to proceed. This stage was reached when the extra weight of construction involved in an increase in wing area was just sufficient to counterbalance the increase in lift. The method of using aerofoils in biplanes is desirable in the first place from the fact that, with a smaller loss in the necessary weight of construction, extra wing area may thus be obtained." Thus, whereas some of the combatants used monoplanes at the start of World War I [e.g., the Morane-Saulnier type N (France) and the Fokker series of E-type fighters (E for Eindecker; Germany)], most of the planes in service at the end of the war were biplanes [e.g., the SE5a (United Kingdom), the Fokker D-VII (Germany) and the SPAD XIII (France)] to carry the increased weight of the engine and of the payload.

Although the serious design of biplanes continued until the late 1930s, with the Fiat C.R. 42 (Italy) making its maiden flight in 1939, the improved performance of monoplane designs brought them to the front. Various methods of changing the wing geometry in flight were proposed in the 1920s and 1930s. Based on a concept proposed by test pilot V. V. Shevchenko, Soviet designer V. V. Nikitin developed a fighter that could translate from a biplane to a monoplane, or vice versa, at the will of the pilot [Air International (1975)]. In the design of Nikitin (known as the *IS*-2), the inboard sections of the lower wing were hinged at their roots, folding upward into recesses in the fuselage sides. The sections outboard of the main undercarriage attachment points also were articulated and, rising vertically and inward, occupied recesses in the upper wing. Thus, the one airplane combined the desirable short-field and low-speed characteristics of a lightly loaded biplane with the higher performance offered by a highly loaded monoplane.

The variable-area concepts included the telescoping wing, an example of which is illustrated in Fig. 13.1. The example is the MAK-10, built in France to the design of an expatriate Russian, Ivan Makhonine. The wing outer panel telescoped into the inner panel to reduce span and wing area for high-speed flight and could be extended for economic cruise and landing.

Trailing-edge flaps, such as the Fowler flap, which extend beyond the normal wing-surface area when deployed, are modern examples of design features that increase the wing area for landing. (Aerodynamic data for these flaps are discussed in the section on multielement airfoils in Chapter 6 and later in this chapter.) The increase in the effective wing area offered by typical multielement, high-lift configurations is illustrated in Fig. 13.2. The area increases available from using a plain (or aileron)

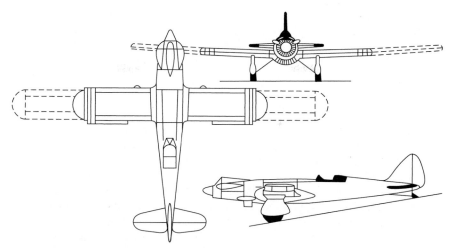

Figure 13.1 Makhonine MAK-10 variable-geometry (telescoping wing) aircraft. [From Air International (1975).]

type flap, a circular motion flap similar to that used on the Boeing 707, and the extended Fowler flap used on the Boeing 737 are presented in Fig. 13.3. The large increase in area for the 737-type flap is the sum of (1) the aft motion of the entire flap, (2) the aft motion of the main flap from the fore flap, (3) the motion of the auxiliary (aft) flap, and (4) the movement of the leading-edge devices.

13.1.2   Increasing the Lift Coefficient

The progress in developing equivalent straight-wing, nonpropulsive high-lift systems is illustrated in Fig. 13.4. Note the relatively high values obtained by experimental aircraft such as the L-19 of Mississippi State University and the MA4 of Cambridge University. Both of these aircraft use distributed suction on the wing so that the flow field approximates that for inviscid flow.

A companion figure from the work of Cleveland (1970) has been included for the interested reader. The parasite drag coefficient, which includes interference drag but does not include induced or compressibility drag, is presented for several airplanes in Fig. 13.5.

The Chance-Vought F8U Crusader offers an interesting design approach for obtaining a sufficiently high lift coefficient for low-speed flight while maintaining good visibility for the pilot during landing on the restricted space of an aircraft carrier deck. As shown in Fig. 13.6, the entire wing could be pivoted about its rear spar to increase its incidence by 7° during takeoff and landing. Thus, while the wing is at a relatively high incidence, the fuselage is nearly horizontal and the pilot has excellent visibility. Furthermore, when the wing is raised, the protruding center section also serves as a large speed brake.

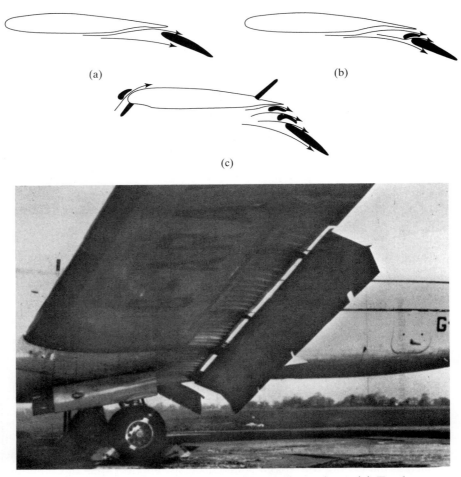

Figure 13.2 Multielement, high-lift configurations: (a) Fowler flap; (b) double-slotted flap; (c) leading-edge slat, Krueger leading-edge flap, spoiler, and triple-slotted flaps (representative of Boeing 727 wing section). (d) Fowler flaps on the HS 748. (Courtesy of British Aerospace.)

13.1.3 Flap Systems

Olason and Norton (1966) note that "if a clean flaps-up wing did not stall, a flap system would not be needed, except perhaps to reduce nose-up attitude (more correctly, angle of attack) in low-speed flight." Thus, a basic goal of the flap system design is to attain the highest possible L/D ratio at the highest possible lift coefficient, as illustrated in Fig. 13.7. A flap system (1) increases the effective wing area, (2) increases the camber of the airfoil section (thereby increasing the lift produced at a given angle of attack), (3) can provide leading-edge camber to help prevent leading-edge stall, and (4) can include slots which affect the boundary layer and its separation characteristics.

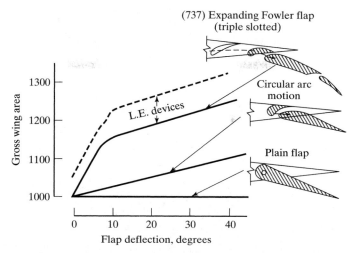

Figure 13.3 Use of flaps to increase the wing area. [From Olason and Norton (1966).]

Symbol: Plane

A:	Wright Flyer	*H*:	749
B:	Spirit of St. Louis	*I*:	1049
C:	C-47 (DC-3)	*J*:	C-130
D:	NACA 23012 Airfoil	*K*:	MA4
E:	B-32	*L*:	L-19
F:	C-54 (DC-6)	*M*:	Boeing 727
G:	C-124	*N*:	C-5A

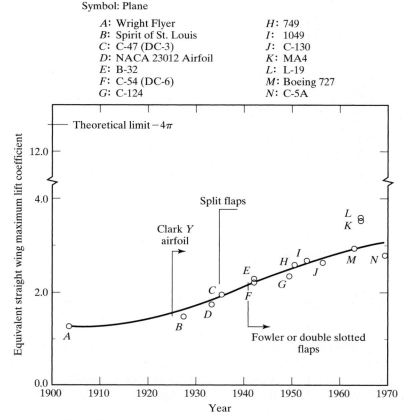

Figure 13.4 History of nonaugmented maximum lift coefficient. [From Cleveland (1970).]

Symbol: Plane

A:	Wright Flyer	*H:*	Bf (Me) 109	*O:*	C-130
B:	WWI Bomber	*I:*	B-29	*P:*	P-51
C:	WWI Fighter	*J:*	B-17	*Q:*	Comet
D:	Spirit of St. Louis	*K:*	Bf (Me) 108	*R:*	Jetstar
E:	Lockheed Vega	*L:*	Me 262	*S:*	C-141A
F:	Curtiss Navy Fighter	*M:*	XB-19	*T:*	Boeing 747
G:	Piper Cub	*N:*	F-80	*U:*	C-5A

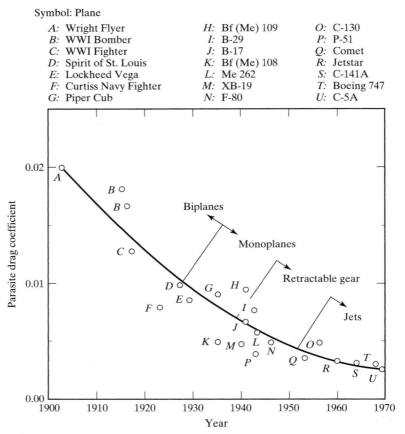

Figure 13.5 History of parasite drag coefficient, where the drag coefficient is based on total surface area. [From Cleveland (1970).]

Figure 13.6 Chance-Vought F8U showing incidence for the wing during landing (or takeoff). (Courtesy of Vought Corp.)

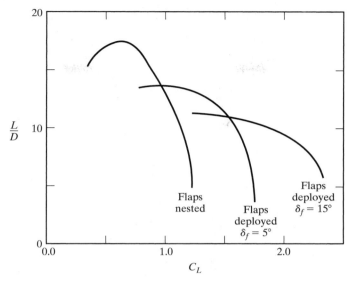

Figure 13.7 Effect of flap deployment on the aerodynamic forces.

Significant increases in the lift coefficient (and in the drag coefficient) can be obtained by increasing the camber of the airfoil section. The effect of deploying a split flap, which is essentially a plate deflected from the lower surface of the airfoil, is illustrated in the pressure distributions of Fig. 13.8. Deployment of the split flap not only causes an increase in the pressure acting on the lower surface upstream of the flap but also causes a pressure reduction on the upper surface of the airfoil. Thus, the deployment of the split flap produces a marked increase in the circulation around the section and, therefore, increases the lift. The relatively low pressure in the separated region in the wake downstream of the deployed plate causes the drag to be relatively high. The effect is so pronounced that it affects the pressure at the trailing edge of the upper surface (Fig. 13.8). The relatively high drag may not be a disadvantage if the application requires relatively steep landing approaches over obstacles or requires higher power from the engine during approach in order to minimize engine acceleration time in the event of wave-off.

The effect of the flap deflection angle on the lift coefficient is presented in Fig. 13.9. Data are presented both for a plain flap and for a split flap. Both flaps were $0.2c$ in length. The split flap produces a slightly greater increase in $C_{l,\max}$ than does the plain flap.

Simple hinge systems as on a plain flap, even though sealed, can have a significant adverse effect on the separation point and hence on the lift and the drag. The adverse effect of the break at the hinge line is indicated in data presented in Stevens et al. (1971). Pressure distributions are compared in this reference for a plain flap, which is $0.25c$ in length and is deflected 25°, and for a variable camber flap whose centerline is a circular arc having a final slope of 25°. Although separation occurs for both flaps, it occurs nearer the trailing edge for the variable camber shape, which "turns out to be better only because of its drastic reduction in the suction peak" [The quotes are from Stevens et al. (1971)].

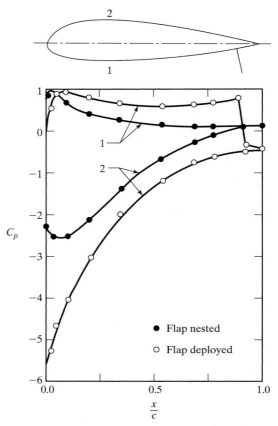

Figure 13.8 Pressure distribution for an airfoil with a split flap. [Data from Schlichting and Truckenbrodt (1969).]

13.1.4 Multielement Airfoils

As noted, the location of separation has a significant effect on the lift, drag, and moment acting on the airfoil section. It has long been recognized that gaps between the main section and the leading edge of the flap can cause a significant increase in $C_{l,\max}$ over that for a split flap or a plain flap. Furthermore, the drag for the slotted flap configurations is reduced. Sketches of airfoil sections with leading-edge slats or with slotted flaps are presented in Fig. 13.2.

Smith (1975) notes that the air through the slot cannot really be called high-energy air, since all the air outside the boundary layer has the same total pressure. Smith states that "There appear to be five primary effects of gaps, and here we speak of properly designed aerodynamic slots.

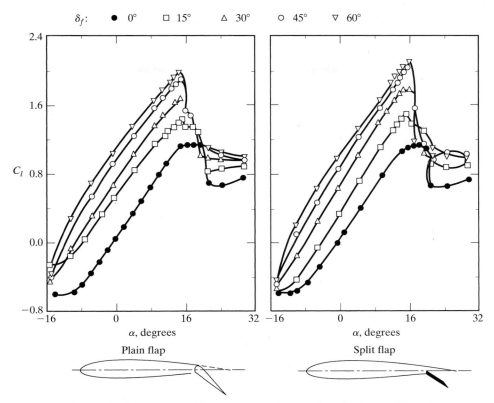

Figure 13.9 Effect of flap angle on the sectional lift coefficient for a NACA 23012 airfoil section, $Re_c = 6 \times 10^5$. [Data from Schlichting and Truckenbrodt (1969).]

1. *Slat effect.* In the vicinity of the leading edge of a downstream element, the velocities due to circulation on a forward element (e.g., a slat) run counter to the velocities on the downstream element and so reduce pressure peaks on the downstream element.

2. *Circulation effect.* In turn, the downstream element causes the trailing edge of the adjacent upstream element to be in a region of high velocity that is inclined to the mean line at the rear of the forward element. Such flow inclination induces considerably greater circulation on the forward element.

3. *Dumping effect.* Because the trailing edge of a forward element is in a region of velocity appreciably higher than free stream, the boundary layer 'dumps' at a high velocity. The higher discharge velocity relieves the pressure rise impressed on the boundary layer, thus alleviating separation problems or permitting increased lift.

4. *Off-the-surface pressure recovery.* The boundary layer from forward elements is dumped at velocities appreciably higher than free stream. The final deceleration

to free-stream velocity is done in an efficient manner. The deceleration of the wake occurs out of contact with a wall. Such a method is more effective than the best possible deceleration in contact with a wall.

5. *Fresh-boundary-layer effect.*   Each new element starts out with a fresh boundary layer at its leading edge. Thin boundary layers can withstand stronger adverse gradients than thick ones."

Since the viscous boundary layer is a dominant factor in determining the aerodynamic performance of a high-lift multielement airfoil, inviscid theory is not sufficient. Typical theoretical methods iteratively couple potential-flow solutions with boundary-layer solutions. The potential-flow methods used to determine the velocity at specified locations on the surface of the airfoil usually employ singularity-distribution methods. As discussed in Chapters 3 and 6, singularity-distribution methods, which have been widely used since the advent of the high-speed, high-capacity digital computers needed to solve the large systems of simultaneous equations, can handle arbitrarily shaped airfoils at "any" orientation relative to the free stream. (The word "any" is in quotes since there are limits to the validity of the numerical simulation of the actual flow.) For singularity-distribution methods, either source, sink, or vortex singularities are distributed on the surface of the airfoil and integral equations formulated to determine the resultant velocity induced at a point by the singularities. The airfoil surface is divided into N segments with the boundary condition that the inviscid flow is tangent to the surface at the control point of each and every segment. The integral equations can be approximated by a corresponding system of $N - 1$ simultaneous equations. By satisfying the Kutta condition at the trailing edge of the airfoil, the Nth equation can be formulated and then the singularity strengths determined with a matrix-inversion technique. As noted previously, the Kutta condition which is usually employed is that the velocities at the upper and lower surface trailing edge be tangent to the surface and equal in magnitude. The various investigators use diverse combinations of integral and finite-difference techniques to generate solutions to the laminar, the transitional, and the turbulent boundary layers. The interested reader is referred to the brief discussion in Chapter 6 or to one of the numerous analyses of the multielement airfoil problem [e.g., Stevens et al. (1971)], Morgan (1975), Olsen and Dvorak (1976), and Anon. (1993)].

The lift coefficients for a GA(W)-1 airfoil with a Fowler-type, single-slotted flap which are taken from Wentz and Seetharam (1974), are presented as a function of the angle of attack in Fig. 13.10. Note the large increases in $C_{l,\max}$ which are obtained with the slotted Fowler flap. As shown in the sketch of the airfoil section, the deflected flap segment is moved aft along a set of tracks that increases the chord and the effective wing area. Thus, the Fowler flap is characterized by large increases in $C_{l,\max}$ with minimum changes in drag. The ability of numerical techniques to predict the pressure distribution is illustrated in Fig. 13.11. Data are presented for the GA(W)-1 airfoil at an angle of attack of 5°, with a flap deflection of 30°.

In this section we have seen how the use of mechanical changes in camber can be used to increase the lift generated by airfoils and wings, independent of the change in angle of attack. Typically, one could expect maximum values of C_L of 2.5 to 3.5 using mechanical systems on wings of commercial and military transports. In addition to the obvious mechanical complexity (e.g., tracks, brackets, and actuators) and weight, mechanical systems are limited aerodynamically in the maximum lift they can generate.

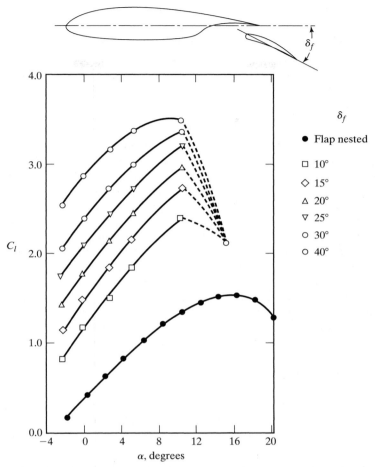

Figure 13.10 Experimental lift coefficient for a GA(W)-1 airfoil with a slotted Fowler flap. [From Wentz and Seetharam (1974).]

One limitation, the Kutta condition, essentially fixes airfoil circulation (and thus the lift) to the value where the free-system flow leaves the airfoil at the trailing edge. A second limitation, viscous effects, usually reduces the attainable lift to less than that value, because the flow separates from the highly curved flap upper surface before the trailing edge is reached. Slotted flaps, of course, reduce the effects of separation. Nevertheless, concepts employing jet-engine bleed gases are often used to generate increased lift.

13.1.5 Power-Augmented Lift

An additional factor to consider in the comparison of flap types is the aerodynamic moment created by deployment of the flap. Positive camber produces a nose-down pitching moment, which is especially great when applied well aft on the chord, and produces

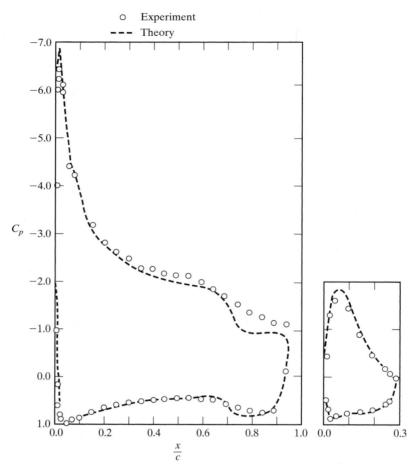

Figure 13.11 Comparison of the theoretical and the experimental pressure distribution for a GA(W)-1 airfoil with a slotted flap, $\alpha = 5°, \delta_f = 30°$. [From Wentz and Seetharam (1974).]

twisting loads on the structure. The pitching moments must be controlled with the horizontal tail. Unfortunately, the flap types which produce the greatest increase in $C_{l,\max}$ usually produce the largest moments. Thus, as shown in the sketches of the MiG 21s presented in Fig. 13.12, the Fowler flap with its extended guides and fairing plates is replaced by blown flaps for some applications. Separation from the surface of the flap is prevented by discharging fluid from the interior of the main airfoil section. The fluid injected tangentially imparts additional energy to the fluid particles in the boundary layer so that the boundary layer remains attached.

The internally blown flap is compared with two other techniques which use engine power to achieve very high lift in Fig. 13.13. The corresponding drag polars [as taken from Goodmanson and Gratzer (1973)] are included. The externally blown flap (EBF) spreads

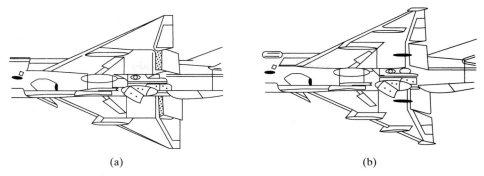

(a)   (b)

Figure 13.12 Fowler-type flaps used on early series MiG 21s and the blown flaps used on later series: (a) Fowler-type flaps employed on the MiG-21 PF; (b) blown flaps employed on the MiG-21 MF. [From Air Enthusiast International (1974).]

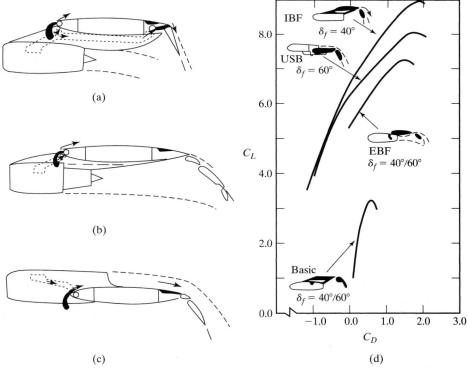

(a)

(b)

(c)   (d)

Figure 13.13 Power-augmented high-lift configurations: (a) internally blown flap (IBF); (b) externally blown flap (EBF); (c) upper-surface blowing (USB); (d) drag polars for four-engine configuration. $cj = 2.0$ for blown configuration. [From Goodmanson and Gratzer (1973).]

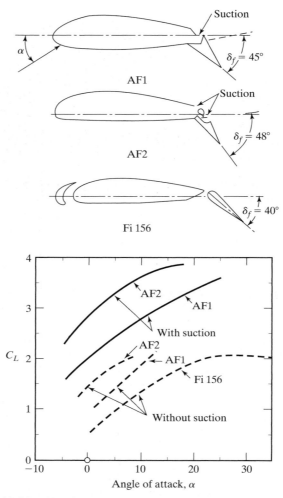

Figure 13.14 Lift of three STOL airplanes for full landing flap deflection (power off). AF1 and AF2 are boundary layer control airplanes of the Aerodynamische Versuchsanstalt, Fi 156 is the Fieseler Storch. [Data from Schlichting (1960).]

and turns the jet exhaust directed at the trailing-edge flap. A portion of the flow emerging through the flap slots maintains attachment of the boundary layer over the flap's upper surface. The upper-surface-blowing (USB) concept resembles the externally blown flaps. However, the data indicate "better performance than the externally blown flap if the air-turning process is executed properly. Also, the path of the engine exhaust permits a certain amount of acoustic shielding by the wing, and consequently a significant reduction in noise." The quote is that of the authors of Goodmanson and Gratzer (1973), who work for the Boeing Company, which designed the YC-14 AMST (Advanced Medium STOL Transport), an aircraft that employs USB.

Figure 13.15 Fieseler Storch, Fi 156, showing fixed leading-edge slots and hinged trailing edge. (Courtesy of Jay Miller, "Aerophile.")

Using suction to remove the decelerated fluid particles from the boundary layer before they separate from the surface in the presence of the adverse pressure gradient is another means of increasing the maximum lift. The "new" boundary layer which is formed downstream of the suction slot can overcome a relatively large adverse pressure gradient without separating. Flight data obtained in the late 1930s at the Aerodynamische Versuchsanstalt at Göttingen are reproduced in Fig. 13.14. The data demonstrate that the application of suction through a slit between the wing and the flap can prevent separation. Since the flaps can operate at relatively large deflection angles and the airfoil at relatively high angles of attack without separation, large increases in lift can be obtained. The maximum lift coefficients for the airplanes equipped with suction are almost twice that for the Fieseler Storch (Fi 156), a famous short takeoff and landing (STOL) airplane of the World War II period, which is shown in Fig. 13.15. The entire trailing edge of the Storch wing was hinged; the outer portions acting as statically balanced and slotted ailerons, and the inner portions as slotted camber-changing flaps. A fixed slot occupied the entire leading edge. Initial flight tests showed the speed range of the Fieseler Storch to be 51 to 174 km/h (32 to 108 mi/h) and that the landing run in a 13 km/h (8 mi/h) wind using brakes is 16 m. The interested reader is referred to Green (1970) for more details.

13.2 CIRCULATION CONTROL WING

As discussed by Englar (1987), the circulation control wing (CCW) concept avoids the problems of mechanical and blown flaps by replacing the sharp trailing edge with a fixed nondeflecting, round or near-round surface, such as shown in Fig. 13.16. The tangential blowing jet remains attached to the curved surface by creating a balance between the subambient pressure and the centrifugal force. Thus, as shown in Fig. 13.16, at low values for the blowing momentum coefficient, it serves as a boundary-layer control (BLC)

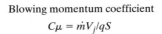

Blowing momentum coefficient

$$C\mu = \dot{m}V_j/qS$$

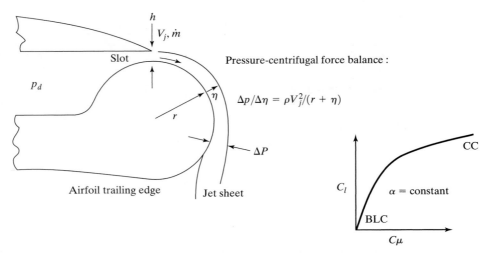

Figure 13.16 Basic principles of circulation control aerodynamics. [From Englar (1987).]

device to entrain the flow field and prevent it from separating. Once the flow is returned to the inviscid condition, the jet continues to turn around the trailing edge. It thus entrains and deflects the flow field, providing pneumatic deflection of the streamlines (equivalent to camber) and resulting in supercirculation lift. In the circulation control region (CC), very high lift can be generated since the jet turning is not limited by a sharp trailing edge. As shown in Fig. 13.17, relatively high lift coefficients can be generated for two-dimensional CCW airfoils with much less blowing than that typical of mechanical blown flaps.

13.3 DESIGN CONSIDERATIONS FOR TACTICAL MILITARY AIRCRAFT[*]

As noted by Bradley (1981), the designer of a tactical military aircraft is faced with a multitude of design points throughout the subsonic/supersonic flow regimes plus many off-design constraints. The design goals for a tactical weapons system may include efficient cruise at both subsonic and supersonic Mach numbers, superior maneuverability at both subsonic and supersonic Mach numbers, and rapid acceleration.

Bradley states, "The multiple design point requirement turns out to be the major driver for the designer of fighter aircraft. The aerodynamic requirements for each of

---

[*]Quotations and data presented in this section are reproduced by permission from General Dynamics Corporation.

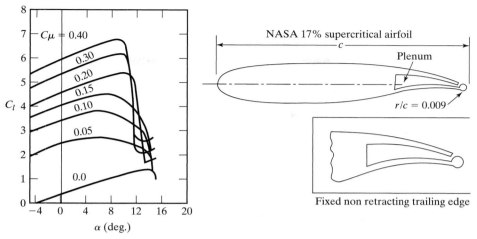

Figure 13.17 Sectional lift characteristics for a 17% thick CCW/-supercritical airfoil. [From Englar (1987).]

the design points often present conflicting requirements. For example, the need for rapid acceleration to supersonic flight and efficient supersonic cruise calls for thin wing sections with relatively high sweep and with camber that is designed to trim out the moments resulting from aft ac movement at supersonic flight. However, these requirements are contrary to those requirements for efficient transonic maneuver, where the designer would prefer to have thicker wing sections designed with camber for high C_L operation and a high-aspect-ratio planform to provide a good transonic drag polar. Designers are thus faced with a situation of compromise. These conflicting requirements suggest the obvious solution of variable geometry, that is, variable sweep wing and/or variable camber. Although this is a satisfactory aerodynamic solution, in many cases the resultant weight increases to a configuration can be prohibitive.

"A typical performance spectrum corresponding to mission requirements is illustrated in Fig. 13.18, which presents a map of lift coefficient versus Mach number. The low Mach end of the spectrum throughout the C_L range is typical of takeoff and landing for the configuration. The subsonic cruise and supersonic cruise portions are noted in the moderate lift range. Acceleration to high supersonic speeds occurs at the low lift coefficients. Sustained maneuver takes place in the C_L range of less than one for most fighter configurations. Above this lift coefficient, the aircraft is in the instantaneous maneuver regime. Drag rise occurs, depending on the wing geometry, in the range of 0.8 to 1.2 Mach.

"The particular flow conditions that correspond to the flow map are shown in Fig. 13.19. At the cruise and acceleration points, the aircraft designer is dealing primarily with attached flow, and his design objective is to maintain attached flow for maximum efficiency. At the higher C_L's, corresponding to instantaneous maneuver, separated flow becomes the dominant feature. Current designs take advantage of the separated flow by forming vortex flows in this range. Intermediate C_L's corresponding to sustained maneuver are usually a mixture of separated and attached flow. Consequently, if the aircraft is designed with camber to minimize separation in the maneuver regime,

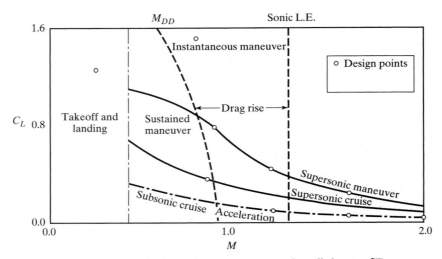

Figure 13.18 Typical performance map for fighters. [From Bradley (1981).]

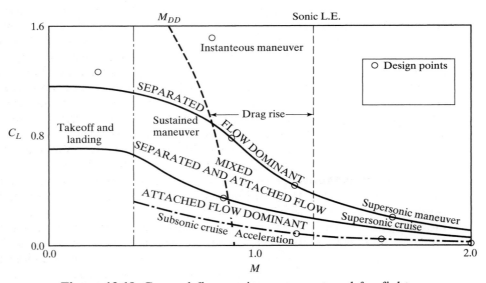

Figure 13.19 General flow regimes encountered for fighters. [From Bradley (1981).]

the configuration will have camber drag and may have lower surface separation, which increases drag at the low C_L's needed for acceleration.

"Thus, the aircraft design is a compromise to achieve an optimal flow efficiency considering the numerous design points associated with the mission objectives. It is easily seen that the transonic flow problems that must be addressed in fighter design are driven to a very large extent by the constraints imposed at the other design points—supersonic and subsonic."

Bradley (1981) notes that "supersonic design enjoys relatively precise computation and optimization thanks to wide applicability of linearized theory." Further, it is noted that "the design of efficient transonic configurations may proceed from two conceptual schools of logic. One acknowledges that the optimum low-drag flow must accelerate rapidly over the airfoil to supercritical flow and decelerate in a nearly isentropic manner, avoiding strong shocks and/or steep gradients that can lead to significant regions of separation. This approach sets attached flow or near fully attached flow as an intuitive design goal and is typically used for aircraft that permit strong emphasis to be placed on transonic cruise or sustained maneuver design points." The design of airfoils for transonic cruise applications was discussed extensively in Chapter 9.

"The second school of thought recognizes the inevitability of significant flow separations at design conditions and adopts a philosophy of controlling certain regions of separation through vortex flows to complement other regions of attached supercritical flow. This approach is appropriate for configurations constrained by multiple design points that emphasize added supersonic requirements. Current tactical fighters that rely on high wing loadings for transonic performance are good examples. The F-16 and F-18 employ a combination of controlled vortex flow and variable camber to achieve maneuverability."

Even though a designer has worked carefully to design a wing for attached flow, the flow fields associated with sustained and instantaneous maneuvers involve predominantly separated flows. The manner in which the flow separates and how the separation develops over the configuration will strongly affect the vehicle's drag and its controllability at the higher C_L's. Aircraft, such as the F-16 and the F-18, employ strakes or leading edge extension devices to provide a controlled separated flow. Controlled vortex flow can then be integrated with variable camber devices on the wing surface to provide satisfactory high-lift, stability and control, and buffet characteristics. The resulting flow field is a complex one, combining attached flows over portions of the wing with the vortex flow from the strake.

"Vortex control devices can improve the transonic drag polar, but highly integrated designs are required to avoid other aerodynamic problems and also to minimize the potential weight penalties. The F-16 forebody strake vortex system is clearly visible as it interacts with the wing flow field in Fig. 13.20. Forebody strakes and canards have similar aerodynamic effects—both good and bad. For example, the F-16 forebody strake design required extensive integration with the wing variable leading edge flap system and empennage to achieve significant aerodynamic improvements. The aerodynamic improvements achieved with the forebody strake/variable leading edge flap combination are illustrated in Fig. 13.21 for the YF-16 aircraft. The dashed curve depicts the lift curve and drag polar for the wing without the strake and variable leading edge flaps. The strake with no leading edge flap significantly improves the static aerodynamic characteristics. Further improvement is seen with the variable leading edge flap in combination with the strake configuration."

Bradley (1981) states: "We should not pass from the separated flow wing design discussion without mentioning a new concept of controlled vortex flow for designing military aircraft having emphasis on supersonic configurations. Wing planforms for supersonic cruise have higher leading edge sweep and generally lower aspect ratios; these planforms develop vortex flows at relatively low angle of attack. As a result, the transonic drag characteristics are lacking in the maneuver regime since drag polars

Figure 13.20 Strake vortex system on an F-16. (Photograph provided by General Dynamics Corporation, Fort Worth Division.)

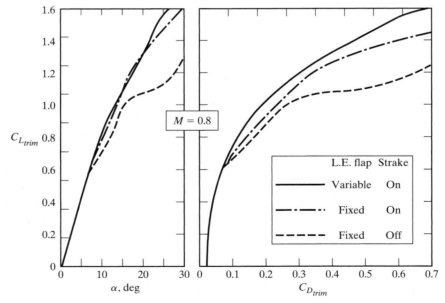

Figure 13.21 Aerodynamic benefits of combined forebody-strake/-variable-camber system. [From Bradley (1981).]

generally reflect very little leading edge suction recovery. Recently, wings of this type have been designed to take advantage of the separated vortex flows rather than to try to maintain attached flow to higher C_L values."

13.4 DRAG REDUCTION

Possible fuel shortages combined with sharp price increases and the requirements of high performance over a wide-speed range emphasize the need for reducing the drag on a vehicle and, therefore, improving the aerodynamic efficiency. Of the various drag reduction concepts, we will discuss

1. Variable-twist, variable-camber wings
2. Laminar-flow control (LFC)
3. Wingtip devices
4. Wing planform

13.4.1 Variable-Twist, Variable-Camber Wings

Survivability and mission effectiveness of a supersonic-cruise military aircraft require relatively high lift/drag ratios while retaining adequate maneuverability. The performance of a moderate-aspect-ratio, thin swept wing is significantly degraded at high lift coefficients at high subsonic Mach numbers because of shock-induced boundary-layer separation and, at higher angles of attack, because of leading-edge separation and wing stall. The resulting degradation in handling qualities significantly reduces the combat effectiveness of such airplanes. There are several techniques to counter leading-edge stall, including leading-edge flaps, slats, and boundary-layer control by suction or by blowing. These techniques, along with trailing-edge flaps, have been used effectively to increase the maximum usable lift coefficient for low-speed landing and for higher subsonic speeds.

Low-thickness-ratio wings incorporating variable camber and twist appear to offer higher performance for fighters with a fixed-wing planform [Meyer and Fields (1978)], since the camber can be reduced or reflexed for the supersonic mission and increased to provide the high lift coefficients required for transonic and subsonic maneuverability.

A test program was conducted to determine the effect of variable-twist, variable-camber on the aerodynamic characteristics of a low-thickness-ratio wing [Ferris (1977)]. The basic wing was planar with a NACA 65A005 airfoil at the root and a NACA 65A004 airfoil at the tip (i.e., there was no camber and no twist). Section camber was varied using four leading-edge segments and four trailing-edge segments, all with spanwise hinge lines. Variable twist was achieved since the leading-edge (or trailing-edge) segments were parallel and were swept more than the leading edge (or trailing edge). Camber and twist could be applied to the wing as shown in Fig. 13.22. Deploying the trailing-edge segments near the root creates a cambered section whose effective chord is at an increased incidence. Similarly, deploying the leading-edge segments near the wing tip creates a cambered section whose local incidence is decreased. Thus, the modified wing could have an effective twist of approximately 8° washout. As noted in Ferris (1977), use of leading-edge camber lowers the drag substantially for lift coefficients up

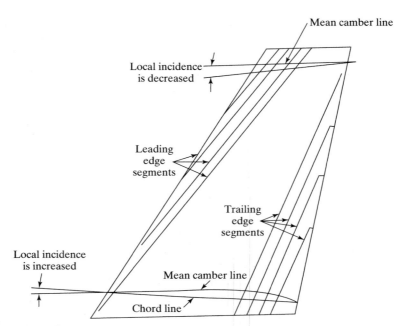

Figure 13.22 Use of leading-edge segments and trailing-edge segments to produce camber and twist on a basic planar wing.

to 0.4. Furthermore, use of leading-edge camber significantly increased the maximum lift/drag ratio over a Mach number range of 0.6 to 0.9. At the higher lift coefficients ($\geq$0.5), the combination of twist and camber achieved using both leading-edge segments and trailing-edge segments was effective in reducing the drag. Trailing-edge camber causes very large increments in C_L with substantial negative shifts in the pitching moment coefficients.

The effectiveness of leading-edge segments and trailing-edge segments in increasing the lift coefficient and in reducing the drag coefficient at these relatively high lift coefficients is illustrated in the data presented in Fig. 13.23. As a result, the maximum lift/drag ratio for this particular configuration at $M_\infty = 0.80$ is 18 and it occurs when $C_L = 0.4$.

13.4.2 Laminar-Flow Control

In previous chapters we have seen that the skin-friction component of the drag is markedly higher when the boundary layer is turbulent. Thus, in an effort to reduce skin friction, which is a major part of the airplane's "parasitic" drag, attempts have been made to maintain laminar flow over substantial portions of the aircraft's surface. Attempts at delaying transition by appropriately shaping the airfoil section geometry were discussed in Chapter 5. However, a natural boundary layer cannot withstand even very small disturbances at the higher Reynolds numbers, making transition difficult to avoid. Theoretical solutions reveal that removing the innermost part of the boundary

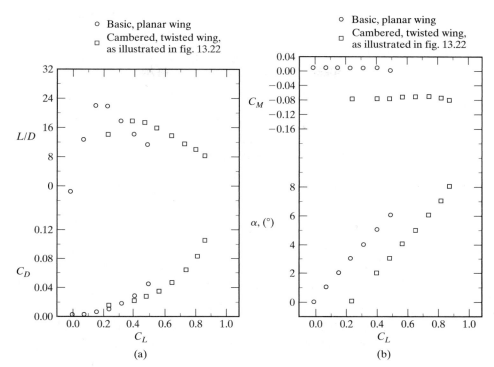

Figure 13.23 Effect of twist and camber on the longitudinal aerodynamic characteristics $M_\infty = 0.80$, $Re_c = 7.4 \times 10^6$: (a) lift-to-drag ratio and the drag polar, (b) pitching moment and lift coefficients. [From Ferris (1977).]

layer using even very small amounts of suction substantially increases the stability of a laminar boundary layer. Maintaining a laminar profile by suction is termed *laminar-flow control* (LFC).

The aerodynamic analysis of a LFC surface is divided into three parts: (1) the prediction of the inviscid flow field, (2) the calculation of the natural development of the boundary layer, and (3) the suction system analysis. In this approach (which may in reality require an iterative procedure), the first step is to determine the pressure and the velocity distribution of the inviscid flow at the edge of the boundary layer. The second step is to calculate the three-dimensional boundary layer, including both the velocity profiles and the integral thicknesses. It might be noted that because of cross flow, the boundary layer on a swept wing may be more unstable than that on an unswept wing. Finally, the suction required to stabilize the boundary layer must be calculated and the suction system designed.

In 1960, two WB-66 aircraft were adapted to a 30° swept wing with an aspect ratio of 7 and a thickness ratio of approximately 10%. The modified aircraft were designated X-21A. A suction system consisting of turbocompressor units removed boundary-layer air from the wing through many narrowly spaced LFC suction slots.

With suction-inflow velocities varying from $0.0001U_\infty$ in regions of negligible pressure gradient to $0.0010U_\infty$ near the wing leading edge, full-chord laminar flows were obtained up to a maximum Reynolds number of 45.7×10^6 [Kosin (1965)]. It was concluded that laminar-flow control significantly reduced the wake drag on the wing.

Using the propulsion, structural, flight controls and system technologies predicted for 1985, Jobe et al. (1978) predicted fuel savings from 27 to 30% by applying LFC to the design of large subsonic military transports. Jobe et al. assumed that the LFC system used in their design would maintain a laminar boundary layer to $0.70c$, even though full-chord laminarization of a wing with trailing-edge controls is technically feasible. The optimum wing planform for the minimum-fuel airplane has the highest aspect ratio, the lowest thickness-chord ratio, and a quarter-chord sweep of about 12°. The cruise Mach number for this aircraft design is 0.78. As noted in Table 13.1, their sensitivity analysis showed that a high aspect ratio is the most important parameter for minimizing fuel consumption, wing thickness is of secondary importance, and sweep is relatively unimportant. However, since productivity varies linearly with the cruise speed, a maximum productivity airplane requires a relatively high sweep, a maximum aspect ratio, and a low thickness ratio for the section. The resultant aircraft cruises at a Mach number of 0.85. The sensitivity analysis of Jobe et al. (1978) indicates that a low thickness ratio is most important to the design of the wing for a maximum productivity airplane, followed by aspect ratio and sweep.

TABLE 13.1 Desirable Laminar-Flow Control Wing Planform Characteristics

	Wing design parameter		
Figure of merit	Aspect ratio	Thickness ratio	Sweep
Performance			
Minimum fuel	High	Low	NMC[a]
Minimum takeoff gross weight	High	NMC	Low
Maximum $\dfrac{\text{maximum payload}}{\text{takeoff gross weight}}$	High	Low	NMC
Ease of laminarization			
Low chord Reynolds number	High	NMC	NMC
Low unit Reynolds number	NMC	NMC	NMC
Minimize cross flow	NMC	Low	Low
Minimize leading-edge contamination	High	Low	Low

[a]NMC, not a major consideration.
Source: Jobe et al. (1978).

13.4.3 Wingtip Devices

As discussed in Chapter 7, one of the ways of decreasing the induced drag is by increasing the aspect ratio. Although increased wing span provides improved lift/drag ratios, the higher bending moments at the wing root create the need for a stronger wing structure. Furthermore, there are problems in maneuvering and parking once on the ground. As noted by Thomas (1985), using large-aspect-ratio wings will reduce the induced drag because the tip vortices will be further separated, reducing the strength of the average induced flow between them. However, as Thomas points out, as the aspect ratio is increased for the same chord, there is a weight penalty that may offset the drag reduction. Thus, as noted in Thomas (1985), "optimal wing aspect ratio for a transport aircraft varies from 7.5 for minimum acquisition cost, to 9.8 for minimum gross weight, to 12.0 for minimum direct operating cost, and to 15.2 for minimum fuel. At present aspect ratios as large as 15.2 are not structurally feasible but the importance of aspect ratio is clear."

A possible means of reducing the drag is by the use of fixed winglets. As illustrated in Fig. 13.24, winglets are used on the Gates Learjet Model 28/29, the Longhorn. The drag polars at $M_\infty = 0.7$ and at $M_\infty = 0.8$ for the M28/29 are compared with those for the M25D/F in Fig. 13.25. As can be seen in these data, the greatest improvement is at the lower Mach number. However, this is of no concern for this design application, since the normal cruise speed and long-range cruise speed of this airplane are always less than $M_\infty = 0.8$.

To generate an optimum winglet for a particular flight condition, we must calculate the flow field for the complex wing. The subsonic aerodynamic load distributions for a lifting surface with winglets can be calculated using the vortex lattice method discussed in Chapter 7. The theoretical lift-curve slopes for a swept wing with end plates which were calculated using a distribution of vortices, such as illustrated in Fig. 13.26, were in good agreement with experimentally determined data [see Blackwell (1969)]. The lifting-surface geometry shown in the sketch indicates that the technique can be

Figure 13.24 Gates Learjet Model 28/29, the Longhorn, illustrating use of winglets. (Courtesy of Gates Learjet Corp.)

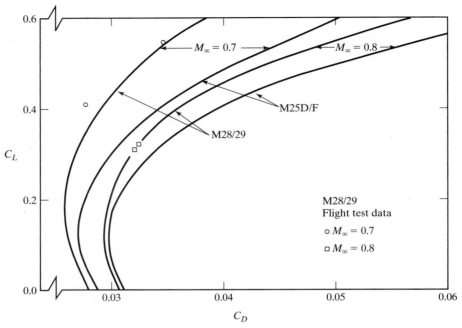

Figure 13.25 Comparison of the drag polars for the Gates Learjet M28/29 with those for the M25D/F. (Unpublished data provided by Gates Learjet Corp.)

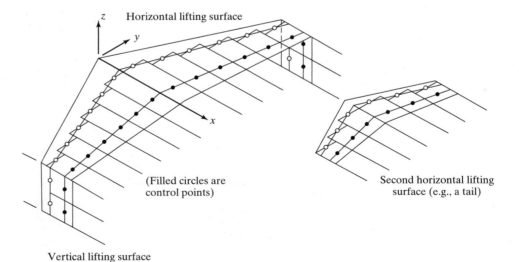

Figure 13.26 Distribution of vortices which can be used to calculate the aerodynamic load distribution for a combination of lifting surfaces. [From Blackwell (1969).]

used to calculate the aerodynamic load distribution for lifting surfaces involving a non-planar wing, a wing with end plates, and/or a wing and empennage.

For best performance, the proper design of winglets includes the following [Thomas (1985)].

1. For supercritical performance, the winglet should be tapered and swept aft. It should be mounted behind the region of lowest pressure of the main wing to minimize interference effects.

2. Smooth fillets should be used between the wing tip and the winglet or smaller drag reduction benefits might result.

3. Some toe-out of the winglet is needed due to the inflow angles at the wing tip. This is also desirable, since it reduces the likelihood of winglet stall during sideslip.

4. Although the drag reduction increases with winglet span, the decrease is less than linear. Therefore, the optimal winglet height involves a trade-off between improved aerodynamics and the increased moments due to longer moment arms.

Various additional wing-tip devices, as shown in Fig. 13.27, have been used in recent years in an attempt to reduce drag, especially on aircraft that fly long distances in cruise configuration where the drag reduction would have an appreciable impact on the range and fuel efficiency of the aircraft. Some of the wing-tip concepts that have been used include winglets, blended winglets, split winglets, raked wing-tips, and spiroids [McLean (2005)].

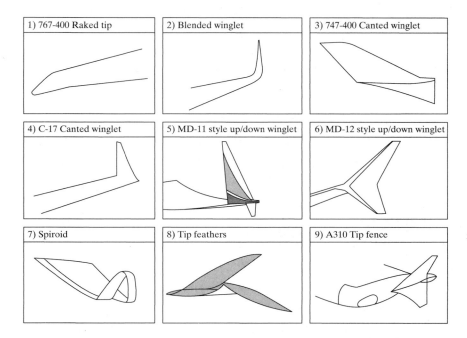

Figure 13.27 An assortment of wing-tip device configurations. [From McLean (2005).]

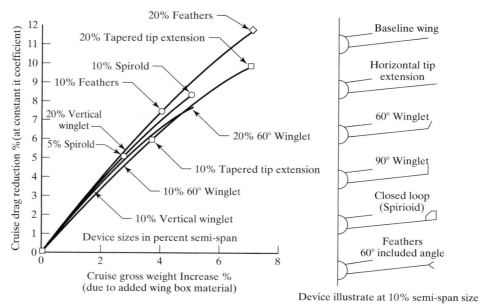

Figure 13.28 CFD-based drag prediction and flight test for blended winglets on a candidate configuration. [From McLean (2005).]

Whitcomb (1976) was one of the first to fully realize the benefit of wing-tip devices (such as winglets) and to realize that, while the goal of using a winglet was to reduce induced drag, there were several offsetting impacts that needed to be optimized in order to make any given wing-tip device worthwhile. Specifically, wing-tip devices have the positive impact of reducing induced drag at takeoff and during cruise, but they also add to profile drag of the wing by increasing the planform area and adding junctions to the wing-tip [McLean (2005)]. In addition, the weight and structural requirements of the wing also are changed by the addition of a wing-tip device. As with any aerodynamic concept, there are positive and negative aspects to the design feature. Figure 13.28 shows the cruise drag reduction for various wing-tip concepts on a generic aircraft and the resulting increase in wing structural weight. While all of the concepts add similar weight penalties for small amount of drag reduction, several of the devices perform better while achieving higher levels of drag reduction (from 4 to 12%).

13.4.4 Wing Planform

Since the induced drag of an airplane is determined by the total shed vorticity from the wing (not just the wing-tip shape), some researchers have chosen to modify the wing-planform shape as another method for reducing induced drag. One of the more interesting approaches along these lines is the lunate or crescent wing, which mimics wing shapes

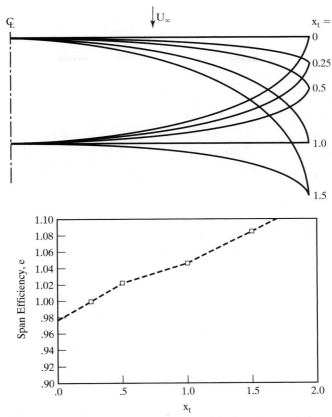

Figure 13.29 Crescent shaped wings and their corresponding span efficiency factors. [From van Dam (1987).]

often found in nature (see Fig. 13.29). Some of these shapes were found to reduce induced drag (or increase the span-efficiency factor) by as much as 8% in wind-tunnel tests [van Dam (1991)] and computational simulations [van Dam (1987)]. Other numerical simulations did not find the drag savings to be quite that high, showing drag reduction closer to 1% in some cases [Smith and Kroo (1993)]. Van Dam (1987) hypothesized that the decrease in drag was due to favorable interactions of the rolled-up wake of the wing.

Results such as these have led to an evolution of wing design in recent years, although the extreme crescent shapes have not been used on commercial aircraft (probably due to manufacturing limitations and costs). Wing-tip devices, such as the raked wing-tip used on the Boeing 767–400 (Fig. 13.30a), originally were done as a wing-tip extension rather than as a planform modification. However, the geometric impact of the rake is to make the planform more "crescent-like," and when that is coupled with wing sweep, the use of trailing-edge extensions (referred to as "Yehudis") and highly flexible structures, the advanced wing design for aircraft like the Boeing 787 (see Fig. 13.30b) are certainly similar to earlier "lunate-like" shapes.

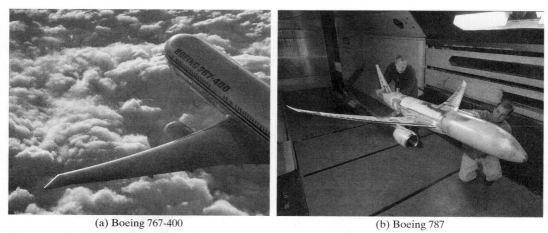

(a) Boeing 767-400   (b) Boeing 787

Figure 13.30 (a) The Boeing 767–400 raked wing tip and (b) the Boeing 787 in the Boeing Transonic Wind Tunnel. (Courtesy of The Boeing Company.)

13.5 DEVELOPMENT OF AN AIRFRAME MODIFICATION TO IMPROVE THE MISSION EFFECTIVENESS OF AN EXISTING AIRPLANE

13.5.1 The EA-6B

The EA-6B is a four-place, subsonic, twin-jet, electronic countermeasures airplane designed for land and carrier-based operations. As noted in Hanley (1987), the EA-6B airplane was difficult to handle when the pilot attempted to maneuver the aircraft at low airspeed (approximately 250 KIAS). These problems relate to the fact that the airplane through its evolution had grown to gross weights above 54,000 lb. To compound the problem further, both the lateral and directional moments become unstable near the stall angle of attack (approximately 16°). These characteristics contribute to the sharp roll-off and directional nose slice that is inherent in the EA-6B at stall.

As noted by Gato and Masiello (1987), the EA-6B Prowler is 10,000 lb heavier than the A-6 Intruder. In addition to additional ECM equipment and structural weight, the EA-6B was provided with more powerful engines. However, in the cruise configuration, there were no aerodynamic changes to account for the extra weight. Since the two aircraft have the same maximum lift capability, when maneuvering at low speeds and/or high altitude, the heavier EA-6B flies much closer to the stall than its predecessor, the A-6. This is illustrated in Fig. 13.31 "where the lift coefficient required to execute a banked 2g, 60 degree banked turn at 250 knots and the angle of attack margin to stall are indicated for an A-6, a current production EA-6B, and a future ADVCAP EA-6B at representative landing weights. This is a typical maneuver for these aircraft, always performed while decelerating prior to entering the landing pattern and initiating landing gear, slat, and flap extension. Incidentally, the stall for these aircraft is defined by loss of lateral control and subsequent roll-off. Further penetration of the stall results in a directional departure which may develop into a spin."

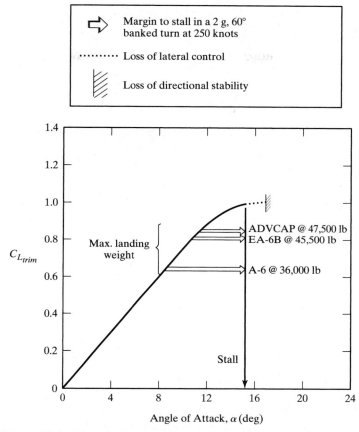

Figure 13.31 EA-6B/A-6 cruise configuration lift. [From Gato and Masiello (1987).]

   In order to diagnose the flow mechanism contributing to the directional instability near stall, flow visualization studies were conducted [Jordan et al. (1987)]. "Results from these studies showed that a pair of vortices are generated at the fuselage-wing junctures. These vortices trail behind the wing, close to the fuselage and below the tail at low angles of attack (Fig. 13.32a). As angle of attack is increased, wing downwash maintains the vortex system at the same relative location—that is, low with respect to the vertical tail. At stall angles of attack, flow separation on the wing and consequent downwash breakdown cause the vortices to rise to the level of the vertical tail (Fig. 13.32b). In sideslip, the vortex generated on the windward side of the airplane drifts leeward such that, as angle of attack is increased through stall, the vertical tail becomes immersed in the windward vortex flow field (Fig. 13.32c). Because of the rotational sense of the vortex system, the bottom portion of the vertical tail first becomes immersed in a region of proverse (stabilizing) sidewash—the top portion of the windward vortex. As angle of attack is increased and the vortex system rises further, the vertical tail becomes immersed

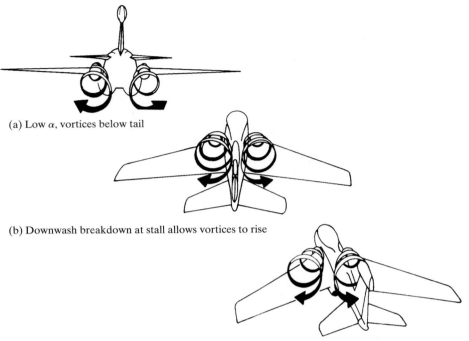

(a) Low α, vortices below tail

(b) Downwash breakdown at stall allows vortices to rise

(c) In sideslip, windward wing vortex creates adverse/destabilizing flow field

Figure 13.32 Directional destabilizing vortex system. [From Jordan et al. (1987).]

in the lower portion of the windward vortex where a condition of adverse (destabilizing) sidewash exists. Clearly, the abrupt changes in sidewash that occur as the windward vortex traverses the span of the vertical tail have a direct impact on directional stability. In fact, it is this phenomenon that causes the directional instability the airplane configuration experiences near stall."

To eliminate these dangers and to provide the EA-6B with electronics growth capability, the Navy undertook a comprehensive Maneuvering Improvement Program. The ground rules imposed by the Navy were that no major changes were allowed to the main wing or airframe structure. The project involved personnel from the Navy, Grumman Aircraft Systems Division, and the Langley Research Center (NASA). The results of the various tasks are described in Gato and Masiello (1987), Hanley (1987), Jordan et al. (1987), Sewall et al. (1987), Waggener and Allison (1987).

As a result of these efforts, an integrated mix of computational and experimental investigations, an aerodynamic upgrade package was proposed for the EA-6B Prowler, as shown in Fig. 13.33. The upgrade package includes the following features:

1. Drooped inboard leading edges (add-on to inboard slat elements)
2. Wing-glove strakes
3. A fin extension
4. Wing-tip speed-brake/ailerons

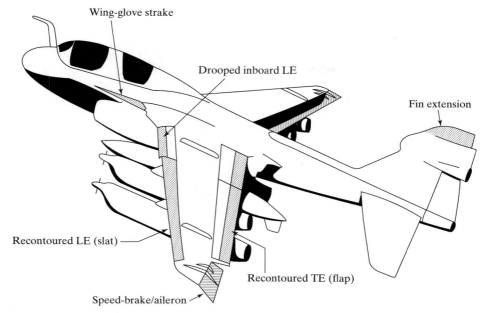

Figure 13.33 Proposed aerodynamics upgrade package for the EA-6B Prowler. [From Gato and Masiello (1987).]

The combined effect of fin extension, strakes, and drooped inboard leading edges on directional stability is illustrated in Fig. 13.34. The unstable directional characteristic was shifted to higher angles of attack, by a total of 6°. The level of directional stability at low angles of attack was also significantly increased as a result of the additional vertical tail area. In fact, the fin extension was sized to provide the EA-6B with the same low angle of attack directional stability as the A-6, thereby compensating for the destabilizing effect of the longer fuselage.

Besides improving the lateral-directional stability, the strakes and the drooped inboard leading edges increased the maximum lift and the stall angle of attack, as shown in Figs. 13.35 and 13.36. This combination of aerodynamic modifications/devices provides significant improvements in the maximum lift coefficient. As shown in Fig. 13.35, the increase in cruise configuration maximum usable lift is 22% at low Mach numbers and 30% at higher Mach numbers. Consequently, the aerodynamically upgraded EA-6B will have essentially the same stall/maneuver margins of the lighter A-6. Improved lateral-directional stability and positive lateral control beyond stall allow full use of this increment in maximum lift.

13.5.2 The Evolution of the F-16

As stated by Bradley (1981), "The design of tactical military aircraft presents quite a challenge to the aerodynamicist because of the vast spectrum of operational requirements encompassed by today's military scenario. The designer is faced with a multitude

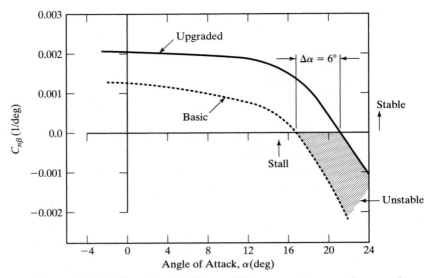

Figure 13.34 Combined effect of fin extension, strakes, and drooped inboard leading edges on lateral-directional stability. [From Gato and Masiello (1987).]

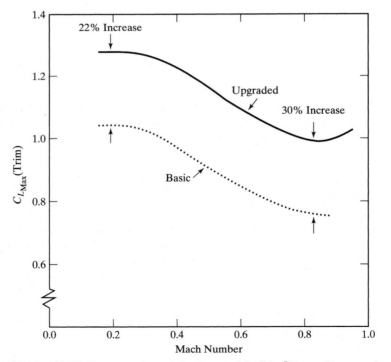

Figure 13.35 Increase in maximum usable lift. [From Gato and Masiello (1987).]

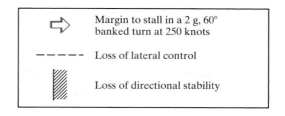

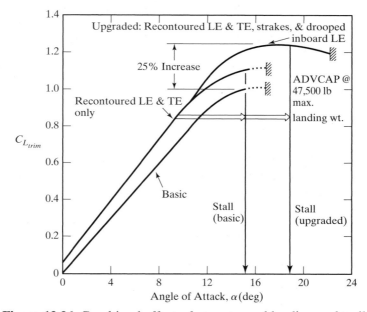

Figure 13.36 Combined effect of recontoured leading and trailing edges, strakes, and drooped inboard leading edges on cruise configuration lift. [From Gato and Masiello (1987).]

of design points throughout the subsonic-supersonic flow regimes plus many off-design constraints that call for imaginative approaches and compromises. Transonic design objectives are often made more difficult by restraints imposed on by subsonic and supersonic requirements. For example, wings designed for efficient transonic cruise and maneuver must also have the capability to accelerate rapidly to supersonic speeds and exhibit efficient performance in that regime."

Bradley continues, "The design problem is further complicated by the fact that the weapon systems of today are required to fill multiple roles. For example, an aircraft designed to fill the basic air superiority role is often used for air-to-ground support, strike penetration, or intercept missions. Thus, carriage and delivery of ordnance and carriage of external fuel present additional key considerations for the aerodynamicist. The resulting aircraft flowfield environment encompasses a complex mixture of interacting flows."

"For example, the need for rapid acceleration to supersonic flight and efficient supersonic cruise calls for thin wing sections with relatively high sweep and with camber that is designed to trim out the aft ac movement at supersonic flight. However, these requirements are contrary to those requirements for efficient transonic maneuver, where the designer would prefer to have thicker wing sections designed with camber for high C_L operation and a high-aspect-ratio planform to provide a good transonic drag polar."

Harry Hillaker (1997) who was the Chief Project Engineer on the YF-16 and Vice President and Deputy Program Director of the F-16XL Program, discussed early studies that provided the technology base from which the YF-16 design was developed. Examination of results from air-to-air combat over Southeast Asia/Vietnam revealed that aircraft in the U.S. inventory had only marginal success over their opponents. From 1965 to 1968, engineers at General Dynamics in Fort Worth examined the data from the Southeast Asian conflict to determine what parameters provided an edge in air-to-air combat. Wing loadings, thrust loadings, control issues, g-tolerances for the pilot, and empty-weight fraction were identified as key parameters in air-to-air combat. Since these studies were focused on technology, there were no configuration designs.

In 1997 Hillaker continued, "From 1969–1971, an extensive wind-tunnel program was conducted in which data were obtained on a wide variety of configurations over a range of free-stream test conditions and angles of attack." As shown in the sketches of Fig. 13.37, configuration variables included wing planforms, airfoil sections, wing-body relationships, inlet locations, horizontal and vertical tail configurations, and forebody

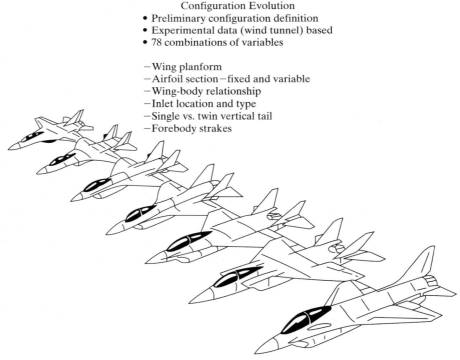

Configuration Evolution
- Preliminary configuration definition
- Experimental data (wind tunnel) based
- 78 combinations of variables

−Wing planform
−Airfoil section−fixed and variable
−Wing-body relationship
−Inlet location and type
−Single vs. twin vertical tail
−Forebody strakes

Figure 13.37 Configuration evolution for the F-16.

strakes. Buckner et al. (1974) reported, "The 1969–1970 period produced technology studies in four areas very important to air combat maneuvering fighter design:

1. Project Tailormate (1969, 1970) — an experimental study of a wide variety of inlet types and locations on typical fighter designs with the goal of maintaining low distortion and high-pressure recovery over wide ranges of angle of attack and sideslip. The YF-16 inlet location is largely a result of the experience gained in this work.

2. Wing mounted Roll Control Devices for Transonic, High-Light (sic) Conditions (1969) — a study of a variety of leading-and trailing-edge devices for roll control in the combat Mach number and high-angle-of-attack range.

3. Aerodynamic Contouring of a Wing-Body Design for an Advanced Air-Superiority Fighter(1970) — an add-on to the roll-control study, which produced analytical information on a blended wing-body design. This experience was helpful in the later development of the YF-16 overall planform and blended cross-section concept.

4. Buffet Studies (1969–1970) — a series of efforts producing new knowledge and methodology on the phenomenon of increased buffet with increasing angle of attack. The methodology, in turn, allowed the YF-16 to be designed for buffet intensity to be mild enough to permit tracking at essentially any angle of attack the pilot can command in the combat arena.

The above studies produced knowledge that was brought to bear on the configuration studies, specifically for the development of a lightweight low-cost fighter, accomplished in 1970 and 1971."

The lightweight, low-cost fighter configuration studies intensified when a new set of guidelines were defined for application to combat scenarios in Europe. Hillaker (1997) noted, "These mission rules included a 525 nautical mile (nm) radius, four turns at 0.9M at 30,000 feet, accelerate from 0.8M to 1.6M, three turns at 1.2M at 30,000 feet, and pull a four-g sustained turn at 0.8M at 40,000 feet. Furthermore, the aircraft had to demonstrate operability both for the U.S. Air Force and its NATO allies."

As noted by Buckner et al. (1974): "The YF-16 really got its start, then, as a result of the 'in-house' studies initiated in late 1970 in response to the new mission rules . . . Major emphasis was placed on achieving flight and configuration characteristics that would contribute directly to the air-to-air kill potential in the combat arena — specifically maximizing maneuver/energy potential and eliminating aerodynamic anomalies up to the maneuver angle-of-attack limits." Buckner and Webb (1974) reported, "Examples of the 'design to cost' in the case of the YF-16 aerodynamic features are (1) a single engine, eliminating the complex question of what to do between the nozzles, (2) an empennage/nozzle integrated design, devoid of adverse interference, (3) a single vertical tail tucked in safely between the forebody vortices, (4) a simple underslung, open-nosed inlet with no complex moving parts, (5) a thin-wing airfoil with only slight camber, minimizing the question of Reynolds number effects on transonic shock locations, and (6) simple trailing-edge ailerons. All of these features reduced the cost through virtual elimination of design changes in the refinement stage after contract go-ahead and through simplification of the task required to fully define the vehicle aerodynamics." In 1972,

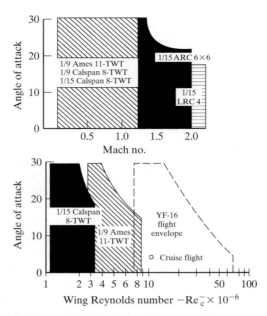

Figure 13.38 Angle of attack, Mach number, Reynolds number coverage for YF-16 design tests. [From Buckner et al. (1974).]

aerodynamic data were obtained over a wide range of Mach number, of Reynolds number, and of angle of attack. The ranges of these variables, as taken from Buckner and Webb (1974), are reproduced in Fig. 13.38.

Buckner, et al. (1974) reported that "from these data it is obvious that the taper ratio (λ) should be as low as practical, consistent with reasonable tip-chord structural thickness and early tip-stall configurations. The final selection was a taper ratio of 0.227. The start combat weight is lowest at a wing aspect ratio of 3.0, the final selected value, and is relatively independent of wing sweep in the range of 35 to 40 degrees. The greater wing sweep is beneficial to the supersonic performance, giving reduced acceleration time and increased turn rate. Some penalty in aircraft size is noted as the wing sweep is increased to 45 degrees, and the higher-sweep wings are more prone to have a aileron reversal because of aeroelastic effects. As a result, wind tunnel tests were limited to sweep angles between 35 and 45 degrees. Perturbation of wing thickness ratio t/c indicates a lighter weight airplane results with a thicker wing, but supersonic maneuverability improves with thinner wings. The desire to achieve a balance in subsonic and supersonic maneuver capability dictated selection of as thin a wing as practical ($t/c = 0.04$), consistent with flutter and aileron reversal considerations."

Bradley (1981) notes that: "The tactical military aircraft design problem is made more difficult by the supersonic acceleration requirement. The wings must be as thin as structurally feasible to reduce drag, but fixed camber suitable for optimum transonic maneuver is not practical because of supersonic camber drag. An obvious solution is a smoothly varying wing camber design. However, structural and actuation system weights prove to be prohibitive for thin wings. Simple leading edge and trailing edge flaps often prove to be the most practical compromise for high-performance, multiple design point configurations."

"The extreme possibilities are apparent. One may design a wing with optimized transonic maneuver camber and twist and attempt to decamber the wing with simple flaps for supersonic flight. On the other hand, one may design the wing with no camber or with a mild supersonic camber and attempt to obtain transonic maneuver with simple flaps."

Even though the designer has worked carefully to design a wing for attached flow and to optimize its performance, there are points in the sustained and instantaneous maneuver regimes where the flow field contains extensive regions of separated flow. The manner in which the flow separates will strongly affect the vehicle's drag and its controllability at the higher values of the lift coefficient. For many designs, strakes or leading-edge extension devices are used to provide a controlled separated flow. Controlled vortex flow can then be integrated with the variable camber devices on the wing surface to provide satisfactory high-lift, stability and control, and buffet characteristics. The resulting flowfield is a complex one, combining attached flows over portions of the wing with the vortex flow from the strake. See Fig. 13.20.

One of the questions addressed by the designers of the Lightweight Fighter was single vertical tail versus twin-tailed configurations. See Fig. 13.37. NASA data available at this time indicated advantages of the twin-tailed configurations. However, as noted by Hillaker (1997), the NASA data were limited to angles of attack of 15°, or less. Buckner et al. (1974) noted, "To the dismay of the design team, however, the directional stability characteristics of the twin-vertical-tailed 401F-0 configuration were not as expected. In fact, a severe loss of directional stability occurred at moderate-to-high angles of attack.... Analysis of oil flow visualization photographs led to the belief that forebody flow separations and the interaction of the resulting vortices with the wing and vertical tail flow fields were major causes of the stability problem."

Modifications were made to delay forebody separation to higher angles of attack. Buckner et al. (1974) reported, "It was more difficult to make the twin-tail configurations satisfactory compared to the single vertical tail configurations (in addition, some combinations of angle of attack and sideslip produced visible buffeting of the twin tails). Beneficial effects on the directionally stability derivatives were noted when relatively small highly swept 'vortex generators' (strakes) were located on the maximum half-breadth of the forebody."

"At this point, NASA Langley Research Center aerodynamicists were consulted and they suggested that the lift of the wide forebody could be increased by sharpening the leading edge to strengthen the vortices rather than weaken them as our earlier attempts had done. The point was that forebody separation is inevitable at very high angle of attack; therefore, the lift advantages offered by sharp leading edges should be exploited. This also would allow the forebody vortices to dominate and stabilize the high-angle-of-attack flow field over the entire aircraft, improving, even, the flow over the outboard wing panels."

Once the YF-16 was in the flight-test program with the YF-17, it was no longer called the Lightweight Fighter. It was called the Air Combat Fighter (ACF). Orders for the ACF, the F-16, came in 1975. The first operational units were formed in 1978. In

Figure 13.39 An F-105D piloted by Major E. R. Bracken on a mission during Vietnam War carrying ten stores externally. (Photo provided by M. Gen. E. R. Bracken, USAF Ret.)

the late 1970s, production versions of the F-16 were quickly modified to add a full radar capability, to add hard points to accommodate the ability to handle air-to-ground capability, and to increase the combat radius to 725 nautical miles. Thus, the aircraft had become the Multi-Role Fighter (MRF).

13.5.3   External Carriage of Stores

Starting in the 1950s, the air-to-ground role was often performed by military aircraft carrying the ordnance externally. In a photo provided by Bracken (2000), an F-105D is shown on a mission during the Vietnam War. The F-105D (Fig. 13.39) carries externally two 500-pound bombs (one outboard on each wing), two 450-gallon fuel tanks (one on each wing), and six 750-pound bombs (clustered near the centerline of the aircraft).

Whitford (1991) reported large improvements in the load-carrying ability from the designs of the 1950s to more recent designs. As shown in Fig. 13.40, the Hawk 200 is able to lift an external load equal to 85% of its empty weight.

Bradley (1981) notes, "Perhaps the greatest irony for the tactical aircraft designer results from the fact that an aircraft designed to be the ultimate in aerodynamic efficiency throughout a performance spectrum is often used as a 'truck' to deliver armaments. Aircraft that are designed in a clean configuration are often used operationally to carry an assortment of pylons, racks, missiles, fuel tanks, bombs, designator seeker pods, launchers, dispensers, and antennaes that are attached to the configuration at any conceivable location."

Although the F-16 was originally conceived as a lightweight fighter (see Fig. 7.40), it has assumed a multirole capability to meet the demands of widely differing operational requirements, including air superiority, air intercept, battlefield support, precision-strike/interdiction, defense suppression, maritime interdiction, (airborne) forward air controller, and reconnaissance. Weapons that can be carried by the F-16 are presented

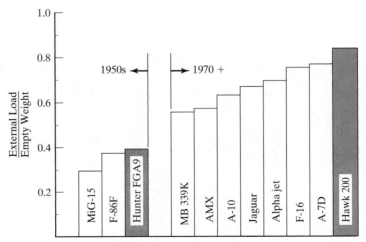

Figure 13.40 Capability of aircraft to carry stores externally. [As taken from Whitford (1991).]

AIM-132 Asraam

MICA

AIM-120 Amraam

AIM-7 Sparrow/Skyflash

ALQ-131 ECM Pod

LAU-3A/5003 Rocket Pod

CRV-7 Rocket

MK-82 LDGP (500 lb)

MK-82 Snakeye (500 lb)

MK-36 Destructor (500 lb)

20 mm

30 mm

Magic II

AGM-84 Harpoon

ALQ-119/184 ECM Pod

BLU-107/B Durandal

AAS-35 Pave Penny
Laser Spot Tracker Pod

Rubis Navigation Pod

ASQ-213 Harm Targeting
system Pod

AAQ-20 Pathfinder
Navigation Pod

Atlas II Targeting Pod

AGM-65 Maverick

AGM-119 Penguin

AAQ-14 Lantirn/Sharpshooter
Targeting Pod

Atlantic
Navigation Pod

AAQ-13 Lantirn Navigation Pod

GPU-5/A 30 mm GUN Pod

BSU-49 "BALOOT" (500 lb)

AS30L

AGM-88 Harm

AIM-9 Sidewinder

AGM-45 Shrike

AMCI Pod

Reconnaissance Pods

Hydra-70

MK-106 BDU-33
Practice Practice

SUU-20 Practice

MXU-648
Cargo Pod

AXQ-14 Data Link Pod

N/A 37U-36 Aerial Gunnery Target

MK-84 LDGP (2000 lb)

BSU-50 "BALOOT" (2000 lb)

300-GAL
Centerline Tank

370-GAL Tank

600-GAL Tank

MK-20 Mod-4 Rockeye

GBU-15 (2000 lb)

BLU-109 (2000 lb)

GBU-24 Paveway III
(2000 lb)

GBU-10 Paveway II
(2000 lb)

GBU-12 Paveway II
(500 lb)

CBU-52/58/71

CBU-87/89/97 TMD

BL 755

Autonomous free-flight
Dispenser system

Figure 13.41 Weapons that can be carried by the F-16. (Figure provided by Lockheed Martin Aeronautics Company.)

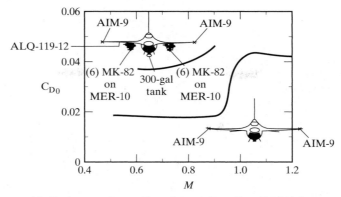

Figure 13.42 Stores drag effect [From Bradley (1981).]

in Fig. 13.41, which was provided by M. J. Nipper (1996). The figure, which illustrates the certified stores capability, clearly exhibits the F-16's versatility of weapons carriage. A number of new weapons are being developed that will be certified on the F-16 by 2001.

According to Bradley (1981), "The carriage drag of the stores is often of the same order of magnitude as the total minimum drag of the aircraft itself. For example, the minimum drag of the F-16 aircraft is compared in Fig. 13.42 with and without the air-to-ground weapons load. It is readily seen that the store drags themselves present as large a problem to the aircraft designer as the drag of the clean configuration. Store carriage on modern technical aircraft is extremely important, particularly as one approaches the transonic regime, where the interference effect of the stores and pylons are highest and most detrimental to performance."

Bradley (1981) continues, "Not only must external stores be carried efficiently by tactical aircraft, but they must release cleanly and follow a predictable trajectory through the vehicle's flowfield. The store trajectory is governed by the highly unsteady forces and moments acting on the store produced by the nonuniform flowfield about the configuration and the aerodynamic characteristics and motions of the store itself. The problem is complicated by realistic combat requirements for jettison or launch at maneuver conditions and multiple release conditions where the weapons must not 'fly' into one another."

Military aircraft of the future will have added emphasis on store carriage and release early in the design process. Some possible concepts for weapons carriage are contrasted in Table 13.2.

Bradley (1981) presented wind-tunnel measurements that compared the drag for conformal carriage versus conventional installations of MK-82 bombs. As shown in Fig. 13.43, conformal carriage permits one to carry 14 MK-82 bombs at substantially less drag than a 12 MK-82 pylon/multiple bomb rack mounting. Bradley noted (1981), "Significant benefits for the conformal carriage approach, in addition to increased range, are realized: increased number of weapons and carriage flexibility; increased penetration speed; higher maneuver limits; and improved supersonic persistence. Lateral directional stability is actually improved with the weapons on."

TABLE 13.2   Weapon Carriage concepts [Bradley (1981).]

Store carriage concepts	Advantages	Disadvantages
1. Wing pylon carriage	Most flexible carriage mode—large payloads, inefficient store shapes	High drag High radar cross section
2. Internal carriage	Low drag	Limited weapon flexibility
	Low radar cross section	Increased fuselage volume
3. Semisub-merged carriage	Low drag	"Holes" must be covered up after weapons drop
	Low radar cross section	
		Severely restricted payload flexibility
4. Conformal carriage	Most flexible of low drag carriage concepts	Size restrained

Hillaker (1997) noted that in 1974, General Dynamics embarked on a Supersonic Cruise and Maneuver Program (SCAMP) to develop a supersonic cruise derivative of the F-16. The "supercruiser" concept envisioned optimization at supersonic cruise lift conditions so that sustained cruise speeds on dry power (nonafterburner power) could be achieved in the Mach 1.2 to 1.3 speed range. Trade-offs to the aerodynamics required for supersonic cruise, subsonic cruise, and maneuvering flight were explored. The goal was to arrive at a design that would offer at least a 50% increase in the supersonic lift-to-drag ratio (L/D) that would retain a high subsonic L/D ratio and that would provide the level of maneuverability of a fighter.

Hillaker (1997) noted that, by 1977 to 1978, it was clear that the F-16 was to serve both the air-to-air mission and the air-to-ground mission (with the air-to-ground mission

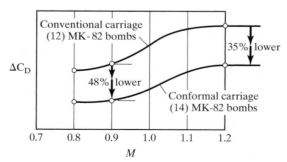

Figure 13.43 Drag comparison—conformal vs. conventional carriage. [From Bradley (1981).]

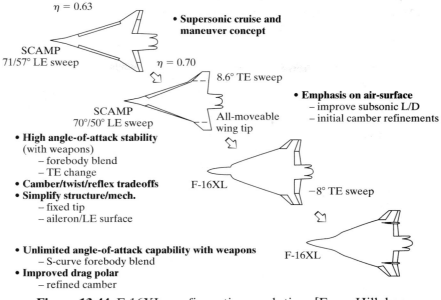

Figure 13.44 F-16XL configuration evolution. [From Hillaker (1982).]

dominating). Thus, engineers sought to develop a design that was a straight-forward modification of the F-16. The approach was to build on the technology developed for the original SCAMP configuration adding a maneuver requirement to the supersonic cruise capability. From 1974 to 1982, the SCAMP/F-16XL configurations underwent significant refinements. The configuration evolution, as taken from a paper by Hillaker (1982), is reproduced in Fig. 13.44. Hillaker (1982) reported, "The planform requirements included a forebody blend (strake area) for high angle-of-attack stability, an inboard trailing-edge extension for pitching moment improvement, and fixed wing tips with ailerons and leading-edge device to enhance the flow over the aileron at high angles of attack. The combination of ailerons and a leading-edge device (flaps for flow control, not lift) was developed to resolve the mechanical and structural complexities of the all-movable wing tip on the previous configuration. All-movable wing tips provided high roll rates and adverse yaw for resistance to yaw divergence at high angles of attack. The aileron-leading-edge device combination provided the same aerodynamic advantages and simplified wing structure, reduced weight, increased fuel volume, and allowed tip missile carriage."

The benefits of the F-16XL configuration include 40% to 85% lower drag with the integrated weapons carriage, 50% better supersonic lift-to-drag ratio with no subsonic penalty, 17% lower wave drag with drag with 83% more internal fuel, and increased lift with expanded maneuver angle of attack.

The first two demonstrator aircraft (modified from two F-16A aircraft) took to the air at Carswell Air Force Base in Fort Worth, Texas on July 3, 1982. They are still being used by NASA for flight research.

TABLE 13.3 Summary of main design features required to achieve a variety of operational requirements [Taken from Whitford (1991)]

Operational Requirements	Primary Design Features
Short takeoff	High afterburning (A/B) thrust/weight (T/W) ratio, high flap lift, low wing loading (W/S), thrust vectoring
Economical transit/loiter	Low throttled specific fuel consumption (sfc), high cruise lift-to-drag (L/D) ratio
High penetration speed	High dry thrust/weight (T/W) ratio, low store drag, good ride quality
High combat agility	High dry thrust/weight (T/W) ratio, high control power, good lift-to-drag (L/D) ratio at high g, high usable lift, thrust vectoring
High combat persistence	Low combat specific fuel consumption (sfc), good lift-to-drag (L/D) ratio at high g, versatility of weapon carriage
High combat mobility	High afterburning (A/B) thrust/weight (T/W) ratio, low drag at all Mach numbers
High survivability	Stealth (to give first shoot/first kill), threat awareness and countermeasures, robustness/redundancy
Short landing	High flap lift, low wing loading (W/S), flareless, good retardation

13.5.4 Additional Comments

The ability to modify a design to accommodate the increased weights allows an aircraft to satisfy added missions, handle more ordnance, be fitted with more powerful engines, and so on. The Supermarine Spitfire, a single-engine pursuit aircraft of World War II, grew by the weight of 32 airline passengers and their luggage during its evolution from 1936 through 1946.

The examples discussed in this section illustrate the philosophy represented by the comment made by Montulli (1986): "When establishing airplane performance requirements, allow for potential changes in operational requirements. Do not allow a point design." Table 13.3, which is taken from Whitford (1991), identifies the main design features that allow a military aircraft to achieve a variety of operational requirements.

13.6 CONSIDERATIONS FOR WING/CANARD, WING/TAIL, AND TAILLESS CONFIGURATIONS

The relative merits of wing/canard, wing/tail, and tailless configurations for high-performance aircraft have been the subject of numerous investigations. The use of canards usually allows a shorter fuselage, which results in less skin friction drag, and a more optimum cross-sectional area distribution, which results in less wave drag. However, although aft-tail configurations need a longer fuselage, they usually have smaller wing area, which also results in lower skin-friction drag and in lower wave drag.

The stall progression for a forward swept wing (FSW) is compared with that for an aft swept wing (ASW) in Fig. 13.45, which is taken from Weeks (1986). As depicted in Fig. 13.45a, the stall pattern for a forward swept wing proceeds from the root to the tip, as opposed to the normal behavior for aft swept wings, in which the stall pattern starts at the tip and proceeds inboard toward the root. See the discussion in

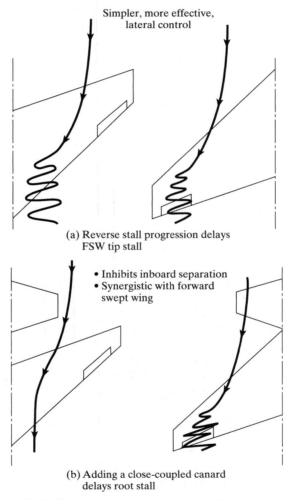

Simpler, more effective,
lateral control

(a) Reverse stall progression delays
FSW tip stall

• Inhibits inboard separation
• Synergistic with forward
  swept wing

(b) Adding a close-coupled canard
delays root stall

Figure 13.45 Stall-progression patterns for a forward swept wing (FSW) as compared with those for an aft swept wing. [From Weeks (1986).]

Section 7.3.5. Both cases would result in pitch-up tendencies. Weeks states, "By placing a full authority canard ahead of the wing root region (Figure 13.45b), the FSW configuration avoids pitch-up and achieves delayed stall providing full utilization of the inboard, large lift contributing portions of the wing to high angles-of-attack. Lateral control can then be maintained with simple (light weight) tip region located ailerons. By further integrating negative subsonic stall margin and nearly neutral supersonic margin, positive lift to trim can be maintained over the entire envelope. Then, by further introduction of variable camber and three-surface control, airframe drag can be minimized."

In 1982 and 1983, General Dynamics and the Langley Research Center (NASA) conducted a wind-tunnel test program to generate a data base needed for a more general understanding of the aerodynamic performance comparisons between a wing/canard configuration, a wing/tail configuration, and a tailless configuration. Data were obtained for two sets of configurations: one employed a 60-degree leading-edge-sweep delta wing and the other employed a 44-degree leading-edge-sweep trapezoidal wing. Data for the 44-degree leading-edge-sweep trapezoidal wing, as presented in Nicholas et al. (1984), will be reproduced here. A description of the models is presented in Fig. 13.46. The leading and trailing-edge flaps were optimally scheduled to minimize the drag

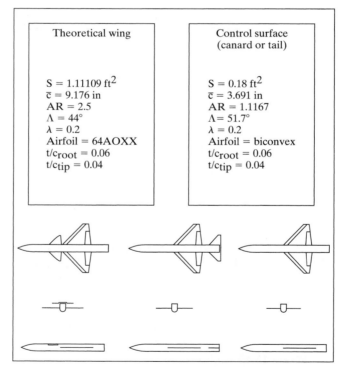

Figure 13.46 Description of models for wing/canard, wing/tail, and tailless configurations comparisons. [From Nicholas et al. (1984).]

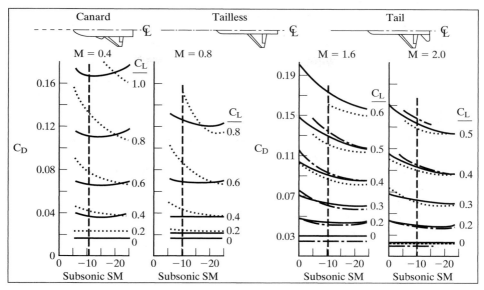

Figure 13.47 Summary of trimmed drag comparison of canard, tail, and tailless arrangements with the 44-degree sweep trapezoidal wing. [From Nicholas et al. (1984).]

for the wing/canard configuration and for the wing/tail configuration. For the tailless configuration, the subsonic data were insufficient to define the trimmed polars. At supersonic speeds, trailing-edge flap deflections were governed by the trim requirements rather than by the drag optimization.

Referring to the data presented in Fig. 13.47, Nicholas et al. (1984) noted, "At subsonic speeds, large negative static margins are required to achieve small polar benefits for the wing-canard as compared to the wing-tail. The wing-tail drag is optimized with subsonic static margins in the range of -10 to $-15\%\bar{c}$. This nominal level of instability is considered achievable because of satisfactory high-angle-of-attack stability and control characteristics observed on the transonic fighter model. Furthermore, the drag penalties for slightly increased stability on the wing-tail are not severe. The wing-canard subsonic drag polar appears to be approaching optimum with a subsonic static margin of $-25\%\bar{c}$, and increased stability is accompanied by severe drag penalties. The risk associated with an aircraft designed for this level of instability is significant, and the potential drag benefits appear to be small. It is also apparent that the subsonic polar shapes of both the canard and tailless arrangements are far more sensitive to subsonic static margin variations than those of the wing-tail."

Nicholas et al. (1984) continued, "At supersonic speeds, static margin sensitivity is roughly similar for the canard, tail, and tailless arrangements. Here, the optimization comparison between canard and tail is somewhat reversed, with the subsonic static margin being optimum at $-15\%\bar{c}$ to $-20\%\bar{c}$ for the wing-canard versus approaching

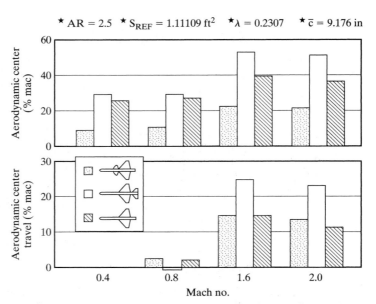

Figure 13.48 Variation of aerodynamic center with Mach number for canard, tail, and tailless arrangements with the 44-degree sweep trapezoidal wing. [From Nicholas et al. (1984).]

optimum at $-25\%\bar{c}$ for the wing-tail. The reason for this reversal is associated with aerodynamic center (defined with all surfaces fixed at zero deflection) travel from subsonic to supersonic speeds, which is greater for the wing-tail than for the wing-canard, as shown in Fig. 13.48. This happens because the fraction of the total lift carried by the tail increases significantly as the wing downwash field decreases from subsonic to supersonic speeds. However, because of its forward location, the canard experiences relatively little variation in the fraction of total lift that it carries between subsonic and supersonic speeds. Therefore, for a fixed level of subsonic stability, the wing-body is more stable at supersonic speeds for the wing-tail than it is for the wing-canard. It is this supersonic wing-body stability that determines canard or tail trim requirements. Both the wing-canard and the wing-tail optimize with approximately 12% of the total lift carried in the control surface, as shown for Mach 1.6 in Fig. 13.49. This optimum control-surface/wing lift ratio is achieved with a subsonic static margin of nominally $-21\%\bar{c}$ for the wing-canard versus $-27\%\bar{c}$ for the wing-tail. The net result is that, at supersonic speeds, the wing-canard provides a small drag advantage over the wing-tail with a reasonable level of subsonic static margin (i.e., -10 to $-15\%\bar{c}$)."

Both for wing-canard and wing-tail configurations, the optimum polars are achieved by carrying a very small fraction of the total lift in the control surface.

From the results of their studies, Nicholas et al. (1984) reached the following conclusions:

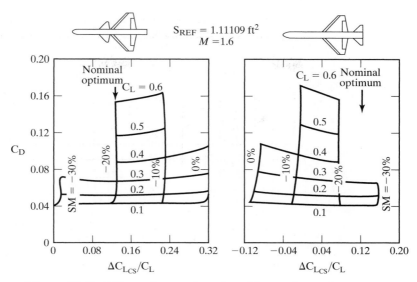

Figure 13.49 Effect of control surface lift on trimmed supersonic drag for canard and tail arrangements with the 44-degree sweep trapezoidal wing. [From Nicholas et al. (1984).]

1. For highly efficient, variable-camber wings, large negative stability levels are required to achieve small subsonic polar shape benefits for wing-canard arrangements, as compared to wing-tail arrangements. However, these large negative stability levels are accompanied by reduced maximum lift for canard arrangements along with potential stability and control problems at high angles of attack. The need for large negative stability levels with a wing-canard diminishes as the main wing efficiency is decreased.

2. Subsonic polars for canard and tailless arrangements are more sensitive to subsonic static margin than those of wing-tail arrangements.

3. At supersonic speeds, the canard arrangements show some advantage because their polar shapes optimize at higher subsonic stability levels than wing-tail or tailless arrangements.

4. The minimum drag and weight advantages of tailless delta arrangements can overcome polar shape deficiency to provide a TOGW advantage for typical advanced fighter mission/performance requirements.

5. Static margin limit is a critical issue in control surface (canard, tail, tailless) selection."

13.7 COMMENTS ON THE F-15 DESIGN

A planform view of an F-15 in flight is presented in the photograph of Fig. 13.50. Let us discuss two features evident in the photograph: (1) the notch or "snag" in the leading edge of the stabilator and (2) the clipped tip of the wing. The following information on these two features was provided by Peters, Roos, and Graber (2001):

Figure 13.50 Plan view of F-15. (Copyrighted image reproduced courtesy of The Boeing Company.)

"The notch, or snag, in the leading edge of the F-15 stabilator solved a flutter-margin issue. Moving the inboard leading-edge aft moved the center of pressure aft and also removed mass from the leading edge. This allowed the F-15 stabilator to meet an 800 KCAS (Knot Calibrated Air Speed) +15% margin requirement. The snag design was one of several options that were investigated for improving the flutter margin of the empennage. The other options (structural beef-up, mass balance, tip pods, etc.) introduced more weight and/or had aero performance penalties.

"During the 80% flight-loads testing of the F-15, it was discovered that the maximum bending moment was 3% to 4% higher than the estimates based on the wind-tunnel results. To reduce the bending moments, the wing tips were cut off on the spot and were replaced with raked tips made out of mahogany. The mahogany wing tips flew until ship number 4 was retired many years later. With the modified wing tips in place, the flight-loads testing was completed successfully."

13.8 THE DESIGN OF THE F-22

The F-22 design is the result of more than twenty years of work to build an Advanced Tactical Fighter (ATF). As noted by Hehs (1998) "the term *advanced tactical fighter* and its abbreviation ATF, however, appeared in a general operational requirements document issued to contractors . . . in 1972." However, 1981 is usually cited as the beginning of the ATF program. The basic challenge of the ATF design was to pack stealth, supercruise, highly integrated avionics, and agility into an airplane with an operating range that bettered the F-15, the aircraft it was to replace.

Skow (1992) noted that "Whereas there is universal agreement that agility in air combat is valuable, many other aircraft attributes such as acceleration, speed, maneuverability, and payload/range performance have a value also. Each of these attributes has a value and a cost."

Mullin (1992) reported that "The configuration design evolution was primarily driven by the requirement to obtain excellent subsonic and supersonic aerodynamics and low observability. With the requirement for internal weapons carriage and all mission fuel internal (except for the ferry mission), the design challenge was formidable." All configurations were premised on the availability of the following technologies:

1. External aircraft geometry
   (a) Low observable
   (b) Low supersonic drag
   (c) Unrestricted maneuverability
2. Propulsion
   (a) Low observable supersonic inlet
   (b) Low observable augmenter/thrust vectoring nozzle
3. Avionics
   (a) Common modules
   (b) Liquid cooling for reliability
   (c) Low observable apertures (radar, infrared, communications/navigation, electronics warfare)
4. Other subsystems and equipment
   (a) Low observable air data systems
   (b) Low observable canopy

Unlike stealth and supercruise, high maneuverability is more often used as a defensive tactic rather than an offensive one. As reported by Hehs (1998), "Maneuverability was quantified in the 1960s by John R. Boyd in his energy-maneuverability theory. Boyd's ideas were used on the F-15, but the F-16 was the first airplane to be designed specifically to emphasize the principles established by his theory. The most common measures of merit for energy maneuverability are sustained g capability (the ability to turn hard without losing airspeed and altitude); instantaneous g (the ability to turn the nose without regard to the effect on speed); and specific excess power (a measure of an aircraft's potential to climb, accelerate, or turn at any flight condition). Another parameter of interest is transonic acceleration time (for example, the time needed to go from Mach 0.8 to Mach 1.2). Comparing these characteristics for two fighters shows which one should have the tactical advantage in a maneuvering engagement."

Skow (1992) noted that "in an aggressive engagement against a highly maneuverable adversary, maximum load factor or maximum lift maneuvering may be required for survivability. The high values of turn rate that are achieved at these maximum conditions come at the expense of energy. Since maximum rate maneuvering bleeds energy rapidly, it can only be continued for short durations. When the pilot decides to terminate a maximum rate maneuver that has caused his airspeed to be bled to a low value, he needs to accelerate quickly. ... Two factors influence the ability of the aircraft to accelerate rapidly: thrust minus drag and gravity. ... For a high thrust/weight fighter, the rate at which thrust increases after a movement of the throttle can have an important effect on the energy addition achieved during a short acceleration. However, the rate at which drag is reduced can have a substantially greater effect."

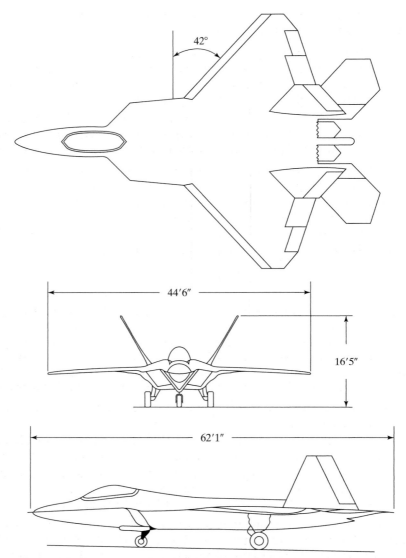

Figure 13.51 External configuration of the F-22. [As taken from Mullin (1992).]

Hehs (1998) wrote that Dick Hardy (the ATF Program Director for Boeing) said, "One problem we typically face when trying to stuff everything inside an airplane is that everything wants to be at the center of gravity. ... The weapons want to be at the center of gravity so that when they drop, the airplane doesn't change its stability modes.... The fuel volume wants to be at the center of gravity, so the center of gravity doesn't shift as the fuel tanks empty. Having the center of gravity move as fuel burns reduces stability and control."

Mullin (1992) wrote, "Although they are really inseparable, we faced two major areas of issues during the ATF Dem/Val program:

1. System engineering issues, based on performance requirements versus flyaway cost and weight. Many initial 'requirements' which seemed to be no problem turned out to be major drivers of weight and/or cost, and, as a result, were substantially changed by the Air Force, based on our trade study data.

2. Design engineering issues, based on conflicting requirements, such as supersonic drag versus internal fuel and weapons load, supersonic and subsonic maneuverability, installed propulsion performance versus low observability, and many others."

Lockheed, Boeing, and General Dynamics teamed to develop a demonstration/-validation (dem/val) design phase for the ATF. From 1986 to 1989 the team accumulated 20,000 hours in the wind tunnel. The resultant external configuration of the F-22 is shown Fig. 13.51. Note that the modified diamond wing has a reference area of 840 square feet with a wing span of 44.6 feet, compatible with existing aircraft shelters. The aspect ratio is 2.37. The F-22 has constant chord full leading-edge flaps, ailerons, and flaperons.

13.9 THE DESIGN OF THE F-35

The F-35 design is the product of the Joint Strike Fighter (JSF) program and represents one of the most ambitious military aircraft designs pursued in many years. The program is also the largest aerospace defense program in history. The resulting aircraft (now named the Lightning II) has to accomplish diverse missions, while at the same time being affordable. In fact, in spite of numerous difficult design requirements, the primary goal of the program is affordability. The F-35 was designed to fulfill the following roles:

- A strike fighter to complement the F/A-18E/F for the U.S. Navy (designated CV for Carrier Variant).
- A multirole aircraft to replace the F-16 and A-10 and to complement the F-22 for the U.S. Air Force (designated CTOL for Conventional Take-Off and Landing).
- A strike fighter with short-takeoff and vertical landing capability to replace the AV-8B and F/A-18 for the U.S. Marines and the Harriers of the U.K. Royal Navy (designated STOVL for Short Take-Off and Vertical Landing).

Designing a single airframe that can accomplish all of these missions while also being affordable is a necessarily difficult task. The approach taken by Lockheed Martin is to manufacture a single airframe with differing systems for the CTOL, STOVL, and CV versions (the CV version also has a larger wing). This has proven to be a cost-effective approach.

In addition to the multirole and low-cost nature of the aircraft, there were numerous other technical requirements that have to be fulfilled by the F-35, including [Hehs (1998a, 1998b)]:

- Overall performance similar to the F-16 and F-18
- Low observable
- Electronic countermeasures
- Advanced avionics, including adverse-weather precision targeting
- Increased range with internal fuel and weapons storage
- State-of-the-art "health" management for maintainability

The design of the F-35 involved significant use of computational fluid dynamics (CFD) for developing the overall configuration and for the addition of unique features to the aircraft. The CFD analysis was used to design and improve the outer shape of the fuselage to maximize internal fuel-storage volume, design the vertical-tail cant and rotation, and incorporate the diverterless supersonic inlet (Wooden, 2006). Of special interest, the diverterless supersonic inlet [Hehs (2000)] replaces the boundary-layer diverter present for most inlets and also provides the compression necessary for operating the engine at supersonic speeds (see Figure 13.52). This feature reduces drag and also is favorable for low observability.

Extensive wind-tunnel testing of all variants of the F-35 has been performed at a number of wind tunnels in both the U.S. and U.K, especially the U.S. Air Force wind tunnels at the Arnold Engineering Development Center (AEDC). The AEDC tests were performed using an integrated test and evaluation approach that takes advantage of the rapid growth in CFD capabilities over the past twenty years. "As computational power has increased, so has the capability of computational analysis to contribute to aircraft design." [See Skelley et al. (2007).] Specifically, a typical full-aircraft CFD simulation from 1988 could be represented by simulating the F-15E with an Euler equation calculation on a grid with approximately one million grid points and requiring a turn-around time of four weeks (grid generation, solution, and post-processing time). Skelley stated in 2007 that their current CFD capability has given them a nearly four order of magnitude improvement over the 1988 capability. They could now simulate the JSF with the Navier-Stokes equations on a grid with 25 million grid points and a turnaround time of two weeks. This rapid increase in CFD capability led to full integration of the ground testing and computational analysis throughout the design phase of the F-35, including the analysis of inlet performance, carriage loads of stores, store separation, and free-flight data.

The STOVL variant of the F-35 (shown in Fig. 13.53) uses a unique lift system during take-offs and landings. The Harrier aircraft uses direct lift by rotating the jet exhaust downward and creating vertical thrust, while the F-35 uses a combination of direct lift

Figure 13.52 The JSF diverterless supersonic inlet. (Photo courtesy of Lockheed Martin.)

Figure 13.53 The Lockheed Martin X-35B approaching landing in STOVL configuration. [From Buccholz (2002); photo courtesy of Lockheed Martin.]

and a unique lift fan system. The Pratt & Whitney 119-611 engine is used to create three very different lift and control devices: (a) the core nozzle with a swivel duct at the end of the aircraft provides direct lift, (b) a drive shaft and gearbox from the engine rotates a lift fan located directly behind the cockpit, and (c) bleed air from the engine is used for roll control with roll posts and ducts on the wing of the aircraft [Hehs (2001)].

Specifically, the lift system operates in the following way: "The conversion from CTOL to STOVL begins when the pilot pulls back on the thrust vector lever. Doors open above and below the lift fan. Doors behind the lift fan intake open for an auxiliary engine intake. The engine nozzle twists and vectors downward. The clutch engages and transfers energy from the spinning shaft via the gearbox to the lift fan. Control valves open to divert bypass air from the engine to the roll posts. All of these various changes occur in seconds in precise, computer-controlled order." [See Hehs (2001).] While this approach may seem unnecessarily complex, the goal of this lifting approach is multi-fold, including the avoidance of ingesting hot gasses into the engine intake (which can cause the engine damage or failure), avoiding hot-exhaust damage to runway surfaces, and aiding the aircraft in accomplishing the difficult task of landing vertically. Hehs (2001) notes: "The lift fan approach was chosen for its many attributes. It extracts power from the engine, thus reducing exhaust temperatures from the engine by about 200 degrees compared to exhaust temperatures of direct-lift systems. It significantly reduces exhaust velocity as well. Engine exhaust air combines with low-temperature and low-velocity air from the lift fan to produce a more benign ground environment. Cool exhaust air from the fan prevents hotter [engine] exhaust from being reingested into the intakes. Hot gas reingestion, a common problem on legacy Harrier-type approaches, causes compressor stalls, and other severe engine performance degradations. Most importantly, the lift fan system was chosen because it does not detract from the up-and-away performance of the JSF 119-611 engine." This approach was verified with significant wind-tunnel testing and flight testing, including defining the transition capabilities of the F-35 as it changes from horizontal to vertical flight [Buchholz (2002)].

Lockheed Martin, Northrop Grumman, and BAE Systems teamed to develop the F-35 during the concept development phase (CDP) from 1997 through 2001 and the system development and demonstration (SDD) period starting in 2001. Wind-tunnel testing during CDP included 3500 hours at AEDC, as well as significant wind-tunnel testing of the STOVL variant at BAE Systems, NASA Ames, NASA Langley, and DNW, the German-Dutch wind tunnels [Buchholz (2002)]. The first flight of the F-35 took place in 2006, and flight testing is ongoing, with the first delivery scheduled for 2009. Approximately 23 aircraft will be built and tested during SDD [Hehs (2007)] as the flight envelope is expanded, including testing of the STOVL capabilities of the aircraft.

PROBLEMS

13.1. Discuss the variation of C_L as a low-speed aircraft consumes fuel during a constant altitude, constant airspeed cruise. What is the variation of α?

13.2. Based on the results of Problem 5.3, discuss ways to increase $(L/D)_{max}$.

13.3. Discuss the desirability of adding winglets to
   (a) a transport aircraft that operates for long periods in cruise flight
   (b) a stunt or aerobatic-type airplane

13.4. You are a "spy" who has taken a position near one of the runways of an important military air base of your opponent. You plan to observe the enemy aircraft as they fly over and use the projected view of the aircraft and the trailing vortices to determine the capabilities of the aircraft.

   Consider the planform views of the airplanes presented in Figures P13.4(a) through P13.4(f). For each of the six planforms, predict the corresponding value of the "design" Mach number. Also, what is the aspect ratio of the wing and the approximate value of $(L/D)_{max}$? You get extra credit if you can identify the aircraft by its name and the company that originally built it.

M_∞ _____

AR _____
$(L/D)_{max}$ _____

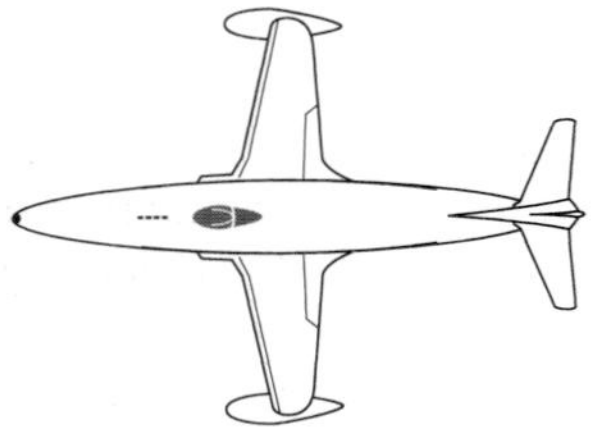

Figure P13.4(a)

M_∞ _____

AR _____
$(L/D)_{max}$ _____

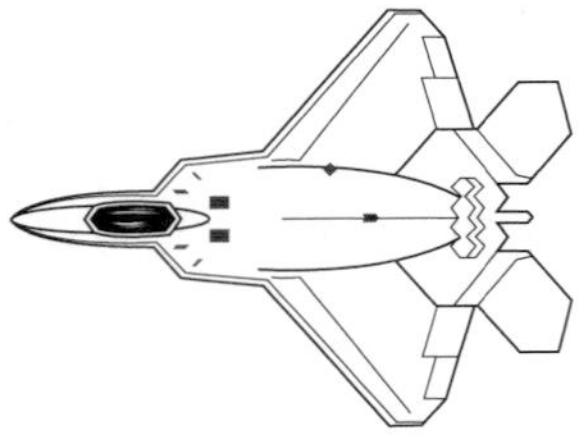

Figure P13.4(b)

M_∞ _____

AR _____

$(L/D)_{max}$ _____

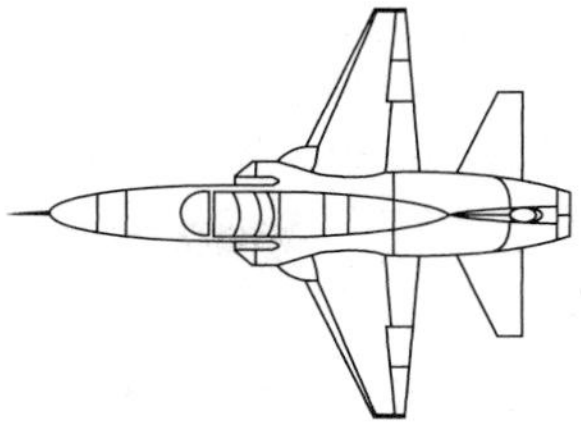

Figure P13.4(c)

M_∞ _____

AR _____

$(L/D)_{max}$ _____

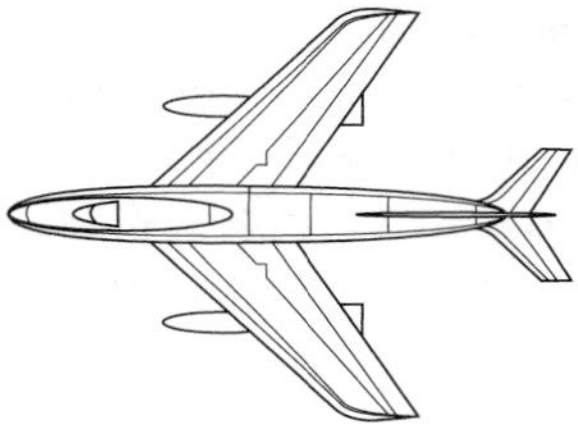

Figure P13.4(d)

M_∞ _____

AR _____

$(L/D)_{max}$ _____

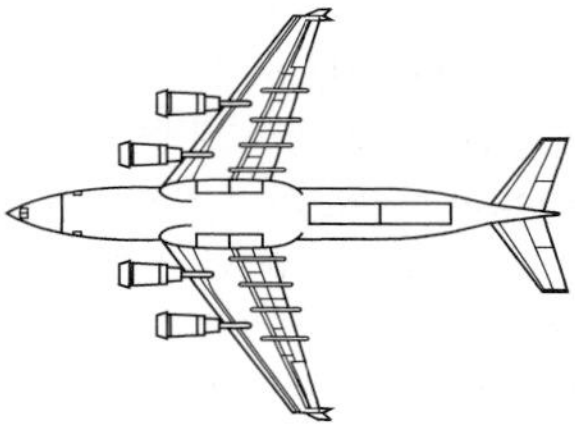

Figure P13.4(e)

M_∞ _____

AR _____

$(L/D)_{max}$ _____

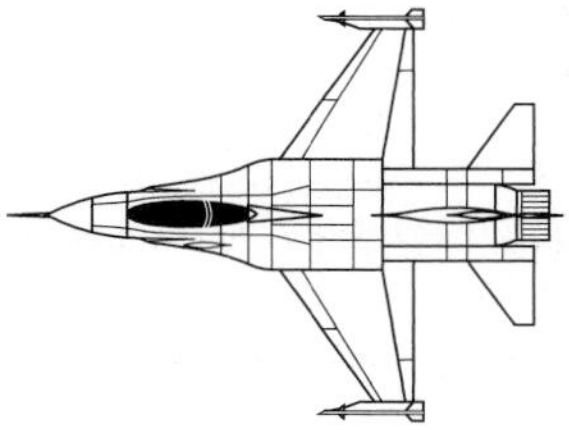

Figure P13.4(f)

13.5. A T-41 (Cessna 172) is cruising at an airspeed of 125 knots at an altitude of 10,000 *ft*. Let us treat the airfoil cross-section as a flat-plat at zero angle-of-attack. The wing planform is essentially unswept and rectangular with a span of 35 *ft* and a chord of 7 *ft*.

In order to obtain "very approximate" estimates of the location of boundary-layer transition, we need a transition criteria and a method for calculating the values of the required parameters. We will assume that the local Reynolds number is obtained by using the free-stream values in place of the parameters at the edge of the boundary layer (which is true for an x-coordinate):

$$\text{Re}_x = \frac{\rho_\infty U_\infty x}{\mu_\infty}$$

If the boundary-layer transition Reynolds number is 500,000,

$$x_{tr} = \frac{\text{Re}_{x_{tr}} \mu_\infty}{\rho_\infty U_\infty} = \frac{500{,}000 \, \mu_\infty}{\rho_\infty U_\infty}$$

Does the boundary layer for the T-41 remain laminar for the entire airfoil section. If it does not remain laminar, where does boundary-layer transition occur?

REFERENCES

Air International. 1993. High-lift system aerodynamics. *AGARD CP-515*

Air International. 1974. Two decades of the 'Twenty-One'. *Air Enthusiast Internatonal* 6(5):226–232

Air International. 1975. The annals of the polymorph, a short history of V-G. *Air International* 8(3):134–140

Blackwell JA. 1969. A finite-step method for calculation of theoretical load distributions for arbitrary lifting surface arrangements at subsonic speeds. *NASA Tech. Note D-5335*

Bracken ER. 2000. Private Communication

Bradley RG. 1981. *Practical aerodynamic problems—military aircraft.* Presented at Transonic Perspective, NASA Ames Research Center, CA

Buchholz MD. 2002. *Highlights of the JSF X-35 STOVL jet effects test effort.* Presented at Biennial International Powered Lift Conf. AIAA Pap. 2002–5962, Williamsburg, VA

Buckner JK, Hill PW, Benepe D. 1974. *Aerodynamic design evolution of the YF-16.* Presented at Aircraft Design, Flight Test, and Operations Meet., 6th, AIAA Pap. 74–935, Los Angeles, CA

Buckner JK, Webb JB. 1974. *Selected results from the YF-16 wind-tunnel test program.* Presented at Aerodynamic Testing Conference, 8th, AIAA Pap. 74–619, Bethesda, MD

Cleveland FA. 1970. Size effects in conventional aircraft design. *J. Aircraft* 7:483–512

Cowley WL, Levy H. 1918. *Aeronautics in Theory and Experiment.* London: Edward Arnold, Publisher

Englar RJ. 1987. Circulation control technology for powered-lift STOL aircraft. *Lockheed Horizons* 24:44–55

Ferris JC. 1977. Wind-tunnel investigation of a variable camber and twist wing. *NASA Tech. Note D-8457*

Gato W, Masiello MF. 1987. *Innovative aerodynamics: the sensible way of restoring the growth capability of the EA-6B Prowler.* Presented at Appl. Aerodynamics Conf., 5th, AIAA Pap. 87–2362, Monterey, CA

Goodmanson LT, Gratzer LB. 1973. Recent advances in aerodynamics for transport aircraft. *Aeronaut. Astronaut.* 11(12):30–45

Green W. 1970. *The Warplanes of the Third Reich.* Garden City, NJ: Doubleday & Company

Hanley RJ. 1987. Development of an airplane modification to improve the mission effectiveness of the EA-6B airplane. Presented at Appl. Aerodynamics Conf., 5th, AIAA Pap. 87–2358, Monterey, CA

Hehs E. 1998a. F-22 design evolution. *Code One Magazine* 13(2)

Hehs E. 1998b. F-22 design evolution, part II. *Code One Magazine* 13(4)

Hehs E. 2000. JSF diverterless supersonic inlet. *Code One Magazine* 15(3)

Hehs E. 2001. Going vertical—X-35B flight testing. *Code One Magazine* 16(3)

Hehs E. 2007. F-35 Lightning II flight tests. *Code One Magazine* 22(2)

Hillaker HJ. 1982. A supersonic cruise fighter evolution. Paper 14

Hillaker HJ. 1997. Private Communication

Jobe CE, Kulfan RM, Vachal JC. 1978. *Application of laminar flow control to large subsonic military transport airplanes*. Presented at Aerosp. Sci. Meet., 16th, AIAA Pap. 78–0095, Huntsville, AL

Jordan FL, Hahne DE, Masiello MF, Gato W. 1987. *High-angle-of-attack stability and control improvements for the EA-6B Prowler*. Presented at Appl. Aerodynamcis Conf., 5th, AIAA Pap. 87–2361, Monterey, CA

Kosin RE. 1965. Laminar flow control by suction as applied to the X-21A aircraft. *J. Aircraft.* 2:384–390

McLean D. 2005. Wingtip devices: what they do and how they do it. Presented at Boeing Performance and Flight Operations Engineering Conf., Article 4, St. Louis, MO

McMasters JH, Kroo IM. 1998. Advanced configurations for very large transport airplanes. *Aircraft Design* 1:217–242

Meyer RC, Fields WD. 1978. *Configuration development of a supersonic cruise strike-fighter*. Presented at Aerosp. Sci. Meet., 16th, AIAA Pap. 78–0148, Huntsville, AL

Montulli LT. 1986. *Lessons learned from the B-52 program evolution: past, present, and future*. Presented at Aircraft Systems, Design, and Tech. Meet., AIAA Pap. 86–2639, Dayton, OH

Morgan HL. 1975. A computer program for the analysis of multielement airfoils in two-dimensional subsonic, viscous flow. *NASA Spec. Pub. 347*

Mullin SN. 1992. *The evolution of the F-22 Advanced Tactical Fighter*. Presented at Aircraft Design Systems Meet., AIAA Pap. 92–4188, Hilton Head Island, SC

Nicholas WU, Naville GL, Hoffschwelle JE, Huffman JK, Covell PF. 1984. *An evaluation of wing-canard, tail-canard, and tailless arrangements for advanced fighter applications*. Presented at Congress of the Intern. Council of the Aeronaut. Sci., ICAS Pap. 84–2.7.3, Tolouse, France

Nipper MJ. 1996. Private Communication

Olason ML, Norton DA. 1966. Aerodynamic design philosophy of the Boeing 737. *J. Aircraft* 3:524–528

Olsen LE, Dvorak FA. 1976. *Viscous /potential flow about multi-element two-dimensional and infinite-span swept wings: theory and experiment*. Presented at the Aerosp. Sci. Meet., 14th, AIAA Pap. 76–18, Washington, DC

Peters CE, Roos HN, Graber GL. 2001. Private Communication

Schlichting H. 1960. Some developments in boundary layer research in the past thirty years. *J. Roy. Aeronaut. Soc.* 64(590):4–79

Schlichting H, Truckenbrodt E. 1969. *Aerodynamik des Flugzeuges*. Berlin: Springer Verlag

Sewall WG, McGhee RJ, Ferris JC. 1987. *Wind-tunnel test results of airfoil modifications for the EA-6B*. Presented at the Appl. Aerodynamics Conf., 5th, AIAA Pap. 87–2359, Monterey, CA

Skelley ML, Langham TF, Peters WL, Frantz BG. 2007. *Lessons learned during the Joint Strike Fighter ground testing and evaluation at AEDC*. Presented at USAF T&E Days. AIAA Pap. 2007–1635, Destin, FL

Skow AM. 1992. Agility as a contributor to design balance. *J. Aircraft* 29:34–46

Smith AMO. 1975. High-lift aerodynamics. *J. Aircraft* 12:501–530

Smith SC, Kroo IM. 1993. Computation of induced drag for elliptical and crescent-shaped wings. *J. Aircraft* 30: 446–452

Stevens WA, Goradia SH, Braden JA. 1971. Mathematical model for two-dimensional multi-component airfoils in viscous flow. *NASA Cont. Rep. 1843*

Thomas ASW. 1985. Aircraft drag reduction technology. *AGARD Report 723* (Quotations from this reference are reproduced by permission of AGARD and ASW Thomas, Lockheed Aeronautical Systems Company, Georgia Division)

van Dam CP. 1987. Induced-drag characteristics of crescent-moon-shaped wings. *J. Aircraft* 24: 115–119

van Dam CP, Vijgen PMHW, Holmes BJ. 1991. Experimental investigation on the effect of crescent planform on lift and drag. *J. Aircraft* 28: 713–730

Waggoner EG, Allison DO. 1987. *EA-6B high-lift wing modifications*. Presented at the Appl. Aerodynamics Conf., 5th, AIAA Pap. 87–2360, Monterey, CA

Weeks TM. 1986. *Advanced technology integration for tomorrow's fighter aircraft*. Presented at the Aircraft Systems, Design, and Tech. Meet., AIAA Pap. 86–2613, Dayton, OH

Wentz WH, Seetharam FC. 1974. Development of a Fowler flap system for a high performance general aviation airfoil. *NASA Cont. Rep. 2443*

Whitcomb RT. 1976. A design approach and selected wind tunnel results at high subsonic speeds for wing-tip mounted winglets. NASA TN-D-8260

Whitford R. 1991. Four decades of transonic fighter design. *J. Aircraft* 28:805–811

Whitford R. 1996. Fundamentals of fighter design, part I – requirements. *Air International* 52(1):37–43

Wooden PA, Azevedo JJ. 2006. *Use of CFD in developing the JSF F-35 outer mold lines*. Presented at Appl. Aerodyn. Conf., 24th, AIAA Pap. 2006–3663, San Francisco, CA

14 TOOLS FOR DEFINING THE AERODYNAMIC ENVIRONMENT

Aircraft designers have a wide variety of tools available to them. We will divide the tools into two categories:

1. Analytical tools, which include exact analytical solutions, empirical-based conceptual design codes, and computational fluid dynamics (CFD) codes
2. Experimental programs, which employ either ground-based test facilities (e.g., wind tunnels) or flight tests

Most aircraft designers use both analytical tools and experimental programs to define the configurations that could meet their mission requirements (range, payload, costs, maintainability, etc.). The division between the usage of analytical tools and of experimental programs depends on the organization's history of design practices (i.e., experience), on the facilities (computational and experimental) available to the designers, and on the personnel resources of the organization.

Kafyeke and Mavriplis (1997) noted: "The approach to aircraft design has traditionally been based on wind tunnel testing with flight testing being used for final validation. CFD emerged in the late 1960s. Its role in aircraft design increased steadily as speed and memory of computers increased. Today CFD is a principal aerodynamic technology along with wind tunnel testing and flight testing. State-of-the-art capabilities in each of these technologies are needed to achieve superior performance with reduced risk and low cost." This last sentence recommending that state-of-the-art capabilities are needed in each of these three areas to achieve superior performance with reduced risk and low cost is a key to the designers' success.

The assumptions that one can make for a steady, low-speed flow of a viscous fluid in an infinitely long, two-dimensional channel of height h or in an infinitely long circular pipe allow one to obtain exact solutions to the simplified governing equations. Examples of these flows, which are called Poiseuille flow, Couette flow, etc., appear in the homework problems of Chapter 2. Such applications provide the student with important practice in developing the governing equations, in specifying the boundary conditions, and in solving viscous flow problems. However, they have limited practical use for aircraft design applications.

Rapid advances in computer hardware and in computer software have enabled the designer to use relatively sophisticated CFD codes to generate flow-field solutions for realistic aerodynamic configurations. However, CFD is not purely theoretical analysis. CFD changes the fundamental nature of the analysis from calculus to arithmetic and from the continuum domain to the discrete domain so that the problem of interest can be solved using a computer. The existing mathematical theory for numerical solutions of nonlinear partial differential equations is continually being improved. The computational fluid dynamicist often relies on the mathematical analysis of simpler, linearized formulations, and on heuristic reasoning, physical intuition, data from experimental programs, and trial-and-error procedures. Furthermore, the numerical algorithms that are chosen depend heavily on the dominant physics of the specific application.

Thus, even the most rigorous CFD codes employ models and approximations both for physical properties and for fluid dynamic processes. The approximations range from the model used to generate numerical values for the absolute viscosity as a function of pressure and of temperature to the numerical algorithm used to model the turbulent boundary layer. It is important to keep in mind that turbulence models are not universal. One turbulence model may provide reasonable values for the engineering parameters for a particular class of flows. See Chapter 4 for a brief discussion of turbulence models.

As noted by Shang (1995): "As an analog, the ground testing facilities can easily measure the global aerodynamic force and moment exerted by the flow field on the tested model. Once a scaled model is installed in the test section, the data-generating process is the most efficient among all simulation techniques. Therefore, for a large class of design problems, the ground testing method is preferred over others. However, like CFD, experimental simulation does not necessarily reproduce accurate results in flight. The inherent limitation is derived from the principle of dynamic similarity—the scaling rule. Even if a perfect match to flight conditions is reached in dimensionless similarity parameters of Mach, Reynolds, and Eckert numbers, the small-scale model still may not describe the fine-scale surface features. If the flow fields under study are strongly influenced by fine-scale turbulence and laminar-turbulent transition, the accuracy of simulations to flow physics is uncertain."

Although testing in ground-based facilities can deliver prodigious amounts of quality data, the ability to extrapolate the information obtained to the flight environment may be affected by shortcomings in the simulation of the flow. Data obtained in ground-based test facilities may be affected by model scale. As a result, one fails to match critical simulation parameters (e.g., the Reynolds number). Only a full-scale vehicle flying at the desired velocity, the desired altitude, and the desired attitude (i.e., its orientation with respect to the free-stream velocity vector) provides information about the aerodynamic parameters in the true environment. However, flight-test experiments present considerable challenges to define the free-stream conditions with suitable accuracy (especially for high-altitude flights at hypersonic speeds). Furthermore, because of their size, their weight, and their fragile nature, many of the types of instrumentation available to the wind-tunnel-test engineer are not available to the flight-test engineer.

14.1 CFD TOOLS

There are a wide variety of analytical/computational tools that incorporate flow models of varying degrees of rigor available to the aircraft designer. Many readers of this text will participate in the development of their organization's CFD codes. A large fraction of the readers will have had no role in the development of the CFD codes that they use. Often multiple organizations participate for several years in the development of a single code. Whether you are developing your own code or using a code that was developed elsewhere, it is important that you understand (1) the grid scheme that is used to represent the body and the grid scheme that is used in the solution of the flow field; (2) the numerical algorithms used to obtain the flow-field solution; and (3) the models used to represent fluid mechanic phenomena, thermochemical phenomena, and flow properties. In this section, we will discuss these analytical/computational tools in approximately the increasing order of sophistication.

14.1.1 Semiempirical Methods

There is a large number of codes that employ semiempirical, data-based methods that are used for preliminary design purposes. An example of a semiempirical code, which uses a combination of theoretical methods with nonlinear corrections for the body and an extensive experimental database for wing and tail fin loads, is the MISL3 code, developed by Lesieutre et al. (1989). As Mendenhall et al. (1990) discuss, "The data base inherently includes viscous and compressibility effects as well as fin-body gap effects.... Mutual interference between control surfaces is also considered in the data base." The MISL3 code is limited to applications of Mach 5 or less.

There is a variety of conceptual design codes for predicting the aerothermodynamic environment for configurations in hypersonic flows (i.e., Mach numbers of approximately 5 or greater). One such code is the Supersonic/Hypersonic Arbitrary Body Program (S/HABP) [Gentry et al. (1973)]. In the first step, the vehicle geometry is divided into flat elemental panels. Frequently used techniques for computing the inviscid pressure acting on a panel employ one of the many simple techniques based on impact or on expansion methods. Once the pressure distribution has been determined, boundary-layer correlations are used to estimate the skin friction and the heat transfer acting at the vehicle surface. The local pressures and the local skin friction forces can be integrated to compute estimates of the resultant forces and moments acting on the vehicle.

14.1.2   Surface Panel Methods for Inviscid Flows

As the level of complexity of the flow increases, more detailed information is needed. Therefore, numerical codes using panel surface methods, or surface singularity methods, have been under development since the 1960s. These panel surface methods use the linearized/potential flow equations. Panel surface method codes for subsonic flows are discussed in Chapter 7 and for supersonic flows are discussed in Chapter 11.

For subsonic flows, vortex lattice methods (VLMs) combine the basic building blocks of the constant source panel and the vortex lattice methods, representing thickness effects with source panels and lift effects with vortex panels. Every panel has some form of source singularity imposed upon it, whether it is a lifting or nonlifting component. Lifting components have vortex singularities imposed upon them or interior to them. There is an infinite number of combinations of vortex singularity variations that will provide lift and satisfy the boundary conditions. Therefore, a numerical technique must be developed to generate a unique solution. This is done by prescribing an assumed chordwise variation and spanwise variation of vortex strength, then solving both for the source strengths and for the vortex strengths, subject to the Kutta condition. The particular variation of the vortex singularity chosen impacts the resulting magnitudes of the source strengths. An improper choice can lead to large source gradients and inaccuracies in the solution.

For each of the control points in the lattice, the velocities induced at that control point by each of the panels of the configuration are summed, resulting in a set of linear algebraic equations which express the exact boundary condition of flow tangency on the surface. The local velocities are then used to compute the static pressure acting on each of the panels. These pressures are then integrated to obtain the forces and the moments.

The PAN AIR code was developed to model both subsonic and supersonic linear potential flows about arbitrary configurations [Towne et al. (1983)]. The Prandtl-Glauert equations [the reader should refer to equation (9.13)] represent inviscid, compressible, subsonic/supersonic, irrotational, isentropic, small perturbation flows. PAN AIR uses a linear source distribution and a quadratic doublet distribution on each panel to reduce the number of panels needed to attain a given accuracy. The flexibility of the PAN AIR code provides the user with a great deal of freedom and capability in modeling linear potential flow problems.

Everson et al. (1987) noted that "panel methods have long been able to handle complex configurations and boundary conditions, but they are limited to linear flows." Thus, TRANAIR was developed to solve the full potential equation about arbitrary configurations. The key to the ability to handle complex configurations is the use of a rectangular grid rather than a surface fitted grid. Therefore, the grid is defined by a hybrid method employing both *surface panels* and a grid containing points away from the surface where solutions to the flow field parameters are generated.

Since the VLM, the PAN AIR code, and the TRANAIR code all are inviscid codes, they do not model the boundary layer. Thus, these codes do not provide estimates for the skin friction component of the drag coefficient. Furthermore, they do not model the shock-wave/boundary-layer interactions that produce the separation-induced-drag component of the drag coefficient at transonic speeds. However, these codes provide reasonable estimates of the wave drag for supersonic flows. The lift coefficient and the pitching moment coefficient generated using these codes are best for low angles of attack (i.e., cruise applications). Furthermore, because only a surface discretization scheme is required, very complex configurations can be modeled in which the panel density is adjustable to the desired accuracy. Therefore, since the computational intensity for panel methods is very low, the cost of computing the lift and the pitching moments is relatively inexpensive for those conditions where these methods are applicable.

14.1.3   Euler Codes for Inviscid Flow Fields

Neglecting the viscous terms in the Navier-Stokes equations (see Appendix A) provides the analyst with the Euler equations to model the flow field. Since the Euler equations neglect all viscous terms, the solutions cannot be used to compute either the shear forces or the heat transfer to the surface of the vehicle. However, they can provide solutions for the unsteady, inviscid flow field over the configuration in either subsonic, transonic, or supersonic streams. Thus, Euler codes can be used to compute engineering estimates of the lift and of the pitching moments. In comparison with panel methods, one of the difficulties of numerically solving the Euler equations involves generating a body-fitted, discrete grid about the configuration geometry (i.e., a grid that is curved to follow the contours of the body). The job of creating a surface definition for a complicated three-dimensional geometry that is needed for panel methods is much easier than the generation of a body-fitted grid that is needed for an Euler-based code.

Rapid developments of grid-generation tools support the growing use of Euler codes. Thus, Euler codes are becoming available to even the smallest design groups due to the ever-improving capabilities of computational hardware and software. Thus, since the costs for use continue to decrease, more and more organizations are using Euler codes to generate the lift and the pitching moments for those applications where these aerodynamic coefficients were formerly computed using the surface panel methods described in the previous paragraphs.

14.1.4   Two-Layer Flow Models

If one wishes to estimate the shear forces and the heat transfer to the surface of the vehicle, one can use computations based on a two-layer flow model. As noted in the prologue for Chapter 4, for many high Reynolds number flows, the flow field may be divided into two regions: (1) a viscous boundary layer adjacent to the surface, and (2) the essentially inviscid flow outside of the boundary layer. Thus, in the two-layer flow model, the Euler technique or a panel method can be used to generate the inviscid portion of the flow field and, therefore, the boundary conditions at the outer edge of the boundary layer. The two velocity boundary conditions at the wall are defined by the no-slip condition ($u = 0$) and by the assumption that either there is no flow through the wall ($v = 0$) or a transpiration boundary condition. The temperature boundary condition at the wall that is required for high-speed flows is either that the wall temperature (T_W) is given or that there is no heat transfer to the wall [i.e., the adiabatic-wall assumption ($\partial T / \partial y = 0$)]. A solution for the boundary layer is computed subject to these boundary conditions and the transition/turbulence models. An iterative procedure is used to compute the flow field using the two-layer model. To start the second iteration, the inviscid flow field is recalculated, replacing the actual configuration by the *effective configuration*. The effective configuration is determined by adding the displacement thickness of the boundary layer as computed during the first iteration to the surface coordinate of the actual configuration. The boundary layer is recalculated using the second-iterate inviscid flow field as the boundary condition.

Two-layer flow models can be used for applications where the viscous boundary layer near the surface is thin and does not significantly alter the inviscid-region flow field. Thus, this procedure would not apply to flow fields for which there are shock-wave/boundary-layer interactions or significant regions of separated flow. Thus, the two-layer method would not be the proper tool to generate solutions for a transonic flow, where shock-wave/boundary-layer interactions cause the boundary layer to separate or for flows where the vehicle is at large angles of attack, so that there are extensive regions of separated flow.

Because the flow field can be computed in a relatively short time using the two-layer methods, they remain popular computational tools at many organizations.

14.1.5 Computational Techniques that Treat the Entire Flow Field in a Unified Fashion

Because of the limitations described in the previous paragraphs, the ultimate computational tool treats the entire flow field (both the viscous regions and the inviscid regions) in a unified fashion. Starting with the Navier-Stokes equations (see Appendix A), one can develop a code that treats the entire flow field in a unified fashion. Li (1989) stated, "The CFD code development and application may follow seven steps:

1. Select the physical processes to be considered.
2. Decide upon the mathematical and topographical models.
3. Build body geometry and space grid.
4. Develop a numerical solution method.
5. Incorporate the above into a computer code.
6. Calibrate and validate the code against benchmark data.
7. Predict aerodynamic coefficients, flow properties, and aeroheatings."

Every code that is used to compute the flow field requires a subprogram to define the geometric characteristics of the vehicle and a grid scheme to identify points or volumes within the flow field. There are a number of techniques that have been developed for generating the computational grids that are required in the finite-difference, the finite-volume, and the finite-element solutions of the equations for arbitrary regions.

Grids may be structured or unstructured. There are no a priori requirements on how grids are to be oriented. However, in some cases, the manner in which the flow-modeling information is formulated may influence the grid structure. For instance, since turbulence models are often formulated in terms of the distance normal to the surface, the grid schemes utilized for these turbulent boundary-layer models employ surface-oriented coordinates where one of the coordinate axes is locally perpendicular to the body surface.

The use of adaptive grid techniques is a grid-generation strategy that can be used both with structured and with unstructured grids. Adaptive grid techniques might be applied to portions of the flow field where accurate numerical solutions require the resolution of events within extremely small distances. Often such resolution requirements lead to the use of very fine grids and lengthy computations. Aftosmis and Baron (1989) state, "Adaptive grid embedding provides a promising alternative to more traditional clustering techniques. This method locally refines the computational mesh by sub-dividing existing computational cells based on information from developing solutions. By responding to the resolution demands of chemical relaxation, viscous transport, or other features, adaptation provides additional mesh refinement only where actually required by the developing solution." Grid-clustering strategies also provide a valuable tool, when shock-capturing formulations are employed.

A variety of assumptions and simplifications may be introduced into the computational models that are used to treat the flow field in a unified fashion. For instance as noted by Deiwert et al. (1988), when there is no flow reversal and when the inviscid portion of the flow field is supersonic in the streamwise direction, the Navier-Stokes equations can be simplified by neglecting the streamwise viscous terms. By neglecting the unsteady terms and the streamwise viscous derivative terms in the full Navier-Stokes equations, one obtains the *parabolized* Navier-Stokes (PNS) equations. These are reasonable assumptions for large Reynolds number flows over bodies which do not experience severe geometric variations in the streamwise direction. Thus, a practical limitation for the PNS method is that there is no streamwise separation. Crossflow separation is allowed. These PNS equations are used to compute the shock-layer flow field for certain high-Mach-number flows. If it is assumed that the viscous, streamwise derivatives are small compared with the viscous, normal, and circumferential derivatives, a tremendous reduction in computing time and in storage requirements is possible over that required for the time-dependent approaches. Since the equations are parabolic in the streamwise direction, a spatial marching-type numerical-solution technique can be used. A PNS code requires a starting solution to generate the initial conditions on a surface where the inviscid portion of the flow is supersonic. [Mendenhall et al. (1990)] used an Euler code to generate the starting conditions for the region of the flow field modeled by the PNS formulation.

14.1.6 Integrating the Diverse CFD Tools

The aerodynamic design of the Pegasus™ was based on proven techniques developed during the design of existing vehicles. No wind-tunnel tests were included in the program. However, readily available computational codes were used for all aerodynamic analyses. As shown in Fig. 14.1, all levels of codes, ranging in complexity from empirical data-based methods to three-dimensional Navier-Stokes codes, were used in the program to develop the aerodynamic design of the Pegasus™ [Mendenhall et al. (1990)]. A PNS code was used to predict fuselage pressure distributions; an axisymmetric Navier-Stokes code was used to explore the possibility of rocket-plume-induced separation near the tail control surfaces; and a three-dimensional code was used to compute the complete flow field for critical conditions to check details of the flow, which may have been missed by

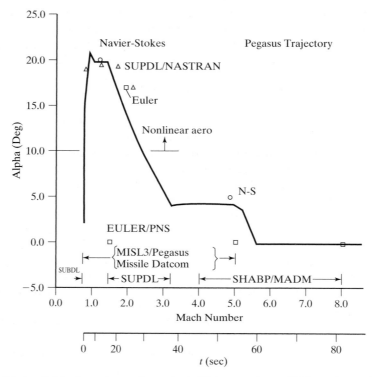

Figure 14.1 Aerodynamic analysis flight envelope; CFD tools used for α / M_∞ space for the Pegasus^{TM}. [Taken from Mendenhall et al. (1990).]

the simpler methods. The CFD results were used to compute the aerodynamic heating environment in the design of the thermal protection system.

14.2 ESTABLISHING THE CREDIBILITY OF CFD SIMULATIONS

Neumann (1988) notes, "The codes are NOT an end in themselves. . . . They represent engineering tools; tools that require engineering to use and critical appraisal to understand. Hypersonics must be dominated by an increased understanding of fluid mechanic reality and an appreciation between reality and the modeling of that reality. CFD represents the framework for that modeling study and experiments represent the technique for introducing physical reality into the modeling process. Finally, classical analytical theory and the trend information produced by theory gives us the direction with which to assemble the point data from these numerical solutions in an efficient and meaningful way."

Shang (1995) notes: "Poor numerical approximations to physical phenomena can result from solving over-simplified governing equations. Common mistakes have been made in using Euler equations to investigate viscous dominated flows, and employing the thin-layer approximation to Navier-Stokes equations for flowfield containing catastrophic separation. Under these circumstances, no meaningful quantification of errors for the numerical procedure can be achieved. The physically correct value and the implementation of initial and/or boundary conditions are another major source of error in numerical procedures in which the appropriate placement and type of boundary/initial conditions have a determining effect on numerical accuracy."

As suggested by Barber (1996), we will "focus on the issues of accuracy and achieving reduced variability (robustness), wherein a range of physical processes and a variety of desired outcomes can occur. Most organizations seek to achieve this goal by performing validation or certification studies. When the end-user is a research group, the metrics for evaluating the accuracy of the code (the desired outcomes) are typically fundamental flow variables (streamlines, velocity profiles, etc.). However, when the end-user is a member of a design team, the metrics are more performance oriented, e.g. lift and drag coefficients, system efficiency, etc. The central basis of the validation process is benchmarking, whereby a limited number of numerical predictions are made and compared to experimental data. Even though one successfully performs such calculations, the choice of cases frequently is not appropriate for minimizing the risk of faulty data for a given design process. Significant risks can be introduced by, for example, poorly defined design requirements. This can occur from a failure to choose the test cases (validation/calibration) from an operational or end-user perspective, but instead choosing them from a research perspective."

Bradley (1988) defined the concepts of *code validation* and *code calibration* as follows.

"CFD code validation implies detailed surface and flow field comparisons with experimental data to verify the code's ability to accurately model the critical physics of the flow. Validation can occur only when the accuracy and limitations of the experimental data are known and thoroughly understood and when the accuracy and limitations of the code's numerical algorithms, grid-density effects, and physical basis are equally known and understood over a range of specified parameters.

"CFD code calibration implies the comparison of CFD code results with experimental data for realistic geometries that are similar to the ones of design interest, made in order to provide a measure of the code's ability to predict specific parameters that are of importance to the design objectives without necessarily verifying that all the features of the flow are correctly modeled."

To the two definitions for *validation* and *calibration*, Barber (1996) and Rizzi and Vos (1996) add the terms *verification* and *certification*. Verification establishes the ability of a computer code to generate numerical solutions for the specific set of governing equations and boundary conditions. Through the verification process, the accuracy of the solution and the sensitivity of the results to parameters appearing in the numerical formulation are established through purely numerical experiments. These numerical experiments include both grid-refinement studies and comparison with solutions to problems that have exact analytical solutions.

Certification relates to programming issues (e.g., logic checks, programming style, documentation, and quality assurance issues). Thus, certification test cases are run before a new version of the code is released in order to be certain that no new errors have been introduced into the previously certified version.

Rizzi and Vos (1996) note that the code developers are responsible for building a credible code and verifying and certifying that code. Experts with a strong background in developing numerical models to represent physical processes carry out the tasks of validating and of calibrating the code. The code is then passed on to the users, who are neither experts in code development nor in numerical modeling.

Cosner (1995) suggests that "a range of test cases is analyzed which illustrates the various outcomes which can be expected across the design range of interest. Also, a set of standard practices usually are defined, to be used throughout the validation study. The skill level of the user should be representative of the engineering (user) environment. Therefore, it is preferable that this validation should be performed by representative engineers from the user community, not by the experts in the code or technology which is being tested."

Cosner (1995) notes that "Computational Fluid Dynamics technology, as a basis for design decisions, is rapidly gaining acceptance in the aerospace industry. The pace of acceptance is set by the advancing confidence of design team leaders that reliance on CFD can improve the quality of their end product, and reduce the schedule, costs and risks in developing that product." Barber (1996) concluded that "Risk reduction to an engineering design team is rooted in (sic) a determination to deliver *reliable* engineering data having a specified level of *accuracy*, in a specified *time*, and for a specified *cost*."

Bradley (1995) has said, "Engineers have always been able to use less than perfect tools coupled with experiences and calibration to known physical quantities to provide design guidance. Calibration and validation should not be confused. Calibration provides an error band or correction factor to enhance the ability of a particular code to predict specific parameters that are important to the design objectives for a particular design without verifying that all other features of the flow are modeled accurately. For example, one might calibrate a code's ability to predict shock location and lift and moment on a wing without any assurance that the flowfield off the surface and the wake behind the wing are properly modeled. Or one may calibrate a code's ability to compute the gross pressure loss through a supersonic inlet-duct combination without concern for the distortion distribution at the compressor face. Although the use of calibrated CFD solutions is dangerous because of the subtle viscous interactions that are extremely sensitive to geometry and flowfield, skilled engineers can often obtain useful design information and guidance from relatively immature codes."

14.3   GROUND-BASED TEST PROGRAMS

A large fraction of the relevant experimental information about flow fields of interest to the aerodynamicist is obtained in ground-based test facilities. Since complete simulations of the flow field are seldom obtained in a ground-based facility, the first and most important step in planning a ground-based test program is establishing the test objectives. As stated by Matthews et al. (1985), "A precisely defined test objective coupled with comprehensive pretest planning are essential for a successful test program."

There are many reasons for conducting ground-based test programs. The test objectives include the following.

1. Obtain data defining the aerodynamic forces and moments and/or heat-transfer distributions (especially for complete configurations whose complex flow fields resist computational modeling).

2. Use partial configurations to obtain data defining local flow phenomena, such as shock-wave/boundary-layer interactions, using a fin or a wing mounted on a plane surface.

3. Determine the effects of specific design features on the overall aerodynamic coefficients of the vehicle (e.g., the drag increment due to protuberances of the full-scale aircraft).

4. Certify airbreathing engines in ground-based test programs.

5. Obtain detailed flow-field data to be used in developing flow models for use in a computational algorithm (i.e., tests to obtain code-validation data).

6. Obtain measurements of parameters, such as the heat transfer or the total drag, to be used in comparison with computed flow field solutions over a range of configuration geometries and of flow conditions (i.e., tests to obtain code-calibration data).

As noted in Chapter 13, Lockheed, Boeing, and General Dynamics used 20,000 hours in the wind tunnel during the demonstration/validation design phase for the Advanced Technology Fighter. Thus, even for a relevantly recent design activity, considerable resources were spent on wind-tunnel programs for objective 1 in the preceding list.

Niewald and Parker (2000) noted that the drag increment due to the protuberances on the full-scale aircraft is significant and must be included during the database development phase. "An extensive effort was made to identify and estimate the drag of all protuberance items. The design process included continuous tracking of all outer moldline protuberances to facilitate trade studies. There were 94 types of protuberances, resulting in a total of 386 individual protuberance items. The items included external fasteners of various types, which numbered 88,340, and skin panel gaps totaling 2,180 ft in length. Outer moldline surveys of production F/A-18C aircraft were used to determine the percentage of gaps that would be forward or aft facing. The large model scale permitted testing of the most significant protuberance items (see Fig. 14.2). Seventy-one percent of the F/A-18E protuberance drag was based on wind-tunnel test data." When using Fig. 14.2, the reader should note that *one drag count* corresponds to a drag coefficient of 0.0001.

Niewald and Parker (2000) concluded, "Development of a credible preflight database for accurate aircraft predictions requires commitment and resources. The success of the F/A-18E/F wind-tunnel program was a direct result of both of these. The commitment was made at the outset to develop and implement test techniques that would properly account for each item impacting aircraft performance either by wind-tunnel testing or by estimation. Adequate resources allowed the development of high-fidelity models, use of large, interference-free wind-tunnels, comprehensive test programs, and integration of CFD methods to ensure first-time quality test results. Front loading the project resources to the wind-tunnel program were beneficial to the subsequent performance flight test program. The excellent agreement between wind-tunnel and flight

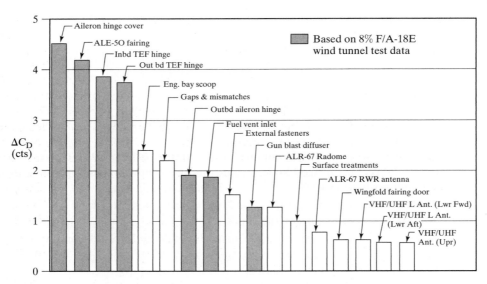

Figure 14.2 Protuberance drag for the F/A-18E. [Taken from Niewald and Parker (2000).]

results allowed the performance flight evaluation plan to be reduced by 60 flights and eliminated aircraft development flight testing for drag reduction as a result of optimistic predictions."

The planner of a ground-based test program should consider the following parameters:

1. the free-stream Mach number,
2. the free-stream Reynolds number (and its influence on the character of the boundary layer),
3. the free-stream velocity,
4. the pressure altitude,
5. the wall-to-total temperature ratio,
6. the total enthalpy of the flow,
7. the density ratio across the shock wave,
8. the test gas, and
9. the thermochemistry of the flow field.

Some of these nine parameters are important only for hypersonic flows (e.g., the density ratio across the shock wave, the test gas, and the thermochemistry of the flow field).

Note that some parameters are interrelated (e.g., the free-stream velocity, the free-stream Mach number, and the free-stream Reynolds number). The reader should review the discussion relating to equation (5.27) in Chapter 5. Most often, the boundary layer is turbulent over the majority of the aircraft in flight. However, wind-tunnel models are usually of relatively small scale. In such cases, the Reynolds number is relatively low for conventional wind tunnels and the boundary layer may be laminar or transitional. This may not be critical if the objective of the test is to generate lift and/or pitching moment data at low angles of attack. However, for determining the onset of stall, simulation of the Reynolds number is much more important.

Laster et al. (1998) note: "Experience has shown that lift and pitching moments are usually not too sensitive to Reynolds number up to the onset of buffet; but, buffet boundary, maximum lift, drag, and drag rise are usually very sensitive to Reynolds number. Therefore, the aircraft developer is faced with accounting for Reynolds number effects with these parameters as best he/she can. The usual practice is to use a combination of test techniques and empirical corrections. Because of little sensitivity of lift and pitching moment to Reynolds number below buffet onset, in most cases, the engineer has been able to directly use low Reynolds number wind tunnel measurements of lift and pitching moment in his/her design without having to resort to Reynolds number corrections. However, this is not necessarily true for wings with high aft loading. Test technique plays an important role in the determination of drag from wind tunnel data. Because the boundary layer is mostly turbulent in flight, experience has shown that forcing the model boundary layer to be turbulent in the wind tunnel makes the task easier in accounting for Reynolds number effects."

Laster et al. (1998) continue, "For accurate calculation of the forces (especially drag), the location of transition of the model boundary layer from laminar to turbulent must be accurately predicted, measured, or fixed. Otherwise, in the case of transonic flow, the shock/boundary-layer interaction cannot be properly modeled. For transition fixing, the prediction of the untripped transition location is important to assure that the boundary-layer trips are placed ahead of the location where transition occurs 'naturally' in the test facility. The transition location is dependent on the pressure gradient, surface roughness, turbulence and/or noise, and instabilities associated with the three-dimensionality of the flow. Therefore, the location of transition is both model- and flow-field dependent."

Complete simulations of the flow field are rarely obtained in any one ground-based facility. In a statement attributed to Potter, Trimmer et al. (1986) noted: "Aerodynamic modeling is the art of partial simulation." Thus, one must decide which parameters are critical to accomplishing the objectives of the test program. In fact, during the development of a particular vehicle, the designers will most likely utilize many different facilities with the run schedule, the model, the instrumentation, and the test conditions for each program tailored to answer specific questions.

14.4 FLIGHT-TEST PROGRAMS

Neumann (1986) suggests a variety of reasons for conducting flight tests. To the four reasons suggested by Draper et al. (1983), which are

1. To demonstrate interactive technologies and to identify unanticipated problems;
2. To form a catalyst (or a focus) for technology;
3. To gain knowledge not only from the flights but also from the process of development; and
4. To demonstrate technology in flight so that it is credible for larger-scale applications;

Neumann added three reasons of his own:

5. To verify ground-test data and/or to understand the bridge between ground-test simulations and actual flight;

6. To validate the overall performance of the system; and

7. To generate information not available on the ground.

Saltzman and Ayers (1982) wrote, "Although aircraft designers must depend heavily upon model data and theory, their confidence in each should occasionally be bolstered by a flight demonstration to evaluate whether ground-based tools can indeed simulate real-world aerodynamic phenomena. Over the years as the increments of improvement in performance have become smaller and aircraft development costs have risen, casual model-to-flight drag comparisons have sometimes given way to very comprehensive correlation efforts involving even more precise sensors, the careful control of variables, and great attention to detail on behalf of both tunnel experimenters and their flight counterparts."

Lift and drag characteristics for the F/A-18E have been determined throughout the flight envelope via in-flight thrust determination techniques, as reported by Niewald and Parker (1999). Three maneuver types were employed during the flight-test program to establish the aerodynamic database: steady-state maneuvers, quasi-steady-state maneuvers, and dynamic maneuvers. Steady-state maneuvers (e.g., cruise at constant Mach number and constant altitude) yield the most accurate data but require a large amount of flight time and airspace. It takes one minute of stabilization and three minutes of data-acquisition time to define one point on the drag polar using cruise. Quasi-steady maneuvers, such as constant-Mach-number climbs, provide more points for the drag polar definition. However, uncertainty is increased relative to steady-state maneuvers. Dynamic maneuvers, such as the pitch up/pitch down maneuvers, also known as the roller coaster, allow the estimation of aerodynamic characteristics covering a range of angle of attack that cannot be achieved through steady-state maneuvers or quasi-steady-state maneuvers. However, highly accurate instrumentation and data synchronization are required to reduce the potentially large data scatter that may result. As a result of the difficulties with dynamic maneuvers, they were not used in the F/A-18E flight-test program.

14.5 INTEGRATION OF EXPERIMENTAL AND COMPUTATIONAL TOOLS: THE AERODYNAMIC DESIGN PHILOSOPHY

The vehicle design integration process should integrate experimental data obtained from ground-based facilities with computed flow-field solutions and with data from flight tests to define the aerodynamic parameters and the heating environments (the combination of aerodynamic parameters and the heating environments is termed the *aerothermodynamic environments*) to which the vehicle is subjected during its mission.

At the time of this writing, flight tests are relatively expensive. One test point in a store separation flight-test program might cost $1M. Therefore, an extensive flight-test program of store separation and deployment might cost $50M. One Research and Development flight test to develop the technology base for a Mach 10 vehicle might cost $50M itself. Sophisticated wind-tunnel programs for a high-performance vehicle may cost $250K to $900K for the model and for the tunnel occupancy costs. To run a code is relatively cheap, say $50K to generate solutions for a number of flow conditions. And this is the cost once the code has been developed. Furthermore, as has been noted, it might take several years to develop the code to the point where it will deliver the desired computations. In addition, generating grids for the vehicle surface and for the flow field may take months. The grid generation process is a key element in generating suitable numerical solutions, when using Euler codes or Navier-Stokes codes. Thus, a very crude order of magnitude estimate for the costs might be as follows:

1. A flight-test program is two orders-of-magnitude more expensive than a wind-tunnel program, while

2. A wind-tunnel-test program is an order-of-magnitude more expensive than a CFD program.

The reader is cautioned that these numbers are very approximate.

The advances in numerical tools have led some designers of recent programs to state at the outset that their design would be developed without wind-tunnel tests and without flight tests. The aerodynamic design and analysis of the Pegasus™ vehicle were conducted without the benefit of wind-tunnel and subscale model testing using only computational aerodynamic and fluid dynamic methods. Mendenhall, et al. (1990) noted that "All levels of codes, ranging in complexity from empirical data-base methods to three-dimensional Navier-Stokes codes, were used in the design. . . . Most of the aerodynamic design is based on proven, existing vehicles; therefore, no wind tunnel tests were included in the program, and readily available computational codes were used for all aerodynamic analyses." For many other programs where the designers stated at the outset that the design would be accomplished using only CFD tools, the concepts never reached the hardware stage.

Mullin (1992) noted, "The aeronautical technology changes over 87-years have been enormous; almost beyond description, but the most important factor in successful aeronautical engineering has not changed: making technical decisions based on analysis, test data, and good engineering judgement. Then and now, the paramount obligation of aeronautical engineers is to make critical technical decisions. The increasingly dominant role of digital computers in aircraft design and analysis has tended to confuse some of our engineering colleagues, but the truth of the matter is inescapable."

All of the tools available to the aircraft designer, whether analytical/computational or experimental, require that he or she bring judgment born of experience to the application of these tools in the design process. *Good judgment comes from experience; experience comes from bad judgment.*

REFERENCES

Aftosmis MJ, Baron JR. 1989. *Adaptive grid embedding in nonequilibrium hypersonic flow.* Presented at Thermophysics Conference, 24th, AIAA Pap. 89–1652, Buffalo, NY

Barber TJ. 1996. *The role of code validation and certification in the design environment.* Presented at Fluid Dynamics Conference, 27th, AIAA Pap. 96–2033, New Orleans, LA

Bradley RG. 1988. *CFD validation philosophy*. Presented at Validation of Computational Fluid Dynamics Conf., AGARD Conf. Proc. 437, Pap. 1, Lisbon, Portugal

Bradley RG. 1995. Private Transmission

Cosner RR. 1995. *Validation requirements for technology transition*. Presented at Fluid Dynamics Conf., 26th, AIAA Pap. 95–2227, San Diego, CA

Deiwert GS, Strawa AW, Sharma SP, Park C. 1988. *Experimental program for real gas flow code validation at NASA Ames Research Center*. Presented at Validation of Computational Fluid Dynamics Conf., AGARD Conf. Proc. 437, Pap. 20, Lisbon, Portugal

Draper AC, Buck ML, Selegan DR. 1983. *Aerospace technology demonstrators/research and operational options*. Presented at Aircraft Prototype and Technology Demonstrator Symp., AIAA Pap. 83–1054, Dayton, OH

Everson BL, Bussoletti JE, Johnson FT, Samant SS. 1987. *TRANAIR and its NAS implementation*. Presented at the NASA Conf. on Supercomputing in Aerosp., Moffett Field, CA

Gentry AE, Smyth DN, Oliver WR. 1973. The Mark IV supersonic-hypersonic arbitrary-body program. *AFFDL-TR-73–159*

Kafyeke F, Mavriplis F. 1997. *CFD for the design of Bombardier's Global Express^R high performance jet*. Presented at the Appl. Aerodynamics Conf., 15th, AIAA Pap. 97–2269, Atlanta, GA

Laster M, Stanewsky E, Sinclair FW, Sickles WL. 1998. *Reynolds number scaling at transonic speeds*. Presented at Advanced Measurement and Ground Testing Tech. Conf., 20th, AIAA Pap. 98–2878, Albuquerque, NM

Lesieutre DJ, Dillenius MFE, Mendenhall MR, Torres TO. 1989. Aerodynamic analysis program MISL3 for conventional missiles with cruciform fin sections. *Nielsen Engineering and Research TR 404*

Li CP. 1987. Computations of Hypersonic Flow Fields. In J. J. Bertin, R. Glowinski, and J. Periaux (eds.) *Hypersonics, Volume II: Computation and Measurement of Hypersonic Flows*, Boston: Birkhauser Boston

Matthews RK, Nutt KW, Wannenwetsch GD, Kidd CT, Boudreau AH. 1985. *Developments in aerothermal test techniques at the AEDC supersonic-hypersonic wind tunnels*. Presented at Thermophysics Conf., 20th, AIAA Pap. 85–1003, Snowmass, CO

Mendenhall MR, Lesieutre DJ, Caruso SC, Dillenius MFE, Kuhn GD. 1990. Aerodynamic design of Pegasus[TM], concept to flight with CFD. In *Missile Aerodynamics*, AGARD Conf. Proc. No. 493, Symposium of the Fluid Dynamics Panel, Friedrichshafen, Germany

Mullin SN. 1992. *The evolution of the F-22 Advanced Tactical Fighter*. Presented at Aircraft Design Systems Meet., AIAA Pap. 92–4188, Hilton Head Island, SC

Neumann RD. 1986. *Designing a flight test program*. Presented at University of Texas Short Course on Hypersonics, Austin, TX

Neumann RD. 1988. *Missions and requirements*. Presented at Special Course on Aerothermodynamics of Hypersonic Vehicles, AGARD Report 761, Neuilly sur Seine, France

Niewald PW, Parker SL. 1999. *Flight test techniques employed successfully to verify F/A-18E in-flight lift and drag*. Presented at Aerosp. Sci. Meet., 37th, AIAA Pap. 99–0768, Reno, NV

Niewald PW, Parker SL. 2000. Wind-tunnel techniques to successfully predict F/A-18E in-flight lift and drag. *J. Aircraft* 37:9–14

Rizzi A, Vos J. 1996. *Towards establishing credibility in CFD simulations*. Presented at Fluid Dyn. Conf., 27th, AIAA Pap. 96–2029, New Orleans, LA

Saltzman EJ, Ayers TG. 1982. Review of flight-to-wind-tunnel drag correlations. *J. Aircraft* 19:801–811.

Shang JS. 1995. Assessment of technology for aircraft development. *J. Aircraft* 32:611–617

Towne MC, Strande SM, Erickson LL, Kroo IM, Enomoto FY, Carmichael RL, McPherson KF. 1983. *PAN AIR modeling studies*. Presented at Appl. Aerodyn. Conf., AIAA Pap. 83–1830, Danvers, MA

Trimmer LL, Cary A, Voisinet RL. 1986. *The optimum hypersonic wind tunnel*. Presented at Aerodyn. Test. Conf., AIAA Pap. 86–0739, West Palm Beach, FL

A THE EQUATIONS OF MOTION WRITTEN IN CONSERVATION FORM

In the main text the basic equations of motion were developed and were used in the nonconservative form, e.g., the momentum equation (equation (2.12)] and the energy equation (equation (2.32)]. However, the reader who pursues advanced applications of fluid mechanics often encounters the basic equations in conservation form for use in computational fluid dynamics. The conservation form of the basic equations is used for the following reasons (among others). First, it is easier to derive a noniterative, second-order, implicit algorithm if the nonlinear equations are in conservation form. Second, when shock waves and shear layers are expected in the flow field, it is essential that conservative difference approximations be used if one uses a "shock-capturing" method (Ref. A.1). *Shock capturing* is defined as ignoring the presence of embedded discontinuities in the sense that they are not treated as internal boundaries in the difference algorithm.

Let us establish the following nomenclature for the general form of the equations of motion in conservation form:

$$\frac{\partial U}{\partial t} + \frac{\partial (E_i - E_v)}{\partial x} + \frac{\partial (F_i - F_v)}{\partial y} + \frac{\partial (G_i - G_v)}{\partial z} = 0 \qquad \textbf{(A.1)}$$

where the subscript i denotes the terms that are included in the equations of motion for an inviscid flow and the subscript v denotes the terms that are unique to the equations of motion when the viscous and heat-transfer effects are included.

Recall that the continuity equation (2.1) is

$$\frac{\partial \rho}{\partial t} + \frac{\partial(\rho u)}{\partial x} + \frac{\partial(\rho v)}{\partial y} + \frac{\partial(\rho w)}{\partial z} = 0 \qquad \textbf{(A.2)}$$

Note that this equation is already in conservation form. It is valid for an inviscid flow as well as a viscous flow. Therefore, comparing equations (A.1) and (A.2), we obtain

$$U = \rho \qquad E_i = \rho u \qquad F_i = \rho v \qquad G_i = \rho w \qquad \textbf{(A.3a)–(A.3d)}$$

$$E_v = F_v = G_v = 0 \qquad \textbf{(A.3e)–(A.3g)}$$

Consider the x component of the momentum equation, as given by equation (2.11a). Assuming that the body forces are negligible, as is commonly done for gas flows,

$$\rho\frac{\partial u}{\partial t} + \rho u\frac{\partial u}{\partial x} + \rho v\frac{\partial u}{\partial y} + \rho w\frac{\partial u}{\partial z} = -\frac{\partial p}{\partial x} + \frac{\partial(\tau'_{xx})}{\partial x} + \frac{\partial(\tau_{yx})}{\partial y} + \frac{\partial(\tau_{zx})}{\partial z} \qquad \textbf{(A.4)}$$

To obtain equation (A.4), we have divided the expression for the normal shear stress τ_{xx}, as defined in Chapter 2, into the sum of the pressure p which is present in the equations for an inviscid flow and a term relating to the viscous flow τ'_{xx}. This is done to satisfy more easily the format of equation (A.1). Thus,

$$\tau'_{xx} = -\frac{2}{3}\mu\nabla\cdot\vec{V} + 2\mu\frac{\partial u}{\partial x}$$

The definitions for τ_{yx} and τ_{zx} remain as defined in Chapter 2. If we multiply the continuity equation (A.2) by u, we obtain

$$u\frac{\partial \rho}{\partial t} + u\frac{\partial(\rho u)}{\partial x} + u\frac{\partial(\rho v)}{\partial y} + u\frac{\partial(\rho w)}{\partial z} = 0 \qquad \textbf{(A.5)}$$

Adding equations (A.4) and (A.5) and rearranging, we obtain

$$\frac{\partial(\rho u)}{\partial t} + \frac{\partial}{\partial x}[(p + \rho u^2) - (\tau'_{xx})]$$

$$+ \frac{\partial}{\partial y}[(\rho uv) - (\tau_{yx})] + \frac{\partial}{\partial z}[(\rho uw) - (\tau_{zx})] = 0 \qquad \textbf{(A.6)}$$

Comparing equation (A.6) to the general form of the equations of motion in conservation form, equation (A.1), it is clear that

$$U = \rho u \qquad E_i = p + \rho u^2 \qquad F_i = \rho uv \qquad G_i = \rho uw \quad \textbf{(A.7a)–(A.7d)}$$

$$E_v = \tau'_{xx} \qquad F_v = \tau_{yx} \qquad G_v = \tau_{zx} \qquad \textbf{(A.7e)–(A.7g)}$$

Similar manipulations of the y momentum will yield

$$U = \rho v \qquad E_i = \rho vu \qquad F_i = p + \rho v^2 \qquad G_i = \rho vw \quad \textbf{(A.8a)–(A.8d)}$$

$$E_v = \tau_{xy} \qquad F_v = \tau'_{yy} \qquad G_v = \tau_{zy} \qquad \textbf{(A.8e)–(A.8g)}$$

where

$$\tau'_{yy} = -\frac{2}{3}\mu\nabla\cdot\vec{V} + 2\mu\frac{\partial v}{\partial y}$$

Similarly, the z-momentum equation can be written in conservation form, if

$$U = \rho w \quad E_i = \rho wu \quad F_i = \rho wv \quad G_i = p + \rho w^2 \quad \text{(A.9a)–(A.9d)}$$

$$E_v = \tau_{xz} \quad F_v = \tau_{yz} \quad G_v = \tau'_{zz} \quad \text{(A.9e)–(A.9g)}$$

where

$$\tau'_{zz} = -\frac{2}{3}\mu \nabla \cdot \vec{V} + 2\mu \frac{\partial w}{\partial z}$$

The energy equation (2.25) is

$$\rho \dot{q} - \rho \dot{w} = \rho \frac{d}{dt}(ke) + \rho \frac{d}{dt}(pe) + \rho \frac{d}{dt}(u_e) \quad \text{(2.25)}$$

Neglecting the changes in potential energy and using the definitions for $\rho \dot{q}$ and $\rho \dot{w}$, we obtain

$$\frac{\partial}{\partial x}\left(k\frac{\partial T}{\partial x}\right) + \frac{\partial}{\partial y}\left(k\frac{\partial T}{\partial y}\right) + \frac{\partial}{\partial z}\left(k\frac{\partial T}{\partial z}\right)$$

$$-\frac{\partial}{\partial x}(up) + \frac{\partial}{\partial x}(u\tau'_{xx}) + \frac{\partial}{\partial x}(v\tau_{xy}) + \frac{\partial}{\partial x}(w\tau_{xz})$$

$$+\frac{\partial}{\partial y}(u\tau_{yx}) - \frac{\partial}{\partial y}(vp) + \frac{\partial}{\partial y}(v\tau'_{yy}) + \frac{\partial}{\partial y}(w\tau_{yz})$$

$$+\frac{\partial}{\partial z}(u\tau_{zx}) + \frac{\partial}{\partial z}(v\tau_{zy}) - \frac{\partial}{\partial z}(wp) + \frac{\partial}{\partial z}(w\tau'_{zz})$$

$$= \rho \frac{de_t}{dt} = \rho \frac{\partial e_t}{\partial t} + \rho u \frac{\partial e_t}{\partial x} + \rho v \frac{\partial e_t}{\partial y} + \rho w \frac{\partial e_t}{\partial z} \quad \text{(A.10)}$$

where e_t, the specific total energy of the flow (neglecting the potential energy), is given by

$$e_t = u_e + \tfrac{1}{2}(u^2 + v^2 + w^2) \quad \text{(A.11)}$$

The continuity equation multiplied by e_t is

$$e_t \frac{\partial \rho}{\partial t} + e_t \frac{\partial(\rho u)}{\partial x} + e_t \frac{\partial(\rho v)}{\partial y} + e_t \frac{\partial(\rho w)}{\partial z} = 0 \quad \text{(A.12)}$$

Adding equations (A.10) and (A.12) and rearranging, we obtain

$$\frac{\partial}{\partial t}(\rho e_t) + \frac{\partial}{\partial x}[(\rho e_t + p)u - (u\tau'_{xx} + v\tau_{xy} + w\tau_{xz} + \dot{q}_x)]$$

$$+ \frac{\partial}{\partial y}[(\rho e_t + p)v - (u\tau_{yx} + v\tau'_{yy} + w\tau_{yz} + \dot{q}_y)]$$

$$+ \frac{\partial}{\partial z}[(\rho e_t + p)w - (u\tau_{zx} + v\tau_{zy} + w\tau'_{zz} + \dot{q}_z)] \quad \text{(A.13)}$$

Equation (A.13) is the energy equation in the conservation form. Comparing equation (A.13) to the general form, equation (A.1), it is clear that

$$U = \rho e_t \qquad E_i = (\rho e_t + p)u \qquad F_i = (\rho e_t + p)v$$

$$G_i = (\rho e_t + p)w \tag{A.14a)-(A.14d)}$$

$$E_v = u\tau'_{xx} + v\tau_{xy} + w\tau_{xz} + \dot{q}_x \tag{A.14e}$$

$$F_v = u\tau_{yx} + v\tau'_{yy} + w\tau_{yz} + \dot{q}_y \tag{A.14f}$$

$$G_v = u\tau_{zx} + v\tau_{zy} + w\tau'_{zz} + \dot{q}_z \tag{A.14g}$$

When the fundamental equations governing the unsteady flow of gas, without body forces or external heat addition, are written in conservation form [i.e., equation (A.1)], the terms $U, E_i, E_v, F_i, F_v, G_i$, and G_v can be represented by the following vectors.

$$U = \begin{bmatrix} \rho \\ \rho u \\ \rho v \\ \rho w \\ \rho e_t \end{bmatrix}$$

$$E_i = \begin{bmatrix} \rho u \\ p + \rho u^2 \\ \rho u v \\ \rho u w \\ (\rho e_t + p)u \end{bmatrix} \qquad E_v = \begin{bmatrix} 0 \\ \tau'_{xx} \\ \tau_{xy} \\ \tau_{xz} \\ (u\tau'_{xx} + v\tau_{xy} + w\tau_{xz} + \dot{q}_x) \end{bmatrix}$$

$$F_i = \begin{bmatrix} \rho v \\ \rho v u \\ p + \rho v^2 \\ \rho v w \\ (\rho e_t + p)v \end{bmatrix} \qquad F_v = \begin{bmatrix} 0 \\ \tau_{yx} \\ \tau'_{yy} \\ \tau_{yz} \\ (u\tau_{yx} + v\tau'_{yy} + w\tau_{yz} + \dot{q}_y) \end{bmatrix}$$

$$G_i = \begin{bmatrix} \rho w \\ \rho w u \\ \rho w v \\ p + \rho w^2 \\ (\rho e_t + p)w \end{bmatrix} \qquad G_v = \begin{bmatrix} 0 \\ \tau_{zx} \\ \tau_{zy} \\ \tau'_{zz} \\ (u\tau_{zx} + v\tau_{zy} + w\tau'_{zz} + \dot{q}_z) \end{bmatrix}$$

The successive lines of these vectors are the continuity equation, the x-momentum equation, the y-momentum equation, the z-momentum equation, and the energy equation. Note that considering only U and the inviscid terms (E_i, F_i, and G_i) of lines 2 through 4 yields Euler's equations in conservation form.

Repeating the information of Chapter 2 (with the modifications noted previously),

$$\tau'_{xx} = 2\mu\frac{\partial u}{\partial x} - \frac{2}{3}\mu\nabla \cdot \vec{V}$$

$$\tau'_{yy} = 2\mu\frac{\partial v}{\partial y} - \frac{2}{3}\mu\nabla \cdot \vec{V}$$

$$\tau'_{zz} = 2\mu\frac{\partial w}{\partial z} - \frac{2}{3}\mu\nabla \cdot \vec{V}$$

$$\tau_{xy} = \tau_{yx} = \mu\left(\frac{\partial u}{\partial y} + \frac{\partial v}{\partial x}\right)$$

$$\tau_{xz} = \tau_{zx} = \mu\left(\frac{\partial u}{\partial z} + \frac{\partial w}{\partial x}\right)$$

$$\tau_{yz} = \tau_{zy} = \mu\left(\frac{\partial v}{\partial z} + \frac{\partial w}{\partial y}\right)$$

$$\dot{q}_x = k\frac{\partial T}{\partial x} \qquad \dot{q}_y = k\frac{\partial T}{\partial y} \qquad \dot{q}_z = k\frac{\partial T}{\partial z}$$

Similar procedures can be followed to develop the conservative form of the governing equations for flow in cylindrical coordinates.

$$\frac{\partial U}{\partial t} + \frac{\partial(F_i - F_v)}{\partial z} + \frac{\partial(G_i - G_v)}{\partial r} + \frac{\partial(H_i - H_v)}{\partial \theta} + (R_i - R_v) = 0$$

$$U = \begin{bmatrix} \rho \\ \rho v_z \\ \rho v_r \\ \rho v_\theta \\ \rho e_t \end{bmatrix}$$

$$F_i = \begin{bmatrix} \rho v_z \\ p + \rho v_z^2 \\ \rho v_z v_r \\ \rho v_z v_\theta \\ (\rho e_t + p)v_z \end{bmatrix} \qquad F_v = \begin{bmatrix} 0 \\ \tau_{zz} \\ \tau_{zr} \\ \tau_{z\theta} \\ (v_z\tau_{zz} + v_r\tau_{zr} + v_\theta\tau_{z\theta} + \dot{q}_z) \end{bmatrix}$$

$$G_i = \begin{bmatrix} \rho v_r \\ \rho v_r v_z \\ p + \rho v_r^2 \\ \rho v_r v_\theta \\ (\rho e_t + p)v_r \end{bmatrix} \qquad G_v = \begin{bmatrix} 0 \\ \tau_{rz} \\ \tau_{rr} \\ \tau_{r\theta} \\ (v_z\tau_{rz} + v_r\tau_{rr} + v_\theta\tau_{r\theta} + \dot{q}_r) \end{bmatrix}$$

$$H_i = \frac{1}{r} \begin{bmatrix} \rho v_\theta \\ \rho v_\theta v_z \\ \rho v_\theta v_r \\ p + \rho v_\theta^2 \\ (\rho e_t + p) v_\theta \end{bmatrix} \qquad H_v = \frac{1}{r} \begin{bmatrix} 0 \\ \tau_{\theta z} \\ \tau_{\theta r} \\ \tau_{\theta \theta} \\ (v_z \tau_{z\theta} + v_r \tau_{r\theta} + v_\theta \tau_{\theta\theta} + \dot{q}_\theta) \end{bmatrix}$$

$$R_i = \frac{1}{r} \begin{bmatrix} \rho v_r \\ \rho v_r v_z \\ \rho v_r^2 - \rho v_\theta^2 \\ 2\rho v_r v_\theta \\ (\rho e_t + p) v_r \end{bmatrix} \qquad R_v = \frac{1}{r} \begin{bmatrix} 0 \\ \tau_{rz} \\ \tau_{rr} - \tau_{\theta\theta} \\ 2\tau_{r\theta} \\ (v_z \tau_{zr} + v_r \tau_{rr} + v_\theta \tau_{\theta r} + \dot{q}_r) \end{bmatrix} \qquad \textbf{(A.15)}$$

where

$$\tau_{rr} = 2\mu \frac{\partial v_r}{\partial r} - \frac{2}{3} \mu \nabla \cdot \vec{V}$$

$$\tau_{\theta\theta} = 2\mu \left(\frac{1}{r} \frac{\partial v_\theta}{\partial \theta} + \frac{v_r}{r} \right) - \frac{2}{3} \mu \nabla \cdot \vec{V}$$

$$\tau_{zz} = 2\mu \frac{\partial v_z}{\partial z} - \frac{2}{3} \mu \nabla \cdot \vec{V}$$

$$\tau_{\theta z} = \tau_{z\theta} = \mu \left(\frac{1}{r} \frac{\partial v_z}{\partial \theta} + \frac{\partial v_\theta}{\partial z} \right)$$

$$\tau_{\theta r} = \tau_{r\theta} = \mu \left(\frac{\partial v_\theta}{\partial r} - \frac{v_\theta}{r} + \frac{1}{r} \frac{\partial v_r}{\partial \theta} \right)$$

$$\tau_{zr} = \tau_{rz} = \mu \left(\frac{\partial v_r}{\partial z} + \frac{\partial v_z}{\partial r} \right)$$

$$q_z = k \frac{\partial T}{\partial z} \qquad q_r = k \frac{\partial T}{\partial r} \qquad q_\theta = \frac{k}{r} \frac{\partial T}{\partial \theta}$$

REFERENCES

Warming RF, Beam RM. 1978. On the construction and application of implicit factored schemes for conservation laws. *SIAM-AMS Proceedings* 11:85–129

B A COLLECTION OF OFTEN USED TABLES

TABLE 1.2A U.S. Standard Atmosphere, 1976 SI Units

Geometric Altitude (km)	Pressure (p/p_{SL})	Temperature (K)	Density (ρ/ρ_{SL})	Viscosity (μ/μ_{SL})	Speed of Sound (m/s)
0	1.0000 E+00	288.150	1.0000 E+00	1.00000	340.29
1	8.8700 E−01	281.651	9.0748 E−01	0.98237	336.43
2	7.8461 E−01	275.154	8.2168 E−01	0.96456	332.53
3	6.9204 E−01	268.659	7.4225 E−01	0.94656	328.58
4	6.0854 E−01	262.166	6.6885 E−01	0.92836	324.59
5	5.3341 E−01	255.676	6.0117 E−01	0.90995	320.55
6	4.6600 E−01	249.187	5.3887 E−01	0.89133	316.45
7	4.0567 E−01	242.700	4.8165 E−01	0.87249	312.31
8	3.5185 E−01	236.215	4.2921 E−01	0.85343	308.11
9	3.0397 E−01	229.733	3.8128 E−01	0.83414	303.85
10	2.6153 E−01	223.252	3.3756 E−01	0.81461	299.53
11	2.2403 E−01	216.774	2.9780 E−01	0.79485	295.15
12	1.9145 E−01	216.650	2.5464 E−01	0.79447	295.07
13	1.6362 E−01	216.650	2.1763 E−01	0.79447	295.07
14	1.3985 E−01	216.650	1.8601 E−01	0.79447	295.07
15	1.1953 E−01	216.650	1.5898 E−01	0.79447	295.07

(continued on next page)

TABLE 1.2A   (*Continued*)

Geometric Altitude (km)	Pressure (p/p_SL)	Temperature (K)	Density (ρ/ρ_SL)	Viscosity (μ/μ_SL)	Speed of Sound (m/s)
16	1.0217 E−01	216.650	1.3589 E−01	0.79447	295.07
17	8.7340 E−02	216.650	1.1616 E−01	0.79447	295.07
18	7.4663 E−02	216.650	9.9304 E−02	0.79447	295.07
19	6.3829 E−02	216.650	8.4894 E−02	0.79447	295.07
20	5.4570 E−02	216.650	7.2580 E−02	0.79447	295.07
21	4.6671 E−02	217.581	6.1808 E−02	0.79732	295.70
22	3.9945 E−02	218.574	5.2661 E−02	0.80037	296.38
23	3.4215 E−02	219.567	4.4903 E−02	0.80340	297.05
24	2.9328 E−02	220.560	3.8317 E−02	0.80643	297.72
25	2.5158 E−02	221.552	3.2722 E−02	0.80945	298.39
26	2.1597 E−02	222.544	2.7965 E−02	0.81247	299.06
27	1.8553 E−02	223.536	2.3917 E−02	0.81547	299.72
28	1.5950 E−02	224.527	2.0470 E−02	0.81847	300.39
29	1.3722 E−02	225.518	1.7533 E−02	0.82147	301.05
30	1.1813 E−02	226.509	1.5029 E−02	0.82446	301.71

Reference values: $p_{SL} = 1.01325 \times 10^5 \, N/m^2$; $T_{SL} = 288.150 \, K$

$\rho_{SL} = 1.2250 \, kg/m^3$; $\mu_{SL} = 1.7894 \times 10^{-5} \, kg/s \cdot m$

TABLE 1.2B   U.S. Standard Atmosphere, 1976 English Units

Geometric Altitude (kft)	Pressure (p/p_SL)	Temperature (°R)	Density (ρ/ρ_SL)	Viscosity (μ/μ_SL)	Speed of Sound (ft/s)
0	1.0000 E+00	518.67	1.0000 E+00	1.00000 E+00	1116.44
2	9.2981 E−01	511.54	9.4278 E−01	9.8928 E−01	1108.76
4	8.6368 E−01	504.41	8.8811 E−01	9.7849 E−01	1100.98
6	8.0142 E−01	497.28	8.3590 E−01	9.6763 E−01	1093.18
8	7.4286 E−01	490.15	7.8609 E−01	9.5670 E−01	1085.33
10	6.8783 E−01	483.02	7.3859 E−01	9.4569 E−01	1077.40
12	6.3615 E−01	475.90	6.9333 E−01	9.3461 E−01	1069.42
14	5.8767 E−01	468.78	6.5022 E−01	9.2346 E−01	1061.38
16	5.4224 E−01	461.66	6.0921 E−01	9.1223 E−01	1053.31
18	4.9970 E−01	454.53	5.7021 E−01	9.0092 E−01	1045.14
20	4.5991 E−01	447.42	5.3316 E−01	8.8953 E−01	1036.94
22	4.2273 E−01	440.30	4.9798 E−01	8.7806 E−01	1028.64
24	3.8803 E−01	433.18	4.6462 E−01	8.6650 E−01	1020.31
26	3.5568 E−01	426.07	4.3300 E−01	8.5487 E−01	1011.88
28	3.2556 E−01	418.95	4.0305 E−01	8.4315 E−01	1003.41
30	2.9754 E−01	411.84	3.7473 E−01	8.3134 E−01	994.85

(*continued on next page*)

TABLE 1.2B (Continued)

Geometric Altitude (kft)	Pressure (p/p_{SL})	Temperature (°R)	Density (ρ/ρ_{SL})	Viscosity (μ/μ_{SL})	Speed of Sound (ft/s)
32	2.7151 E−01	404.73	3.4795 E−01	8.1945 E−01	986.22
34	2.4736 E−01	397.62	3.2267 E−01	8.0746 E−01	977.53
36	2.2498 E−01	390.51	2.9883 E−01	7.9539 E−01	968.73
38	2.0443 E−01	389.97	2.7191 E−01	7.9447 E−01	968.08
40	1.8576 E−01	389.97	2.4708 E−01	7.9447 E−01	968.08
42	1.6880 E−01	389.97	2.2452 E−01	7.9447 E−01	968.08
44	1.5339 E−01	389.97	2.0402 E−01	7.9447 E−01	968.08
46	1.3939 E−01	389.97	1.8540 E−01	7.9447 E−01	968.08
48	1.2667 E−01	389.97	1.6848 E−01	7.9447 E−01	968.08
50	1.1511 E−01	389.97	1.5311 E−01	7.9447 E−01	968.08
52	1.0461 E−01	389.97	1.3914 E−01	7.9447 E−01	968.08
54	9.5072 E−02	389.97	1.2645 E−01	7.9447 E−01	968.08
56	8.6402 E−02	389.97	1.1492 E−01	7.9447 E−01	968.08
58	7.8524 E−02	389.97	1.0444 E−01	7.9447 E−01	968.08
60	7.1366 E−02	389.97	9.4919 E−02	7.9447 E−01	968.08
62	6.4861 E−02	389.97	8.6268 E−02	7.9447 E−01	968.08
64	5.8951 E−02	389.97	7.8407 E−02	7.9447 E−01	968.08
66	5.3580 E−02	390.07	7.1246 E−02	7.9463 E−01	968.21
68	4.8707 E−02	391.16	6.4585 E−02	7.9649 E−01	969.55
70	4.4289 E−02	392.25	5.8565 E−02	7.9835 E−01	970.90
72	4.0284 E−02	393.34	5.3121 E−02	8.0020 E−01	972.24
74	3.6651 E−02	394.43	4.8197 E−02	8.0205 E−01	973.59
76	3.3355 E−02	395.52	4.3742 E−02	8.0390 E−01	974.93
78	3.0364 E−02	396.60	3.9710 E−02	8.0575 E−01	976.28
80	2.7649 E−02	397.69	3.6060 E−02	8.0759 E−01	977.62
82	2.5183 E−02	398.78	3.2755 E−02	8.0943 E−01	978.94
84	2.2943 E−02	399.87	2.9761 E−02	8.1127 E−01	980.28
86	2.0909 E−02	400.96	2.7048 E−02	8.1311 E−01	981.63
88	1.9060 E−02	402.05	2.4589 E−02	8.1494 E−01	982.94
90	1.7379 E−02	403.14	2.2360 E−02	8.1677 E−01	984.28
92	1.5850 E−02	404.22	2.0339 E−02	8.1860 E−01	985.60
94	1.4460 E−02	405.31	1.8505 E−02	8.2043 E−01	986.94
96	1.3195 E−02	406.40	1.6841 E−02	8.2225 E−01	988.25
98	1.2044 E−02	407.49	1.5331 E−02	8.2407 E−01	989.57
100	1.0997 E−02	408.57	1.3960 E−02	8.2589 E−01	990.91

Reference values: $p_{SL} = 2116.22$ lbf/ft^2
$T_{SL} = 518.67$°R
$\rho_{SL} = 0.002377$ slugs/ft^3
$\mu_{SL} = 1.2024 \times 10^{-5}$ lbm/ft·s
$= 3.740 \times 10^{-7}$ lbf·s/ft^2

TABLE 8.1 Correlations for a One-Dimensional, Isentropic Flow of Perfect Air ($\gamma = 1.4$)

M	$\dfrac{A}{A^*}$	$\dfrac{p}{p_{t1}}$	$\dfrac{\rho}{\rho_{t1}}$	$\dfrac{T}{T_t}$	$\dfrac{A}{A^*}\dfrac{p}{p_{t1}}$
0	/	1.00000	1.00000	1.00000	/
0.05	11.592	0.99825	0.99875	0.99950	11.571
0.10	5.8218	0.99303	0.99502	0.99800	5.7812
0.15	3.9103	0.98441	0.98884	0.99552	3.8493
0.20	2.9635	0.97250	0.98027	0.99206	2.8820
0.25	2.4027	0.95745	0.96942	0.98765	2.3005
0.30	2.0351	0.93947	0.95638	0.98232	1.9119
0.35	1.7780	0.91877	0.94128	0.97608	1.6336
0.40	1.5901	0.89562	0.92428	0.96899	1.4241
0.45	1.4487	0.87027	0.90552	0.96108	1.2607
0.50	1.3398	0.84302	0.88517	0.95238	1.12951
0.55	1.2550	0.81416	0.86342	0.94295	1.02174
0.60	1.1882	0.78400	0.84045	0.93284	0.93155
0.65	1.1356	0.75283	0.81644	0.92208	0.85493
0.70	1.09437	0.72092	0.79158	0.91075	0.78896
0.75	1.06242	0.68857	0.76603	0.89888	0.73155
0.80	1.03823	0.65602	0.74000	0.88652	0.68110
0.85	1.02067	0.62351	0.71361	0.87374	0.63640
0.90	1.00886	0.59126	0.68704	0.86058	0.59650
0.95	1.00214	0.55946	0.66044	0.84710	0.56066
1.00	1.00000	0.52828	0.63394	0.83333	0.52828
1.05	1.00202	0.49787	0.60765	0.81933	0.49888
1.10	1.00793	0.46835	0.58169	0.80515	0.47206
1.15	1.01746	0.43983	0.55616	0.79083	0.44751
1.20	1.03044	0.41238	0.53114	0.77640	0.42493
1.25	1.04676	0.38606	0.50670	0.76190	0.40411
1.30	1.06631	0.36092	0.48291	0.74738	0.38484
1.35	1.08904	0.33697	0.45980	0.73287	0.36697
1.40	1.1149	0.31424	0.43742	0.71839	0.35036
1.45	1.1440	0.29272	0.41581	0.70397	0.33486
1.50	1.1762	0.27240	0.39498	0.68965	0.32039
1.55	1.2115	0.25326	0.37496	0.67545	0.30685
1.60	1.2502	0.23527	0.35573	0.66138	0.29414
1.65	1.2922	0.21839	0.33731	0.64746	0.28221
1.70	1.3376	0.20259	0.31969	0.63372	0.27099
1.75	1.3865	0.18782	0.30287	0.62016	0.26042
1.80	1.4390	0.17404	0.28682	0.60680	0.25044
1.85	1.4952	0.16120	0.27153	0.59365	0.24102
1.90	1.5555	0.14924	0.25699	0.58072	0.23211
1.95	1.6193	0.13813	0.24317	0.56802	0.22367
2.00	1.6875	0.12780	0.23005	0.55556	0.21567
2.05	1.7600	0.11823	0.21760	0.54333	0.20808
2.10	1.8369	0.10935	0.20580	0.53135	0.20087
2.15	1.9185	0.10113	0.19463	0.51962	0.19403
2.20	2.0050	0.09352	0.18405	0.50813	0.18751

(continued on next page)

TABLE 8.1 (*Continued*)

M	$\dfrac{A}{A^*}$	$\dfrac{p}{p_{t1}}$	$\dfrac{\rho}{\rho_{t1}}$	$\dfrac{T}{T_t}$	$\dfrac{A}{A^*}\dfrac{p}{p_{t1}}$
2.25	2.0964	0.08648	0.17404	0.49689	0.18130
2.30	2.1931	0.07997	0.16458	0.48591	0.17539
2.35	2.2953	0.07396	0.15564	0.47517	0.16975
2.40	2.4031	0.06840	0.14720	0.46468	0.16437
2.45	2.5168	0.06327	0.13922	0.45444	0.15923
2.50	2.6367	0.05853	0.13169	0.44444	0.15432
2.55	2.7630	0.05415	0.12458	0.43469	0.14963
2.60	2.8960	0.05012	0.11787	0.42517	0.14513
2.65	3.0359	0.04639	0.11154	0.41589	0.14083
2.70	3.1830	0.04295	0.10557	0.40684	0.13671
2.75	3.3376	0.03977	0.09994	0.39801	0.13276
2.80	3.5001	0.03685	0.09462	0.38941	0.12897
2.85	3.6707	0.03415	0.08962	0.38102	0.12534
2.90	3.8498	0.03165	0.08489	0.37286	0.12185
2.95	4.0376	0.02935	0.08043	0.36490	0.11850
3.00	4.2346	0.02722	0.07623	0.35714	0.11527
3.50	6.7896	0.01311	0.04523	0.28986	0.08902
4.00	10.719	0.00658	0.02766	0.23810	0.07059
4.50	16.562	0.00346	0.01745	0.19802	0.05723
5.00	25.000	$189(10)^{-5}$	0.01134	0.16667	0.04725
6.00	53.189	$633(10)^{-6}$	0.00519	0.12195	0.03368
7.00	104.143	$242(10)^{-6}$	0.00261	0.09259	0.02516
8.00	190.109	$102(10)^{-6}$	0.00141	0.07246	0.01947
9.00	327.189	$474(10)^{-7}$	0.000815	0.05814	0.01550
10.00	535.938	$236(10)^{-7}$	0.000495	0.04762	0.01263
∞	∞	0	0	0	0

TABLE 8.2   Mach Number and Mach Angle as a Function of Prandt-Meyer Angle

ν (deg)	M	μ (deg)	ν (deg)	M	μ (deg)
0.0	1.000	90.000	7.5	1.348	47.896
0.5	1.051	72.099	8.0	1.366	47.082
1.0	1.082	67.574	8.5	1.383	46.306
1.5	1.108	64.451	9.0	1.400	45.566
2.0	1.133	61.997	9.5	1.418	44.857
2.5	1.155	59.950	10.0	1.435	44.177
3.0	1.177	58.180	10.5	1.452	43.523
3.5	1.198	56.614	11.0	1.469	42.894
4.0	1.218	55.205	11.5	1.486	42.287
4.5	1.237	53.920	12.0	1.503	41.701
5.0	1.256	52.738	12.5	1.520	41.134
5.5	1.275	51.642	13.0	1.537	40.585
6.0	1.294	50.619	13.5	1.554	40.053
6.5	1.312	49.658	14.0	1.571	39.537
7.0	1.330	48.753	14.5	1.588	39.035

(*continued on next page*)

TABLE 8.2   (*Continued*)

ν (deg)	M	μ (deg)	ν (deg)	M	μ (deg)
15.0	1.605	38.547	40.0	2.538	23.206
15.5	1.622	38.073	40.5	2.560	22.997
16.0	1.639	37.611	41.0	2.582	22.790
16.5	1.655	37.160	41.5	2.604	22.585
17.0	1.672	36.721	42.0	2.626	22.382
17.5	1.689	36.293	42.5	2.649	22.182
18.0	1.706	35.874	43.0	2.671	21.983
18.5	1.724	35.465	43.5	2.694	21.786
19.0	1.741	35.065	44.0	2.718	21.591
19.5	1.758	34.673	44.5	2.741	21.398
20.0	1.775	34.290	45.0	2.764	21.207
20.5	1.792	33.915	45.5	2.788	21.017
21.0	1.810	33.548	46.0	2.812	20.830
21.5	1.827	33.188	46.5	2.836	20.644
22.0	1.844	32.834	47.0	2.861	20.459
22.5	1.862	32.488	47.5	2.886	20.277
23.0	1.879	32.148	48.0	2.910	20.096
23.5	1.897	31.814	48.5	2.936	19.916
24.0	1.915	31.486	49.0	2.961	19.738
24.5	1.932	31.164	49.5	2.987	15.561
25.0	1.950	30.847	50.0	3.013	19.386
25.5	1.968	30.536	50.5	3.039	19.213
26.0	1.986	30.229	51.0	3.065	19.041
26.5	2.004	29.928	51.5	3.092	18.870
27.0	2.023	29.632	52.0	3.119	18.701
27.5	2.041	29.340	52.5	3.146	18.532
28.0	2.059	29.052	53.0	3.174	18.366
28.5	2.078	28.769	53.5	3.202	18.200
29.0	2.096	28.491	54.0	3.230	18.036
29.5	2.115	28.216	54.5	3.258	17.873
30.0	2.134	27.945	55.0	3.287	17.711
30.5	2.153	27.678	55.5	3.316	17.551
31.0	2.172	27.415	56.0	3.346	17.391
31.5	2.191	27.155	56.5	3.375	17.233
32.0	2.210	26.899	57.0	3.406	17.076
32.5	2.230	26.646	57.5	3.436	16.920
33.0	2.249	26.397	58.0	3.467	16.765
33.5	2.269	26.151	58.5	3.498	16.611
34.0	2.289	25.908	59.0	3.530	16.458
34.5	2.309	25.668	59.5	3.562	16.306
35.0	2.329	25.430	60.0	3.594	16.155
35.5	2.349	25.196	60.5	3.627	16.006
36.0	2.369	24.965	61.0	3.660	15.856
36.5	2.390	24.736	61.5	3.694	15.708
37.0	2.410	24.510	62.0	3.728	15.561
37.5	2.431	24.287	62.5	3.762	15.415
38.0	2.452	24.066	63.0	3.797	15.270
38.5	2.473	23.847	63.5	3.832	15.126
39.0	2.495	23.631	64.0	3.868	14.983
39.5	2.516	23.418	64.5	3.904	14.840

(*continued on next page*)

TABLE 8.2   (*Continued*)

ν (deg)	M	μ (deg)	ν (deg)	M	μ (deg)
65.0	3.941	14.698	85.5	6.080	9.467
65.5	3.979	14.557	86.0	6.155	9.350
66.0	4.016	14.417	86.5	6.232	9.234
66.5	4.055	14.278	87.0	6.310	9.119
67.0	4.094	14.140	87.5	6.390	9.003
67.5	4.133	14.002	88.0	6.472	8.888
68.0	4.173	13.865	88.5	6.556	8.774
68.5	4.214	13.729	89.0	6.642	8.660
69.0	4.255	13.593	89.5	6.729	8.546
69.5	4.297	13.459	90.0	6.819	8.433
70.0	4.339	13.325	90.5	6.911	8.320
70.5	4.382	13.191	91.0	7.005	8.207
71.0	4.426	13.059	91.5	7.102	8.095
71.5	4.470	12.927	92.0	7.201	7.983
72.0	4.515	12.795	92.5	7.302	7.871
72.5	4.561	12.665	93.0	7.406	7.760
73.0	4.608	12.535	93.5	7.513	7.649
73.5	4.655	12.406	94.0	7.623	7.538
74.0	4.703	12.277	94.5	7.735	7.428
74.5	4.752	12.149	94.5	5.935	9.701
75.0	4.801	12.021	95.0	7.851	7.318
75.5	4.852	11.894	95.5	7.970	7.208
76.0	4.903	11.768	96.0	8.092	7.099
76.5	4.955	11.642	96.5	8.218	6.989
77.0	5.009	11.517	97.0	8.347	6.881
77.5	5.063	11.392	97.5	8.480	6.772
78.0	5.118	11.268	98.0	8.618	6.664
78.5	5.175	11.145	98.5	8.759	6.556
79.0	5.231	11.022	99.0	8.905	6.448
79.5	5.289	10.899	99.5	9.055	6.340
80.0	5.348	10.777	100.0	9.210	6.233
80.5	5.408	10.656	100.5	9.371	6.126
81.0	5.470	10.535	101.0	9.536	6.019
81.5	5.532	10.414	101.5	9.708	5.913
82.0	5.596	10.294	102.0	9.885	5.806
82.5	5.661	10.175			
83.0	5.727	10.056			
83.5	5.795	9.937			
84.0	5.864	9.819			
84.5	5.935	9.701			
85.0	6.006	9.584			

TABLE 8.3   Correlation of Flow Properties Across a Normal Shock Wave as a Function of the Upstream Mach Number for Air, $\gamma = 1.4$

M_1	M_2	$\dfrac{p_2}{p_2}$	$\dfrac{p_2}{p_1}$	$\dfrac{T_2}{T_1}$	$\dfrac{p_{t2}}{p_{t1}}$
1.00	1.00000	1.00000	1.00000	1.00000	1.00000
1.05	0.95312	1.1196	1.08398	1.03284	0.99987
1.10	0.91177	1.2450	1.1691	1.06494	0.99892

(*continued on next page*)

TABLE 8.3 (*Continued*)

M_1	M_2	$\dfrac{p_2}{p_2}$	$\dfrac{\rho_2}{\rho_1}$	$\dfrac{T_2}{T_1}$	$\dfrac{p_{t2}}{p_{t1}}$
1.15	0.87502	1.3762	1.2550	1.09657	0.99669
1.20	0.84217	1.5133	1.3416	1.1280	0.99280
1.25	0.81264	1.6562	1.4286	1.1594	0.98706
1.30	0.78596	1.8050	1.5157	1.1909	0.97935
1.35	0.76175	1.9596	1.6027	1.2226	0.96972
1.40	0.73971	2.1200	1.6896	1.2547	0.95819
1.45	0.71956	2.2862	1.7761	1.2872	0.94483
1.50	0.70109	2.4583	1.8621	1.3202	0.92978
1.55	0.68410	2.6363	1.9473	1.3538	0.91319
1.60	0.66844	2.8201	2.0317	1.3880	0.89520
1.65	0.65396	3.0096	2.1152	1.4228	0.87598
1.70	0.64055	3.2050	2.1977	1.4583	0.85573
1.75	0.62809	3.4062	2.2781	1.4946	0.83456
1.80	0.61650	3.6133	2.3592	1.5316	0.81268
1.85	0.60570	3.8262	2.4381	1.5694	0.79021
1.90	0.59562	4.0450	2.5157	1.6079	0.76735
1.95	0.58618	4.2696	2.5919	1.6473	0.74418
2.00	0.57735	4.5000	2.6666	1.6875	0.72088
2.05	0.56907	4.7363	2.7400	1.7286	0.69752
2.10	0.56128	4.9784	2.8119	1.7704	0.67422
2.15	0.55395	5.2262	2.8823	1.8132	0.65105
2.20	0.54706	5.4800	2.9512	1.8569	0.62812
2.25	0.54055	5.7396	3.0186	1.9014	0.60554
2.30	0.53441	6.0050	3.0846	1.9468	0.58331
2.35	0.52861	6.2762	3.1490	1.9931	0.56148
2.40	0.52312	6.5533	3.2119	2.0403	0.54015
2.45	0.51792	6.8362	3.2733	2.0885	0.51932
2.50	0.51299	7.1250	3.3333	2.1375	0.49902
2.55	0.50831	7.4196	3.3918	2.1875	0.47927
2.60	0.50387	7.7200	3.4489	2.2383	0.46012
2.65	0.49965	8.0262	3.5047	2.2901	0.44155
2.70	0.49563	8.3383	3.5590	2.3429	0.42359
2.75	0.49181	8.6562	3.6119	2.3966	0.40622
2.80	0.48817	8.9800	3.6635	2.4512	0.38946
2.85	0.48470	9.3096	3.7139	2.5067	0.37330
2.90	0.48138	9.6450	3.7629	2.5632	0.35773
2.95	0.47821	9.986	3.8106	2.6206	0.34275
3.00	0.47519	10.333	3.8571	2.6790	0.32834
3.50	0.45115	14.125	4.2608	3.3150	0.21295
4.00	0.43496	18.500	4.5714	4.0469	0.13876
4.50	0.42355	23.458	4.8119	4.8761	0.09170
5.00	0.41523	29.000	5.0000	5.8000	0.06172
6.00	0.40416	41.833	5.2683	7.941	0.02965
7.00	0.39736	57.000	5.4444	10.469	0.01535
8.00	0.39289	74.500	5.5652	13.387	0.00849
9.00	0.38980	94.333	5.6512	16.693	0.00496
10.00	0.38757	116.50	5.1743	20.388	0.00304
∞	0.37796	∞	6.000	∞	0

Index